PROGRESS IN BRAIN RESEARCH

VOLUME 143

BRAIN MECHANISMS FOR THE INTEGRATION OF
POSTURE AND MOVEMENT

Other volumes in PROGRESS IN BRAIN RESEARCH

Volume 115: Brain Function in Hot Environment, by H.S. Sharma and J. Westman (Eds.) – 1998, ISBN 0-444-82377-8.
Volume 116: The Glutamate Synapse as a Therapeutical Target: Molecular Organization and Pathology of the Glutamate Synapse, by O.P. Ottersen, I.A. Langmoen and L. Gjerstad (Eds.) – 1998, ISBN 0-444-82754-4.
Volume 117: Neuronal Degeneration and Regeneration: From Basic Mechanisms to Prospects for Therapy, by F.W. van Leeuwen, A. Salehi, R.J. Giger, A.J.G.D. Holtmaat and J. Verhaagen (Eds.) – 1998, ISBN 0-444-82817-6.
Volume 118: Nitric Oxide in Brain Development, Plasticity and Disease, by R.R. Mize, T.M. Dawson, V.L. Dawson and M.J. Friedlander (Eds.) – 1998, ISBN 0-444-82885-0.
Volume 119: Advances in Brain Vasopressin, by I.J.A. Urban, J.P.H. Burbach and D. De Wied (Eds.) – 1999, ISBN 0-444-50080-4.
Volume 120: Nucleotides and their Receptors in the Nervous System, by P. Illes and H. Zimmermann (Eds.) – 1999, ISBN 0-444-50082-0.
Volume 121: Disorders of Brain, Behavior and Cognition: The Neurocomputational Perspective, by J.A. Reggia, E. Ruppin and D. Glanzman (Eds.) – 1999, ISBN 0-444-50175-4.
Volume 122: The Biological Basis for Mind Body Interactions, by E.A. Mayer and C.B. Saper (Eds.) – 1999, ISBN 0-444-50049-9.
Volume 123: Peripheral and Spinal Mechanisms in the Neural Control of Movement, by M.D. Binder (Ed.) – 1999, ISBN 0-444-50288-2.
Volume 124: Cerebellar Modules: Molecules, Morphology and Function, by N.M. Gerrits, T.J.H. Ruigrok and C.E. De Zeeuw (Eds.) – 2000, ISBN 0-444-50108-8.
Volume 125: Volume Transmission Revisited, by L.F. Agnati, K. Fuxe, C. Nicholson and E. Syková (Eds.) – 2000, ISBN 0-444-50314-5.
Volume 126: Cognition, Emotion and Autonomic Responses: the Integrative Role of the Prefrontal Cortex and Limbic Structures, by H.B.M. Uylings, C.G. Van Eden, J.P.C. De Bruin, M.G.P. Feenstra and C.M.A. Pennartz (Eds.) – 2000, ISBN 0-444-50332-3.
Volume 127: Neural Transplantation II. Novel Cell Therapies for CNS Disorders, by S.B. Dunnett and A. Björklund (Eds.) – 2000, ISBN 0-444-50109-6.
Volume 128: Neural Plasticity and Regeneration, by F.J. Seil (Ed.) – 2000, ISBN 0-444-50209-2.
Volume 129: Nervous System Plasticity and Chronic Pain, by J. Sandkühler, B. Bromm and G.F. Gebhart (Eds.) – 2000, ISBN 0-444-50509-1.
Volume 130: Advances in Neural Population Coding, by M.A.L. Nicolelis (Ed.) – 2001, ISBN 0-444-50110-X.
Volume 131: Concepts and Challenges in Retinal Biology, by H. Kolb, H. Ripps, and S. Wu (Eds.), – 2001, ISBN 0-444-50677-2.
Volume 132: Glial Cell Function, by B. Castellano López and M. Nieto-Sampedro (Eds.) – 2001. ISBN 0-444-50508-3.
Volume 133: The Maternal Brain. Neurobiological and neuroendocrine adaptation and disorders in pregnancy and post partum, by J.A. Russell, A.J. Douglas, R.J. Windle and C.D. Ingram (Eds.) – 2001, ISBN 0-444-50548-2.
Volume 134: Vision: From Neurons to Cognition, by C. Casanova and M. Ptito (Eds.) – 2001, ISBN 0-444-50586-5.
Volume 135: Do Seizures Damage the Brain, by A. Pitkanen and T. Sutula (Eds.) – 2002, ISBN 0-444-50814-7.
Volume 136: Changing Views of Cajal's Neuron, by E.C. Azmitia, J. DeFelipe, E.G. Jones, P Rakic and C.E. Ribak (Eds.) – 2002, ISBN 0-444-50815-5.
Volume 137: Spinal Cord Trauma: Regeneration, Neural Repair and Functional Recovery, by L. McKerracher, G. Doucet and S. Rossignol (Eds.) – 2002, ISBN 0-444-50817-1.
Volume 138: Plasticity in the Adult Brain: From Genes to Neurotherapy, by M.A. Hofman, G.J. Boer, A.J.G.D. Holtmaat, E.J.W. Van Someren, J. Verhaagen and D.F. Swaab (Eds.)-2002, ISBN 0-444-50981-X.
Volume 139: Vasopressin and Oxytocin: From Genes to Clinical Applications, by D. Poulain, S. Oliet and D. Theodosis (Eds.) – 2002. ISBN 0-444-50982-8.
Volume 140: The Brain's Eye, by J. Hybna, D.P. Munoz. W. Heide and R. Radach (Eds.) – 2002, ISBN 0-444-51097-4.
Volume 141: Gonadotrophin-Releasing Hormone: Molecules and Receptors, by I.S. Parhar (Ed.) – 2002, ISBN 0-444-50979-8.
Volume 142: Neural Control of Space Coding and Action Production, by C. Prablanc, D. Pélisson and Y. Rossetti (Eds.) – 2003, ISBN 0-444-509771.

PROGRESS IN BRAIN RESEARCH

VOLUME 143

BRAIN MECHANISMS FOR THE INTEGRATION OF POSTURE AND MOVEMENT

EDITED BY

SHIGEMI MORI

Department of Biological Control System, National Institute for Physiological Sciences, Okazaki 444-85-885-8, Japan

DOUGLAS G. STUART

Department of Physiology, University of Arizona College of Medicine, Tucson, AZ 85724, USA

MARIO WIESENDANGER

Laboratory of Motor Systems, Deparmtent of Neurology, University of Berne, CH-3010-12 Berne, Switzerland

TECHNICAL EDITOR

PATRICIA A. PIERCE

Department of Physiology, University of Arizona College of Medicine, Tucson, AZ 85724, USA

ELSEVIER

AMSTERDAM – BOSTON – HEIDELBERG – LONDON – NEW YORK – OXFORD
PARIS – SAN DIEGO – SAN FRANCISCO – SINGAPORE – SYDNEY – TOKYO
2004

ELSEVIER B.V.
Sara Burgerhartstraat 25
P.O. Box 211, 1000 AE Amsterdam, The Netherlands

© 2004 Elsevier B.V. All rights reserved.

This work is protected under copyright by Elsevier and the following terms and conditions apply to its use:

Photocopying
Single photocopies of single chapters may be made for personal use as allowed by national copyright laws. Permission of the Publisher and payment of a fee is required for all other photocopying, including multiple or systematic copying, copying for advertising or promotional purposes, resale, and all forms of document delivery. Special rates are available for educational institutions that wish to make photocopies for non-profit educational classroom use.

Permissions may be sought directly from Elsevier's Science & Technology Rights Department in Oxford, UK: phone: (+44) 1865 843830, fax: (+44) 1865 853333, e-mail: permissions@elsevier.com. You may also complete your request on-line via the Elsevier Science homepage (http://www.elsevier.com), by selecting 'Customer Support' and then 'Obtaining Permissions'.

In the USA, users may clear permissions and make payments through the Copyright Clearance Center, Inc., 222 Rosewood Drive, Danvers, MA 01923, USA; phone: (+1) (978) 7508400, fax: (+1) (978) 7504744, and in the UK through the Copyright Licensing Agency Rapid Clearance Service (CLARCS), 90 Tottenham Court Road, London W1P 0LP, UK; phone: (+44) 207 631 5555; fax: (+44) 207 631 5500. Other countries may have a local reprographic rights agency for payments.

Derivative Works
Tables of contents may be reproduced for internal circulation, but permission of Elsevier is required for external resale or distribution of such material.
Permission of the Publisher is required for all other derivative works, including compilations and translations.

Electronic Storage or Usage
Permission of the Publisher is required to store or use electronically any material contained in this work, including any chapter or part of a chapter.

Except as outlined above, no part of this work may be reproduced, stored in a retrieval system or transmitted in any form or by any means, electronic, mechanical, photocopying, recording or otherwise, without prior written permission of the Publisher.
Address permissions requests to: Elsevier's Science & Technology Rights Department, at the phone, fax and e-mail addresses noted above.

Notice
No responsibility is assumed by the Publisher for any injury and/or damage to persons or property as a matter of products liability, negligence or otherwise, or from any use or operation of any methods, products, instructions or ideas contained in the material herein. Because of rapid advances in the medical sciences, in particular, independent verification of diagnoses and drug dosages should be made.

First edition 2004

Library of Congress Cataloging in Publication Data
A catalog record from the Library of Congress has been applied for.

British Library Cataloguing in Publication Data
Brain mechanisms for the integration of posture and
  movement. – (Progress in brain research ; v. 143)
  1.Neurophysiology 2.Posture 3.Human locomotion
  I. Mori, Shigemi II. Pierce, Patricia A.
  612.8

ISBN:	0-444-51389-2 (volume)
ISBN:	0-444-80104-9 (series)
ISSN:	0079-6123

∞ The paper used in this publication meets the requirements of ANSI/NISO Z39.48-1992 (Permanence of Paper).
Printed in The Netherlands.

List of Contributors

A. Alexandrov, Institute of Higher Nervous Activity and Neurophysiology, Russian Academy of Science, Moscow 117865, Russia

M. Aoki, Department of Physiology, Sapporo Medical University, School of Medicine, Sapporo 060-8556, Japan

T. Asahara, Department of Physiology, Faculty of Medicine, Mie University, Tsu 514-8507, Japan

D. Barthélemy, Center for Research in Neurological Sciences, Department of Physiology, University of Montreal, Faculty of Medicine, Montreal, QC H3C 3J7, Canada

A.J. Bastian, Departments of Neurology, Neuroscience and Physical Medicine and Rehabilitation, Johns Hopkins Medical School, and the Kennedy Krieger Institute, Baltimore, MD 21205, USA

J.R. Bloedel, Department of Health and Human Performance and Biomedical Sciences, Iowa State University, Ames, IA 50013, USA

L. Bouyer, Center for Research in Neurological Sciences, Department of Physiology, University of Montreal, Faculty of Medicine, Montreal, QC H3C 3J7, Canada

V. Bracha, Department of Biomedical Sciences, Iowa State University College of Veterinary Medicine, Ames, IA 50011, USA

F. Brocard, Centre De Recherches en Sciences Neurologiques, C.P. 6128, Succ Centre-Ville, Montréal, QC H3C 3J7, Canada

R.B. Brownstone, Division of Neurosurgery, Dalhousie University, Halifax, NS B3H 4R2, Canada

E. Brustein, Center for Research in Neurological Sciences, Department of Physiology, University of Montreal, Faculty of Medicine, Montreal, QC H3C 3J7, Canada

C. Chau, Center for Research in Neurological Sciences, Department of Physiology, University of Montreal, Faculty of Medicine, Montreal, QC H3C 3J7, Canada

F. Clarac, Institut de Neurosciences Physiologiques et Cognitives – INPC, CNRS, Aix-Marseille II, 13402 Marseille Cedex 20, France

P.J. Cordo, Neurological Sciences Institute, Oregon Health and Science University, Portland, OR 97201, USA

N.V. Dounskaia, Motor Control Laboratory, Arizona State University, Tempe, AZ 85257-0404, USA

T. Drew, Department of Physiology, University of Montreal Faculty of Medicine, Montreal, QC H3C 3J7, Canada

K. Ezure, Department of Neurobiology, Tokyo Metropolitan Institute for Neuroscience, Tokyo 183-8526, Japan

T.W. Ford, Sobell Department for Motor Neuroscience and Movement Disorders, Institute of Neurology, University College London, Queen Square, London, WC1N 3BG, UK

A. Frolov, Institute of Higher Nervous Activity and Neurophysiology, Russian Academy of Science, Moscow 117865, Russia

J. Fukushima, Department of Physiology, Hokkaido University School of Medicine, Sapporo 060-8638, Japan

K. Fukushima, Department of Physiology, Hokkaido University School of Medicine, Sapporo 060-8638, Japan

H. Fukuyama, Human Brain Research Center, Kyoto University Graduate School of Medicine, Kyoto 606-8507, Japan

E. Garcia-Rill, Department of Anatomy and Neurobiology, University of Arkansas Medical Sciences, Little Rock, AR 72205, USA

J.R. Georgiadis, Department of Anatomy and Embryology, Groningen Medical School, University of Groningen, Groningen 9700 AD, The Netherlands

N. Giroux, Center for Research in Neurological Sciences, Department of Physiology, University of Montreal, Faculty of Medicine, Montreal, QC H3C 3J7, Canada

J.-P. Gossard, Department of Physiology, University of Montreal, Montreal, QC H3C 3J7, Canada

A. Grantyn, Laboratoire de Physiologie de la Perception et de l'Action, CNRS-Collège de France, 75005 Paris, France

S. Grillner, Nobel Institute for Neurophysiology, Department of Neuroscience, Karolinska Institutet, SE-17177 Stockholm, Sweden

V.S. Gurfinkel, Neurological Sciences Institute, Oregon Health and Science University, Portland, OR 97201, USA

T. Habaguchi, Department of Physiology, Asahikawa Medical College, Asahikawa 078-8510, Japan

T. Hanakawa, Human Brain Research Center, Kyoto University Graduate School of Medicine, Kyoto 606-8507, Japan

J. He, Department of Bioengineering, Arizona State University, Tempe, AZ 85287, USA

G. Holstege, Department of Anatomy and Embryology, Groningen Medical School, University of Groningen, Groningen 9713 AV, The Netherlands

Y. Homma, Department of Anatomy and Neurobiology, University of Arkansas Medical Sciences, Little Rock, AR 72205, USA

E. Hoshi, Department of Physiology, Tohoku University School of Medicine, Sendai 980-8575, Japan

H. Hultborn, Department of Medical Physiology, The Panum Institute, University of Copenhagen, Copenhagen DK-2200, Denmark

Y. Inoue, Department of Integrative Physiology, National Institute for Physiological Sciences, Okazaki 444-8585, Japan

T. Isa, Department of Integrative Physiology, National Institute for Physiological Sciences, Okazaki 444-8585, Japan

T. Iseda, Department of Integrative Brain Science, Kyoto University Graduate School of Medicine, Kyoto 606-8501, Japan

Y. Izawa, Department of Systems Neurophysiology, Graduate School of Medicine, Tokyo Medical and Dental University, Tokyo 113-8519, Japan

I. Kagan, Department of Biomedical Engineering, Technion, Haifa 32000, Israel

S. Kakei, Department of Systems Neurophysiology, Graduate School of Medicine, Tokyo Medical and Dental University, Tokyo 113-8519, Japan

K. Kanda, Department of Central Nervous System, Tokyo Metropolitan Institute of Gerontology, Tokyo 173-0015, Japan

N. Katakura, Department of Gerontology, Graduate School, Tokyo Medical and Dental University, Tokyo 113-8549, Japan

S. Kawaguchi, Department of Integrative Brain Science, Kyoto University Graduate School of Medicine, Kyoto 606-8501, Japan

R.M. Kelly, Johnson and Johnson Pharmaceutical Research and Development, San Diego, CA 92121, USA

C.J. Ketcham, Motor Control Laboratory, Arizona State University, Tempe, AZ 85287-0404, USA

B.-H. Kim, ARL Division of Neural Systems, Memory and Aging, and Department of Psychology, University of Arizona, Tucson, AZ 85721, USA

P.A. Kirkwood, Sobell Department for Motor Neuroscience and Movement Disorders, Institute of Neurology, University College London, Queen Square, London WC1N 3BG, UK

T. Kitama, Department of Physiology, Yamanashi Medical University, Tamaho, Yamanashi 409-38, Japan

Y. Kobayashi, Neuroscience Laboratories, Visual Neuroscience Group, Osaka University Graduate School of Frontier Biosciences, 1-3 Machikaneyama, Toyomaka-shi, Osaka 560-8531, Japan

N. Kudo, Department of Physiology, Institute of Basic Medical Sciences, University of Tsukuba, Tsukuba, Ibaraki 305-8575, Japan

S. Kurkin, Department of Physiology, Hokkaido University School of Medicine, Sapporo 060-8638, Japan

C. Langlet, Center for Research in Neurological Sciences, Department of Physiology, University of Montreal, Faculty of Medicine, Montreal, QC H3C 3J7, Canada

H. Leblond, Center for Research in Neurological Sciences, Department of Physiology, University of Montreal, Faculty of Medicine, Montreal, QC H3C 3J7, Canada

R.N. Lemon, Sobell Department for Motor Neuroscience and Movement Disorders, Institute of Neurology, University College London, Queen Square, London WC1N 3BG, UK

M. Lin, Department of Physiology, Faculty of Medicine, Mie University, Tsu 514-8507, Japan

J. Liu, Stomatology Hospital, Nanjin Medical University, Nanjin 210029, China

Y.-Y. Ma, Kunming Institute of Zoology, Kunming, China

M.A. Maier, INSERM U483, Université Pierre et Marie Curie, 75005 Paris, France

J. Marcoux, Center for Research in Neurological Sciences, Department of Physiology, University of Montreal, Faculty of Medicine, Montreal, QC H3C 3J7, Canada

J. Massion, Laboratoire Parole et Langage, Université de Provence, 13621 Aix-en-Provence, France

T. Matsuura, Department of Physiology, Faculty of Medicine, Mie University, Tsu 514-8507, Japan

K. Matsuyama, Department of Physiology, Sapporo Medical University, Sapporo 060-8556, Japan

F. Mori, Department of Biological Control System, National Institute for Physiological Sciences, Okazaki 444-8585, Japan

M. Mori, Department of Biological Control System, National Institute for Physiological Sciences, Okazaki 444-8585, Japan

S. Mori, Department of Biological Control System, National Institute for Physiological Sciences, Okazaki 444-8585, Japan

A.K. Moschovakis, Institute of Applied and Computational Mathematics, F.O.R.T.H., Heraklion, Crete, Greece

M. Murray, Department of Neurobiology and Anatomy, Drexel University College of Medicine, Philadelphia, PA 19129, USA

K. Naito, Department of Neurophysiology, Tokyo Metropolitan Institute for Neuroscience, Tokyo 183-8526, Japan

K. Nakajima, Department of Physiology, Kinki University, School of Medicine, Osaka-Sayama 589-8511, Japan

M. Nakajima, Clinical Research Administration Department, Clinical Development Center, Head Office, Toray Industries, Inc., Tokyo 103-8666, Japan and Department of Biological Control System, National Institute for Physiological Sciences, Okazaki 444-8585, Japan

Y. Nakamura, Department of Welfare and Information, Faculty of Informatics, Teikyo Heisei University, Ichihara 290-0193, Japan

K. Nakayama, Department of Physiology, Institute of Basic Medical Sciences, University of Tsukuba, Tsukuba, Ibaraki 305-8575, Japan

A. Nambu, Department of System Neuroscience, Tokyo Metropolitan Institute for Neuroscience, Tokyo Metropolitan Organization for Medical Research, Tokyo 183-8526, Japan

P. Nathan, The National Hospital for Neurology and Neurosurgery, Queen Square, London WC1N 3BG, UK (In Memoriam)

H. Nishimaru, Department of Physiology, Institute of Basic Medical Sciences, University of Tsukuba, Tsukuba, Ibaraki 305-8575, Japan

Y. Nishimura, Department of Physiology, Faculty of Medicine, Mie University, Tsu 514-8507, Japan

T. Nishio, Department of Integrative Brain Science, Kyoto University Graduate School of Medicine, Kyoto 606-8501, Japan

E. Olivier, Laboratory of Neurophysiology, Faculty of Medicine, Université Catholique de Louvain, B-1200 Brussels, Belgium

J. Oohinata-Sugimoto, Department of Physiology, Asahikawa Medical College, Asahikawa 078-8510, Japan

K.G. Pearson, Department of Physiology, University of Alberta, Edmonton, AB T6G 2H7, Canada

B.W. Peterson, Department of Physiology, The Feinberg Medical School, Northwestern University, Chicago, IL 60611, USA

S. Prentice, Department of Kinesiology, University of Waterloo, Waterloo, ON N2L 3G1, Canada

T.A. Reader, Center for Research in Neurological Sciences, Department of Physiology, University of Montreal, Faculty of Medicine, Montreal, QC H3C 3J7, Canada (In Memoriam)

S. Rossignol, Center for Research in Neurological Sciences, Department of Physiology, University of Montreal, Faculty of Medicine, Montreal, QC H3C 3J7, Canada

E.M. Rouiller, Institute of Physiology, University of Fribourg, Fribourg CH-1700, Switzerland

J.-W. Ryou, ARL Division of Neural Systems, Memory and Aging, and Department of Psychology, University of Arizona, Tucson, AZ 85721, USA

K. Saitoh, Department of Physiology, Asahikawa Medical College, Asahikawa 078-8510, Japan

S. Sasaki, Department of Neurophysiology, Tokyo Metropolitan Institute for Neuroscience, Tokyo 183-8526, Japan

B. Schepens, Unité De Physiologie et Biomécanique de la Locomotion, Département d'Éducation Physique et de Réadaptation, Université Catholique de Louvain, 1348 Louvain-La-Neuve, Belgium

D.J. Serrien, Sobell Department of Neurophysiology, Institute of Neurology, Queen Square, London WC1N 3BG, UK

H. Shibasaki, Department of Neurology, Kyoto University Graduate School of Medicine, Kyoto 606-8507, Japan

H. Shibuya, Department of Physiology, Faculty of Medicine, Mie University, Tsu 514-8507, Japan

M.L. Shik, Department of Zoology, Tel Aviv University, Tel Aviv 69978, Israel

Y. Shinmei, Department of Physiology, Hokkaido University School of Medicine, Sapporo 060-8638, Japan

Y. Shinoda, Department of Systems Neurophysiology, Graduate School of Medicine, Tokyo Medical and Dental University, Tokyo 113-8519, Japan

R.D. Skinner, Department of Anatomy and Neurobiology, University of Arkansas Medical Sciences, Little Rock, AR 72205, USA

G.E. Stelmach, Motor Control Laboratory, Arizona State University, Tempe, AZ 85287-0404, USA

P.L. Strick, Department of Neurobiology, Neurological Surgery and Psychiatry, University of Pittsburgh, Pittsburgh, PA 15261, USA

D.G. Stuart, Department of Physiology, University of Arizona, College of Medicine, Tucson, AZ 85724, USA

Y. Sugiuchi, Department of Systems Neurophysiology, Graduate School of Medicine, Tokyo Medical and Dental University, Tokyo 113-8519, Japan

A. Tachibana, Department of Biological Control System, National Institute for Physiological Sciences, Okazaki 444-8585, Japan

K. Takakusaki, Department of Physiology, Asahikawa Medical College, Asahikawa 078-8510, Japan

C. Takasu, Department of Biological Control System, National Institute for Physiological Sciences, Okazaki 444-8585, Japan

J. Tanji, Department of Physiology, Tohoku University School of Medicine, Sendai 980-8575, Japan

A. Tessler, Department of Neurobiology and Anatomy, Drexel University College of Medicine, Philadelphia, PA 19129, USA

W.T. Thach, Department of Anatomy and Neurobiology, Program in Physical Therapy, Neurology and Neurological Surgery, and the Irene Walter Johnson Rehabilitation Research Institute, Washington University School of Medicine, St. Louis, MO 63110, USA

T.I. Toth, Physiology Unit, School of Biosciences, Cardiff University, Cardiff CF10 3US, UK

T. Tsujimoto, Center for Brain Experiment, National Institute for Physiological Sciences, Okazaki 444-8585, Japan

H. Tsukada, PET Center, Central Research Laboratory, Hamamatsu Photonics KK, Hamakita 434-8601, Japan

Y. Uchino, Department of Physiology, Tokyo Medical University, 6-1-1 Shinjuku, Shinjuku-ku, Tokyo 160-8402, Japan

L. Vinay, Dévelopement et Pathologie du Mouvement, CNRS, Aix-Marseille II, 13402 Marseille, Cedex 20, France

N. Wada, Department of Veterinary Physiology, Yamaguchi University, Yamaguchi 753-8515, Japan

P. Wallén, Nobel Institute for Neurophysiology, Department of Neuroscience, Karolinska Institutet, SE-17177 Stockholm, Sweden

D.J. Weber, Department of Bioengineering, Arizona State University, Tempe, AZ 85287, USA

M. Wiesendanger, Laboratory of Motor Systems, Department of Neurology, University of Berne, CH-3010 Berne, Switzerland

F.A.W. Wilson, Department of Neurology and ARL Division of Neural Systems, University of Arizona, Tucson, AZ 85721, USA

T. Yamaguchi, Graduate School of Engineering, Yamagata University, Yamagata 992-8510, Japan

T. Yamamoto, Department of Physiology, Faculty of Medicine, Mie University, Tsu 514-8507, Japan

T. Yamanobe, Department of Physiology, Hokkaido University School of Medicine, Sapporo 060-8638, Japan

K. Yoshimura, Department of Neurophysiology, Tokyo Metropolitan Institute for Neuroscience, Tokyo 183-8526, Japan

Preface

This volume of Progress in Brain Research is dedicated to the scientific mentors of Shigemi Mori: Bunichi Fujimori (1910–86), John Brookhart (1913–85), and Victor Gurfinkel (1922–present).

The volume describes the current state of our knowledge on the role of parallel and distributed neuronal systems in the integration of posture and movement. Our charge to the authors of the various chapters was twofold: to provide a conceptual overview of the topic that could serve as a balanced reference text for the next generation of movement neuroscientists; and, to stimulate further experimental and theoretical work in the field.

Key issues are addressed in ten interrelated sections: perspectives on the overall issues; three aspects of brainstem–spinal cord interactions (developmental and comparative; motoneuron properties, pattern generation, and sensory feedback; adaptive mechanisms issues; biomechanical and imaging approaches; descending command issues; supraspinal sensorimotor interactions; cerebellar interactions and control mechanisms; eye–head–neck coordination; and, higher control from the basal ganglia, sensorimotor cortex, and frontal lobe). Relevant chapters are cross-referenced, but no attempt has been made to adjudicate current disparities between our authors' results and their interpretations. These differences bring out the lively state of current work in the aspects of movement neuroscience addressed in this volume.

The above topics illustrate the need to understand early-appearing movements (e.g., chewing and similar rhythms) and later-developed skilled movements (like prehension and speech) in a *behavioral, goal-oriented context*. It is now evident that in addition to the local dedicated neural systems that control specific movements, neural networks are implicated in parallel to smoothly steer a highly coordinated motor apparatus in its execution of goal-directed behavior. Attention, object-identification, motivation, and reward expectation are all examples of the multifold ingredients that comprise such behavior. The underlying neural control mechanisms are widespread throughout the brain and spinal cord, and they are recruited in patterns that are both task- and context-dependent.

It is fitting to recognize the contributions of Fujimori, Brookhart, and Gurfinkel because each of them contributed importantly to the above areas: Fujimori for his contributions to integrative systems neuroscience and rebuilding Japanese movement neuroscience after WW II (particularly with Toshihiko Tokizane, 1909–1973); Brookhart for encouraging young neuroscientists and applying bioengineering principles to movement neuroscience; and Gurfinkel, for extending Nicolai Bernstein's ideas into modern movement neuroscience. All three have also contributed importantly to the international, indeed transnational, cooperation and goodwill that pervade this field of endeavor.

Shigemi Mori
Douglas G. Stuart
Mario Wiesendanger

Acknowledgments

We greatly appreciate the dedication, effort, and talent that our technical editor, Patricia Pierce, brought to bear on bringing this volume to successful fruition.

Much of the work in this volume was presented by the authors of most of the chapters at an International Symposium organized and directed by Shigemi Mori, *Higher Nervous Control of Posture and Locomotion: Parallel and Centralized Control Mechanisms*, and held at the National Institute for Physiological Sciences (NIPS), Okazaki Conference center, Okazaki National Research Institutes, Okazaki, Japan on March 18–22, 2001. The oral presentations in Okazaki were made by 18 invited scientists from Japan and 26 from other countries (Canada, 4; Denmark, 1; France, 3; Israel, 1; Sweden, 1; Switzerland, 2; The Netherlands, 1; UK, 2; USA, 11). In addition, there were 20 poster presentations by other invited scientists from Hungary (1), Japan (15), Poland (2), Romania (1), and the USA (2). We would like to emphasize that the poster presentations, and their informal discussion, contributed importantly to the overall success of the symposium because one of its key themes was to encourage the next generation of movement neuroscientists in Japan and abroad.

NIPS and the Ministry of Education, Culture, Sports, Science and Technology, Japan largely supported the expenses of the symposium and the visiting scientists. Additional support was kindly provided by the Uehara Memorial Foundation, the Sapporo Ko-Jinkai Medical Hospital, and several commercial companies including, Ad Instruments Japan, Bio Research Center, Canama Scientific, Eicom, Eicou, Herotec, Inter Medical, Kinoshita Rika, Medical Agent, Nac Imaging Technology, Nihon Kohden, Nihon Poladigital K.K., Ozawa Science, Shinko Camera, Shoshin Em, and Yasahisa Biomechanics.

We would also like to thank the local organization committee for their considerable efforts to ensure the success of the symposium. This fine group included Drs. Futoshi Mori, Katsumi Nakajima, Tetsu Okumura and Atsumichi Tachibana, Ms. Chijiko Takasu, and Mr. Masahiro Mori. Finally, we would like to recognize the ever-present and much-appreciated support of the former (1991–97) and the present (1997–2003). Director-General of NIPS, Professors Kiyoshi Hama and Kazuo Sasaki.

The Editors

Contents

List of Contributors v

Preface xi

Acknowledgments xiii

Section I. Perspectives

1. Innate versus learned movements—a false dichotomy?
   S. Grillner and P. Wallén (Stockholm, Sweden) 3

2. Why and how are posture and movement coordinated?
   J. Massion, A. Alexandrov and A. Frolov (Aix-en-Provence,
   France and Moscow, Russia) 13

3. Motor coordination can be fully understood only by studying complex movements
   P.J. Cordo and V.S. Gurfinkel (Portland, OR, USA) 29

4. The emotional brain: neural correlates of cat sexual behavior and human male ejaculation
   G. Holstege and J.R. Georgiadis (Groningen,
   The Netherlands) 39

Section II. Spinal cord and brainstem: developmental and comparative issues

5. Developmental changes in rhythmic spinal neuronal activity in the rat fetus
   N. Kudo, H. Nishimaru and K. Nakayama (Tsukuba, Japan) 49

6. The maturation of locomotor networks
   F. Clarac, F. Brocard and L. Vinay (Marseille, France) 57

7. Reflections on respiratory rhythm generation
   K. Ezure (Tokyo, Japan) 67

Section III. Spinal cord and brainstem: motoneurons, pattern generation and sensory feedback

8. Key mechanisms for setting the input–output gain across the motoneuron pool
   H. Hultborn, R.B. Brownstone, T.I. Toth and J.-P. Gossard
   (Copenhagen, Denmark, Halifax, NS and Montreal, QC, Canada
   and Cardiff, UK) 77

9. Rhythm generation for food-ingestive movements
   Y. Nakamura, N. Katakura, M. Nakajima and J. Liu
   (Ichihara and Tokyo, Japan) 97

10. Do respiratory neurons control female receptive behavior: a suggested
    role for a medullary central pattern generator?
    P.A. Kirkwood and T.W. Ford (London, UK) 105

11. The central pattern generator for forelimb locomotion in the cat
    T. Yamaguchi (Yamagata, Japan) 115

12. Generating the walking gait: role of sensory feedback
    K.G. Pearson (Edmonton, AB, Canada) 123

Section IV. Spinal cord and brainstem: adaptive mechanisms

13. Cellular transplants: steps toward restoration of function in spinal injured animals
    M. Murray (Philadelphia, PA, USA) 133

14. Neurotrophic effects on dorsal root regeneration into the spinal cord
    A. Tessler (Philadelphia, PA, USA) 147

15. Effects of an embryonic repair graft on recovery from spinal cord injury
    S. Kawaguchi, T. Iseda and T. Nishio (Kyoto, Japan) 155

16. Determinants of locomotor recovery after spinal injury in the cat
    S. Rossignol, L. Bouyer, C. Langlet, D. Barthélemy, C. Chau,
    N. Giroux, E. Brustein, J. Marcoux, H. Leblond and
    T.A. Reader (Montreal, QC, Canada) 163

Section V. Biomechanical and imaging approaches in movement neuroscience

17. Trunk movements and EMG activity in the cat: level versus upslope walking
    N. Wada and K. Kanda (Yamaguchi and Tokyo, Japan) 175

18. Biomechanical constraints in hindlimb joints during the quadrupedal
    versus bipedal locomotion of *M. fuscata*
    K. Nakajima, F. Mori, C. Takasu, M. Mori, K. Matsuyama and
    S. Mori (Osaka, Okazaki and Sapporo, Japan) 183

19. Reactive and anticipatory control of posture and bipedal locomotion in a nonhuman primate
F. Mori, K. Nakajima, A. Tachibana, C. Takasu, M. Mori, T. Tsujimoto, H. Tsukada and S. Mori (Okazaki and Hamikata, Japan) 191

20. Neural control mechanisms for normal versus Parkinsonian gait
H. Shibasaki, H. Fukuyama and T. Hanakawa (Kyoto, Japan) 199

21. Multijoint movement control: the importance of interactive torques
C.J. Ketcham, N.V. Dounskaia and G.E. Stelmach (Tempe, AZ, USA) 207

Section VI. Descending command issues

22. How the mesencephalic locomotor region recruits hindbrain neurons
I. Kagan and M.L. Shik (Tel Aviv and Haifa, Israel) 221

23. Role of basal ganglia–brainstem systems in the control of postural muscle tone and locomotion
K. Takakusaki, J. Oohinata-Sugimoto, K. Saitoh and T. Habaguchi (Asahikawa, Japan) 231

24. Locomotor role of the corticoreticular–reticulospinal–spinal interneuronal system
K. Matsuyama, F. Mori, K. Nakajima, T. Drew, M. Aoki and S. Mori (Sapporo and Okazaki, Japan and Montreal, QC, Canada) 239

25. Cortical and brainstem control of locomotion
T. Drew, S. Prentice and B. Schepens (Montreal, QC and Waterloo, ON, Canada and Louvain-La-Neuve, Belgium) 251

26. Direct and indirect pathways for corticospinal control of upper limb motoneurons in the primate
R.N. Lemon, P.A. Kirkwood, M.A. Maier, K. Nakajima and P. Nathan (London, UK, Paris, France and Osaka, Japan) 263

Section VII. Supraspinal sensorimotor interactions

27. Arousal mechanisms related to posture and locomotion: 1. Descending modulation
E. Garcia-Rill, Y. Homma and R.D. Skinner (Little Rock, AR, USA) 283

28. Arousal mechanisms related to posture and locomotion: 2. Ascending modulation
R.D. Skinner, Y. Homma and E. Garcia-Rill (Little Rock, AR, USA) 291

29. Switching between cortical and subcortical sensorimotor pathways
T. Isa and Y. Kobayashi (Okazaki, Japan) 299

Section VIII. Cerebellar interactions and control mechanisms

30. Cerebellar activation of cortical motor regions: comparisons across mammals
T. Yamamoto, Y. Nishimura, T. Matsuura, H. Shibuya, M. Lin and T. Asahara (Tsu, Japan) 309

31. Task-dependent role of the cerebellum in motor learning
J.R. Bloedel (Ames, IA, USA) 319

32. Role of the cerebellum in eyeblink conditioning
V. Bracha (Ames, IA, USA) 331

33. Integration of multiple motor segments for the elaboration of locomotion: role of the fastigial nucleus of the cerebellum
S. Mori, K. Nakajima, F. Mori and K. Matusyama (Okazaki, Japan) 341

34. Role of the cerebellum in the control and adaptation of gait in health and disease
W.T. Thach and A.J. Bastian (St. Louis, MO and Baltimore, MD, USA) 353

Section IX. Eye–head–neck coordination

35. Current approaches and future directions to understanding control of head movement
B.W. Peterson (Chicago, IL, USA) 369

36. The neural control of orienting: role of multiple-branching reticulospinal neurons
S. Sasaki, K. Yoshimura and K. Naito (Tokyo, Japan) 383

37. Role of the frontal eye fields in smooth-gaze tracking
K. Fukushima, T. Yamanobe, Y. Shinmei, J. Fukushima and S. Kurkin (Sapporo, Japan) 391

38. Role of cross-striolar and commissural inhibition in the vestibulocollic reflex
Y. Uchino (Tokyo, Japan) 403

39. Functional synergies among neck muscles revealed by branching patterns of single long descending motor-tract axons
Y. Sugiuchi, S. Kakei, Y. Izawa and Y. Shinoda (Tokyo, Japan) ... 411

40. Control of orienting movements: role of multiple tectal projections to the lower brainstem
    A. Grantyn, A.K. Moschovakis and T. Kitama (Paris, France, Heraklion, Greece and Yamanashi, Japan) 423

41. Pedunculo-pontine control of visually guided saccades
    Y. Kobayashi, Y. Inoue and T. Isa (Okazaki, Japan) 439

Section X. Higher control mechanisms: basal ganglia, sensorimotor cortex and frontal lobe

42. Macro-architecture of basal ganglia loops with the cerebral cortex: use of rabies virus to reveal multisynaptic circuits
    R.M. Kelly and P.L. Strick (San Diego, CA and Pittsburgh, PA, USA) 449

43. A new dynamic model of the cortico-basal ganglia loop
    A. Nambu (Tokyo, Japan) 461

44. Functional recovery after lesions of the primary motor cortex
    E.M. Rouiller and E. Olivier (Fribourg, Switzerland and Brussels, Belgium) 467

45. Adaptive behavior of cortical neurons during a perturbed arm-reaching movement in a nonhuman primate
    D.J. Weber and J. He (Tempe, AZ, USA) 477

46. The quest to understand bimanual coordination
    M. Wiesendanger and D.J. Serrien (Berne, Switzerland) 491

47. Functional specialization in dorsal and ventral premotor areas
    E. Hoshi and J. Tanji (Sendai and Kawaguchi, Japan) 507

48. Spatially directed movement and neuronal activity in freely moving monkey
    Y.-Y. Ma, J.-W. Ryou, B.-H. Kim and F.A.W. Wilson (Kunming, China and Tucson, AZ, USA) 513

Subject Index 521

SECTION I

Perspectives

CHAPTER 1

Innate versus learned movements—a false dichotomy?

Sten Grillner* and Peter Wallén

Nobel Institute for Neurophysiology, Department of Neuroscience, Karolinska Institutet, SE-17177 Stockholm, Sweden

Abstract: It is argued that the nervous systems of vertebrates are equipped with a "motor infrastructure," which enables them to perform the full extent of the motor repertoire characteristic of their particular species. In the human, it extends from the networks/circuits underlying locomotion and feeding to sound production in speech and arm-hand-finger coordination. Contrary to current opinion, these diverse motor patterns should be labeled as *voluntary*, because they can be recruited *at will*. Moreover, most, if not all, of the motor patterns available at birth are subject to maturation and are modified substantially through learning. We thus argue that the all-too-common distinction between learned and innate movements is based on a fundamental misconception about the neural control of the vertebrate motor system.

Introduction

When vertebrates are born, the capabilities of their motor systems vary markedly. Some hoofed mammals are able to stand and even run a few minutes after birth, and chickens are able to walk on two legs directly after hatching. Others, like human babies, are quite immature, and have an extended period of maturation, lasting until after puberty.

Animals, in general, are born with a certain 'motor infrastructure' that enables them to perform a variety of vital tasks like breathing, eating and responding to a variety of stimuli. Some species can also immediately stand and locomote, as mentioned previously. After birth, a maturation process takes place. In addition, a learning process occurs that utilizes the different motor programs available at any given stage. At birth, the human is able to reach out to different objects, and even grasp forcefully with the hands (the grasping reflex). Progressively, the reaching movements acquire a finer precision, and the coordination of the fingers matures to enable the manipulation of objects with remarkable spatial and temporal precision. At an age of ~1 year, the human can perform certain additional motor tasks, like standing, walking, reaching and grasping objects. Regardless of how much the child trains at this age, however, it is unable to tie shoelaces or brush the teeth. These motor tasks only become possible several years later. At the age of ~5 years, a child may be able to swim using a crude crawl stroke, but is still unable to coordinate arms and legs in a normal breaststroke manner (i.e., like a frog's swimming). With regard to the categories of motor tasks/programs that can be mastered at a certain age, the human can learn to further modify and improve a particular type of coordination but not learn patterns of coordination not yet matured. An adaptation/calibration of the motor patterns takes place as the child changes its body dimensions (a type of learning). The child starts walking by will, utilizing a standard motor program for locomotion, but somewhat later also learns to modify the motor program for walking, e.g., as in a Charlie Chaplin movie. The developing human also becomes able to adapt the locomotor pattern to different conditions, like carrying a suitcase or backpack, or carrying a child.

*Corresponding author: Tel.: +46-8-7286900;
Fax: +46-8-349544; E-mail: sten.grillner@neuro.ki.se

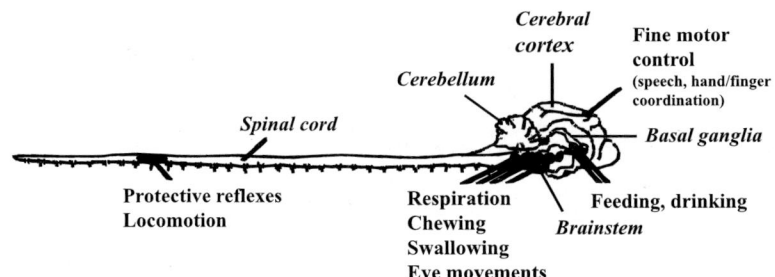

Fig. 1. Different patterns of motor behavior are controlled by a library of different networks located in the various parts of the CNS. See text for further details.

Further, we briefly review the different patterns of coordination that humans and other species are able to recruit, such that a multitude of different movements can be executed at will. We will argue that these patterns of motor behavior can also be modified at will and that species learn to modify or sequence them in a reproducible way. Moreover, most, if not all, innate patterns of motor behavior can be modified by learning.

The motor infrastructure: networks involved in the control of movement

Along the entire neuraxis, there are a number of neuronal networks that are designed to generate different aspects of the motor-repertoire characteristic of a given vertebrate species. These motor patterns can be recruited at will and they can be referred to as the 'motor infrastructure'. Figure 1 illustrates the location of some of these networks.

Locomotion

In the spinal cord, there are networks that mediate protective withdrawal reflexes, and also those that are responsible for the segmental activation of different muscles during locomotion (Grillner, 1981, 2002; Rossignol, 1996; Orlovsky et al., 1999). Figure 2 shows that the locomotor central pattern generators (CPGs) in the spinal cord can be activated from the brainstem via reticulospinal pathways. These, in turn, are controlled from a mesencephalic (mesopontine; MLR) and a diencephalic (DLR) locomotor center (Orlovsky and Shik, 1976; Grillner et al., 1997; Orlovsky et al., 1999).

The locomotor centers in Fig. 2A are conserved throughout vertebrate evolution (Eidelberg et al., 1981; S. Mori et al. and K. Takakusaki et al., chapters 33 and 23 of this volume). When, for example, the MLR is stimulated at low strength in the rat, cat and monkey, the animal will start walking, and at progressively higher stimulus strength, it will walk faster, then trot and finally gallop (Fig. 2B). A bird will start walking and, at higher stimulus strength, start flapping its wings (Steeves et al., 1987). Similarly, a lamprey or fish will swim at a progressively faster speed (McClellan and Grillner, 1984; Di Prisco et al., 2000). The brain is thus controlling a complex motor pattern with hundreds of different muscles by way of a simple, graded control signal. The details of the motor pattern are handled at the spinal level by CPGs (Yamaguchi, chapter 11 of this volume), which also become adapted to external events through a sensory control that adapts the burst duration (Grillner and Rossignol, 1978; Duysens and Pearson, 1980; Pearson, this volume) and, at slow speeds, the level of EMG activity (Stein et al., 2000). Bernstein (1947, 1967) has emphasized the need to use synergies to reduce the degrees of freedom in the control system.

Posture and control of body orientation during movement

Intimately linked with locomotion is the control system responsible for maintaining the body's orientation during this form of movement. At will, a number of stable postures can be assumed, e.g., standing and also more complex configurations with the trunk and limbs oriented in a multitude of ways.

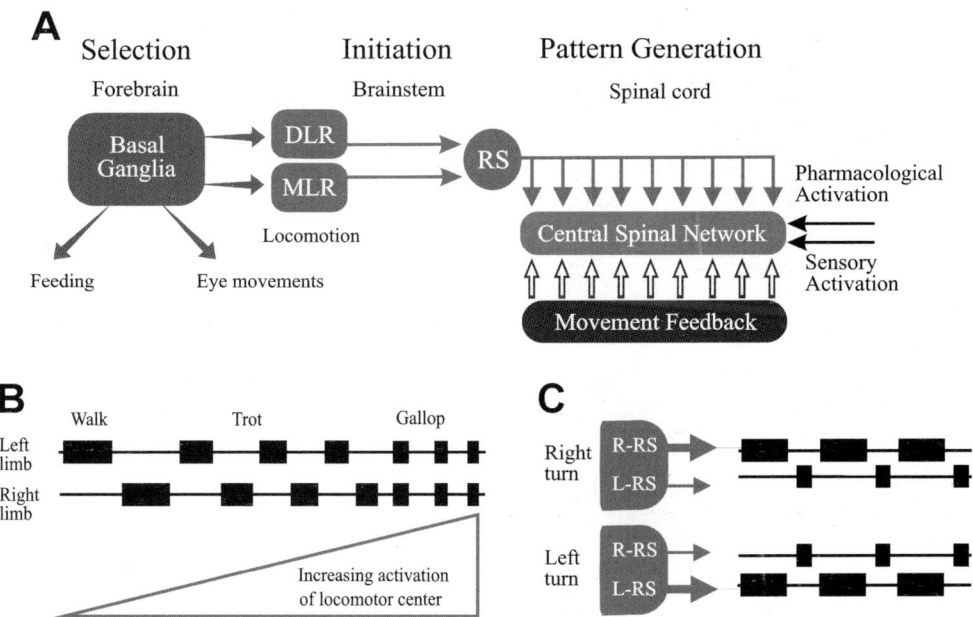

Fig. 2. Overview of the control of vertebrate locomotion. (A) General control strategy. The locomotor CPGs in the spinal cord are turned on from the brainstem via reticulospinal pathways. Disinhibition of the basal ganglia's input to the mesencephalic (mesopontine, MLR) and diencephalic (DLR) locomotor centers results in increased activity in reticulospinal neurons (RS), which, in cooperation with sensory feedback, activate the central spinal network, which, in turn, produces the locomotor pattern. The basal ganglia exert a tonic inhibitory influence on different motor centers. Once a pattern of motor behavior is selected, this inhibition is released, allowing, in this case, the locomotor centers to be activated. (B) With increased activation of the locomotor centers, the speed of locomotion increases. In quadrupeds, this also leads to a shift in interlimb coordination, from walk to trot and, finally, to gallop. Experimentally, locomotion can also be elicited pharmacologically by administration of excitatory amino acid agonists combined with sensory input. (C) An asymmetric activation of RS neurons gives rise to an asymmetric output on the left (L) and right (R) sides. This will result in a turning movement to one side or the other. (Reproduced, in part, from Grillner and Wallén 1999 with permission from Elsevier Science.)

The essential requirement is that the body be in a biomechanically stable position throughout movement, with the projection of the center of gravity falling between the different points of support. Figure 3 shows an example of a voluntary posture assumed at will in two different species. During movement, there may be a dynamic equilibrium moving the body from one side to another as seen during walking and even more pronouncedly, in skating.

Respiration, chewing and feeding

In the brainstem, there are CPG circuits for respiration operating throughout life. They are modulated by metabolic demands at different levels of activity. The respiratory CPGs are also influenced by sensory input from the lungs that help to regulate the amplitude of each breath (von Euler, 1983; Gray et al., 2001). In addition, the airflow is modulated during vocalization in the animals, and during speech and singing in the human. The respiratory CPGs are thus subject to modulation in a variety of voluntary tasks. There are also networks in the brainstem which coordinate swallowing, and sucking in the case of newborn mammals (Lund, 1991). In mammals, there are also networks coordinating chewing movements, whereas other vertebrates only tear up the food into appropriate portions to bite and swallow, as in birds and crocodiles.

Vocalization and expression of emotions

In the brainstem, there are also specific areas that can elicit vocalization. In birds, specific warning

Fig. 3. Voluntary posture in human and animal.

calls are produced that communicate a species-specific message to their peers. When a baby is born, it uses crying to effectively communicate when it is uneasy. This is most likely a motor pattern generated at the brainstem level. As the baby grows older, it progressively becomes able to express a greater variety of emotions, like smiling, laughing and distress. Altogether, at least six or seven basic, distinct facial patterns of coordination are used to express emotions in humans (Ekman, 1973; Russell and Bullock, 1985). Early on, Darwin (1872) noted in a seminal publication that children born blind could smile as beautifully as seeing children. This clearly demonstrates that this pattern of coordination is innate. He and other authors have emphasized that the motor patterns underlying these different patterns of coordination are innate and present in humans in all parts of the world. On the other hand, what the human laughs about is clearly dependent on the cultural environment and idiosyncratic (personal) traits. The human, and possibly apes, can modify by will these innate motor patterns to imitate, for instance, the smile of another individual. Moreover, the human can recruit these patterns of coordination (e.g., smiling) by will (i.e., without an emotional context), if the intent is to please somebody like a photographer or a customer!

Saccadic eye movements and the superior colliculus

The circuitry underlying saccadic eye movements, and the accompanying slower head movements, is located in the superior colliculus (Sparks, 1997; Hikosaka et al., 2000; Isa and Saito, 2001; see also Section X of this volume). In this case, there is a fast central command (appropriate direction and amplitude) that rapidly brings the eye (fovea) to the spot of visual interest, and a slower counter-rotation of the eye, as the 'slow' head is rotating to face the object. This is required to maintain the same spot projected onto the retina within the fovea. This is accommodated by the fast vestibulo-ocular reflex. In the superior colliculus, there is a well-designed organization from medial to lateral that generates eye movements at different angles (produced by a combination of eye muscles) and a rostro-caudal organization that determines the amplitude of the movement. In mammals, the frontal eye fields within the cerebral cortex act via the superior colliculus, whereas in some other vertebrates, saccades appear to be controlled directly and in interaction with the basal ganglia (Isa, 2002).

Forebrain centers control integrated motor behavior

In the forebrain (including the hypothalamus), there are centers controlling a variety of integrated patterns of motor behavior. Aggressive attack behavior can be induced by stimulation of some areas, and in others, drinking and feeding behavior (Hess, 1949). If the hypothalamic area that senses the osmolality of the extracellular fluid in the mammal is stimulated by implanted electrodes, the animal will perceive thirst by presumably subjective means. The animal will also find a place where water is to be found and then activate the locomotor

CPGs to arrive at this place. It will then position itself so that it can start drinking. The animal will continue to ingest water as long as the stimulation continues, even if this leads to 'water intoxication'. Thus, a sequence of different motor programs becomes active and, together, they ensure that the animal finds and ingests water.

If the centers for aggressive behavior are activated, the behavior of the animal will change dramatically. Rather than being a friendly interacting being, for example, the cat will display all outer signs of aggression and be prepared to viciously attack surrounding objects and subjects. At the termination of the stimulus, the cat may sit down and look surprised and puzzled, again displaying its previous friendly behavior (Hess, 1949; S. Grillner, personal observation). This stimulus-induced aggressive behavior utilizes a variety of integrated motor acts including attacking, biting, locomotion and posture. The attack behavior is directed toward different objects and subjects in an environment that were previously (normally) considered innocuous. Thus, this behavior is adaptable. The only difference is that the cat would normally not react with aggression to such stimuli.

It is also worth noting that the different patterns of motor behavior discussed earlier can be displayed in animals lacking a cerebral cortex, but with the rest of the forebrain intact. Such decorticate mammals can be maintained chronically. They display a surprisingly normal behavioral repertoire, which actually looks quite normal to at least the casual observer. The animals seemingly search for food and water, maintain appropriate food and fluid intake, and undergo phases of 'sleep' and 'wakefulness' (Bjursten et al., 1976). In a laboratory environment, they can thus survive, although they most likely would not in the wild. These decorticate cats are even able to learn some visual discrimination tasks related to where food is to be found (Norrsell and Norrsell, 1970), presumably by utilizing visual input directed to the midbrain. The decorticate cat often reacts with sham rage toward other individuals, and it appears unable to interpret more subtle signals in the environment. In a colony of cats, the 'wild type' avoids decorticate ones, which thereby become social 'outcasts' (Norrsell and Norrsell, 1970).

Visuomotor coordination during locomotion: a corticospinal precursor of independent reaching movements?

When a quadrupedal mammal (e.g., cat) walks over an even ground like a floor, corticospinal neurons are not modulated significantly in relation to the step cycle. The situation changes dramatically if the cat is required to place its feet very accurately in each step: i.e., precision walking (Beloozerova and Sirota, 1993; Drew, 1993; Drew et al., this volume), as when walking on a very complex terrain. Under these conditions, a cortically mediated visuomotor control signal is superimposed on the spinally generated basic locomotor pattern. There is a major projection to the motor areas from the visual areas and the related dorsal stream for movement perception. These projections are the most likely responsible ones for this type of visuomotor coordination, which is critically important for the animal in the wild. If the pyramidal tracts (and thereby the corticospinal contribution) are transected, the animal walks comfortably over a floor, but is unable to perform precision walking, e.g., up a ladder (Liddell and Phillips, 1944).

Some time ago, we pointed out that the precise positioning of the limbs during fast locomotion on uneven terrain is a very demanding task, which requires a corticospinal contribution (Georgopoulos and Grillner, 1989; see also Grillner et al., 1997). This task is obviously more demanding than just reaching out toward an object without concurrent locomotion (as in Fig. 4). It has recently been shown (Drew, 1993; Drew et al., Chapter 25 of this volume) that the *same* corticospinal neurons are active in the cat during *both* locomotion and an arm-reaching task without concurrent locomotion. This clearly demonstrates that the same corticospinal circuits are at least partially used during both locomotion and reaching.

It was also pointed out earlier that in arboreal primates that swing from one tree branch to another, there has been a powerful evolutionary pressure to reach out and grasp each branch with great precision. Individuals that fail to successfully target a branch will fall, and may not even survive (see Georgopoulos and Grillner, 1989). Thus, this precision locomotor system may also form the basis for fast accurate

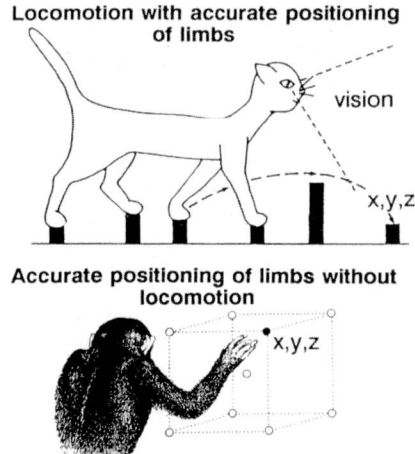

Fig. 4. Precision movements during locomotion and reaching. During normal locomotion on a complex terrain, the limbs need to be positioned with great accuracy (upper drawing). This requires integration between general locomotor commands and corticospinal visuomotor coordination, thereby leading to an accurate placement of the foot in each step. This aspect of visuomotor coordination has been shown to depend on a corticospinal contribution in the cat (see text), and it presumably applies to other species, as well. Arboreal primates move from branch to branch, requiring very accurate reaching and grasping in practically every step cycle. Positioning the limb accurately in relation to the environment, as the body is continuously moving, is, in fact, a more difficult task than a normal reaching one (lower drawing). In the latter case, the body is kept semistationary while the arm and hand reach out to different points in the surrounding space. It is therefore likely that the neuronal systems used for accurate positioning of the limb during locomotion involve circuits overlapping with those used during reaching, as exemplified in the lower drawing of a monkey. (Reproduced from Grillner et al. 1997 with permission from MIT Press.)

reaching–grasping movements. This neural system, however, is presumably not identical to that responsible for the independent fine finger movements underlying the skilled manipulation that humans display, like handwriting and the fine work of a silversmith.

Cortical contribution to motor control and visuomotor coordination

Clearly, the different motor cortices, and associated frontal areas, play an important role in the motor coordination underlying speech (Broca's area) and skilled hand and finger movements, like those used when playing the piano or executing other motor patterns that require independent finger movements (Lawrence and Kuypers, 1968a,b). In the nonhuman primate, in particular, the ability to imitate different movements performed by others seem to be linked to specific areas within the prefrontal areas (Rizzolatti et al., 2001). Also, in this animal, selected neurons in the motor cortex (M1) respond during and even just before a reaching movement. Their discharge is in a preferred direction such that the actual direction of a movement can be predicted from the response of this population of M1 neurons (Georgopoulos, 1996a, 1999).

Fast- and slow-conducting axons from the cortical motor areas project to the motor systems of the brainstem where they connect with a variety of reticulospinal neurons and the rubrospinal tract. Many corticospinal axons also project to the spinal cord, where they synapse mainly on different classes of propriospinal and segmental interneurones in the gray matter. Only a very small proportion of such axons (largely from M1) make direct monosynaptic connections with motoneurons, and these are largely with cells controlling distal muscles (Brodal, 1981; Alstermark et al., 1999; Maier et al., 2002). It is thus important to realize that only a very limited part of the corticospinal contribution to motor control is exerted via the direct monosynaptic pathway to motoneurons. Furthermore, this direct pathway is present in primates but not carnivores and other mammalian and nonmammalian vertebrates.

It is unfortunate that most current studies on the cortical control of movements include interpretations of results that reason as if there is little or no processing taking place between the cortex and the motoneuronal level (cf. Georgopoulos, 1996a; Prut et al., 2001). This conceptual obstacle must be overcome if substantial progress is to be made. It is far more likely that cortical control systems utilize the interneuronal systems at the brainstem/spinal cord level that are designed to control parts of integrated motor acts (see Grillner, 1981, 1985; Alstermark et al., 1999).

Most motor patterns performed at will are not labeled as voluntary

Most motor control scientists working on human and nonhuman primates refer to hand–arm–finger movements as voluntary and consider them as something fundamentally different from other motor acts performed by will, like chewing, breathing, standing, swallowing, walking and so forth. This is therefore a confusing terminology (see also Prochazka et al., 2000). The motor patterns mentioned earlier can be utilized, controlled and modified by will. Contrary to current dogma, these motor patterns are indeed voluntary/volitional. Note also that the abovementioned results of Drew demonstrate that the *same* corticospinal neurons can contribute to coordination during *both* locomotion and a reaching task, with only the latter traditionally termed voluntary. We propose that it is far more logical to consider that *all* the available motor patterns are part and parcel of the same general motor infrastructure that can be used to generate all 'voluntary' motor tasks in a wide variety of vertebrates.

Motor learning, adaptation and the basic motor infrastructure

We now consider in more detail three types of motor coordination (locomotion, saccadic eye movements, reaching) and how they adapt to changes in body dimensions and how they are modified by explicit learning.

Locomotion

As emphasized earlier, a spinal network coordinates the basic motor pattern of locomotion, and it can be recruited at will. In some species, it is operative and reasonably calibrated within minutes after birth. In others, it matures over a longer time. Clearly, the degree of activation of different muscles will change over time as an individual grows. Moreover, the motor pattern becomes perfected as the animal matures. Thus, it most likely that a progressive adaptation of the motor pattern represents a type of learning. Moreover, explicit learning of new aspects of a motor pattern can be achieved. A horse can be trained to trot or gallop in particular ways to please the owner or a jury evaluating the performance. All humans have a characteristic pattern of walking with a heel strike, and a somewhat different pattern during running. Each individual also tends to add a personal touch to the walking pattern, however, and can learn to imitate the walk-pattern of others. Moreover, the human can learn to modify the walking pattern to walk on high heels or in slalom boots, which both require learning. Following a pulled muscle, etc., a limping gait can be instigated instantly in order to minimize pain. After the pain has abated, the limp may continue out of habit until the normal pattern is reestablished. Skating and cross-country skiing are other examples of a restructuring of the locomotor and postural motor programs.

Saccadic eye movements

Saccadic eye movements are followed by a head movement, which requires an accurate counter-rotation to maintain the object of interest within the fovea (see Section IX, this volume). As mentioned earlier, this is achieved through the vestibulo-ocular reflex, which senses accurately the details of the head movement, and activates with minimal delay the eye muscles, which produce the necessary counter-rotation. A 'family' of vestibulo-ocular reflexes is actually needed, and it must operate continuously and in accurate fashion. An error signal occurs if the visual object 'slips' over the retina. This causes a long-lasting plastic change of the gain of the reflex, which is required to maintain accuracy. The gain of the reflex changes as the individual grows and changes body dimensions. Experimentally inappropriate vestibulo-ocular reflexes can be produced in man and animals by introducing magnifying lenses or reversing prisms. Although the world may then appear strange during the first few hours, particularly with reversing prisms, the vestibulo-ocular reflex then undergoes substantial adaptation over a broad range and finally reverses in sign. This adaptation/learning of the vestibulo-ocular reflex takes place through plastic changes occurring in both the cerebellar cortex and subcortical structures (Raymond and Lisberger, 1997; Ito, 2001).

Reaching

When a baby is born, it will immediately extend the arms toward different objects. These movements are very imprecise but not completely random. As the child develops, the precision increases, this being due to not only a maturation of the nervous system but also to a better calibration of the reaching movement, which is acquired through learning. Within certain limits, the young child at a given age can be trained to improve performance. The reaching movements can soon be specified as to the direction and amplitude of the movement, and be directed to a specific target in space. The reach may also have a more complicated trajectory that may take a curve-shaped form and be subdivided into different smaller components (Georgopoulos, 1996b). Just as the machinery for saccadic visuomotor coordination is innate, and then calibrated through learning, the basic machinery for reaching, too, is most likely a basic motor infrastructure of the brain. The fundamental patterns of coordination are thus available in a standard form, i.e., a set of neuronal networks that allow specification in terms of direction, amplitude and temporal characteristics. As different complex movements are learned, they presumably utilize and sequence elements of the innate machinery to achieve a complex, efficient and reproducible set of trajectories.

It is noteworthy that a basic memory trace of a given type of movement appears to be stored in the form of its general shape. Bernstein (1967) pointed out that in writing, the human learns the specific shape of, for example, the letter 'a'. If this letter is then written with a pencil on paper, a chalk on the blackboard or a pen taped onto the foot, its general shape remains the same, even though the precision differs. A completely different set of muscles is used in each case, of course. In these different cases, the motor command is thus directed to different sets of neurons to achieve the same general form, but of different dimensions.

Concluding thoughts: innate versus learned movements represents a false dichotomy

The purpose of this chapter is to argue that the vertebrate nervous system contains a basic motor infrastructure that provides a very extensive and flexible movement repertoire, which can express all manner of movements, including, for example, chewing, eye movements, grasping, locomotion, posture, reaching and speech. The basic machinery underlying these patterns of coordination is available in at least rudimentary form at birth, after which the patterns mature at different rates, and to different degree. They are all controlled at will, and thus voluntary in the true meaning of the word. Moreover, each pattern can be instantly modified at will and remembered, the latter being a type of learning. As emphasized earlier, the organism can initiate a variety of walking styles. Similarly, the multitude of different basic patterns of arm–hand–finger coordination can be expressed and recombined through learning. The speech motor control apparatus (e.g., airflow, shape of oral cavity, timing) allows the production of different phonemes. As speech is learned, the human picks up the appropriate sounds, and becomes able to reproduce them, while suppressing others (including perception) that do not occur in the local language (e.g., compare 'r' and 'l' elaborated by a native Chinese speaker). The human learns as a child to communicate through imitation of words spoken in the environment, and to sequence different phonemes and words while testing the efficacy of verbal interactions with other individuals. The ability of primates (in particular, humans) to imitate others is of fundamental importance for rapidly learning new patterns of motor coordination, including speech (Fogassi et al., 1992; Rizzolatti et al., 2001).

Certain motor patterns are relatively standard, and can be modified only to a limited degree, like volitional swallowing. Chewing, posture and walking are clearly more adaptive and modifiable through learning. Nonetheless, they, too, usually involve relatively stereotyped motor patterns. In contrast, arm–hand–finger movements and speech allow for far greater flexibility, and they utilize a much greater number of movement–pattern combinations.

In final summary, we conclude that practically all motor patterns are flexible and subject to learning. The all-too-common view that movements can be divided into innate and learned categories is thus incorrect and it should be abandoned. Similarly, virtually all these motor patterns can be generated at will and should thus be considered voluntary.

Acknowledgments

Support from the Swedish Medical Research Council (3026), the Science Research Council, the Wallenberg

Foundation and an EU Grant is gratefully acknowledged, as are the comments on the manuscript by Drs. Abdeljabbar El Manira, Ole Kiehn and Grigori N. Orlovsky.

Abbreviations

CNS central nervous system
CPG central pattern generator
DLR diencephalic locomotor region
L   left
MLR mesencephalic locomotor region
R   right
RS   reticulospinal

References

Alstermark, B., Isa, T., Ohki, Y. and Saito, Y. (1999). Disynaptic pyramidal excitation in forelimb motoneurons mediated via C(3)-C(4) propriospinal neurons in the Macaca fuscata. J. Neurophysiol., 82: 3580–3585.

Beloozerova, I.N. and Sirota, M.G. (1993). The role of the motor cortex in the control of accuracy of locomotor movements in the cat. J. Physiol. (Lond.), 461: 1–25.

Bernstein, N. (1947). On the Construction of Movements. Medgiz, Moscow (in Russian).

Bernstein, N. (1967). The Co-ordination and Regulation of Movements. Pergamon Press Ltd, London.

Bjursten, L.M., Norrsell, K. and Norrsell, U. (1976). Behavioural repertory of cats without cerebral cortex from infancy. Exp. Brain Res., 25: 115–130.

Brodal, A. (1981). Neurological Anatomy in Relation to Clinical Medicine, 3rd ed., Oxford University Press, New York, Oxford.

Darwin, C. (1872). The Expression of the Emotions in Man and Animals. The University of Chicago Press (1965), Chicago and London.

Di Prisco, G.V., Pearlstein, E., Le Ray, D., Robitaille, R. and Dubuc, R. (2000). A cellular mechanism for the transformation of a sensory input into a motor command. J. Neurosci., 20: 8169–8176.

Drew, T. (1993). Motor cortical activity during voluntary gait modifications in the cat. I. Cells related to the forelimbs. J. Neurophysiol., 70: 179–199.

Duysens, J. and Pearson, K.G. (1980). Inhibition of flexor burst generation by loading ankle extensor muscles in walking cats. Brain Res., 187: 321–332.

Eidelberg, E., Walden, J.G. and Nguyen, L.H. (1981). Locomotor control in macaque monkeys. Brain, 104: 647–663.

Ekman, P. (1973). Cross-cultural studies of facial expressions. In: Ekman P. (Ed.), Darwin and Facial Expressions. Academic Press, New York, pp. 169–222.

Fogassi, L., Gallese, V., di Pellegrino, G., Fadiga, L., Gentilucci, M., Luppino, G., Matelli, M., Pedotti, A. and Rizzolatti, G. (1992). Space coding by premotor cortex. Exp. Brain Res., 89: 686–690.

Georgopoulos, A.P. (1996a). On the translation of directional motor cortical commands to activation of muscles via spinal interneuronal systems. Brain Res. Cogn. Brain Res., 3: 151–155.

Georgopoulos, A.P. (1996b). Arm movements in monkeys: behavior and neurophysiology. J. Comp. Physiol., 179: 603–612.

Georgopoulos, A.P. (1999). News in motor cortical physiology. News Physiol. Sci., 14: 64–68.

Georgopoulos, A.P. and Grillner, S. (1989). Visuomotor coordination in reaching and locomotion. Science, 245: 1209–1210.

Gray, P.A., Janczewski, W.A., Mellen, N., McCrimmon, D.R. and Feldman, J.L. (2001). Normal breathing requires preBotzinger complex neurokinin-1 receptor-expressing neurons. Nat. Neurosci., 4: 927–930.

Grillner, S. (1981). Control of locomotion in bipeds, tetrapods and fish. In: Brooks V.B. (Ed.), Handbook of Physiology, Sec. 1. The Nervous System II. Motor Control. American Physiological Society, Bethesda, pp. 1179–1236.

Grillner, S. (1985). Neurobiological bases of rhythmic motor acts in vertebrates. Science, 228: 143–149.

Grillner, S. (2002). Fundamentals of motor systems. In: Squire L.R., Bloom F.E., McConnell S.K., Roberts J.L., Spitzer N.C. and Zigmond M.J., (Eds.), Fundamental Neuroscience. Academic Press, San Diego, pp. 753–766.

Grillner, S. and Rossignol, S. (1978). On the initiation of the swing phase of locomotion in chronic spinal cats. Brain Res., 146: 269–277.

Grillner, S. and Wallén, P. (1999). On the cellular bases of vertebrate locomotion. Prog. Brain Res., 123: 297–309.

Grillner, S., Georgopoulos, A.P. and Jordan, L. (1997). Selection and initiation of motor behavior. In: Stein P.S.G., Grillner S., Selverston A.I. and Stuart D.G., (Eds.), Neurons, Networks, and Motor Behavior. MIT Press, Cambridge, pp. 3–19.

Hess, W.R. (1949). Das zwischenhirn. Benno Schwabe, Basel.

Hikosaka, O., Takikawa, Y. and Kawagoe, R. (2000). Role of the basal ganglia in the control of purposive saccadic eye movements. Physiol. Rev., 80: 953–978.

Isa, T. (2002) Intrinsic processing in the mammalian superior colliculus. Curr. Opin. Neurobiol., 12: 668–677.

Isa, T. and Saito, Y. (2001). The direct visuo-motor pathway in mammalian superior colliculus; novel perspective on the interlaminar connection. Neurosci. Res., 41: 107–113.

Ito, M. (2001). Cerebellar long-term depression: characterization, signal transduction, and functional roles. Physiol. Rev., 81: 1143–1195.

Lawrence, D.G. and Kuypers, H.G. (1968a). The functional organization of the motor system in the monkey. I. The effects of bilateral pyramidal lesions. Brain, 91: 1–14.

Lawrence, D.G. and Kuypers, H.G. (1968b). The functional organization of the motor system in the monkey. II. The effects of lesions of the descending brain-stem pathways. Brain, 91: 15–36.

Liddell, E.G.T. and Phillips, C.G. (1944). Pyramidal section in the cat. Brain Behav. Evol., 67: 1–9.

Lund, J.P. (1991). Mastication and its control by the brain stem. Crit. Rev. Oral Biol. Med., 2: 33–64.

Maier, M.A., Armand, J., Kirkwood, P.A., Yang, H.W., Davis, J.N. and Lemon, R.N. (2002). Differences in the corticospinal projection from primary motor cortex and supplementary motor area to macaque upper limb motoneurons: an anatomical and electrophysiological study. Cereb. Cortex, 12: 281–296.

McClellan, A.D. and Grillner, S. (1984). Activation of 'fictive swimming' by electrical microstimulation of brainstem locomotor regions in an in vitro preparation of the lamprey central nervous system. Brain Res., 300: 357–361.

Norrsell, K. and Norrsell, U. (1970). Visually guided behaviour after total neonatal removal or the neocortex in the cat. Acta Physiol. Scand., 80: 14A–15A.

Orlovsky, G.N. and Shik, M.L. (1976). Control of locomotion: a neurophysiological analysis of the cat locomotor system. Int. Rev. Physiol., Neurophysiol., 10: 281–317.

Orlovsky, G.N., Deliagina, T.G. and Grillner, S. (1999). Neuronal Control of Locomotion. From Mollusc to Man. Oxford University Press, New York.

Prochazka, A., Clarac, F., Loeb, G.E., Rothwell, J.C. and Wolpaw, J.R. (2000). What do reflex and voluntary mean? Modern views on an ancient debate. Exp. Brain Res., 130: 417–432.

Prut, Y., Perlmutter, S.I. and Fetz, E.E. (2001). Distributed processing in the motor system: spinal cord perspective. Prog. Brain Res., 130: 267–278.

Raymond, J.L. and Lisberger, S.G. (1997). Multiple subclasses of purkinje cells in the primate floccular complex provide similar signals to guide learning in the vestibulo-ocular reflex. Learn. Mem., 3: 503–518.

Rizzolatti, G., Fogassi, L. and Gallese, V. (2001). Neurophysiological mechanisms underlying the understanding and imitation of action. Nat. Rev. Neurosci., 2: 661–670.

Rossignol, S. (1996). Neural Control of Stereotypic Limb Movements. Oxford University Press, New York, pp. 173–216.

Russell, J. and Bullock, M. (1985). Multidimensional scaling of emotional facial expressions: similarity from preschoolers to adults. J. Pers. Soc. Psychol., 48: 1290–1298.

Sparks, D.L. (1997). The role of population coding in the control of movement. In: Stein P.S.G., Grillner S., Selverston A.I. and Stuart D.G., (Eds.), Neurons, Networks, and Motor Behavior. MIT Press, Cambridge, pp. 21–32.

Steeves, J.D., Sholomenko, G.N. and Webster, D.M. (1987). Stimulation of the pontomedullary reticular formation initiates locomotion in decerebrate birds. Brain Res., 401: 205–212.

Stein, R.B., Misiaszek, J.E. and Pearson, K.G. (2000). Functional role of muscle reflexes for force generation in the decerebrate walking cat. J. Physiol. (Lond.), 525 Pt 3: 781–791.

von Euler, C. (1983). On the central pattern generator for the basic breathing rhythmicity. J. Appl. Physiol., 55: 1647–1659.

CHAPTER 2

Why and how are posture and movement coordinated?

Jean Massion[1,*], Alexei Alexandrov[2] and Alexander Frolov[2]

[1]*Laboratoire Parole et Langage, Université de Provence, 13621 Aix-en-Provence, France*
[2]*Institute of Higher Nervous Activity and Neurophysiology, Russian Academy of Science, Moscow 117865, Russia*

Abstract: In most motor acts, posture and movement must be coordinated in order to achieve the goal of the task. The focus of this chapter is on why and how this coordination takes place. First, the nature of posture is discussed. Two of its general functions are recognized; an antigravity role, and a role in interfacing the body with its environment such that perception and action can ensue. Next addressed is how posture is controlled centrally. Two models are presented and evaluated; a genetic and a hierarchical one. The latter has two levels; internal representation and execution. Finally, we consider how central control processes might achieve an effective coordination between posture and movement. Is a single central control process responsible for both movement and its associated posture? Alternatively, is there a dual coordinated control system: one for movement, and the other for posture? We provide evidence for the latter, in the form of a biomechanical analysis that features the use of eigenmovement approach.

Introduction

During the execution of voluntary movements, many aspects of the same task are coordinated and performed simultaneously. Gaze orientation (toward a target) results from the parallel control of eye and head movements (Bizzi et al., 1977). Coordination between posture and limb movement is another example seen in daily life. Often, two goals must be achieved at the same time. There is need for an accurate performance of a goal-directed movement on the one hand, and the maintenance of equilibrium and an appropriate posture, or set of postures, on the other (for review, Massion, 1992, 1994, 1998; Horak and Macpherson, 1996). Similarly, the elaboration of a locomotor gait requires coordination between a postural function (support of the body segments against gravity; forward acceleration of the center of mass, CM) and the rhythmic movements of locomotion (Mori, 1989; Mori et al., 1999). In order to understand why and how this coordination occurs, it is important to precisely define posture and explore how it differs from movement.

A simple, coarse-grain definition is that posture (also called 'attitude') is the body segments' configuration at any given time (Thomas, 1940). This definition is purely descriptive and it does not refer to the many functions subserved by posture. To this end, several approaches have been proposed in the literature. One refers to stance as a genetically determined reference posture, with characteristics which are species dependent. This view was championed in the classical works of Sherrington (1906), Magnus (1924) and Rademaker (1931). They all emphasized that the body's orientation with respect to gravity was determined by a group of reflexes that sets the body segments' collective orientation and stabilizes this orientation against external disturbances.

A second approach is related to the concept of support. Kuypers (1981) proposed it on the basis of anatomical and functional considerations. In this concept, axial and proximal body segments, and their corresponding musculature, serve as a support for the

*Corresponding author: Tel.: +33-4-4257-1228;
Fax: +33-4-4257-1228; E-mail: jean.massion@lpl.univ-aix.fr

distal musculature, such as the hands for reaching and grasping. This view was actually proposed previously by Hess (1943). He made a distinction between the 'eiresmatic' (supporting function) aspect of a motor act and the 'teleokinetic' (goal-directed movement) aspect. Hess proposed that missing the goal resulted from the lack of appropriate support for the movement (see Stuart et al., Chapter 1). More recently, Bouisset et al. (1992) proposed a modified version of the support concept, 'posturo-kinetic capacity'. It involves the capacity of the supporting segments to assist the ongoing movement in terms of their speed and forcefulness.

A third approach is to distinguish relatively sharply between posture and movement. Its proponents argue that the maintenance of a posture means that the position of one or several body segments are fixed with respect to other segments (also termed 'joint fixation') or with respect to space (i.e., the external environment). This position is stabilized against external disturbances. In contrast, 'movement' to these workers means a change of posture (or position), i.e., the initial posture is destabilized to achieve a new position, which is reached through a trajectory (see also S. Mori et al., Chapter 33). Feldman's (1966) equilibrium point hypothesis (for an introductory review: Latash, 1993) is in line with this view.

Posture has two primary functions

It is clear that the current definitions of posture are multiple, each resulting in different experimental and theoretical approaches. One way to identify a unifying model is to ask the question "what are the underlying postural functions common to these different views?". From this perspective, two primary functions emerge: *antigravity control*, which requires the buildup against gravity of the body segments' collective, overall configuration, and the *interface between perception and action* (i.e., controlling the relationship between the external world and the body).

Antigravity control

Two related processes accomplish antigravity control. First, the various segments of the kinematic chain from feet to head should be able to support the body's weight against gravity forces and resulting ground reaction forces. In addition, the kinematic chain should be able to provide a dynamic support to the moving segments when they are involved in a goal-directed action.

The second process is equilibrium control. It ensures that under static conditions, the center of pressure and the projection of the center of gravity (CG) should remain inside the support surface (i.e., the support's contact with the environment).

Interface between perception and action

Each body segment position can be defined with respect to the other body segments. An *egocentric reference frame* is used to calculate each body segment's position with respect to the axis of a given reference segment, such as the trunk or head. When a target is moving in external space, its position can be coded into a coordinate system external to the body (*allocentric reference frame*). The latter usually uses the vertical gravity axis as a reference (Berthoz, 1991; Paillard, 1991). In order to match the egocentric reference frame to the external world, the orientation of the trunk or head (the referent in the egocentric system) is calculated with respect to the vertical gravity axis (the referent for the external world). This calculation is based primarily on a set of sensors: the otoliths, and body graviceptors (in the soles of the feet and pelvis, possibly also muscle Golgi tendon organs), which monitor the gravity axis (Horak and Macpherson, 1996; Massion, 1998). The known orientation of the head and/or trunk with respect to the gravity axis can then be utilized both for the calculation of the other body segments' orientation with respect to the external world, and the external target's position with respect to the body (Assaiante and Amblard, 1993; Mouchnino et al., 1993; Pozzo et al., 1995). The reference is for the perception of body's movements with respect to the environment and the control of balance. Body segment orientation with respect to the vertical axis also serves as a reference frame for calculation of a target's position in space and the trajectories required to reach this target (Wise et al., 1991).

Synthesis of the two roles

Antigravity control and *interface between perception and action* are closely related to the execution of movement. For example, when raising a leg laterally, equilibrium is maintained by shifting the CG toward the supporting leg (Mouchnino et al., 1992). This is a form of antigravity control. During the same task, it can be shown that the orientation of body segments with respect to space, which serves as an interface between perception and action, is also preserved. In untrained subjects, the head axis remains vertical throughout the movement, whereas the trunk is inclined toward the supporting leg. In expert dancers, however, the trunk axis also remains vertical. Mouchnino et al. (1992, 1993) have shown that the trunk axis serves as a reference frame for the calculation of leg angle (egocentric reference frame). While the trunk axis remains vertical throughout the movements of an elite dancer, the calculation of leg angle relative to the trunk (egocentric coordinates) becomes equivalent to the calculation of leg angle relative to the vertical gravity axis. The latter is the reference axis for the external space coordinating system. Thus, for the expert dancer, keeping the trunk axis vertical throughout the task clearly simplifies the calculation of leg position in space.

Central organization and control of posture

Each species has its own genetic patrimony, which provides throughout fetal development, birth and ontogeny a central and reflex organization supporting basic behaviors, which are critical for survival. There is consensus that posture and locomotion are examples of such basic, genetically determined behaviors. There has been much controversy these past three decades, however, concerning the relative contribution of genetic versus learned networks to the central organization and control of posture (Prochazka et al., 2000; Grillner and Wallén, Chapter 1).

Genetic model of posture

Three main functions are identified in the genetic model of posture: the body segments' orientation with respect to gravity, their support against gravity and the adaptation of posture to ongoing movement. These three functions are closely related to the primary ones of *antigravity control* and *interface between perception and action.*

Body orientation with respect to gravity

In mammals, with a body which is segmented into head, trunk and leg sections, the otoliths and vision provide sensory input for controlling the head's orientation with respect to gravity. The investigations on righting reflexes described by Rademaker (1931) illustrate this claim. Note that a cat falling from an upside-down initial posture first reorients the head along the horizontal plane, then the body along the head plane, and finally, the legs adopt a vertical position that makes the animal ready for landing. There are also a series of reflexes aimed at stabilizing the segmental orientation with respect to gravity. For example, Vestibulo-collic reflexes stabilize head orientation in all manners of movements. Other reflexes help orient the foot with respect to the support (e.g., placing reactions) or the leg with respect to the gravity axis (hopping reactions).

Antigravity function

There is need to support the body segments against the contact forces exerted by gravity forces acting upon individual segments and the body as a whole. Postural muscle tone (i.e., the level of EMG activity in antigravity muscles) is a key means of meeting this need. It depends in large part on operation of the myotatic (stretch reflex) loop, predominantly in antigravity extensor muscles. Such muscle activity is exaggerated in decerebrate animals, which exhibit the well known, albeit mistermed, 'decerebrate rigidity' (i.e., it is really a form of spasticity). These postural reactions, which occur in the presence of an external perturbation of stance, are determined, in part, by both genetic endowment and learning throughout ontogeny (Forssberg, 1999).

Adaptation of the body's segments to ongoing movement

Genetic patrimony is also fundamentally concerned with the coordination between posture and movement. The distribution of tone throughout the body's musculature depends on the orientation of the head in space and in relation to the trunk, as achieved in part by labyrinthine and neck reflexes. Note also the role of lumbar reflexes, which are concerned with the orientation of the trunk with respect to the pelvis (Tokizane et al., 1951). These latter reflexes adapt tone in muscles supplying the upper and lower segments to the ongoing movement.

In summary, the genetic model of posture proposes that its purpose is to support the body against gravity, preserve balance, orient the body with respect to gravity and also adapt the body's segments to the ongoing movement.

Limitations of the genetic model

There are several challenges to the idea that the genetic model of posture can be the sole basis for the central organization and control of posture. Two key issues concern the flexibility of postural reactions, and anticipatory postural adjustments, of which the latter ones are involved in the coordination between posture and movement (see also F. Mori et al., S. Mori et al., and Takakusaki et al., Chapters 19, 33 and 23 of this volume).

Flexibility of postural reactions

Postural reactions induced by external disturbances are now known to be remarkably flexible (for review, Macpherson, 1991; Horak and Macpherson, 1996). For example, when a balance disturbance occurs while standing, the leg muscles are the main ones involved in the correction. If the subject simultaneously grasps a support with the hands, however, the postural reaction will mainly involve the arm muscles (Nashner and McCollum, 1985). Thus, the postural reactions to the same stimulus depend, in terms of their spatial distribution, on external constraints, such as the conditions of support. This marked flexibility is, however, a feature of proprioceptive reflex organization (Forssberg et al., 1975; Stuart, 2002). Segmental reflexes are both task and context dependent, and under a higher level of control for selecting appropriate reflex actions (gate and gain control).

Anticipatory postural adjustments

Analysis of the coordination between posture and movement argues against a purely genetically based reflex organization of posture. During the performance of most voluntary movements, the primary functions of posture must be preserved in order to accomplish the goal in an accurate fashion. Preserving equilibrium is a necessary condition for an accurate performance and preserving the orientation of the body's segments with respect to gravity are essential because they provide a reference value for planning the trajectory of the movement toward its goal. In addition, the supporting function of the kinematic chain from the ground to the moving body segments has to be preserved during the dynamic conditions of movement execution.

The coordination between posture and movement must accommodate a major problem. The performance of a movement is a source of disturbance of posture for two reasons. First, it changes the body's geometry and, as a consequence, the CG's position with respect to the support surface (often the ground). Second, the movement is initiated by internal muscle forces, which are associated with reaction forces acting on the supporting segments and the ground. The resulting dynamic interactions between body segments must be accommodated. Otherwise, there would be a deviation from the planned trajectory and misreaching would occur. This is prevented by anticipatory postural adjustments. These were first described by Belenkiy et al. (1967) for an arm-raising task and they have since been observed in a wide variety of voluntary movements (for review, Massion, 1992; Horak and Macpherson, 1996; see also Bouisset and Zattara, 1987). Figure 1 models how anticipatory adjustments correct in advance for perturbations of posture and equilibrium that are associated with the execution of movement.

Anticipation involves prediction of a forthcoming perturbation. It implies that due to the effect of repeated experience and learning on adaptive central neural networks, memorized models are developed in the central nervous system (CNS) and used during the performance of a task. Such models accommodate the external world, the body's biomechanical characteristics and their interactions. This idea of a memorized representation (or model) was first proposed for the

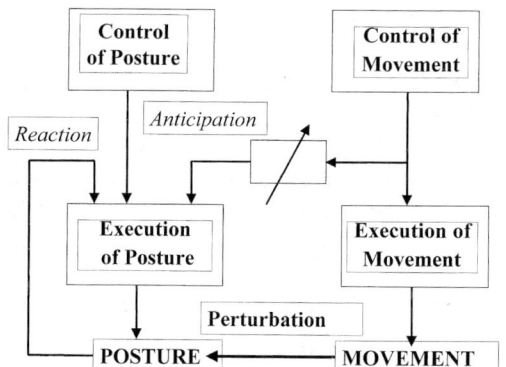

Fig. 1. Schematic representation of the integrated control of posture and movement. Note the need for a postural control system that integrates interactions between postural reactions and anticipatory adjustments. The execution of a movement provokes a perturbation of posture (internal perturbation), which is compensated for by a postural reaction. This occurs with a delay, however. It is comprised of the time for the perturbation to be detected plus that for transmission through the corrective pathway(s). In contrast, anticipatory postural adjustments correct in advance for the forthcoming perturbation. Adaptive neural networks, whose development requires learning, control this anticipatory process.

whole somatic system by Bernstein on the basis of his analysis of motor learning (for the evolution of his ideas, see his posthumous 1967 volume). The idea was further extended into the postural domain by Clément et al. (1984), Gurfinkel et al. (1988) and Gurfinkel and Levik (1991). It has also appeared in the area of oculomotor control, where it was initially termed 'internal models' (see Robinson, 1986).

The genetic model of posture cannot account for the important role of anticipation in helping achieve an appropriate coordination between posture and movement. The key reason is that such anticipation is based on learning.

It thus appears that in addition to the genetic model of posture, which nonetheless remains valid for many basic postural controls, a higher level of postural organization is present. It makes use of experience and learning, and provides a finely tuned adaptation of postural functions to the movement needs of daily life.

Hierarchical model of posture

The flexibility of postural reactions, and anticipatory postural adjustments associated with voluntary movements, are much better accounted for by a hierarchical model of posture. This approach is derived from an analogous model of movement organization proposed initially by Bernstein. (The evolution of this idea, too, is presented in Bernstein's 1967 volume.) It includes both inborn genetically endowed reactions (Gurfinkel and Shik, 1973) and those built up by learning (Clément et al., 1984; Gurfinkel et al., 1988; Gurfinkel and Levik, 1991). This model proposes that two types of CNS processing are required for the control of posture. A higher-order one accomplishes an internal representation of body posture (the so called 'body schema'). In parallel, lower-order control brings about the kinematics and force (kinetics), which are required for implementing postural functions as a necessary aspect of the task. In reality, this proposal is not basically different from the body schema model proposed by Head (1920) on neurological grounds, or the concept of internal representations (or models) required for the organization of movements (Wise et al., 1991; Wolpert et al., 1995, 1998). For example, in an arm target-reaching task, one may identify a set of control processes, which depend on representational mechanisms, such as the coding of target position in head and trunk coordinates, the coding of hand position in trunk coordinates and planning the desired trajectory. All three processes require internal models. For the execution of an arm-reach trajectory, both static and dynamic models, and direct and inverse models, are used for implementing the command in terms of muscle force (Wise et al., 1991; Wolpert et al., 1998).

Higher-order processing of body representation

Higher-order processing has been shown in experiments that made use of artificial or biased sensory inputs. These have included a visual-moving scene (Dichgans et al., 1972; Lestienne et al., 1977), galvanic stimulation of the labyrinth (Lund and Broberg, 1983; Gurfinkel et al., 1988; Hlavacka et al., 1995) and tendon vibration (the latter stimulus activates largely muscle sensory Ia axons; Roll and Roll, 1988; Roll et al., 1993, 1998). These studies showed that the internal CNS representation of the body's posture (or body schema) includes at least three components. First is body kinematics, i.e., the kinematic chain from feet to head. These make use of muscle proprioceptive Ia afferent input and/or an efferent copy of the motor command. The latter mainly concerns the position of the eyes in their respective orbits (Roll et al., 1998). Second are the body segments' mass and inertia, which are derived from sensory inputs not yet clearly identified. Finally, there are the positions of the body's individual segments and their weight, both with respect to the external world. The vertical axis related to the gravity vector is assessed by sensory inputs from multiple sources, including vestibular graviceptors, body graviceptors (pelvis—Mittelstaedt, 1998; plantar sole—Kavounoudias et al., 1998, 1999) and Golgi tendon organ receptors in extensor muscles (Dietz et al., 1989, 1992). In addition, visual input provides information about the vertical axis (Dichgans et al., 1972; Lestienne et al., 1977; Dijkstra et al., 1994). Finally, haptic (tactile) contacts of the hands on external supports, like a wall or a cane in contact with the ground, serve to adjust the body's posture relative to the external support (Jeka and Lackner, 1995).

According to Mergner and Rosemeier (1998), the representation of the body segments' positions with respect to space is accomplished in two ways. A top-down one uses the head as a reference frame with respect to space. The position of the various body segments can then be calculated with respect to the head. A bottom-up representation uses the basis of support as a reference frame. This serves mainly to calculate the CG position with respect to the ground.

Evidence for this internal representation of the proprioceptive kinematic chain is illustrated by experiments which noted the effects of vibration of the biceps' tendon when a human subject was holding the nose or skull with a hand (recall that biceps vibration stimulates Ia biceps afferents, thereby mimicking his muscle's stretch during arm extension). Surprisingly, the resulting illusory perception is an elongation of either a finger or the skull (Lackner, 1988). This finding shows that Ia afferent input is processed by the CNS in such manner as to provide the most likely interpretation according to the subject's previous experience.

Lower-order implementation of posture

The higher-order component of the hierarchical model of posture is rather stable even when the constraints are drastically changed. This occurs, for example, when the subject is subjected to microgravity conditions (Gurfinkel et al., 1993a,b; Massion et al., 1998). In contrast, the lower-order control of implementation (kinematics and kinetics) is quite flexible and adaptable to the constraints.

Experiments performed in microgravity have illustrated this initially surprising claim (Vernazza-Martin et al., 2000; Baroni et al., 2001). For example, the performance of trunk bending in normogravity is characterized by opposite movements of upper and lower body segments. These minimize a shift of the CM's horizontal position with respect to the feet, and thereby preserve equilibrium during the movement. In a short-term exposure to microgravity, such as a parabolic flight, the kinematics of axial movements during trunk bending is preserved just as well as in normogravity. Similarly, the close covariation between ankle, knee and hip joint movements is retained (Vernazza-Martin et al., 2000). Also, the CM's displacement during movement is minimized in microgravity to the same extent as in normogravity (Massion et al., 1997; Baroni et al., 2001). All these findings suggest that in normogravity, a movement is planned kinematically on the basis of the internal representation of body kinematics (higher-order processing). This representation is well preserved in microgravity. In contrast, the EMG patterns of the limbs' musculature, which correspond to the implementation of kinematic synergies, are markedly changed in microgravity. This shows that muscle forces adapt to new constraints in order to maintain an invariant kinematic synergy.

Summary on models of posture

Both the genetic and hierarchical models of posture make use of several identical mechanisms for achieving postural control. These include: (1) joint stiffness, which restricts the body's deviation from the desired position while standing; (2) postural reactions, synergies and strategies, which reduce the effect of postural and balance disturbances; and

(3) anticipatory postural adjustments, which are associated with movement performance in order to preserve the body segments' orientation and equilibrium, and provide appropriate postural support for an accurate movement performance. Despite these similarities, the genetic model alone cannot explain the flexibility of postural reactions and anticipatory postural adjustments. The hierarchical model of posture, with its higher-order processing of the body schema and lower-order processing of implementation, is far more appropriate for accommodating the experimental findings in the domain of posture and posturo-kinetic coordination. It is not exclusive of the genetic model, however, because it uses many genetically prewired circuits for accomplishing postural tasks.

Integrated CNS control of posture and movement

In daily life, postural control is most intimately associated with the execution of movements (reaching and grasping tasks, locomotion, etc.). As such, it is one key aspect of the overall control of motor tasks, which usually involve *both* postures and movements.

A major difficulty in postural control, especially in humans, concerns the 'multijoint kinematic chain' from the feet to the head. It includes body segments with quite different mass and inertia, which are linked by muscles with their own idiosyncratic viscoelastic characteristics. Their totality (for the whole body) is responsible for the production of force and kinematics. For example, Eng et al. (1992) showed that for an arm-reaching task, each single joint movement involves dynamic interactions with the other segments of the kinematic chain, and this becomes a source of postural disturbance. In the human, the relatively high level of the CG's position, and the narrow support surface for the feet, are other sources of instability, which further confound postural control during the execution of movement.

It bears emphasis that the multijoint chain from the feet to the head is not specific to posture. Movement of segments of this chain accompanies all manner of bodily movements during daily life, e.g., trunk bending forwards and backwards. As such, the control of axial movement involves issues not basically different from those for other multijoint movements, such as arm-reaching ones. The unique feature of posture in the execution of any movement involves the need to control balance and/or the body's segmental orientation during the motor act. Thus, two primary problems of postural control in a given movement task require resolution. The first is how to simplify control by reducing the number of degrees of freedom (Bernstein, 1967). The second problem is how to counteract the disturbing effects of movement performance on posture in terms of body segment orientation and equilibrium. It is interesting to see how far the various hypotheses proposed for movement control can accommodate the above two requirements. Three such concepts are presented and evaluated below, using the guidelines that (1) internal representations of body kinematics and dynamics provide a basis for planning movement trajectory (internal 'direct' model), and (2) during the execution of movement, dynamic interactions between segments perturb the movement's trajectory and the body's equilibrium.

'Referent posture'

An early 20th century concept in motor control was that EMG patterns were sufficient to reveal the characteristics of a CNS-operated central pattern generator (Brown, 1911; see also Pearson's present Chapter 12). This concept was first put into question by Wachholder (1928). He introduced the idea that the neural pattern also depended on the biomechanical characteristics of the segment to be moved. These include inertia, weight and the dynamic characteristics of the task, such as velocity. Similarly, Weiss (1961) insisted on the fact that the EMG pattern of individual muscles in a given goal-oriented task varied from trial to trial without changes in the characteristics of the overall performance. He introduced the concept that the performance resulted from the action of *the whole system* involved in the task (for review, Paillard, 1960; Wiesendanger, 1997; Sternad and Corcos, 2001). The lamba model of motor control was proposed by Feldman (for its history, see Feldman and Levin, 1995). This idea is in keeping with Bernstein's (1967) views on the need for a systemic approach. Feldman has argued that EMG patterns are not directly programmed by a central controller, rather, they emerge from the interaction between the central command and pattern-generating signals, reflex signals and the biomechanical system's components. Furthermore, central control must also accommodate the biomechanical system's interactions with the environment (see also Hasan and Enoka, 1985; Latash, 1993).

The concept of a 'referent posture' (Feldman et al., 1999) is related to Feldman's lamba model. (The equilibrium point hypothesis or alpha model, which is based on other premises, also proposes that the final posture is programmed by the CNS; see Bizzi et al., 1992.) Feldman's referent posture is based on the idea that the CNS controls the joints' positions by adjusting the spring-like properties of the muscles. These are determined, in part, by the muscles' 'threshold' for their own myotatic reflex. This spring-like adjustment results in an equilibrium position of the joints being reached, thereby defining a final position of the limb. A movement results from a change in the equilibrium position of the joints, which also produces a new referent postural configuration. The actual body configuration differs from that of the referent one, however, due to the action of external forces, such as gravity.

Feldman's concept is attractive due to its relative simplicity. A single control process results in achieving both the movement trajectory and preserving postural functions. There is thus a reduction of the controlled degrees of freedom. The main controversy about this concept concerns the control of dynamic interactions between the body's segments, which disturb movement trajectory, posture and equilibrium. For example, is muscle stiffness along the kinematic chain able on its own to efficiently minimize the disturbing effect on the joints' movements on the dynamic interactions between body segments? This question is central to evaluation of Feldman's ideas (Gomi and Kawato, 1996, 1997).

Direct and inverse dynamics

Direct and inverse dynamic models of the control of posture and movement are based on the assumption that the CNS has used learning to create an internal model of the dynamics of the motor system. The direct dynamic model is used to assess on-line the dynamic state of the body's biomechanical system. The inverse dynamic model represents the central command which takes into account all of the dynamic properties of the system required to perform a given task. These concepts have been used by several authors (Kawato et al., 1987; Atkeson, 1989; Miall et al., 1993; Wolpert et al., 1995).

Gomi and Kawato (1996, 1997) claim that Feldman's equilibrium point model cannot account for dynamical interactions between segments. Thus, it fails to describe the accurate performance of a goal-directed task. They have opined that a direct dynamic model and/or an inverse dynamic control is more likely, i.e., one which emphasizes that the CNS has a control system for dynamic interactions between body segments during the accurate performance of a task. In keeping with Ito (1984), they have proposed that the cerebellum plays a critical role in building and storing the direct and/or inverse models appropriate for each motor task (cf. Wolpert et al., 1998).

In summary, direct and inverse dynamic models account for both balance control and movement performance by the use of a *single* central control system.

Concept of dual, coordinated control

The above two concepts feature motor actions being controlled as a whole, i.e., no specific separate control processes for posture and movement. The third concept in vogue is that two independent, yet coordinated, control processes are used for the execution of motor acts. One is for the prime (focal) movement, and the other for the associated postural movement (in terms of equilibrium preservation and appropriate orientations of the body segments). The suggestion that any movement comprises these two components was first claimed in the works of Hess (1943) and Belenkiy et al. (1967). More recently, Cordo and Nashner (1982) proposed that the same repertoire of synergies could be utilized for both reactive and anticipatory postural adjustments associated with a voluntary movement. They demonstrated that the postural component often started earlier than the prime one in order to minimize the perturbation of equilibrium caused by the expected prime movement component (Cordo and Nashner, 1982). Bouisset and Zattara (1987) have argued that these anticipatory postural adjustments are preprogrammed and integral parts of the central motor program for the movement in question.

This concept of a dual-coordinated control for posture and movement has been questioned in several studies. Some of them have involved an analysis of trunk bending. In this task, the movement involves motion of the upper trunk. Such motion imposes a large dynamic perturbation between body segments along the kinematic chain from the ground to the head. The task requires a precise coordination between movement and posture. The size of the base of support is limited by the relatively small length of the feet, and the CM is located at a relatively large distance above the support. On the other hand, decomposition of trunk bending into its prime movement and postural components is not so evident as in the case of an arm-reaching task. Trunk bending is accompanied by opposite movements of the lower segments (Babinski, 1899; Oddsson and Thorstensson, 1986; Crenna et al., 1987; Pedotti et al., 1989). The identification of a postural versus movement component cannot be made on the basis of movement kinematics. The motions of the various joints along the kinematic chain (hip, knee, ankle) are too highly synchronized. This suggests that a single central control system is responsible for all of the kinematics (Alexandrov et al., 1998). As a result, a different approach becomes necessary, like the one presented in the next section.

The eigenmovement approach

The main difficulty in the control of a multijoint mechanical system, such as the human body, is the dynamical interaction between its different segments. The movement of any one segment influences the movement of all the other segments of the kinematic chain. As a result, the control of a given single segment cannot be performed without simultaneous control of the neighboring ones. For such multijoint systems, however, a set of dynamically independent multijoint movements (multijoint motion units) can be described by using the eigenmovement approach.

The eigenmovements represent movements along the eigenvectors of the motion equation. Each eigenvector defines a set of ratios between the relevant joint angles and joint torques. The number of eigenvectors is equal to the number of degrees of freedom in the multijoint system. For example, for the three joints primarily involved in a trunk-bending task (hip, knee, ankle), there are three degrees of freedom. Three eigenvectors define for each joint a given set of ratios between the three joints' angles and their corresponding torques. An *eigenmovement is a unique movement in which the ratios between the changes in joint angle are fixed and, simultaneously, the changes in joint torque are fixed.* Since in each eigenmovement, all joint angular changes are synchronized, all angular changes in each eigenmovement can be described by the time course of a single eigenmovement's 'kinematic amplitude'. Similarly, all joint torques changes can be described by the time course of a single eigenmovement's 'dynamic amplitude'. The relationship between kinematic and dynamic amplitudes of an individual eigenmovement is described by the simple motion equation of an inverted pendulum with its individual inertia. Any given multijoint movement can be represented as a superimposition of the eigenmovements of the relevant joints. Due to the eigenmovements' dynamical independence from each other, and due to the simplicity of their motion equations, the eigenmovement approach is a standard and useful method for the analysis of the dynamics of *any* multijoint system, be it living or inanimate. Recently, we applied this method to the investigation of human trunk bending in the sagittal plane (Alexandrov et al., 2001a,b). For this, a three-joint (hip, knee, ankle) biomechanical model of the human body was used.

The fixed ratios between joint angles and their corresponding joint torques, and the inertia in each eigenmovement, are determined by the structure of the multijoint chain (from feet to head in the present case) and by the inertia of its segments. Figure 2 represents the ratios between joint angles (stick diagrams) and joint torques (sizes of circles) for each eigenmovement of our three-joint human body model, using standard anthropometric parameters (Winter, 1990).

In Fig. 2, the eigenmovements are termed according to their predominant joint. For our

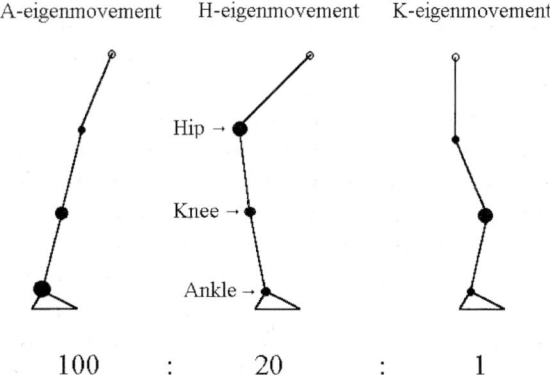

Fig. 2. Application of the eigenmovement approach to a human trunk-bending task. Kinematic (stick diagrams) and kinetic (circles at joints) patterns of ankle (A-), knee (K-), and hip (H-) eigenmovements during voluntary trunk bending. The term, 'eigenmovement' is defined in the text. In each eigenmovement, the sticks' inclinations and the sizes of circles correspond to the ratios between the changes in joint angles and joint torques, respectively. The values at the bottom correspond to the normalized ratios between the three eigenmovement's inertias (i.e., relative to a value of 1 for the K-eigenmovement).

trunk-bending task, ankle (A-) eigenmovement involved ankle dorsiflexion combined with knee extension and hip flexion. Hip (H-) eigenmovement consisted of hip flexion combined with knee and ankle extension. Finally, knee (K-) eigenmovement was composed of knee and hip flexion, and ankle dorsiflexion. The inertia of the A-eigenmovement was five times greater than that of the H-eigenmovement. In turn, the inertia of the H-eigenmovement was 20 times greater than that of the K-eigenmovement. In our three-joint model, any arbitrary movement in the sagittal plane could be represented, in terms of its kinematics and kinetics, as a superimposition of these three eigenmovements.

We found previously that the K-eigenmovement (again, knee and hip flexion with ankle dorsiflexion) during human forward upper trunk bending was negligibly small (Alexandrov et al., 2001b). Therefore, we could decompose the task into only two (H- and A-) of our three eigenmovements. We found that the H-eigenmovement (hip flexion with knee and ankle extension) was responsible for the main part of the movement kinematics (i.e., it described more than 95% of the total angular variance at the hip, knee and ankle). Due to the ankle extension accompanying hip flexion, the H-eigenmovement was the main reason for a backward shift of the CM (to about 4.5 cm), which perturbed equilibrium. The A-eigenmovement (ankle dorsiflexion with knee extension and hip flexion) was responsible for only 5% of total angular variance at the hip, knee and ankle. Its contribution to the total CM displacement was comparable to that of the H-eigenmovement, however. The changes in ankle angle produced a forward CM shift of about 5 cm, which compensated for the backward CM shift caused by the H-eigenmovement. Therefore, the A-eigenmovement was concerned mainly with equilibrium control during the task.

The above analysis suggests that in a trunk-bending task, the CNS uses a dual-control process; one for bending, per se (the H-eigenmovement), and the other for equilibrium maintenance (the A-eigenmovement). The arguments in favor of the idea that the A- and H-eigenmovements are controlled in the task as entire (involving all three joints) motion units were developed previously (Alexandrov et al., 2001a,b).

An important aspect of the control of H- and A-eigenmovements during trunk bending is their relative timing. Due to the large inertia of the A-eigenmovement (five times greater than that of H-eigenmovement), it must start earlier and last longer than the H-eigenmovement. Our previous modeling experiments on forward bending on an extremely narrow support (Alexandrov et al., 2001a) showed that when both eigenmovements started simultaneously, the A-eigenmovement, which produces forward acceleration of the CM, could not compensate for the backward CM movement associated with the H-eigenmovement. As a result, the body experienced a fall backward.

Fig. 3 shows that during our present trunk-bending task, the A-eigenmovement actually started 70 ± 20 ms earlier and lasted longer (for 820 ± 200 ms) than did the H-eigenmovement (510 ± 80 ms). The initial forward CM acceleration in the A-eigenmovement was presumably produced by activation of ankle dorsiflexors. This interpretation clarifies the functional meaning of the early EMG (activation) pattern of tibialis anterior, which is usually observed in trunk-bending tasks (Oddsson and Thorstensson, 1986; Crenna et al., 1987; Pedotti et al., 1989).

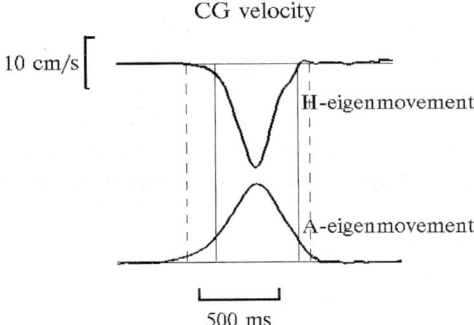

Fig. 3. Calculated changes in the CG's velocity for A- and H-eigenmovements during voluntary forward trunk bending by a human subject. Note that a positive (upward) velocity corresponds to forward CG displacement, and vice versa. The thin vertical lines show the onset and offset of the CG velocity due to the H-eigenmovement. The vertical broken lines show the onset and offset of the CG velocity due to the A-eigenmovement (threshold effect at 5% of the maximal velocity). Note that the onset of the CG velocity changes due to the A-eigenmovement started earlier and ended later than did that due to the H-eigenmovement. Note also that the velocity changes because these two eigenmovements are in an opposite direction. Abbreviation: CG, center of gravity. See text for further information.

The results of our eigenmovement approach suggest that the CNS needs an internal model of the inertia of the mentioned multijoint motion units. It can use this model to adapt the timing of its neural control mechanisms to the task in question, i.e., movement performance (trunk bending in our experiments) and equilibrium maintenance.

Summary on eigenmovements

We believe that the eigenmovements identified by our analysis correspond not only to purely mathematical formalisms, but also to *real components* of the CNS's control system. First, in each eigenmovement, joint angles and joint torques change synchronously in all three (hip, knee, ankle) joints. In our opinion, this is the simplest scaling procedure to be controlled by the CNS. Second, the two goal components of the task (i.e., movement and appropriate postures) can be explained by each one having its own single eigenmovement; the H-eigenmovement for the prime movement of bending per se, and the A-eigenmovement for postural control by forward acceleration of the CM and minimizing its perturbation by the prime mover (i.e., the H-eigenmovement). Third, recall the experiments of Horak and Nashner (1986), in which they studied postural reactions to an external disturbance of the erect posture. They proposed that two strategies, *hip* and *ankle* ones, are the main mechanisms for restoring balance control in response to sudden perturbations. They showed that an *ankle strategy* (rotation of the whole body around the ankle joint) was of much higher inertia than a *hip strategy* (rotation of the trunk around the hip). They argued that the two strategies were task dependent, with the *ankle strategy* used to compensate for small and slow perturbations, whereas the *hip strategy* accommodated large and fast ones.

The ankle and hip strategies of Horak and Nashner (1986) can be identified in our A- and H-eigenmovements. For example, in the A-eigenmovement (ankle dorsiflexion with knee extension and hip flexion), the ankle dominates, just as it does in their *ankle strategy*. Similarly, the kinematic pattern of H-eigenmovement (hip flexion with knee and ankle extension) is close to that of their *hip strategy* (see also Kuo and Zajac, 1993). Note further that the high inertial A-eigenmovement (again like the *ankle strategy* in the experiments of Horak and Nashner) is efficient in slow CM displacements, whereas the five times less inertial H-eigenmovement (as in their *hip strategy*) is efficient when the perturbation is fast.

In summary, our biomechanical analysis of human trunk bending suggests that hip and ankle strategies (best quantified as H- and A-eigenmovements) represent elementary multijoint motion units of the biomechanical system. They can be independently controlled by the CNS to achieve different behavioral goals within the same task (trunk bending using H-eigenmovement control, and equilibrium maintenance using A-eigenmovement control). They act conjointly to prevent a potentially forthcoming disturbance of posture and equilibrium provoked by the movement performance. The same motion units are also presumed by us to operate either separately or in conjunction with other tasks, e.g., when initiating gait or rising on tiptoe (Crenna and Frigo, 1991); or, during low or high amplitude body oscillations during fixating on an oscillating visual target (Bardy et al., 1999).

How these two motion units are controlled by the CNS, including their relative timing, is still an open question. Is an internal model of the biomechanical characteristics (mass, inertia) of each of these multijoint motion units stored at the cerebellar level? This would be in keeping with the direct or inverse dynamic model of Gomi and Kawato (1996), Gomi and Kawato (1997) and Wolpert et al. (1998). Is the control of these multijoint motion units compatible with the equilibrium point hypothesis and, more specifically, with the concept of referent posture as proposed by Feldman and Levin (1995) and Feldman et al. (1999)? The answering of these questions remains open in the absence of special analytical techniques, like those of Domen et al. (1999), or a model of the neuromuscular apparatus similar to those developed for the human arm (Flanagan et al., 1993; Gribble et al., 1998; Frolov et al., 2000).

Summary and conclusions

The coordination between posture and movement results from two parallel controls. They operate on multijoint motion units (eigenmovements) of the overall biomechanical system responsible for both the movement and its associated postural adjustment(s). An example is ankle and hip eigenmovements during voluntary forward bending of the trunk. In this case, a critical aspect for the coordination of the two eigenmovements is their timing, which is a function of their respective inertias. It is suggested that this type of control might be a general rule for co-ordinating posture and movement. It is likely that the cerebellum plays a critical role in this coordination, particularly in terms of the timing of eigenmovements. Recall, for example, that it has long been known that the coordination between movement and posture is impaired in cerebellar patients (Babinski, 1899).

Acknowledgments

We thank Dr. Becky Farley, The University of Arizona, for her very effective help in reviewing an early draft of this chapter. The work was supported by the Russian Foundation For Basic Research (projects 02-04-48410a and 01-04-48924a) and the Russian Foundation of Humanity (project 00-06-00242a).

Abbreviations

A	ankle
CM	center of mass
CNS	central nervous system
CG	center of gravity
H	hip
K	knee

References

Alexandrov, A.V., Frolov, A.A. and Massion, J. (1998). Axial synergies during human upper trunk bending. Exp. Brain Res., 118: 210–220.

Alexandrov, A.V., Frolov, A.A. and Massion, J. (2001a). Biomechanical analysis of movement strategies in human forward bending. I. Modeling. Biol. Cybern., 84: 425–434.

Alexandrov, A.V., Frolov, A.A. and Massion, J. (2001b). Biomechanical analysis of movement strategies in human forward bending. II. Experimental study. Biol. Cybern., 84: 435–443.

Assaiante, C. and Amblard, B. (1993). Ontogenesis of head stabilization in space during locomotion in children: influence of visual cues. Exp. Brain Res., 93: 499–515.

Atkeson, C.G. (1989). Learning arm kinematics and dynamics. Annu. Rev. Neurosci., 12: 157–183.

Babinski, J. (1899). De l'asynergie cérébelleuse. Rev. Neurol., 7: 806–816.

Bardy, B.G., Marin, L., Stoffregen, T.A. and Bootsma, R.J. (1999). Postural coordination modes considered as emergent phenomena. J. Exp. Psychol. Hum. Percept. Perform., 25: 1284–1301.

Baroni, G., Pedrocchi, A., Ferrigno, G., Massion, J. and Pedotti, A. (2001). Static and dynamic postural control in long-term microgravity: evidence of dual adaptation. Am. J. Physiol., 90: 205–215.

Belenkiy, V.E., Gurfinkel, V.S. and Paltsev, E.I. (1967). On elements of control of voluntary movements. Biofizica, 12: 135–141.

Bernstein, N.A. (1967). The Coordination and Regulation of Movements. Pergamon Press, New York.

Berthoz, A. (1991). Reference frames for the perception and control of movement. In: Paillard J. (Ed.), Brain and Space. Oxford University Press, Oxford, pp. 81–111.

Bizzi, E., Kalil, R.E. and Tagliasco, V. (1977). Eye-head co-ordination in monkeys: evidence for centrally patterned organization. Science, 1: 452–454.

Bizzi, E., Hogan, N., Mussa-Ivaldi, F.A. and Giszter, S. (1992). Does the nervous system use equlibrium-point control to guide single and multiple joint movements?. Behav. Brain Sci., 15: 603–613.

Bouisset, S. and Zattara, M. (1987). Biomechanical study of the programming of anticipatory postural adjustments associated with voluntary movement. J. Biomech., 20: 735–742.

Bouisset, S., Do, M.C. and Zattara, M. (1992). Posturo-kinetic capacity assessed in paraplegics and parkinsonians. In: Woollacott M. and Horak F. (Eds.), Posture and Gait: Control Mechanisms, Vol. II, University of Oregon Press, Eugene, pp. 19–22.

Brown, T.G. (1911). The intrinsic factors in the act of progression in the mammal. Proc. R. Soc. Lond. B. Biol. Sci., 84: 308–319.

Clément, G., Gurfinkel, V.S., Lestienne, F., Lipshits, M.I. and Popov, K.E. (1984). Adaptation of postural control to weightlessness. Exp. Brain Res., 57: 61–72.

Cordo, P.J. and Nashner, L.M. (1982). Properties of postural adjustments associated with rapid arm movements. J. Neurophysiol., 47: 287–302.

Crenna, P. and Frigo, C. (1991). A motor program for the initiation of forward oriented movement in man. J. Physiol. (Lond.), 437: 635–653.

Crenna, P., Frigo, C., Massion, J. and Pedotti, A. (1987). Forward and backward axial synergies in man. Exp. Brain Res., 65: 538–548.

Dichgans, J., Held, R., Young, L. and Brandt, T. (1972). Moving visual scenes influence the apparent direction of gravity. Science, 178: 1217–1219.

Dietz, V., Horstmann, G.A., Trippel, M. and Gollhofer, A. (1989). Human postural reflexes and gravity—an under water simulation. Neurosci. Lett., 106: 350–355.

Dietz, V., Gollhofer, A., Kleiber, M. and Trippel, M. (1992). Regulation of bipedal stance: dependency on 'load' receptors. Exp. Brain Res., 89: 229–231.

Dijkstra, T.M.H., Schöner, G. and Gielen, C.C.A.M. (1994). Temporal stability of the action-perception cycle for postural control in a moving visual environment. Exp. Brain Res., 97: 477–486.

Domen, K., Latash, M.L. and Zatsiorsky, V.M. (1999). Reconstruction of equilibrium trajectories during whole-body movements. Biol. Cybern., 80: 195–204.

Eng, J.J., Winter, D.A., Colum, D., MacKinnon, D. and Patla, A.E. (1992). Interaction of the reactive moments and center of mass displacement for postural control during voluntary arm movements. Neurosci. Res. Commun., 11: 73–80.

Feldman, A.G. (1966). Functional tuning of the nervous system with control of movement or maintenance of a steady posture. II. Controllable paramets of the muscle. Biophysics, 11: 565–578.

Feldman, A.G. and Levin, M.F. (1995). The origin and use of positional frames of reference in motor control. Behav. Brain Sci., 18: 723–806.

Feldman, A.G., Archambault, P. and Lestienne, F.G. (1999). Multimuscle control is based on the specification of referent body image. In: Gantchev G.S., Mori S. and Massion J. (Eds.), Motor Control Today and Tomorrow. Academic Publishing House 'Prof. Marin Drinov', Sofia, pp. 163–179.

Flanagan, A.G., Ostry, D.J. and Feldman, A.G. (1993). Control of trajectory modification in target-directed reaching. J. Mot. Behav., 25: 140–152.

Forssberg, H. (1999). Neural control of human development. Curr. Opin. Neurobiol., 9: 676–682.

Forssberg, H., Grillner, S. and Rossignol, S. (1975). Phase dependant reflex reversal during walking in chronic spinal cats. Brain Res., 85: 103–107.

Frolov, A.A., Dufosse, M., Rizek, S. and Kaladjian, A. (2000). On the possibility of linear modelling the human arm neuromuscular apparatus. Biol. Cybern., 82: 499–515.

Gomi, H. and Kawato, M. (1996). Equilibrium-point control hypothesis examined by measured arm stiffness during multijoint movement. Science, 272: 117–120.

Gomi, H. and Kawato, M. (1997). Human arm stiffness and equilibrium-point trajectory during multi-joint movement. Biol. Cybern., 76: 163–171.

Gribble, P.L., Ostry, D.J., Sanguineti, V. and Laboissiere, R. (1998). Are complex control signals required for human arm movement?. J. Neurophysiol., 79: 1409–1424.

Gurfinkel, V.S. and Levik, Y.S. (1991). Perceptual and automatic aspects of the postural body scheme. In: Paillard J. (Ed.), Brain and Space. Oxford University Press, Oxford, pp. 147–162.

Gurfinkel, V.S. and Shik, M.L. (1973). The control of posture and locomotion. In: Gydikov A.A., Tankov N.T. and Kosarov D.S. (Eds.), Motor Control. Plenum Press, New York, pp. 217–234.

Gurfinkel, V.S., Levik, Yu.S., Popov, K.E., Smetanin, B.N. and Shlikov, Y. (1988). Body scheme in the control of postural activity. In: Gurfinkel V.S., Ioffé M.E., Massion J. and Roll J.-P. (Eds.), Stance and Motion: Facts and Concepts. Plenum Press, New York, pp. 185–193.

Gurfinkel, V.S., Lestienne, F., Levik, Yu.S. and Popov, K.E. (1993a). Egocentric references and human spatial orientation in microgravity. I. Perception of spatial tactile stimuli. Exp. Brain Res., 95: 339–342.

Gurfinkel, V.S., Lestienne, F., Levik, Yu.S., Popov, K.E. and Lefort, L. (1993b). Egocentric references and human spatial orientation in microgravity. II. Body-centred coordinates in the task of drawing ellipses with prescribed orientation. Exp. Brain Res., 95: 343–348.

Hasan, Z. and Enoka, R.M. (1985). Isometric torque-ankle relationship and movement-related activity of human elbow

flexors: implications for the equilibrium-point hypothesis. Exp. Brain Res., 59: 441–450.
Head, H. (1920). Studies in Neurology. Hodder and Stoughton, London.
Hess, W.R. (1943). Teleokinetisches und erreismatisches Kräftesystem in der Biomotorik?. Helv. Physiol. Pharmac. Acta, 1: C62–C63.
Hlavacka, F., Kriskova, M. and Horak, F.B. (1995). Modification of human postural response to leg muscle vibration by electrical vestibular stimulation. Neurosci. Lett., 189: 9–12.
Horak, F.B. and Macpherson, J.M. (1996). Postural orientation and equilibrium. In: Towell L.B. and Shepherd J.T. (Eds.), Handbook on Integration of Motor Circulatory, Respiratory, and Metabolic Control During Exercise. American Physiological Society, Bethesda, pp. 255–292.
Horak, F.B. and Nashner, L.M. (1986). Central programming of postural movements: adaptation to altered support-surface configurations. J. Neurophysiol., 55: 1369–1381.
Ito, M. (1984). The Cerebellum and Neural Control. Raven Press, New York, pp. 580.
Jeka, J.J. and Lackner, J.R. (1995). The role of haptic cues from rough and slippery surfaces in human postural control. Exp. Brain Res., 103: 267–276.
Kavounoudias, A., Roll, R. and Roll, J.P. (1998). The plantar sole is a 'dynamometric map' for human balance control. Neuroreport, 9: 3247–3252.
Kavounoudias, A., Gilhodes, J.C., Roll, R. and Roll, J.P. (1999). From balance regulation to body orientation: two goals for muscle proprioceptive information processing?. Exp. Brain Res., 124: 80–88.
Kawato, M., Furukawa, K. and Suzuki, R. (1987). A hierarchial neural-network model for control and learning of voluntary movement. Biol. Cybern., 69: 169–185.
Kuo, A.D. and Zajac, F.E. (1993). A biomechanical analysis of muscle strength as a limiting factor in standing posture. J. Biomech., 26 Suppl. 1: 137–150.
Kuypers, H.G.J.M. (1981). Anatomy of the descending pathways. In: Brooks V.B. (Ed.), Handbook of Physiology, Sec. 1, The Nervous System, Vol. II, Motor Control,. American Physiology Society, Bethesda, pp. 597–666 Part 1.
Lackner, J.R. (1988). Some proprioceptive influences on the perceptual representation and orientation. Brain, 111: 281–297.
Latash, M. (1993) Control of Human Movement. Human Kinetics, Champaign, IL, USA.
Lestienne, F., Soechting, J. and Berthoz, A. (1977). Postural readjustment induced by linear motion of visual scenes. Exp. Brain Res., 28: 363–384.
Lund, S. and Broberg, C. (1983). Effects of different head positions on postural sway in man induced by a reproducible vestibular error signal. Acta Physiol. Scand., 117: 307–309.

Macpherson, J.M. (1991). How flexible are muscle synergies?. In: Humphrey D.R. and Freund H.J. (Eds.), Motor Control: Concepts and Issues. John Wiley, Chichester, pp. 33–37.
Magnus, R. (1924). Der Korperstellung. Springer Verlag, Berlin.
Massion, J. (1992). Movement, posture and equilibrium: interaction and co-ordination. Prog. Neurobiol., 38: 35–56.
Massion, J. (1994). Postural control system. Curr. Opin. Neurobiol., 4: 877–887.
Massion, J. (1998). Postural control systems in developmental perspective. Neurosci. Biobehav. Rev., 22: 465–472.
Massion, J., Popov, K., Fabre, J.C., Rage, P. and Gurfinkel, V. (1997). Is the erect posture in microgravity based on the control of trunk orientation or center of mass position?. Exp. Brain Res., 114: 384–389.
Massion, J., Amblard, B., Assaiante, C., Mouchnino, L. and Vernazza, S. (1998). Body orientation and control of coordinated movements in microgravity. Brain Res. Rev., 28: 83–91.
Mergner, T. and Rosemeier, T. (1998). Interaction of vestibular, somatosensory and visual signals for postural control and motion perception under terrestrial and microgravity conditions—a conceptual model. Brain Res. Rev., 28: 118–135.
Miall, R.C., Weir, D.J., Wolpert, D.M. and Stein, J.F. (1993). Is the cerebellum a Smith predictor?. J. Mot. Behav., 25: 203–216.
Mittelstaedt, H. (1998). Origin and processing of postural information. Neurosci. Biobehav. Rev., 22: 473–478.
Mori, S. (1989). Contribution of postural muscle tone to full expression of posture and locomotion movements: multifaceted analyses of its setting brainstem-spinal cord mechanisms in the cat. Jpn. J. Physiol., 39: 785–809.
Mori, S., Matsuyama, K., Kuze, B. and Mori, F. (1999). Features of the fastigio-reticulo-spinal system involved in the control of posture and locomotion in the cat. In: Gantchev G.N., Mori S. and Massion J. (Eds.), Motor Control Today and Tomorrow. Academic Publishing House 'Prof. M. Drinov', Sofia, pp. 31–43.
Mouchnino, L., Aurenty, R., Massion, J. and Pedotti, A. (1992). Coordination between equilibrium and head-trunk orientation during leg movement: a new strategy built up by training. J. Neurophysiol., 67: 1587–1598.
Mouchnino, L., Aurenty, R., Massion, J. and Pedotti, A. (1993). Is the trunk a reference frame for calculating leg position?. Neuroreport, 4: 125–127.
Nashner, L.M. and McCollum, G. (1985). The organization of human postural movements: a formal basis and experimental synthesis. Behav. Brain Sci., 8: 135–172.
Oddsson, L. and Thorstensson, A. (1986). Fast voluntary trunk flexion movements in standing: primary movements and associated postural adjsutments. Acta Physiol. Scand., 128: 341–349.

Paillard, J. (1960). The patterning of skilled movements. In: Field J., Magoun H.W. and Hall V.E. (Eds.), Handbook of Physiology, Neurophysiology, Vol. III, American Physiological Society, Bethesda, pp. 1679–1708.

Paillard, J. (1991). Motor and representational framing of space. In: Paillard J. (Ed.), Brain and Space. Oxford University Press, Oxford, pp. 163–184.

Pedotti, A., Crenna, P., Frigo, C. and Massion, J. (1989). Postural synergies in axial movements: short and long-term adaptation. Exp. Brain Res., 74: 3–11.

Pozzo, T., Levik, Y. and Berthoz, A. (1995). Head and trunk movements in the frontal plane during complex dynamic equilibrium tasks in humans. Exp. Brain Res., 106: 327–338.

Prochazka, A., Clarac, F., Loeb, G., Rothwell, J.C. and Wolpaw, J.R. (2000). What do reflex and voluntary mean? Exp. Brain Res., 130: 417–432.

Rademaker, G.G.J. (1931). Das Stehen: Statische Reactionen, Gleichwichtsreaktionen und Muskeltonus unter Besondere Berucksichtung Ihres Verhaltens Bei Kleinhirnlosen Tieren. Springer Verlag, Berlin.

Robinson, D.A. (1986). The systems approach to the oculomotor system. Vision Res., 1: 91–99.

Roll, J.P. and Roll, R. (1988). From eye to foot: a proprioceptive chain involved in postural control. In: Amblard B., Berthoz A. and Clarac F. (Eds.), Posture and Gait: Development Adaptation and Modulation. Elsevier, Amsterdam, pp. 155–164.

Roll, J.-P., Popov, K., Gurfinkel, V., Lipshits, M., André-Deshays, C., Gilhodes, J.C. and Quoniam, C. (1993). Sensorimotor and perceptual function of muscle proprioception in microgravity. J. Vestibul. Res., 3: 259–273.

Roll, R., Gilhodes, J.C., Roll, J.P., Popov, K., Charade, O. and Gurfinkel, V. (1998). Proprioceptive information processing in weightlessness. Exp. Brain Res., 122: 393–402.

Sherrington, C.S. (1906). The Integrative Action of the Nervous System. Constable, London.

Sternad, D. and Corcos, D. (2001). Effect of task and instruction on patterns of muscle activation: wacholder and beyond. Motor Control, 5: 307–336.

Stuart, D.G. (2002). Reflections on spinal reflexes. Adv. Exp. Biol. Med., 508: 249–257.

Thomas, A. (1940). Equilibre et Equilibration. Masson, Paris.

Tokizane, T., Murao, M., Ogata, T. and Kondo, T. (1951). Electromyographic studies on tonic neck, lumbar and labyrinthine reflexes in normal persons. Jpn. J. Physiol., 2: 130–146.

Vernazza-Martin, S., Martin, N. and Massion, J. (2000). Kinematic synergy adaptation to microgravity during forward trunk movement. J. Neurophysiol., 83: 453–464.

Wachholder, K. (1928). Willkürliche Haltung und Bewegung. Ergeb. Physiol., 26: 568–775.

Weiss, P. (1961). Self differentiation of basic patterns of coordination. Comp. Psychol. Monogr., 17: 1–96.

Wiesendanger, M. (1997). Paths of discovery in human motor control. In: Hepp-Reymond M.-C. and Marini G. (Eds.), Perspectives of Motor Behavior and Its Neural Basis. Karger, Basel, pp. 103–134.

Winter, DA. (1990). Biomechanics and Motor Control in Human Movement, 2nd ed., John Wiley and Sons, New York.

Wise, S.P., Alexander, G.E., Altman, J.S., Brooks, V.B., Freund, H.-J., Fromm, C.J., Humphrey, D.R., Sasaki, K., Strick, P.L., Tanji, J., Vogel, S., Wiesendanger, M. (1991). What are the specific functions of the different motor areas? Humphrey D.R. and Freund H.-J. (Eds.), Motor Control: Concepts and Issues. John Wiley & Sons, New York, pp. 463–485.

Wolpert, D.M., Ghahramani, Z. and Jordan, M.A. (1995). An internal model for sensori-motor integration. Science, 269: 1880–1882.

Wolpert, D.M., Miall, R.C. and Kawato, M. (1998). Internal models in the cerebellum. Trends Cogn. Sci., 2: 338–347.

CHAPTER 3

Motor coordination can be fully understood only by studying complex movements

Paul J. Cordo* and Victor S. Gurfinkel

Neurological Sciences Institute, Oregon Health and Science University, Portland, OR 97201, USA

Abstract: In this chapter, we use the sit-up to illustrate the complexity of coordination in movements that involve many muscles, joints, degrees of freedom, and high levels of muscle activity. Complex movements often involve the body axis. In addition to the intentional, focal part of any voluntary movement, complex movements also include "associated movements" that are not consciously controlled, but are necessary for the movement to succeed. Some associated movements serve a purpose, and others may not. During sitting up, the leg-lift is a purposive associated movement, whereas three-joint flexion is a non-purposive associated movement. The control of complex movements is also likely to be complex and, we argue, is hierarchically controlled. Associated movements may, themselves, be hierarchically organized and triggered by lower brain structures, local changes in neuronal excitability, and sensory feedback. Complex movements typically involve a high level of mobility. Because this mobility can lead to instability, anticipatory postural adjustments, a type of purposive associated movement, are commonly used to regulate posture. Thus, a number of important aspects of motor coordination can only be revealed by the study of complex movements.

Introduction

The ability to reduce everything to simple fundamental laws does not imply the ability to start from those laws and reconstruct the universe (Anderson, 1972). Up to the mid-20th century, much of what was known about the central nervous system (CNS) originated from studies of motor function. The organization of simple circuits, especially those within the spinal cord, was investigated in studies of spinal reflexes such as the withdrawal (flexion) reflex (Eccles, 1953) and the myotatic (stretch) reflex (Sherrington, 1906). Much of the early knowledge of brainstem physiology developed from research on posture control (Magnus, 1924). The study of higher brain centers, especially the cerebellum (Babinski, 1899; Dow and Moruzzi, 1958), developed largely from investigation of motor disorders. Thus, early studies of motor control laid the groundwork for much of modern neuroscience.

The study of motor control has also influenced the field of robotics. The skeletomotor structure of humans and other mammals is complex. This structure involves scores of joints with multiple degrees of freedom and an even larger number of muscles. Attempts to develop anthropomorphic robots, such as those that engage in bipedal locomotion, have been singularly unsuccessful due to the required sophistication of their control mechanisms. More recently, researchers in the field of robotics have looked to naturally occurring control mechanisms, for example those used by the human CNS (Benati et al., 1980), to provide insight into how complex artificial mechanical systems might be controlled. Conversely, researchers of natural motor control systems have borrowed from robotics and other areas of engineering to make sense of neuromuscular behavior (Houk, 1979; Robinson, 1981; Whitney, 1987). Because of the bewildering

*Corresponding author: Tel.: +503-418-2520;
Fax: +503-418-2501; E-mail: cordop@ohsu.edu

complexity of the skeletomotor structure of higher animals, most attempts to elucidate motor control mechanisms have been reductionist. These studies have focused largely on relatively simple reflexes and on voluntary movements involving only one or a few joints.

The reductionist approach is limited, however. To paraphrase Bernstein (1967), complex movements cannot be reconstructed from the sum of simple motor activities. Many simple motor activities that have been studied to this point represent little more than laboratory analogs of natural movement. Although this perspective may seem hyperbolic, reductionism does indeed restrict one's view of the broad landscape of motor control. Accordingly, the goal of this chapter is to present some reasons why the control of natural movement can be fully understood only by studying complex movements. No unified theory of motor control is proposed. Rather, we focus attention on several aspects of control peculiar to complex motor activities.

Coordination—what does it mean?

An often-used keyword in the motor control literature is 'coordination'. Attempts to define the term are relatively rare, and they have lead to diverse views on what constitutes coordination. Most of us have an intuitive, subjective appreciation of what constitutes a coordinated motor act, but this appreciation derives mostly from the negative. We can easily tell when a motor action is clumsily performed, and we are acutely aware of abnormal motor behavior. For example, an unusual gait pattern is obvious to the eye, although just what is different about the gait is sometimes difficult to describe in words.

The most obvious evidence of coordination in a motor activity is that intent is achieved, be it dodging an obstacle, hitting a target or flawlessly executing a triple axel. Strictly defined, 'coordination' implies the registration of two or more 'things' into a 'harmonious' relationship (Guralnik, 1979). In movement, these 'things' might equally well be sensory feedback and motor output, movements of different joints or bursts of activity in different muscles. Thus, coordination can be defined at the structural, kinematic or muscular levels, and it may be expressed in relation to space as well as time.

Coordination of complex movements

A common goal of motor control researchers is to understand how the CNS produces natural, coordinated movements, from simple ones involving just a few joints (Hepp-Reymond et al., 1996) to complex movements engaging most or all of the body. We define complex movements as naturally occurring motor activities with high dimensionality. Relevant dimensions include the number of joints, degrees of freedom and muscles, the levels of force generated, the extent of extrapersonal space traversed and the number of possible trajectories. Also, complex movements often involve voluntary control of the body axis.

Multijoint movements have been investigated previously, but with a primary focus on movements of the extremities. Such studies include, for example, work on: the legs during locomotion (Shik et al., 1966; Engberg and Lundberg, 1969; Grillner, 1985; Mori, 1987; Rossignol, 1996), the arms during space-oriented movement (Biguer et al., 1984; Soechting et al., 1986) and the wrist and fingers in the performance of a precision grip (Johansson and Westling, 1987; Hepp-Reymond et al., 1996). Locomotor movements in animals and humans are highly automatic, whereas arm and hand movements are more focally controlled. To investigate the coordination of complex movements, we chose the sit-up, a many degrees of freedom movement that incorporates both focal and automatic forms of control. The sit-up also involves the axial and proximal muscles at relatively high levels of activation.

As with rising from a chair (Riley et al., 1991; Schultz et al., 1992), the goal of the sit-up cannot be accomplished directly because the initial distribution of body mass and its configuration do not provide the appropriate leverage. Therefore, there are two distinct parts to the sit-up. The first changes the configuration of the trunk and the second rotates the pelvis, trunk and head in the direction of the feet. The arrow labeled 'T' in Fig. 1 identifies the transition point from the first (dashed lines) to the second part (solid lines) of the sit-up. As shown in

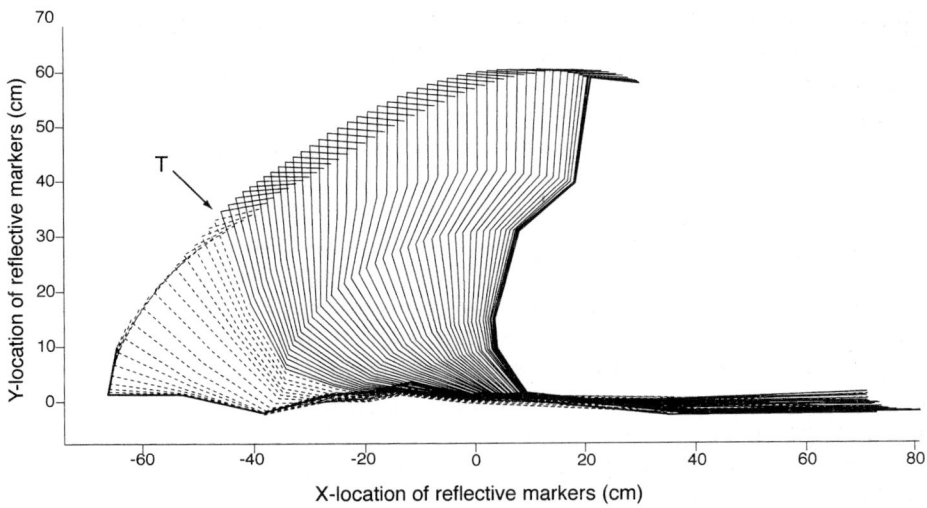

Fig. 1. Stick figures showing smooth transition from supine to a sitting position. The distance between successive figures is 20 ms. The dashed lines represent the first part of the sit-up, a transition occurs at 'T', and the solid lines represent the second part of the sit-up. Marker points: nose, temple, shoulder, 8th rib, 10th rib, 12th rib, anterior–superior iliac crest, trochanter major, knee and ankle.

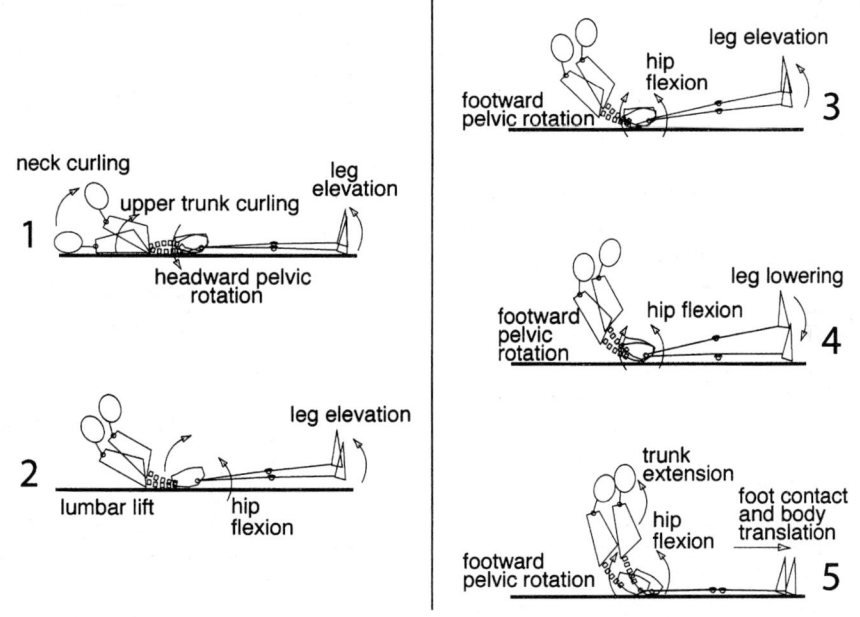

Fig. 2. Two parts of the sit-up with many purposive-associated movements. Left column shows trunk curling (1) and lumbar lift (2), and right column shows headward pelvic rotation (3–5). Footward pelvic rotation is broken down into periods when the legs continue to move up (3), when they move down (4) and when they contact the support surface (5).

Fig. 2, the change in trunk configuration during the first part of the sit-up (left column) includes an initial *head and upper trunk curling* (Fig. 2-1), which is then followed by a *lumbar lift* (Fig. 2-2). These changes in trunk configuration reduce the static torque at the point of contact between the trunk and the support surface (see also Ketcham et al., Chapter 21 of this volume).

During head and trunk curling, the contact point shifts gradually toward the pelvis, and during lumbar lift, it shifts incrementally. In the supine individual, the static torque on the upper body (i.e., above the pelvis) is twice that on the lower body. Reducing the static torque on the upper body is therefore a precondition for the second part of the sit-up, in which the pelvis and upper body rotate in the direction of the feet.

The second part of the sit-up, which encompasses more than half of the trajectory and movement duration, begins when the center of gravity passes through the point of pelvic rotation ('T' in Fig. 1), and the net passive torque rotates the body toward the feet (Fig. 2, right column). Despite the apparently passive nature of the second part of the sit-up, active contraction of the hip flexors contributes in several important ways. First, hip-flexor activity continues the process of hip flexion, thereby delaying the moment of contact of the legs with the support surface. This allows the subject to 'rest' while gravity pulls the body partway toward the sitting posture (Fig. 2-3 and 2-4). Second, once leg contact occurs, the hip flexors actively rotate the pelvis and upper body into their final posture (Fig. 2-5). Despite the duration and distance of the second part of the sit-up, as well as the contributions of hip-flexor activity, the coordination involved in the first part of the sit-up appears to be much more complex than that involved in the second part.

Purposive-associated movements—leg elevation in the sit-up

In addition to the two main parts of the sit-up, a number of additional actions support the intended movement. One of these is leg elevation. Loss of contact of the legs with the support surface, which begins near the beginning of the sit-up (Fig. 2), assists trunk curling and the lumbar lift. Leg elevation can be considered as a purposive-associated movement (Gahery, 1987) insofar as it is not intentionally produced, yet it is an essential part of the sit-up. Leg elevation mobilizes the static torque of the legs to counteract the static torque on the upper body while the upper body is being lifted from the support surface. Leg elevation may also stabilize the pelvis and keep the abdominal flexors from over-shortening while they are contracting near-maximally during trunk curling (RecA and EObl in Fig. 3). In addition to leg elevation, many other movements are involved in the sit-up (Fig. 2), some essential for successful execution of the movement.

We hypothesize that leg elevation, rather than being generated as part of the central motor program, is actually part of a muscle synergy associated with trunk curling. Therefore, leg elevation is not under conscious control. When the head and shoulders are lifted from the support surface, but the sit-up stops at

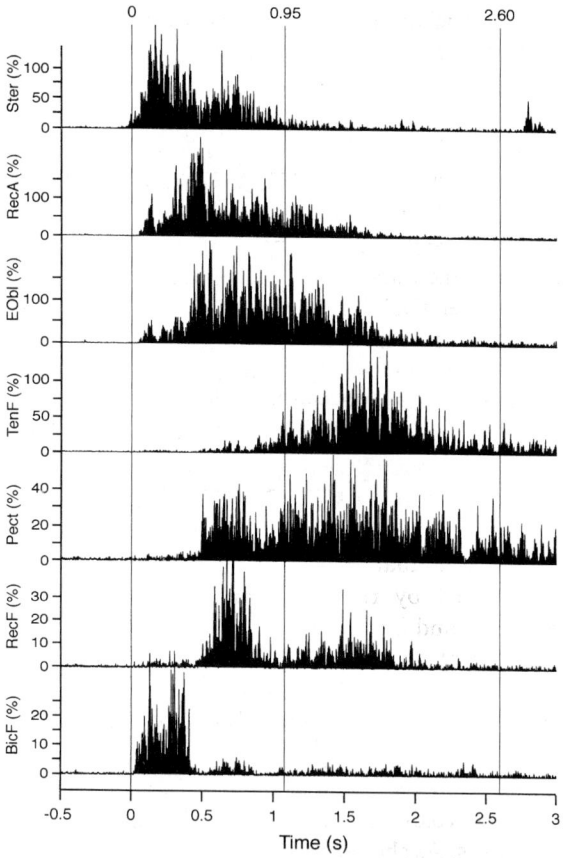

Fig. 3. Rectified, raw EMG activity associated with a sit-up. The start, end (2.6 s) and transition point between the two parts of the sit-up (0.96 s) is indicated by vertical lines. The amplitude of EMG activity is scaled to the maximum, although true maximal contractions in axial muscles were difficult to produce under isometric conditions. The tested muscles include: sternocleidomastoidus (Ster), rectus abdominis (RecA), external obliquus (EObl), tensor fascia latae (TenF), pectineus (Pect), rectus femoris (RecF) and biceps femoris (BicF).

this point, the legs still rise. The elevation so achieved by the legs is highly correlated with the elevation of the head and shoulders. During trunk curling in a full-blown sit-up, the elevation of the legs is also correlated with head and shoulder elevation, as well as with the magnitude of hip-flexor activity. These observations strongly suggest that leg elevation is an active movement resulting from a muscle synergy. This idea is not new, having been previously alluded to by Babinski (1899) in his description of two types of 'asynergies' associated with cerebellar degeneration: (1) hypersynergy in the sit-up task, wherein some unspecified disorganization of muscle activity causes cerebellar patients to elevate the legs instead of the trunk (Dusser de Barenne, 1924; Tomas, 1925); and (2) hyposynergy in a backward trunk-bending task, wherein an absence of knee flexion and lower leg-forward inclination leads to loss of balance. Similarly, Rademaker (1931) showed that when the head and upper trunk of a decerebellated dog are flexed passively, the legs move actively into ventral flexion. In the intact dog, ventral flexion also occurs, but it is relatively small in magnitude. One more observation in support of the hypothesized synergic interrelation between abdominal and hip flexors comes from studies of breathing in standing humans. Breathing involves a tightly linked counterphase movement in the upper trunk and hip joint, termed the 'respiratory synergy of vertical posture' (Gurfinkel et al., 1971). Hodges et al. (2002) demonstrated that this counter-phase movement is accompanied by reciprocal muscle activity in the abdominal and hip flexors. In sum, these clinical and experimental observations support the hypothesis that leg elevation during the sit-up is connected with trunk curling and the abdominal-flexor activity that produces it.

With increasing movement complexity come additional roles for muscles not found in simpler movements. As shown in Fig. 3, the pattern of muscle activity in a sit-up is complex. Recordings are shown from seven different muscles of the body axis, each of which is activated at a distinct time and with a distinct profile and duration. The most obvious role of a muscle is as prime mover. In both simple and complex movements, the prime mover can usually be identified unambiguously, because the known action of the muscle relates closely to the overt goal of the movement, and the prime mover is often the most strongly activated muscle. Of the additional roles available to muscles in simple movements, an antagonist muscle may contract toward the end of a movement to slow it down. Also, late agonist muscle activity can dampen oscillation of a joint about the endpoint of a movement. These three roles—prime mover, antagonist and dampener—make up the complete repertoire of muscle actions in single joint movements.

In movements involving several joints, interaction forces become significant (Sainberg et al., 1993; Dounskaia et al., 2002). This requires joint stabilization, a role usually served by proximal joints. In even more complex movements, muscles can act to reposition joints not directly associated with the goal of the movement in order to focus internal and external forces at localized parts of the body. For more global functions, this process helps to maintain balance. Muscles, with their indwelling sensory receptors, may also serve as motion detectors in complex movements. This role does not necessarily require the generation of active force. Thus, the available roles for muscles expand with movement complexity. In complex movements, the role served by a particular muscle may even change partway through a movement. For example, in the sit-up, the hip flexors start as pelvic stabilizers, and then switch to become the prime movers (Cordo et al., 2003). To make the available choices almost limitless, a given muscle can even serve as prime mover and, simultaneously, produce an associated movement. An example is the role of abdominal flexors in the sit-up, which lift the trunk, but also rotate the pelvis toward the head (Fig. 2-1).

Nonpurposive-associated movements—three-joint flexion

There exist other associated movements during the sit-up, which appear to serve no functional purpose ('nonpurposive-associated movements'; Gahery, 1987). At the beginning of the sit-up, the hips, knees and ankles all flex (Fig. 4A). This is a characteristic of the 'flexion-reflex synergy', which is so called because all flexor muscles in an extremity are activated. In the sit-up, initial knee flexion and

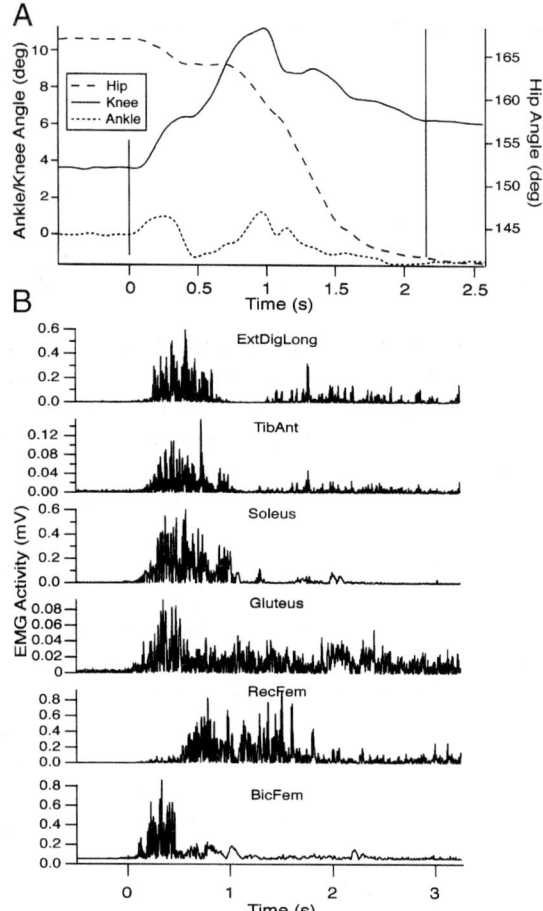

Fig. 4. Nonpurposive-associated movements. (A) Joint angle trajectories are shown for hip, knee and ankle in a 2.1-s duration sit-up. Note initial flexion (up for ankle and knee, and down for hip) in all three joints. (B) Muscle activity from six muscles of the legs is shown.

ankle dorsiflexion and plantarflexion have no clear purpose. Most mammals have a strong flexion reflex as a defense mechanism, in which the limbs, especially the legs, are automatically retracted on painful stimulation. A reflex response involving flexion of the three main joints of the leg (hip, knee, ankle) has also been reported in patients afflicted by stroke or other types of brain and spinal cord disorders. In these cases, three-joint flexion is generated in the absence of a noxious stimulus (Sherrington, 1906; Hallett et al., 1975). This is an example of a pathological manifestation of a normal motor behavior, which is possibly due to interruption of normal descending inhibitory tone. In the case of the normal sit-up, we hypothesize that three-joint flexion is an overt and natural manifestation of the flexion reflex, which is provoked by strong descending signals to the abdominal muscles. In the sit-up, other muscles (Fig. 4B) produce toe extension and foot inversion, which can also be classified as nonpurposive-associated movements.

Like the sit-up, forward bending from the standing position is associated with upper trunk flexion and with activation of abdominal flexors and hip flexors as the prime movers. Therefore, it should not be surprising that in fast-forward trunk bending, an early and nonpurposive flexion of the knee occurs, and is accompanied by ankle and hip flexion. The knee flexion then reverses to extension, which serves to shift the pelvis backwards and maintain balance. In slow-forward trunk bending, which involves lower levels of muscle contraction, a reflex-like combination of joint flexions does not occur (Crenna et al., 1987; Oddsson and Thorsstenson, 1987; Oddsson and Thorsstenson, 1990). Thus, it appears that complex movements involving the axis and extremities, which also involve high levels of muscle activity, transiently activate reflexive muscle synergies with no functional purpose. As a result, any comprehensive description of coordination should include such nonpurposive-associated movements. It is important to understand how nonpurposive movements are compensated for, in order that they do not interfere with the main goal of the movement.

Control of complex movements

Given all of the possible functional roles of focal and associated movements and the individual muscles producing these movements, the CNS control of complex motor tasks is likely to also be complex. Formally, purposive movement originates from a central motor program that is voluntary, conscious and focused on a goal (see also Grillner and Wallén and Ma et al., Chapter 1 and 48 of this volume). It is difficult, however, to understand how a single descending motor pathway coordinates all the associated movements in a complex motor task. We suggest that a more plausible means of control is through a hierarchy of pathways (Weiss, 1936; Bernstein, 1947).

At the top of this hierarchy is a primary motor command to the prime movers. This control is focused on the overt goal of the task, for example, transporting the body from one location to another or changing body posture from one configuration to another. The prime movers are most often muscles involved in propulsive actions. In complex movements, the primary motor command represents a rough pattern of muscle activity required to achieve the overt goal. It may not include, however, all of the control signals required to accomplish the task. For example, the primary motor command may not include signals to generate adequate joint torque, in the most efficient body configuration, while also sending signals to maintain a stable state of the body. In addition to the primary motor command, other command mechanisms must exist, which generate associated movements. Some of the latter movements are closely related to the primary goal and others are not. Also, some associated movements are synchronized to the onset of movement while others are triggered later.

We hypothesize that associated movements are, in and of themselves, hierarchically organized. Those closely related to the primary goal may be loaded and triggered at the same time as the primary motor command, or they may be triggered after a preplanned delay. While associated movements are executed subconsciously, they are often necessary to create the proper conditions, internally and externally, for intention to be achieved.

On a more fundamental level, complex movements may engage highly automatic processes such as pattern generation and reflex activity. How the CNS triggers these lower-level-associated movements is unclear. A strong possibility is that they are triggered by local changes in state (e.g., excitability, gating) in the CNS, particularly in the spinal cord. Neural structures in the brainstem are known to regulate some changes in state (Burke, 1999), although overt manifestations of these state-changes are most often described as pathological and long lasting. Other more transient changes in state occur, however, under quite normal conditions. They sometimes trigger unintentional movements with no obvious function. In adolescents, handwriting is often associated with unintentional motions of the contralateral limb and the lips. Similarly, with the onset of fatigue, even simple weight-lifting movements of a single finger can evoke associated movements in the proximal limb, shoulders, body axis and contralateral limb. These unintentional movements are likely to be caused by changes in state, perhaps evoked by corollary discharge in different parts of the CNS, and they must be accounted for in any complete description of motor coordination.

Another means of triggering associated movements in complex motor tasks is via sensory feedback. Historically, sensory feedback in movement coordination has been thought of in terms of servocontrol, to correct for externally generated (Morton, 1953; Marsden et al., 1976) and internally generated (Houk, 1979) errors. Sensory feedback may also be used, however, to trigger movement fragments (Cordo et al., 1994, 1995) in complex sequences of actions such as locomotion (Andersson and Grillner, 1983; Conway et al., 1987; Wolf and Pearson, 1988). In the sit-to-stand maneuver, for example, sensory feedback generated by trunk flexion may trigger leg extension when the center of gravity has shifted far enough forward for the body to rise vertically. Similarly, in the sit-up, feedback related to the center of gravity can trigger hip-flexor activity when weight distribution permits these muscles to lift the trunk instead of the legs (Cordo et al., 1996). One implication of sensory-based triggering is that the site at which the feedback is generated might be remote to the site at which the triggered movement is produced.

Mobility versus stability

Another task of associated movements is to maintain postural stability during movements with high mobility. Such associated movements have been termed 'anticipatory postural adjustments', although activity associated with these adjustments both precede and follow the onset of the focal movement. Anticipatory postural adjustments commonly serve to maintain a stable standing posture during focal activities that shift the body's center of pressure (Belen'kii et al., 1967; Cordo and Nashner, 1982; Hodges et al., 2002). Anticipatory postural activity may be required when a large mass is moved, or when a focal movement involves high acceleration. On a smaller scale, anticipatory postural adjustments can function to maintain posture in only a limited part of the body. For example, they provide the ability to hold a tray steady with one hand while the other hand lifts a glass of water from it (Hugon et al., 1982). As with other associated movements, postural control is automatically and subconsciously regulated during focal movement.

One example of postural regulation by associated movement occurs during forward and backward trunk bending in standing individuals (Babinski, 1899; Crenna et al., 1987). During forward trunk bending, the pelvis automatically moves in the posterior direction to prevent the body's center of mass from wandering outside a stable base of support. This process of mass redistribution is highly efficient, with a net shift in the whole-body center of gravity of < 1 cm. The compensatory pelvic shift is performed subconsciously but it is well coordinated with the intentional trunk motion. Therefore, pelvic movement may be a functionally useful associated activity, which is distinct from the focal motor program to bend the trunk forward.

Babinski (1899) observed that patients with cerebellar degeneration are unable to bend the trunk without losing balance, thereby suggesting that coordination between focal and associated movements is impaired in these individuals. Similarly, breathing movements in the standing individual, though small, tend to destabilize balance (Gurfinkel et al., 1971). The musculature of the lower body acts automatically to counter these destabilizing movements (Hodges et al., 2002). Thus, in complex motor tasks, the mobility provided by many joints, degrees of freedom and muscles leads to destabilization. To maintain balance, associated movements that are coordinated with the primary motor program must compensate for such mobility.

Conclusions

Historically, coordination has been thought of as a process of bringing together a number of submovements into a single whole movement (Novak et al., 2002). There are many reasons to think of coordination in this way. For example,

Zackowski et al. (2002) showed that in cerebellar patients, two-joint movements of the arm are not well coordinated. The same individuals produce well-controlled movements in the individual joints, however, as long as the joints are moved one at a time. Our own observations on sit-ups performed by normal individuals support this point of view (Cordo et al., 2003). We also need to understand, however, mechanisms underlying the coordination of *all* movements, both simple and complex. We are at an early stage of grasping how direct control, synergies and automatisms are combined to produce mobility, while also preserving stability. To gain this understanding will require, in very large part, the study of complex movements.

Acknowledgments

The work described in this chapter was supported by USPHS NIH grant R01 AR46007. The authors also thank Sandra Oster, Ph.D., for her careful reading and editing of the manuscript.

Abbreviations

BicF	biceps femoris
CNS	central nervous system
EObl	external obliquus
Pect	pectineus
RecA	rectus abdominis
RecF	rectus femoris
Ster	sternocleidomastoidus
TenF	tensor fascia latae

References

Anderson, P. (1972). More is different. Science, 177: 393–396.

Andersson, O. and Grillner, S. (1983). Peripheral control of the cat's step cycle. II. Entrainment of the central pattern generator by sinusoidal hip movements during 'fictive' locomotion. Acta Physiol. Scand., 118: 229–239.

Babinski, J. (1899). De l'asynergie cerebelleuse. Rev. Neurol., 7: 806–816.

Belen'kii, V.Ye., Gurfinkel, V.S. and Pal'tsev, Ye.I. (1967). Elements of control of voluntary movements. Biofizika, 12: 135–141.

Benati, M., Gaglio, S., Morasso, P., Tagliasco, V. and Zaccaria, R. (1980). Anthropomorphic robotics. I. Representing mechanical complexity. Biol. Cybern., 38: 125–140.

Bernstein, N. (1947). On the Construction of Movements. Medgiz, Moscow (in Russian).

Bernstein, N. (1967). The Coordination and Regulation of Movements. Pergamon, New York.

Biguer, B., Prablanc, C. and Jeannerod, M. (1984). The contribution of coordinated eye and head movements in hand pointing accuracy. Exp. Brain Res., 55: 426–469.

Burke, R.E. (1999). The use of state-dependent modulation of spinal reflexes as a tool to investigate the organization of spinal interneurons. Exp. Brain Res., 128: 263–277.

Conway, B.A., Hultborn, H. and Kiehn, O. (1987). Proprioceptive input resets central locomotor rhythm in the spinal cat. Exp. Brain Res., 68: 643–656.

Cordo, P. and Nashner, L.M. (1982). Properties of postural adjustments associated with rapid arm movements. J. Neurophysiol., 47: 287–302.

Cordo, P.J., Carlton, L., Bevan, L., Carlton, M.J. and Kerr, G. (1994). Proprioceptive coordination of movement sequences: role of velocity and position information. J. Neurophysiol., 71: 1848–1861.

Cordo, P.J., Gurfinkel, V.S., Bevan, L. and Kerr, G.K. (1995). Proprioceptive consequences of tendon vibration during movement. J. Neurophysiol., 74: 1675–1688.

Cordo, P., Gurfinkel, V., Verschueren, S., Smith, T. and Collins, J.J. (1996). Proprioceptive coordination of axial movement. Soc. Neurosci. Abstr., 22: 129.

Cordo, P., Gurfinkel, V.S., Smith, T.S., Hodges, P.W., Verschueren, S.M.P. and Brumagne, S. (2003) The sit-up: complex kinematics and muscle activity in a voluntary axial movement. J. Electromyogr. Kinesiol. 12: 239–252.

Crenna, P., Frigo, C., Massion, J. and Pedotti, A. (1987). Forward and backward axial synergies in man. Exp. Brain Res., 65: 538–548.

Dounskaia, N., Ketcham, C.J. and Stelmach, G.E. (2002). Commonalities and differences in control of various drawing movements. Exp. Brain Res., 146: 11–25.

Dow, R.S. and Moruzzi, G. (1958). Physiology and Pathology of the Cerebellum. University of Minnesota Press, Minneapolis.

Dusser de Barenne, J.G. (1924). Die Funktionen des Kleinhirns. In: Alexander G. and Marburg O. (Eds.), Handbuch der Neurologie des Ohres. Urban und Schwarzenberg, Berlin.

Eccles, J.C. (1953). The Neurophysiological Basis of Mind. The Principles of Neurophysiology. The Claredon Press, Oxford.

Engberg, I. and Lundberg, A. (1969). An electromyographic analysis of muscular activity in the hindlimb of the cat during locomotion. Acta Physiol. Scand., 75: 614–630.

Gahery, Y. (1987). Associated movement. Postural adjustments and synergie: some comments about the history and significance of three motor concepts. Arch. Ital. Biol., 125: 345–360.

Grillner, S. (1985). Neurobiological bases of rhythmic motor acts in vertebrate. Science, 228: 143–149.

Guralnik D.B. (Ed.) (1979). Webster's New World Dictionary of the American Language. William Collins Publishers, Inc, Cleveland.

Gurfinkel, V.S., Kots, Ya.M., Pal'tsev, Ye.I. and Feldman, A.G. (1971). The compensation of respiratory disturbances of the erect posture of man as an example of the organization of interarticular interaction. In: Gelfand I.M., Gurfinkel V.S., Fomin S.V. and Tsetlin M.L. (Eds.), Models of the Structural-Functional Organization of Certain Biological Systems. MIT Press, Cambridge, MA, pp. 382–395.

Hallett, M., Shahani, B.T. and Young, R.R. (1975). EMG analysis of stereotyped voluntary movements in man. J. Neurol. Neurosurg. Psychiatry, 38: 1154–1162.

Hepp-Reymond, M.-C., Huesler, E.J. and Maier, M.A. (1996). Precision grip in humans. Temporal and spatial synergies. In: Wing A.M., Haggard P. and Flanagan J.R. (Eds.), Hand and Brain: The Neurophysiology and Psychology of Hand Movements. Academic Press, San Diego, pp. 37–68.

Hodges, P.W., Gurfinkel, V.S., Brumagne, S., Smith, T.C. and Cordo, P.J. (2002). Coexistence of stability and mobility in postural control: evidence from postural compensation for respiration. Exp. Brain Res., 144: 293–302.

Houk, J.C. (1979). Regulation of stiffness by skeletomotor reflexes. Annu. Rev. Physiol., 41: 99–114.

Hugon, M., Massion, J. and Wiesendanger, M. (1982). Anticipatory postural changes induced by active unloading and comparison with passive unloading in man. Pflugers Arch., 393: 292–296.

Johansson, R.S. and Westling, G. (1987). Signals in tactile afferents from the fingers eliciting adaptive motor responses during precision grip. Exp. Brain Res., 66: 141–154.

Magnus, R. (1924). Körperstellung. Julius Springer, Berlin.

Marsden, C.D., Merton, P.A. and Morton, H.B. (1976). Servo action in the human thumb. J. Physiol. (Lond.), 257: 1–44.

Morton, P.A. (1953). Speculations on the servo-control of movement: Ciba Foundation Symposium on the Spinal Cord. Churchill, London, pp. 247–255.

Mori, S. (1987). Integration of posture and locomotion in acute decerebrate cat and in awake freely moving cat. Prog. Neurobiol., 28: 161–195.

Novak, K.E., Miller, L.E. and Houk, J.C. (2002). The use of overlapping submovements in the control of rapid hand movements. Exp. Brain Res., 144: 351–364.

Oddsson, L. (1990). Control of voluntary trunk movements in man. Mechanisms for postural equilibrium during standing. Acta Physiol. Scand. Suppl., 595: 1–60.

Oddsson, L. and Thorstensson, A. (1987). Fast voluntary trunk flexion movements in standing: motor patterns. Acta Physiol. Scand., 129: 93–106.

Rademaker, G.G.J. (1931). Das Stehen. Statische Reaktionen, Gleichgewichtsreaktionen und Muskeltonus unter Besonderer Berucksichtigung Ihres Verhaltens Bei Kleinhirnlosen Tieren. Julius Springer, Berlin.

Riley, P.O., Schenkman, M.L., Mann, R.W. and Hodge, W.A. (1991). Mechanics of constrained chair rise. J. Biomech., 24: 7–85.

Robinson, D.A. (1981). The use of control systems analysis in the neurophysiology of eye movements. Annu. Rev. Neurosci., 4: 463–503.

Rossignol, S. (1996). Neural control of stereotypic limb movements. In: Rowell L.B. and Shepherd J.T. (Eds.), Handbook of Physiology, Sec. 12. Exercise, Regulation and Integration of Multiple Systems. Oxford University Press, New York, pp. 173–216.

Sainberg, R.L., Poizner, H. and Ghez, C. (1993). Loss of proprioception produces deficits in interjoint coordination. J. Neurophysiol., 70: 2136–2147.

Schultz, A.B., Alexander, N.B. and Ashton-Miller, J.A. (1992). Biomechanical analysis of rising from a chair. J. Biomech., 25: 1383–1391.

Sherrington, C.S. (1906). The Integrative Action of the Nervous System. Yale University Press, New Haven.

Shik, M.L., Severin, F.V. and Orlovsky, G.N. (1966). Control of walking and running by means of electrical stimulation of the mid-brain. Biophysics, 11: 756–765.

Soechting, J.F., Lacquaniti, F. and Terzuolo, C.A. (1986). Coordination of arm movements in three-dimensional space. Sensorimotor mapping during drawing movmement. Neuroscience, 17: 295–311.

Tomas, A. (1925). In Roger G.H., Widal F. and Teissier P.J. (Eds.), Pathologie du Cervelet. Nouveau Traité de Méd, Vol. 39, Masson et Cie, Paris.

Weiss, P. (1936). Selectivity controlling the central-peripheral relations in the nervous system. Biol. Rev., 11: 494–531.

Whitney, D.E. (1987). Neuronal coding and robotics. Science, 237: 300–302.

Wolf, H. and Pearson, K.G. (1988). Proprioceptive input patterns elevator activity in the locust flight system. J. Neurophysiol., 59: 1831–1853.

Zackowski, K.M., Thach, W.T. and Bastian, A.J. (2002). Cerebellar subjects show impaired coupling of reach and grasp movements. Exp. Brain Res., 146: 511–522.

CHAPTER 4

The emotional brain: neural correlates of cat sexual behavior and human male ejaculation

Gert Holstege* and Janniko R. Georgiadis

Department of Anatomy and Embryology, University of Groningen, Groningen 9713 AV, The Netherlands

Abstract: The organization of virtually all basic survival mechanisms in the central nervous system (CNS) is within the most central regions of the mesencephalon and the rostrally adjoining diencephalon; in particular, the mesencephalic periaqueductal gray (PAG) and hypothalamus. The PAG sends specific pathways to the caudal brainstem where neurons are located that, in turn, control nociception, blood pressure, heart rate, and micturition. Via projections to the nucleus retroambiguus (NRA) in the most caudal part of the medulla, the PAG controls the intra-abdominal pressure associated with vocalization, vomiting, and parturition. In cats, the PAG also controls sexual posture via NRA projections to motoneurons in the lumbosacral cord. These NRA-lumbosacral motoneuronal pathways are almost nine times stronger in the estrous vs. non-estrous female cat. While neuronal activity in specific CNS pathways is now known to control sexual behavior in the cat, how is it organized in the human? PET-scan results on human ejaculation have revealed that the meso-diencephalic transition zone is particularly and strongly activated. This region includes the so-called ventral tegmental area that is also known as a "reward area." For example, it is also activated during a heroin rush. Other strongly activated structures during sexual activity include the cerebellum and lateral part of the corpus striatum. At the level of the cerebral cortex, areas in the prefrontal and parietal cortex are also activated, but exclusively on the right side. Further study of these structures should certainly lead to better insight into human sexual behavior and provide the possibility to improve sexual activity in those who suffer from problems in this area.

Introduction

The two tasks of the central nervous system (CNS) are to ensure survival of the individual and survival of the species. The organization of these tasks is found at different levels of the neuraxis, depending on their complexity. The caudal brainstem controls the most basic mechanisms, such as regulation of blood pressure, heart rate, respiration and micturition. Defensive mechanisms, such as aggression and freezing, are organized in the mesencephalon, while hunger, thirst and temperature regulations are controlled from the caudal diencephalon. All these functions are strongly influenced by more rostrally located structures at diencephalic and telencephalic levels, which are part of the so-called limbic system. The reproductive function, the second task of the CNS, is thought to be mainly organized at diencephalic levels, and sexual functions in animals have been studied mostly at the level of the hypothalamus and preoptic region.

How do these diencephalic structures control the motor output necessary for reproductive behavior? Specific somatic and autonomic motoneuron (MN) cell groups in the spinal cord generate the final motor output, but how the diencephalic structures control these MNs is not clear. A short survey will be given of descending pathways possibly involved in the reproductive behavior of cats.

Similar studies about the anatomy and physiology of the CNS organization of sexual behavior in humans do not exist. On the other hand, the recent introduction of powerful neuroimaging techniques allows

*Corresponding author: Tel.: +31-50-363-2460;
Fax: +31-50-363-2461; E-mail: g.holstege@med.rug.nl

investigators to uncover the parts of the CNS that are involved in human reproductive behavior. The first positron emission tomography (PET) scan results regarding which brain structures play a role in male ejaculation will be presented at the end of this chapter.

How does the brain control the MNs that generate ejaculation?

Studies in rats (Baum and Everitt, 1992; Coolen et al., 1996) and gerbils (Heeb and Yahr, 1996) revealed that there are four regions that express C-Fos immunoreactivity after ejaculation: the medial preoptic area (MPOA), the medial nucleus of the amygdala (MeA), the bed nucleus of the stria terminalis (BNST) and the midbrain lateral central tegmental field (LCTF)/subparafascicular nucleus (SPF). Baum and Everitt (1992) suggested that genital and olfactory vomeronasal input activates the LCTF/SPF and MeA, respectively, and that these regions interact to activate the MPOA and BNST.

Lesions in the posterodorsal preoptic nucleus and the posterodorsal part of the MeA of the gerbil resulted in delay of ejaculation, but lesions in the SPF had no such effect (Heeb and Yahr, 2000). Brackett and Edwards (1984) concluded that the connection between the MPOA and the LCTF is essential for copulation. This was because bilateral lesions in the area of the LCTF eliminated mating behavior in male rats and reproductive behavior was also eliminated after a MPOA lesion on one side, combined with a lesion in the LCTF on the other.

The C-Fos studies, however, because of their low temporal resolution, could not unravel the precise function of the respective regions in ejaculation. Moreover, C-Fos is thought to be expressed mainly in neurons belonging to the somatosensory rather than the motor system. Lesion studies could not provide any insight, either, because it was unclear which systems became dysfunctional as a result of the respective lesions. The inescapable conclusion is that there is *no* current concept regarding which motor systems play a key role in ejaculation.

Pathways involved in reproductive behavior in cats

In the cat a descending motor system has been demonstrated that is thought to play a role in mating behavior. It originates in the nucleus retroambiguus (NRA), a cell group located laterally in the caudal medulla (see also Kirkwood and Ford, Chapter 10 of this volume). The NRA is especially known for its role in respiration and vocalization. It sends projections to MNs of intercostal and abdominal muscles in the thoracic and upper lumbar cord (Holstege, 1989). In addition, the NRA projects to a distinct set of MN cell groups in the lumbosacral cord (VanderHorst and Holstege, 1995) that participate in producing the female (lordosis) and male posture necessary for successful mating. The strength of this NRA-MN projection appeared to depend strongly on the estrous cycle, and it was almost nine times greater in estrous versus nonestrous females (Fig. 1; VanderHorst and Holstege, 1997a).

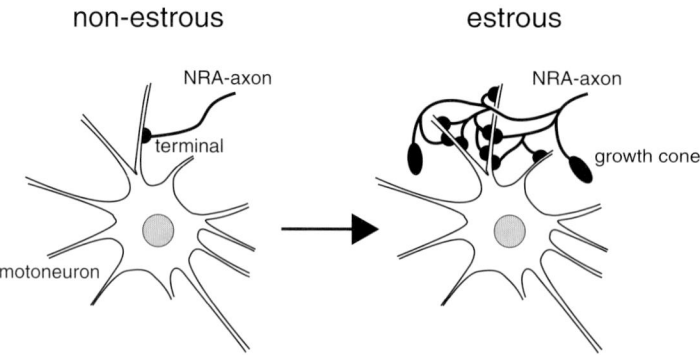

Fig. 1. Schematic illustration of estrogen-induced axonal sprouting of NRA fibers to lumbosacral (semimembranosus) MNs.

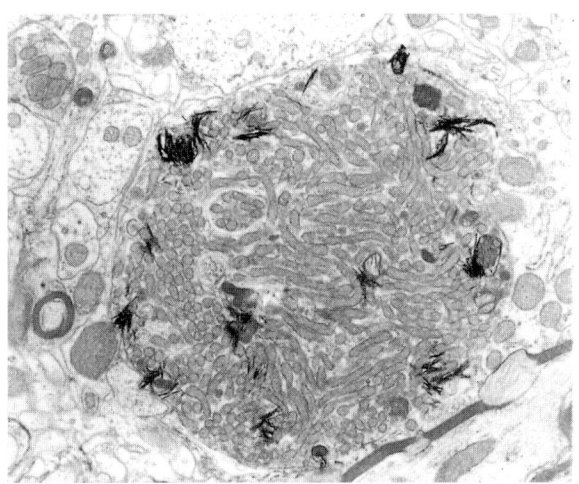

Fig. 2. Electronmicrograph representing a labeled growth cone in the semimembranosus MN cell group in the sixth lumbar spinal segment of an estrous cat following an injection of wheatgerm agglutinin horseradish peroxidase (WGA-HRP) into the NRA. Note the tetramethylbenzidine (TMB) reaction products in mainly the peripheral parts of the cone.

This process of a greater strength of the NRA-MN pathway during estrous is caused by growth of the terminal NRA fibers, since electronmicroscopically, it was found that in estrous cats, growth cones developed in the NRA terminals on lumbosacral MNs (Fig. 2; VanderHorst and Holstege, 1997a). In all likelihood, estrogen induces this increase of NRA terminals on MNs, because a similar rise of NRA terminals was found in the ovariectomized female cat intrathecally injected with estradiol.

The NRA-lumbosacral pathway in males involves slightly different MN cell groups than in females, possibly because during mating, males adopt a somewhat different posture than females. The NRA-lumbosacral MN pathway in males is stronger than in nonestrous females, but weaker than in estrous females (VanderHorst and Holstege, 1997b). Any plasticity in the male, NRA-lumbosacral MN pathway has not been demonstrated.

Afferent projections to the nucleus retroambiguus

The finding that the NRA might regulate the postures necessary for sexual behavior in the cat brings up the question of which structures, in turn, control the NRA. First of all, because of its important role in respiration and respiration-related activities like vocalization and vomiting, the NRA receives strong afferent projections from respiration-related cell groups in the lateral medulla oblongata. The question is whether these afferents play a role in the control of mating behavior. No specific ascending projections to the NRA have been demonstrated from the lumbosacral cord, but it receives a very strong projection from the caudal periaqueductal gray (PAG; Holstege, 1989) and from a cell group in the ventromedial tegmentum of the upper medulla (Gerrits and Holstege, 1996). Both of these cell groups, in turn, receive strong afferent projections from the hypothalamus and preoptic region, as well as from the central nucleus of the amygdala and BNST. All of these forebrain regions do not project to the NRA, but they may use the PAG, and the ventromedial and lateral tegmentum of the medulla oblongata, as a relay to gain access to the NRA.

The spino-brainstem-spinal circuit for the mating reflex

Many sensory afferents from the bladder, colon, vagina, pelvic floor, perineal skin and related structures terminate in the lateral dorsal horn and lateral intermediate zone of the L7–S1–S2 spinal segments of the cat. In this same region, a large cell group is located that sends fibers to the central part of the lateral PAG (VanderHorst et al., 1996; Fig. 3). As indicated previously, the PAG has direct access to the NRA in the caudal medulla which, in turn, projects to those MNs in the lumbosacral cord that produce the mating posture (Fig. 4). Although further behavioral studies are needed to confirm our hypothesis, we propose that at least in the cat, the sacral cord-PAG-NRA-mating MN pathway generates the posture necessary for mating. Obviously, other parts of the limbic brain that have access to the PAG can influence this spino-brainstem-spinal circuit. It seems likely that comparable systems exist in other mammals, too, because in the female hamster, which exhibits a less complicated but very pronounced mating or lordosis posture, a similar PAG-NRA-axial muscle MN pathway has been found (Gerrits and Holstege, 1999).

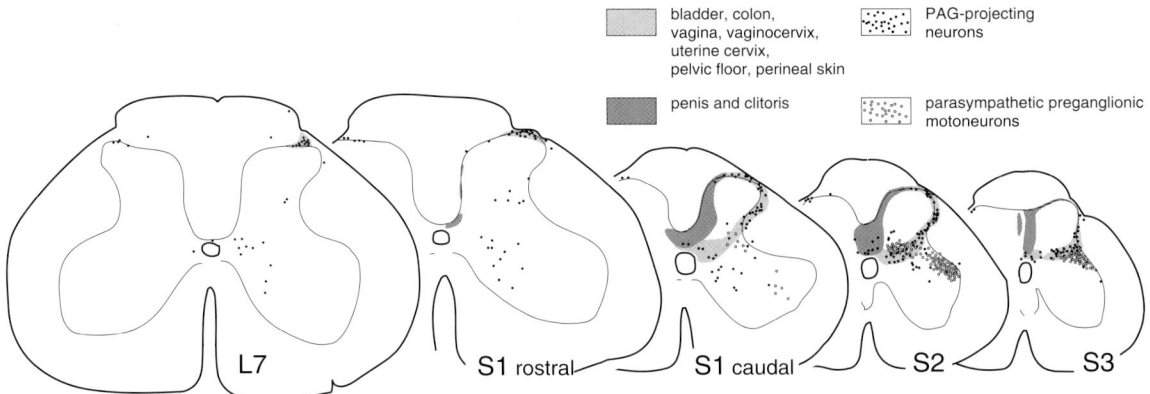

Fig. 3. Schematic overview of where the urogenital afferent projections terminate in the sacral cord and where the neurons are located that project to the PAG.

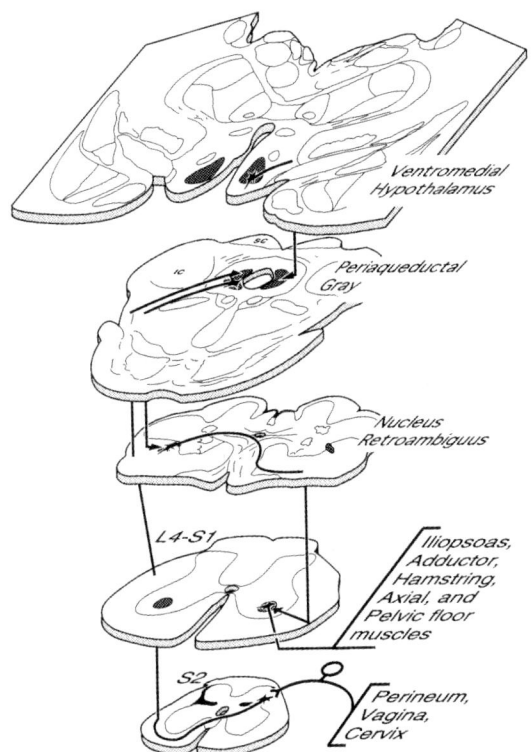

Fig. 4. Schematic overview of the ascending and descending pathways that play a role in the mating reflex circuit in female cats.

Mating behavior in the human

A largely unresolved issue is how mating is organized in the human. Our species does not have a very distinct mating posture like that in the cat and hamster. It can be assumed, however, that in the human, too, the motor output related to mating is controlled by specific descending pathways. Techniques used in experimental animals to investigate these pathways are not applicable to the human. Modern neuroimaging techniques, however, supply the means to reveal which human brain structures are involved in mating, and especially ejaculation and orgasm.

Recently, we undertook a PET study in which heterosexual male volunteers were asked to ejaculate in response to manual stimulation by their female partner (Holstege et al., 2003). The PET technique using radioactive water ($H_2^{15}O$) shows increases or decreases in blood flow in distinct parts of the brain, which represent increases or decreases, respectively, in the activation of neurons in these areas (see also F. Mori et al. and H. Shibasaki et al., this volume). Although the resolution of PET is relatively limited, the results of our ejaculation/orgasm studies were remarkable. During ejaculation, compared to manual stimulation of the penis, the strongest activation was in the so-called meso-diencephalic region. It includes the ventral tegmental area (VTA; known as the 'reward' area), SPF, ventromedial posterior thalamic nucleus, intralaminar nuclei, and the LCTF (Fig. 5). A similar region in the meso-diencephalic transition zone was activated during a cocaine (Breiter et al., 1997), and especially heroin, rush (Sell et al., 1999), thereby suggesting that these experiences are similar to that in a sexual orgasm. This finding might also explain why substances such as cocaine and heroin are so addictive.

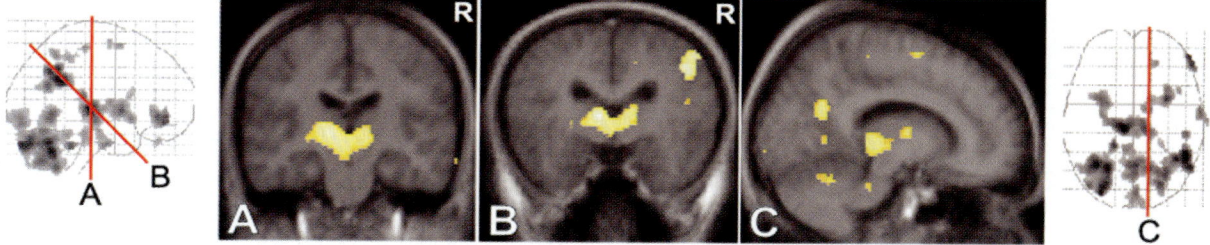

Fig. 5. Significant activations for ejaculation minus penile stimulation ($P < 0.01$; corrected for multiple comparisons), superimposed on the average T1-weighted MRI of the subjects, in the meso-diencephalic junction in a coronal (A), oblique (B) and sagittal (C) section. The red lines on the glass brains (SPMs) indicate the orientation of the sections.

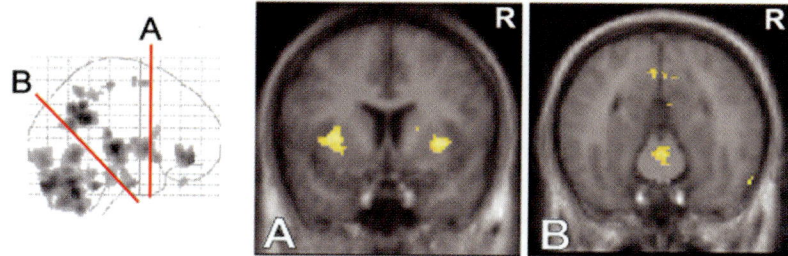

Fig. 6. Significant activations for ejaculation minus penile stimulation ($P < 0.01$; corrected for multiple comparisons). In (A), a coronal section through the striatum shows strong activation in the putamen/claustrum. In (B), an oblique section through the pons and caudal inferior colliculus reveals increased blood flow in the pontine medial tegmentum.

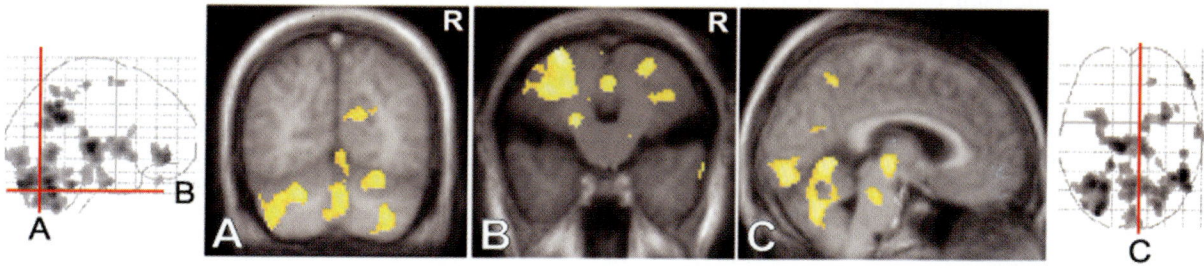

Fig. 7. Significant activations for ejaculation minus penile stimulation ($P < 0.01$; corrected for multiple comparisons). Coronal (A), horizontal (B) and sagittal (C) sections depict large regions of increased blood flow in the cerebellum. The sagittal section (C) also shows activation in the occipital cortex (BA 18), the medial ponto-mesencephalic tegmentum, and the meso-diencephalic junction.

Other findings in our ejaculation/orgasm studies included demonstration of an increased activation of the lateral putamen (Fig. 6A), selected parts of the prefrontal, temporal, parietal and insular cortex, and a very strong activation of the cerebellum (Fig. 7). Strong activation was also found in the medial pontine tegmentum (Fig. 6B). Surprisingly, no activation was found in the hypothalamus or preoptic area (Fig. 8a). One might speculate that these regions play a role in creating the conditions in which mating can take place (e.g., estrous vs. nonestrous) but that they are not involved in controlling the motor act, itself.

An important decrease of activation was also found. The medial parts of the amygdala (Fig. 8b) were deactivated, not only during ejaculation and orgasm, but also during sexual stimulation and erection. A similar decrease of amygdala activity was also found during romantic love (Bartels and Zeki, 2000). An active amygdala is crucial for

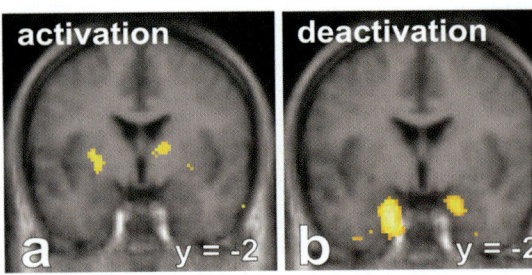

Fig. 8. (a) A coronal section 2-mm caudal to the anterior commissure shows significant activations for ejaculation minus stimulation ($P < 0.01$; corrected for multiple comparisons). Activation is found in the putamen/claustrum and anterior thalamus but not in the hypothalamus. (b) PET-scan results showing significant amygdala deactivations ($P < 0.001$; uncorrected) comparing ejaculation minus rest.

individual survival by its constant monitoring of environmental stimuli. In case of possibly hazardous events, the amygdala elicits a fear response to protect the organism from harm. In the context of sexual behavior, however, such vigilance could easily block the sexual act, thereby leading to unsuccessful reproduction. It would seem that in order to prevent such a disruption, brain structures involved in sexual behavior decrease vigilance by inhibiting amygdala activity. Although deactivation of the amygdala may make the organism less concerned of potential danger, it enhances the odds of species survival, because reproduction is more likely to succeed.

One important conclusion that one might draw from the PET-scan results is that sexual behavior is brought about by a limited number of structures within the CNS. One might expect that these regions play a role in sexual motor activity and, as such, they should be considered as CNS control components for posture and movement. According to the concept of Holstege (1997), sexual motor activity is controlled by the 'emotional' motor system. Whether these PET-identified structures control MNs in the same fashion as is known for the cat, however, remains to be elucidated. Further imaging studies are now needed to delineate the most crucial brain regions for sexual activity in the human. Finally, such studies will certainly advance understanding of human sexual behavior and provide the possibility to improve sexual activity in those who suffer from problems in this area.

Abbreviations

BNST   bed nucleus of the stria terminalis
CNS   central nervous system
LCTF   lateral central tegmental field
MeA   medial nucleus of the amygdala
MPOA   medial preoptic area
NRA   nucleus retroambiguus
PAG   periaqueductal gray
PET   positron emission tomography
SPF   subparafascicular nucleus

References

Bartels, A. and Zeki, S. (2000). The neural basis of romantic love. Neuroreport, 11: 3829–3834.

Baum, M.J. and Everitt, B.J. (1992). Increased expression of c-fos in the medial preoptic area after mating in male rats: role of afferent inputs from the medial amygdala and central tegmental field. Neuroscience, 50: 627–646.

Brackett, N.L. and Edwards, D.A. (1984). Medial preoptic connections with the midbrain tegmentum are essential for male sexual behavior. Physiol. Behav., 32: 79–84.

Breiter, H.C., Gollub, R.L., Weisskoff, R.M., Kennedy, D.N., Makris, N., Berke, J.D., Goodman, J.M., Kantor, H.L., Gastfriend, D.R., Riorden, J.P., Mathew, R.T., Rosen, B.R., Hyman, S.E. (1997). Acute effects of cocaine on human brain activity and emotion. Neuron, 19: 591–611.

Coolen, L.M., Peters, H.J. and Veening, J.G. (1996). Fos immunoreactivity in the rat brain following consummatory elements of sexual behavior: a sex comparison. Brain Res., 738: 67–82.

Gerrits, P.O. and Holstege, G. (1996). Pontine and medullary projections to the nucleus retroambiguus: A WGA-HRP and autoradiographic tracing study in the cat. J. Comp. Neurol., 373: 173–185.

Gerrits, P.O. and Holstege, G. (1999). Descending projections from the nucleus retroambiguus to the iliopsoas motoneuronal cell groups in the female golden hamster: possible role in reproductive behavior. J. Comp. Neurol., 403: 219–228.

Heeb, M.M. and Yahr, P. (1996). C-fos immunoreactivity in the sexually dimorphic area of the hypothalamus and related brain regions of male gerbils after exposure to sex-related stimuli or performance of specific sexual behaviors. Neuroscience, 72: 1049–1071.

Heeb, M.M. and Yahr, P. (2000). Cell-body lesions of the posterodorsal preoptic nucleus or posterodorsal medial amygdala, but not the parvicellar subparafascicular thalamus, disrupt mating in male gerbils. Physiol. Behav., 68: 317–331.

Holstege, G. (1989). An anatomical study of the final common pathway for vocalization in the cat. J. Comp. Neurol., 284: 242–252.

Holstege, G., Georgiadis, J.R., Paans, A.M.J., Meiners, L.C., van der Graaf, F.H.C.E. and Reinders, A.A.T.S. (2003) Brain activation during human male ejaculation. J. Neurosci., in press.

Sell, L.A., Morris, J., Bearn, J., Frackowiak, R.S., Friston, K.J. and Dolan, R.J. (1999). Activation of reward circuitry in human opiate addicts. Eur. J. Neurosci., 11: 1042–1048.

VanderHorst, V.G.J.M. and Holstege, G. (1995). Caudal medullary pathways to lumbosacral motoneuronal cell groups in the cat; evidence for direct projections possibly representing the final common pathway for lordosis. J. Comp. Neurol., 359: 457–475.

VanderHorst, V.G.J.M. and Holstege, G. (1997a). Estrogen induces axonal outgrowth in the nucleus retroambiguus-lumbosacral motoneuronal pathway in the adult female cat. J. Neurosci., 17: 1122–1136.

VanderHorst, V.G.J.M. and Holstege, G. (1997b). Nucleus retroambiguus projections to lumbosacral motoneuronal cell groups in the male cat. J. Comp. Neurol., 382: 77–88.

VanderHorst, V.G.J.M., Mouton, L.J., Blok, B.F.M. and Holstege, G. (1996). Somatotopical organization of input from the lumbosacral cord to the periaqueductal gray in the cat; possible implications for aggressive and defensive behavior, micturition, and lordosis. J. Comp. Neurol., 376: 361–385.

SECTION II

Spinal cord and brainstem: developmental and comparative issues

CHAPTER 5

Developmental changes in rhythmic spinal neuronal activity in the rat fetus

Norio Kudo*, Hiroshi Nishimaru and Kiyomi Nakayama

Department of Physiology, Institute of Basic Medical Sciences, University of Tsukuba, Tsukuba, Ibaraki 305-8575, Japan

Abstract: In the developing rat spinal cord, formation and differentiation of the central pattern generator for locomotion occur during the prenatal period. Early on, excitatory synaptic transmission mediated by glycine receptors plays a leading role for rhythmogenesis, at a later stage, followed by glutamate-receptor-mediated synaptic transmission becoming dominant. The maturation of inhibitory circuitry in the spinal cord, mediated largely by glycinergic synapses, is crucial for the generation of alternating activity between left/right limbs and flexor/extensor muscles. Formation of left/right alternation is presumably due to developmental changes in the properties of the postsynaptic neurons, themselves, whereas flexor/extensor alternation requires the additional emergence of inhibitory synaptic functions in the spinal cord.

Introduction

Neuronal networks located in the spinal cord can generate rhythmic motor activities, such as swimming and walking (for review, see Grillner, 1975, 1985). In the rat, behavioral studies have shown that interactive, locomotor-like movements of the hindlimb appear by the last stage of gestation (Bekoff and Lau, 1980; Suzue, 1992). Much is to be learned about the prenatal development of the neuronal network than can produce such movements in the fetus. In this chapter, we focus on our recent studies on this topic, using an in vitro spinal cord preparation of the fetal rat. There is emphasis on developmental changes in glycinergic transmission within the neuronal networks that produce rhythmic motor activity (see also Clarac et al. and Ezure, Chapters 6 and 7 of this volume).

Background: locomotor activity in an in vitro preparation in the neonatal rat

Locomotor-like muscle activity in an in vitro preparation was first demonstrated in a spinal cord–hindlimb muscle preparation taken from the neonatal rat (Kudo and Yamada, 1987). Alternate EMG burst activity in ipsilateral ankle flexor and extensor muscles was induced by bath application of an appropriate N-methyl-D-aspartate (NMDA) receptor agonist. Such activity was also observed in homonymous bilateral muscles just as it occurs during regular postnatal hindlimb locomotion (Fig. 1A). Subsequently, we showed that it was possible to record locomotor activity by use of ventral-root (neurogram) recordings at different lumbar segments in an in vitro spinal cord preparation of the neonatal rat. For example, Fig. 1B shows the relationship between EMG activity and neurogram discharges in this preparation. Alternate rhythmic neurogram bursts were recorded from the central stump of the L2 and L5 ventral roots ipsilateral to the EMG recordings. A morphological study (Nicolopoulos-Stournaras and Iles, 1983)

*Corresponding author: Tel.: +81-29-853-3080; Fax: +81-29-853-3495; E-mail: kudo@md.tsukuba.ac.jp

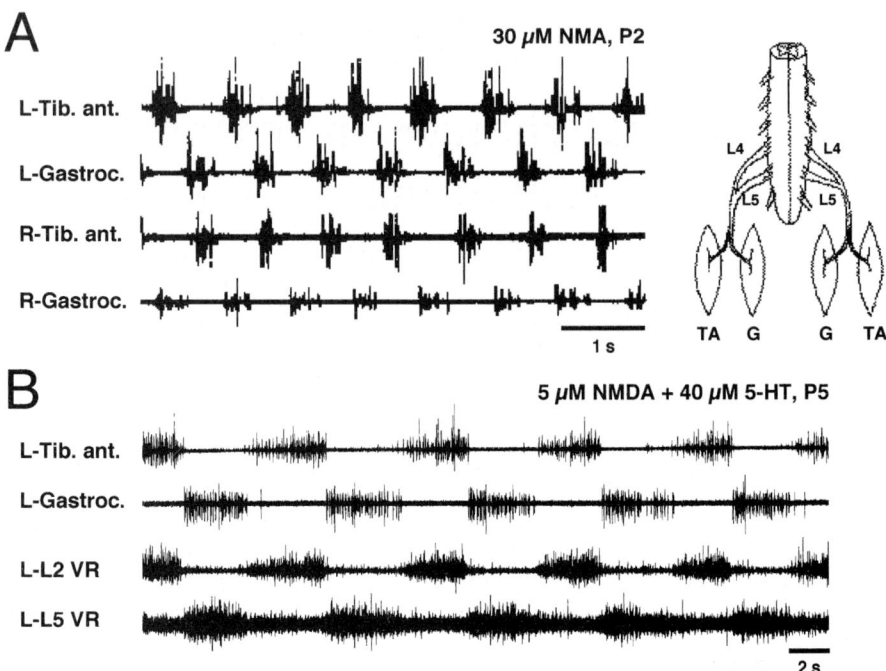

Fig. 1. Locomotor-like activity recorded in a hindlimb-attached spinal cord preparation of the neonatal rat following: (A) bath application of N-methyl-D,L-aspartate (NMA); and (B) application of NMA plus 5-HT. (A) Simultaneous EMG recordings from the left (L) and right (R) tibialis anterior (Tib. ant.) and gastrocnemius (Gastroc.) muscles at postnatal (P) day 2. The right-side schematic shows the anatomical arrangement. (B) Simultaneous EMG recordings from L-Tib. ant. and L. Gastroc., and neurograms recorded from the L-L2 and L-L5 ventral roots (VR) on P5. Note that the activity in L-L2 VR corresponds to activity in the ipsilateral flexor (tib. ant.) muscle and that of L-L5 to the ipsilateral extensor (Gastroc.).

revealed that both of these spinal segments contain flexor and extensor motoneurons. Note, however, that Fig. 1B shows that the bursts in the L2 neurogram were in phase with the bursts of flexor muscle EMG activity. In contrast, burst activity in the L5 neurogram mainly represented extensor muscle activity (Kiehn and Kjærulff, 1996; Iizuka et al., 1997). This result might reflect a large preponderance of one type of motoneuron population in a given spinal segment and/or different excitability levels within a given segment for the motoneurons supplying flexor versus extensor muscles.

It has been shown to this point that in isolated spinal cord or spinal cord–hindlimb preparations of the neonatal rat, locomotor-like EMG activity in hindlimb muscles, or alternate rhythmic neurogram bursts in bilateral ventral roots of a given lumbar segment, can be induced by bath application of any one of several neuroactive substances, including NMDA, 5-hydroxytryptamine (5-HT), acetylcholine, dopamine or noradrenarine (Kudo and Yamada, 1987; Smith and Feldman, 1987; Cazalets et al., 1992; Cowley and Schmidt, 1994; Kiehn and Kjærulff, 1996; Kiehn et al., 1999). Such in vitro preparations have been employed extensively in studies of motor rhythmogenesis, and their use has provided much valuable information on the neuronal mechanisms operating within the central pattern generator (CPG) for mammalian locomotion (for recent reviews, see Cazalets et al., 1998; Kiehn and Kjærulff, 1998; Schmidt et al., 1998).

Fetal motor activity: developmental changes and their mechanisms

Until quite recently, little was known about the prenatal development of neuronal networks generating locomotor activity in the mammal. This deficit was due largely to the technical difficulty of

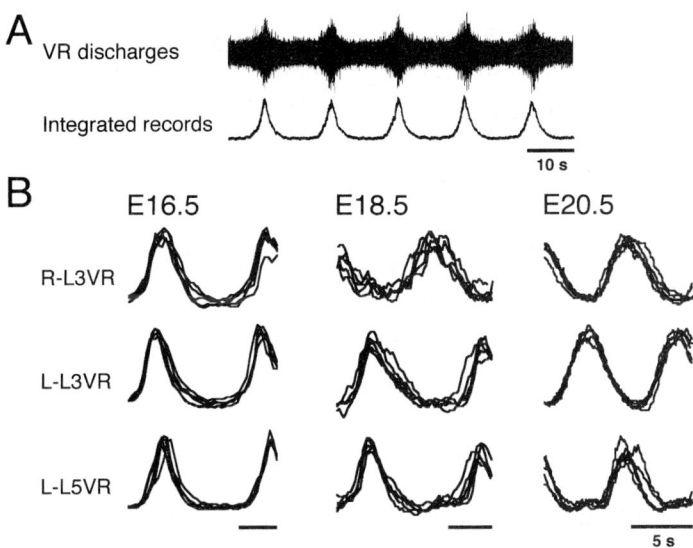

Fig. 2. Neurogram recordings of a 5-HT-induced rhythm at E16.5, E18.5 and E20.5. Characteristic features of the spatial pattern are illustrated by superimposing integrated records of the raw activity in selected VRs. (A) An example of rhythmic VR discharges (upper trace) and their integrals (lower trace). (B) Neurograms recorded simultaneously from the R-L3 (upper), L-L3 (middle) and L-L5 (lower) VRs at E16.5, E18.5 and E20.5.

maintaining stable recording from spinal neurons, and the lack of a suitable method for evoking sustained rhythmic motor activity. Nonetheless, it was necessary to resolve these problems, because an in vitro experimental system is extremely useful for studying the formation and differentiation of the locomotor CPG during the mammalian fetal stage (Greer et al., 1992; Nishimaru et al., 1996).

Embryonic day (E) 14.5 is the first at which 5-HT-induced rhythmic motor activity appears in the lumbar ventral roots, i.e., 8 days before birth (Nakayama et al., 2001). The spatial pattern of this activity is stage dependent. For example, Fig. 2 shows the 5-HT-induced rhythm recorded from various ventral roots at E16.5, E18.5 and E20.5. In E16.5 preparations, the rhythmic bursts occur synchronously in bilateral roots at the same segmental level. At E18.5, an alternating rhythm is observed between the left- and right-side roots. Motor activity alternating between L3 and L5 segments first appears at E20.5, i.e., 2 days later than the onset of the left/right alternation (Iizuka et al., 1998; Nishimaru and Kudo, 2000).

An NMDA- or 5-HT-induced alternate (locomotor-like) spinal activity pattern is changed to a synchronous output by the concomitant application of strychnine, a glycine-receptor antagonist. This occurs in both the neonatal rat (Cowley and Schmidt, 1995) and the rat fetus at and after E18.5 (Ozaki et al., 1996; Iizuka et al., 1998). Moreover, intracellular recordings from the neonatal rat spinal cord have shown that motoneurons receive a mainly glycinergic rhythmic synaptic inhibition from the contralateral half of the locomotor network (Kjaerulff and Kiehn, 1997). Thus, the left/right alternation of motor output from the rat spinal cord is likely to be mediated by glycinergic inhibitory synaptic inputs, just as it is in the tadpole (Soffe, 1987) and lamprey (for review, see Grillner and Matsushima, 1991).

The transition from a synchronous to an alternate pattern of rhythmic spinal motor activity seems to occur near E17.5 (Ozaki et al., 1996; Nishimaru and Kudo, 2000). At this stage, the pattern of rhythmic bursting on the left- versus right-side ventral roots is variable (i.e., alternate vs. synchronized), even within a single preparation. Furthermore, the rhythm is frequently at a different frequency on the two sides. Such variable phase relationship and burst frequency at E17.5 may be due to the presence of immature inhibitory interactions between the neuronal circuits on the two sides of the spinal cord. Immature interconnections might impede, to a variable extent, the phasic rhythm induced on one side by the activity of the other side. Moreover, it is noteworthy that E17.5 is exactly the age at which glycine-induced inhibitory synaptic transmission becomes manifest in the neuronal networks of the spinal cord (Wu et al., 1992; Nishimaru et al., 1996).

Glycinergic inhibition also plays an important role in the alternation of activity observed at different segmental levels. For example, in prenatal preparations, an alternate 5-HT-induced rhythmic pattern at different segmental levels can be changed to a synchronous one by the concomitant application of strychnine (Iizuka et al., 1998). A similar strychnine effect has also been shown in the neonatal rat (Cowley and Schmidt, 1995), i.e., neurographic recordings from the tibial (flexor) and peroneal (extensor) branch of the sciatic nerve revealed that a reciprocal pattern of the 5-HT-induced rhythmic activity was changed to a synchronous one in the presence of strychnine.

Glycinergic and glutamatergic synaptic transmission in rhythmogenesis

Glutamatergic transmission is important in generating, or at least mediating, the rhythmic activity of interneuronal circuits which is impressed on motoneurons in the lamprey (Dale and Grillner, 1986), tadpole (Dale and Roberts, 1985), neonatal rat (Cazalets et al., 1992) and cat (Brownstone et al., 1994). Both NMDA and non-NMDA receptors are likely to be involved in this process during locomotor-like activity of the neonatal rat (Beato et al., 1997).

In the fetal spinal cord of the rat, kynurenate, a broad-spectrum glutamate-receptor antagonist, completely blocks 5-HT-induced rhythmic activity at and after E18.5, whereas, at this stage, strychnine provides no such blockage. These findings indicate that generation of rhythmic activity at and after E18.5 is mediated by glutamate receptors, with glycine-mediated inhibitory synaptic transmission not essential for rhythmogenesis at these late fetal stages (Nakayama et al., 2001).

Interestingly, at the earlier E14.5 stage, bath application of kynurenate causes little decrease in the frequency and amplitude of rhythmic bursts. However, it partially suppresses them at E16.5 (Nakayama et al., 2001). In contrast, strychnine drastically reduces such rhythmic motor activity at both E14.5 and E15.5. This suggests that glycine plays its important role as an excitatory transmitter in 5-HT-induced rhythmic activity at the earlier fetal stages. Indeed, at these stages, exogenously applied glycine excites both interneurons and motoneurons, and induces burst discharges in the ventral roots (Nishimaru et al., 1996). At and shortly after the onset of rhythmogenesis, activity is driven mainly by glycinergic versus glutamatergic excitatory synaptic inputs. It is noteworthy that the simultaneous application of kynurenate and strychnine completely blocks 5-HT-induced rhythmic activity at E16.5 (Nakayama et al., 2001), thereby suggesting that at this stage, both glutamatergic and glycinergic excitations are implicated in rhythmic motor activity.

Fig. 3 illustrates the idea that the future (postnatal) inhibitory circuit emerges earlier in the fetus than the future excitatory circuit. First, during the early fetal period, the future inhibitory synaptic connections meditated by glycine and GABA emerge in the spinal cord. Owing to the high intracellular Cl^- concentration at this stage, these amino acids excite fetal spinal neurons via excitatory synaptic connections. It has been shown that glycine and GABA elevate Ca^{2+} in immature spinal neurons (Reichling et al., 1994; Kulik et al., 2000). This elevation, brought on by glycine-mediated depolarization, seems to control the outgrowth of the neurons' dendrites and axons (Kater et al., 1988; Metzger et al., 1998; Collins et al., 1991). Moreover, it has also been shown recently that such Ca^{2+} elevation is required for the clustering of postsynaptic glycine receptors in cultured embryonic spinal neurons (Kirsch and Betz, 1998). Such a transient excitatory action of glycine may be important for synaptogenesis in the spinal cord, as well.

The amplitude of depolarization induced by glycine and GABA decreases as birth approaches (Wu et al., 1992). In all likelihood, this developmental decrease in the magnitude of the GABA- and glycine-induced depolarization is associated with the development of an outwardly directed Cl^- pump, as occurs in developing pyramidal neurons of the rat hippocampus (Rivera et al., 1999). Moreover, it has been reported recently that GABA itself activates the expression of the Cl^- transporter, KCC2, and thereby promotes the developmental switch of

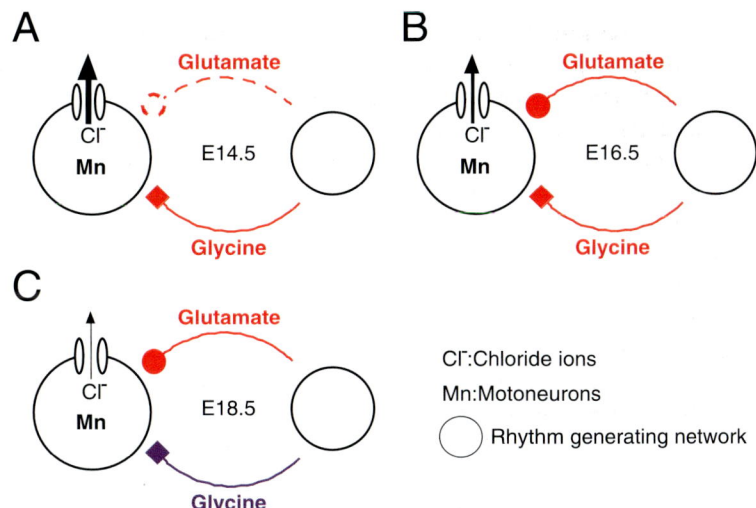

Fig. 3. Scheme of the development of the neuronal network that generates locomotor activity in the rat spinal cord. (A) At E14.5, excitatory synaptic transmission (indicated with a filled-circle terminal) from the network to produce rhythmic motoneuron (Mn) activity is mediated largely by glycine, with a weak effect from glutamate mainly convey rhythmic activity to motoneurons. This excitatory action is due to high intracellular Cl$^-$ concentration in the Mns. (B) By E16.5, glutamatergic excitatory synaptic transmission is clear-cut and it becomes the predominant effect by E18.5 (C) at which time the glycine effect becomes inhibitory (indicated by a filled-circle terminal).

neuronal GABAergic responses from excitation to inhibition in cultured immature hippocampal neurons (Ganguly et al., 2001). During the fetal period, GABA, and presumably glycine as well, could have similar effects on the developing spinal neurons.

Developmental changes of the intersegmental connection of the CPG

In the previous section, we described the development of interlimb coordination during locomotor activity, but intermuscular (e.g., flexor–extensor) coordination is equally important in mammalian locomotion. At the earlier fetal stages, the function and operation of neuromuscular junctions are still immature. This makes it difficult to record EMG activity in individual muscles. Recall, however, that Fig. 1B shows that in the neonatal rat, a ventral-root neurogram at a given segmental level can represent predominant flexor or extensor motor activity. This arrangement can also be used in the rat fetus to examine developmental changes in pattern generating spinal networks (Ozaki et al., 1996; Iizuka et al., 1998; Kudo and Nishimaru, 1998).

Our previous studies revealed that motor activity alternating between different spinal segments, probably represents alternating flexor and extensor activity. It first appears at E20.5, 2 days later than the onset of left/right alternation (Fig. 2). At E20.5, there is already an inhibitory coupling between the left and right sides. This suggests that an L3 (or L2)–L5 alternation is not likely to be initiated by a simple change in the effect of glycine and GABA on motoneurons from excitatory to inhibitory, such is likely to cause left–right alternation. The most likely explanation is that inhibitory connections between L2/L3 and L5 neurons are formed after E18.5, and they then start to function at E20.5. Note also that the amplitude of spinal-reflex responses evoked by dorsal-root stimulation is not affected by strychnine before E19.5. At and after E19.5, however, strychnine enhances such reflex discharges, especially intersegmental ones (Saito, 1979). It is likely that a similar delayed emergence of an intersegmental inhibitory system may be a feature of the development of the locomotor CPG.

An alternative possibility is that the rhythmic motor activity seen at different segmental levels may represent near-exclusive flexor motor activity before E20.5, because the extensor system is immature at early developmental stages. For example, it has been reported that during the early neonatal period, activity in extensor muscles develops later than in flexors (Westerga and Gramsbergen, 1993). This difference is unlikely to be the main factor for L2/L3–L5 synchronicity before E20.5, however, because a behavioral study has shown that spontaneous movements of alternating flexion and extension can be observed in the hindlimb even before E20.5 (Suzue, 1992). Furthermore, a stretch reflex of the triceps surae (ankle extensor) muscle has been observed as early as E19.5 (Kudo and Yamada, 1985).

Concluding remarks

Formation and differentiation of the CPG for locomotion occurs during the prenatal period in the rat. Glycinergic transmission appears to play the leading role for rhythmogenesis at the earlier developmental stages, with glutamatergic transmission becoming more dominant at later stages. The generation of alternate contralateral activity as well as ipsilateral alternation between flexor and extensor muscles require the appearance of a glycinergic inhibitory system in the spinal cord. The development of contralateral alternation may be due to developmental changes in the properties of the postsynaptic spinal neurons, themselves, while ipsilateral flexor/extensor alternation may require the additional emergence of inhibitory synaptic functions in the spinal cord.

Acknowledgments

We would like to thank Ms. A. Ohgami for excellent technical assistance. This work was supported in part by a Grant in Aid for Scientific Research from the Ministry of Education, Science and of Japan.

Abbreviations

5-HT	5-hydroxytryptamine
CPG	central pattern generator
E	embryonic day
EMG	electromyogram
GABA	gamma-aminobutyric acid
KCC2	neuronal Cl^--extruding K^+/Cl^- cotransporter
L2	second lumbar segment
L3	third lumbar segment
L5	fifth lumbar segment
NMA	N-methyl-D,L-aspartate
NMDA	N-methyl-D-aspartate
VR	ventral root

References

Beato, M., Bracci, E. and Nistri, A. (1997). Contribution of NMDA and non-NMDA glutamate receptors to locomotor pattern generation in the neonatal rat spinal cord. Proc. R. Soc. Lond. B Biol. Sci., 264: 877–884.

Bekoff, A. and Lau, B. (1980). Interlimb coordination in 20-day-old rat fetuses. J. Exp. Zool., 214: 173–175.

Brownstone, R.M., Gossard, J.P. and Hultborn, H. (1994). Voltage-dependent excitation of motoneurones from spinal locomotor centres in the cat. Exp. Brain Res., 102: 34–44.

Cazalets, J.R., Sqalli-Houssaini, Y. and Clarac, F. (1992). Activation of the CPGs for locomotion by serotonin and excitatory amino acids in neonatal rat. J. Physiol. (Lond.), 455: 187–204.

Cazalets, J.R., Bertrand, S., Sqalli-Houssaini, Y. and Clarac, F. (1998). GABAergic control of spinal locomotor networks in the neonatal rat. Ann. N.Y. Acad. Sci., 860: 168–180.

Collins, F., Schmidt, M.F., Guthrie, P.B. and Kater, S.B. (1991). Sustained increase in intracellular calcium promotes neuronal survival. J. Neurosci., 11: 2582–2587.

Cowley, K.C. and Schmidt, B.J. (1994). A comparison of motor patterns induced by N-methyl-D-aspartate, acetylcholine and serotonin in the in vitro neonatal rat spinal cord. Neurosci. Lett., 171: 147–150.

Cowley, K.C. and Schmidt, B.J. (1995). Effects of inhibitory amino acid antagonists on reciprocal inhibitory interactions during rhythmic motor activity in the in vitro neonatal rat spinal cord. J. Neurophysiol., 74: 1109–1117.

Dale, N. and Grillner, S. (1986). Dual-component synaptic potentials in the lamprey mediated by excitatory amino acid receptors. J. Neurosci., 6: 2653–2661.

Dale, N. and Roberts, A. (1985). Dual-component amino-acid-mediated synaptic potentials: excitatory drive for swimming in Xenopus embryos. J. Physiol. (Lond.), 363: 35–59.

Ganguly, K., Schinder, A.F., Wong, S.T. and Poo, M. (2001). GABA itself promotes the developmental switch of neuronal GABAergic responses from excitation to inhibition. Cell, 105: 521–532.

Greer, J.J., Smith, J.C. and Feldman, J.L. (1992). Respiratory and locomotor patterns generated in the fetal rat brain stem-spinal cord in vitro. J. Neurophysiol., 67: 996–999.

Grillner, S. (1975). Locomotion in vertebrates: central mechanisms and reflex interaction. Physiol. Rev., 55: 247–304.

Grillner, S. (1985). Neurobiological bases of rhythmic motor acts in vertebrates. Science, 228: 143–149.

Grillner, S. and Matsushima, T. (1991). The neural network underlying locomotion in lamprey. Neuron, 7: 1–15.

Iizuka, M., Kiehn, O. and Kudo, N. (1997). Development in neonatal rats of the sensory resetting of the locomotor rhythm induced by NMDA and 5-HT. Exp. Brain Res., 114: 193–204.

Iizuka, M., Nishimaru, H. and Kudo, N. (1998). Development of the spatial pattern of 5-HT-induced locomotor rhythm in the lumbar spinal cord of rat fetuses in vitro. Neurosci. Res., 31: 107–111.

Kater, S.B., Mattson, M.P., Cohan, C. and Connor, J. (1988). Calcium regulation of the neuronal growth cone. Trends Neurosci., 11: 315–321.

Kiehn, O. and Kjærulff, O. (1996). Spatiotemporal characteristics of 5-HT and dopamine-induced rhythmic hindlimb activity in the in vitro neonatal rat. J. Neurophysiol., 75: 1472–1482.

Kiehn, O. and Kjærulff, O. (1998). Distribution of CPGs for rhythmic motor outputs in the spinal cord of limbed vertebrates. Ann. N.Y. Acad. Sci., 860: 110–129.

Kiehn, O., Sillar, K.T., Kjaerulff, O. and McDearmid, J.R. (1999). Effects of noradrenaline on locomotor rhythm-generating networks in the isolated neonatal rat spinal cord. J. Neurophysiol., 82: 741–746.

Kirsch, J. and Betz, H. (1998). Glycine-receptor activation is required for receptor clustering in spinal neurons. Nature, 392: 717–720.

Kjaerulff, O. and Kiehn, O. (1997). Crossed rhythmic synaptic input to motoneurons during selective activation of the contralateral spinal locomotor network. J. Neurosci., 17: 9433–9447.

Kudo, N. and Nishimaru, H. (1998). Reorganization of locomotor activity during development in the prenatal rat. Ann. N.Y. Acad. Sci., 860: 306–317.

Kudo, N. and Yamada, T. (1985). Development of the monosynaptic stretch reflex in the rat: an in vitro study. J. Physiol. (Lond.), 369: 127–144.

Kudo, N. and Yamada, T. (1987). N-Methyl-D,L-aspartate-induced locomotor activity in a spinal cord-hindlimb muscles preparation of the newborn rat studied in vitro. Neurosci. Lett., 75: 43–48.

Kulik, A., Nishimaru, H. and Ballanyi, K. (2000). Role of bicarbonate and chloride in GABA- and glycine-induced depolarization and $[Ca^{2+}]_i$ rise in fetal rat motoneurons in situ. J. Neurosci., 20: 7905–7913.

Metzger, F., Wiese, S. and Sendtner, M. (1998). Effect of glutamate on dendritic growth in embryonic rat motoneurons. J. Neurosci., 18: 1735–1742.

Nakayama, K., Nishimaru, H. and Kudo, N. (2001). Developmental changes in 5-HT-induced rhythmic activity in the spinal cord of rat fetuses in vitro. Neurosci. Lett., 307: 1–4.

Nicolopoulos-Stournaras, S. and Iles, J.F. (1983). Motor neuron columns in the lumbar spinal cord of the rat. J. Comp. Neurol., 217: 75–85.

Nishimaru, H. and Kudo, N. (2000). Formation of the CPG for locomotion in the rat and mouse. Brain Res. Bull., 53: 661–669.

Nishimaru, H., Iizuka, M., Ozaki, S. and Kudo, N. (1996). Spontaneous motoneuronal activity mediated by glycine and GABA in the spinal cord of rat fetuses in vitro. J. Physiol. (Lond.), 497: 131–143.

Ozaki, S., Yamada, T., Iizuka, M., Nishimaru, H. and Kudo, N. (1996). Development of locomotor activity induced by NMDA receptor activation in the lumbar spinal cord of the rat fetus studied in vitro. Brain Res. Dev. Brain Res., 97: 118–125.

Reichling, D.B., Kyrozis, A., Wang, J. and MacDermott, A.B. (1994). Mechanisms of GABA and glycine depolarization-induced calcium transients in rat dorsal horn neurons. J. Physiol. (Lond.), 476: 411–421.

Rivera, C., Voipio, J., Payne, J.A., Ruusuvuori, E., Lahtinen, H., Lamsa, K., Pirvola, U., Saarma, M. and Kaila, K. (1999). The K^+/Cl^- co-transporter KCC2 renders GABA hyperpolarizing during neuronal maturation. Nature, 397: 251–255.

Saito, K. (1979). Development of spinal reflexes in the rat fetus studied in vitro. J. Physiol. (Lond.), 294: 581–594.

Schmidt, B.J., Hochman, S. and MacLean, J.N. (1998). NMDA receptor-mediated oscillatory properties: potential role in rhythm generation in the mammalian spinal cord. Ann. N.Y. Acad. Sci., 860: 189–202.

Smith, J.C. and Feldman, J.L. (1987). In vitro brainstem-spinal cord preparations for study of motor systems for mammalian respiration and locomotion. J. Neurosci. Methods, 21: 321–333.

Soffe, S.R. (1987). Ionic and pharmacological properties of reciprocal inhibition in Xenopus embryo motoneurones. J. Physiol. (Lond.), 382: 463–473.

Suzue, T. (1992). Physiological activities of late-gestation rat fetuses in vitro. Neurosci. Res., 14: 145–157.

Westerga, J. and Gramsbergen, A. (1993). Changes in the electromyogram of two major hindlimb muscles during locomotor development in the rat. Exp. Brain Res., 92: 479–488.

Wu, W.L., Ziskind-Conhaim, L. and Sweet, M.A. (1992). Early development of glycine- and GABA-mediated synapses in rat spinal cord. J. Neurosci., 12: 3935–3945.

CHAPTER 6

The maturation of locomotor networks

Francois Clarac*, Frédéric Brocard and Laurent Vinay

Institut de Neurosciences Physiologiques et Cognitives – INPC, CNRS, Aix-Marseille II, 13402 Marseille Cedex 20, France

Abstract: In both vertebrates and invertebrates, the elaboration of locomotion, and its neural control by the central nervous system, are extremely flexible. This is due not only to the network properties of relevant sets of central neurons, but also to the active participation of mutually co-operative central and peripheral loops of neural projections and activity. In this chapter, we describe experiments in which the above concepts have been advanced by comparing locomotor properties in the adult vs. neonatal rat preparation. Data obtained from the in vivo vs. in vitro preparation, and swimming vs. walking behavior, suggest that the locomotor pattern progressively exhibited after birth corresponds to successive steps in the maturation of locomotor networks. Our work emphasises that during the late pre- and early postnatal period, three distinct neural entities—segmental sensory input, descending pathways, and motoneurons—play a key role in the maturation of locomotion and its neural control. We propose that the neonatal rat preparation is an excellent model for studying the conversion from immature to adult locomotion. Some neural controls are more clearly demonstrable in the developing animal preparation than in the adult because the latter exhibits an array of complex and redundant adaptive mechanisms.

Introduction

The neural control of locomotion has been extensively studied these last four decades and it continues to be a pace-setting topic for movement neuroscientists. It is a complex, coordinated behavior whose most basic feature is its pattern of muscle contractions. Kinematically, locomotion consists of an assemblage of motor sequences with several levels of coordination (intramuscular, intrajoint, interjoint, inter-bodypart, etc.). The extensive literature available on locomotor patterns is due to the extreme diversity of both the species studied (from mollusc to human) and the preparations used (from in vitro tissue to freely moving animals; Orlovsky et al., 1999). In some extremely valuable, albeit still exceptional cases, it is now possible to "... bridge the gap from ion channels to networks and behavior" (Grillner, 1999).

At first glance, locomotor organization and control by the CNS are remarkably similar from one species to another. It is generally accepted that the fundamental rhythm of locomotion originates within relatively low-order central circuits which produce alternation between functionally antagonistic (usually flexor and extensor) muscles. This program (often termed a central pattern generator or CPG) is triggered into activity by an intrinsic command and/or an extrinsic sensory input. It is regulated on a moment-to-moment basis by central and peripheral feedback loops which are necessary for the organism's adaptation to its external environment. The central neural patterns are extremely flexible as is the behavior, itself. Pearson (2000) has emphasized distinctions between rapid adaptations, wherein motor programs are able to switch immediately from one pattern to another, and much longer adaptations brought on by anatomical and morphological changes due to training or traumatic injuries. During maturation, both adaptive effects are present, with the organism able to both

*Corresponding author: Tel.: +33-491-164-139;
Fax: +33-491-775-084; E-mail: clarac@dpm.cnrs-mrs.fr

adapt quickly to relevant surrounding events and integrate ontogenetic processes on a much slower timescale.

In recent years, the in vitro isolated spinal cord of a neonatal rat in a manipulable bathing medium has become a widely used preparation (Cazalets et al., 1992; Kiehn and Kjaerulff, 1996). Its locomotor pattern can be studied for several hours with intracellular and extracellular recording techniques before central myelination has occurred (postnatal days 0–12; P_{0-12}). In parallel, ethologists have studied a variety of motor behaviors over the same period (Westerga and Gramsbergen, 1990; Jamon and Clarac, 1998). The close phylogenetic association between the rat and mouse, and the great potential of transgenic models of the latter, has increased even further the general interest of movement neuroscientists in rodents.

In this chapter, we proceed from the flexibility of locomotor behavior in adult species to comparison of locomotor patterns in adult versus neonatal rat preparations and finally, to the maturation of the fundamental locomotor pattern in the rat embryo (see also Kudo et al., this volume). There is an emphasis on comparing data obtained from in vivo versus in vitro preparations (cf. Ezure, Chapter 7 of this volume) and swimming versus walking behavior. All in all, the results to be presented suggest that the locomotor patterns progressively exhibited after birth correspond to the successive steps in the maturation of the locomotor networks in the embryo.

Flexibility of locomotor behavior in adult species

Most animals possess multifarious modes of locomotion, including varying combinations of the gait used (walk, trot, gallop), the direction followed (forward, backward, lateral) and even the medium traversed (terrestrial, aquatic, aerial). In some species and situations, different locomotor behaviors make use of specialized appendages, with each engaged in a particular task. In other cases, the same motor apparatus can produce different patterns. In such latter instances, how is flexibility achieved? It may have different origins: the CPG, itself, which is composed of a predefined assemblage of neurons; or the sensory pathways, which can reinforce or decrease the CPG's synaptic connections; or the central pathways, like those descending from suprasegmental structures. Furthermore, chemical modulators can modify both the CPG's synaptic connections and the intrinsic membrane properties of the neurons comprising the generator (Marder, 2000).

For a simple example of locomotor flexibility, consider the forward and backward walking of a decapod crustacean, the crayfish. Using an in vitro preparation of this species, Chrachri and Clarac (1989) have identified different types of coordinating interneurons corresponding to the swing versus stance phase of both forward and backward walking. It appears that both groups of interneurons inhibit each other, i.e., when the forward swing-phase interneurons are operating, the forward stance-phase interneurons are inhibited, and vice versa. Similarly, Lieske et al. (2000) showed in an in vitro brainstem slice of rodents that the medullary interneurons that generate different forms of respiration are found in the same network, and are active in different breathing behaviors such as eupnea, sighs and gasps.

Such network flexibility appears to be a common feature in vertebrates and invertebrates. Often, the activity of quite complex sets of pattern-generating neurons can be completely rearranged by sensory inputs operating either independent of, or in conjunction with, CNS activity. A striking example has been presented by Wendler et al. (1985) for the hemiptere, Nepa rubra, which both walks and swims. When walking, this animal exhibits classical alternating tripod coordination with alternation between its six limbs. When swimming, all these limbs are in phase, and the swing phase is significantly longer than the stance phase. This difference in locomotor pattern seems due to the difference in frictional load for walking versus swimming. These authors also studied situations wherein the load was intermediate between ground and water conditions, i.e., when walking on mercury or a slippery surface. In the latter case, the swing and stance phases are equal in duration and both in-phase and out-of-phase locomotor patterns are used in alternation.

Work on the crab has also demonstrated network flexibility (Bevengut et al., 1986). This animal walks and swims with the same pairs of limbs. Its exoskeleton contains some contact and force sensitive mechanoreceptors, which are only stimulated by ground contact during walking. Behavioral observations have shown that swimming occurs only in the absence of such contact. During walking, these mechanoreceptors control the duration of the stance phase and the switch into the swing phase. Swimming sequences can be triggered by tilting the head of the animal downward. In the latter case, selective electrical stimulation of the mechanoreceptors' sensory nerve mimics a ground contact and suppresses the activity of the muscles used for swimming. This finding demonstrates that a given sensory receptor involved in a specific locomotor pattern can also inhibit an entirely different motor pattern.

For vertebrates, it is well known that constant-rate electrical stimulation of the mesencephalic locomotor region (MLR) can produce different locomotor gaits in the high-decerebrate cat (controlled locomotion; Orlovsky et al., 1999). Low-strength stimulation induces slow walking and progressively stronger stimulation produces fast walking, then trotting and finally, galloping. Similarly, Steeves et al. (1986) showed in birds that analogous stimulation of the MLR at low voltage induces walking whereas stronger stimulation evokes flying (at three times the walking threshold). The movement control exerted by the MLR is in fact a dual one: for the elaboration of both locomotion and its associated dynamically changing postures. It is particularly appropriate to comment in this volume on a remarkable finding of Mori (1987). He defined two pathways operating in opposition and controlling the muscle tone of limb-extensor muscles during the controlled treadmill locomotion of the high-decerebrate cat. His results show that for various gaits, the level of extensor muscle tone in the hindlimbs can be increased by stimulation of the ventral tegmental field (VTF) and decreased by the dorsal tegmental field (DTF). The two pathways finely tune postural muscle tone to precisely fit the level required for a given gait. His work emphasizes that the activity of a brainstem-spinal generator for locomotion is always linked with a brainstem setting mechanism for posture. In the same preparation, Mori also analyzed another crucial descending pathway for posture, the vestibulospinal motor system. He found significant changes in the unitary discharge of identified Dieters' neurons during different gaits (slow walk, fast walk, trot, gallop). During slow walking, 75% of his neuron sample discharged tonically whereas during fast walking and trotting, 80% of the same population discharged in phase with the activity of the ipsilateral or contralateral gastrocnemius muscle. This shows that the activity of Dieters' neurons differs according to the ongoing motor program, contributing to the control of both the static equilibrium posture and bilateral dynamic locomotor equilibrium.

All of the above examples demonstrate that the changes from one posturo-locomotor pattern to another is due not only to the network properties of relevant sets of central neurons but also to an active participation of mutually cooperative central and peripheral loops.

Locomotor patterns of the adult versus neonatal rat

Figure 1 provides a schematic representation of many of the results reported in the following sections.

The adult

The two primary modes of locomotion of the adult rat, stepping and swimming, have been studied in intact (Gruner and Altman, 1980; De Leon et al., 1999), acute thalamic (Goudard et al., 1992) and decerebrate (Nicolopoulos-Stournaras and Iles, 1984) preparations. Kinematic locomotor patterns are quite similar in these three models, and rather like those of the cat and dog. One difference, however, is that the rat has a more crouched stepping posture, e.g., its mean angular range of the femur displacement relative to the horizontal is 20–80° versus 55–105° for the cat and dog.

During the rat step cycle, the duration of the power-stroke (extension or retraction) phase, which is the longer one (particularly at slow speed), is highly correlated with the cycle's total duration. The duration of the return-stroke (flexion or protraction) phase is shorter and more-or-less independent of forward speed. The situation is different for swimming, during which the return stroke is identical or even longer in duration than the power stroke.

There are some differences in muscle EMG bursts during stepping versus swimming. The generalized characteristic asymmetric EMG waveform of

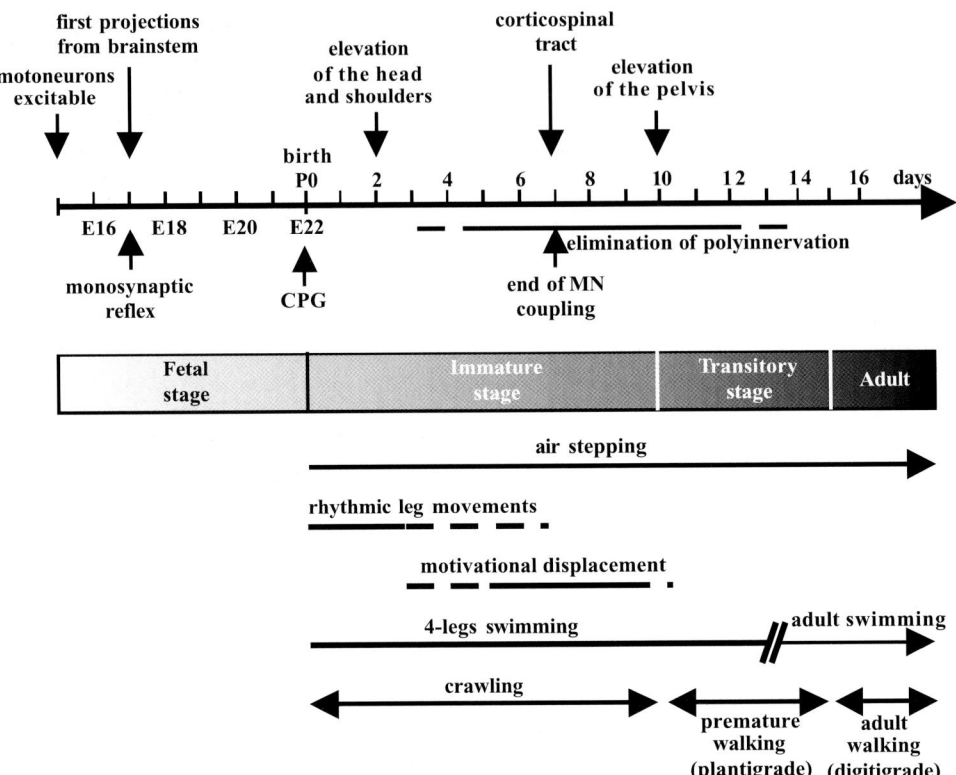

Fig. 1. Different stages in the development of locomotion in the rat. The maturation of locomotion is nonlinear and can be divided into three stages before the adult. The fetal stage starts at ~E_{15} when MNs become excitable and CPGs begin to operate. The immature stage starts at birth when the CPG can produce a rhythmic-alternating pattern even though postural structures are still deficient. Maturation of posture occurs along a rostro-caudal gradient from P_0 to P_{10}. During this period, only behaviors without complex postural regulations are available. The transitory stage, from P_{10} to P_{15}, corresponds to eye opening and the animal switches from immature to adult motor behavior (for further details, see the text). From P_{15} onwards, motor behaviors operate like those in the adult, even though maturation is still proceeding.

stepping is reduced to a simple bell-shaped curve in swimming (Gruner and Altman, 1980). The timing of these bursts is also different. During stepping, some muscles always behave as flexors (e.g., iliopsoas, gluteus superficialis, tibialis anterior) and others always act as extensors (hip adductors, semimembranous, gastrocnemius, soleus). Some bifunctional muscles have a double EMG burst: one in the flexor and one in the extensor phase (biceps femoris, semitendinosus, rectus femoris, vastus lateralis). The EMGs of the two latter muscles have been compared by De Leon et al. (1994). They observed a double burst during stepping and a single, extensor-phase one during swimming. The overall EMG pattern during swimming seemed more simple to them, except for semitendinosus, which had two bursts.

In the acute thalamic and decerebrate rat, the EMG patterns they recorded during controlled locomotion were very similar to those obtained during the unrestrained (free) walking of the intact rat. This similarity of patterns among different preparations confirms that they are centrally organized. The effects of sensory input on EMG patterns during controlled locomotion have been studied in decerebrate rats for three conditions: loaded walking, normal walking and swimming (Iles and Coles, 1991). The animals were locomoting on a freely moving wheel to which frictional loads could be added, thereby increasing the duration of extensor bursts by 10–40%. For muscles with double bursts, such as the semitendinosus, the extensor-phase activity was prolonged during loaded walking. In contrast, when the gravitational load was reduced by performing similar experiments in a tank of water, the duration of the extensor bursts was reduced by 35%, while flexor bursts increased by 60%. In summary, Iles and Coles' (1991) study showed that the duration and amplitude of motor bursts are closely related to the characteristics of the flexor and extensor muscle torques at the joints.

The neonate

The earliest form of neonatal locomotor behavior is crawling, which occurs at P_0. Its main characteristic is that the belly, which supports most of the animal's weight, is dragged along. This structure can be partially or completely lifted up for short periods of time, however. The respective role of the limbs have not yet been defined: all four limbs are indeed involved, but the sluggish hindlimbs often fail to keep up with the forelimbs, being dragged along in an extended position. At P_2, the pup is able to make some pivoting movements of the head and initiate some crawling sequences. During the first postnatal week, only the front part of the body exhibits a consistent postural adaptation (elevation of the shoulders) but there are some sporadic postural reactions of the hindlimbs (Brocard et al., 1999a). At P_5, the pup can raise its head, sniff in the horizontal plane even though it remains prone, and point the head downward. Apparently, there is a rostro-caudal gradient of motor maturation. During the second postnatal week, the pelvic region begins to support the bodyweight and by P_{10}, the pup can assume a raised quadruped posture. This period involves a sudden acceleration in the functional maturation of the hindlimbs. During P_{0-4}, Fady et al. (1998) have shown that a pup disposed on a support is able to move its limbs in alternation if it is stimulated by the odor of the mother. This behavior, which is akin to air stepping, demonstrates that at birth, olfactory stimulation plays a particularly important role in motor behavior. Note further that the olfactory stimulus used by Fady et al. (1998) can elicit plantigrade walking at P_{3-9} (Jamon and Clarac, 1998).

At P_{10-12}, the first form of walking occurs: limb movements are still quite slow, however, with poor control of the trunk. The locomotion is plantigrade (Westerga and Gramsbergen, 1990) but due to its immaturity, it can be considered as close to swimming (Jamon and Clarac, 1998).

At P_{15}, more versatile walking occurs, and rearing appears, with the neonate able to sit on its hindlimbs and use its forepaws for manipulation. These more complex motor behaviors are now smoother and much more efficient. Walking becomes digitigrade. Westerga and Gramsbergen (1990) have quantified some locomotor parameters during free walking at this stage. Step-cycle duration which is > 1400 ms between P_{11} and P_{14}, suddenly reduces to ~ 600 ms at P_{15-16}. Immediately thereafter, complex motor behaviors are exhibited, including walking on a turning rod, traversing paths of different length and width, ascending and descending on a ladder, and climbing up a rope.

The above-described delay in the onset of adult-like stepping is mainly due to an absence of postural regulation at P_{0-10}. At P_0, only locomotion with a weak postural contribution is evident. Both air stepping (with the rat suspended) and swimming (with the animal immersed in cool water) can be evoked, however (Cazalets et al., 1990; McEwen et al., 1997). During both these latter behaviors, the limbs are active in alternation at cycle duration of ~ 1 s. This time decreases over the next 3 weeks to < 300 ms. Dopa-induced air stepping is always obtained at P_0. During swimming at P_0, only the forelimbs are active. Within the next 24 further hours, however, all four limbs are activated, with an alternation evident between the ipsi- and contralateral limbs, and the diagonal limbs moving in synchrony. The adult swimming pattern, with rhythmic propulsive movements restricted to the hindlimbs, and the forelimbs used only for steering, occurs at P_{12-15}. Dopa-induced free swimming is quite stereotyped, with a very tight synchrony of diagonal limb movements. If Dopa is injected after P_{14}, the immature quadruped swimming pattern is still observed. The adult bipedal pattern is delayed demonstrating that amines or peptides can modulate the locomotor networks. Cazalets et al. (1990) have studied the maturation of the fore- and hindlimb swimming generators. They found that the stability of hindlimb movements increases between P_3, P_8 and P_{15}, whereas no change occurs in the regularity of forelimb movements between at least P_3 and P_8.

For P_{0-4}, Kiehn and Kjaerulff (1996) have characterized the fictive rhythms induced by serotonin (5-HT) and dopamine. Their preparation consisted of an isolated spinal cord with one hindlimb attached. The 5-HT-induced hindlimb rhythm was slower than that obtained with NMDA (2–5 vs. 1 s cycle periods). The dopamine-induced rhythm was more irregular and very slow (2–70 s/cycle period). The ratio between EMG burst duration and cycle period was always more variable than in the adult. This ratio was more consistent in the dopamine-induced rhythm of extensor muscles, but not in the flexors. The timing of muscular activity was more-or-less similar for both 5-HT and dopamine-induced rhythms in the hip flexors, hip extensors and for the ankle and digit muscles. Bifunctional muscles displayed different patterns, however. Biceps femoris and semitendinosus were extensor-like during 5-HT-induced rhythms, but exhibited double EMG bursts during dopamine-induced rhythms. Also, rectus femoris and vastus lateralis were activated during flexion under 5-HT and extension under dopamine. In summary, Kiehn and Kjaerulff (1996) showed that 5-HT has a tendency to facilitate flexor drive whereas dopamine may exaggerate extensor activity. The authors further suggested that the 5-HT-induced rhythm is closer to a swimming-like pattern and the dopamine-induced one to a stepping pattern.

In summary, the above comparison of adult versus neonatal rat locomotion emphasizes that the immature spinal cord is able to produce various motor patterns according to the type of pharmacological stimulation. In the adult, different patterns also exist but they are much more stereotyped.

Developmental processes in locomotor networks

The ontogeny of the locomotor generator in the rat embryo and fetus has been studied in detail by Kudo's Tsukuba group (see Kudo et al., this volume). They have shown that it is operative in its earliest form on embryonic (E) day 15.5 (i.e., about one week prior to birth, which occurs on E_{22}). On P_0, and during the immediately following days, the locomotor program is functioning but in a limited way, with central neuronal generators inducing a specific pattern (crawling), sensory-afferent input resetting an ongoing rhythm and brainstem structures able to trigger the crawling rhythm (Cazalets et al., 1992; Keihn and Kjaerulff, 1996).

During the late pre- and early postnatal period, three different entities—segmental sensory input, descending pathways and motoneurons (MNs)—seem to play a key role in the maturation of locomotion.

Segmental sensory input

At P_0, the somatic system is present but still incomplete. Muscle fiber types are not fully differentiated. At E_{18}, the plantar surface of the hindlimb gives some response to central stimulation, albeit after a long conduction delay. Muscle spindles can be identified at $E_{19.5}$ but the intrafusal fibers are only mature at P_{12}. It seems that the proprioceptive system develops first centrally during the embryonic period, and later peripherally at the level of the muscle spindle, itself (Vejsada et al., 1985; Kudo and Yamada, 1987). The development of dorsal-root projections into the lumbar spinal cord has been studied with morphological and physiological techniques (Kudo and Yamada, 1987). Collaterals from dorsal root fibers start to reach the dorsal horn at $E_{15.5}$, the intermediate region at $E_{16.5}$ and the motor nuclei at $E_{17.5}$, i.e., the percentage of axon collaterals entering the ventral horn increases with age. Dorsal-root stimulation evokes long-latency excitatory postsynaptic potentials at even E_{15}, and a monosynaptic response at E_{17}. The stretch reflex is manifest at E_{19-20}, just before birth. It would seem that afferent inputs are important for strengthening and refining the output of the spinal locomotor generator.

Descending pathways

In general, the first specialized descending pathways from the brainstem to the spinal cord are evident during the week preceding birth. The GABAergic, serotonergic and noradrenergic reticulospinal pathways arrive between E_{15} and E_{18} in the lumbar enlargement. The vestibulospinal pathway arrive at about the same time whereas the rubrospinal pathway develops a little later (around birth) and the corticospinal pathway reaches the cervical and thoracic level at P_3 and the lumbar level at P_6. We have studied the physiological correlates of the development of descending pathways within the ventral funiculus (VF), including those from reticular and vestibular nuclei, and the locus coeruleus. The neurons of these nuclei are identified from E_{11} to E_{15}. Their fibers are detected in the lower thoracic cord at E_{14-15}, and in the lumbar cord before birth. The latency of the response decreased whereas the amplitude of the MN EPSP and the ventral-root potential increased (Brocard et al., 1999b). The gradual arrival of descending fibers in the lumbar enlargement suggests that the supraspinal control at birth is exerted primarily on proximal muscles whereas the control over distal muscles increases during the very first days after birth. This is in agreement with a proximo-distal gradient in the maturation of postural mechanisms (Brocard et al., 1999a).

Lumbar motoneurons (MNs)

We have recently shown that lumbar MNs become excitable at E_{15} and then have a very low threshold until birth. This behavior is associated with the abundant spontaneous motor activity observed in the maturing fetus (Vinay et al., 2000a,b). At E_{15}, sodium-channel activation is evident at a MN threshold at about −31 mV. This value then progressively increases (to about −47 mV at P_{1-3}). The action potentials are characterized by a pronounced afterdepolarization (ADP) which is followed by a prolonged afterhyperpolarization (AHP). The ADP is calcium dependent and in some MNs it is due to both a high-voltage-activated (HVA) calcium conductance and a low-voltage-activated (LVA) calcium conductance. The progressive development of potassium currents, which truncate the calcium current, starts before birth and continues into the first postnatal week. Within this period, the action potential shortens, the AHP develops further, and MN repolarization and repetitive discharge become more robust. Immediately after birth, there is a large increase in noninactivating, delayed-rectifier-type current (I_K) and a calcium-dependent potassium current (I_{KC}) becomes evident.

We also compared the development of repetitive firing in response to a prolonged depolarizing pulse in flexor versus extensor MNs (Vinay et al., 2000a). The fraction of MNs exhibiting sustained firing at birth was higher in flexor versus extensor cells. At P_{3-5}, this percentage reached 100% versus 71% in extensor MNs. We further found that the steady-state frequency of discharge is higher in flexor versus extensor MNs, a difference possibly attributable to the longer AHP of the extensor MNs (Vinay et al., 2000a). Electrical coupling between MNs of the same motor nucleus is present likely from the very beginning of MN formation until P_{7-9}, after which latter stage it became nonfunctional. In summary, the earlier results on the intrinsic properties of spinal MNs suggest that in the earliest stages, motor pools function as a syncitium whereas in adults each MN of a given pool has its own characteristics functionally adapted to the target muscle fibers.

The maturation of neuromuscular connections is also an important factor in the expression of a smoothly executed locomotor behavior. During the second postnatal week, muscle fibers in the rat are still innervated by more than one MN. Over the next week, there is a progressive conversion to unimodal innervation, and the separation of muscle fibers into relatively distinct 'types' (fast-glycolytic, fast-oxidative-glycolytic, slow-oxidative). This elimination of polyneuronal innervation and fiber-type differentiation starts somewhat earlier in flexor than extensor muscles (reviewed in Navarrete and Vrbova, 1993). The need for such postnatal maturation is one of the determinants limiting an earlier expression of full-scale locomotor behavior.

Summary

Below, we summarize the development of rat locomotion, which is clearly nonlinear. It has distinct stages from the onset of CNS differentiation and MN formation until one month after birth.

Embryonic and fetal stage (E_{11}-birth)

During this stage, there is development of different motor nuclei and their MNs, and central and peripheral pathways. The locomotor pattern begins to operate, albeit quite sluggishly.

Immature stage (P_{0-10})

Throughout this period, the early neonate is only able to survive in the restricted environment of the nest. It has very few postural reactions, and these are limited largely to the shoulders and forelimbs. The in vitro preparation is normally studied at this stage (particularly between P_0 and P_6). The immaturity of the CNS and the various behavioral reactions suggest that such fictive activity corresponds to a swimming-like behavior.

Later, the nervous system becomes almost completely myelinated (Roseik and Von Keyserlingk, 1987), thereby stabilizing central and peripheral axonal conduction velocity. Descending pathways and sensory afferents become fully functional, the MNs lose their electrical coupling, and their discharge becomes robust and adult-like.

Transitory stage (P_{10-15})

At this stage, supraspinal structures become more developed and efficient. The cortex starts to develop rapidly with the maximum increase in its connectivity occurring between P_{12} and P_{20} (Super et al., 1998). The main development of the cerebellar cortex occurs between P_{10} and P_{17} (Gramsbergen and Ijkema-Paassen, 1984). Immature swimming behavior switches to its adult form. Cycling activity of the forelimbs begins to be inhibited by corticospinal pathways, and these limbs begin to be mainly used for manipulation and guiding (steering) the direction of motor behaviors. Rhythmic motor patterns are then limited to the hindlimbs, which alone propel the body. The first free plantigrade walking pattern is evident at this stage but it is more akin to swimming. Interestingly, after removal of the neocortex and hippocampal formation in the adult rat, the neonatal pattern of four-limbed swimming reemerges (Vanderwolf et al., 1978).

Adult stage ($>P_{15}$)

By P_{15}, the eyes are open and they, together with auditory perception, allow the animal to explore its surrounding environment. Walking becomes digitigrade and it begins to resemble that of the adult. Several motor repertoires are in place. The detailed maturation of full-scale locomotor behavior is not complete, however, until a further 2 weeks of development.

Concluding thoughts

Results obtained in extensive vivo and in vitro studies have shown that the rat is a particularly valuable model for studying the conversion from immature to adult locomotion. It is often argued that developmental studies are confounded by the simultaneous progressive maturation of the adult nervous system and transient presence of mechanisms which only operate in the embryo, fetus and neonate. In fact, however, these two elements can be clearly dissociated. Furthermore, some facets of the neural control of locomotion are more clearly demonstrable in developing animal preparations than in adult because the latter has an array of complex and redundant adaptive mechanisms.

Acknowledgments

This work was supported in part by the French CNRS and CNES.

Abbreviations

5-HT	5-hydroxytryptamine (serotonin)
ADP	afterdepolarization
AHP	afterhyperpolarization
CNS	central nervous system
CPG	central pattern generator
DTF	dorsal tegmental field
E_0–E_n	embryonic day$_{0-n}$
EMG	electromyogram/electromyography
HVA	high-voltage-activated calcium conductance
LVA	low-voltage-activated calcium conductance
MLR	mesencephalic (midbrain) locomotor region

MN motoneuron
P_0–P_n postnatal day$_{0-n}$
VF ventral funiculus
VTF ventral tegmental field

References

Bevengut, M., Libersat, F. and Clarac, F. (1986). Dual locomotor activity selectively controlled by force- and contact-sensitive mechanoreceptors. Neurosci. Lett., 66: 323–327.

Brocard, F., Vinay, L. and Clarac, F. (1999a). Development of hind-limb postural control during the first post-natal week in the rat. Dev. Brain Res., 117: 81–89.

Brocard, F., Vinay, L. and Clarac, F. (1999b). Gradual development of the ventral funiculus input to lumbar motoneurons in the neonatal rat. Neuroscience, 90: 1543–1554.

Cazalets, J.R., Menard, I., Crémieux, J. and Clarac, F. (1990). Variability as a characteristic of immature motor systems: an electromyographic study of swimming in the newborn rat. Behav. Brain Res., 40: 215–225.

Cazalets, J.R., Sqalli-Houssaini, Y. and Clarac, F. (1992). Activation of the central pattern generators for locomotion by serotonin and excitatory amino acids in neonatal rat. J. Physiol. (Lond.), 455: 187–204.

Chrachri, A. and Clarac, F. (1989). Synaptic connections between motor neurons and interneurons in the fourth thoracic ganglion of the crayfish Procambarus clarkii. J. Neurophysiol., 62: 1237–1250.

Clarac, F., Vinay, L., Cazalets, J.R., Fady, J.C. and Jamon, M. (1998). Role of gravity in the development of posture and locomotion in the neonatal rat. Brain Res. Rev., 28: 35–43.

De Leon, R., Hodgson, J.A., Roy, R.R. and Edgerton, W.R. (1994). Extensor- and flexor-like modulation within motor pools of the rat hindlimb during treadmill locomotion and swimming. Brain Res., 654: 241–250.

Fady, J.C., Jamon, F. and Clarac, F. (1998). Early olfactory-induced rhythmic limb activity in the newborn rat. Dev. Brain Res., 108: 111–123.

Goudard, I., Orsal, D. and Cabelguen, J.M. (1992). An electromyographic study of the hindlimb locomotor movements in the acute thalamic rat. Eur. J. Neurosci., 4: 1130–1139.

Gramsbergen, A. and Ijkema-Paassen, J. (1984). The effects of early hemispherectomy in the rat: behavioral, neuroanatomical, and electrophysiological sequelae. In: Almli C.R. and Finger S. (Eds.), Early Brain Damage, Vol. 2, Academic Press, New York, pp. 155–177.

Grillner, S. (1999). Bridging the gap-from ion channels to networks and behaviour. Curr. Opin. Neurobiol., 6(6): 663–669.

Gruner, J.A. and Altman, J. (1980). Swimming in the rat: analysis of locomotor performance in comparison to stepping. Exp. Brain Res., 40: 374–382.

Iles, J.F. and Coles, S.K. (1991). Effects of loading on muscle activity during locomotion in the rat. In: Armstrong D.M. and Bush B.M.H. (Eds.), Locomotor Neural Mechanisms in Arthropods and Vertebrates. Manchester University Press, New York, pp. 196–201.

Jamon, M. and Clarac, F. (1998). Early walking in the neonatal rat: a kinematic study. Behav. Neurosci., 112: 1218–1228.

Kiehn, O. and Kjaerulff, O. (1996). Spatiotemporal characteristics of 5-HT and dopamine-induced rhythmic hindlimb activity in the in vitro neonatal rat. J. Neurophysiol., 75: 1471–1482.

Kudo, N. and Yamada, T. (1987). Morphological and physiological studies of development of the monosynaptic reflex pathway in the rat lumbar spinal cord. J. Physiol. (Lond.), 389: 441–459.

Lieske, S.P., Thoby-Brisson, M., Telgkamp, P. and Ramirez, J.M. (2000). Reconfiguration of the neural network controlling multiple breathing patterns: eupnea, sighs and gasps. Nat. Neurosci., 3: 600–607.

Marder, E. (2000). Motor pattern generation. Curr. Opin. Neurobiol., 10: 691–698.

McEwen, M.L., Van Hartesveldt, C. and Stehouwer, D.J. (1997). A kinematic comparison of L-DOPA-induced air-stepping and swimming in developing rats. Dev. Psychobiol., 30: 313–327.

Mori, S. (1987). Integration of posture and locomotion in acute decerebrate cats in awake, freely moving cats. Prog. Neurobiol., 28: 161–195.

Navarrete, R. and Vrbova, G. (1993). Activity-dependent interactions between motoneurones and muscles: their role in the development of the motor unit. Prog. Neurobiol., 41(1): 93–124.

Nicolopoulos-Stournaras, S. and Iles, J.F. (1984). Hindlimb muscle activity during locomotion in the rat (Rattus norvegicus) (Rodentia-Muridae). J. Zool. Lond., 203: 427–440.

(1999). In Orlovsky G.N., Deliagina T.G. and Grillner S. (Eds.), Neuronal Control of Locomotion—From Mollusc to Man. Oxford University Press, Oxford, pp. 322.

Pearson, K.G. (2000). Neural adaptation in the generation of rhythmic behavior. Ann. Rev. Physiol., 62: 723–753.

Roseik, C. and Von Keyserlingk, D. (1987). The sequence of myelination in the brainstem of the rat monitored by myelin-basic protein immunohistochemistry. Dev. Brain Res., 35: 183–190.

Steeves, J.D., Sholomenko, G.B. and Webster, D.M.S. (1986). Reticular formation stimulation evokes walking and flying in birds. In: Grillner S., Stein P.S.G., Stuart D.G., Forssberg H. and Herman M.M. (Eds.), Neurobiology of Vertebrate Locomotion. Macmillan Press, Basingstoke, pp. 51–54.

Super, H., Soriano, E. and Uylings, H.B. (1998). The functions of the preplate in development and evolution of the neocortex and hippocampus. Brain Res. Rev., 27(1): 40–64.

Vanderwolf, C.H., Kolb, B. and Cooley, R.K. (1978). Behavior of the rat after removal of the neocortex and hippocampal functions. J. Comp. Physiol. Psychol., 92: 156–175.

Vejsada, R., Hnik, P., Payne, R., Ujec, E. and Palecek, J. (1985). The postnatal functional development of muscle stretch receptors in the rat. Somatosens. Res., 2: 205–222.

Vinay, L., Brocard, F. and Clarac, F. (2000a). Differential maturation of motoneurons innervating ankle flexor and extensor muscles in the neonatal rat. Eur. J. Neurosci., 12: 4562–4566.

Vinay, L., Brocard, F., Pfieger, J.F., Simeoni-Alias, H. and Clarac, F. (2000b). Perinatal development of lumbar motoneurons and their inputs in the rat. Brain Res. Bull., 53: 635–648.

Wendler, G., Teuber, H. and Jander, J.P. (1985). Walking, swimming and intermediate locomotion in Nepra rubra. In: Gewecke M. and Wendler G. (Eds.), Insect Locomotion. Parey, Berlin, pp. 103–110.

Westerga, J. and Gramsbergen, A. (1990). The development of locomotion in the rat. Dev. Brain Res., 57: 164–174.

CHAPTER 7

Reflections on respiratory rhythm generation

Kazuhisa Ezure*

Department of Neurobiology, Tokyo Metropolitan Institute for Neuroscience, Tokyo 183-8526, Japan

Abstract: Knowledge about neuronal mechanisms that control respiration is being advanced rapidly by studies that make use of both mature in vivo animals and in vitro neonates. The available data suggest that particular types of neurons within selected networks of the ventrolateral medulla are essential for respiratory rhythm generation. There are many uncertainties, however, about the correspondence between neurons identified by the above two approaches, because there are virtually no studies that have combined them. In this chapter, I propose a hypothesis that shows how neonatal respiratory neurons, with either retained or modified intrinsic cellular properties, develop into mature, well-characterized respiratory neurons located in medullary areas called the Bötzinger and pre-Bötzinger complex. Currently, the most plausible models of respiratory rhythmogenesis are hybrid ones that include both intrinsic cellular and network properties.

Introduction

Respiratory movement, ventilation of the lungs with air, is characterized by its automatic rhythmicity and finely regulated contraction of respiration-related muscles that accommodate moment-to-moment chemical needs and mechanical conditions. The basic question of how the automatic rhythmicity is produced by the CNS has attracted much attention from physiologists. Recently, this field has made substantial progress using two different but complementary approaches. One is the in vivo approach on surgically reduced mature whole animals, whose use, together with refined electrophysiological and neuroanatomical methods at the level of single neurons, has greatly increased our knowledge about the overall respiratory system and its network mechanisms (Feldman, 1986; Bianchi et al., 1995; Ezure, 1996). The other is the in vitro approach, which involves recording from whole amounts of portions of brainstem-spinal cord or thin brainstem slices in neonates (Suzue, 1984; Onimaru and Homma, 1987; Smith et al., 1990). These much-more-reduced in vitro preparations allow precise control of the extracellular and intracellular environment of single neurons (e.g., by using patch-clamp electrodes), and have thereby revealed many microproperties of the respiratory system at the cellular and subcellular level (e.g., Kudo et al. and Nakamura et al., Chapters 5 and 9 of this volume). The combined data from mature in vivo and neonatal in vitro preparations now suggest that particular types of neurons within particular parts of the ventrolateral medulla and their networks are essential for respiratory rhythm generation (see also Kirkwood and Ford, Chapter 10 of this volume).

There are incongruities, however, in the knowledge obtained by the use of these two approaches. Because there are virtually no studies that have combined them, there is much current uncertainty about information obtained from neonatal in vitro

*Corresponding author: Tel.: +81-42-325-3881; Fax: +81-42-321-8678; E-mail: ezurek@tmin.ac.jp

preparations versus mature in vivo preparations. The purpose of this chapter is to address this issue, with special emphasis on the genesis of the respiratory rhythm.

Key questions

In vivo studies using adult mammals have identified a wide variety of respiratory neurons (defined below), their efferent and afferent properties, and their interconnections (Cohen, 1979; Feldman, 1986; Ezure, 1990; Bianchi et al., 1995). In the brainstem's network of respiratory neurons, excitatory and inhibitory synaptic connections play a crucial role in the generation of the respiratory rhythm. In particular, glycinergic and GABAergic inhibitory mechanisms are essential in adult animals (Hayashi and Lipski, 1992; Paton and Richter, 1995; Pierrefiche et al., 1998). On the other hand, in vitro studies on largely neonatal rats have revealed pacemaker-like neurons. These, together with their excitatory (not inhibitory) interconnections, have been put forward as the fundamental kernel for respiratory rhythm generation (Onimaru and Homma, 1987; Ballanyi et al., 1999; Gray et al., 1999; Johnson et al., 2001).

To what extent are the mechanisms of respiratory rhythm generation inferred in neonatal animals comparable with those inferred in mature mammals? To answer this question, it is first necessary to determine how pacemaker-like neurons found in neonatal in vitro preparations correspond to or develop into neurons identified in mature in vivo animals. Next, resolution is required of the long-standing issue: the relative importance for rhythm generation of network versus pacemaker properties. Which of the two is essential for the rhythm, or are both critical?

Subtle aspects of respiratory center terminology

Various neurons in the central nervous system fire in synchrony with respiration and are called respiration-related neurons, or simply, respiratory neurons. Of these, the ones primarily responsible for respiratory control constitute the so-called 'respiratory center'. It is composed of three major cell groups in the medulla: the dorsal respiratory group (DRG), the ventral respiratory group (VRG) and the Bötzinger complex (BOT). These groups, and other cells of the overall respiratory center, include neurons with different functions. Some excite or inhibit other respiratory cells. Others are bulbospinal neurons that transmit respiratory activity to the spinal cord. Still others are motoneurons that innervate upper airway muscles. A variety of other neurons exchange information between the respiratory center and other neuronal centers and networks. Finally, chemo-receptor neurons, which may or may not have a respiratory rhythm, help sustain the activity of rhythm-generating neurons. These neurons, as a whole, comprise the respiratory center whose overall operation is to produce automatic rhythmicity, receive chemical and mechanical information, and integrate this information for the subsequent generation of motor outputs that result in optimal gas exchange within the lungs.

Some terms require careful definition. For example, 'rostral ventrolateral medulla (RVLM)' was originally used to demarcate cardiovascular-related neurons (Ross et al., 1984). Anatomically, it refers to the medullary region located immediately caudal to the facial nucleus, extending from the compact formation of the nucleus ambiguus to the ventral medullary surface. The BOT, which is functionally defined as the region containing E-AUG (see the next section) neurons (Merrill et al., 1983), is included within the RVLM. Thus, the RVLM includes both the BOT and the region containing cardiovascular presympathetic neurons (Kanjhan et al., 1995). The pre-Bötzinger complex (pre-BOT) is a region of respiratory neurons caudal to the BOT (Smith et al., 1991). Its boundaries are reasonably well defined, albeit not with absolute precision (Pilowsky and Feldman, 2001). Anatomically, the pre-BOT is reported to be situated caudal to the compact formation of the nucleus ambiguus, i.e., just caudal to the RVLM (Johnson et al., 1994). Functionally, it corresponds to the transitional zone between the BOT and the rostral VRG (Ezure, 1990, 1996; Connelly et al., 1992; Johnson et al., 1994). In this chapter, the term pre-BOT/RVLM refers collectively to the RVLM and the pre-BOT region.

In vivo studies reveal network properties

Respiratory neurons that fire during the inspiratory (I) phase of breathing are called I neurons, whereas those that fire during the expiratory (E) phase are termed E neurons. Phase-spanning neurons fire in both phases (Cohen, 1979). In both phases, respiratory neurons exhibit characteristic firing patterns, e.g., there are augmenting (AUG), decrementing (DEC) and constant (CON) types of discharge (Ezure, 1990; Duffin et al., 1995).

A large number of neurons in the respiratory center are motoneurons that supply upper airway muscles. They do not contribute to rhythm generation, however. Some bulbospinal neurons, such as the E-AUG neurons of the caudal VRG, have no medullary collaterals. Therefore, they, too, do not contribute to rhythm generation (Arita et al., 1987). On the other hand, many other of the respiratory neurons identified to date have medullary projections and can indeed contribute to rhythm generation. For example, bulbospinal I-AUG neurons of the DRG and VRG excite not only spinal I neurons but also medullary I neurons via their medullary axon collaterals (Ezure and Manabe, 1989a). E-AUG neurons of the BOT, whose distribution actually defines this region, arborize extensively within the respiratory center and maintain the E phase by widely inhibiting I neurons (Merrill et al., 1983; Otake et al., 1987; Ezure, 1990; Jiang and Lipski, 1990; Bryant et al., 1993). E-DEC, I-CON and I-DEC neurons of the BOT and VRG are propriobulbar neurons with extensive medullary projections (Ezure et al., 1989b; Otake et al., 1990). E-DEC neurons contribute to the transition from the I to E phase by inhibiting I neurons (Ezure and Manabe, 1988). This transition may be assisted by the firing of I/E 'phase-spanning' neurons under some conditions, albeit this function has not yet been established unequivocally. I-DEC neurons begin firing at the transition from the E to I phase, and they inhibit not only E but also I neurons (Ezure, 1990). I-CON neurons exhibit a burst of discharge in the I phase, and make extensive excitatory connections with various types of I neuron (Ezure et al., 1989b).

Of the above, I-CON, I-AUG, I-DEC and E-AUG (of the BOT) neurons are of particular interest in relation to neurons studied in neonatal in vitro preparations. Those are located within the pre-BOT/RVLM region, where they span the BOT and rostral part of the VRG.

In vitro analysis of the RVLM/pre-BOT

The pre-BOT of neonates was proposed by Feldman and colleagues to contain a rhythm-generating kernel (Smith et al., 1991; Rekling and Feldman, 1998). Alternatively, Onimaru and colleagues stress that a more rostral region, the RVLM, contains such a kernel (Onimaru et al., 1997; Ballanyi et al., 1999). In both of these in-vitro-derived hypotheses, the rhythm-generating kernel comprises a network of synaptically coupled excitatory neurons that possess pacemaker-like properties. Their hypotheses share four findings. (1) In the pre-BOT/RVLM of neonatal rats, neurons possessing some properties of pacemaker-like activity have been identified by two independent groups (Onimaru and Homma, 1987; Smith et al., 1991). (2) A medullary slice in either the transverse or sagittal plane can produce a respiratory rhythm if it includes the pre-BOT region (Smith et al., 1991; Paton et al., 1994). (3) Rhythmic activity persists even after blockage of inhibitory synaptic transmission (Feldman and Smith, 1989; Onimaru et al., 1990), thereby indicating that glycinergic and/or GABAergic inhibition is not necessary for basic rhythm generation. (4) Chemical and mechanical manipulation of the pre-BOT/RVLM region strongly modulates and can even abolish the respiratory rhythm (Smith et al., 1991; Funk et al., 1993; Gray et al., 1999).

The RVLM pacemaker-like neurons identified by Onimaru and colleagues are called 'pre-I neurons'. They start firing before the I phase, receive inhibition during this phase, and fire again after it (post-I phase), i.e., they exhibit characteristic biphasic E activity. Some (20%) of them do not receive inhibition in the I phase. When their synaptic connections are blocked by a low Ca^{2+}/high Mg^{2+} bathing medium, these neurons still continue to discharge in bursts. On the other hand, the pacemaker-like neurons found by Feldman and colleagues within the pre-BOT are called conditional busters. They exhibit bursting activity during blockage of synaptic transmission when their membrane potential voltage is within a selected range. Many I neurons exhibit this property in in vitro preparations (Rekling and Feldman 1998; Butera et al., 1999). Pre-I neurons of the RVLM also display conditional bursting (Ballanyi et al., 1999). Although the two pacemaker types are clearly different in properties and location, motor output is dependent on their activity. The two regions seem to produce rhythmic activity either independently or jointly (Mellen and Feldman, 2001). Thus, a substantial amount of evidence supports the conclusion that at least in reduced neonatal preparations, a kernel for rhythm generation comprises pacemaker-like neurons and their excitatory connections in the pre-BOT/RVLM region.

Some ideas on the missing link

Figure 1 provides a schematic representation of the ideas presented in the following sections.

Correspondence between neurons studied in in vivo versus in vitro preparations

Currently, there is uniform consensus that the pre-BOT/RVLM is essential for rhythm generation, as shown in *both* in vivo and in vitro studies. In vivo studies on adult cats have revealed critical excitatory and inhibitory respiratory neurons in this region. First, I-CON and I-DEC neurons, which are fundamental excitatory and inhibitory neurons, are

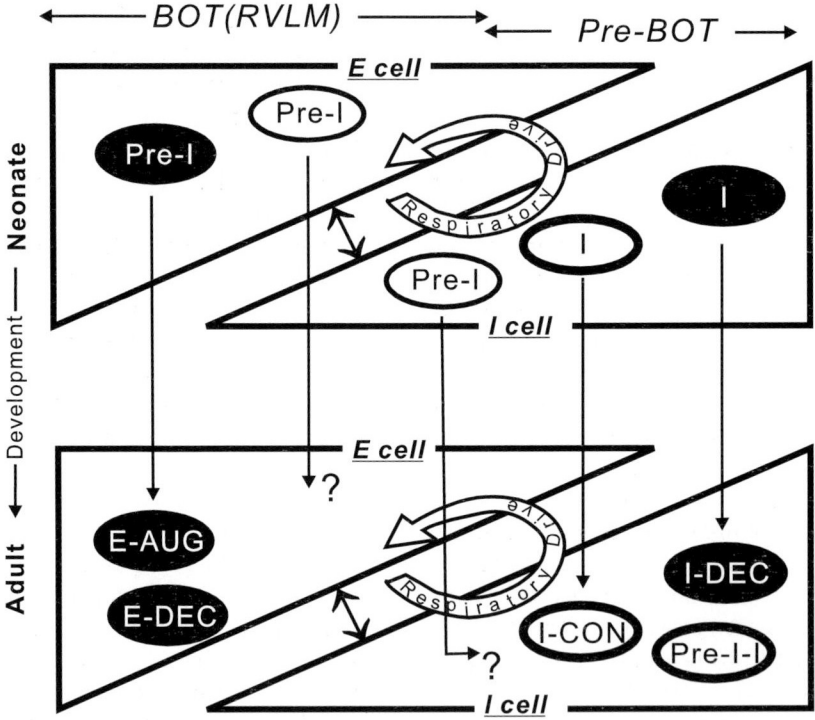

Fig. 1. Hypothetical correspondence between neurons identified in neonatal in vitro versus mature in vivo preparations. Studies in both preparations agree that respiration-related neurons in the pre-BOT/RVLM region and their interconnections are essential for rhythm generation. Many of the neonatal respiratory neurons have pacemaker or conditional bursting properties. These properties together with excitatory connections may produce the basic respiratory rhythm, with the inspiratory and the expiratory phases differentiated by mutual inhibition between the respective neuron groups. Neonatal respiratory neurons, while either retaining or modifying their intrinsic cellular properties, may develop into mature respiratory neurons. These are proposed to correspond to well-characterized respiratory neurons of the mature pre-BOT/RVLM. Neurons with filled and open ellipsoids are inhibitory and excitatory, respectively.

concentrated in the region caudal to BOT and rostral to the rVRG. This region presumably corresponds to the pre-BOT (Ezure, 1990, 1996; Connelly et al., 1992). Second, E-AUG and E-DEC neurons, which are fundamental inhibitory neurons, are distributed in the BOT, and a part of the RVLM. It is parsimonious to propose that these neurons are the mature equivalent of pre-BOT/RVLM neurons found in neonatal animals.

The author further proposes that I-CON and I-DEC neurons found in mature in vivo preparations correspond to the pre-BOT pacemaker-like neurons found in neonates. I-CON (rather than I-DEC) neurons are more likely to make excitatory connections in neonates. They have strong and widely distributed excitatory connections in adult cats, and are virtually the only known excitatory neurons in the pre-BOT region. Excitatory synaptic connections between the I-CON neurons, themselves, are feasible, because they project extensively to both the pre-BOT and several other medullary regions (Otake et al., 1990). Although I-AUG neurons make excitatory synaptic connections with I neurons (Ezure and Manabe, 1989a), the former are not pre-BOT neurons in the strict sense of the definition. Thus, the equivalent of the I-CON neuron in both mature rat and cat is likely to correspond to a pacemaker neuron with excitatory connections in the neonatal rat. Furthermore, the I-DEC neuron, which has inhibitory connections and is thereby not found in excitatory networks, may also be the equivalent of a pacemaker-like neuron in neonatal preparations.

On the other hand, Feldman and colleagues hypothesize that the E–I phase-spanning neurons found in the pre-BOT region of adult (and neonatal) rodents have pacemaker properties and contribute primarily to rhythm generation (Smith et al., 1990; Funk and Feldman, 1995; Sun et al., 1998; Pilowsky and Feldman, 2001; Wang et al., 2001). These neurons have discharge that begins in the late E phase and continues into the I phase. They are also called 'pre-I neurons'. Here they are called 'pre-I–I neurons' to discriminate them from Onimaru's pre-I neurons of the RVLM. These pre-I–I neurons are reported to comprise a conspicuous population in the pre-BOT region in rodents (Paton, 1997; Sun et al., 1998; Wang et al., 2001). They are found also in adult cats (Connelly et al., 1992; Schwarzacher et al., 1995), albeit in a smaller-sized population. It remains unclear whether pre-I–I neurons constitute a special group and whether they are equivalent to I-CON and I-DEC neurons of cats. Indeed, most I-CON and I-DEC neurons and some I-AUG neurons in the cat initiate their firing before the onset of the I phase, and thereby exhibit more-or-less preinspiratory firing. It is also not currently known if pre-I–I neurons have excitatory or inhibitory connections, or both. Such a determination is eagerly awaited (Wang et al., 2001).

The final hypothesis is that the pre-I neurons of the neonatal RVLM, biphasic E neurons, develop into E-AUG or E-DEC neurons of the mature BOT. At a first glance, this hypothesis may seem far-fetched, but it is plausible for several reasons. One of the most characteristic properties of neonatal pacemaker-like neurons is the synaptic inhibition they receive in the I phase. Therefore, they should almost certainly behave like E neurons in mature animals. Next, by definition, the region of these pre-I neurons in the RVLM is the BOT. Although some cardiovascular-related neurons of the RVLM have a respiratory rhythm in mature preparations (Häbler et al., 1994; Sun, 1995), such a rhythm is faint, and it is unlikely that such cells correspond to pre-I neurons. Therefore, only E-AUG and E-DEC neurons in the adult RVLM correspond to neonatal pre-I neurons. Furthermore, the inhibition of pre-I neurons in the I phase has an equivalent in connectivity observed in adult cats (Ezure, 1990), i.e., I-DEC neurons that are located slightly caudal to the RVLM and inhibit E-AUG neurons.

In summary, immature biphasic E neurons (pre-I neurons of the neonatal RVLM) are potentially analogous to mature E-AUG (and/or E-DEC) neurons, which can indeed exhibit biphasic E firing under some conditions. About this hypothesis, however, there are two reservations. First, E-AUG neurons already exist in immature preparations (Smith et al., 1990; Arata et al., 1998). However, it is not certain whether they are interneurons or cranial motoneurons. It is assumed that these neurons may not necessarily develop into mature E-AUG neurons, since the former are possibly inhibited by the pre-I neurons (Arata et al., 1998) that change their own firing during development. Second, some pre-I neurons seem excitatory (Onimaru et al., 1997) and may not correspond to mature E-AUG neurons, which are inhibitory. A group of pre-I neurons that are not inhibited at the I phase are I neurons from the beginning and it is assumed that they may develop into mature I neurons, such as I-CON neurons. If biphasic type pre-I neurons are excitatory, their corresponding neurons in mature animals cannot be specified. Here, it should be emphasized that any expiratory neurons that excite E-AUG neurons of the mature BOT have not yet been identified. Such excitatory neurons may not exist. E-AUG (and/or E-DEC) neurons of the BOT may receive tonic (possibly chemical) drives that modulate their intrinsic membrane properties as well as inhibitory inputs from I neurons. That is, these BOT neurons are called E neurons because they are inhibited by the other group of neurons that are called I neurons (see the next section for related discussion).

Network versus intrinsic cellular properties

It may not be meaningful to ask which is essential for the rhythm generation—network or pacemaker properties. Evidence obtained from both mature in vivo and neonatal in vitro preparations has shown that rhythm generation requires synaptic connections between relevant neurons. Without excitatory synaptic connections, synchronized motor activity cannot be produced in neonatal animals, even if each pacemaker neuron in the circuit maintains its own independent rhythmicity. One major difference is that inhibitory mechanisms are absolutely essential for rhythm generation in mature but not neonatal preparations. In the latter, inhibitory connections are nonetheless important. Although their blockage does not completely stop the primary rhythm, it does perturb both qualitative and quantitative features of the overall rhythmicity (Funk and Feldman, 1995; Ballanyi et al., 1999). For example, Iizuka (1999) showed that after such blockage, the basic alternation between I and E activity disappeared, and both activities synchronized. His work had the important implication that primary respiratory activity is neither inspiratory nor expiratory. Rather, the two phases are differentiated only by the presence of reciprocal inhibition. It is evident, therefore, that both excitatory and inhibitory connections are essential for eupnea in both mature and neonatal animals. Admittedly, synaptic inhibition seems more significant for rhythm generation in mature versus neonatal animals. The fact that even 22-day-old mice can produce a respiratory rhythm after blocking glycinergic inhibition (Ramirez et al., 1996) suggests, however, that the role of inhibitory connections is not so qualitatively different in the two preparations.

A more appropriate question is whether the pacemaker mechanisms that are essential for the neonatal rhythm are also involved in the mature rhythm. The author is that intrinsic cellular properties, which may not necessarily be called pacemaker properties, are crucially involved in rhythm generation in adult animals. This idea is in keeping with various current hybrid models that combine active membrane properties like conditional bursting with network circuitry into the rhythm-generating kernel (Bianchi et al., 1995; Funk and Feldman, 1995; Ramirez and Richter, 1996; Butera et al., 1999). It does not seem, however, that autonomic rhythmicity can be produced by respiratory neurons, simply on the basis of their passive and active membrane properties and their excitatory and inhibitory synaptic connections. For example, it is quite difficult to interpret the firing patterns of respiratory neurons such as I-DEC and I-CON neurons without taking into account their intrinsic cellular properties. There seem to be no respiratory neurons whose synaptic connections to I-DEC neurons alone make their firing pattern decremental. It is also hard to account for the sudden start and stop of I-CON neuron firing solely on the basis of their synaptic bombardment. Therefore, it is reasonable to propose that the intrinsic and neuromodulated ionic mechanisms of I-CON and I-DEC neurons play a crucial role during generation of the respiratory rhythm.

Concluding thoughts

Knowledge about neuronal mechanisms of respiration is rapidly increasing by studies that use both mature in vivo animals and in vitro neonates. Although there are missing links between the two approaches, respiratory physiologists are well positioned to model the mechanisms of respiratory rhythm generation. The available data suggest that the neuronal mechanisms for respiratory rhythm generation may be basically similar in both neonates and adults. Currently, hybrid models of respiratory rhythmogenesis that include both intrinsic cellular properties and network properties seem the most plausible ones. Pacemaker-like neurons in neonates may change throughout development, but there is indication that they still exist and function in the mature pre-BOT/RVLM region. In other words, respiratory neurons of adults give indication of their neonatal ancestry. Respiratory neurons in the pre-BOT/RVLM region with intrinsic properties like spike-frequency adaptation and neuromodulated ionic properties like conditional bursting, together with their excitatory and inhibitory interconnections, may be primarily responsible for the respiratory rhythm.

Abbreviations

BOT	Bötzinger complex
CNS	central nervous system
DRG	dorsal respiratory group
E	expiratory
E-AUG	augmenting expiratory
E-DEC	decrementing expiratory
I	inspiratory
I-CON	constant inspiratory
I-DEC	decrementing inspiratory
pre-BOT	pre-Bötzinger complex
RVLM	rostral ventrolateral medulla
VRG	ventral respiratory group

References

Arata, A., Onimaru, H. and Homma, I. (1998). Possible synaptic connections of excitatory neurons in the medulla of newborn rat in vitro. Neuroreport, 9: 743–746.

Arita, H., Kogo, N. and Koshiya, N. (1987). Morphological and physiological properties of caudal medullary expiratory neurons of the cat. Brain Res., 401: 258–266.

Ballanyi, K., Onimaru, H. and Homma, I. (1999). Respiratory network function in the isolated brainstem-spinal cord of newborn rats. Prog. Neurobiol., 59: 583–634.

Bianchi, A.L., Denavit-Saubie, M. and Champagnat, J. (1995). Central control of breathing in mammals: neuronal circuitry, membrane properties, and neurotransmitters. Physiol. Rev., 75: 1–45.

Bryant, T.H., Yoshida, S., de Castro, D. and Lipski, J. (1993). Expiratory neurons of the Bötzinger complex in the rat: a morphological study following intracellular labeling with biocytin. J. Comp. Neurol., 335: 267–282.

Butera, R.J., Rinzel, J. and Smith, J.C. (1999). Models of respiratory rhythm generation in the pre-Bötzinger complex. I. Bursting pacemaker neurons. J. Neurophysiol., 82: 383–397.

Cohen, M.I. (1979). Neurogenesis of respiratory rhythm in the mammal. Physiol. Rev., 59: 1105–1173.

Connelly, C.A., Dobbins, E.G. and Feldman, J.F. (1992). Pre-Bötzinger complex in cats: respiratory neuronal discharge patterns. Brain Res., 590: 337–340.

Duffin, J., Ezure, K. and Lipski, J. (1995). Breathing rhythm generation: focus on the rostral ventrolateral medulla. News Physiol. Sci., 10: 133–140.

Ezure, K. (1990). Synaptic connections between medullary respiratory neurons and considerations on the genesis of respiratory rhythm. Prog. Neurobiol., 35: 429–450.

Ezure, K. (1996). Respiratory control. In: Yetes B.J. and Miller A.D. (Eds.), Vestibular Autonomic Regulation. CRC Press, Boca Raton, pp. 53–84.

Ezure, K. and Manabe, M. (1988). Decrementing expiratory neurons of the Bötzinger complex. II. Direct inhibitory synaptic linkage with ventral respiratory group neurons. Exp. Brain Res., 72: 159–166.

Ezure, K. and Manabe, M. (1989). Monosynaptic excitation of medullary inspiratory neurons by bulbospinal inspiratory neurons of the ventral respiratory group in the cat. Exp. Brain Res., 74: 501–511.

Ezure, K., Manabe, M. and Otake, K. (1989). Excitation and inhibition of medullary inspiratory neurons by two types of burst inspiratory neurons in the cat. Neurosci. Lett., 104: 303–308.

Feldman, J.L. (1986). Neurophysiology of breathing in mammals. In: Bloom F.E. (Ed.), Handbook of Physiology. The Nervous System. Intrinsic Regulatory System in the Brain, Vol. IV, American Physiological Society, Bethesda, pp. 463–524 Sec. 1.

Feldman, J.L. and Smith, J.C. (1989). Cellular mechanisms underlying modulation of breathing pattern in mammals. Ann. N.Y. Acad. Sci., 563: 114–130.

Funk, G.D. and Feldman, J.L. (1995). Generation of respiratory rhythm and pattern in mammals: insights from developmental studies. Curr. Opin. Neurobiol., 5: 778–785.

Funk, G.D., Smith, J.C. and Feldman, J.L. (1993). Generation and transmission of respiratory oscillation in medullary slices: role of excitatory amino acid. J. Neurophysiol., 70: 1497–1515.

Gray, P.A., Rekling, J.C., Bocchiaro, C.M. and Feldman, J.L. (1999). Modulation of respiratory frequency by peptidergic input to rhythmogenic neurons in the preBötzinger complex. Science, 286: 1566–1568.

Häbler, H.-J., Jänig, W. and Michaelis, M. (1994). Respiratory modulation in the activity of sympathetic neurones. Prog. Neurobiol., 43: 567–606.

Hayashi, F. and Lipski, J. (1992). The role of inhibitory aminoacids in control of respiratory motor output in an arterially perfused rat. Respir. Physiol., 89: 47–63.

Iizuka, M. (1999). Intercostal expiratory activity in an in vitro brainstem-spinal cord-rib preparation from the neonatal rat. J. Physiol. (Lond.), 520: 293–302.

Jiang, C. and Lipski, J. (1990). Extensive monosynaptic inhibition of ventral respiratory group neurons by augmenting neurons in the Bötzinger complex in the cat. Exp. Brain Res., 81: 631–648.

Johnson, S.M., Smith, J.C., Funk, G.D. and Feldman, J.L. (1994). Pacemaker behavior of respiratory neurons in medullary slices from neonatal rat. J. Neurophysiol., 72: 2598–2608.

Johnson, S.M., Koshiya, N. and Smith, J.C. (2001). Isolation of the kernel for respiratory rhythm generation in a novel preparation: the pre-Bötzinger complex 'island'. J. Neurophysiol., 85: 1772–1776.

Kanjhan, R., Lipski, J., Kruszewska, B. and Rong, W. (1995). A comparative study of pre-sympathetic and Bötzinger neurons in the rostral ventrolateral medulla (RVLM) of the rat. Brain Res., 699: 19–32.

Mellen, N.M. and Feldman, J.L. (2001). Evidence for multi rhythmogenic networks in the rat en bloc preparation. Respir. Res., 2: S21.

Merrill, E.G., Lipski, J., Kubin, L. and Fedorko, L. (1983). Origin of the expiratory inhibition of the nucleus tractus solitarius inspiratory neurones. Brain Res., 263: 43–50.

Onimaru, H. and Homma, I. (1987). Respiratory rhythm generator neurons in medulla of brainstem-spinal cord preparation from newborn rat. Brain Res., 403: 380–384.

Onimaru, H., Arata, A. and Homma, I. (1990). Inhibitory synaptic inputs to the respiratory rhythm generator in the medulla isolated from newborn rats. Pflügers Arch., 417: 425–432.

Onimaru, H., Arata, A. and Homma, I. (1997). Neuronal mechanisms of respiratory rhythm generation: an approach using in vitro preparation. Jpn. J. Physiol., 47: 385–403.

Otake, K., Sasaki, H., Mannen, H. and Ezure, K. (1987). Morphology of expiratory neurons of the Bötzinger complex: an HRP study in the cat. J. Comp. Neurol., 258: 565–579.

Otake, K., Sasaki, H., Ezure, K. and Manabe, M. (1990). Medullary projection of non-augmenting inspiratory neurons of the ventrolateral medulla in the cat. J. Comp. Neurol., 302: 485–499.

Paton, J.F.R. (1997). Rhythmic bursting of pre- and post-inspiratory neurones during central apnoea in mature mice. J. Physiol. (Lond.), 502: 623–639.

Paton, J.F.R. and Richter, D.W. (1995). Role of fast inhibitory synaptic mechanisms in respiratory rhythm generation in the maturing mouse. J. Physiol. (Lond.), 484: 505–521.

Paton, J.F.R., Ramirez, J-M. and Richter, D.W. (1994). Functionally intact in vitro preparation generating respiratory activity in neonatal and mature mammals. Pflügers Arch., 428: 250–260.

Pierrefiche, O., Schwarzacher, S.W., Bischoff, A.M. and Richter, D.W. (1998). Blockage of synaptic inhibition within the pre-Bötzinger complex in the cat suppresses respiratory rhythm generation in vivo. J. Physiol. (Lond.), 509: 245–254.

Pilowsky, P.M. and Feldman, J.L. (2001). Identifying neurons in the preBötzinger complex that generate respiratory rhythm: visualizing the ghost in the machine. J. Comp. Neurol., 434: 125–127.

Ramirez, J.M. and Richter, D.W. (1996). The neuronal mechanisms of respiratory rhythm generation. Curr. Opin. Neurobiol., 6: 817–825.

Ramirez, J.M., Quellmalz, U.J.A. and Richter, D.W. (1996). Postnatal changes in the mammalian respiratory network as revealed by the transverse brainstem slice of mice. J. Physiol. (Lond.), 491: 799–812.

Rekling, J.C. and Feldman, J.L. (1998). PreBötzinger complex and pacemaker neurons: hypothesized site and kernel for respiratory rhythm generation. Ann. Rev. Physiol., 60: 385–405.

Ross, C.A., Ruggiero, D.A., Joh, T.H., Park, D.H. and Reis, D.J. (1984). Rostral ventrolateral medulla: selective projections to the thoracic autonomic cell column from the region containing C1 adrenaline neurons. J. Comp. Neurol., 228: 168–185.

Schwarzacher, S.W., Smith, J.C. and Richter, D.W. (1995). Pre-Bötzinger complex in the cat. J. Neurophysiol., 73: 1452–1461.

Smith, J.C., Greer, J.J., Liu, G. and Feldman, J.L. (1990). Neural mechanisms generating respiratory pattern in mammalian brain stem-spinal cord in vitro. I. Spatiotemporal patterns of motor and medullary neuron activity. J. Neurophysiol., 64: 1149–1169.

Smith, J.C., Ellenberger, H.H., Ballanyi, K., Richter, D.W. and Feldman, J.L. (1991). Pre-Bötzinger complex: a brainstem region that may generate respiratory rhythm in mammals. Science, 254: 726–729.

Sun, M.-K. (1995). Central neural organization and control of sympathetic nervous system in mammals. Prog. Neurobiol., 47: 157–233.

Sun, Q.-J., Goodchild, A.K., Chalmers, J.P. and Pilowsky, P.M. (1998). The pre-Bötzinger complex and phase-spanning neurons in the adult rat. Brain Res., 809: 204–213.

Suzue, T. (1984). Respiratory rhythm generation in the in vitro brain stem-spinal cord preparation of the neonatal rat. J. Physiol. (Lond.), 354: 173–183.

Wang, H., Stornetta, R.L., Rosin, D.L. and Guyenet, P.G. (2001). Neurokinin-1 receptor immunoreactive neurons of the ventral respiratory group in the rat. J. Comp. Neurol., 434: 128–146.

SECTION III

Spinal cord and brainstem: motoneurons, pattern generation and sensory feedback

CHAPTER 8

Key mechanisms for setting the input–output gain across the motoneuron pool

Hans Hultborn[1]*, Robert B. Brownstone[2], Tibor I. Toth[3] and Jean-Pierre Gossard[4]

[1]*Department of Medical Physiology, The Panum Institute, University of Copenhagen, Copenhagen DK-2200, Denmark*
[2]*Division of Neurosurgery, Dalhousie University, Halifax, NS B3H 4R2, Canada*
[3]*Physiology Unit, School of Biosciences, Cardiff University, Cardiff CF10 3US, UK*
[4]*Department of Physiology, University of Montreal, Montreal, QC H3C 3J7, Canada*

Abstract: This chapter summarizes a number of factors that control the "input–output" function across the motoneurons (MNs) comprising a single spinal motor nucleus. The main focus is on intrinsic properties of individual MNs that can be controlled by neuromodulators. These include: (1) amplification of the synaptic input at the cell's dendritic level by voltage-gated, persistent inward currents (plateau potentials); and (2) transduction of the net synaptic excitation into a frequency code (the MN's stimulus current-spike frequency relation) at the cell's soma/initial segment. Two other aspects of the synaptic control of MNs, which may affect their input–output gain, are also discussed. They include the hypotheses that: (1) a non-uniform distribution of synaptic effects to low- and high-threshold motor units causes a change in recruitment gain; and (2) recurrent inhibition, via motor axon collaterals and Renshaw cells, functions as a variable gain regulator of MN discharge.

Introduction

During the last 30 years, much information has been gathered on the activity of individual neurons in supraspinal motor centers during voluntary movement (Porter and Lemon, 1993). It has been possible to correlate the activity of neurons in these centers with movement and muscle activity. There have also been important developments both in the investigation of spinal centers that mediate the supraspinal command signals of voluntary movement to relevant motoneurons (MNs; Baldissera et al., 1981; Jankowska, 1992) and in understanding the function of the MNs, themselves (Powers and Binder, 2001; Hornby et al., 2002). It is nevertheless obvious that present knowledge is far from sufficient to provide an understanding of the transfer functions that operate from the activity of 'upper motoneurons' to activation of muscles. In this chapter, we present a short overview on four factors of importance for the input–output relations across the MNs of a single mammalian spinal motor nucleus (pool). They include: (1) amplification by persistent inward currents of the synaptic excitation at the dendritic level of MN operation; (2) transduction of the net synaptic excitation to a frequency code (the I–f relation) at the MN's soma/initial segment; (3) recruitment gain of MNs in a single spinal pool; and (4) negative feedback by recurrent inhibition to MNs via motor axon collaterals and Renshaw cells. All four of these mechanisms have their own idiosyncratic-controlled variables. This suggests that the MN pool is designed to handle many more external as well as internal challenges than have hitherto been considered.

Plateau potentials and amplification of synaptic input

The classical view of the mammalian spinal MN, which emerged from the laboratories of Eccles and

*Corresponding author: Tel.: +45-3532-7461;
Fax: +45-3532-7499; E-mail: h.hultborn@mfi.ku.dk

Granit in the 1950s–1960s, was that its membrane was essentially passive in its areas of synaptic contact (i.e., largely in the dendrites). As a result, the relation between a MN's synaptic excitation and its firing frequency was initially thought to be determined solely by the cell's postspike afterhyperpolarization (AHP; see Granit et al., 1963; Kernell, 1965a,b; Granit et al., 1966a,b). When Schwindt and Crill (1977) discovered that MNs could display prolonged plateau potentials and self-sustained firing, it certainly opened new horizons in the exploration of MN function (see also their 1984 review). They demonstrated that plateau potentials were due to voltage-dependent, noninactivating Ca^{2+} currents (Schwindt and Crill, 1980c). These inward currents were large enough as compared to steady-state outward currents to produce a maintained shift in the membrane potential, i.e., the plateau potential. In their experiments, penicillin (Schwindt and Crill, 1980a) or tetraethylammonium (Schwindt and Crill, 1980b) were used to block outward Cl^- and K^+ currents, respectively, thereby 'uncovering' the inward Ca^{2+} current. A few years later, the Copenhagen group continued the analysis of self-sustained MN discharge, including its mechanisms and function. The phenomenon, and its dependence on physiological transmitters (monoamines), was first analyzed in the in vivo cat (Conway et al., 1988; Hounsgaard et al., 1988a). Subsequent analysis of the cellular mechanisms of such self-sustained discharge was mainly performed in a newly developed in vitro preparation of the turtle spinal cord (Hounsgaard and Mintz, 1988; Hounsgaard et al., 1988b). It is now established that a number of physiological transmitters can facilitate the appearance of plateau properties. Such actions may reduce outward currents (Hounsgaard and Kiehn, 1989) and/or augment the persistent inward Ca^{2+} current (Delgado-Lezama et al., 1997; Svirskis and Hounsgaard, 1998; reviewed by Hultborn, 1999; Delgado-Lezama and Hounsgaard, 1999; Alaburda et al., 2002).

The presence of a plateau potential (due to a persistent inward current) is clearly reflected in the firing pattern of the MN during injection of current pulses through the recording microelectrode (see also Kirkwood and Ford, Chapter 10 of this volume). For example, the recordings in Fig. 1 were made in the unanesthetized decerebrate cat preparation, in which condition plateau properties are enhanced (Hounsgaard et al., 1988a).

Fig. 1A shows self-sustained firing. Note that it is initiated and terminated by depolarizing and hyperpolarizing current pulses, respectively. When plateau potentials were present in this study, there was a pronounced 'frequency acceleration' during the stimulus current pulse (see arrow in Fig. 1B), thereby indicating the delay in onset and gradual subsequent development of a persistent inward current. The initiation of the plateau could also be seen in the MN's response to long-lasting triangular current pulses (Fig. 1C–D). When the injected current reached the cell's firing threshold, the spike frequency increased linearly with an f/I slope of 2.0 imp s^{-1} nA^{-1} (the so-called 'primary range' of Kernell, 1965a,b). Subsequently, a critical spike-frequency transition point was reached, at which the f/I slope increased abruptly to ~ 6.4 imp s^{-1} nA^{-1}, thereby corresponding to the 'secondary range' of Kernell (1965b). This steep rise in spike frequency indicated the initiation of a persistent inward current, because a reduction in injected current after the spike-frequency 'jump' did not bring the firing rate back to the prejump, matched-current level. This behavior yields a counter-clockwise hysteresis in the I–f plot (Fig. 1D).

In summary, when using classical current-clamp recording conditions, a persistent inward current is revealed by (1) self-sustained MN discharge, (2) spike-frequency acceleration during either constant-current or triangular-current injection, and (3) a pronounced hysteresis in the firing frequency as the injected current is reduced. A persistent inward current is visualized most directly in the form of a plateau potential after inactivation of the MN's Na^+ spikes. This can occur either spontaneously or by injecting into the MN the Na^+ channel blocker, QX314 (Hounsgaard et al., 1988a; Brownstone et al., 1994; Bennett et al., 1998a).

Why did plateau potential properties go unnoticed for so many years after the first intensive investigation of spinal MNs in the early 1950s? The answer is that most such initial work made use of anesthetized and/or spinal preparations in which plateau potentials do not occur spontaneously. Rather, in MNs, the plateau potential is a latent property, which can only become manifest under appropriate facilitation (neuromodulation). For example, it appears in the

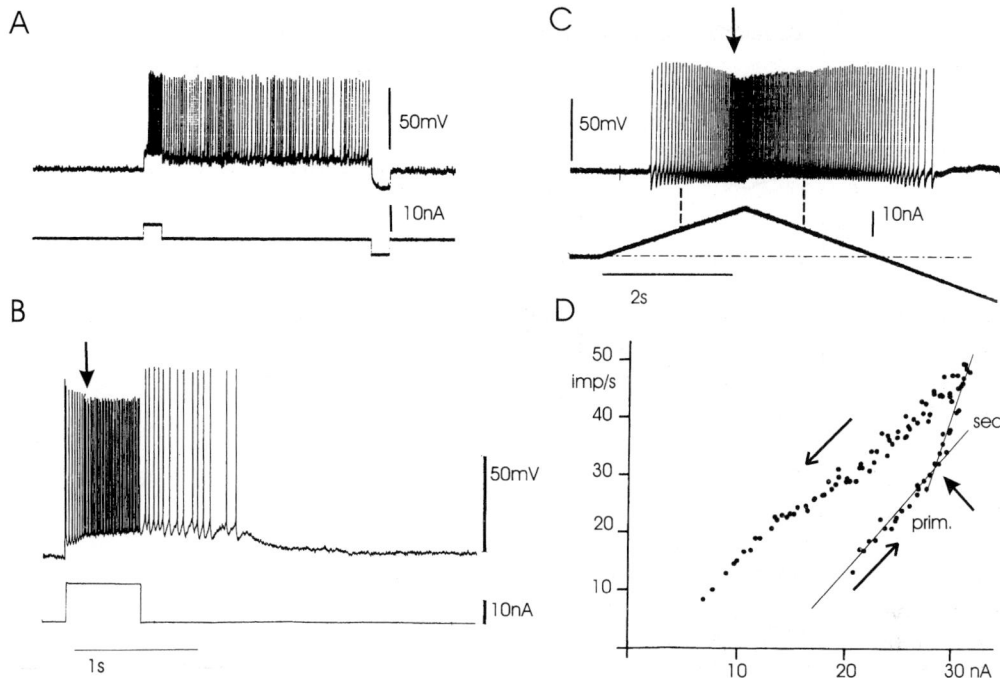

Fig. 1. Examples and signs of plateau potentials recorded in spinal MNs. (A–B) Sustained shifts in excitability triggered by depolarizing and hyperpolarizing currents injected intracellularly into cat spinal MNs in a decerebrate, unanesthetized preparation. Upper traces are intracellular recordings. Lower traces indicate the amount and timing of the injected current. (A) Sustained repetitive firing in a MN, which was initiated by a short depolarizing current pulse and terminated by a short hyperpolarizing pulse. (Reproduced, in part, from Hounsgaard et al., 1984, with permission from Springer-Verlag.) (B) Record from another MN during the rectangular stimulus current pulse. Note the spike-frequency acceleration (at arrow), which was a sign of the onset of a plateau current. (Reproduced, in part, from Hounsgaard et al., 1988a, with permission from Cambridge University Press.) (C–D) Firing pattern and membrane potentials during injection of triangular current pulses into cat MNs in an unanesthetized, decerebrate cat preparation. (C) Spike activity in a MN (upper trace) and triangular profile of its injected current (second trace). The interrupted vertical lines point to the difference in firing frequency at similar current intensities during the ascending and descending phase. The interrupted horizontal line shows the zero level for injected current. (D) Plot of spike frequency versus injected current for the C data. The thin lines approximate the f/I slopes for Kernell's primary (prim.) and secondary (sec.) range of firing. Note the counter-clockwise hysteresis. (C–D reproduced, in part, from Hounsgaard et al., 1984, with permission from Springer-Verlag.)

unanesthetized decerebrate cat by virtue of facilitatory activity in descending monoaminergic projections (Crone et al., 1988; Hounsgaard et al., 1988a; for further history, see Hornby et al., 2002). Fig. 2A–B illustrates that no counter-clockwise hysteresis occurs in the MN in response to injection of triangular current pulses following a spinal transection. It returns, however (Fig. 2C), after intravenous injection of the serotonin precursor, 5-HTP (Hounsgaard et al., 1988a). Noradrenergic precursors are also effective (Conway et al., 1988). At the time of these late 1980s' studies, it was argued that monoamines enabled plateau properties primarily by reducing outward K^+ currents (Hounsgaard and Kiehn, 1989). Subsequent studies by Hounsgaard and colleagues in the turtle, however, have shown that several other transmitters may be involved. They also demonstrated that part of the plateau-enhancing effect was due to a facilitation of the L-type Ca^{2+} channels (Delgado-Lezama et al., 1997; Svirskis and Hounsgaard, 1998; for review, see Powers and Binder, 2001; Alaburda et al., 2002; Hornby et al., 2002).

Fig. 2D–E illustrates results from a barbiturate-anesthetized cat with an intact central nervous system (CNS). Triangular current pulses did not produce counter-clockwise hysteresis (Fig. 2E). During continuous low-frequency stimulation of the descending raphe-spinal (serotonergic) projection to the spinal

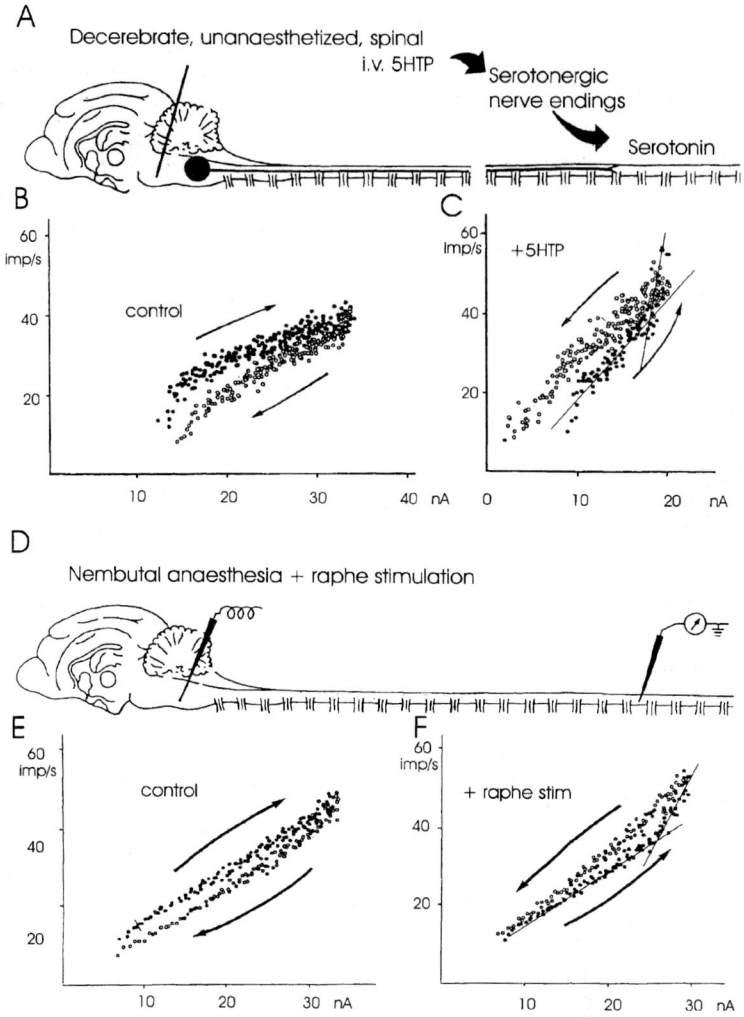

Fig. 2. Response properties of MNs to triangular current-pulse injection. (A–C) MN responses to a triangular current-pulse injection in (A) a decerebrate, unanesthetized acute spinal cat. (B–C) Stimulus current–spike frequency plot of the cell before (B) and after (C) i.v. administration of 5-HTP. Note the counter-clockwise hysteresis before, and clockwise hysteresis after drug application. Thin lines in C as in Fig. 1D. Note the steeper slopes and lower thresholds for repetitive firing following 5-HTP application. (D–F) Response properties of a MN to a triangular current-pulse injection in (D) an intact cat preparation under barbiturate anesthesia. (E–F) Stimulus current–spike frequency plot of the cell before (E, control) and during (F) a train of 20-Hz stimuli to the descending raphe-spinal projection. Note the stimulus-evoked clockwise (E) versus counter-clockwise (F) hysteresis. (A–C reproduced, in part, from Hounsgaard et al., 1988a with permission from Cambridge University Press; D–F are unpublished results of R.M.B., T.I.T., J.-P.G. and H.H.)

cord, however (Fig. 2F), the same triangular stimulus current pulse resulted in a spike-frequency acceleration and a counter-clockwise hysteresis (an unpublished result of R.M.B., T.I.T., J.-P.G. and H.H.). The lack of plateau properties during barbiturate anesthesia of the cat was first interpreted as a sign of a drug-induced depression of spontaneous activity in descending monoaminergic pathways (Crone et al., 1988). In turtle MNs, however, it was later demonstrated that barbiturates also suppress the L-type Ca^{2+} channel (Guertin and Hounsgaard, 1999).

The plateau current is predominantly of dendritic origin. This was first predicted on the basis of a theoretical analysis (Gutman, 1991) of the data of Schwindt and Crill. This dendritic origin was subsequently demonstrated in experiments on turtle MNs in which the soma and dendrites were differentially polarized by application of extracellular electrical fields (Hounsgaard and Kiehn, 1993). This result was also supported by an analysis of the dynamics of the hysteresis in the plateau current (Svirskis et al., 2001). Further electrophysiological and pharmacological analyses in the mouse spinal cord demonstrated that a special L-subtype of Ca^{2+} channels (the Cav 1.3, which expresses the α1D subunit; see Ertel et al., 2002 for nomenclature) is likely mediating the plateau current in MNs (Carlin et al., 2000b). In this latter study, the conclusion as to the mainly dendritic origin of the plateau currents was strongly reinforced by direct visualization of these channels on MN dendrites. Furthermore, it was shown that the hysteresis properties of this current could only be explained by a dendritic localization of the Ca^{2+} channels in question (Carlin et al., 2000b).

A dendritic localization of the plateau current suggests that uniformly distributed synaptic currents should be more efficient in activating the plateau current than preferential depolarization of the soma region by currents injected through a recording microelectrode. This proposition stimulated Bennett et al. (1998a) to systematically investigate the threshold of the plateau potential to intracellular current injections during tonic (subthreshold) synaptic excitation and inhibition. In this study, synaptic excitation was evoked by maintained stretch of the triceps surae muscle, and synaptic inhibition was achieved by trains of stimuli to a nerve supplying antagonist muscles (i.e., reciprocal inhibition). The main finding was that the plateau threshold decreased substantially during synaptic excitation. (It was recognized, of course, that less microelectrode-delivered current was needed, because synaptic current was also provided.) The truly novel finding is shown in Fig. 3. Intracellular current ramps initiated the plateau at significantly lower frequencies of spike discharge.

Bennett et al. (1998a) also showed that during tonic synaptic excitation, the plateau occurred at a more negative membrane potential [i.e., as recorded at the soma; with the action potentials (APs) blocked by QX314]. The reduction in plateau threshold was graded according to the amount of excitation. During larger muscle stretches, the plateau threshold was often lowered to near the MN's initial recruitment level. Figure 3 also shows that synaptic inhibition had the opposite effect by increasing the plateau's threshold. The important conclusion from this study is that during synaptic excitation of a MN, the thresholds for initiating the plateau potential and recruiting the MN into activity are almost the same. On several occasions, the plateau potential was initiated at an even lower threshold than that of the APs, thereby directly contributing to the MN's recruitment (Bennett et al., 1998a).

In the Copenhagen studies described earlier, the experimental paradigm was designed to investigate the effect of synaptic excitation and inhibition on the threshold of the plateau potential elicited by current injection. In actuality, most of our experimental paradigms employed current injection through a recording electrode that was presumably positioned in the cell's soma. This had a tendency to over-emphasize the 'all-or-none' character of the current producing the plateau potential. It seems functionally more realistic to imagine that a local synaptic excitation (predominately in the dendrites) would activate a persistent inward current only in its immediate environs. As such, it would be insufficient to generate a full-blown plateau potential involving the whole neuron. This possibility raises a series of questions: (1) is plateau-induced amplification of synaptic excitation a graded event; (2) if so, how large is such amplification; (3) could amplification of inputs differ for different dendrites; and (4) over what range of spike frequencies does such amplification occur?

Figure 4 shows that our ongoing experiments clearly demonstrate a graded, strong facilitation of synaptic inputs by the plateau potential (Hultborn and Nielsen, 2002; also the unpublished results of H.H., J.B. Nielsen, M.E. Denton, J. Wienecke). In these experiments on the decerebrate cat, we injected slowly rising triangular currents into triceps surae MNs, while simultaneously noting the increment in the cell's spike frequency produced by sinusoidal stretches of the homonymous muscles. We found that the additional firing caused by muscle

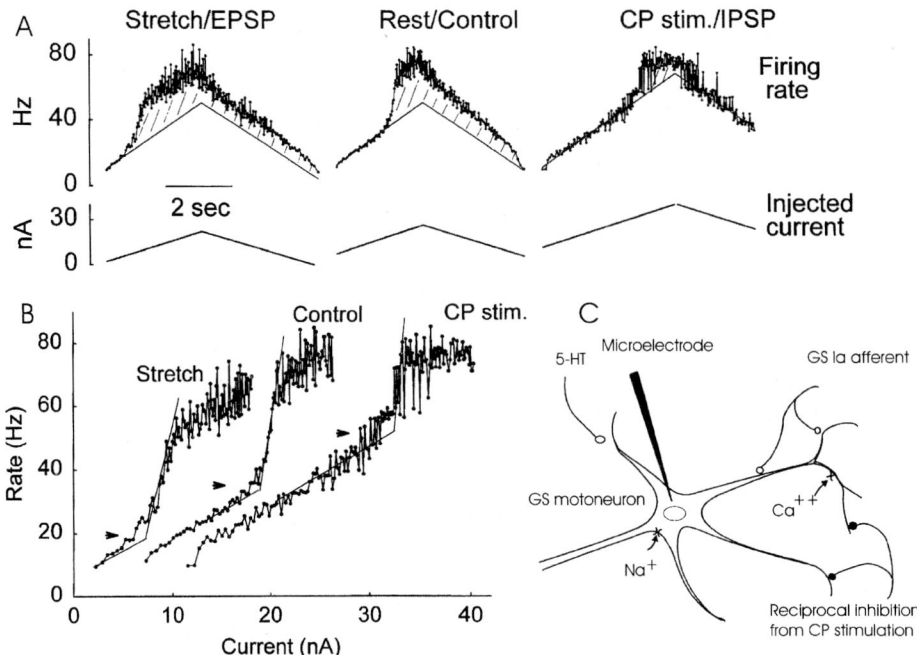

Fig. 3. Demonstration that synaptic input changes the threshold at which intracellular current injection into a MN generates a plateau potential. Such activation was studied with and without a steady (tonic) peripheral synaptic input (Ia afferent excitation and reciprocal Ia inhibition; see the C sketch). (A) In each panel, the upper trace shows the intracellularly- recorded spike-frequency response of the MN. The middle trace (thin line) shows the expected firing frequency by injected current as extrapolated from the initial primary-range slope. The cross-hatched area shows the contribution to the firing rate of the plateau potential. The lower trace shows the triangular profile of the injected current. Middle panel, control situation without added tonic synaptic input. Left-side panel, effect of tonic muscle stretch applied throughout the triangular current. Before current injection, the tonic, stretch-evoked synaptic excitation was adjusted to be subthreshold, i.e., below the MN's recruitment level. Right-side panel, effect of tonic inhibitory nerve stimulation, i.e., IPSPs produced in the MN by stimulation of common peroneal nerve (CP). (B) Composite of results from the A measurements showing MN responses during the ascending phase of the ramp stimulus plotted against stimulus current. Note that relative to the control situation, the firing frequency at which the plateau current was activated was lower during tonic synaptic excitation and higher during tonic synaptic inhibition. (Reproduced, in part, from Bennett et al., 1998a with permission from The American Physiological Society.)

stretch increased stepwise as the firing frequency slowly increased in response to current injection into the MN. The stretch (synaptically)-induced increase in spike frequency revealed a many-fold amplification of the output to a given synaptic input even before the current injection, itself, triggered a plateau potential. These results provide evidence of for a powerful nonlinear facilitation of synaptic inputs by voltage-dependent, noninactivating inward currents. Interestingly, the design of these experiments was almost identical to that of Granit et al. (1966a,b), who also used synaptic input superimposed on repetitive MN firing evoked by injected current. Our conceptual framework,

however, was substantially different to that of Granit, thereby leading to a different interpretation of the results.

The possibility of a localized presence of, and control of, plateau potentials in different parts of the dendritic tree in an individual MN is still a very open possibility. It has been demonstrated that only a fraction of the dendrites is already sufficient to produce plateau potentials by extracellularly applied electric fields (Hounsgaard and Kiehn, 1993). Sequential, stepwise activation of dendrites were also seen following intracellular current pulses in young mouse MNs (Carlin et al., 2000a). Furthermore, synaptic input shown to facilitate the

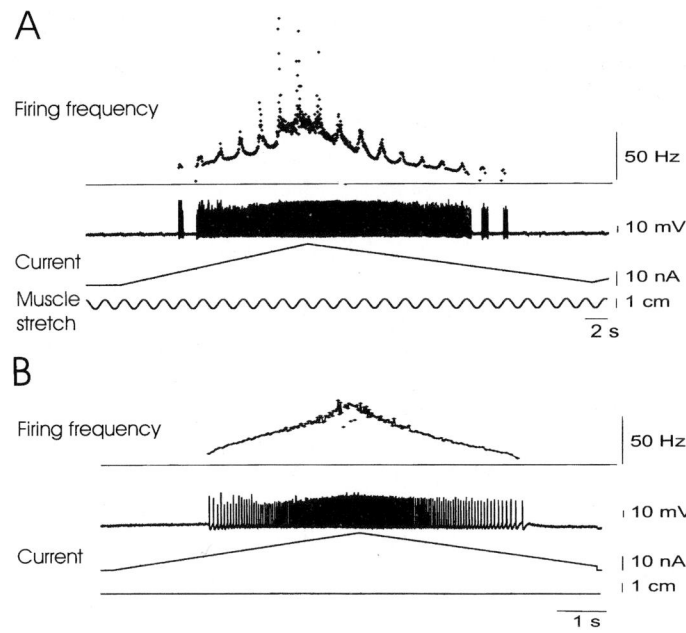

Fig. 4. Amplification of the MN's spike-frequency response to muscle stretch by gradually increasing depolarizing current. The response of a triceps surae MN in a decerebrate unanesthetized cat is shown to triangular current-pulse injections. (A–B) Traces, from top to bottom, show the cell's firing frequency, spike train (AC-filtered), injected current and the homonymous muscle stretch (at $0.5\,\text{s}^{-1}$ in A; no stretch in B). Muscle stretch alone was subthreshold for cell discharge. The records in B show that in the absence of muscle stretch, the cell's response to a triangular current pulse was quite linear. In contrast, in the presence of muscle stretch (A), there were substantially higher rates of discharge, thereby demonstrating a nonlinear response to the dual input. (Unpublished results of H.H., J.B. Nielsen, M.E. Denton and J. Wienecke; see also abstract of Hultborn and Nielsen, 2002.)

production of plateau potentials (neuromodulation) could be spatially segregated to the medial or lateral dendrites in turtle MNs (Delgado-Lezama et al., 1999).

Ever since their discovery, it has been proposed with varying nuances that plateau potentials serve to amplify synaptic inputs to MNs (for review, see Hultborn, 1999; Powers and Binder, 2001; Hornby et al., 2002). Over the years, it has been firmly established that the exogenous administration of neurotransmitters can greatly enhance the amplification of synaptic inputs—at least six-fold according to a recent quantitative analysis by Lee and Heckman (2000) under voltage-clamp conditions in cat spinal MNs. The experiments that the Copenhagen group has in progress, exemplified by Fig. 4, demonstrate that even with a constant monoaminergic drive, the dendritic amplification will vary considerably, as dependent on the level of the MN's activation. Figure 4 clearly illustrates the graded character of this amplification within the physiological range of AP discharge.

Transduction of the summated synaptic current into a spike-frequency code by postspike AHP

While much attention has been devoted in recent years to complex dendritic processing in MNs (for review, see Powers and Binder, 2001), it is important to remember that for all neurons with a relatively long axon, the cell's global output is encoded as a spike frequency. Postspike AHP has a key function in transducing the summed result of processed synaptic input into a variable spike frequency. In the case of sensory axons, Adrian and Zotterman concluded as early as 1926 that the "... regular rhythm could be explained as the natural consequence of the refractory period and the subsequent return of excitability." It was known even then that stronger excitation of a sensory neuron overcomes its 'refractoriness' to result in a higher spike frequency. In 1929, Adrian and Bronk extended the observations to motor axons and wrote "... The likeness between the impulse discharge from a sense organ and that from a motor nerve cell suggests that both are due to some property common to axon or dendritic terminations in general." The direct demonstration of refractoriness in MNs did not occur until many years later, when it was uncovered by intracellular recording as the AHP (Brock et al., 1952; Eccles et al., 1958). Eccles et al. (1958) compared the duration of the AHP for fast versus slow motor units and suggested that "... since it seems likely that the frequency of repetitive discharge from motoneurons is controlled by the AHP, durations of the AHPs are matched to the twitch durations of muscle." To further advance understanding of this MN transduction process, Granit et al. (1963) used current injection through the recording microelectrode to mimic synaptic currents. Their I–f relation values typically had a slope of $\sim 1-2$ imp s^{-1} nA^{-1} for a relatively long linear segment, which was subsequently named the primary range of cell discharge (Kernell, 1965a,b). In some cases, large current injections reached a level where a steeper linear slope was seen. This was called the secondary range (Kernell, 1965b). It was typically seen when the spike frequency reached ~ 50 imp s^{-1}. The time course and properties of the AHP as seen after individual APs can directly explain much of the I–f relation. Additional mechanisms are probably involved, however, especially for elaboration of the secondary range, one possibly being a persistent inward current (Schwindt and Crill, 1982; see Powers and Binder, 2001 for alternative interpretations). In 1966, Granit et al. (1966a,b) reported on the superimposition of synaptic excitation or inhibition to MN discharge evoked by current pulses that were injected through the recording microelectrode. Their basic assumption was that the discharge due to the two sources would add or subtract 'algebraically' (linearly). This would validate the idea that current injection through the recording microelectrode could be used to mimic synaptic inputs. Their results did indeed seem to support this hypothesis as long as the MNs fired within the primary range. Within the secondary range, however, the increase in spike firing by increased excitation and the decrease by increased inhibition was larger than expected. According to our present understanding, the secondary range (during steady-state discharge) arises due to the recruitment of a persistent inward current in the MN's dendrites.

Experimentally, a persistent inward current is activated by injecting depolarizing current through the recording microelectrode into the cell's soma. As a result, a highly nonlinear interaction with synaptic inputs is to be expected. Because the Granit group's experiments were performed on preparations under light barbiturate anesthesia, the dendritic inward currents producing the plateau potential were presumably rather depressed. Consequently, secondary-range firing was seen only occasionally. Results of our recent experiments on decerebrate preparations without anesthesia, however, show that a dendritic persistent inward current is activated much more easily (Fig. 4). Here, as described earlier, synaptic excitation caused more than algebraic summation—and incrementally so—and at firing frequencies corresponding to Kernell's (1965a and 1965b) primary range.

The AHP is due to a Ca^{2+}-dependent K^+-conductance (for review, see Powers and Binder, 2001). A reduction of this AHP current, by blocking either Ca^{2+} entry (by Mn^{2+}; Walton and Fulton, 1986) or the AHP's K^+ conductance (by apamin; Hounsgaard and Mintz, 1988), increases the slope of the I–f relation. It has been demonstrated that serotonin reduces the AHP in MNs in the lamprey (Van Dongen et al., 1986), turtle (Hounsgaard and Kiehn, 1989), rat (Wu et al., 1991) and cat (White and Fung, 1989; for further reports, see Powers and Binder, 2001). Initially, it was thought that serotonin directly reduced the K^+ conductance underlying the AHP (Hounsgaard and Kiehn, 1989). Recent experiments on the lamprey spinal cord (Svensson et al., 2001) have, however, demonstrated that serotonin inhibits N-type Ca^{2+} channels and Ca^{2+} entry during the AP, and thereby indirectly reduces the opening of the KCa channels. Moreover, Fig. 5 illustrates a depression of the AHP following repetitive stimulation of the caudal raphe nuclei in a cat when under barbiturate anesthesia (unpublished results of R.M.B., T.I.T., J.-P.G. and H.H.).

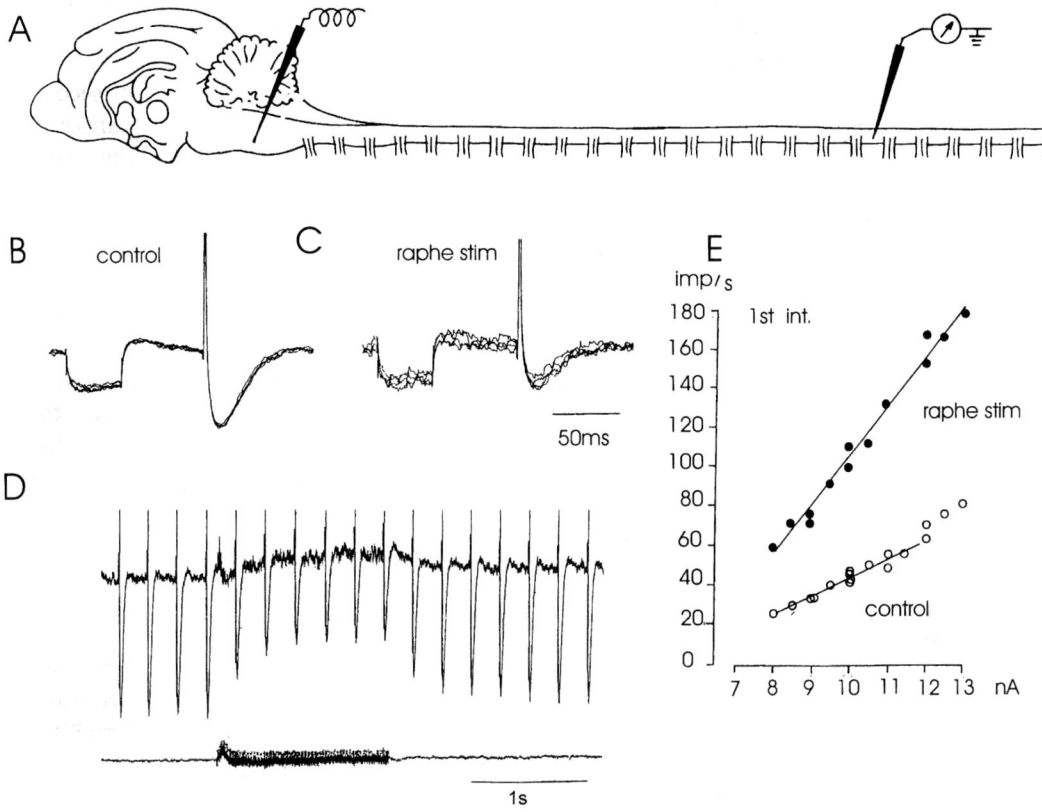

Fig. 5. Depression of a MN's postspike AHP and f/I slope during stimulation of the raphe nuclei. The MN was recorded in an intact CNS cat preparation under barbiturate anesthesia. (A) Intracellular current pulses were used to measure the input resistance of spinal MNs and to evoke their APs (spike + AHP). (B–D) AP spikes are truncated to focus on the AHP. (B) Control response to a brief injected current pulse. (C) Response to the same injected pulse during tetanic (20 s^{-1}) stimulation of the caudal raphe nuclei. (D) Upper trace, APs elicited by injected current pulses at a frequency of 5 Hz. Lower trace, time marker for a short train of stimuli to the raphe nuclei. Note the relatively fast onset and disappearance of the raphe-evoked depression of the AHP. (E) The MN's I–f relation for the first interspike interval when tested with graded current injection (1 s square pulses) in the control versus raphe-stimulated state. Note the extent to which raphe stimulation steepened the f/I slope. (Unpublished results of R.M.B., T.I.T., J.-P.G. and H.H.)

Figure 5 shows (B vs. C) that without a significant change in input resistance, raphe stimulation caused a rather dramatic depression of AHP amplitude. This depression cannot be explained by a membrane hyperpolarization, because raphe stimulation resulted in a moderate depolarization (Fig. 5D). Note also that a reduction in the AHP began rather quickly after onset of the repetitive stimulus train, and the depressive effect continued to grow for almost a second. This gradual increase in AHP depression, and similar delay in restoration of the control AHP after termination of raphe stimulation, can possibly be explained by the kinetics of activation of protein kinase C. This occurs in the presence of Ca^{2+}, diacylglycerol and arachidonic acid, which are all involved in controlling K^+ channels (Hellgren Koteleski et al., 2003).

In the experiment shown in Fig. 5, the spike discharge evoked by graded, rectangular current pulses was tested in the same MN in which AHP depression was observed. The cell's I–f relation was first determined without raphe stimulation (control) and then during repetitive stimulation of the raphe nuclei. As anticipated, Fig. 5E shows that the I–f relation became much steeper during raphe stimulation.

In summary, the spike-frequency code of the global output of the MN is a controlled variable, and monoaminergic innervation of the MN is directly involved in this control.

Effects of the distribution of synaptic excitation and inhibition on recruitment gain in a spinal MN pool

A commonly held notion is that MN size is a key factor underlying the orderly recruitment of motor units according to their contraction force. The basic hypothesis of this 'size principle' (Henneman and Mendell, 1981), is that MN excitability is a function of MN size. If synaptic density and distribution are the same for small and large MNs, however, size per se cannot totally explain orderly recruitment. For example, Gustafsson and Pinter (1984, 1985) demonstrated systematic intrinsic threshold differences (higher resistivity) in small versus large MNs. This built-in feature in the MN pool could account for motor units being recruited in a force-related manner under conditions of constant MN synaptic density and efficacy. This finding does not exclude the possibility, however, that variation in the distribution of synaptic inputs among the MNs of a given pool play an additional role in the orderly recruitment phenomenon. Exceptions to such recruitment are well known, of course. For example, synaptic inputs to MNs from skin afferents and the rubrospinal tract produce inhibition in MNs supplying slow-twitch motor units and excitation to those innervating fast-twitch ones (Burke, 1981; Powers and Binder, 2001). There has been much discussion concerning to what extent (or even whether) this deviant projection pattern might change orderly motor-unit recruitment (Burke, 1981; Clark et al., 1993; Cope and Clark, 1995). If so, what functional consequences might ensue? In the authors' opinion, such a deviant distribution of synaptic inputs will not only change the recruitment order of MNs, but also the ease of grading their recruitment, i.e., the MNs' recruitment gain. This argument is illustrated in Fig. 6A–D (Kernell and Hultborn, 1990).

A MN may be said to have an intrinsic excitability, which we have defined operationally as the amount of injected current required to discharge the MN in the absence of background synaptic activity. At any given moment, the ability of a MN to generate an AP will depend on both its intrinsic excitability and the sum of its synaptic inputs. Kernell and Hultborn (1990) pointed out that synaptic input systems with a nonuniform distribution ('pool bias' in Fig. 6A–C) might expand or compress the range of functional thresholds among the MNs of a single pool. This changes the ease with which a uniformly distributed excitatory system ('pool drive' in Fig. 6A–C) will activate the pool. More precisely, the term, recruitment gain, was used to indicate the relation between the intensity of the synaptic input (drive) to a MN pool and the output from that pool (i.e., the number of active cells). It predicted that a biasing input (with distributions to low- and high-threshold motor units different from that of the drive input) would indeed change this gain (Fig. 6D).

To evaluate the above hypothesis experimentally, Nielsen et al. (1990) investigated the effect of a conditioning stimulation of the caudal cutaneous sural nerve on the input/output relation of medial gastrocnemius versus soleus MNs. Figures 6E–G show that for a test stimulus, graded stimulation of the dorsal roots was used. The size of the EPSP in the tested medial gastrocnemius versus soleus MNs was used as a measure of the input (pool drive) to each of these two sets of MNs. The size (in % of maximum response) of the simultaneously recorded monosynaptic reflex (i.e., the neurogram response of the nerve to medial gastrocnemius vs. soleus) was each pool's output. Because the relation between the monosynaptic reflex and the EPSP became steeper for medial gastrocnemius MNs following stimulation of the sural nerve, they concluded that the input/output relation of these cells' pool (i.e., the recruitment gain) had indeed increased. Fig. 6F shows that small reflexes (engaging slow motor units) were inhibited, whereas large reflexes (dominated by fast units) were facilitated. In contrast, Fig. 6G shows that monosynaptic reflexes of the soleus MNs (virtually all supplying slow motor units) were inhibited independent of reflex size, and the slopes for the input/output relation of the tested MNs were the same for the unconditioned and conditioned monosynaptic reflexes. Further experiments along these lines have shown that a similar increase in recruitment gain can occur in humans (Nielsen and Kagamihara, 1993). In this latter case, sural nerve stimulation inhibited low-threshold motor units and facilitated high-threshold

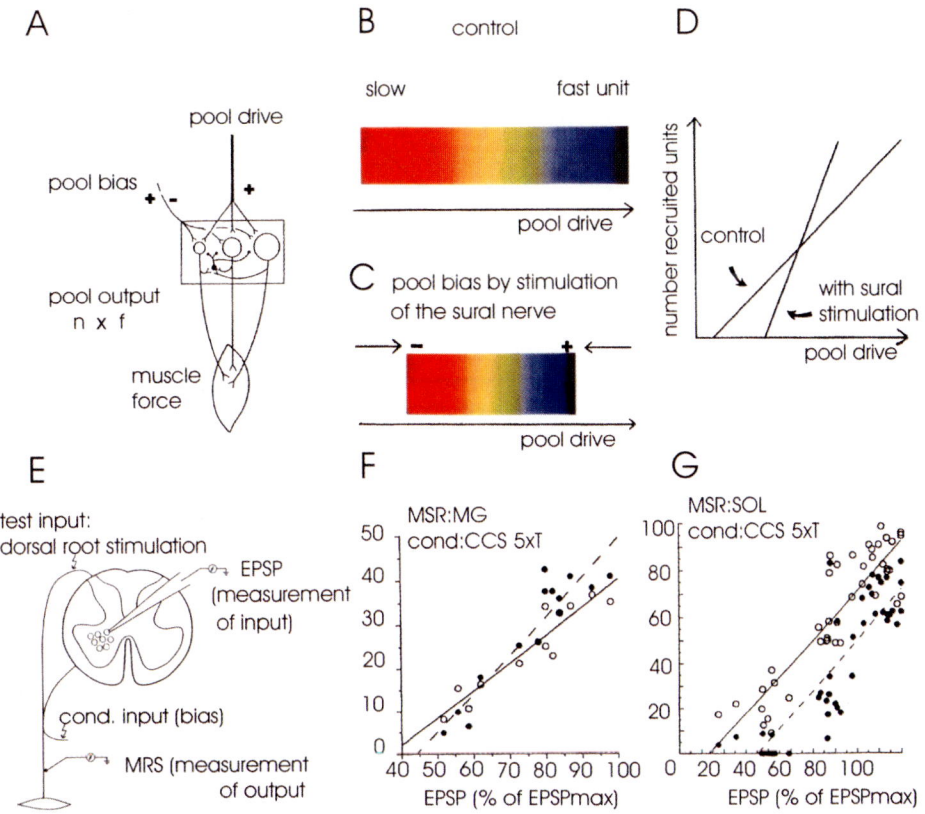

Fig. 6. Testing the recruitment gain hypothesis. (A–C) The hypothesis concerns changes in recruitment gain produced by conditioning volleys with a nonuniform distribution to the pool (pool bias), which differs from the orderly distribution of the test volley (pool drive). (A) Diagram showing inputs (pool drive, pool bias) to a MN pool with early- and late-recruited MNs shown as small and large, respectively. (B) The test volley recruits the MNs in the normal orderly manner from slow (low-threshold) to fast (high-threshold) units. (C) When a conditioning volley (pool bias) is added that inhibits slow motor units, and facilitates fast ones, the test volley still recruits MNs in the same order, but within a smaller range of the test volley. This is seen as a compression of the excitability spectrum. The recruitment gain is increased because a smaller increment of the test volley is needed to recruit successive MNs. (D) Diagram illustrating the relation between the test volley and the number of recruited MNs in the control situation (as in B) versus sural nerve stimulation (i.e., a pool bias conditioning stimulus as in C). (E–F) Experimental evidence of changes in recruitment gain. (E) The experimental arrangement. Decerebrate, unanesthetized cats were used to obtain large monosynaptic reflexes recruiting both slow (early-recruited) and fast (late-recruited) motor units. Graded stimulation of dorsal roots L7-S1 activated Ia afferents (among other afferents) to produce monosynaptic Ia EPSPs in MNs (test input) and evoke monosynaptic (test) reflexes (MSR). The latter were recorded from the nerves to medial gastrocnemius (MG) and soleus (SOL) as a measure of the output. Conditioning volleys were delivered to skin nerves, especially the caudal sural nerve (CCS) and the cutaneous part of the superficial peroneal nerve (SP). The size of a total activation of the motor pool was estimated from the response obtained by supramaximal stimulation of the ventral roots, and the MSRs were expressed in % of this motor response. (F–G) The effect of conditioning stimulation to the CCS on the input/output relation of MG (E) versus SOL (F) MNs. The test stimulation was given to the L7-S1 dorsal roots. It was graded from 1.1 to $5.0 \times T$ (threshold). The resulting input (abscissa) to each MN was expressed as the cell's monosynaptic (Ia) EPSP in % of its maximal EPSP size. The simultaneous pool output is shown on the ordinate (in % of maximum MSR). The conditioning (single shock) stimulation preceded the test stimulation by 2.2 ms (which was too short a time for presynaptic inhibition of the test volley). In E (for MG MNs), the slope for the curve obtained in the conditioned situation (interrupted line, filled circles) was significantly steeper than the slope of the control one (continuous line, open circles). In contrast, in F (for soleus MNs), there was no significant difference between the two slopes. Rather, there was a parallel shift. (Unpublished material from Nielsen et al., 1990.)

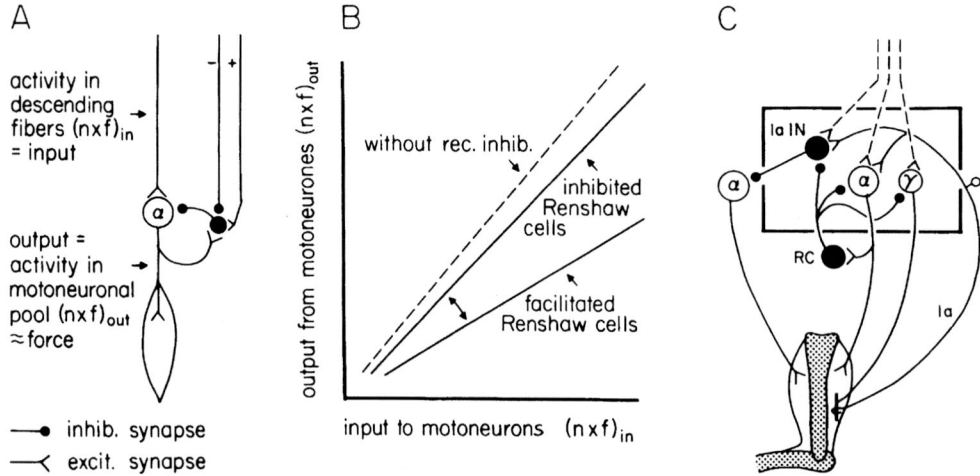

Fig. 7. Recurrent inhibition as variable gain regulator at the MN level. (A) Input–output connections of α MNs and Renshaw cells. (B) Simplified diagram of input–output relations of a single MN pool in three different situations. The expression, $(n \times f)$ denotes the number of motor axons (n) and their firing frequencies (f) of the input (in) to and output (out) from the MN pool. See text for further explanations. (C) Concept of the spinal motor output stage. Neurons constituting this stage are framed by thick lines. Abbreviations: α, alpha MNs; γ, gamma MNs; RC, Renshaw cells; IaIN, Ia inhibitory interneurons. (Reproduced, in part, from Hultborn et al., 1979 with permission from Springer-Verlag.)

motor units in the tibialis anterior muscle, thereby compressing the range of functional thresholds in the pool. It was also possible to demonstrate a steeper input/output relation for the monosynaptic reflex following such conditioning stimulation.

In summary, it has been demonstrated experimentally that variations in the distribution of synaptic inputs to low- versus high-threshold motor units can indeed change the recruitment gain. There are still no experiments, however, which have evaluated if this occurs under more natural circumstances, and, if so, whether the change would have a significant effect on the gradation of force production.

The Renshaw system as a variable gain regulator at the spinal motor output stage

Hultborn et al. (1979) proposed that recurrent inhibition may contribute to setting the input/output relation across a single spinal MN pool, which is otherwise largely determined by the intrinsic properties of MNs and, at least to some extent, their other synaptic input (see the previous section). With recurrent inhibition in operation, a smaller total output would be expected for any given input. Because Renshaw cells receive both strong excitatory and inhibitory control from many motor centers (Fig. 7A), it was proposed that the CNS presumably use the recurrent inhibitory pathway to regulate gain at the spinal output stage (Hultborn et al., 1979).

Fig. 7B shows that a low-gain condition during facilitation of Renshaw cells would allow force-generating circuits in the CNS to operate over a considerable part of their range and cause only small changes in muscular force. In contrast, a high-gain state (during inhibition of Renshaw cells) would allow the central command to generate larger forces for a given drive. Such a variable gain control at the MN level may thus adjust motor output during weak as well as strong contractions. Experiments on humans (Hultborn and Pierrot-Deseilligny, 1979a) have indeed suggested that the excitability of Renshaw cells changes in relation to the strength of a graded voluntary contraction. This result is consistent with the variable gain-regulator hypothesis, with inhibition of Renshaw cells during strong contractions (increased slope of the input/output relation across the motor nucleus) and facilitation during weak ones (decreased slope).

Fig. 7C shows that the general idea of a gain control for the input/output relation across a single

MN pool can be extended to cover the whole motor output stage of the spinal cord. This concept evolved from the findings that α and γ MNs of a single spinal motor pool receive excitatory and inhibitory inputs from descending and segmental sources in a pattern that is surprisingly similar to that received by the pool's corresponding Ia inhibitory interneurons (i.e., those with the same Ia connections). For example, recurrent inhibition is distributed quite similarly to the pool's MNs and Ia inhibitory interneurons (Baldissera et al., 1981). The output-stage concept was thus introduced to describe the idealized basic neuronal organization subserving a coordinated activation and relaxation of antagonist muscle pairs. Within this framework, a central position was attributed to the recurrent Renshaw system.

The above hypothesis has met with several objections. First, as pointed out by Windhorst (1996), it is essential to prove that inhibition and facilitation of Renshaw cells are not causing just a bias, but also a change in the gain (slope) of their input/output relation. A change in bias would merely cause a parallel shift of the idealized relations for the input/output relations across the motor pool shown in Fig. 7B, but it would not change their slope. In an investigation on the input/output relations of Renshaw cells in the cat, Hultborn and Pierrot-Deseilligny (1979b) did provide evidence for a distinct nonlinear facilitation (especially with phasic inputs during tonic facilitations), but whether this operates within the normal working range remains totally unknown.

A second objection to the Renshaw variable gain-regulator hypothesis is the traditional view that high-threshold motor units excite Renshaw cells, which in turn inhibit low-threshold units (Windhorst, 1996). Such a unidirectional distribution of Renshaw inhibition within a motor pool would not be able to bring about the slope changes shown earlier in Fig. 7B. Indeed, it has been demonstrated both anatomically (Cullheim and Kellerth, 1978) and physiologically (Hultborn et al., 1988a) that the relative contribution of individual slow (S), fast fatigue resistant (FR) and fast fatiguing (FF) motor units to the excitation of Renshaw cells is approximately 1:2:4. Because the range of force output for S, FR and FF motor units is much larger (1:4:14), however, even low forces (that require many early recruited motor units) must contribute to a sizable excitation of Renshaw cells (Hultborn et al., 1988a). It is also well established that the average size of recurrent inhibition is larger in MNs to slow versus faster motor units (i.e., in the order $S > FR > FF$) when measured at resting membrane potential. When the amplitude was measured at threshold level, however, and the synaptic current was calculated (by taking input resistance into account), there was no longer any significant difference between the recurrent inhibition of MNs to fast versus slow motor units (Hultborn et al., 1988b; Lindsay and Binder, 1991). Thus, the distribution of recurrent inhibition within the motor nucleus does indeed comply with the requirements of the gain-control hypothesis

The third major objection to the hypothesis is that the average magnitude of recurrent Renshaw inhibition is too small to be functionally significant (Lindsay and Binder, 1991). The peak amplitude of Renshaw inhibition in MNs when measured at firing threshold range from 1.4 to 3.1 mV (fast and slow units, respectively), which corresponds to a calculated synaptic current of 1.6–1.8 nA (with no significant differences between MNs to fast and slow units; Hultborn et al., 1988b). At an f/I slope of 1–2 imp s^{-1} nA^{-1}, the regulation of firing rate by recurrent inhibition would be marginal at best. When strong amplification of synaptic input by the dendritic-persistent inward current is added, however, the real situation may be very different. Indeed, there is evidence that recurrent inhibition is very effective in reducing synaptic-induced MN firing rate, as compared to current-injected discharge (unpublished results of H.H., J.B. Nielsen, M.E. Denton and J. Wienecke).

In summary, the functional significance of recurrent inhibition has remained speculative, with several ideas put forward over the past 50 years (for a comprehensive review, see Windhorst, 1996). Among these, we would argue that a reasonable one is still the hypothesis that recurrent inhibition serves as variable gain regulator at the spinal motor output stage. Several key assumptions of this hypothesis regarding factors affecting the input/output slope of Renshaw cells need, however, to be tested directly by experiments.

Comments on normal function and pathophysiology

A considerable part of this chapter has been devoted to mechanisms underlying amplification of the net synaptic excitation of MNs by way of currents through L-type Ca^{2+} channels in the dendritic (synaptic) region of the cell. Our own experiments were performed mainly in either decerebrate cat preparations—often with addition of monoaminergic transmitters—or in in vitro slices of turtle spinal cord, in which cocktails of neurotransmitters were added to the bathing medium. Does the behavior of MNs under such conditions reflect a true physiological state, or is it an experimental artifact? To investigate this question, Eken and Kiehn (1989) recorded from individual motor units in freely moving rats. Their experimental design allowed simultaneous recording of EMG activity and movement, together with afferent nerve stimulation to produce brief periods of synaptic excitation or inhibition. Their initial study focused on the firing behavior of soleus motor units during quiet standing. They succeeded in reproducing the firing-rate behavior seen previously in MNs of the decerebrate cat preparation. Sudden persistent increases in firing frequency (typically from 10 to 20 Hz) were evoked by trains of low-threshold afferent stimulation of the homonymous nerve (causing a burst of excitation), while maintained decreases in firing rate (typically from 20 to 10 Hz) were triggered by stimulation of cutaneous nerves. The jumps in firing frequency could not be graded, but rather, they appeared in an all-or-none fashion. Furthermore, these jumps occurred in individual motor units without any change in firing frequency of other simultaneously active units. It was therefore suggested that the spike-frequency shifts were related to intrinsic properties of individual MNs rather than to long-lasting changes in the descending drive.

The term, 'bistable firing' was introduced to describe shifts between two rather stable spike frequencies during short periods of synaptic excitation or inhibition. It was recorded both intracellularly from cat MNs (see Fig. 1 in Hounsgaard et al., 1988a) and as single motor-unit spike discharge in the rat soleus muscle (see Figs. 2–3 in Eken and Kiehn, 1989). Such firing, with shifts between two stable spike frequencies, may be expected if the Na^+ spike threshold is lower than that of the plateau potential. Theoretically, it will not occur, however, if the plateau is induced below the spike's threshold, i.e., before the MN is recruited into discharge (Kiehn and Eken, 1998). In the latter case, the plateau activation should boost firing rate at the onset of the cell's recruitment. Furthermore, an abrupt jump in spike frequency should be to a rate significantly higher than the minimal one at the threshold for repetitive discharge. We have supported this hypothesis by observations on the firing behavior of intracellularly recorded cat MNs in which the plateau potential was activated synaptically just below the threshold for the cell's recruitment (Bennett et al., 1998a). A similar discharge pattern was also observed in single motor units recorded in unrestrained rats during both phasic muscle stretch (Gorassini et al., 1999) and locomotion (Gorassini et al., 2000). The original observation of bistable-firing behavior in the rat soleus (Eken and Kiehn, 1989) would thus be compatible with the plateau threshold being higher than for the Na^+ spikes (recruitment level), whereas the later studies by Gorassini et al. (1999, 2000) suggested that plateau potentials were already triggered before recruitment. How can these different results be reconciled? One possibility is a sampling bias in the Eken and Kiehn's (1989) study. They searched for motor units in which the bistable behavior seen previously in cat MNs could be reproduced. Another possibility is the subtly varying conditions that prevail in work on conscious animals. Such differences may be correlated with differences in the dynamic regulation of plateau properties by descending command systems. For example, the activity and spinal effects of the descending serotonergic system are strongly correlated to the behavioral situation (for review, see Jacobs and Fornal, 1997). It is intuitively attractive to reason that a decreasing serotonergic innervation of MNs should cause an increase in their plateau threshold. Such an effect has indeed been formally demonstrated in a compartmental model of vertebrate MNs (Booth et al., 1997).

Bistable firing has also been studied in human motor-unit discharge during tonic voluntary contractions with imposed short-lasting bursts of vibration, which activates largely spindle Ia afferents (Kiehn and Eken, 1997; Gorassini et al., 1998). Although a bistable-firing pattern was never truly demonstrated, short periods of vibration often recruited new motor units into self-sustained/maintained activity. Because this occurred without an increase in the firing frequency of other simultaneously recorded, active motor units, it was argued that the descending drive had remained constant. These results are thus compatible with the idea that activation of plateau potentials is actually part of the normal recruitment process. The paired motor-unit recording paradigm assumes that the lower-threshold (control) unit can monitor changes in common synaptic drive to the whole MN pool, and that it, itself, is continuously firing either below or (more likely) above its plateau threshold. If the test unit has a significantly higher recruitment threshold, it is possible to use its recruitment/derecruitment levels (in terms of the discharge frequency of the control unit) as a provisional indication of its plateau-influenced discharge. This provides the means to explore how intrinsic properties of the test unit contribute to its recruitment and prolonged firing during periods of decreasing excitatory drive to the pool (e.g., during changes in the level of voluntary contraction). Gorassini et al. (2002a,b) have recently considerably extended work on human MNs by quantifying the paradigm of paired motor-unit recording. Repeated voluntary contractions lowered the recruitment level of another higher-threshold (test) motor unit. This finding is analogous to demonstration in animal preparations of the plateau potential 'wind-up' phenomenon (Svirskis and Hounsgaard, 1997; Bennett et al., 1998b). A detailed analysis of the difference in firing frequency of the low-threshold control unit at which the higher-threshold unit was recruited versus derecruited during voluntarily generated triangular wave-contraction forces, was also used to estimate the driving force of the presumed plateau potential (Gorassini et al., 2002b). This work provided strong evidence for the presence, threshold and size of the plateau potential in human MNs. Much further research of this kind must yet be undertaken to establish reliable and valid methods for estimating plateau potentials during normal movements in healthy subjects, as well as under pathophysiological conditions (see also Hornby et al., 2001).

There is much clinical evidence that large spinal lesions in the human lead to an immediate depression of spinal-reflex activity (spinal shock). Subsequently, flexor- and tonic-stretch reflexes return and become exaggerated several weeks and even months later. A similar spastic syndrome is seen in the cat following a complete spinal transection at the low thoracic level (Bailey et al., 1980; see also review by Eken et al., 1989). In a few cases, it has been possible to demonstrate that plateau potentials can again be evoked in the chronic spinal state without adding any neurotransmitter precursors or agonists (unpublished result of T. Eken, H.H., O. Kiehn and T.I.T.; see also Eken et al., 1989). This suggested that plateau properties, returning late after the spinal injury, could play a role in the pathophysiology of spasticity. Recently, Bennett et al. (2001a,b) have developed a rat preparation in which a very low chronic spinal lesion causes a pronounced spasticity of the tail, without interfering with normal hindlimb or bladder function. Results from paired motor-unit recording strongly suggested that their activation was supported by plateau potentials (Bennett et al., 2001a). Because the sacrocaudal spinal cord is very thin, it was possible to study the tail MNs in an in vitro preparation. It then became possible to compare the intrinsic properties of tail MNs after an acute transection (control cells) and after chronic lesions (test cells), in which the rats had developed spasticity and hyperreflexia. It was shown that plateau properties were regularly seen in the chronic test state, but not in the acute control one (Bennett et al., 2001b). These observations have certainly opened new horizons for further investigation of the cellular mechanisms underlying spasticity and other changes following chronic lesions within the CNS.

Concluding thoughts

As stated in the Introduction, we are still far from understanding the transfer function of command signals from the brain to activation of muscles. In all likelihood, this relation may change significantly depending on, for example, the emotional setting ('flight and fight'; the monoaminergic systems), type of motor activity and various kinds of pathology, as, for example, spasticity following CNS lesions. For all mechanisms discussed in this review, it is necessary to establish, or further develop, valid and reliable methods to evaluate their effect in humans: at rest, during voluntary activity, and in various pathological conditions. It is only when all of the above has been achieved that we can hope for a coherent understanding of the 'key mechanisms setting the input–output gain across the motoneuron pool'.

Acknowledgments

We thank Lillian Grondahl for her expert, long-provided technical assistance in our Copenhagen laboratory. The research has been supported by grants from the Danish Medical Research Council, the NOVO Nordisk Foundation, the Ludvig & Sara Elsass Foundation, the Michaelsen Foundation, and the Sofus & Olga Friis Foundation.

Abbreviations

AHP	afterhyperpolarization
AP	action potential
CNS	central nervous system
FR*	fast contracting and fatigue resistant
FF*	fast contracting and fatiguing
I–f	stimulus current–spike frequency
MN	motoneuron
TEA	tetraethylammonium
S*	slow

(*Descriptor for a type of motor unit)

References

Adrian, E.D. and Bronk, D.W. (1929). The discharge of impulses in motor nerve fibres. Part II. The frequency of discharge in reflex and voluntary contractions. J. Physiol. (Lond.), 67: 119–151.

Adrian, E.D. and Zotterman, Y. (1926). The impulses produced by sensory nerve-endings. Part 2. The response of single end-organ. J. Physiol. (Lond.), 61: 151–171.

Alaburda, A., Perrier, J.F. and Hounsgaard, J. (2002). Mechanisms causing plateau potentials in spinal motoneurones. Adv. Exp. Med. Biol., 508: 219–226.

Bailey, C.S., Lieberman, J.S. and Kitchell, R.L. (1980). Response of muscle spindle primary endings to static stretch in acute and chronic spinal cats. Am. J. Vet. Res., 41: 2030–2036.

Baldissera, F., Hultborn, H. and Illert, M. (1981). Integration in spinal neuronal systems. In: Brooks V.B. (Ed.), Handbook of Physiology, Vol. II. Motor Control. The Williams & Wilkins Company, Baltimore, pp. 509–595.

Bennett, D.J., Hultborn, H., Fedirchuk, B. and Gorassini, M. (1998a). Synaptic activation of plateaus in hindlimb motoneurons of decerebrate cats. J. Neurophysiol., 80: 2023–2037.

Bennett, D.J., Hultborn, H., Fedirchuk, B. and Gorassini, M. (1998b). Short-term plasticity in hindlimb motoneurons of decerebrate cats. J. Neurophysiol., 80: 2038–2045.

Bennett, D.J., Li, Y., Harvey, P.J. and Gorassini, M. (2001a). Evidence for plateau potentials in tail motoneurons of awake chronic spinal rats with spasticity. J. Neurophysiol., 86: 1972–1982.

Bennett, D.J., Li, Y. and Siu, M. (2001b). Plateau potentials in sacrocaudal motoneurons of chronic spinal rats, recorded in vitro. J. Neurophysiol., 86: 1955–1971.

Booth, V., Rinzel, J. and Kiehn, O. (1997). Compartmental model of vertebrate motoneurons for Ca^{2+}-dependent spiking and plateau potentials under pharmacological treatment. J. Neurophysiol., 78: 3371–3385.

Brock, L.G., Coombs, J.S. and Eccles, J.C. (1952). The recording of potentials from motoneurones with an intracellular electrode. J. Physiol. (Lond.), 117: 431–460.

Brownstone, R.M., Gossard, J.P. and Hultborn, H. (1994). Voltage-dependent excitation of motoneurones from spinal locomotor centres in the cat. Exp. Brain Res., 102: 34–44.

Burke, R.E. (1981). Motor units: anatomy, physiology, and functional organization. In: Brooks V.B. (Ed.), Handbook of Physiology, Vol. II. Motor Control. The Williams & Wilkins Company, Baltimore, pp. 345–422.

Carlin, K.P., Jones, K.E., Jiang, Z. and Brownstone, R.M. (2000a) Sequential activation of dendrites in the generation of calcium currents in mammalian motoneurones. Soc. Neurosci. Abstr., 257.10.

Carlin, K.P., Jones, K.E., Jiang, Z., Jordan, L.M. and Brownstone, R.M. (2000b). Dendritic L-type calcium currents in mouse spinal motoneurons: implications for bistability. Eur. J. Neurosci., 12: 1635–1646.

Clark, B.D., Dacko, S.M. and Cope, T.C. (1993). Cutaneous stimulation fails to alter motor unit recruitment in the decerebrate cat. J. Neurophysiol., 70: 1433–1439.

Conway, B.A., Hultborn, H., Kiehn, O. and Mintz, I. (1988). Plateau potentials in alpha-motoneurones induced by intravenous injection of L-dopa and clonidine in the spinal cat. J. Physiol. (Lond.), 405: 369–384.

Cope, T.C. and Clark, B.D. (1995). Are there important exceptions to the size principle of α-motoneurone recruitment? In: Taylor A., Gladden M.H. and Durbaba R. (Eds.), Alpha and Gamma Motor Systems. Plenum Press, New York, pp. 71–78.

Crone, C., Hultborn, H., Kiehn, O., Mazieres, L. and Wigstrom, H. (1988). Maintained changes in motoneuronal excitability by short-lasting synaptic inputs in the decerebrate cat. J. Physiol. (Lond.), 405: 321–343.

Cullheim, S. and Kellerth, J.O. (1978). A morphological study of the axons and recurrent axon collaterals of cat alpha-motoneurones supplying different functional types of muscle unit. J. Physiol. (Lond.), 281: 301–313.

Delgado-Lezama, R. and Hounsgaard, J. (1999). Adapting motoneurons for motor behavior. Prog. Brain Res., 123: 57–63.

Delgado-Lezama, R., Perrier, J.F., Nedergaard, S., Svirskis, G. and Hounsgaard, J. (1997). Metabotropic synaptic regulation of intrinsic response properties of turtle spinal motoneurones. J. Physiol. (Lond.), 504: 97–102.

Delgado-Lezama, R., Perrier, J.F. and Hounsgaard, J. (1999). Local facilitation of plateau potentials in dendrites of turtle motoneurones by synaptic activation of metabotropic receptors. J. Physiol. (Lond.), 515: 203–207.

Eccles, J.C., Eccles, R.M. and Lundberg, A. (1958). The action potentials of alpha motoneurones supplying fast and slow muscles. J. Physiol. (Lond.), 142: 275–291.

Eken, T. and Kiehn, O. (1989). Bistable firing properties of soleus motor units in unrestrained rats. Acta Physiol. Scand., 136: 383–394.

Eken, T., Hultborn, H. and Kiehn, O. (1989). Possible functions of transmitter-controlled plateau potentials in alpha motoneurones. Prog. Brain Res., 80: 257–267.

Ertel, E.A., Campbell, K.P., Harpold, M.M., Hofmann, F., Mori, Y., Perez-Reyes, E., Schwartz, A., Snutch, T.P., Tanabe, T., Birnbaumer, L., Tsien, R.W., Catterall, W.A. (2002). Nomenclature of voltage-gated calcium channels. Neuron, 25: 533–535.

Gorassini, M.A., Bennett, D.J. and Yang, J.F. (1998). Self-sustained firing of human motor units. Neurosci. Lett., 247: 13–16.

Gorassini, M., Bennett, D.J., Kiehn, O., Eken, T. and Hultborn, H. (1999). Activation patterns of hindlimb motor units in the awake rat and their relation to motoneuron intrinsic properties. J. Neurophysiol., 82: 709–717.

Gorassini, M., Eken, T., Bennett, D.J., Kiehn, O. and Hultborn, H. (2000). Activity of hindlimb motor units during locomotion in the conscious rat. J. Neurophysiol., 83: 2002–2011.

Gorassini, M., Yang, J.F., Siu, M. and Bennett, D.J. (2002a). Intrinsic activation of human motoneurons: reduction of motor unit recruitment thresholds by repeated contractions. J. Neurophysiol., 87: 1859–1866.

Gorassini, M., Yang, J.F., Siu, M. and Bennett, D.J. (2002b). Intrinsic activation of human motoneurons: possible contribution to motor unit excitation. J. Neurophysiol., 87: 1850–1858.

Granit, R., Kernell, D. and Shortess, G.K. (1963). Quantitative aspects of repetitive firing of mammalian motoneurons, caused by injected currents. J. Physiol. (Lond.), 168: 911–931.

Granit, R., Kernell, D. and Lamarre, Y. (1966a). Algebraical summation in synaptic activation of motoneurones firing within the 'primary range' to injected currents. J. Physiol. (Lond.), 187: 379–399.

Granit, R., Kernell, D. and Lamarre, Y. (1966b). Synaptic stimulation superimposed on motoneurones firing in the 'secondary range' to injected current. J. Physiol. (Lond.), 187: 401–415.

Guertin, P.A. and Hounsgaard, J. (1999). Non-volatile general anaesthetics reduce spinal activity by suppressing plateau potentials. Neuroscience, 88: 353–358.

Gustafsson, B. and Pinter, M.J. (1984). Relations among passive electrical properties of lumbar alpha-motoneurones of the cat. J. Physiol. (Lond.), 356: 401–431.

Gustafsson, B. and Pinter, M.J. (1985). On factors determining orderly recruitment of motor units: a role for intrinsic membrane properties. Trends Neurosci., 8: 431–433.

Gutman, A.M. (1991). Bistability of dendrites. Int. J. Neural Syst., 1: 291–304.

Hellgren Koteleski, J., Lester, D. and Blackwell, K.T. (2003) Subcellular interactions between parallel fibre and climbing fibre signals in Purkinje cells predict sensitivity of classical conditioning to interstimulus interval. In: Steinmetz, J.E. (Ed.), Integrative Physiological and Behavioural Science. Transaction Publishers, Piscataway, in press.

Henneman, E. and Mendell, L.M. (1981). Functional organization of motoneuron pool and its inputs. In: Brooks V.B. (Ed.), Handbook of Physiology, Vol. II. Motor Control. The Williams & Wilkins Company, Baltimore, pp. 423–507.

Hornby, T.G., Stauffer, E.K. and Stuart, D.G. (2001). Open issues on the functional role of the plateau potential in the repetitive discharge of motoneurons in experimental animals and humans. In: Dengler R. and Kossev A. (Eds.), Sensorimotor Control, NATO Science Series. Series 1, Life and Behavioural Sciences 326. IOS Press, Amsterdam, pp. 65–74.

Hornby, T.G., McDonagh, J.C., Reinking, R.M. and Stuart, D.G. (2002). Motoneurons: a preferred firing range across vertebrate species?. Muscle Nerve, 25: 632–648.

Hounsgaard, J. and Kiehn, O. (1989). Serotonin-induced bistability of turtle motoneurones caused by a nifedipine-sensitive calcium plateau potential. J. Physiol. (Lond.), 414: 265–282.

Hounsgaard, J. and Kiehn, O. (1993). Calcium spikes and calcium plateaux evoked by differential polarization in dendrites of turtle motoneurones in vitro. J. Physiol. (Lond.), 468: 245–259.

Hounsgaard, J. and Mintz, I. (1988). Calcium conductance and firing properties of spinal motoneurones in the turtle. J. Physiol. (Lond.), 398: 591–603.

Hounsgaard, J., Hultborn, H., Jespersen, B. and Kiehn, O. (1984). Intrinsic membrane properties causing a bistable behaviour of alpha-motoneurones. Exp. Brain Res., 55: 391–394.

Hounsgaard, J., Hultborn, H., Jespersen, B. and Kiehn, O. (1988a). Bistability of alpha-motoneurones in the decerebrate cat and in the acute spinal cat after intravenous 5-hydroxytryptophan. J. Physiol. (Lond.), 405: 345–367.

Hounsgaard, J., Kiehn, O. and Mintz, I. (1988b). Response properties of motoneurones in a slice preparation of the turtle spinal cord. J. Physiol. (Lond.), 398: 575–589.

Hultborn, H. (1999). Plateau potentials and their role in regulating motoneuronal firing. Prog. Brain Res., 123: 39–48.

Hultborn, H. and Nielsen, J.B. (2002) Plateau potentials and their role in regulating motoneuronal firing. In: Zijdewind, I., Bakels, R., van der Sluis, T. and Boddeke, E. (Eds.), Abstr. Symp.: Motoneurones and Muscles: The Output Machinery. Groningen, The Netherlands, p. 21.

Hultborn, H. and Pierrot-Deseilligny, E. (1979a). Changes in recurrent inhibition during voluntary soleus contractions in man studied by an H-reflex technique. J. Physiol. (Lond.), 297: 229–251.

Hultborn, H. and Pierrot-Deseilligny, E. (1979b). Input-output relations in the pathway of recurrent inhibition to motoneurones in the cat. J. Physiol. (Lond.), 297: 267–287.

Hultborn, H., Lindstrom, S. and Wigstrom, H. (1979). On the function of recurrent inhibition in the spinal cord. Exp. Brain Res., 37: 399–403.

Hultborn, H., Katz, R. and Mackel, R. (1988a). Distribution of recurrent inhibition within a motor nucleus. II. Amount of recurrent inhibition in motoneurones to fast and slow units. Acta Physiol. Scand., 134: 363–374.

Hultborn, H., Lipski, J., Mackel, R. and Wigstrom, H. (1988b). Distribution of recurrent inhibition within a motor nucleus. I. Contribution from slow and fast motor units to the excitation of Renshaw cells. Acta Physiol. Scand., 134: 347–361.

Jacobs, B.L. and Fornal, C.A. (1997). Serotonin and motor activity. Curr. Opin. Neurobiol., 7: 820–825.

Jankowska, E. (1992). Interneuronal relay in spinal pathways from proprioceptors. Prog. Neurobiol., 38: 335–378.

Kernell, D. (1965a). The adaptation and the relation between discharge frequency and current strength of cat lumbosacral motoneurones stimulated by long-lasting injected currents. Acta Physiol. Scand., 65: 65–73.

Kernell, D. (1965b). High-frequency repetitive firing of cat lumbosacral motoneurones stimulated by long-lasting injected currents. Acta Physiol. Scand., 65: 74–86.

Kernell, D. and Hultborn, H. (1990). Synaptic effects on recruitment gain: a mechanism of importance for the input-output relations of motoneurone pools?. Brain Res., 507: 176–179.

Kiehn, O. and Eken, T. (1997). Prolonged firing in motor units: evidence of plateau potentials in human motoneurons? J. Neurophysiol., 78: 3061–3068.

Kiehn, O. and Eken, T. (1998). Functional role of plateau potentials in vertebrate motor neurons. Curr. Opin. Neurobiol., 8: 746–752.

Lee, R.H. and Heckman, C.J. (2000). Adjustable amplification of synaptic input in the dendrites of spinal motoneurons in vivo. J. Neurosci., 20: 6734–6740.

Lindsay, A.D. and Binder, M.D. (1991). Distribution of effective synaptic currents underlying recurrent inhibition in cat triceps surae motoneurons. J. Neurophysiol., 65: 168–177.

Nielsen, J. and Kagamihara, Y. (1993). Differential projection of the sural nerve to early and late recruited human tibialis anterior motor units: change of recruitment gain. Acta Physiol. Scand., 147: 385–401.

Nielsen, J., Hultborn, H. and Gossard, J.-P. (1990). Change in recruitment gain by stimulation of the caudal sural or superficial peroneal nerve in the cat. Eur. J. Neurosci., Suppl. 3: 193.

Porter, R. and Lemon, R. (1993). *Corticospinal Function and Voluntary Movement. Clarendon Press, Oxford, pp. 428.*

Powers, R.K. and Binder, M.D. (2001). Input-output functions of mammalian motoneurons. Rev. Physiol. Biochem. Pharmacol., 143: 137–263.

Schwindt, P. and Crill, W.E. (1977). A persistent negative resistance in cat lumbar motoneurons. Brain Res., 120: 173–178.

Schwindt, P. and Crill, W. (1980a). Role of a persistent inward current in motoneuron bursting during spinal seizures. J. Neurophysiol., 43: 1296–1318.

Schwindt, P.C. and Crill, W.E. (1980b). Properties of a persistent inward current in normal and TEA-injected motoneurons. J. Neurophysiol., 43: 1700–1724.

Schwindt, P.C. and Crill, W.E. (1980c). Effects of barium on cat spinal motoneurons studied by voltage clamp. J. Neurophysiol., 44: 827–846.

Schwindt, P.C. and Crill, W.E. (1982). Factors influencing motoneuron rhythmic firing: results from a voltage-clamp study. J. Neurophysiol., 48: 875–890.

Schwindt, P.C. and Crill, W.E. (1984). Membrane properties of cat spinal motoneurones. In: Davidoff R.A. (Ed.), Handbook of the Spinal Cord, Vol. 2 and 3, Marcel Dekker, New York, pp. 199–246.

Svensson, E., Dewael, Y., Hill, R.H. and Grillner, S. (2001) 5-HT inhibits N-type calcium channels in lamprey spinal neurons. Soc. Neurosci. Abstr., 934.14.

Svirskis, G. and Hounsgaard, J. (1997). Depolarization-induced facilitation of plateau generating currents in ventral horn neurons in the turtle spinal cord. J. Neurophysiol., 78: 1740–1742.

Svirskis, G. and Hounsgaard, J. (1998). Transmitter regulation of plateau properties in turtle motoneurons. J. Neurophysiol., 79: 45–50.

Svirskis, G., Gutman, A. and Hounsgaard, J. (2001). Electrotonic structure of motoneurons in the spinal cord of the turtle: inferences for the mechanisms of bistability. J. Neurophysiol., 85: 391–398.

Van Dongen, P.A., Grillner, S. and Hokfelt, T. (1986). 5-Hydroxytryptamine (serotonin) causes a reduction in the afterhyperpolarization following the action potential in lamprey motoneurons and premotor interneurons. Brain Res., 366: 320–325.

Walton, K. and Fulton, B.P. (1986). Ionic mechanisms underlying the firing properties of rat neonatal motoneurons studied in vitro. Neuroscience, 19: 669–683.

White, S.R. and Fung, S.J. (1989). Serotonin depolarizes cat spinal motoneurons in situ and decreases motoneuron afterhyperpolarizing potentials. Brain Res., 502: 205–213.

Windhorst, U. (1996). The spinal cord and its brain: representations and models. To what extent do forebrain mechanisms appear at brainstem spinal cord levels? Prog. Neurobiol., 49: 381–414.

Wu, S.Y., Wang, M.Y. and Dun, N.J. (1991). Serotonin via presynaptic 5-HT1 receptors attenuates synaptic transmission to immature rat motoneurons in vitro. Brain Res., 554: 111–121.

CHAPTER 9

Rhythm generation for food-ingestive movements

Yoshio Nakamura*[1], Nobuo Katakura, Misuzu Nakajima and Jia Liu

Department of Physiology, Faculty of Dentistry, Tokyo Medical and Dental University, Tokyo 113-8549, Japan

Abstract: In vitro block preparations of the central nervous system (CNS) are particularly valuable for study of central neuronal mechanisms controlling the respiratory and locomotor rhythms. No comparable in vitro preparation has been described previously, however, for analysis of analogous feeding rhythms. In this chapter, we present such a model. It is comprised of an in vitro brainstem-spinal cord preparation isolated from the newborn rat and mouse. Bath application to this preparation of N-methyl-D-aspartate (NMDA) induces rhythmical burst activity in the V, VII and XII nerves, which, collectively, is indicative of feeding behavior. Selected transections of the brainstem reveal that the central sucking rhythm generators for such V, VII and XII activity are separate from one another, and located segmentally in the brainstem at the level of their respective motor nuclei. We believe that use of this in vitro preparation will advance understanding of the central neuronal mechanisms controlling sucking and mastication, and the developmental transition from sucking to mastication.

Introduction

The postnatal development of the conversion from suckling to mastication is a central issue in research on neural mechanisms underlying mammalian food-ingestive movements. It has generally been thought that this conversion is triggered peripherally by the eruption of teeth, because both events coincide in several species, including primates (Bosma, 1967; Dubner et al., 1978). Such tooth eruption may not play a generalized essential role, however. For example, dentition is completed at birth in some precocious animals, like the guinea pig (Ainamo, 1971), and yet their nutrition depends totally on suckling during the initial postnatal period. A plastic change in the CNS seems more likely to play a critical role in this species, because the guinea pig is born with a mature oro-facial structure such that its conversion from suckling to mastication can be accomplished without a sensory-input trigger.

Rhythmicity is a fundamental property of both suckling and masticatory movements. It consists of alternating jaw-closing and jaw-opening movements, which are phase-locked to rhythmical activity of the tongue and facial muscles. The overall rhythm is generated centrally by a selected neuronal population in the lower brainstem, which is termed the central rhythm generator (CRG; for review, see Nakamura and Katakura, 1995; see also Grillner and Wallén, Section II; Yamaguchi, and Pearson, Chapters 11 and 12 of this volume). Strikingly, the central issue concerning the postnatal development of the CRG remains open. Is the suckling CRG the same as the masticatory CRG? If different, how is the suckling CRG reorganized to provide mastication?

Substantial data is available on the masticatory CRG, including its localization, cortical and subcortical sources of activation, and the intercalated neurons projecting its output to trigeminal (V) and hypoglossal (XII) motoneurons (Nakamura and Katakura, 1995). Similarly, there is evidence that the suckling CRG is located in the brainstem, and

---

*Corresponding author.
[1]Present address: Department of Welfare and Information, Faculty of Informatics, Teikyo Heisei University, Ichihara 290-0193, Japan. Tel.: +81-436-74-7137; Fax: +81-45-826-4905; E-mail: yoshio-n@sk9.so-net.ne.jp

DOI: 10.1016/S0079-6123(03)43009-4

that the corticobulbar projection system involved in its activation is reorganized upon the conversion from suckling to mastication in guinea pigs (Iriki et al., 1988). Despite these noteworthy advances, much remains unknown about the neuronal organization of the CRG for both suckling and mastication. Further advancement requires: (1) identifying the complete neuronal circuitry of the CRG, and its neurons' membrane properties; (2) determining the organization and type of synaptic connections among the CRG neurons; and (3) clarifying the effects of transmitters on these neurons. For such analyses, an in vitro CNS preparation is particularly suitable, like that used in physiological and pharmacological studies on the central rhythm generation for respiration and locomotion. Such latter studies have made use of in vitro CNS block preparations, i.e., those that have included the *complete* neuronal circuitry for the selected behavior. One such example of this approach is demonstration that excitatory amino acids can initiate and maintain rhythmical locomotor activity in relevant neurons, particularly following activation of their N-methyl-D-aspartate (NMDA) receptors (Nakamura and Katakura, 1995).

In contrast to work on the respiratory and locomotor CRGs, no in vitro preparation has been used previously for study of the suckling and/or masticatory CRG. Here, we report on the development of such a model, using newborn and young rodents.

NMDA-induction of rhythmical activity in n. XII

Figure 1 shows our use of Suzue's (1984) isolated brainstem-spinal cord preparation of the newborn rat. This preparation exhibits spontaneous rhythmical inspiratory burst activities in n. XII and C5 VRs which can continue for >5 h.

Figure 2A shows that bath application of NMDA-induced rhythmical burst activity in n. XII and the C5 VR, following pronounced tonic activity. Under continuous bath application of NMDA, the rhythmical n. XII activity continued steadily for >1 h (Katakura et al., 1995a). These NMDA effects were completely blocked by the simultaneous bath application of 2-amino-5-phosphonovalerate (AP5), with no obvious effects on this nerve's inspiratory activity (Katakura et al., 1995a). This finding showed that the NMDA effect was mediated by NMDA receptors.

Such NMDA-induced n. XII activity was distinct from this nerve's spontaneous inspiratory activity. It had a shorter cycle length (2.5–6.0 s) than the inspiratory activity (>10 s). Also, the latter featured rapidly rising and slowly falling bursts, whereas, as shown in Fig. 2B, the NMDA-induced rhythmical activity consisted of more gradual increases and decreases, and often 2–3 peaks of activity during each single burst.

Another difference related to the extent of bilateral synchrony of discharge. The right-side versus left-side onset time of the NMDA-induced bursts had a significantly larger coefficient of variation than the analogous inspiratory bursts. A cross-correlation analysis of the NMDA-induced versus inspiratory activity revealed that the latter could reset the former, but not vice versa. In other words, the CRG for NMDA-induced rhythmical n. XII activity was strongly influenced by the activity of the inspiratory CRG (Katakura et al., 1996, 1998a). Despite this effect, NMDA-induced, rhythmical n.

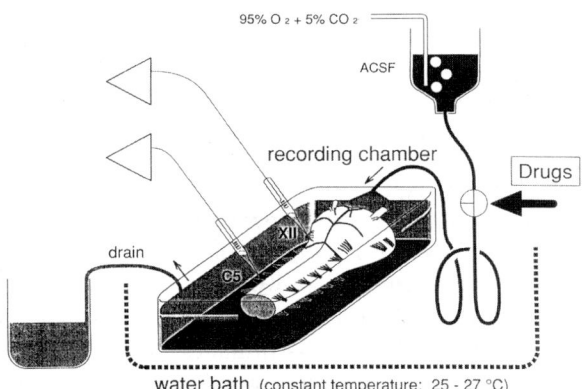

Fig. 1. Schematic representation of the experimental arrangement. Under ether anesthesia, newborn rats (0–3 days old) were decerebrated at the intercollicular level, and the spinal cord was transected between C7 and C8. The brainstem-spinal cord was removed, and placed with the ventral surface upward in a chamber, where it was perfused continuously with an ACSF, which was equilibrated with 95% O_2–5% CO_2, and buffered to pH 7.45. The temperature in the chamber was maintained at 25–27°C. The neural activity of the hypoglossal nerve (n. XII) and the C5 ventral roots (VRs) were recorded with suction electrodes. Inspiratory activity was monitored from the C5 VR containing the phrenic motoneuronal axons. NMDA was dissolved in the ACSF, and applied to the recording chamber through the perfusing system. (Reproduced from Nakamura et al., 1999 with permission from Tokyo Medical and Dental University.)

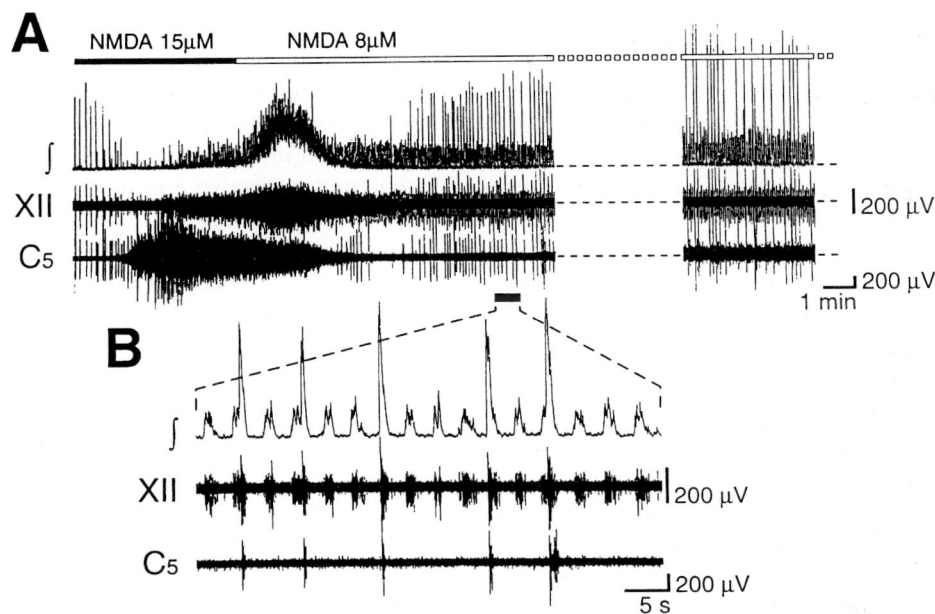

Fig. 2. Spontaneous rhythmical inspiratory activity and NMDA-induced rhythmical activity in n. XII. (A) In this case, 15 μM NMDA was applied to the recording bath for 5 min, followed by the continuous application of 8 μM NMDA. The top trace is an integrated (∫) version of n. XII activity (middle trace). The bottom recording is from a C5 VR. Note that NMDA induced a rhythmical burst of activity in n. XII, following pronounced tonic activity in both n. XII and the C5 VR. The extreme right traces were recorded ~ 1 h after the onset of 8-μM NMDA application. They show that continuous bath application of NMDA resulted in maintained rhythmical burst activity. (B) Expanded records of the period indicated by a bar in A. (Reproduced from Nakamura et al., 1999 with permission from Tokyo Medical and Dental University.)

XII activity persisted after this nerve's inspiratory activity was completely abolished by the bilateral microinjection of kynurenic acid into the rostral ventro-lateral medulla, where the neurons of the respiratory CRG are located. Thus, the CRG for NMDA-induced n. N. XII activity is *separate* from, but *strongly influenced* by the respiratory CRG.

Tongue activity inferred from NMDA-induced rhythmical n. XII activity

Rhythmical tongue movements are observed not only during suckling, but also during respiration, mastication and swallowing. Despite such ubiquity, three lines of evidence indicate that NMDA-induced activity in n. XII is the neural representation of rhythmical tongue movements during suckling. (1) NMDA-induced activity in n. XII is distinct from inspiratory activity in cycle length and temporal pattern. (2) Unlike suckling, mastication is not seen in newborn rats. Rather, like the eruption of teeth, it starts near the 12th postnatal day (Westneat and Hall, 1992). (3) As shown in the next section, NMDA-induced rhythmical n. XII activity is retained after removal of the solitary tract nucleus, which plays a critical role in the elaboration of swallowing (Kessler and Jean, 1985).

To further test the above suckling inference, we studied the pattern of NMDA-induced tongue movements in another kind of in vitro preparation. It involved a similar brainstem-spinal cord preparation to which was added selected oro-facial structures. The maxilla was removed on the one side, but the neural connections of the tongue and jaw muscles with the brainstem were retained bilaterally. This preparation allowed us to phase-relate neural activity with camera- and EMG-recorded movements of the tongue and jaw. During continuous bath application of NMDA, two kinds of rhythmical burst activity were found in the tongue EMG. One featured a rapidly rising and slowly falling pattern, which was synchronous with inspiratory activity in the C5 VR. This pattern was the same as that measured *before* NMDA application, thereby indicating that it resulted from inspiratory CRG activity. Coincident with this EMG activity, the tongue's tip and body were depressed, and the mandible undertook a large opening movement.

The second type of rhythmical activity in the tongue EMG was seen only after adding NMDA to the bath. It was lower in amplitude than the inspiratory burst, and it featured a gradually waxing and waning pattern, which was in phase with the NMDA-induced n. XII activity. Coincident with this EMG activity, the tongue's tip was elevated slightly, its body depressed, and there was a slight opening of the jaw. This pattern was quite similar to that of the tongue during the natural suckling of newborn rats. We believe that this is a strong evidence for the inference that NMDA-induced activity in n. XII reflects rhythmical tongue movements during suckling behavior (Katakura et al., 1996, 1998a).

Localization of the CRG for NMDA-induced rhythmical n. XII activity

It has been shown that after a midline section of the brainstem, NMDA-induced rhythmical n. XII activity persists on both sides, but with a different rhythm, thereby indicating separate CRGs on each side of the brainstem (Katakura et al., 1995b; Liu, 1997). To identify the neurons of this CRG in the Fig. 1 preparation, we applied sulforhodamine 101, a fluorescent dye taken up by neurons in an activity-dependent manner (Lichtman et al., 1985; Keifer et al., 1992), before and after rhythmical n. XII activity was induced by bath application of NMDA. In the control (no-NMDA) of animals, with spontaneous rhythmical respiratory activity in n. XII and the C5 VR, sulforhodamine-labeled neurons were found in the n. XII nucleus, facial nucleus, nucleus ambiguus, spinal V nucleus, solitary tract nucleus, lateral reticular nucleus, bulbar reticular formation and inferior olivary nucleus. In the test group, following NMDA-induced rhythmical suckling-like activity, sulforhodamine-labeled neurons were found in the above structures, and, in addition, the gigantocellular reticular nucleus (Katakura et al., 1996; Katakura and Nakamura, 1997; Liu, 1997).

We next studied which of these sulforhodamine-labeled neuron groups comprise the CRG, by eliminating respective group of these neurons with selective transections of the brainstem. We made coronal, parasagittal and horizontal sections within the brainstem of our Fig. 1 preparation. NMDA-induced rhythmical n. XII activity persisted after: (1) complete transection of the brainstem at both the bulbo-spinal junction, and the coronal level between n. VII (facial) and n. XII; (2) bilateral removal of the lateral part of the brainstem, including the spinal V nucleus; and (3) removal of the dorsal portion of the brainstem, including the solitary tract nucleus. These results showed that the CRG for NMDA-induced rhythmical n. XII activity is located within the ventro-medial medulla including the gigantocellular reticular nucleus (Katakura and Nakamura, 1997; Liu, 1997).

Demonstration of separate CRGs to n. V, VII and XII motoneurons

Both suckling and mastication require coordination of rhythmical jaw, tongue and facial movements. This raises the issue as to whether the rhythmical activities in n. V, VII and XII are generated by a *single* CRG for these three motoneuron groups or by three separate ones. We addressed this problem in our Fig. 1 preparation, using newborn mice rather than rats, because NMDA-induced n. VII activity is more consistent in the former. First, we recorded simultaneously from n. V's motor root, n. VII and n. XII. We found that the cycle length of NMDA-induced rhythmical activity was *different* among the three nerves, thereby suggesting the possibility for *separate* CRGs for these motoneuron groups. To test this possibility, as shown in Fig. 3, we made coronal sections of the brainstem at levels between n. V and VII and between n. VII and XII, such that we could then record from preparations that contained the isolated motor V nucleus, facial nucleus and hypoglossal nucleus, respectively.

Figure 3B shows that after the above transections, bath application of NMDA induced rhythmical

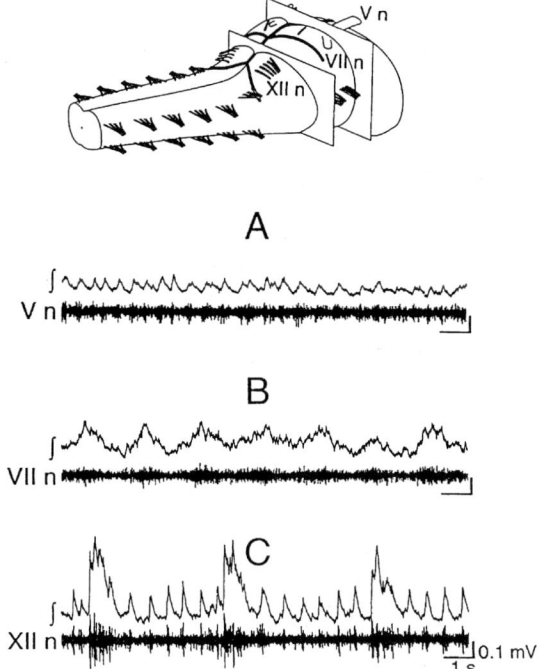

Fig. 3. NMDA-induced rhythmical activity in n. V, VII and XII. The top schematic shows coronal transections of the isolated brainstem of a newborn mouse at levels between n. V and VII and between n. VII and XII. NMDA was bath applied to the transected brainstem. (A–C) NMDA-induced rhythmical activity recorded from n. V (A), VII (B) and XII (C). In each pair of traces, the upper one is an integrated ($\int$) version of the raw recording shown in the lower one. The A, B and C traces were recorded 3, 8 and 4 min, respectively, after a 3-min application of 20 μM NMDA to the recording bath. (Reproduced from Nakamura et al., 1999 with permission from Tokyo Medical and Dental University.)

activities in the three nerves at *different frequencies*. This result suggests that NMDA-induced rhythmical activity in the three motoneuron groups is driven by separate CRGs at appropriate segmental levels of the brainstem (Katakura et al., 1998b; Nakajima, 1999).

The earlier results using newborn mice were then confirmed in analogous newborn rat preparations, with the focus on n. VII and XII. Again, the cycle length differed for these two nerves, the rhythm being significantly faster in n. XII than in n. VII, and again the rhythmical discharge was retained after a brainstem transection between the two nerves (Nakajima et al., 1998).

Rhythmical jaw muscle activity induced by pyramidal tract stimulation in an in vitro preparation isolated from the young adult mouse

To study the CRG for masticatory rather than suckling movements in vitro, we developed a preparation as shown in Fig. 1, but it was isolated from adult rats with well-developed mastication. It is not possible to keep a large-size, adult preparation viable solely by diffusion of oxygen from the perfusing solution. It is possible, however, to use Paton's (1996) technique of supplying oxygen by a combination of diffusion from the perfusing solution and transport via the vascular system. (Furuta et al., 1997; Furuta, 1998). Our approach involved isolating a brainstem-thoracic spinal cord preparation with selected oro-facial structures from the young adult mouse. The preparation was again placed in a chamber continuously perfused with an artificial cerebrospinal fluid (ACSF) as shown in Fig. 1, and infused with the same ACSF including a heparin sodium solution through a cannula inserted into the descending aorta.

Repetitive electrical stimulation of the medullary pyramidal tract induced reciprocal, rhythmical EMG activity at ~4–5 Hz in the digastric and masseter muscles. This pattern was quite similar to that induced by repetitive electrical stimulation of the cortical masticatory area in in vivo preparations, except for the rhythm being slower. Bath and intra-arterial application of kynurenic acid abolished this type of rhythmical EMG activity, thereby implicating glutamate receptors of the relevant neurons.

Even conceding that the necessity to supply oxygen via the vascular system is a crucial disadvantage of the above adult mouse preparation, we hope that its use will provide the means to determine the brainstem structure(s) critically involved in masticatory rhythm generation. Subsequent to this advance, we hope to develop a block or slice preparation which can be kept viable without the aid of oxygen supplied by the cardiovascular system. For example, it was recently reported that bath application of NMDA induced rhythmical membrane potential changes in the identified n. XII motoneurons of a sagittal medial brainstem slice of the newborn mouse. These changes persisted during hyperpolarizing current injection into the test cells, and during the extracellular iontophoretic application of AP5. These results demonstrate that during NMDA-induced rhythmical activity, n. XII motoneurons receive rhythmical inputs from a CRG located in the medial brainstem (Katakura and Ohno, 2000). Progress along these lines in in vitro CNS preparations isolated from newborn and progressively older animals will allow us to elucidate the postnatal development and reorganization of the CRG for food-ingestive movements in mammals.

Acknowledgments

Supported by Grants in Aid for Scientific Research from the Japanese Ministry of Education, Culture, Sports, Sciences and Technology.

Abbreviations

ACSF	artificial cerebrospinal fluid
AP5	2-amino-5-phosphonovalerate
C5	fifth cervical segment
CNS	central nervous system
CRG	central rhythm generator
EMG	electromyogram/electromyographic
n.	nerve
NMDA	N-methyl-D-aspartate
V	trigeminal
VII	facial
VR	ventral root
XII	hypoglossal

References

Ainamo, J. (1971). Prenatal occulsal wear in guinea pig molars. Scand. J. Dent. Res., 79: 69–71.

Bosma, J.F. (1967). Human infant oral function. In: Bosma J.F. (Ed.), Symposium on Oral Sensation and Perception. Thomas, Springfield, pp. 98–110.

Dubner, R., Sessle, B.J. and Storey, A.T. (1978). The Neural Basis of Oral and Facial Function. Plenum, New York, pp. 312–313.

Furuta, S. (1998). Induction of rhythmical activity in digastric muscles in an in vitro brainstem preparation from adult mice. J. Stomatol. Soc. Jpn., 65: 1–10.

Furuta, S., Katakura, N., Hokazono, T., Liu, J., Soma, K. and Nakamura, Y. (1997). Induction of rhythmical activity in digastric muscles in in vitro preparation from adult mice. Jpn. J. Physiol., 47(Suppl. 2): S172.

Iriki, A., Nozaki, S. and Nakamura, Y. (1988). Feeding behavior in mammals: corticobulbar projection is reorganized during conversion from sucking to chewing. Dev. Brain Res., 44: 189–196.

Katakura, N. and Nakamura, Y. (1997) Location of the brainstem neurons related to NMDA-induced sucking-like activity in an isolated brainstem-spinal cord preparation from newborn rats. Abstr. XXXIII Int. Cong. Physiol. Sci., P52.24.

Katakura, N. and Ohno, T. (2000). Rhythmical membrane potential changes of hypoglossal motoneurons during NMDA-induced sucking-like activity in an en bloc brainstem preparations isolated from newborn mice. Neurosci. Res., Suppl. 24: S150.

Katakura, N., Liu, J. and Nakamura, Y. (1995a). NMDA-induced rhythmical activity in XII nerve of isolated CNS from newborn rats. Neuroreport, 6: 601–604.

Katakura, N., Liu, J. and Nakamura, Y. (1995b) Separate oscillators for NMDA-induced rhythmical hypoglossal nerve activities in both halves of the brainstem in an in vitro preparation from newborn rats. Fourth IBRO World Cong. Neurosci. Abstr., 331.

Katakura, N., Liu, J., Furuta, S. and Nakamura, Y. (1996). Influence of inspiratory activity on NMDA-induced suckling-like activity in an in vitro brainstem-spinal cord preparation of rats. Jpn. J. Physiol., Suppl. 46: S141.

Katakura, N., Nakajima, M. and Nakamura, Y. (1998a). Relationship between inspiratory and NMDA-induced sucking-like activities in isolated brainstem from rats. J. Dent. Res., 77: 751 (Special issue B).

Katakura, N., Nakajima, M. and Nakamura, Y. (1998b). Separate oscillators for NMDA-induced rhythmical V, VII and nerve activities along the neuraxis in an in vitro brainstem preparation in mice. Jpn. J. Physiol., Suppl. 48: S153.

Keifer, J., Vyas, D. and Houk, J.C. (1992). Sulforhodamine labeling of neural circuits engaged in motor pattern generation in the in vitro turtle brainstem-cerebellum. J. Neurosci., 12: 3187–3199.

Kessler, J.P. and Jean, A. (1985). Identification of the medullary swallowing regions in the rat. Exp. Brain Res., 57: 256–263.

Lichtman, J.W., Willkinson, R.S. and Rich, M. (1985). Multiple innervation of tonic endplates revealed by activity-dependent uptake of fluorescent probes. Nature, 3(143): 357–359.

Liu, J. (1997). Localization of central rhythm generator for tongue movements in sucking—analysis in isolated brainstem-spinal cord preparation from newborn rats. J. Stomatol. Soc. Jpn., 64: 499–511.

Liu, J., Katakura, N., Furuta, S. and Nakamura, Y. (1997). Brainstem neurons involved in NMDA-induction of sucking-like activity in vitro. J. Dent. Res., 76: 358 (Special issue).

Nakajima, M. (1999). Brainstem segmental arrangement of sucking rhythm generators for trigeminal, facial and hypoglossal motoneurons. J. Stomatol. Soc. Jpn., 66: 88–97.

Nakajima, M., Katakura, N. and Nakamura, Y. (1998). Separate oscillators for NMDA-induced VII and XII activities in isolated rat brainstem. J. Dent. Res., 77: 1005 (Special issue B).

Nakamura, Y. and Katakura, N. (1995). Generation of masticatory rhythm in the brainstem. Neurosci. Res., 23: 1–19.

Nakamura, Y., Katakura, N. and Nakajima, M. (1999). Generation of rhythmical food ingestive activities of the trigeminal, facial and hypoglossal motoneurons in in vitro CNS preparations isolated from rats and mice. J. Med. Dent. Sci., 46: 63–73.

Paton, J.F. (1996). A working heart-brainstem preparation of the mouse. J. Neurosci. Methods, 65: 63–68.

Suzue, T. (1984). Respiratory rhythm generation in the in vitro brain stem-spinal cord preparation of the neonatal rat. J. Physiol. (Lond.), 354: 173–183.

Westneat, M.W. and Hall, W.G. (1992). Ontogeny of feeding motor patterns in infant rats: an electromyographic analysis of suckling and chewing. Behav. Neurosci., 106: 539–554.

CHAPTER 10

Do respiratory neurons control female receptive behavior: a suggested role for a medullary central pattern generator?

Peter A. Kirkwood* and Tim W. Ford

Sobell Department for Motor Neuroscience and Movement Disorders, Institute of Neurology, University College London, Queen Square, London WC1N 3BG, UK

Abstract: Nucleus retroambiguus (NRA) consists of a column of neurons in the caudal medulla with crossed descending axons that terminate in almost all spinal segments. Many of these neurons transmit the drive for expiratory movements to the spinal cord. The same neurons are also known to participate, however, in other motor acts, such as vomiting and abdominal straining, for which it appears that the medullary circuits controlling the respiratory pattern are reconfigured. Plasticity in projections from the NRA to hindlimb motor nuclei provides evidence that some of these projections are involved in yet another motor act, female receptive behavior. Here, we present the hypothesis that the medullary circuits are also reconfigured to act as a central pattern generator for this behavior. In addition, we suggest that during estrus, plasticity is shown not only in spinal cord connections, but also in a selected membrane property of hindlimb motoneurons.

Introduction

Holstege (1991) has identified nucleus retroambiguus (NRA) in the caudal medulla as an important part of the 'emotional motor system'. The NRA was identified as such largely because it receives projections from the periaqueductal gray (PAG) (Holstege, 1989; also Holstege and Georgiadis, Chapter 4 of this volume). One of the vital motor acts that depend on the integrity of the PAG is female receptive behavior (Ogawa et al., 1991). VanderHorst and Holstege (1995) proposed a role for the NRA in the control of this behavior because the pattern of its projections to particular hindlimb motor nuclei in the cat appeared to be appropriate for the characteristic mating posture. This proposition received strong support when one component of this projection, that to the semimembranosus (Sm) motor nucleus, was shown to be nearly nine times stronger during estrus versus nonestrus (VanderHorst and Holstege, 1997).

The question in the title to our chapter arose because, physiologically, the NRA has been identified as the location of the caudal part of the ventral respiratory group. In particular, virtually all of the excitatory expiratory bulbospinal neurons (EBSNs) are found in the caudal part of the NRA (Merrill, 1970, 1974). Now, it is known that respiratory neurons in the medulla can participate in a number of other motor acts, i.e., they are 'multifunctional' (Grélot et al., 1996). Those in which EBSNs participate all involve actions with direct synergies of expiration, such as coughing or vocalization, or related synergies involving raised intra-abdominal pressure such as vomiting or straining. For these actions it seems logical that the EBSNs should be

*Corresponding author: Tel.: +44-(0)20-7837-3611 (Ext. 4189); Fax: +44-(0)20-7813-3107; E-mail: pkirkwoo@ion.ucl.ac.uk

involved. The proposal from VanderHorst and Holstege (1995) concerning the mating posture is different; the principal muscles are those of the hindlimb and no obvious synergy with expiration is present. The proposal is consistent, however, with Holstege's (1989) suggestion that specific groups of NRA neurons with different functions and different projections are present in the NRA. We describe below recent experiments in this laboratory, in which we tested a simpler hypothesis, that the EBSNs actually constitute the projections demonstrated by VanderHorst and Holstege (1995, 1997). Our conclusion is to reject this hypothesis and thus to support the proposal of VanderHorst and Holstege (1995). The observations do not fully support Holstege's (1989) suggestion, however, of completely separate groups of neurons in the NRA. The proposal that we wish to put forward in this chapter is that, while individual groups of neurons adapted for particular motor tasks are present in the NRA and related respiratory nuclei in the medulla, they always act in concert with other such groups in shaping the particular motor output. The motor acts need not be limited to those with close synergies to respiration, such as swallowing and vocalization (Larson et al., 1994). They may also include others for which the projections of the NRA seem appropriate. Thus, we propose that in the sexually receptive female cat during mounting by the male, the output from the PAG transforms circuits within the NRA, and other medullary cell groups, into a pattern generator for receptive behavior. This behavior is therefore another stereotyped motor response brought about by reconfiguration within the NRA, in a manner similar to that which occurs for the EBSNs in vomiting versus respiration (Miller et al., 1987) or straining versus respiration (Fukuda and Fukai, 1988). The proposal is that if evolution has indeed chosen to control female receptive behavior in the cat via the NRA, it is because the NRA operates as part of a pattern generator, and one with extensive output connections to many levels of the neuraxis. Although we are presenting this suggestion specifically with respect to the cat, it should be noted that very similar projections are also present in the hamster and rhesus monkey (Gerrits and Holstege, 1999; Vanderhorst et al., 2000).

Background

NRA projections defined anatomically

The NRA was originally defined in man (Olszewski and Baxter, 1954). In the cat, it consists of a compact column of cells in the ventrolateral medulla extending from about the level of the obex to about 8 mm caudal. Merrill (1970) identified it as an important location of bulbospinal respiratory neurons. The descending projections from the NRA have been well described anatomically by both retrograde and anterograde tracing methods (Holstege and Kuypers, 1982; Feldman et al., 1985; Holstege, 1989, 1991; Iscoe, 1998; Kirkwood et al., 1999). The projections extend throughout the thoracic and lumbar segments, to as far as S2–3. Diffuse terminations are seen bilaterally in the ventral horn, with a contralateral focus in the lateral (intercostal and abdominal) motor nuclei of the thoracic and upper lumbar segments. Strong bilateral projections are also seen in S1, particularly to Onuf's nucleus. Experiments, which included cord hemisection or midline section of the caudal medulla (Miller et al., 1989; VanderHorst and Holstege, 1995), indicate that the descending axons are crossed and that the ipsilateral projections are derived from collaterals which recross the cord at the level of their terminations. A degree of somatotopy may be present (cf. Miller et al., 1985), but it does not appear to be marked. Projections have also been identified to the cranial motor nuclei innervating the pharynx, larynx and soft palate (Holstege, 1989).

The lateral motor nuclei in thoracic and upper lumbar segments receiving NRA projections innervate the abdominal and internal intercostal muscles. Thus, the functional significance of these projections is readily understood in terms of a variety of motor acts that involve raised intra-abdominal and/or intrathoracic pressure. The projections to the sacral cord also fit in with these roles, since these projections are involved in control of the pelvic floor, including the sphincters (Holstege and Tan, 1987; Miller et al., 1995). The projections to the mid and lower lumbar segments, which are those ascribed by VanderHorst and Holstege (1995, 1997) to a role in female receptive behavior, appear in the light microscope to be weaker than those to the thoracic and upper lumbar segments, even when exaggerated during estrus. Both categories of projections are characterized, however, in the electron microscope by direct connections to motoneurons (MNs) (VanderHorst et al., 1997; Boers and Holstege, 1999).

EBSN projections defined physiologically

EBSNs are large cells with fast-conducting axons (19–105 m.s^{-1}; Kirkwood, 1995). They are packed close together throughout the length of the narrow column which comprises the NRA (Merrill, 1970; Miller et al., 1985). They are generally accepted as being responsible for conveying the excitatory expiratory drive to the spinal cord. In the cat, antidromic stimulation shows that virtually all EBSN axons cross at the level of their somata (Merrill, 1974; Arita et al., 1987). The axons form a relatively compact tract at C3, but they are more dispersed in the ventral and ventrolateral funiculi at more caudal levels (Merrill, 1971; Kirkwood, 1995; Kirkwood et al., 1999). These features correspond to those of the anatomically described axons of the NRA. With some relatively minor exceptions (Cohen et al., 1985; Miller et al., 1987), their physiological properties seem rather uniform (Iscoe, 1998), as are their connections. For instance, nearly all of them (perhaps all of them, given the limited sampling) make direct connections to thoracic expiratory MNs (Kirkwood, 1995). In addition, some have axons projecting beyond the thoracic cord, with direct connections being made to abdominal MNs in the upper lumbar cord (Miller et al., 1985; Road and Kirkwood, 1993). Some axons project as far as the sacral cord, giving off at least some collaterals as they pass through the lumbar segments (Sasaki et al., 1994).

Thus, many properties of the EBSNs are consistent with those that characterize the anatomically defined descending projections from the NRA (for further details, see Kirkwood et al., 1999). A priori, this makes EBSNs obvious candidates for those projections from the NRA to the mid and lower lumbar cord that VanderHorst and Holstege (1995) hypothesized to play a role in female receptive behavior.

Multifunctionality

EBSNs are defined operationally as respiratory neurons by their activity in anesthetized or decerebrate preparations. They are also known to participate, however, in a number of other motor acts which can be evoked in these preparations, and which involve activity in thoracic or abdominal MNs. These include coughing (Jakuš et al., 1985), sneezing (Price and Batsel, 1970), vomiting (Miller et al., 1987), vocalization (Katada et al., 1996) and rhythmic straining (Fukuda and Fukai, 1988). Each of these is characterized by a particular stereotyped spatial and temporal pattern of MN discharge, which can be recognized in paralyzed preparations. These essentially fictive motor patterns are centrally generated. The pattern for both *normal* (nonfictive) and fictive vomiting has been extensively investigated (McCarthy and Borison, 1974; Grélot and Miller, 1994) and is illustrated in Fig. 1. It consists of two

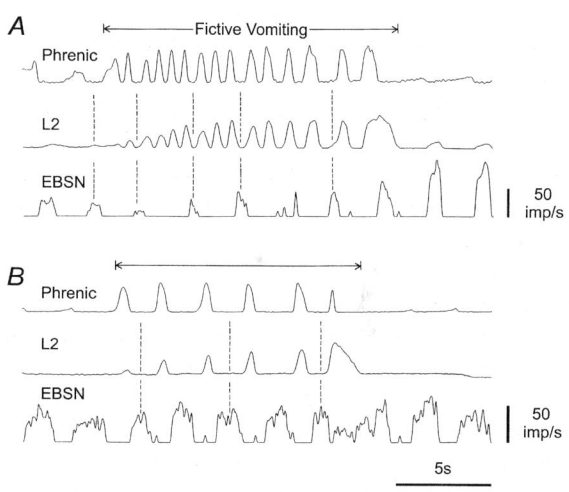

Fig. 1. Activities of two EBSNs during fictive vomiting. The first two traces in A–B show the integrated neurograms of the phrenic and L2 abdominal nerve efferent discharges during fictive vomiting, with about two cycles of respiration at the beginning and end of each panel. Note the coactivation of the two nerve discharges during the repetitive retching phase, followed by the more prolonged discharge in the abdominal nerve during the last burst of the pattern, expulsion. The lowest traces in A–B show the firing pattern of two EBSNs, both of which were of the type that fired between the phrenic/abdominal bursts during vomiting, when the intercostal muscles are most active. (Reproduced from Miller et al., 1987 with permission from The American Physiological Society.)

phases. First there is retching, which involves a series of rhythmic motor discharges, in which the diaphragm and abdominal MNs discharge together but alternate with at least some of the internal intercostal MNs. Retching is followed by expulsion, in which a single burst occurs in diaphragm MNs, together with more prolonged discharge in abdominal MNs. In contrast, during normal respiration, internal intercostal and abdominal MNs act in synergy and alternate with the diaphragm MNs. The reconfiguration of this output pattern to the one for vomiting is accompanied by a reconfiguration of the patterns of discharge of the EBSNs. These divide into two groups, which fire in phase with each of the two groups of MNs, abdominal or intercostal (Miller et al., 1987). Two examples of the latter group of EBSNs are shown in Fig. 1.

As with respiration, it seems clear that EBSNs provide MN excitation during vomiting. Interruption of their axons by a midline section of the caudal medulla abolishes activation of abdominal and pudendal MNs (Miller and Nonaka, 1990; Miller et al., 1995). There does not appear to be a simple correspondence, however, between the EBSNs that fire in one phase of retching and the MNs activated. In respiration, a calculation by simple integration of direct EBSN inputs to thoracic expiratory MNs suggest that excitation from this source alone might be sufficient to bring the MNs to threshold (Kirkwood et al., 1999). This cannot be simply translated to the situation during retching. In the recordings of Miller et al. (1987), the intensity of efferent abdominal nerve discharges increased greatly when retching replaced respiration, yet only a minority of the EBSNs showed a corresponding increase in their discharge frequency. Indeed, because most of the EBSNs fired not during the abdominal phase, but during the intercostal phase (Fig. 1), a large part of the direct input available for the above integral calculation during respiration is not available during the abdominal phase (Kirkwood and Road, 1995). Moreover, inspiratory bulbospinal neurons, which normally supply drive to phrenic MNs, mostly become silent during retching, at least in the cat (Grélot and Miller, 1994). They are also silent in the dog during straining, during which action the diaphragm is also strongly activated (Fukuda and Fukai, 1988). It is likely, then, that during these actions, parallel pathways to the spinal cord are also activated (Miller et al., 1996), as well as interneurons at the spinal level (Nonaka and Miller, 1991; Grélot et al., 1993; Kirkwood and Road, 1995). Also, modulation of direct descending inputs from EBSNs could occur at the MN level.

Recent experiments

It was with the above findings in mind that we began to search for physiological evidence of a functional role for NRA projections to lower lumbar MNs. Our rationale was that whatever the situation during female receptive behavior, if the neurons involved are EBSNs, then their direct connections should create a central respiratory drive potential (CRDP; Sears, 1964) in relevant MNs. Accordingly, intracellular recordings were made in anesthetized, paralyzed female cats either in estrus or nonestrus (protocol of VanderHorst and Holstege, 1997), using a strong CO_2 drive in order to recruit most of the EBSNs. Motoneurons in the Sm motor column [which also contains some semitendinosus (St) MNs] were compared with those in the surrounding motor nuclei, which latter ones are not part of the focus of the NRA projection. The results were very clear (Kirkwood et al., 2002). First, a very high proportion of MNs showed a CRDP. It seemed very unlikely, however, that these CRDPs resulted from the *direct* NRA projection, because the proportions and amplitudes of CRDPs from Sm/St MNs were virtually identical to those from the other, nonrelevant MNs. Also, the time course of most of the CRDPs did not correspond to the firing patterns of EBSNs. Virtually all of the latter have incrementing patterns of discharge during expiration (Miller et al., 1985). Most of the CRDPs showed an expiratory decrementing (E_{dec}) pattern with maximum depolarization near the start of expiration (Fig. 2). Very few of them had a profile that included an expiratory ramp. If present, such a ramp was of quite low amplitude.

The above experiments therefore gave little support to the hypothesis that EBSNs provide the axonal projections investigated by VanderHorst and Holstege (1995, 1997). The few with ramp-like

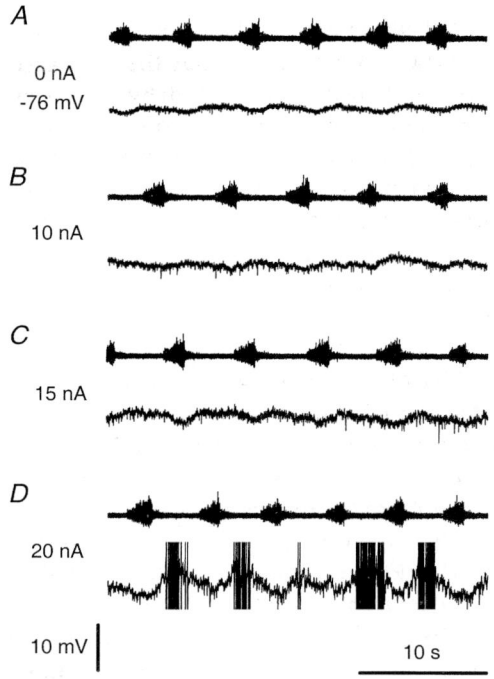

Fig. 2. An example of a CRDP potentiated by depolarizing current. Successive sections of an intracellular recording (about 1 min between each section) from a tibial MN (estrous cat) with the indicated values of current passed via the recording electrode. Upper trace in each panel is the efferent phrenic discharge. Note that the CRDP amplitude increases only slightly from 0 to 15 nA (A–C), but shows a large increase at 20 nA (D). In contrast, the tonic background synaptic noise (dominated by IPSPs) increases steadily. Spikes in D truncated. (Unpublished data from Kirkwood et al., 2002.)

CRDPs may have received some NRA drive (cf. Sasaki et al., 1994). Our experiments also provided some unexpected observations. A number of the CRDPs showed marked variability in amplitude, which frequently covaried with the cell's membrane potential. In some of the tested MNs, their CRDPs could be markedly enhanced (up to ×10–15) by depolarizing currents (Fig. 2). This was attributed (Kirkwood et al., 2002) to the activation of plateau potentials (Hultborn, 1999; also Hultborn et al., Chapter 8 of this volume). In a few MNs, overt plateaus were indeed identified. These observations, in anesthetized animals, are clearly important in their own right, but for the present purposes, the most interesting observation was that the plateau-like effects were observed more frequently during estrus.

Our more recent experiments (Ford et al., 2001) have characterized more directly the NRA neurons that do indeed project to the Sm motor column. Multiunit recordings were made in the NRA of cats in estrus under similar conditions to those described earlier. Antidromically activated units were sought within those populations, using either microstimulation in or around the Sm column or stimulation of the nearby ventrolateral funiculus at L6. Spontaneous discharges (if any) of these units were identified by collisions between spontaneous impulses and those antidromically activated. Units firing in late expiration were identified as EBSNs (Fig. 3). Our current sample of NRA neurons activated from the ventrolateral funiculus contains a reasonable proportion of EBSNs ($\sim$50%), whereas, of those identified antidromically from the Sm motor nucleus, only 1/8 was an EBSN.

We conclude that those NRA neurons that project to the Sm motor nucleus form a distinct population that can be distinguished from EBSNs and therefore are very likely to have a different primary function. Such a function could still be synergistic to expiration. For example, such NRA cells might be excited only subliminally even at high CO_2 levels, but be recruited into activity during more intense abdominal contractions, such as straining or vomiting. Alternatively (or in addition) the primary function could include female receptive behavior, as suggested by the estrous enlargement of the overall NRA projections. We obtained more clues by measurements of the subthreshold excitability modulation of these NRA neurons during the respiratory cycle (technique of Merrill, 1974). For most of the tested EBSNs, a ramp of excitability (of decreasing latency) was detected before the neurons started to fire in late expiration. Very few of the NRA cells projecting to the Sm column displayed this behavior, however. Rather, their latency was mostly constant during expiration (Fig. 3). This argues against the idea of these Sm-projecting NRA cells being even subliminally excited. As such, they cannot contribute to an expiration synergy. The majority of our tested NRA cells, however, both those activated from the funiculus and those activated from the Sm column, showed an increase in their latency, or even a failure of the antidromic impulse to invade the soma during

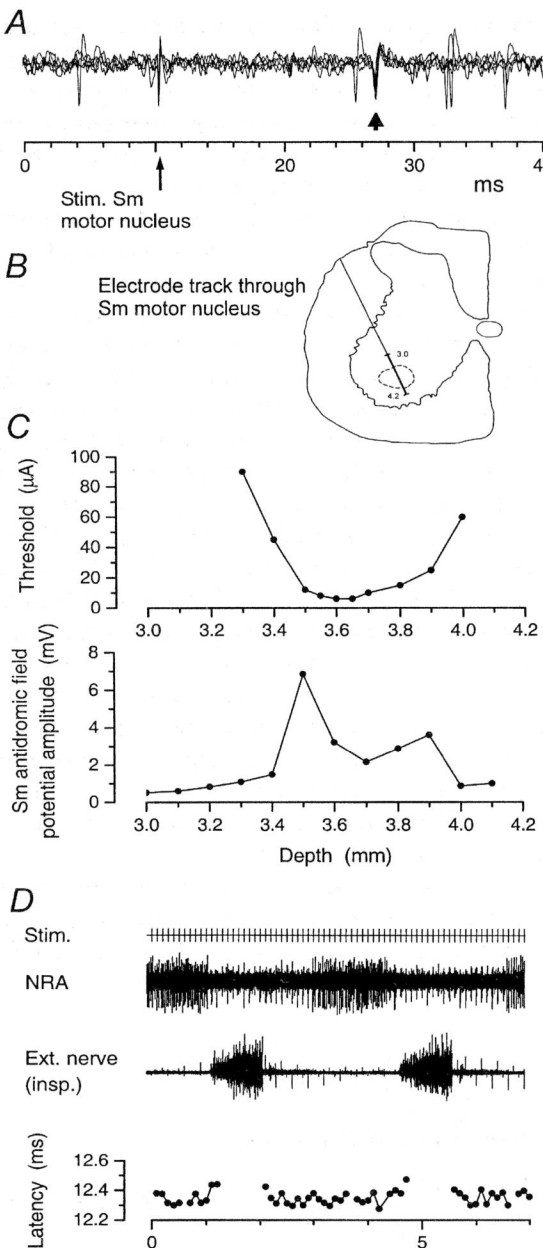

inspiration (Fig. 3). This is highly suggestive of postsynaptic inhibition during inspiration (Merrill, 1974), which is known to occur in EBSNs (Mitchell and Herbert, 1974; Richter et al., 1979). Thus, even if our tested Sm-projecting NRA neurons do have a distinct function, they nevertheless appear to be part of the same overall network as the EBSNs.

Discussion

The results described earlier, which show that the NRA projections to hindlimb MNs involve *very few* EBSNs, are consistent with the proposition that these NRA projections have a role in receptive behavior. Admittedly, our present results do not add *specific* evidence to support this idea. The consistency of our results is sufficient, however, to speculate that the signal these NRA projections transmit to the spinal cord is indeed the output of a medullary pattern generator.

Our suggestion is based partly on teleology. Like other motor actions in which the NRA and the other respiratory cell groups are involved, female receptive behavior is a stereotyped motor act. It involves several movement components controlled from different levels of the neuraxis, including lordosis,

Fig. 3. Some properties of a unit recorded in the NRA and activated antidromically by microstimulation within the Sm motor nucleus. (A) Antidromic activation of unit (broad arrow). Five sweeps superimposed. Note no collisions with the spontaneous spikes: the unit was not firing, so is not an EBSN. (B) Track of stimulating electrode in caudal L6, reconstructed histologically. Dotted outline shows the Sm/St column. Depths indicated in mm. (C) Upper graph, depth-threshold plot for stimulation at points along this track; lower graph, amplitude of MN antidromic field potential from stimulation of Sm nerve at the same depths, used for physiological definition of the Sm motor nucleus. Note that the minimum threshold (6 μA) is within the motor nucleus. The range included on the abscissa is indicated in B by the thick line. (D) Original recordings and corresponding plot of antidromic latency, which shows some dependence on the respiratory phase. Top trace, stimulation times, second trace, multiunit recording in NRA. There are bursts of spontaneous discharges in phase II expiration, which relate to a different unit (or units) from the one concerned, and which indicate that the recording was within the NRA. The unit concerned and/or the stimulus artifact can be seen as the slightly smaller spike at the times of the stimulus. Third trace, efferent discharge in an external intercostal nerve, indicating the timing of inspiration. Stimulus artifacts (amplitude varies due to aliasing) are also visible. Bottom plot, antidromic latency. The antidromic spike was absent (presumably as a result of failure to invade the soma) during most of inspiration. When it was present at the start and end of inspiration, the latency was longer than the rather constant value shown throughout expiration. (Unpublished data from Ford et al., 2001.)

deviation of the tail and rhythmic-treading movements of the hindlimbs (Michael, 1961; VanderHorst and Holstege, 1995; VanderHorst et al., 2000). A medullary pattern generator involving the NRA (especially given its characteristic direct connections to MNs) is ideally placed to control such movements. These movements may not be synergistic with those involved in respiration, as are most of those previously identified as being controlled by these cell groups. Nonetheless, the NRA neurons we have identified do show some respiratory modulation (inhibition during inspiration). Thus, reconfiguration of networks involving these neurons can be regarded as directly analogous to the other reconfigurations listed previously.

Note that our hypothesis does not deny a commanding or controlling role for the PAG in receptive behavior. What we are proposing is that the medullary cell groups, by analogy with other behaviors such as vomiting (Fukuda et al., 1999), have been delegated the function of organizing the spatial and temporal output patterns for receptive behavior. This is a slightly different view to that put forward by some authors with regard to vocalization, which is another function controlled by the PAG (Davis and Zhang, 1991). For this function, the organizing role assigned here to the medulla is largely retained by the PAG. In vocalization, a higher level of control is clearly necessary (Zhang et al., 1995). Vocalization includes one of the most refined examples of skilled movement, the prosodic qualities of human speech or song. In contrast, the extremely stereotyped nature of female receptive behavior in the cat seems no more sophisticated than breathing or vomiting. This viewpoint is in keeping with a medullary site for its pattern generator.

As with vomiting, it is likely that the direct connections from the NRA alone are not sufficient to command the MNs in receptive behavior. The relative weakness of NRA projections to the mid and lower lumbar motor nuclei, as compared to those of the thoracic and upper lumbar levels, suggests that considerable boosting of the command signals is required. Perhaps the E_{dec} CRDPs that we observed in hindlimb MNs represent an additional input. The respiratory signal for these particular CRDPs must come ultimately from the medulla, but via a presently unknown route. One possibility is that the respiratory signal is relayed by thoracic interneurons, some of which have an E_{dec} firing pattern when studied under similar conditions. Moreover, some of these E_{dec} interneurons have long descending axons (Kirkwood et al., 1988; Schmid et al., 1993), which seem appropriate for the proposed purpose.

Alternatively, or in addition, the plateau potential mechanism could provide the necessary amplification. This role is specifically suggested for receptive behavior by our observation of an increase in the occurrence of plateau-like effects during estrus. Perhaps inputs from the NRA are strategically colocated on dendrites along with the channels responsible for the plateau mechanism (Delgado-Lezama et al., 1999). This would be the ideal way for specific stimuli, propagated via relatively few neurons in the NRA, to provide a powerful, sustained, overriding activation of relevant MNs. A similar suggestion was made by Cueva-Rolón et al. (1993) to explain maintained reflex discharges in hindlimb extensor muscles evoked by vaginal stimulation in the decerebrate cat. One of their illustrations is shown in Fig. 4. Of particular interest in this example is the rhythmic excitation, which preceded the sustained reflex discharge. Not only does this suggest a pattern similar to the treading movements in receptive behavior, but it also provides a further analogy with the retching-plus-expulsion pattern of fictive vomiting. In normal vomiting, the retching pattern has been assigned a specific role in moving the gastric

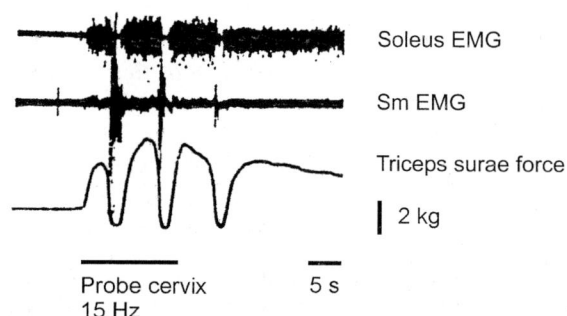

Fig. 4. Activation of hindlimb muscles by coitus-like mechanical stimulation of the cervix in a decerebrate cat. Reciprocal activity in extensor and flexor muscles is followed by a prolonged activation of the extensor. (Reproduced from Cueva-Rolón et al., 1993 with permission from Elsevier Science.)

contents into the esophagus (McCarthy and Borison, 1974; but cf. Grélot and Miller, 1994). For female receptive behavior, however, we can suggest another role for rhythmic excitation. It has frequently been observed that plateaus demonstrate 'warm-up', i.e., they are facilitated by repeated application of a depolarizing stimulus (Bennett et al., 1998). (This was also observed in our above-described experiments). On the assumption that the maintained reflex discharge demonstrated by Cueva-Rolón et al. (1993) is an important component of normal behavior, could it be that one function of the treading movements is to provide just such a warm-up, specifically to promote plateau potentials in MNs for their receptive-behavior discharge?

In addition to our presumed estrus-induced modulation of the plateau mechanism in MNs, it is most likely that additional modulation is supplied by descending monoaminergic systems. The latter are surely activated by stimuli arising from the receptive-inducing actions of the male cat. The descending monaminergic systems are, like the NRA, under control of the PAG. Accordingly, they have also been assigned by Holstege (1991) to the medial component of the emotional motor system (see also Holstege and Georgiadis, Chapter 4 of this volume). Soon after these descending monoaminergic effects on MNs were first described, Kuypers and Huisman (1982) suggested that their gain-setting role could be "... instrumental in providing motivational drive in the execution of movements." (p. 49; see also Holstege and Kuypers, 1982). Subsequently, Conway et al. (1988) showed how this could be performed via the modulation of plateau potentials (see also Hultborn et al., Chapter 8 of this volume).

In final summary, our suggestion is that plateau potentials, acting under the additional modulation of estrogen, provides a substrate for very specific projections from the NRA to command a very particular motor act, and one for which motivation must be as strong as for any.

Acknowledgments

The experimental work was supported by the Wellcome Trust. Claire Meehan is thanked for assistance during one of the experiments and we are grateful to Professor G. Holstege for many useful discussions.

Abbreviations

C3	third cervical segment
CRDP	central respiratory drive potential
EBSN	expiratory bulbospinal neuron
E_{dec}	expiratory decrementing
L6	sixth lumbar segment
MN	motoneuron
NRA	nucleus retroambiguus
PAG	periaqueductal gray
S1	first sacral segment
S2	second sacral segment
S3	third sacral segment
Sm	semimembranosus
St	semitendinosus

References

Arita, H., Kogo, N. and Koshiya, N. (1987). Morphological and physiological properties of caudal medullary expiratory neurons of the cat. Brain Res., 401: 258–266.

Bennett, D.J., Hultborn, H., Fedirchuk, B. and Gorassini, M. (1998). Short-term plasticity in hindlimb motoneurons of decerebrate cats. J. Neurophysiol., 80: 2038–2045.

Boers, J. and Holstege, G. (1999). Evidence for direct projections from the nucleus retroambiguus to the motoneurons of the external oblique abdominal muscle in the female cat. Soc. Neurosci. Abstr., 25: 117.

Cohen, M.I., Feldman, J.L. and Sommer, D. (1985). Caudal medullary expiratory neurone and internal intercostal nerve discharges in the cat: effects of lung inflation. J. Physiol. (Lond.), 368: 147–178.

Conway, B.A., Hultborn, H., Kiehn, O. and Mintz, I. (1988). Plateau potentials in alpha-motoneurones induced by intravenous injection of L-DOPA and clonidine in the spinal cat. J. Physiol. (Lond.), 405: 369–384.

Cueva-Rolón, R., Múñoz-Martínez, E.J., Delgado-Lezama, R., Raya, J.G. and González-Santos, G. (1993). Sustained activation of the triceps surae muscle produced by mechanical stimulation of the genital tract of the female cat. Brain Res., 600: 33–38.

Davis, P.J. and Zhang, S.P. (1991). What is the role of the midbrain periaqueductal gray in respiration and vocalization. In: Depaulis A. and Bandler R. (Eds.), The Midbrain Periaqueductal Gray Matter. Plenum Press, New York, pp. 57–66.

Delgado-Lezama, R., Perrier, J.-F. and Hounsgaard, J. (1999). Local facilitation of plateau potentials in dendrites of turtle motoneurones by synaptic activation of metabotropic receptors. J. Physiol. (Lond.), 515: 203–207.

Feldman, J.L., Loewy, A.D. and Speck, D.F. (1985). Projections from the ventral respiratory group to phrenic and intercostal motoneurons in cat: an autoradiographic study. J. Neurosci., 5: 1993–2000.

Ford, T.W., Meehan, C.F. and Kirkwood, P.A. (2001) Functional roles for nucleus retroambiguus neurones projecting to the lumbosacral spinal cord of the cat. XXXIV Int. Congr. Physiol. Sci., Abstr. 983.

Fukuda, H. and Fukai, K. (1988). Discharges of bulbar respiratory neurons during rhythmic straining evoked by activation of pelvic afferent fibers in dogs. Brain Res., 449: 157–166.

Fukuda, H., Koga, T., Furukawa, N., Nakamura, E. and Shiroshita, Y. (1999). The tachykinin NK_1 receptor antagonist GR205171 abolishes the retching activity of neurons comprising the central pattern generator for vomiting in dogs. Neurosci. Res., 33: 25–32.

Gerrits, P.O. and Holstege, G. (1999) Descending projections from the nucleus retroambiguus to the iliopsoas motoneuronal cell groups in the female golden hamster: possible role in reproductive behavior. J. Comp. Neurol., 403: 291–228.

Grélot, L. and Miller, A.D. (1994). Vomiting—its ins and outs. News Physiol. Sci., 9: 142–147.

Grélot, L., Milano, S., Portillo, F. and Miller, A.D. (1993). Respiratory interneurons of the lower cervical (C4-C5) cord: membrane potential changes during fictive coughing, vomiting, and swallowing in the decerebrate cat. Pflügers Archiv., 425: 313–320.

Grélot, L., Milano, S., Gestreau, C. and Bianchi, A.L. (1996). Are medullary respiratory neurones multipurpose neurons? In: Bostock H., Kirkwood P.A. and Pullen A.H. (Eds.), The Neurobiology of Disease: Contributions from Neuroscience to Clinical Neurology. Cambridge University Press, Cambridge, pp. 299–308.

Holstege, G. (1989). Anatomical study of the final common pathway for vocalization in the cat. J. Comp. Neurol., 284: 242–252.

Holstege, G. (1991). Descending motor pathways and the spinal motor system: limbic and non-limbic components. In: Holstege G. (Ed.), Role of the Forebrain in Sensation and Behaviour, Prog. Brain Res., Vol. 87, Elsevier, Amsterdam, pp. 307–421.

Holstege, G. and Kuypers, H.G.J.M. (1982). The anatomy of brain stem pathways to the spinal cord in cat. A labeled amino acid tracing study. In: Kuypers H.G.J.M. and Martin G.F. (Eds.), Descending Pathways to the Spinal Cord, Prog. Brain Res., Vol. 57, Elsevier, New York, pp. 145–175.

Holstege, G. and Tan, J. (1987). Supraspinal control of motoneurons innervating the striated muscles of the pelvic floor including urethral and anal sphincters in the cat. Brain, 110: 1323–1344.

Hultborn, H. (1999). Plateau potentials and their role in regulating motoneuronal firing. In: Binder M. (Ed.), Peripheral and Spinal Mechanisms in the Neural Control of Movement, Prog. Brain Res., Vol. 123, Elsevier, New York, pp. 39–48.

Iscoe, S. (1998). Control of abdominal muscles. Prog. Neurobiol., 56: 433–506.

Jakuš, J., Tomori, Z. and Stránsky, A. (1985). Activity of bulbar respiratory neurones during cough and other respiratory tract reflexes in cats. Physiol. Bohemoslov., 34: 127–136.

Katada, A., Sugimoto, T., Utsumi, K., Nonaka, S. and Sakamoto, T. (1996). Functional role of ventral respiratory group expiratory neurons during vocalization. Neurosci. Res., 26: 225–233.

Kirkwood, P.A. (1995). Synaptic excitation in the thoracic spinal cord from expiratory bulbospinal neurons in the cat. J. Physiol. (Lond.), 484: 201–225.

Kirkwood, P.A. and Road, J.D. (1995). On the functional significance of long monosynaptic descending pathways to spinal motoneurones. In: Taylor A., Gladden M.H. and Durbaba R. (Eds.), Alpha and Gamma Motor Systems. Plenum Press, London, pp. 589–592.

Kirkwood, P.A., Munson, J.B., Sears, T.A. and Westgaard, R.H. (1988). Respiratory interneurones in the thoracic spinal cord of the cat. J. Physiol. (Lond.), 395: 161–192.

Kirkwood, P.A., Ford, T.W., Donga, R., Saywell, S.A. and Holstege, G. (1999). Assessing the strengths of motoneuron inputs: different anatomical and physiological approaches compared. In: Binder M. (Ed.), Peripheral and Spinal Mechanisms in the Neural Control of Movement, Prog. Brain Res., Vol. 123, Elsevier, New York, pp. 67–82.

Kirkwood, P.A., Lawton, M. and Ford, T.W. (2002). Plateau potentials in hindlimb motoneurones of female cats under anaesthesia. Exp. Brain Res., 146: 399–403.

Kuypers, H.G.J.M. and Huisman, A.M. (1982). The new anatomy of the descending brain pathways. In: Sjölund B. and Björklund A. (Eds.), Brain Stem Control of Spinal Mechanisms. Elsevier, Amsterdam, pp. 29–54.

Larson, C.R., Yajima, Y. and Ko, P. (1994). Modification in activity of medullary respiratory-related neurons for vocalization and swallowing. J. Neurophysiol., 71: 2294–2304.

McCarthy, L.E. and Borison, H.L. (1974). Respiratory mechanics of vomiting in decerebrate cats. Am. J. Physiol., 226: 738–743.

Merrill, E.G. (1970). The lateral respiratory neurones of the medulla: their associations with nucleus ambiguus, nucleus retroambigualis, the spinal accessory nucleus and the spinal cord. Brain Res., 24: 11–28.

Merrill, E.G. (1971). The descending pathways from the lateral respiratory neurones in cats. J. Physiol. (Lond.), 218: 82P–83P.

Merrill, E.G. (1974). Finding a respiratory function for the medullary respiratory neurons. In: Bellairs R. and Gray E.G.

(Eds.), Essays on the Nervous System. Clarendon Press, Oxford, pp. 451–486.

Michael, R.P. (1961). Observations upon the sexual behaviour of the domestic cat (*felis catus* L.) under laboratory conditions. Behaviour, 18: 1–24.

Miller, A.D. and Nonaka, S. (1990). Mechanisms of abdominal muscle activation during vomiting. J. Appl. Physiol., 69: 21–25.

Miller, A.D., Ezure, K. and Suzuki, I. (1985). Control of abdominal muscles by brain stem respiratory neurons in the cat. J. Neurophysiol., 54: 155–167.

Miller, A.D., Tan, L.K. and Suzuki, I. (1987). Control of abdominal and expiratory intercostal muscle activity during vomiting: role of ventral respiratory group expiratory neurons. J. Neurophysiol., 57: 1854–1866.

Miller, A.D., Tan, L.K. and Lakos, S.F. (1989). Brainstem projections to cat's upper lumbar spinal cord: implications for abdominal muscle control. Brain Res., 493: 348–356.

Miller, A.D., Nonaka, S., Siniaia, M.S. and Jakuš, J. (1995). Multifunctional ventral respiratory group: bulbospinal expiratory neurons play a role in pudendal discharge during vomiting. J. Auton. Nerv. Syst., 54: 253–260.

Miller, A.D., Nonaka, S., Jakuš, J. and Yates, B.J. (1996). Modulation of vomiting by the medullary midline. Brain Res., 737: 51–58.

Mitchell, R.A. and Herbert, D.A. (1974). Synchronized high frequency synaptic potentials in medullary respiratory neurons. Brain Res., 75: 350–355.

Nonaka, S. and Miller, A.D. (1991). Behavior of upper cervical inspiratory propriospinal neurons during fictive vomiting. J. Neurophysiol., 65: 1492–1500.

Ogawa, S., Kow, L.-M., McCarthy, M.M., Pfaff, D.W. and Schwartz-Giblin, S. (1991). Midbrain PAG control of female reproductive behavior: in vitro electrophysiological characterization of actions of lordosis-relevant substances. In: Depaulis A. and Bandler R. (Eds.), The Midbrain Periaqueductal Gray Matter. Plenum, New York, pp. 211–235.

Olszewski, J. and Baxter, D. (1954). Cytoarchitecture of the Human Brain Stem. Lippincott, Philadelphia.

Price, W.M. and Batsel, H.L. (1970). Respiratory neurons participating in sneeze and in response to resistance to expiration. Exp. Neurol., 29: 554–570.

Richter, D.W., Camerer, H., Meesmann, M. and Rohrig, N. (1979). Studies on the synaptic interconnection between bulbar respiratory neurones of cats. Pflügers Arch., 380: 245–257.

Road, J.D. and Kirkwood, P.A. (1993) Distribution of monosynaptic connections from expiratory bulbospinal neurones to motoneurones of different expiratory muscles in the cat. Abstr. XXXII Congr. Int. Union Physiol. Sci., 141.39/P.

Sasaki, S.-I., Uchino, H. and Uchino, Y. (1994). Axon branching of medullary expiratory neurons in the lumbar and sacral spinal cord of the cat. Brain Res., 648: 229–238.

Schmid, K., Kirkwood, P.A., Munson, J.B., Shen, E. and Sears, T.A. (1993). Contralateral projections of thoracic respiratory interneurones in the cat. J. Physiol. (Lond.), 461: 647–665.

Sears, T.A. (1964). The slow potentials of thoracic respiratory motoneurones and their relation to breathing. J. Physiol. (Lond.), 175: 404–424.

VanderHorst, V.G.M. and Holstege, G. (1995). Caudal medullary pathways to lumbosacral motoneuronal cell groups in the cat: evidence for direct projections possibly representing the final common pathway for lordosis. J. Comp. Neurol., 359: 457–475.

VanderHorst, V.G.J.M. and Holstege, G. (1997). Estrogen induces axonal outgrowth in the nucleus retroambiguus-lumbosacral motoneuronal pathway in the adult female cat. J. Neurosci., 17: 1122–1136.

VanderHorst, V.G.J.M., de Weerd, H. and Holstege, G. (1997). Evidence for monosynaptic projections from the nucleus retroambiguus to hindlimb motoneurons in the cat. Neurosci. Lett., 224: 33–36.

VanderHorst, V.G.J.M., Terasawa, E.I., Ralston, H.J. and Holstege, G. (2000). Monosynaptic projections from the nucleus retroambiguus to motoneurons supplying the abdominal wall, axial, hindlimb, and pelvic floor muscles in the female rhesus monkey. J. Comp. Neurol., 424: 233–250.

Zhang, S.P., Bandler, R. and Davis, P.J. (1995). Brain stem integration of vocalization: role of the nucleus retroambigualis. J. Neurophysiol., 74: 2500–2512.

CHAPTER 11

The central pattern generator for forelimb locomotion in the cat

Takashi Yamaguchi*

Graduate School of Engineering, Yamagata University, Yamagata 992-8510, Japan

Abstract: This chapter addresses the neuronal organization of the central pattern generator (CPG) for forelimb locomotion in the cat. The focus is on fictive locomotion, as induced by repetitive stimulation of the cervical lateral funiculus (CLF) in the decerebrate preparation. The stimulus time-locked responses of cervical motoneurons (MNs) that follow each current pulse of such repetitive stimulation include disynaptic EPSPs, trisynaptic EPSPs and IPSPs, and polysynaptic PSPs. Their amplitudes are all phase-related to selected aspects of the locomotor rhythm, as determined by the pattern of discharge in nerves to selected muscles. Correlation analysis reveals rhythmic modulation between the responses of extensor and flexor MNs, thereby suggesting mutual interactions between pathways mediating the CLF effects on MNs. It is argued that the shortest CLF pathway to MNs via the CPG is disynaptic: i.e., with only a single intercalated interneuron (IN) between the descending command axons and the MNs. A second proposal is that such intercalated INs contribute to interacting, reverberating IN circuits (i.e., the CPG) through mutual excitation.

Introduction

The spatio-temporal pattern of muscle activity of the forelimbs during locomotion is primarily generated in the cervical enlargement of the spinal cord. This circuitry can be termed the central pattern generator (CPG) for forelimb locomotion. The CPGs for both the forelimbs and the hindlimbs (one for each limb) are controlled in parallel, i.e., ipsilateral or bilateral excitation of the midbrain locomotor region (MLR) elicits a coordinated locomotor rhythm in both sets of limbs (Shik et al., 1966; Jordan et al., 1979; Amemiya and Yamaguchi, 1984; Grillner and Wallén, and Pearson, Chapters 1 and 12 of this volume). Furthermore, the spinal cord pathways activating the locomotor rhythm of both sets of limbs descend through the same region of the lateral funiculus at the upper cervical level of the spinal cord (Steeves and Jordan, 1980; Yamaguchi, 1986). Nevertheless, it is intuitively obvious that the forelimb CPG must have some properties that differ from its hindlimb counterpart because the two sets of limbs subserve different functions during locomotion. The hindlimbs provide the majority of the propulsive force, whereas the forelimbs are used largely for propping, steering, and supporting the head's weighted mass (Vidal et al., 1986; Graf et al., 1997). Accordingly, my focus has been on the neuronal organization of the forelimb CPG in the cat, with the progress to date summarized in the next section.

Overview of the forelimb CPG

The forelimb CPG is activated by tonic locomotor command signals traveling through pathways descending in the lateral funiculus (Yamaguchi, 1986). These pathways presumably have projections into the rostral enlargement (C6–7) of the spinal cord. This was shown by demonstrating that transection of the lateral funiculus just rostral to C6 abolishes MLR-evoked stepping in the decerebrate cat, whereas a

*Corresponding author: Tel.: +81-238-26-3396;
Fax: +81-238-26-3177; E-mail: yamagut@yz.yamagata-u.ac.jp

transection below C7 does not (Hishinuma and Yamaguchi, 1990). The localized projection of the command implies that command-sensitive 'input' (first-order) interneurons (INs) of the forelimb CPG located in the rostral segments of the cervical enlargement receive the command, and bring into play the CPG.

In the immobilized, decerebrate cat, many INs in the cervical enlargement fire rhythmically during fictive forelimb locomotion. This activity produces rhythmic, alternating discharges of motoneurons (MNs) supplying elbow extensors and flexors (Yamaguchi, 1992). Each IN emits action potentials in a specific phase of the step cycle, i.e., the 'active' phase for that particular IN. To date, four major CPG groups of IN have been identified.

(1) INs termed 'f' are active within the flexor phase. Since the flexor phase is short, 'f' neurons are short bursting. Flexor, phase-related neurons with longer active periods are termed '⟨f⟩' ('⟨ ⟩' means expansion of active periods beyond the onset and offset of relevant phases). They are active throughout the entire flexor phase. They start firing in the late extensor phase or just prior to the flexor phase, and cease their discharge in the first one-third ('medium-bursting ⟨f⟩' INs) or middle ('long-bursting') one-third of the immediately following extensor phase.
(2) Those termed 'f + e' start firing in the late flexor phase, and cease their discharge in the following extensor phase. Their 'short', 'medium', and 'long-bursting' subcategories stop firing in the first, second and last one-third of the extensor phase, respectively.
(3) INs termed 'e' discharge from the early extensor phase to the middle one-third ('short-bursting'), last one-third ('medium-bursting') of the extensor phase, or immediately before the transition to the flexor phase ('long-bursting'). Neurons termed '⟨e⟩' are active throughout the entire extensor phase.
(4) INs termed 'e + f' start firing in the extensor phase and stop in the next flexor phase. Their subcategories comprise 'short', 'medium', and 'long-bursting' types, which commence their discharge prior to the flexor phase, in the middle one-third of the extensor phase, and in the first one-third of the extensor phase, respectively.

The above-described nomenclature and activity patterns show that f, ⟨f⟩ and e + f INs are 'flexor-like', while e, ⟨e⟩ and f + e INs are 'extensor-like'. It is assumed that these INs can be first-order, in-between-order, or last-order, the latter exciting MNs to either flexor or extensor muscles.

Figure 1 shows the approximate spinal cord location of these CPG IN types. In the dorsoventral axis, ⟨f⟩ and e + f INs are located rather dorsally (mainly in laminae V–VI), whereas e and f + e INs are located ventrally (mainly in the lamina VII). Along the rostrocaudal axis, e INs are located caudally (mainly C7–T1) and e + f INs are more rostral (mainly in C5–6). The ⟨f⟩ and f + e groups have a bimodal distribution of locations, one rostral (mainly C5–6) and the other caudal (C7–8). As mentioned earlier, the input INs of the CPG are presumably located in the C6–7 segments, but we have not been able to specify the first-order neurons solely on the basis of their active phase, because there are several types of rhythmically discharging INs at the C6–7 level.

The forelimb CPG includes INs that send rhythmic excitation and inhibition to forelimb extensor and flexor MNs (Gödderz et al., 1990). These 'output' INs, the last-order ones of the CPG are, by definition, rhythmically active during fictive locomotion, and they are identifiable among the INs shown in the topographic maps of Fig. 1. Following the identification of such last-order INs by antidromic stimulation of MN pools (i.e., nuclei), their location has been compared to those of the total population of CPG INs (Terakado and Yamaguchi, 1990; Ichikawa et al., 1991).

The last-order INs to the flexor MNs are localized in the more rostral C5–7 segments. Those of f and e + f type are located in the dorsal intermediate zone, whereas those of f + e and e type are located in the ventral intermediate zone. The former (f and e + f) INs are active in virtually the same phase as their target MNs, thereby suggesting that they are excitatory. In contrast, the f + e and e INs are out-of-phase with flexor MNs, thereby suggesting that they have an inhibitory function.

The last-order CPG INs to extensor MNs are distributed in a column throughout the cervical

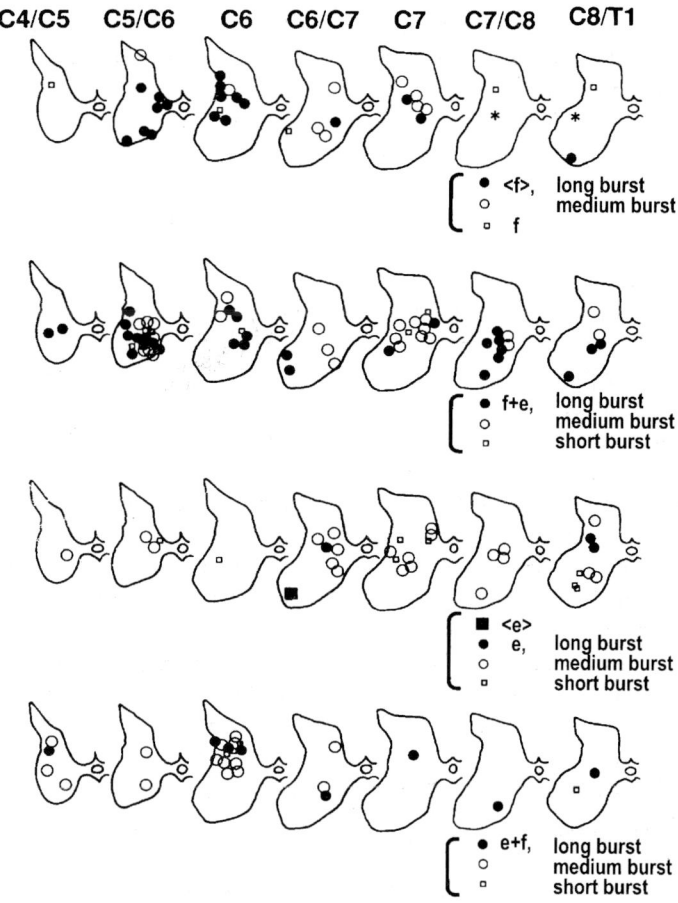

Fig. 1. Spinal cord location of cervical INs contributing to the forelimb CPG for locomotion. Four separate sets of spinal cord cross-section sketches are shown for the four major IN cell types (⟨f⟩, f+e, e, e+f), which, together with their subtypes, are defined in the text. The asterisks (*; uppermost row, C7/8 and C8/T1) are for INs which exhibited a double burst of discharge; the C7/8 neuron showing one burst like ⟨f⟩ INs and the others like short-bursting e INs at the late extensor phase; and, the C8/T1 neuron showing short-bursting e+f and medium-bursting e behavior. Notations with two segments separated by slash mean boundaries between the two. (Reproduced, in part, from Yamaguchi, 1992 with permission from Center for Academic Publications Japan.)

enlargement. The INs of f and e+f type, which are presumably inhibitory (i.e., for the reasons provided earlier) are mainly located more rostrally (C6–7), and those of f+e and e type (presumably excitatory) are more caudal (C7-T1). Note further that potentially excitatory last-order INs, especially those to flexor MNs, are usually located in the same spinal cord segment as their corresponding first-order INs.

In summary, the forelimb CPG receives descending tonic impulses via its first-order input INs at the C6–7 level. Then, the full array of CPG INs develops its locomotor rhythm, and distributes rhythmic excitatory and inhibitory synaptic volleys to MNs by way of various last-order output INs.

The stimulus time-locked-response approach to studying the forelimb CPG

What kinds of signal processing take place between input and output CPG INs? There could be multiple interneuronal paths connecting the two groups of cells. Signal transmission through these paths and mutual interactions among them could conceivably generate a locomotor rhythm. Is it possible to reveal the organization of these interneuronal paths, and delineate the characteristics of their signal transmission? At first glance, one approach would be to analyze MN responses to a single stimulus pulse delivered within a descending pathway involved in the locomotor command. When the CPG is at rest, however, so too are most of its constituent INs. Therefore, a CPG's IN paths can be investigated only when the CPG is active. We use fictive locomotion for this purpose, as evoked by repetitive stimulation of the upper cervical lateral funiculus. Under these conditions, Fig. 2 shows that locomotor discharges in nerves supplying selected flexor and extensor muscles include spikes resembling the action potentials of MNs. These are time-locked to each stimulus, and they are modulated rhythmically (Kinoshita and Yamaguchi, 2001). Such stimulus time-locked discharges presumably result from activity in multiple IN paths within the forelimbs' locomotor CPG (Shefchyk and Jordan, 1985; Degtyarenko et al., 1998). Thus, the CPG network can be investigated indirectly by studying the stimulus time-locked responses of MNs during fictive locomotion.

Analysis and interpretation of stimulus-time-locked responses of MNs

We have made intracellular (IC) recordings in MNs supplying elbow flexor and extensor muscles during fictive locomotion, which was produced as described in the previous section (Kinoshita and Yamaguchi, 2001). Figure 3 shows that under these conditions, both flexor and extensor MNs exhibit stimulus time-locked disynaptic excitatory postsynaptic potentials (EPSPs), trisynaptic inhibitory postsynaptic potentials (IPSPs) and polysynaptic EPSPs. All such postsynaptic potentials (PSPs) are rhythmically modulated with a stereotypic pattern. The disynaptic EPSPs of flexor MNs are facilitated in the flexor phase of locomotion, whereas those of extensor MNs are facilitated mainly at the transition from the flexor to the extensor phase. The depth of such modulation is greater in flexor versus extensor MNs.

The amplitude of trisynaptic IPSPs change in parallel with that of the disynaptic EPSPs of antagonistic MNs. This is a strong indication that the trisynaptic IPSPs of flexor MNs are facilitated in the flexor-to-extensor transition phase, whereas those of extensor MNs are facilitated in the flexor phase.

After transection of the lateral funiculus to abolish disynaptic EPSPs, another type of stimulus time-locked EPSP is found in extensor MNs. It is trisynaptic, and it presumably results from excitation transmitted through the ventral part of the spinal cord. It is clearly facilitated rhythmically in the extensor phase.

The earlier results are summarized in Fig. 4. This schematic diagram proposes that rhythmic modulation of MNs during locomotion is due to the operation of three subsystems of the CPG. These control flexor bursting (F phase of the step), the initiation of extensor bursts (the E1 phase, prior-to-foot-placement phase) and the maintenance of extensor bursts [the E2–3 stance (thrust) phase].

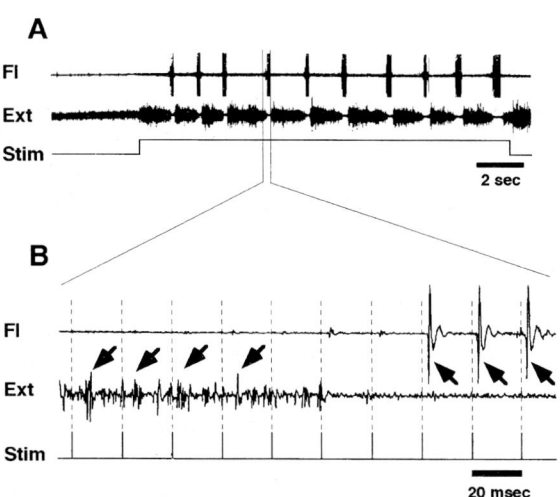

Fig. 2. Stimulus time-locked muscle neurogram discharges during fictive locomotion (A). Forelimb fictive locomotion evoked by 33 Hz stimulation of the spinal cord's lateral funiculus at the C2 level. Such stimulation (Stim) produces rhythmic alternating discharges in nerves supplying flexor (Fl) and extensor (Ext) muscles (B). Expanded timescale records of a transition phase from an extensor to a flexor burst of neurographic discharge. Note that MN action potential-like discharges (at arrows) were induced by each stimulus pulse (at the dotted vertical lines). (Reproduced, in part, from Kinoshita and Yamaguchi, 2001 with permission from Elsevier Science.)

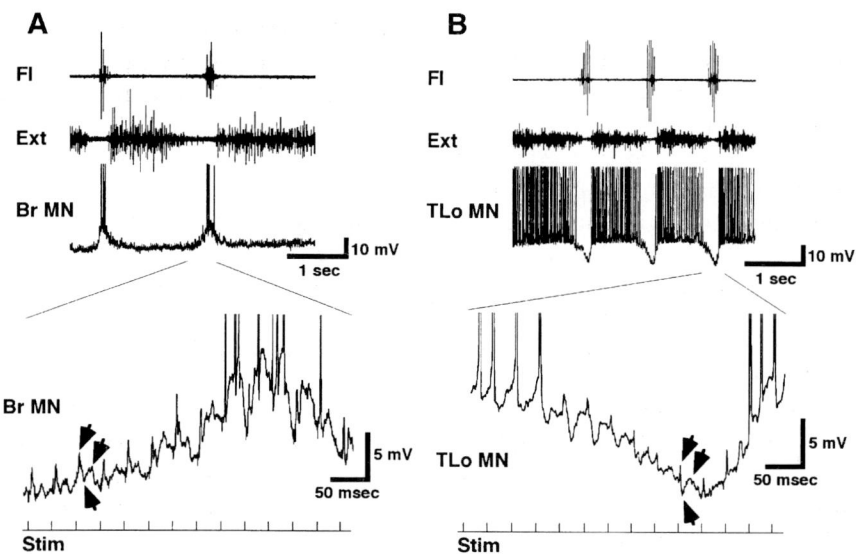

Fig. 3. Exemplary IC recordings made in MNs supplying flexor and extensor muscles during fictive locomotion (A–B). The upper two traces are as in Fig. 2A. The third trace shows IC recordings from MNs supplying the flexor, brachialis (Br) and extensor, triceps brachii, longus (TLo) muscle. The fourth trace shows a portion of the IC-recorded responses on an expanded timescale. The fifth (lowest) trace shows the timing of the stimulus (Stim) pulses. In the two latter traces, note the arrows, which show that each stimulus pulse was followed sequentially by early depolarization, hyperpolarization and late depolarization. (Reproduced, in part, from Kinoshita and Yamaguchi, 2001 with permission from Elsevier Science.)

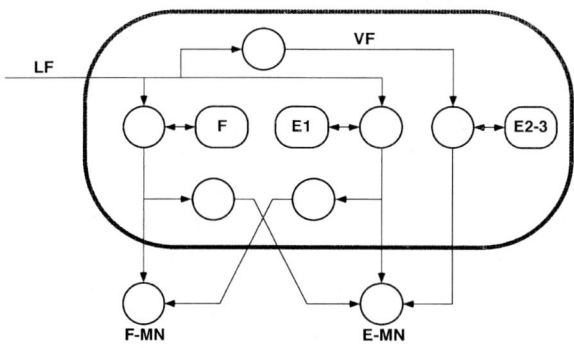

Fig. 4. A proposal on the neuronal mechanisms and pathways underlying the stimulus-time-locked responses of flexor and extensor MNs during fictive forelimb locomotion. F-MN, flexor motoneurons; E-MN, extensor motoneurons; open circles, excitatory INs; closed circles, inhibitory INs. The fictive step phases shown are flexion (F), prior-to-foot-placement (E1), and stance (E2). Other abbreviations: LF, lateral funiculus; VF, ventral funiculus. See the text for further details.

Further thoughts on the stimulus time-locked PSPs of MNs

The above-described repetitive electrical stimulation of the lateral funiculus could excite not only command pathways for locomotion, but other descending pathways, as well. For example, disynaptic excitation of MNs can be evoked by several other descending effects, and these responses could be modulated by the locomotor rhythm if the forelimb CPG is coactivated by the same stimulus. This type of possibility is shown in Fig. 5A.

One such 'nonlocomotor' descending pathway is the corticospinal tract, most of whose axons descend in the dorsolateral funiculus. In fact, stimulation of the medullary pyramid evokes disynaptic EPSPs in forelimb MNs (mainly to flexors), and these are clearly rhythmically modulated during forelimb fictive locomotion (Seki et al., 1997). This suggests that the corticospinal tract and the forelimb CPG share common last-order INs. Similarly, the presence of such interneuronal sharing has been explored for the cervical CPG and the vestibulospinal tract (Kitama et al., 1996); trisynaptic EPSPs evoked from vestibular afferents in forelimb extensor MNs are rhythmically modulated during fictive locomotion. There has been no demonstration to date, however, of these two descending pathways (and the CPG) having their own private last-order INs for connection to MNs. It is well

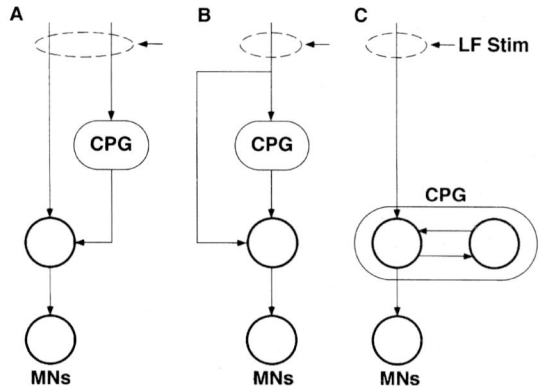

Fig. 5. Schema for the descending control of cervical MNs. The diagrams indicate possible ways that the stimulus-time-locked responses of cervical MNs are subjected to the dual influences of descending effects mediated by stimulation (Stim) of the lateral funiculus (LF) of the spinal cord and the forelimb locomotor CPG, with the latter circuitry also activated by the same descending effect. See the text for further details.

known, however, that at the lumbosacral level of the cat spinal cord, the locomotor hindlimb CPG and its descending and sensory inputs share common last-order INs, and that these come into play for the integration of locomotor and nonlocomotor movements (Edgley and Jankowska, 1987; Edgley et al., 1988; Shefchyk et al., 1990). A similar organization is known at the cervical level for interactions between the locomotor forelimb CPG and cutaneous reflex pathways (Seki and Yamaguchi, 1997).

A second, as yet untested explanation of the basis for stimulus time-locked EPSPs of MNs is that the locomotor command, itself, sends direct parallel excitation to the CPG and its last-order INs (Fig. 5B). Such a feed-forward connection would lower the threshold of the last-order INs to their subsequent rhythmic excitation by the CPG.

If the input INs of the CPG also function as its output INs, then their locomotor discharge must occur in bursts. Such bursts of action potentials can be generated either by network properties, including reverberation (Fig. 5C), intrinsic membrane properties, such as plateau-potential-based discharge (Hounsgaard et al., 1988), or their varying combinations. It is not yet known how these mechanisms relate to the stimulus time-locked EPSPs described earlier. If we assume that a reverberating effect is a feature of the locomotor forelimb CPG, however, we can then propose that the CPG includes three kinds of networks. One is for flexor bursting, with INs mediating stimulus time-locked disynaptic EPSPs of flexor INs. The other two must control the disynaptic and trisynaptic EPSPs of extensor INs, respectively. The former of these two EPSPs is facilitated mainly at the transition from the flexor to extensor phases in the step cycle, i.e., when a reverberating network *initiates* extensor bursting. In contrast, the trisynaptic EPSPs of extensor INs are rhythmically facilitated *throughout* the full extensor phase, this being controlled by a third reverberating network.

The locomotor rhythm and its spatio-temporal pattern of muscle activity could presumably be produced by synaptic interactions among the above three reverberating networks. Their respective inhibitory INs, which produce the trisynaptic IPSPs of flexor and extensor INs, appear to be excited by the axon collaterals of the INs of disynaptic pathways (Fig. 4). Thus, it will be valuable to test for the respective inhibitory INs also projecting axon collaterals to the antagonistic reverberating networks.

It is also necessary to consider how late stimulus time-locked EPSPs might relate to reverberating networks. For example, the late, polysynaptic EPSPs of flexor and extensor MNs are facilitated (increase in amplitude) during the flexor and extensor phases of the step cycle, respectively. If the INs producing the early stimulus time-locked MN EPSPs are components of reverberating networks, they may be reexcited by these networks, such that the late, polysynaptic MN EPSPs are produced by the same INs that produce the early MN EPSPs.

Concluding remarks

Locomotion can result from continuous, repetitive stimulation of various supraspinal structures (Shik et al., 1966, 1968; Orlovsky, 1969; Mori et al., 1977, 1998). This shows that the brain stem sends 'tonic' impulses to the locomotor CPGs of the forelimbs and hindlimbs. These convert the tonic input into a locomotor rhythm, which builds up within the CPG's circuitry. The output INs of each CPG distribute the locomotor rhythm to relevant MN pools. This distribution role may also be possessed by the CPG's input

INs, which receive the tonic command, but such a role has never been shown. For the moment then, it is parsimonious to propose that multiple interneuronal paths should be traceable from input to output INs within the CPG. If so, the identification of such INs and their pathways is critical for advancing understanding of locomotor CPGs. Recall, however, that such interneuronal paths become active (i.e., open) when the CPG, itself, is active. When such pathways are open, their input–output relations and transfer functions can be analyzed indirectly by studying MN responses to stimulus pulses delivered to supraspinal command centers and their descending pathways.

Generation of any central nervous system (CNS) rhythm driven by tonic excitation is a highly nonlinear process. Linear summation of the linear transfer functions of the interneuronal paths alone cannot model such a rhythm. Rather, nonlinear transfer functions and integration are also required. Nonlinear transfer functions and integration are ubiquitous within CPGs, because the single IN, itself, has many nonlinear properties, including its threshold for eliciting action potentials, persistent inward currents (plateau potentials) and so on. As a result, it is unlikely that any single nonlinear process within a CPG will ever be shown to be the predominant mechanism underlying the CPG's capacity to generate a rhythm.

In CPG circuitry consisting of many (indeed thousands of) INs, any tiny nonlinearity can result in large, macroscopic change of state of the CPG. This process is termed self-organization, which, in turn affects elementary processes. Rhythmic modulation of stimulus time-locked MN responses is such an elementary process. Therefore, instead of searching for nonlinear processes within CPG circuitry, it seems much more fruitful to advance our understanding of how the CNS modulates stimulus time-locked MN responses during fictive locomotion. It must be recognized that many questions and points have yet to be investigated, including the study of individual INs within the pathways that comprise the forelimb's locomotor CPG.

Acknowledgments

This work was supported by a Grant-in-Aid for Scientific Research (#10480223 to T.Y.), the Japanese Ministry of Education, Science, Sports and Culture.

Abbreviations

CLF	cervical lateral funiculus
Cn	nth cervical segment
CNS	central nervous system
CPG	central pattern generator
EPSP	excitatory postsynaptic potential
IC	intracellular
IN	interneuron
IPSP	inhibitory postsynaptic potential
MLR	midbrain locomotor region
MN	motoneuron
PSP	postsynaptic potential
Tn	nth thoracic segment

References

Amemiya, M. and Yamaguchi, T. (1984). Fictive locomotion of the forelimb evoked by stimulation of the mesencephalic locomotor region in the decerebrate cat. Neurosci. Lett., 50: 91–96.

Degtyarenko, A.M., Simon, E.S. and Burke, R.E. (1998). Locomotor modulation of disynaptic EPSPs from the mesencephalic locomotor region in cat motoneurons. J. Neurophysiol., 80: 3284–3296.

Edgley, S.A. and Jankowska, E. (1987). An interneuronal relay for group I and II muscle afferents in the midlumbar segments of the cat spinal cord. J. Physiol. (Lond.), 389: 647–674.

Edgley, S.A, Jankowska, E. and Shefchyk, S.J. (1988). Evidence that midlumbar neurones in reflex pathways from group II afferents are involved in locomotion in the cat. J. Physiol. (Lond.), 403: 57–71.

Gödderz, W., Illert, M. and Yamaguchi, T. (1990). Efferent pattern of fictive locomotion in the cat forelimb: with special reference to radial motor nuclei. Eur. J. Neurosci., 2: 663–671.

Graf, W., Keshner, E., Richmond, F.J., Shinoda, Y., Statler, K. and Uchino, Y. (1997). How to construct and move a cat's neck. J. Vest. Res., 7: 219–237.

Hishinuma, M. and Yamaguchi, T. (1990). Axonal projection of descending pathways responsible for eliciting forelimb stepping into the cat cervical spinal cord. Exp. Brain Res., 82: 597–605.

Hounsgaard, J., Hultborn, H., Jespersen, B. and Kiehn, O. (1988). Bistability of alpha-motoneurones in the decerebrate cat and in the acute spinal cat after intravenous 5-hydroxy-tryptophan. J. Physiol. (Lond.), 405: 345–367.

Ichikawa, Y., Terakado, Y. and Yamaguchi, T. (1991). Last-order interneurones controlling activity of elbow extensor motoneurones during forelimb fictive locomotion in the cat. Neurosci. Lett., 121: 37–39.

Jordan, L.M., Pratt, C.A. and Menzies, J.E. (1979). Locomotion evoked by brain stem stimulation: occurrence without phasic segmental afferent input. Brain Res., 177: 204–207.

Kinoshita, M. and Yamaguchi, T. (2001). Stimulus time-locked responses of motoneurons during forelimb fictive locomotion evoked by repetitive stimulation of the lateral funiculus. Brain Res., 904: 31–42.

Kitama, T., Seki, K. and Yamaguchi, T. (1996). Labyrinthine reflex activity during forelimb fictive locomotion in the decerebrate cat. Jpn. J. Physiol., Suppl. 46: S142.

Mori, S., Shik, M.L. and Yagodnitsyn, A.S. (1977). Role of pontine tegmentum for locomotor control in mesencephalic cat. J. Neurophysiol., 40: 284–295.

Mori, S., Matsui, T., Kuze, B., Asanome, M., Nakajima, K. and Matsuyama, K. (1998). Cerebellar-induced locomotion: reticulospinal control of spinal rhythm generating mechanism in cats. Ann. N.Y. Acad. Sci., 860: 94–105.

Orlovsky, G.N. (1969). Spontaneous and induced locomotion of the thalamic cat. Biofizika, 14: 1095–1102.

Seki, K. and Yamaguchi, T. (1997). Cutaneous reflex activity of the cat forelimb during fictive locomotion. Brain Res., 753: 56–62.

Seki, K., Kudo, N., Kolb, F. and Yamaguchi, T. (1997). Effects of pyramidal tract stimulation on forelimb flexor motoneurons during fictive locomotion in cats. Neurosci. Lett., 230: 195–198.

Shefchyk, S.J. and Jordan, L.M. (1985). Excitatory and inhibitory postsynaptic potentials in alpha-motoneurons produced during fictive locomotion by stimulation of the mesencephalic locomotor region. J. Neurophysiol., 53: 1345–1355.

Shefchyk, S.J., McCrea, D., Kriellaars, D., Fortier, P. and Jordan, L.M. (1990). Activity of interneurons within the L4 spinal segment of the cat during brainstem-evoked fictive locomotion. Exp. Brain Res., 80: 290–295.

Shik, M.L., Severin, F.V. and Orlovsky, G.N. (1966). Control of walking and running by means of electrical stimulation of the midbrain. Biophysics, 11: 659–666.

Shik, M.L., Orlovsky, G.N. and Severin, F.V. (1968). Locomotion of the mesencephalic cat evoked by pyramidal stimulation. Biofizika, 13: 127–135.

Steeves, J.D. and Jordan, L.M. (1980). Localization of a descending pathway in the SC which is necessary for controlled treadmill locomotion. Neurosci. Lett., 20: 283–288.

Terakado, Y. and Yamaguchi, T. (1990). Last-order interneurones controlling activity of elbow flexor motoneurones during forelimb fictive locomotion in the cat. Neurosci. Lett., 111: 292–296.

Vidal, P.P., Graf, W. and Berthoz, A. (1986). The orientation of the cervical vertebral column in unrestrained awake animals. I. Resting position. Exp. Brain Res., 61: 549–559.

Yamaguchi, T. (1986). Descending pathways eliciting forelimb stepping in the lateral funiculus: experimental studies with stimulation and lesion of the cervical cord in decerebrate cats. Brain Res., 379: 125–136.

Yamaguchi, T. (1992). Activity of cervical neurons during forelimb fictive locomotion in decerebrate cats. Jpn. J. Physiol., 42: 501–514.

CHAPTER 12

Generating the walking gait: role of sensory feedback

Keir G. Pearson*

Department of Physiology, University of Alberta, Edmonton, AB T6G 2H7, Canada

Abstract: In the walking system of the cat, feedback from muscle proprioceptors establishes the timing of major phase transitions in the motor pattern, contributes to the production of burst activity, generates some features of the motor pattern, and is required for the adaptive modification of the motor pattern in response to alterations in leg mechanics. How proprioceptive signals are integrated into central neuronal networks has not been fully established, largely due to the absence of detailed information on the functional characteristics of central networks in the presence of phasic afferent signals. Nevertheless, it appears likely that afferent signals reorganize the functioning of central networks, and the concept that the generation of the motor pattern can be explained by afferent modulation of a hard-wired central pattern generator may be too simplistic.

Introduction

The notion that central neuronal networks, usually referred to as 'central pattern generators', are primarily responsible for generating the motor patterns associated with rhythmic movements has dominated our thinking over the past 30 years. A common view today is that central pattern generators are functional units within the central nervous system, and that activity in these units is modulated by sensory feedback from peripheral receptors to control the frequency and amplitude of the centrally generated motor pattern (Grillner and Wallén, Chapter 1 of this volume). Although this view is certainly valid for systems in which the elimination of sensory feedback does little to change the normal motor pattern, e.g., lamprey swimming (Grillner et al., 1995), there is less certainty about its validity for those systems in which the motor pattern is substantially altered following deafferentation (Pearson, 1985, 1987). One such system is the walking system of the cat. The centrally generated motor patterns in motoneurons supplying hind leg muscles are enormously variable, and it has proven impossible to define precisely the characteristics of the central pattern generator for each hind leg. For example, the fictive motor pattern in spinal cats treated with L-dihydroxyphenylalanine (L-DOPA) and Nialamide usually consists simply of reciprocal bursts of activity in flexor and extensor motoneurons (Grillner and Zangger, 1979), whereas quite complex patterns can occur in hind leg muscles of decerebrate walking cats following deafferentation (Grillner and Zangger, 1984). Moreover, in immobilized decerebrate preparations, the fictive motor patterns generated either spontaneously or in response to stimulation of the mesencephalic locomotor region show considerable variation from animal to animal (personal observations; Burke et al., 2001), and in the same animal depending on conditions (Perret and Cabelguen, 1980).

In the absence of a well-defined centrally generated pattern, it is difficult to accept the idea that a central pattern generator is the basic functional element in the system generating the motor pattern for normal walking. But obviously there is circuitry in the spinal cord capable of generating burst activity in hind leg motoneurons and ensuring reciprocity of burst activity in certain groups of flexor and extensor motoneurons (see Burke et al., 2001 for a discussion

*Corresponding author: Tel.: +780-492-5628;
Fax: +780-492-8915; E-mail: keir.pearson@ualberta.ca

of the modularity of the central pattern-generating network). Thus, an alternative possibility is that the function of this central circuitry is restructured by afferent signals from leg receptors, and that the timing and intensity characteristics of the normal motor pattern are primarily established by phasic sensory feedback.

Over the past decade considerable evidence has accumulated supporting this alternative position. The purpose of this chapter is to review briefly this evidence. Additional information on the role of sensory feedback in patterning motor activity in rhythmic motor systems can be found in a number of previous reviews (Pearson, 1993, 1995a; Pearson and Ramirez, 1997; Bässler and Büschges, 1998; McCrea, 1998; Pearson et al., 1998; Burke, 1999).

Does sensory feedback establish specific features of the stepping motor program?

An important issue in the analysis of rhythmic motor systems is the extent to which phasic sensory signals are causal factors in the generation of specific features of the motor patterns. If some features are generated by phasic sensory signals, then it follows that the complete motor pattern cannot be simply described by sensory modulation of a central pattern generator. One of the clearest examples of this phenomenon is in the flight system of the locust in which the generation of burst activity in elevator motoneurons depends critically on phasic feedback from sensory receptors (the tegulae) activated by wing depression (Wolf and Pearson, 1988). Specifically, burst generation in elevator motoneurons depends to a large extent on reflex activation of motoneurons via interneurons that are not elements of the central pattern-generating network. A similar mechanism may be responsible for generating the burst of activity in the hind leg semitendinosus muscle that just precedes the onset of stance in trotting cats (Smith et al., 1993). This burst of activity functions to decelerate knee extension in late swing. The primary indication that this burst is generated by afferent feedback is that it is abolished by deafferentation of the hind leg in decerebrate walking cats (compare Figs. 4 and 6 in Grillner and Zangger, 1984). In addition, this burst has never been reported to occur during fictive locomotion in the absence of drugs, although a brief burst sometimes occurs near the end of the flexor activity following the administration of 4-aminopyridine in spinal preparations (see Discussion in Smith et al., 1993). A final piece of evidence for the causal role of a phasic sensory signal in the generation of this burst is that its magnitude is related to the rate of knee extension (Wisleder et al., 1990), thus suggesting that the sensory signal arises from stretch-sensitive receptors in the hamstring (knee flexor) muscles.

Sensory feedback regulates phase transitions during stepping

A common feature of many rhythmic motor systems is that the underlying motor pattern consists of a small number (often two) of distinct phases that are associated with distinct segments of the rhythmic movement (wing elevation and depression in flight, leg swing and stance in walking, and inspiration and expiration in respiration). Although the sequential transition of activity from one phase of the motor pattern to the next can occur in the absence of sensory feedback (demonstrated by the fact that these transitions occur in deafferented or paralyzed preparations), there is now a substantial body of evidence that under normal circumstances many of these transitions are triggered by phasic sensory signals. A simple example is the regulation of the transition from inspiration to expiration in the mammalian respiratory system by feedback from pulmonary stretch receptors (the Hering–Breuer reflex; Feldman, 1986). Another is the transition from wing depression to wing elevation in the locust produced by feedback from the tegulae (Wolf and Pearson, 1988; Pearson and Wolf, 1989). And a third is the transition from stance to swing in the walking systems of mammals and arthropods regulated in part from receptors signaling the loading of the legs (reviewed in Pearson, 1993, 1995a).

It is of some interest to note that the triggering of phase transitions by phasic afferent signals bears a close resemblance to the concept advanced by Charles Sherrington that the chaining of reflexes is responsible for the generation of rhythmic motor activity. Indeed Sherrington's proposal that the swing phase of stepping is initiated in part by sensory signals generated by hip extension (Sherrington, 1910) has been confirmed in a number of contemporary studies (Grillner and Rossignol, 1978; Kriellaars et al., 1994; Hiebert et al., 1996). One aspect of Sherrington's concept of chained reflexes that has proven incorrect, however, is that afferent feedback via reflex pathways is entirely responsible for the generation of the bursts of motoneuronal activity associated with the different phases of the motor pattern. In the majority of cases the generation of these bursts depends primarily on intrinsic properties of neurons and networks in the central nervous system. These intrinsic properties are reflected in the characteristics of centrally generated motor patterns. It now seems obvious why the generation of burst activity, and the switching from one phase of burst activity to the next, cannot rely entirely on phasic afferent feedback, namely, any significant alteration in afferent feedback (caused by an unexpected environmental event, for example) would severely disrupt, or may even stop, the generation of the motor pattern.

Why then are phase transitions normally triggered by phasic sensory signals in many rhythmic motor systems? The answer appears to be to ensure that the motor pattern is appropriately timed according to the biomechanical state of the moving body part. For example, in walking systems it would be maladaptive if swing-phase activity in a leg was initiated when the leg was carrying a significant amount of the body weight. Thus we find that a necessary condition for the swing phase to be initiated is an unloading of the leg (Pearson, 1995a). This occurs naturally during walking as weight is transferred to other legs near the end of the stance phase. Premature unloading of a leg results in an earlier onset of the swing phase (Gorassini et al., 1994) and additional loading prolongs the stance phase (Duysens and Pearson, 1980). Similarly, initiating wing elevation by phasic sensory feedback in the locust flight system limits wing depression to ensure effective aerodynamic force production (Wolf, 1993). The relatively recent finding that spinal stretch receptors in the lamprey are integral elements in the pattern-generating network for swimming (Grillner et al., 1995) also suggests that these receptors function to time the switching of activity from one side of the animal to the other according to an appropriate magnitude of bending of the body.

Sensory feedback contributes to flexor and extensor burst generation during stepping

Another important issue in understanding the generation of rhythmic motor patterns is the extent to which afferent feedback contributes to the generation of ongoing burst activity within a single phase of a rhythmic motor pattern. In other words, to what extent is the magnitude of burst activity dependent on sensory signals from peripheral receptors? Ever since the discovery that the alpha and gamma motoneurons are coactivated during muscle contractions in mammals (Granit and Kaada, 1952), it has been generally accepted that feedback from muscle spindles can reinforce ongoing motoneuronal activity via homonymous activation of alpha motoneurons by group Ia afferents. Strong evidence for this proposal has come from studies of voluntary digit and leg movements in humans (Vallbo et al., 1979; Macefield et al., 1993).

One system in which it is very clear that afferent feedback makes a large direct contribution to the generation of motoneuronal activity is the walking system of the cat. A number of recent studies have demonstrated that afferent feedback during the stance phase more than doubles the magnitude of burst activity in extensor muscles (Gorassini et al., 1994; Hiebert and Pearson, 1999; Stein et al., 2000). A particularly interesting aspect of the afferent contribution to extensor activation is that it is mediated largely via a positive feedback pathway from the force-sensitive Golgi tendon organs (Pearson and Collins, 1993; Gossard et al., 1994; McCrea et al., 1995). This positive feedback pathway functions to regulate automatically the level of motoneuronal activity according to the degree of loading of the extensor muscles. For example, any event that unexpectedly lengthens the extensor muscles initially increases the force in the muscles passively (due to the force-length characteristics of the muscles), then this force increase enhances motoneuronal activity via the positive feedback pathway from the Golgi tendon organs to help restore the muscle towards its original length. The open-loop gain of this positive-feedback pathway has not yet been firmly established, but preliminary studies have yielded values in the range of 0.2–0.5 (Pearson, unpublished). These values are insufficient to cause instability (since they are less than 1.0) but sufficient to contribute substantially to motoneuronal activation. For example, an open-loop gain of 0.5 would yield an afferent contribution to motoneuronal activation equal to the suprathreshold central drive to the motoneurons.

Afferent contribution to the activation of motoneurons via positive feedback pathways has also been found in the walking systems of the cockroach (Pearson, 1972), stick insect (Bässler, 1988), and crayfish and crabs (Sillar et al., 1986; Head and Bush, 1991). Feedback from the campaniform sensilla (cuticular stress detectors) in the legs of cockroaches and stick insects enhances activity in motoneurons extending the leg during stance. As in the cat, this type of positive feedback probably functions to regulate the level of motoneuronal activity according to the load carried by a leg. In addition, positive feedback from other classes of receptors in the legs of insects and crustaceans (chordotonal organs and nonspiking stretch receptors) appears to function to stabilize the velocity of muscle shortening at specific joints (reviewed in Pearson, 1995b).

The extent to which negative feedback from muscle-spindle afferents also contributes to the activation of extensor motoneurons during stance is uncertain. Evidence to date indicates that any spindle contribution is relatively minor. First, primary-spindle endings from the ankle extensors are only modestly activated during the early part of stance as the muscles lengthen (Prochazka et al., 1989), yet afferent signals provide a substantial input to ankle-extensor motoneurons throughout the stance phase (Gorassini et al., 1994; Hiebert and Pearson, 1999). And second, there is little correlation between the level of activity in ankle extensors and either muscle length or the rate of change of muscle length in normal cats walking under different conditions (Pearson, unpublished).

Adaptive modifications of the motor program for walking requires sensory feedback

The production of an accurate movement requires a precise matching of the activation pattern in multiple muscles with the mechanical properties of the muscles and other physical elements (including the properties of the external environment). Since these properties can change in unpredictable ways in the lifetime of an animal, adaptive mechanisms must exist to modify motor patterns to maintain the matching with physical properties of the body and the environment. A number of recent studies have demonstrated this type of adaptive plasticity in the walking system of cats following partial denervation of leg muscles (Carrier et al., 1997; Pearson et al., 1999; Pearson and Misiaszek, 2000; Gritsenko et al., 2001).

The mechanisms underlying the adaptive modifications of the stepping motor pattern in the cat remain largely unknown, although it is clear that a variety of processes may contribute. These include modification of the functioning of neuronal networks in the spinal cord (Carrier et al., 1997), alterations in descending regulation of spinal networks (Carrier et al., 1997), changes in the strength of centrally generated inputs to motoneurons, and modification of the influence of afferent signals on spinal networks (Fouad and Pearson, 1997; Pearson and Misiaszek, 2000). Whatever the mechanisms, it seems likely that the initiation of adaptive responses depends on the utilization of information from sensory receptors in the legs. A priori it is reasonable to expect that some information in sensory signals will be related to motor error, and that this information could be used to induce adaptive modification of the motor pattern. Indeed, consistent with this position is the finding that immobilization of a hind leg of the cat prevents adaptive changes in the motor pattern of an ankle-extensor muscle (medial gastrocnemius) following denervation of its synergists (Pearson et al., 1999; Pearson and Misiaszek, 2000). Adaptive modification of the walking system in spinal cats also depends on signals from leg receptors. Recovery of walking after spinalization depends on use (De Leon et al., 1998) and signals from cutaneous receptors (Rossignol, 2000). Finally, it is likely that alteration in walking behavior in humans after a prolonged period of stepping on a circular treadmill (Gordon et al., 1995) depends on persistent alterations in afferent feedback from leg receptors. If we accept that altered feedback from peripheral receptors is at least partially responsible for initiating adaptive modification of motor programs for walking, then the challenge is to establish the identity of the sensory receptors involved and the mechanisms by which their altered activity changes the functioning of neuronal networks in the central nervous system.

Conclusions

This chapter began by raising the issue of whether the generation of the motor pattern for walking in the cat can be explained by modulation of a central pattern generator by feedback from sensory receptors in the legs. A review of recent findings indicates that this concept may be too simplistic, and that the role of afferent feedback in the generation of the walking motor pattern may be more complex than generally acknowledged over the past two decades. To date we have been unable to precisely identify the temporal characteristics of a central pattern generator for stepping in a leg because a variety of patterns can be generated depending on the state of the preparation, the type of preparation, and the method used to evoke rhythmic activity. Thus, a neural network capable of generating a stereotypic motor pattern resembling the normal motor pattern for walking does not appear to exist. What clearly does exist, however, is shown schematically in Fig. 1.

Mechanisms are available for generating bursts of activity in motoneurons, and for ensuring reciprocity of bursting between specific groups of motoneurons (ankle flexors and extensor, for example). It seems that these elementary central mechanisms are integrated into a complex network, the functioning of which is orchestrated by phasic signals from sensory receptors in the legs. Some of the main findings in support of this position are that the occurrence of specific features of the motor pattern are abolished by removal of afferent input, afferent signals establish the timing of major phase transitions in the motor pattern, afferent feedback contributes to the generation of burst activity in many groups of motoneurons, and afferent information is required for the adaptive modification of the walking motor pattern. Even so, these recent finding do not provide a satisfying conceptual understanding of the mechanisms for motor pattern generation for walking. This will only come when we have much more knowledge of the organization and functional properties of neuronal networks in the spinal cord, and how signals from descending and sensory elements influence the functioning of these networks.

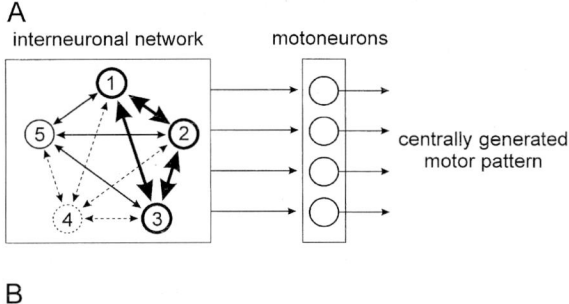

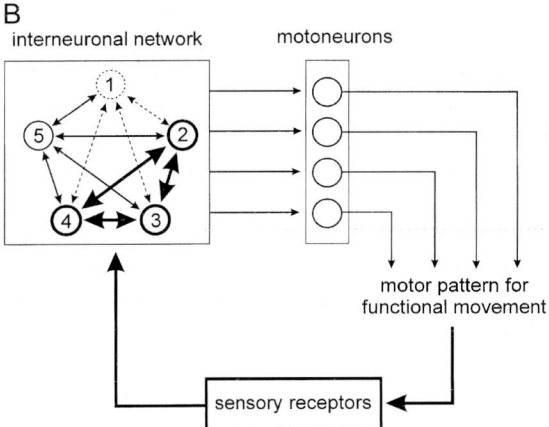

Fig. 1. Schematic diagram illustrating reorganization of a central interneural network by sensory feedback. (A) In the absence of sensory feedback the interneuronal groups 1, 2 and 3 are shown to be primarily responsible for generating a fictive motor pattern (line thickness represents the level of activity and strength of synaptic interaction). Note that in some motor systems (e.g., the walking system of the cat) the functioning of the interneuronal network is variable depending on the preparation and method of activation. This variability is reduced in behaving animals by sensory feedback orchestrating the functioning of the network to generate a motor pattern appropriate for the biomechanical state of the body and limbs. (B) The functional reorganization of the interneuronal network by afferent signals is illustrated by a change in the interneuronal groups primarily involved in motor pattern generation (from groups 1, 2 and 3 to groups 2, 3 and 4).

Acknowledgments

Thanks to Tania Lam for her valuable comments on a draft of this chapter. Support by a grant from the Canadian Institute of Health Research.

Abbreviation

L-DOPA L-dihydroxyphenylalanine

References

Bässler, U. (1988). Functional principles of pattern generation for walking movements of stick insect forelegs: the role of the femoral chordotonal organ afferences. J. Exp. Biol., 136: 125–147.

Bässler, U. and Büschges, A. (1998). Pattern generation for stick insect movements—multisensory control of a locomotor program. Brain Res. Rev., 27: 65–88.

Burke, R.E. (1999). The use of state-dependent modulation of spinal reflexes as a tool to investigate the organization of spinal interneurons. Exp. Brain Res., 128: 263–277.

Burke, R.E., Degtyarenko, A.M. and Simon, E.S. (2001). Patterns of locomotor drive to motoneurons and last-order interneurons: clues to the structure of the CPG. J. Neurophysiol., 86: 447–462.

Carrier, L., Brustein, E. and Rossignol, S. (1997). Locomotion of the hindlimbs after neurectomy of ankle flexors in intact and spinal cats: model for the study of locomotor plasticity. J. Neurophysiol., 77: 1979–1993.

De Leon, R.D., Hodgson, J.A., Roy, R.R. and Edgerton, V.R. (1998). Locomotor capacity attributable to step training versus spontaneous recovery after spinalization in adult cats. J. Neurophysiol., 79: 1329–1340.

Duysens, J.D. and Pearson, K.G. (1980). Inhibition of flexor burst generation by loading ankle extensor muscles in walking cats. Brain Res., 187: 321–332.

Feldman, J.L. (1986). Neurophysiology of breathing in mammals. In: Bloom F.E. (Ed.), Handbook of Physiology. The Nervous System. Intrinsic Regulatory System in the Brain, Vol. IV, American Physiological Society, Bethesda, pp. 463–524, Sec. 1.

Fouad, K. and Pearson, K.G. (1997). Modification of group I field potentials in the intermediate nucleus of the cat spinal cord after chronic axotomy of an extensor nerve. Neurosci. Lett., 236: 9–12.

Gorassini, M., Prochazka, A., Hiebert, G.W. and Gauthier, M.J.A. (1994). Corrective responses to loss of ground support during walking. I. Intact cats. J. Neurophysiol., 71: 603–610.

Gordon, C.R., Fletcher, W.A., Melvill Jones, G. and Block, E.W. (1995). Adaptive plasticity in the control of locomotor trajectory. Exp. Brain Res., 102: 540–545.

Gossard, J.P., Brownstone, R.M., Barajon, I. and Hultborn, H. (1994). Transmission in a locomotor-related group Ib pathway from hindlimb extensor muscles in the cat. Exp. Brain Res., 98: 213–228.

Granit, R. and Kaada, B.R. (1952). Influence of stimulation of central nervous structures on muscle spindles in cat. Acta Physiol. Scand., 27: 130–160.

Grillner, S. and Rossignol, S. (1978). On the initiation of the swing phase of locomotion in chronic spinal cats. Brain Res., 146: 269–277.

Grillner, S. and Zangger, P. (1979). On the central generation of locomotion in the low spinal cat. Exp. Brain Res., 34: 241–261.

Grillner, S. and Zangger, P. (1984). The effect of dorsal root transection of the efferent motor pattern in the cat's hindlimb during locomotion. Acta Physiol. Scand., 120: 393–405.

Grillner, S., Deliagina, T., Ekeberg, O., El Manira, A., Hill, R.H., Lansner, A., Orlovsky, G.N. and Wallen, P. (1995). Neural networks that co-ordinate locomotion and body orientation in lamprey. Trends Neurosci., 18: 270–280.

Gritsenko, V., Mushahwar, V. and Prochazka, A. (2001). Adaptive changes in locomotor control after partial denervation of triceps surae muscles in the cat. J. Physiol. (Lond.), 533: 299–311.

Head, S.I. and Bush, B.M.H. (1991). Proprioceptive reflex interactions with central motor rhythms in the isolated thoracic ganglion of the shore crab. J. Comp. Physiol. [A], 168: 445–459.

Hiebert, G.W. and Pearson, K.G. (1999). The contribution of sensory feedback to the generation of extensor activity during walking in the decerebrate cat. J. Neurophysiol., 81: 758–770.

Hiebert, G.W., Whelan, P.J., Prochazka, A. and Pearson, K.G. (1996). Contributions of hindlimb flexor muscle afferents to the timing of phase transitions in the cat step cycle. J. Neurophysiol., 75: 1126–1137.

Kriellaars, D.J., Brownstone, R.M., Noga, B.R. and Jordan, L.M. (1994). Mechanical entrainment of fictive locomotion in the decerebrate cat. J. Neurophysiol., 71: 2074–2086.

Macefield, V.G., Gandevia, S.C., Bigland-Ritchie, B., Gorman, R.B. and Burke, D. (1993). The firing rates of human motoneurons voluntarily activated in the absence of muscle afferent feedback. J. Physiol. (Lond.), 471: 429–443.

McCrea, D.A. (1998). Neuronal basis of afferent-evoked enhancement of locomotor activity. Ann. N.Y. Acad. Sci., 860: 216–225.

McCrea, D.A., Shefchyk, S.J., Stephens, M.J. and Pearson, K.G. (1995). Disynaptic group I excitation of synergist ankle extensor motoneurones during fictive locomotion. J. Physiol. (Lond.), 487: 527–539.

Pearson, K.G. (1972). Central programming and reflex control of walking in the cockroach. J. Exp. Biol., 56: 321–330.

Pearson, K.G. (1985). Are there central pattern generators for walking and flight in insects?. In: Barnes W.J.P. and Gladden M. (Eds.), Feedback and Motor Control in Invertebrates and Vertebrates. Croom Helm, London, pp. 307–316.

Pearson, K.G. (1987). Central pattern generation: a concept under scrutiny. In: McLennan H., Ledsome J.R., McIntosh C.H.S. and Jones D.R. (Eds.), Advances in Physiological Research. Plenum Press, New York, pp. 167–185.

Pearson, K.G. (1993). Common principles of motor control in vertebrates and invertebrates. Annu. Rev. Neurosci., 16: 265–297.

Pearson, K.G. (1995). Proprioceptive regulation of locomotion. Curr. Opin. Neurobiol., 5: 786–791.

Pearson, K.G. (1995). Reflex reversal in the walking systems of mammals and arthropods. In: Ferrell W.R. and Proske U. (Eds.), Neural Control of Movement. Plenum Press, London, pp. 135–141.

Pearson, K.G. and Collins, D.F. (1993). Reversal of the influence of group Ib afferents from plantaris on activity in medial gastrocnemius muscle during locomotor activity. J. Neurophysiol., 70: 1009–1017.

Pearson, K.G. and Misiaszek, J.E. (2000). Use-dependent gain change in the reflex contribution to extensor activity in walking cats. Brain Res., 883: 131–134.

Pearson, K.G. and Ramirez, J.M. (1997). Sensory modulation of pattern-generating circuits. In: Stein P.S.G., Grillner S., Selverston A.I. and Stuart D.G. (Eds.), Neurons, Networks, and Motor Behavior. MIT Press, Cambridge, MA, pp. 225–236.

Pearson, K.G. and Wolf, H. (1989). Timing of forewing elevator activity during flight in the locust. J. Comp. Physiol. [A], 165: 217–227.

Pearson, K.G., Misiaszek, J.E. and Fouad, K. (1998). Enhancement and resetting of locomotor activity by muscle afferents. Ann. N.Y. Acad. Sci., 860: 203–215.

Pearson, K.G., Fouad, K. and Misiaszek, J.E. (1999). Adaptive changes in motor activity associated with functional recovery following muscle deneration in walking cats. J. Neurophysiol., 82: 370–381.

Perret, C. and Cabelguen, J.M. (1980). Main characteristics of the hindlimb locomotor cycle in the decorticate cat with special reference to bifunctional muscles. Brain Res., 187: 333–352.

Prochazka, A., Trend, P., Hulliger, M. and Vincent, S. (1989). Ensemble proprioceptive activity in the cat step cycle: towards a representative look-up chart. Prog. Brain Res., 80: 61–74.

Rossignol, S. (2000). Locomotion and its recovery after spinal injury. Curr. Opin. Neurobiol., 10: 708–716.

Sherrington, C.S. (1910). Flexor-reflex of the limb, crossed extension reflex and reflex stepping and standing (cat and dog). J. Physiol. (Lond.), 40: 28–121.

Sillar, K.T., Skorupski, P., Elson, R.C. and Bush, B.M.H. (1986). Two identified afferent neurones entrain a central locomotor rhythm generator. Nature, 323: 440–443.

Smith, J.L., Chung, S.H. and Zernicke, R.F. (1993). Gait-related motor patterns and hindlimb kinetics for the cat trot and gallop. Exp. Brain Res., 94: 308–322.

Stein, R.B., Misiaszek, J.E. and Pearson, K.G. (2000). Functional role of muscle reflexes for force generation in the decerebrate walking cat. J. Physiol. (Lond.), 525: 781–791.

Vallbo, A.B., Hagbarth, K.E., Torebjork, H.E. and Wallin, B.G. (1979). Somatosensory, proprioceptive, and sympathetic activity in human peripheral nerve. Physiol. Rev., 59: 919–957.

Wisleder, D., Zernicke, R.F. and Smith, J.L. (1990). Speed-related effects on intersegmental dynamics during the swing phase of cat locomotion. Exp. Brain Res., 79: 651–660.

Wolf, H. (1993). The locust tegula: significance for flight rhythm generation, wing movement control and aerodynamic force production. J. Exp. Biol., 182: 229–253.

Wolf, H. and Pearson, K.G. (1988). Proprioceptive input patterns elevator activity in the locust flight system. J. Neurophysiol., 59: 1831–1853.

SECTION IV

Spinal cord and brainstem: adaptive mechanisms

CHAPTER 13

Cellular transplants: steps toward restoration of function in spinal injured animals

Marion Murray*

Department of Neurobiology and Anatomy, Drexel University College of Medicine, Philadelphia, PA 19129, USA

Abstract: Severe spinal cord injury results in severe, persisting deficits with little hope for substantial recovery. Recent developments in transplantation protocols, gene therapy, and methods of evaluation now offer hope of developing treatments that will lead to better prognoses. This review discusses the consequences of spinal injury, animal models used to study injury and recovery, types of cellular transplants, selection of behavioral and physiological tests of recovery, and ways to test the efficacy of the interventions and to improve transplant-mediated recovery.

Spinal cord injury—what are the problems, what are the solutions?

Until recently there was little hope for recovery after severe spinal cord injury (SCI). Results from research in a variety of fields have now converged, however, to kindle new optimism in the search for treatments that will restore function. Cellular transplants into the site of the injury are particularly promising as a way of improving recovery of motor function. Transplantation protocols are therefore likely to be one component of a multifaceted treatment for SCI.

Failure of axonal regeneration and retrograde degeneration of axotomized neurons result in loss of function after SCI

The diminished ability of adult central neurons to regenerate cut axons and the retrograde degeneration and death of damaged neurons contribute to the devastating consequences of central nervous system (CNS) injury. The failure of axons to regenerate within the CNS is due to the intrinsic properties of the CNS neuron itself, with its limited capacity to regenerate, and to the CNS environment, which lacks sufficient trophic and tropic factors and contains inhibitors. Also, scarring and cyst formations at the lesion site provide additional impediments to axonal growth. Thus far, treatments that correct one or more of these conditions have promoted only a limited amount of regeneration. Nevertheless, it is important to remember that some degree of useful function can be restored with only a modest amount of regeneration (Grill et al., 1997; Miya et al., 1997; Liu et al., 1999a) or sparing (Blight and DeCrescito, 1987) of axons. The retrograde degeneration of axotomized neurons is at least partially an apoptotic process; strategies that interrupt the apoptotic pathway have been successful in rescuing axotomized neurons (Himes and Tessler, 2001). While we do not know how functional these rescued neurons are, they are at least potentially available to participate in neural circuits and to regenerate their axons. Finally, the CNS environment can be modified to make it more permissive to growth by neutralizing the inhibitory molecules or ameliorating physical barriers (Stichel et al., 1999; Moon et al., 2001; Schwab, 2001). The search for treatments that may restore function after CNS injury has been enormously energized by these findings.

*Corresponding author: Tel.: +215-991-8308; Fax: +215-843-9082; E-mail: murray@drexel.edu

Loss of function due to deafferentation

Central nervous system injury may also eliminate afferent input required for function. An example is the loss of descending input resulting from spinal transection that partially deafferents the interneurons that contribute to the central pattern generator. This network of neurons may remain largely intact after a transection, but in the absence of supraspinal input the loss of voluntary locomotor control renders the individual paraplegic even though hindlimb locomotor patterns can still be elicited by appropriate stimulation. Axons that regenerate to reinnervate these networks could restore function. Activation of the neural networks intrinsic to the spinal cord by pharmacological agents and locomotor training are alternative therapies that have improved locomotor performance in spinalized cats and rats (Robinson and Goldberger, 1986b; Kim et al., 1999, 2001a; Rossignol et al., 2001). Pharmacological agents can also act synergistically with transplants that promote regeneration to effect further improvement (Kim et al., 1999, 2001a).

Gain of function due to deafferentation

Denervation also has other effects. Partially deafferented neurons may show upregulation of receptors associated with the lost systems (Kim et al., 1999; Edgerton et al., 2001; Rossignol et al., 2001) and spared axons may also sprout to reinnervate partially denervated neurons (Bregman, 1987; Zhang et al., 1993; Krenz and Weaver, 1998). Both of these events can modify the functional circuitry of the denervated spinal cord.

Spinal cord injury is accompanied by exaggeration of spinally mediated functions which may reflect these changes in sensory input or efficacy. Spasticity and chronic pain are common and serious consequences of spinal injury. Spasticity is associated with a hyperactive stretch reflex. The flexor reflex evoked by noxious stimulation is also hyperactive in spinal adults (Weng and Schouenberg, 1996). Weight-supported stepping in paralyzed rats spinalized as adults can reliably be evoked by noxious stimulation produced by intense perineal stimulation or tail pinch. This suggests that the increased sensitivity to peripheral stimulation can engage the locomotor generator and evoke motor responses in animals spinalized as adults (Gimenez and Ribotta et al., 2000). Evidence of hyperalgesia after spinal hemisection has also been reported (Christensen et al., 1996). It may be ameliorated by transplants (Shumsky et al., 2003). These clinically important sequelae of spinal injury reflect functional alteration of central pathways through loss of inhibitory descending projections, modifications in sensory input, and changes in the sensitivity of target neurons to some transmitters.

Models for the study of recovery from SCI

Contusions or surgical lesions?

Contusion injuries are widely used experimentally because of their resemblance to many clinical injuries. These lesions destroy gray matter at the level of impact and varying amounts of white matter depending on the severity of the impact. In most cases, some white matter is spared and this model thus produces incomplete lesions. Surgical lesions (transections, hemisections, funiculus and tract lesions) allow greater precision in determining which pathways are destroyed. They are therefore more useful for analysis of regeneration and the deficits and recovery of function associated with specific pathways.

Different types of lesions produce not only different deficits but also different types of compensations. Sprouting from spared systems is likely to occur after incomplete lesions and differentiating sprouting from regeneration elicited by an intervention may be difficult. Incomplete lesions may also spare pathways that contribute to a function. The recovery that ensues due to compensatory responses by the spared pathway is likely to differ in details from the normal function and may require subtle measures for its evaluation (Goldberger et al., 1993). Regeneration can be more unequivocally demonstrated in a complete injury model (transection). The severity of the induced deficits limits the amount of residual function, however, and demonstration of recovery after complete lesions may also require subtle tests. Transplants placed in both contusion and surgical lesions have, however, elicited regeneration and been associated with some increased functional recovery (Reier et al., 1992; Itoh et al., 1996; Grill et al., 1997; Diener and Bregman, 1998b; Liu et al., 1999a; Chopp et al., 2000).

Why study neonates?

Much of the work on recovery of function has been done in animals injured as neonates. The most compelling reason to study neonatally operated rats is that they achieve better function than do adults with similar injuries (Robinson and Goldberger, 1986a; Leonard and Goldberger, 1987a; Bregman et al., 1993; Howland et al., 1995; Miya et al., 1997; Stackhouse et al., 1997; Diener and Bregman, 1998a,b; Giszter et al., 1998a; Edgerton et al., 2001). Neonatally injured rats studied as adults provide a model of how the spinal cord, deprived of descending input, can be controlled to enable, for example, weight-supported locomotion. An underlying hypothesis is that neonatal spinal cord contains the neural circuitry that is necessary but not sufficient for function to recover in adults. This is because some inhibitory systems develop postnatally (Robinson and Goldberger, 1986b) and greater compensatory anatomical reorganization occurs after neonatal injuries (Bregman, 1987; Leonard and Goldberger, 1987b; Thallmair et al., 1998). There is also a greater likelihood of the compensatory use of novel or alternative strategies (Leonard and Goldberger, 1987a,b; Diener and Bregman, 1998b; Giszter et al., 1998a; Z'Graggen et al., 2000). Neonatal spinal cord contains higher levels of neurotrophins than adult cord, which are further elevated by injury. This may contribute to greater anatomical reorganization in neonates (Nakamura and Bregman, 2001). Lesion-induced changes in spinal circuitry may, of course, result in qualitative as well as quantitative differences in rats operated as neonates versus adults. It is important to identify the differences and similarities. For example, identification of differences in function between rats operated as neonates versus adults will indicate whether different strategies are used. Also, differences in afferent input or descending monoaminergic innervation between neonates and adults will suggest anatomical pathways that support greater function.

Activation of spinal circuits by reinnervation (through regeneration or sprouting), receptor stimulation, activity-dependent changes, or more likely, a combination of these mechanisms may be required to restore useful function in adults. The data from neonatally operated animals may suggest strategies for improving function in adults. For example, rats receiving complete spinal transections, either as neonates or adults, show long-lasting upregulation of serotonergic receptors in lumbar spinal (Croul et al., 1996; Kim et al., 1999). Similarly, serotonergic agonists can improve hindlimb motor function in animals operated on both as neonates and adults (Kim et al., 1999 and unpublished results). Weight support in neonatal operates can be increased by administration of serotonergic agonists. Preliminary data from our laboratory indicate that rats spinalized as adults also show more extensive hindlimb movements after administration of serotonergic agonists.

Most of the transplant studies in neonates have used fetal tissue grafts. Greater function is attained in animals receiving fetal transplants as neonates versus adults (Kunkel-Bagden and Bregman, 1990; Bregman et al., 1993; Iwashita et al., 1994; Howland et al., 1995; Miya et al., 1997; Stackhouse et al., 1997). This is paralleled by the more extensive growth of host axons elicited by fetal tissue grafts in neonates versus adults (Bregman, 1987; Jakeman and Reier, 1991; Miya et al., 1997; Mori et al., 1997; Diener and Bregman, 1998a). Transplanted fetal cells can, however, extend axons into adult as well as neonatal host tissue (Feraboli-Lohnherr et al., 1997; Miya et al., 1997; Mori et al., 1997). Opportunities for developing novel relays therefore exist in animals receiving transplants, both as neonates and adults that are not present in lesioned animals without transplants.

How to test for recovery of function?

Adequate evaluation of the efficacy of interventions that are neuroprotective or might promote regeneration or other forms of plasticity after injury needs careful assessment to distinguish among spontaneous recovery, compensatory strategies and recovery dependent on the intervention. Testing function after a second lesion that removes the graft, or after antagonist experiments that eliminate the effect of the intervention, may be useful. This is not always conclusive, however, for making the requisite distinctions.

A range of behaviors should therefore be tested to evaluate the consequences of a spinal injury and efforts are being made to develop such sets of tests (Diener and Bregman, 1998a; Metz et al., 2000; Kim et al., 2001b). They should be reliably sensitive to the location and extent of the injury, the degree to which the intervention improves the integrity of the injured region and, because recovery of function is usually incomplete, the tests should also assess residual impairment. Challenging tests may be needed to reveal subtle improvements. Reducing stress during behavioral testing is crucial for acquiring reliable data. Food- and water-deprivation procedures or aversive motivation are required for some tests, but these manipulations may have profound effects on the functional outcome.

Test selection

The use of quite different tests, or a limited number of tests or different endpoints, makes it difficult to compare reports from different laboratories. A number of convenient sensorimotor tests have been developed that are not altered by repeated testing, require little training, aversive motivation or food deprivation and are applicable to established models of SCI. Behavioral methods may not reveal whether the repair process has enabled the CNS to use novel circuits created by the interventions or whether the animal has learned to substitute other strategies or use other pathways to achieve a goal. It is therefore important when possible to use quantitative methods that can provide some insight into physiological mechanisms that underlie recovery and/or compensation.

Spontaneous behaviors

BBB [Basso, Bresnahan and Beattie (1995; i.e., their rating scale for quality of overground locomotion)]. The most widely used test is the Locomotor Rating Scale or BBB test (Basso et al., 1995), which was devised specifically to evaluate hindlimb function after thoracic contusion injuries in rats. This open-field test rates spontaneous locomotor movements made by the animal according to a 21-point scale. A major and laudable goal in the development of this test was to permit different laboratories to compare levels of recovery from contusion lesions after different interventions. Its popularity has resulted in using the scale to test effects of other types of injuries (e.g., cervical spinal injuries) and in other species (e.g., mice) for which it was not originally designed. Efforts are underway to modify the scale to make it more appropriate as a measure of motor function following other lesions and in other species.

The cylinder test. This is an example of a test that evaluates the use of forelimbs in exploratory behavior. A unilateral lateral funiculus lesion, which shows only modest deficits on the BBB test, virtually abolishes the animal's spontaneous use of its affected limb in vertical exploration. A transplant of genetically modified fibroblasts placed within the lesion site, however, will partially restore the animal's use of that limb (Liu et al., 1999a). Interpretation of these results needs to take into account the fact that animals with impaired control of a limb will often neglect that limb. Function may then be revealed, however, by restraining the nonimpaired limb, which forces the animal to use the impaired limb (Schallert et al., 1997).

Triggered behaviors

These complex behaviors are initiated by a stimulus. Contact placing is an example of a triggered behavior, which is sensitive to lesions of the corticospinal tract. The tactile-placing response is elicited by low-intensity, cutaneous stimulation. It is important, but difficult, to differentiate tactile placing from proprioceptive placing, which latter action is elicited by joint displacement and mediated by spinal mechanisms. Fetal transplants into neonatal but not adult spinal cord lesions permit development of contact placing in rats (Bregman et al., 1993).

Trained behaviors can be more revealing

Trained behaviors tend to show less variability in execution and reveal more clearly the deficits, the recovery and the efficacy of the intervention. Thus, the training can bring out a behavior that otherwise would have appeared to have been lost. Animals will spontaneously locomote in an unconstrained environment, of course. If they are preoperatively trained to walk on a platform or treadmill, however, analysis of locomotion that persists after injury and interventions is greatly facilitated. Several characteristics of such constrained gait have been studied in lesioned and transplanted animals (Bregman et al., 1993; Miya et al., 1997; Giszter et al., 1998b; Kim et al., 1999). A locomotor task can be made more challenging by changing the substrate. Requiring the animal to walk on a narrow beam or horizontal rope challenges foot placement and balance. Walking on a grid tests sensorimotor integration since the animal uses sensory input to direct foot placement. Skilled reaching to a target is another complex motor function. Reaching can be accomplished, after several types of injury, using a variety of compensatory mechanisms, which rely upon spared systems. Nevertheless, careful analyses (Whishaw et al., 1998) suggest that certain aspects of the movement are impaired by specific lesions. These observations provide a baseline upon which to recognize recovery mediated by interventions (Li et al., 1997; Diener and Bregman, 1998a).

Sensory tests

In studies of recovery after SCI, motor functions have received more attention than sensory ones. This is partly because there is less evidence that commonly used interventions will support regenerative growth of long, ascending pathways. Nevertheless, sensory impairments are serious concomitants of spinal injury. In particular, central pain states are among the most devastating consequences and it is important to determine whether specific interventions contribute to, or ameliorate, allodynia or hyperalgesia (Christensen et al., 1996; Shumsky et al., 2003). Because most tests of sensory function rely upon motor responses, it may be difficult to distinguish between a change in sensitivity to a stimulus and hyperreflexia resulting from the lesion.

Forces and movements

Kinematic and kinetic analyses of motor function are particularly useful for indicating how certain movements or postures are achieved (Rossignol, 1996). Treadmill locomotion is a form of reflex locomotion that can be examined in spinal cats and in rats spinalized as neonates. Kinematic and kinetic (ground reaction forces) analysis of treadmill (or overground) locomotion can supplement other forms of gait analysis and provide quantitative descriptions about how movement takes place (Leonard and Goldberger, 1987a; De Leon et al., 1998; Giszter et al., 1998b; Rossignol, 2000). Such studies have provided insight into strategies used for recovery. These analyses tend to be quite labor intensive and/or demand highly specialized equipment, however, such as to limit their widespread application.

Can recovery be related to specific pathways?

It is tempting to try to relate specific behaviors to specific pathways. Most behaviorally important movements are mediated by cooperative interactions among a number of sensory and motor pathways, and thus a limited lesion may not abolish the movement.

Skilled reaching is impaired, but not abolished, by lesions to rubrospinal, corticospinal, tectospinal or reticulospinal pathways (Iwaniuk and Whishaw, 2000) because animals use a variety of compensatory movements to achieve the goal. The fact that an animal with damage to both rubrospinal and corticospinal tracts or to the dorsal columns can still obtain an object by reaching indicates the redundancy of mechanisms that underlie this important function (Whishaw et al., 1998; McKenna and Whishaw, 1999). Using successful grasping of an object as an endpoint is thus not necessarily a helpful measure. Aspects of this complex behavior may, however, be dependent on an intact corticospinal system (limb guidance) or an intact rubrospinal system (arpeggio movements) or intact dorsal columns (rotary limb motions).

How to test for the efficacy of local regeneration and sprouting?

Anatomical reorganization in response to injury and interventions, particularly in adults, tends to be spatially restricted (Goldberger et al., 1993). Long-distance regeneration has been achieved only by a minority of axons. Short-distance growth may contribute to the formation of novel circuits through development of relays. These relays may be sufficient to provide some degree of descending input to partially denervated structures below the level of the lesion. Most of the behavioral and physiological tests currently in use are not likely to evaluate directly these local modifications. An absence of behavioral recovery in the presence of demonstrable anatomical reorganization may reflect the insensitivity of the behavioral methods rather than the nonfunctionality of the reorganized circuitry. Sensitive physiological studies may be needed to determine the functional significance of short-distance axonal growth (Chen et al., 2001).

How to test for the efficacy of a transplant?

Transplants may have a quantitative effect, improving the performance on a test that, for example, operated control rats are also able to accomplish. It is important to test the role of the transplant as directly as possible. We have used a relesion strategy designed to eliminate the effect of axons that regenerate or sprout into the rostral portion of the transplant. The considerable reorganization of both cut and spared axonal systems, coupled with preservation of injured neurons and their continued contribution to spinal circuitry, may modify the intrinsic circuitry of the spinal cord in transplant recipients. This can occur in unexpected ways, thereby making it difficult to predict the effect of a second lesion. In addition, the second lesion inevitably enlarges the area of injury and may result in additional retrograde cell loss (Houle and Ye, 1999). Nevertheless, a loss of recovered function to a level comparable to that shown by operated controls, as in the Cylinder test (Liu et al., 1999a), will suggest that the recovered function requires the presence of these newly grown axons. Sprouting of intact axons, or development of novel pathways by regenerating axons elicited by growth factors provided by the graft, might not be affected by the second lesion. If some of these anatomical modifications persist after a lesion that eliminates input to the graft, recovered performance dependent on the persisting changes may be less affected by the lesion. This can occur even though the anatomical changes are induced by, and are thus dependent on, the presence of the acutely grafted tissue (Miya et al., 1997; Thallmair et al., 1998). It is also possible that the spinal cord will become dependent on such modifications. In this case, a second lesion will have more profound effects on function (Kim et al., 2001b).

What to transplant?

Various types of intraspinal cellular transplants have been shown to stimulate regeneration and sprouting, provide neuroprotection and thereby contribute to changes in neuronal circuitry. These effects are presumably due to factors secreted by the transplanted cells, which are therefore acting as biological pumps. When placed in the site of the lesion, cellular transplants can prevent cyst formation and may ameliorate scar formation (Houle and Reier, 1988). When using cells of neural origin, or some types of stem cell, there is also a possibility of replacement of lost neurons or glial cells. An appropriately designed cellular transplant therefore has considerable potential for effecting repair and thereby promoting recovery. Both fetal tissue and cell lines from neural and nonneural sources have been used as donor tissue. Grafting of cell lines is likely to be more useful for promoting repair because they can be prepared in large numbers, stored, and may be amenable to genetic modification. Also, they are not subject to many of the political and ethical issues that surround the use of fetal tissue. A great many cell types are presently being explored as sources of transplant tissue for spinal cord repair. They include, for example, embryonic and adult neural stem cells, restricted precursor cells, Schwann cells, marrow stromal cells (MSCs), activated macrophages and olfactory cells. This chapter focuses on cells whose transplantation into the site of SCI has already been associated with some recovery of function. We can anticipate rapid progress, however, in the experimental evaluation of the properties of additional cellular candidates to arrive at a rational selection of the most efficacious donor cells.

Nonneural transplants

Nonneural cell lines are capable of filling the lesion cavity, thus preventing cyst formation, diminishing scar formation, and providing a permissive environment for regenerating axons. They, however, will not provide targets upon which such axons might synapse.

Genetically modified fibroblasts

Some of the most dramatic examples of enhancement of axon growth, including long-distance regeneration and recovery have been achieved with genetically modified fibroblasts. These cells are easily expanded and genetically modified. At least in principle, they permit autologous grafting. Dorsal roots and descending axons grow into grafts of fibroblasts that have been modified to express neurotrophic factors (Tuszynski et al., 1994; Grill et al., 1997; Liu et al., 1999). For example, fibroblasts genetically modified to synthesize neurotrophin-3 (NT-3) rescue axotomized Clarke's nucleus neurons. Similarly, those modified to synthesize brain-derived neurotrophic factor (BDNF) rescue red nucleus neurons that would otherwise undergo retrograde cell death (Himes and Tessler, 2001; Liu et al., 2002).

Transplants of fibroblasts engineered to express NT-3 into dorsal hemisection lesions were shown to promote regeneration of corticospinal axons and recovery on a grid-walking task (Grill et al., 1997). In another study, BDNF-expressing fibroblasts elicited regeneration of rubrospinal axons around, into and through a transplant placed in a cervical lateral fasciculus lesion in adult rats (Liu et al., 1999a). The graft contributed to recovery of forelimb and hindlimb motor function and the recovered function was abolished by a second lesion placed rostral to the transplant (Kim et al., 2001b; Liu et al., 2002).

Activated macrophages

A rationale for using activated macrophages from peripheral nerve is that these cells can phagocytose degenerating axons and myelin. Activated macrophages also secrete molecules, some of which may be beneficial to the injured spinal cord. These cells can be obtained in large quantities and they permit autologous transplantation. Rapalino et al. (1998) implanted activated macrophages into transected spinal cords of adult rats and reported regeneration of descending axons and recovery of motor function. The latter was assessed behaviorally by hindlimb movement and hindquarter weight support, and physiologically by cortically evoked hindlimb movement. Recovered function was abolished in retransection experiments, thereby suggesting that regeneration or sprouting contributed to the recovery.

Bone marrow stromal cells

The bone marrow contains hematopoietic stem cells and another population of stem cells, termed bone marrow stem cells that produce a variety of mesenchymal tissue, including bone, cartilage and fat. Human bone MSCs are readily obtained from patients, expandable in vitro and already in use in a variety of clinical settings. They are therefore attractive candidates for transplantation into SCI sites. These cells may contribute to a permissive environment when transplanted into such sites because of the complement of trophic factors, cytokines and other bioactive molecules that they secrete. Marrow stem cells have been shown to survive when transplanted into spinal contusion sites where they promote regeneration associated with recovery of function (Himes et al., 1999; Chopp et al., 2000). Greater recovery of hindlimb function, compared to lesion-only controls, was shown by the BBB test (Chopp et al., 2000). There is also evidence that MSCs may differentiate into neurons (Sanchez-Ramos et al., 2000; Deng et al., 2001) or glia (Kopen et al., 1999; Sasaki et al., 2001), thereby suggesting an additional benefit of their use in the injured spinal cord.

Nonneuronal transplants

Work from Aguayo's laboratory (David and Aguayo, 1981), demonstrating that peripheral nerve grafted into the CNS elicits regeneration from CNS neurons, has stimulated much of the recent progress and enthusiasm. Their work showed that the environment provided by the Schwann cells was permissive to axonal growth. Schwann cells, however, do not appear to integrate well within the CNS, and axons that have grown into peripheral nerve grafts do not readily reenter the host tissue (Xu et al., 1997). Grafts of other cells that support neurons, i.e., astrocytes, oligodendrocytes, ependymal cells or olfactory ensheathing cells (OECs), also have the potential to modify the immediate axonal environment and thus enhance regenerative potential. Of these, the most compelling results have come from grafting olfactory ensheathing glia.

OECs are CNS supporting cells that share some of the features of Schwann cells but are more readily integrated into the CNS. They have been reported to promote long-distance regeneration of serotonergic and corticospinal axons when added to guidance channels filled with Schwann cells and placed into a partial cervical SCI (Ramon-Cueto et al., 1998). In addition, they support recovery of motor function when placed into complete transection sites (Ramon-Cueto et al., 2000; Lu et al., 2001). In another model, OECs placed into a corticospinal tract lesion in cervical spinal cord can form continuous bridges across the lesion that supports regeneration of corticospinal axons. In cases where bridges are formed, the OEC recipients showed evidence of recovery of forelimb reaching (Li et al., 1997). In yet another model, transplantation of OECs into a dorsal column lesion cavity promoted both regeneration and myelination, and, in addition, recording studies revealed improved conduction velocities in the regenerating axons (Imaizumi et al., 2000).

Neuronal transplants

Neuronal transplants offer the possibility of replacement of lost neurons and the formation of novel relay pathways.

Fetal tissue transplants

Fetal spinal cord survives when transplanted into neonatal or adult spinal cord. It then integrates with the host tissue, and differentiates into neurons and glial cells. Axotomized neurons destined to undergo retrograde death are rescued by fetal tissue transplanted into both neonate and adult hosts (Himes and Tessler, 2001). In neonates, CNS axons will grow through fetal spinal cord transplants and into caudal host spinal cord where they contribute to the recovery of locomotor function (Howland et al., 1995; Miya et al., 1997; Diener and Bregman, 1998a,b). In adult hosts, regeneration of CNS axons derives almost exclusively from neurons close to the transplants. These axons do not extend into host spinal cord (Jakeman and Reier, 1991), nor is there measurable recovery of locomotor function (cf. however, Kawaguchi et al., Chapter 15 of this volume). Exogenous administration of neurotrophic factors, however, when combined with a fetal graft, increases growth of supraspinal axons into and through the transplant. It is associated with recovery of hindlimb motor function in the adult (Bregman et al., 1997).

Embryonic stem cells

McDonald et al. (1999) transplanted cells differentiated from embryonic mouse stem cells into a contusion lesion. These cells survived and expressed markers associated with oligodendroglia, astrocytes and, in small population, neurons. Behavioral measures with the BBB test indicated a rapid improvement in hindlimb function in stem-cell recipients. Regeneration was not examined in this study but since most of these stem cells differentiated into oligodendrocytes, the authors suggested that the beneficial results might have resulted from remyelination of injured axons.

Neural stem cells and precursors

Multipotential neural stem cells have been prepared from both embryonic (Kalyani et al., 1997) and adult spinal cord (Weiss et al., 1996), and shown to differentiate in vitro into a variety of neurons and glial cells. In addition, restricted precursors committed to a neuronal or glial lineage have been isolated from the developing spinal cord and characterized in terms of their phenotypes in culture (Rao, 1999). The challenge for developing cell replacement therapies for SCI is to find ways to control and direct the fate of the grafted stem cells in the environment of the nonneurogenic adult CNS. For example, transplantation of multipotential stem cells into the adult spinal cord resulted in their differentiation into glial cells but not neurons (Cao et al., 2001). Although this protocol can be utilized to induce remyelination (Akiyama et al., 2001), the potential for neuronal replacement seems limited. Strategies to overcome these difficulties include the differentiation of stem cells prior to grafting, transplantation of neuronal-restricted precursors (Han et al., 2002) or comparable cell lines (Vescovi and Snyder, 1999) and the use of transgenes to control the phenotypic fate of the grafted cells (Liu et al., 1999b).

How to improve transplant-mediated recovery?

Developing more effective transplant paradigms that further enhance regeneration, sprouting and neuroprotection can be expected to improve the extent and quality of recovery. Grafting of cells into spinal injury sites may not be sufficient to support the level of desired recovery. Additional approaches that depend upon different mechanisms, and that can thus act synergistically with cellular transplants, may be needed to permit greater repair and greater recovery.

Prevention of secondary damage

It is well known that the toxic environment that develops at the site of CNS injury will often enlarge the initial injury. This can account for much of the loss of function that develops. Preventing such secondary damage is thus likely to improve outcome. Methylprednisolone is widely used clinically to minimize secondary damage but it appears to be only modestly effective. Blocking apoptosis by providing growth factors (Shibayama et al., 1998; Liu et al., 2002) or Bcl-2 (Shibata et al., 2000), an antiapoptotic molecule, rescues neurons whose axons have been severed. This may also be useful in diminishing the amount of apoptosis locally. Development of additional methods to contain the expansion of a spinal lesion is of great current interest and we can anticipate that pharmacological agents will soon be discovered for this purpose (Sanner et al., 1994; Madsen et al., 1998; Yu et al., 2000; Yoles et al., 2001).

Pharmacological stimulation of denervated circuits

Studies of the circuitry of the central pattern generator for locomotion have identified many of the transmitters that are involved in production of alternating rhythmic movements of the hindlimbs. Agonists and antagonists of some of these transmitters have been used to stimulate movement in animals with complete and incomplete injuries (Barbeau et al., 1999; Edgerton et al., 2001; Rossignol et al., 2001). They have been shown to be even more effective in animals with transplants (Feraboli-Lohnherr et al., 1997; Kim et al., 1999). We can hope that the appropriate mix of pharmacological agents will provide adjunct therapy, which will act synergistically with transplantation protocols to enhance function.

Neuroprostheses

Direct electrophysiological activation of interneuronal networks can potentially provide coordinated control of many muscles (Barbeau et al., 1999). These approaches blend the work on CPGs with that on peripheral functional electrical stimulation with the goal of harnessing the spinal circuitry to produce useful function. These methods can be used to restore some kinds of centrally patterned movements.

Rehabilitative training

Certain kinds of rehabilitative therapies, including partial-weight-supported treadmill locomotion, constraint-induced therapy, and exposure to an enriched environment, have brought out remarkable improvement in function after CNS injuries (Schallert et al., 1997; Barbeau et al., 1999; Wernig et al., 1999; Edgerton et al., 2001). The role of activity in promoting anatomical reorganization is becoming increasingly apparent and some of the mechanisms are being elucidated (Schallert et al., 1997; Dupont-Versteegden et al., 2000; Carro et al., 2001; Chen et al., 2001; Gomez-Pinilla et al., 2001; Raineteau and Schwab, 2001). Development of additional rehabilitation training paradigms based on these studies will become part of a combined treatment for spinal injury.

Concluding thoughts

The seemingly insurmountable problems encountered in trying to repair the injured spinal cord are being reduced to practical issues. It is clear that cellular transplantation into spinal cord lesion sites can ameliorate some motor and sensory deficits arising from the injury. We do not yet have a good understanding, however, of how transplants work to restore function. We know that some injured neurons are rescued but we do not know whether the rescued neurons are functional. We know that some cut axons regenerate and that some spared axons sprout but we do not know the extent to which regenerating and sprouting axons mediate recovery. We need to learn which elements of the novel circuitry contribute to improved function so that we can design interventions that will lead to further improvements. Behavioral tests can identify some kinds of transplant-mediated recovery but it is difficult to assess adequately functional changes arising from modifications in local circuitry. There is therefore the danger that potentially useful interventions may be inadequately assessed functionally. Behavioral tests, however, do suggest directions for the potentially more powerful physiological tests, which may be required for future experiments designed to evaluate mechanisms of recovery.

The problems to be addressed by the next generation of experiments will include identifying the optimal parameters for interventions that will restore function, and developing the appropriate combinations of treatments that can target the deleterious events accompanying SCI.

Acknowledgments

We gratefully acknowledge support from NIH (NS24707), International Spinal Cord Trust, Eastern Paralyzed Veterans Association, Christopher Reeve Paralysis Foundation, and a Center of Excellence grant from MCP Hahnemann University. The contributions to our work made by our Japanese friends and collaborators, Drs. Futoshi Mori, Yasunobu Itoh, Taku Sugawara, Motohide Shibayama, Kosei Takahashi, and Masuta Shibata, are warmly acknowledged. Thanks also to Drs. Itzhak Fischer and Tim Schallert for helpful comments on the manuscript.

Abbreviations

BBB	Basso, Bresnahan and Beattie (1995; i.e., their rating scale for quality of overground locomotion)
BDNF	brain-derived neurotrophic factor
CNS	central nervous system
MSC	marrow stromal cells
NT-3	neurotrophin-3
OEC	olfactory ensheathing cells
SCI	spinal cord injury

References

Akiyama, Y., Honmou, O., Kato, T., Uede, T., Hashi, K. and Kocsis, J.D. (2001). Transplantation of clonal neural precursor cells deprived from adult human brain establishes functional peripheral myelin in the rat spinal cord. Exp. Neurol., 167: 27–39.

Barbeau, H., McCrea, D.A., O'Donovan, M.J., Rossignol, S., Grill, W.M. and Lemay, M.A. (1999). Tapping into spinal circuits to restore function. Brain Res. Rev., 30: 27–51.

Basso, D.M., Beattie, M.S. and Bresnahan, J.C. (1995). A sensitive and reliable locomotor rating scale for open field testing in rats. J. Neurotrauma, 12: 1–21.

Blight, A.R. and DeCrescito, V. (1987). Morphometric analysis of experimental spinal cord injury in the cat. Neuroscience, 19: 331–341.

Bregman, B.S. (1987). Development of serotonin immunoreactivity in the rat spinal cord and its plasticity after neonatal spinal cord lesions. Dev. Brain Res., 34: 245–263.

Bregman, B.S., Kunkel-Bagden, E., Reier, P.J., Dai, H.N., McAtee, M. and Gao, D. (1993). Recovery of function after spinal cord injury: mechanisms underlying transplant mediated recovery of function differ after spinal cord injury in newborn and adult rats. Exp. Neurol., 123: 3–16.

Bregman, B.S., McAtee, M., Dai, H.N. and Kuhn, P.L. (1997). Neurotrophic factors increase axonal growth after spinal cord injury and transplantation in the adult rat. Exp. Neurol., 148: 475–494.

Cao, Q.L., Zhang, Y.P., Howard, R.M., Walters, W.M., Tsoulfas, P. and Whittemore, S.R. (2001). Pluripotent stem cells engrafted into the normal or lesioned adult rat spinal

cord are restricted to a glial lineage. Exp. Neurol., 167: 48–58.
Carro, E., Trejo, J.L., Busiguina, S. and Torres-Aleman, I. (2001). Circulating insulin-like growth factor 1 mediates the protective effects of physical exercise against brain insults of different etiology and anatomy. J. Neurosci., 21: 5678–5684.
Chen, X., Feng-Chen, K.C., Chen, L., Stark, D.M. and Wolpaw, J.R. (2001). Short term and medium term effects of spinal cord tract transections on soleus H-reflex in freely moving rats. J. Neurotrauma, 18: 313–327.
Chopp, M., Zhang, X.H., Li, Y., Wang, L., Chen, J., Lu, D., Lu, M. and Rosenblum, M. (2000). Spinal cord injury in rat: treatment with bone marrow stromal cell transplantation. Neuroreport, 11: 3001–3005.
Christensen, M.D., Everhart, A.W., Pickleman, J.T. and Hulsebosch, C.E. (1996). Mechanical and thermal allodynia in chronic central pain following spinal cord injury. Pain, 68: 97–107.
Croul, S., Radzievsky, A. and Murray, M. (1996). Neurotransmitter receptor modulation caudal to complete spinal transection. Soc. Neurosci. Abstr., 22: 1844.
David, S. and Aguayo, A.J. (1981). Axonal elongation into peripheral nervous system 'bridges' after central nervous system injury in adult rats. Science, 214: 931–933.
De Leon, R.D., Hodgson, J.A., Roy, R.R. and Edgerton, V.R. (1998). Locomotor capacity attributable to step training versus spontaneous recovery after spinalization in adult cats. J. Neurophysiol., 79: 1329–1340.
Deng, W., Obrocka, M., Fischer, I. and Prockop, D.J. (2001). In vitro differentiation of human marrow stromal cells into early progenitors of neural cells by conditions that increase intracellular cyclic AMP. Biochem. Biophys. Res. Commun., 282: 148–152.
Diener, P.S. and Bregman, B.S. (1998a). Fetal spinal cord transplants support the development of target reaching and coordinated postural adjustments after neonatal cervical spinal cord injury. J. Neurosci., 18: 763–778.
Diener, P.S. and Bregman, B.S. (1998b). Fetal spinal cord transplants support growth of supraspinal and segmental projections after cervical spinal cord hemisection in neonatal rat. J. Neurosci., 18: 779–806.
Dupont-Versteegden, E.E., Murphy, R.J., Houle, J.D., Gurley, C.M. and Peterson, C.A. (2000). Mechanisms leading to restoration of muscle size with exercise and transplants after spinal cord injury. Am. J. Physiol. Cell. Physiol., 279: 1677–1684.
Edgerton, V.R., Leon, R.D., Harkema, S.J., Hodgson, J.A., London, N., Reinkensmeyer, D.J., Roy, R.R., Talmadge, R.J., Tillakaratne, N.J., Timoszyk, W., Tobin, A. (2001). Retraining the injured spinal cord. J. Physiol. (Lond.), 533: 15–21.
Feraboli-Lohnherr, D., Orsal, D., Yakovleff, A., Gimenez, M. and Privat, A. (1997). Recovery of locomotor activity in the adult chronic spinal rat after sublesional transplantation of embryonic nervous cells: specific role of serotonergic neurons. Exp. Brain Res., 113: 443–445.
Gimenez y Ribotta, M., Provencher, J., Feraboli-Lhonherr, D., Rossignol, S., Privat, A. and Orsal, D. (2000). Activation of locomotion in adult chronic spinal rats is achieved by transplantation of embryonic raphe cells reinnervating a precise lumbar level. J. Neurosci., 20: 5144–5152.
Giszter, S.F., Kargo, W., Shibayama, M. and Davies, M.R. (1998a). Fetal transplants placed into neonatal spinal transections in rats rescue axial muscle representations in adult motor cortex and improve recovery of locomotion. J. Neurophysiol., 80: 3021–3030.
Giszter, S., Graziani, V., Kargo, W., Hockensmith, G., Davies, M.R., Smeraski, C.S. and Murray, M. (1998b). Pattern generators and cortical maps in locomotion of spinal injured rats. Ann. N.Y. Acad. Sci., 860: 554–556.
Goldberger, M., Murray, M. and Tessler, A. (1993). Sprouting and regeneration in the spinal cord: their roles in recovery of function after spinal injury. In: Gorio A. (Ed.), Neuroregeneration. Raven Press, Ltd, New York, pp. 241–264.
Gomez-Pinilla, F., Ying, Z., Opazo, P., Roy, R.R. and Edgerton, V.R. (2001). Differential regulation by exercise of BDNF and NT-3 in rat spinal cord and skeletal muscle. Eur. J. Neurosci., 13: 1078–1084.
Grill, R., Murai, K., Blesch, A., Gage, F.H. and Tuszynski, M.H. (1997). Cellular delivery of neurotrophin-3 promotes corticospinal axonal growth and partial functional recovery after spinal cord injury. J. Neurosci., 17: 5560–5572.
Han, S.S.W., Kang, D.Y., Mujtanba, T., Rao, M.S. and Fischer, I. (2002). Grafted lineage-restricted precursors differentiate exclusively into neurons in the adult spinal cord. Exp. Neurol., 177: 360–375.
Himes, B.T. and Tessler, A. (2001). Neuroprotection from cell death following axotomy. In: Ingoglia N. and Murray M. (Eds.), Nerve Regeneration. Marcel Dekker, New York, pp. 477–503.
Himes, B.T., Chow, S.Y., Jin, H., Prockop, D.J., Tessler, A. and Fischer, I. (1999). Grafting human bone marrow stromal cells into injured spinal cord of adult rats. Soc. Neurosci. Abstr., 25: 213.
Houle, J.D. and Reier, P.J. (1988). Transplantation of fetal spinal cord tissue into the chronically injured adult rat spinal cord. J. Comp. Neurol., 269: 535–547.
Houle, J.D. and Ye, H. (1999). Survival of chronically injured neurons can be prolonged by treatment with neurotrophic factors. Neuroscience, 94: 929–936.
Howland, D.R., Bregman, B.S., Tessler, A. and Goldberger, M.E. (1995). Transplants enhance locomotion in neonatal kittens whose spinal cords are transected: a behavioral and anatomical study. Exp. Neurol., 135: 123–145.
Imaizumi, T., Lankford, K.L. and Kocsis, J.D. (2000). Transplantation of olfactory ensheathing cells or Schwann cells restores rapid and secure conduction across the transected spinal cord. Brain Res., 854: 70–78.

Itoh, Y., Waldeck, R.F., Tessler, A. and Pinter, M.J. (1996). Regenerated dorsal root fibers form functional synapses in embryonic spinal cord transplants. J. Neurophysiol., 76: 1236–1245.

Iwaniuk, A.N. and Whishaw, I.Q. (2000). On the origin of skilled forelimb movement. Trends Neurosci., 23: 372–376.

Iwashita, Y., Kawaguchi, S. and Murata, M. (1994). Restoration of function by replacement of spinal cord segments in the rat. Nature, 367: 167–170.

Jakeman, L.B. and Reier, P.J. (1991). Axonal projections between fetal spinal cord transplants and the adult rat spinal cord: a neuroanatomical tracing study of local interactions. J. Comp. Neurol., 307: 311–334.

Kalyani, A., Hobson, K. and Rao, M.S. (1997). Neuroepithelial stem cells from the embryonic spinal cord. Dev. Biol., 186: 202–223.

Kim, D., Adipudi, Y., Shibayama, M., Giszter, S., Tessler, A., Murray, M. and Simansky, K.J. (1999). Direct agonists for serotonin (5-HT2) receptors enhance locomotor function in rats that received neural transplants after neonatal spinal transection. J. Neurosci., 19: 6213–6224.

Kim, D., Murray, M. and Simansky, K.J. (2001a). The serotonergic 5-HT2C agonist m-chlorophenylpiperazine increases weight-supported locomotion without development of tolerance in rats with spinal transections. Exp. Neurol., 169: 496–500.

Kim, D., Schallert, T., Liu, Y., Browarak, T., Nayeri, N., Tessler, A., Fischer, I. and Murray, M. (2001b). Transplantation of genetically modified fibroblasts expressing BDNF in adult rats with a subtotal hemisection improve some specific motor and recovery function. Neurorehabil. Neural Repair, 15: 141–150.

Kopen, G.C., Prockop, D.J. and Phinney, D.G. (1999). Marrow stromal cells migrate throughout forebrain and cerebellum, and they differentiate into astrocytes after injection into neonatal mouse brains. Proc. Natl. Acad. Sci. USA, 96: 10711–10716.

Krenz, N.R. and Weaver, L.C. (1998). Sprouting of primary afferents after spinal cord transection in the rat. Neuroscience, 85: 443–458.

Kunkel-Bagden, E. and Bregman, B.S. (1990). Spinal cord transplants enhance the recovery of locomotor function after spinal cord injury at birth. Exp. Brain Res., 81: 25–34.

Leonard, C.T. and Goldberger, M.E. (1987a). Consequences of damage to the sensorimotor cortex in neonatal and adult cats. I. Sparing and recovery of function. Dev. Brain Res., 32: 1–14.

Leonard, C.T. and Goldberger, M.E. (1987b). Consequences of damage to the sensorimotor cortex in neonatal and adult cats. II. Maintenance of exuberant projections. Dev. Brain Res., 32: 15–30.

Li, Y., Field, P.M. and Raisman, G. (1997). Repair of adult rat corticospinal tract by transplants of olfactory ensheathing cells. Science, 277: 2000–2002.

Liu, Y., Kim, D., Himes, B.T., Chow, S.Y., Schallert, T., Murray, M., Tessler, A. and Fischer, I. (1999a). Transplants of fibroblasts genetically modified to express BDNF promote regeneration of adult rat rubrospinal axons and recovery of forelimb function. J. Neurosci., 19: 4370–4387.

Liu, Y., Himes, B.T., Solowska, J., Moul, J., Chow, S.Y., Park, K.I., Tessler, A., Murray, M., Snyder, E.Y. and Fischer I. (1999b). Instrapinal delivery of neurtrophin-3 using neural stem cells genetically modified by recombinant retrovirus. Exp. Neurol., 158(1): 9–26.

Liu, Y., Himes, B.T., Murray, M., Tessler, A. and Fischer, I. (2002). Grafts of BDNF-producing fibroblasts rescue axotomized rubrospinal neurons and prevent their atrophy. Exp. Neurol., 178: 150–164.

Lu, J., Feron, F., Ho, S.M., Mackay-Sim, A. and Waite, P.M.E. (2001). Transplantation of nasal olfactory tissue promotes partial recovery in paraplegic adult rats. Brain Res., 889: 344–357.

Madsen, J.R., MacDonald, P., Irwin, N., Goldberg, D.E., Yao, G.L., Meiri, K.F., Rimm, I.J., Stieg, P.E. and Benowitz, L.I. (1998). Tacrolimus (FK506) increases neuronal expression of GAP-43 and improves functional recovery after spinal cord injury in rats. Exp. Neurol., 154: 673–683.

McDonald, J.W., Liu, X.Z., Qu, Y., Liu, S., Mickey, S.K., Turetsky, D., Gottleib, D.I. and Choi, D.W. (1999). Transplanted embryonic stem cells survive, differentiate and promote recovery in injured rat spinal cord. Nat. Med., 5: 1410–1412.

McKenna, J.E. and Whishaw, I.Q. (1999). Complete compensation in skilled reaching success with associated impairments in limb synergies, after dorsal column lesion in the rat. J. Neurosci., 19: 1885–1894.

Metz, G.A., Merkler, D., Dietz, V., Schwab, M.E. and Fouad, K. (2000). Efficient testing of motor function in spinal cord injured rats. Brain Res., 883: 165–177.

Miya, D., Giszter, S., Mori, F., Adipudi, V., Tessler, A. and Murray, M. (1997). Fetal transplants alter the development of function after spinal cord transection in newborn rats. J. Neurosci., 17: 4856–4872.

Moon, L.D.F., Asher, R.A., Rhodes, K.E. and Fawcett, J.W. (2001). Regeneration of CNS axons back to their target following treatment of adult rat brain with chondroitinase ABC. Nat. Neurosci., 4: 465–466.

Mori, F., Himes, B.T., Kowada, M., Murray, M. and Tessler, A. (1997). Fetal spinal cord transplants rescue some axotomized rubrospinal neurons from retrograde cell death in adult rats. Exp. Neurol., 143: 45–60.

Nakamura, M. and Bregman, B.S. (2001). Differences in neurotrophic factor gene expression profiles between neonate and adult rat spinal cord after injury. Exp. Neurol., 169: 407–415.

Raineteau, O. and Schwab, M.E. (2001). Plasticity of motor systems after incomplete spinal cord injury. Nat. Rev., 2: 263–273.

Ramon-Cueto, A., Plant, G.W., Avila, J. and Bunge, M.B. (1998). Long distance axonal regeneration in the transected adult rat spinal cord is promoted by olfactory ensheathing glial transplants. J. Neurosci., 18: 3803–3815.

Ramon-Cueto, A., Cordero, M.I., Santos-Benito, F.F. and Avila, J. (2000). Functional recovery of paraplegic rats and motor axon regeneration in their spinal cords by olfactory ensheathing glia. Neuron, 25: 425–435.

Rao, M.S. (1999). Multipotent and restricted precursors in the central nervous system. Anat. Rec., 257: 137–148.

Rapalino, O., Lazarov-Spiegler, O., Agranov, E., Velan, G.J., Yoles, E., Fraidakis, M., Solomon, A., Gepstein, R., Katz, A., Belkin, M., Hadani, M., Schwartz, M. (1998). Implantation of stimulated homologous macrophages results in partial recovery of paraplegic rats. Nat. Med., 7: 814–821.

Reier, P.J., Anderson, D.K., Thompson, F.J. and Stokes, B.T. (1992). Neural tissue transplantation and CNS trauma. J. Neurotrauma, 9: 223–248.

Robinson, G.A. and Goldberger, M.E. (1986a). The development and recovery of motor function in spinal cats. I. The infant lesion effect. Exp. Brain Res., 62: 373–386.

Robinson, G.A. and Goldberger, M.E. (1986b). The development and recovery of motor function in spinal cats. II. Pharmacological enhancement of recovery. Exp. Brain Res., 62: 387–400.

Rossignol, S. (1996). Neural control of stereotypic limb movements. In: Rowell L.B. and Shepherd J.T. (Eds.), Handbook of Physiology, Exercise: Regulation and Integration of Multiple Systems. American Physiological Society, Bethesda, pp. 173–216 Sec. 12.

Rossignol, S. (2000). Locomotion and its recovery after spinal injury. Curr. Opin. Neurobiol., 10: 708–716.

Rossignol, S., Giroux, N., Chau, C., Marcoux, J., Brustein, E. and Reader, T.A. (2001). Pharmacological aids to locomotor training after spinal injury in the cat. J. Physiol. (Lond.), 533: 65–74.

Sanchez-Ramos, J., Song, S., Cardozo-Pelaez, F., Hazzi, C., Stedeford, T., Willing, A., Freeman, T.B., Saporta, S., Janssen, W., Patel, N., Cooper, D.R., Sanberg, P.R. (2000). Adult bone marrow stromal cells differentiate into neural cells in vitro. Exp. Neurol., 164: 247–256.

Sanner, C.A., Cunningham, T.J. and Goldberger, M. (1994). NMDA receptor blockade rescues Clarke's and red nucleus neurons after spinal hemisection. J. Neurosci., 14: 6472–6480.

Sasaki, M., Honmou, O., Akiyama, Y., Uede, T., Hashi, K. and Kocsis, J.D. (2001). Transplantation of an acutely isolated bone marrow fraction repairs demyelinated adult rat spinal cord axons. Glia, 35: 26–34.

Schallert, T., Kozlowski, D.A., Humm, J.L. and Cocke, R.R. (1997). Use-dependent structural events in recovery of function. Adv. Neurol., 73: 229–238.

Schwab, M.E. (2001). Inhibition of axonal growth by the myelin associated inhibitory proteins NI-35/250. In: Nogo A., Ingoglia N. and Murray M. (Eds.), Axonal Regeneration in the Central Nervous System. Marcel Dekker, New York, pp. 411–424.

Shibata, M., Murray, M., Tessler, A., Ljubetic, C., Connors, T. and Saavedra, R.A. (2000). Single injections of a DNA plasmid that contains the human Bcl-2 gene prevent loss and atrophy of distinct neuronal populations after spinal cord injury in adult rats. Neurorehabil. Neural Repair, 14: 319–330.

Shibayama, M., Hattori, S., Himes, B.T., Murray, M. and Tessler, A. (1998). Neurotrophin-3 prevents death of axotomized Clarke's nucleus neurons in adult rat. J. Comp. Neurol., 390: 102–111.

Shumsky, J.S., Tobias, C.A., Tumolo, M., Long, W.D., Giszter, S.F. and Murray, M. (2003). Delayed transplantation of fibroblasts genetically modified to secrete BDNF and NT-3 into a spinal cord injury site is associated with limited recovery of function. Exp. Neurol. In press.

Stackhouse, S., Shibayama, M., Bowes, M. and Murray, M. (1997). Fetal tissue transplants improve hindlimb function in adult spinal rats. Soc. Neurosci. Abstr., 27: 906.

Stichel, C.C., Hermanns, S., Luhmann, H.J., Lausberg, F., Niermann, H., D'Urso, D., Servos, G., Hartwig, H.G. and Muller, H.W. (1999). Inhibition of collagen IV deposition promotes regeneration of injured CNS axons. Eur. J. Neurosci., 11: 632–646.

Thallmair, M., Metz, G.A., Z'Graggen, W.J., Raineteau, O., Kartje, G.L. and Schwab, M.E. (1998). Neurite growth inhibitors restrict plasticity and functional recovery following corticospinal tract lesions. Nat. Neurosci., 1: 124–131.

Tuszynski, M.H., Peterson, D.A., Ray, Y., Baird, A., Nakahara, Y. and Gage, F.H. (1994). Fibroblasts genetically modified to produce nerve growth factor induce robust neuritic growth after grafting to the spinal cord. Exp. Neurol., 126: 1–14.

Vescovi, A.L. and Snyder, E.Y. (1999). Establishment and properties of neural stem cell clones: plasticity in vitro and in vivo. Brain Pathol.: 9569–9598.

Weidner, N., Ner, A., Salimi, N. and Tuszynski, M.H. (2001). Spontaneous corticospinal axonal plasticity and functional recovery after adult central nervous system injury. Proc. Natl. Acad. Sci. USA, 98: 3513–3518.

Weiss, A., Dunne, C., Hewson, J., Wohl, C., Wheatley, M., Peterson, A.C. and Reynolds, B.A. (1996). Multipotent CNS stem cells are present in the adult mammalian spinal cord and ventricular neuroaxis. J. Neurosci., 16: 7599–7609.

Weng, H.R. and Schouenberg, J. (1996). Nociceptive inhibition of withdrawal reflex response increases over time in spinalized rats. Neuroreport, 7: 1310–1314.

Wernig, A., Nanassy, A. and Muller, S. (1999). Laufband (treadmill) therapy in incomplete paraplegia and tetraplegia. J. Neurotrauma, 16: 719–726.

Whishaw, I.Q., Gorny, B. and Sarna, J. (1998). Paw and limb use in skilled spontaneous reaching after pyramidal tract, red nucleus and combined lesions in the rat. Behav. Brain Res., 92: 167–183.

Xu, X.M., Chen, A., Guenard, V., Kleitman, N. and Bunge, M.B. (1997). Bridging Schwann cell transplants promote axonal regeneration from both rostral-caudal stumps of transected adult rat spinal cord. J. Neurocytol., 26: 1–16.

Yoles, E., Hauben, E., Palgi, O., Agranov, E., Gothilf, A., Cohen, A., Kuchroo, V., Cohen, I.R., Weiner, H., Schwartz, M. (2001). Protective autoimmunity is a physiological response to CNS trauma. J. Neurosci., 21: 3740–3748.

Yu, C.G., Marcillo, A.E., Fairbanks, C.A., Wilcox, G.L. and Yezierski, R.P. (2000). Agmatine improves locomotor function and reduces tissue damage following spinal cord injury. Neuroreport, 11: 3203–3207.

Z'Graggen, W.J., Fouad, K., Raineteau, O., Metz, G.A.S., Schwab, M.E. and Kartje, L. (2000). Compensatory sprouting and impulse rerouting after unilateral pyramidal tract lesion in neonatal rats. J. Neurosci., 20: 6561–6569.

Zhang, B., Goldberger, M.E. and Murray, M. (1993). Proliferation of SP and 5HT containing terminals in Lamina II of the rat spinal cord following dorsal rhizotomy: quantitative EM immunocytochemical studies. Exp. Neurol., 123: 51–63.

CHAPTER 14

Neurotrophic effects on dorsal root regeneration into the spinal cord

Alan Tessler[1,2]*

[1]*Department of Veterans Affairs Hospital, Philadelphia, PA 19104, USA and* [2]*Department of Neurobiology and Anatomy, Drexel University College of Medicine, Philadelphia, PA 19129, USA*

Abstract: Dorsal root ganglion neurons exhibit a robust and generally successful regenerative response following injury of their peripheral processes. Regeneration fails, however, after section of their central processes in the dorsal roots or dorsal columns. Experiments characterizing the attenuated response of these neurons to injury, and the inhibition of regeneration exerted by astrocytes and oligodendrocytes within the dorsal root entry zone and spinal cord, have contributed important insights into the failure of regeneration after injury to the central nervous system (CNS). Interventions that have enhanced the metabolic response of injured dorsal root ganglion neurons, and altered the inhospitable environment, have increased sensory afferent regeneration and recovery. There is reason to expect that these strategies will help to develop clinically applicable treatments of CNS injuries.

Introduction

Axon regeneration fails after spinal cord injury because the intrinsic neuronal response is ineffective and because the CNS environment is not permissive for axon growth (see also Murray, this volume). The injured CNS environment does not support regeneration because it contains relatively low levels of growth factors and because inhibitory molecules are present in the glial scar formed at the injury site and in rostral and caudal spinal cord. The list of potentially inhibitory influences is expanding. They include components of myelin such as myelin-associated glycoprotein (MAG), oligodendrocyte-myelin glycoprotein (OMgp) and NI250/Nogo, sulfated proteoglycans produced by astrocytes, oligodendrocytes and oligodendrocyte precursor cells. There are also additional substances, whose short- and long-range chemorepellent influence is thought to guide growing axons in the developing nervous system. They include, for example, semaphorins, ephrins, netrins and slits (for review, see Founier and Strittmatter, 2001). Some of these molecules are likely to be present in gray matter (Messersmith et al., 1995).

Dorsal root ganglia provide advantages for studying regeneration

Several features of dorsal root ganglion (DRG) neurons, and their central projections in the dorsal roots and dorsal columns, make them an attractive system for clarifying our understanding of the intrinsic and extrinsic contributions to CNS regeneration and for testing treatments to enhance regeneration. CNS astrocytes are juxtaposed to peripheral nervous system (PNS) Schwann cells as the dorsal root passes through the transitional dorsal root entry zone (DREZ) to enter the spinal cord. This allows dorsal root growth to be directly contrasted in permissive and nonpermissive environments. The reaction of glial cells at the DREZ and within the spinal cord to dorsal root injury resembles that of

*Corresponding author: Tel.: +215-991-8310;
Fax: +215-843-9082; E-mail: atessler@drexel.edu

intraspinal glial cells after spinal cord trauma. These reactions are more easily studied, however, because they are uncontaminated by vascular injury and because DRG neurons do not die after section of their central processes. Regenerating primary afferent axons can be identified by morphological and physiological criteria and their projections are readily labeled and traced within the spinal cord. Because subsets of DRG neurons mediate specific sensory functions, regeneration can be correlated with physiological and behavioral recovery, and the response of specific populations of cells to injury and treatment can be compared. Finally, sensory neurons are unique because the regenerative response to injury of their central processes can be enhanced by a prior injury to their peripheral processes. This conditioning lesion effect allows the neuron's intrinsic response to axotomy to be modulated and assessed directly.

DRG neurons regenerate ineffectively after injury to their central processes

Although DRG neurons mount a robust and largely effective regenerative response to injury of their peripheral processes, they react much less vigorously to injury of their central processes. After dorsal root injury, for example, they show at most a small increase in their expression of the growth-associated protein, GAP-43 (Schreyer and Skene, 1993); the transcription factor, c-jun (Jenkins et al., 1993); the cytoskeletal protein, tubulin (Wong and Oblinger, 1990); or the cell adhesion molecule, CHL1 (Zhang et al., 2000). As a consequence of these attenuated metabolic changes, only 30% of DRG neurons extend axons beyond a crush site within the dorsal root (Andersen and Schreyer, 1999), and few regenerate to the end of a peripheral nerve graft placed in a cut dorsal root (Chong et al., 1996). Regeneration after dorsal column injury, where the environment includes astrocytes and oligodendrocytes rather than Schwann cells, is even less effective.

Astrocytes at the DREZ as well as spinal cord are not permissive for regenerating axons

If the dorsal roots of newborn rats are injured before the end of the first postnatal week, a small number of axons regenerate across the PNS–CNS interface. They reestablish a sparse projection within the dorsal horn but longer projections are not formed within the cord (Carlstedt et al., 1987). The limited extent of regeneration might be due to the poor capacity for growth of DRG neurons after a central injury, a lack of trophic support, or the presence of inhibitory influences even in neonatal spinal cord. Because regeneration into the cord is even more meager after this critical period, however, the failure highlights the role played by astrocytes, whose maturation within the DREZ coincides with the end of the critical period for regeneration. Dorsal rhizotomy induces microglia and oligodendrocyte precursor cells to proliferate within the DREZ (Liu et al., 1998). This causes astrocytes to hypertrophy and send processes into the dorsal root (Aldskogius and Kozlova, 1998). Some regenerating dorsal root axons establish synapse-like profiles with these astrocytes and stop elongating (Carlstedt, 1985a), whereas others turn and grow back down the dorsal root (Zhang et al., 2001). Even the axons of grafted human fetal DRG neurons cannot penetrate the DREZ in adult rats although they enter the cord along blood vessels and grow for a short distance before ramifying within the gray matter (Kozlova et al., 1994). This failure attests to the powerful inhibitory influence of the DREZ glia, because the processes of fetal human neurons grow for remarkable distances when they are grafted into the brain of adult rats (Wictorin et al., 1990).

The molecular basis for the stop and repulsion signals is unknown. Extracellular matrix molecules expressed by astrocytes such as the chondroitin sulfate proteoglycans (CSPGs) and tenascin-C may play a role (Pindzola et al., 1993), as may tenascin-R, which is produced by oligodendrocytes, and NG2, a CSPG expressed by oligodendrocyte precursor cells and associated with actively phagocytic macrophages (Zhang et al., 2001). The absence from this area of cell adhesion and growth-promoting molecules, such as L1 and CHL1, may also contribute to the failure of dorsal root regeneration (Zhang et al., 2001).

The degeneration of sensory root axons and terminals that follows dorsal rhizotomy rapidly establishes an environment within the spinal cord that is inhospitable to axon growth. Astrocytes proliferate and hypertrophy in the gray and white matter, and a densely packed scar forms in the dorsal funiculus (Liu et al., 2000). The onset of degeneration within the white matter during the second week after dorsal rhizotomy exposes inhibitory molecules associated with myelin, including MAG and NI250/Nogo. Their presence correlates temporally with the inability of dorsal root axons to regenerate within the white matter, even when they have been stimulated by the intrathecal administration of NT-3 to cross the DREZ (Ramer et al., 2001). Clearance of myelin debris is much more inefficient and prolonged after dorsal root or CNS injury than after peripheral nerve injury (George and Griffin, 1994). The relative contribution of individual molecules to the regenerative failure is unknown. When ascending DRG axons are cut in the dorsal columns, however, their regeneration appears to be relatively insensitive to the inhibitory effects of NI250/Nogo (Oudega et al., 2000). This is because regeneration does not increase following application of blocking antibodies, which promote regrowth of adult corticospinal axons. Regenerative failure of sensory afferents ascending in the dorsal columns may be more closely related to the expression of other classes of inhibitory factors, including Semaphorin 3A (Pasterkamp et al., 2001). Although the precise role played by individual molecules in the failure of sensory axon regeneration remains unknown, a great deal of evidence emphasizes the importance of the nonpermissive, extraneuronal environment both at the DREZ and within the spinal cord.

Modifying the terrain improves dorsal root regeneration

Various strategies have been used to render the terrain more permissive to dorsal root regeneration. For example, grafts of fetal spinal cord test the capacity of DRG neurons to regenerate under optimal conditions. The donor tissue is taken at an age during normal development [embryonic day (E)14 in the rat] when dorsal roots are actively growing into the spinal cord. Levels of neurotrophic factors are high, the DREZ has not yet formed, myelin is not yet present, and astrocytes generally support rather than inhibit axon growth. Cut dorsal roots (Tessler et al., 1988), but not their collateral axons ascending in the dorsal columns (Dent et al., 1996), respond to these favorable conditions by regenerating into transplants of fetal spinal cord. They arborize extensively and establish synapses on donor neurons. Most of the synapses are axodendritic and morphologically similar to those formed by dorsal roots in the normal dorsal horn (Itoh and Tessler, 1990a). Intracellular and extracellular recording from transplant neurons in response to electrical stimulation of implanted dorsal roots suggests that at least some of these synapses are functional and have conventional properties (Houle et al., 1996; Itoh et al., 1996). Dorsal roots also regenerate into the normally inappropriate targets provided by transplants of embryonic brain, although the ingrowth is less extensive and synapses are less abundant than in embryonic spinal cord grafts (Itoh and Tessler, 1990b). Some of the regenerating axons grow through areas of spinal cord transplants that contain extracellular matrix molecules such as CSPG and tenascin (Sugawara et al., 1999a), which are thought to inhibit growth in the CNS (Fitch and Silver, 1999). In spite of the generally favorable conditions provided by embryonic CNS grafts, regenerating dorsal roots fill only a portion of the transplants and do not traverse them to enter host spinal cord. The explanation for the limited growth within fetal transplants is unknown, but inhibitory factors present in transplants and the ineffective neuronal regenerative response may contribute.

Limited intraspinal regeneration of dorsal root axons has also been observed after irradiation of the newborn spinal cord (Sims and Gilmore, 1994a,b) and in adult rats that received daily injections of the immunosuppressants cyclosporine A (CsA) or FK 506 (Sugawara et al., 1999b). The mechanism of action of immunosuppressants is unknown. Irradiation makes the adult spinal cord more conducive to regeneration, however, by reducing the number of oligodendrocytes and astrocytes and preventing the formation of a glial scar at the DREZ. Destruction of the glial limitans also allows Schwann cells to invade the spinal cord and establish an environment more like that found in the PNS. Although there is regeneration into the irradiated cord, such growth is circumscribed. This suggests that the remaining glial elements continue to exert a robust inhibitory influence and that the growth potential of DRG neurons after dorsal root section is inadequate.

Another approach for enhancing the permissiveness of the CNS terrain is to transplant glial cells that can support growth. Olfactory endothelial glia (OEG) share phenotypic features with both astrocytes and Schwann cells (Franklin and Barnett, 2000). When transplanted into spinal cord stumps rostral and caudal to a complete transection, they have been reported to restore locomotor function by promoting regeneration of corticospinal and bulbospinal axons across the transection, where they then extend for long distances in caudal host spinal cord (Ramon-Cueto et al., 2000). They have also been observed to remyelinate and improve conduction across demyelinated dorsal column axons (Imaizumi et al., 1998). When transplanted into the DREZ after dorsal root transection, OEG migrated into the ipsilateral and contralateral dorsal horn and became associated with regenerated dorsal roots (Ramon-Cueto and Nieto-Sampedro, 1994). The incomplete number of dorsal root axons that regenerated did not enter ventral horn or white matter. Rats that received OEG transplants into the DREZ following L3–L6 dorsal root transections were, however, far more likely than controls to recover H-reflex and withdrawal responses. The amplitude of both reflexes was reduced, however, as compared to normal animals (Navarro et al., 1999). Several modifications of the extraneuronal environment have therefore promoted dorsal root regeneration, but the extent of the growth has been restricted. This suggests that more robust regeneration will require enhancing the neuron's intrinsic response to injury.

Enhancing the regenerative response improves dorsal root regeneration

Successful regeneration depends not only on a supportive terrain but also on a robust intrinsic response by the injured neurons. DRG neurons provide an especially vivid example of the importance of intrinsic neuronal regenerative capacity. Studies performed over many years have established that a preceding, or conditioning lesion, of the peripheral process of sensory neurons can enhance regeneration after a second, or test lesion of the central process either in the dorsal root or in the dorsal columns.

Richardson and Issa (1984), for example, reported that a simultaneous or preceding sciatic nerve injury increased by more than 100-fold the number of lumbar DRG neurons whose axons regenerated into a peripheral nerve graft placed in the dorsal columns at the cervical level. Later studies have demonstrated that transecting the sciatic nerve 1–2 weeks before cutting the dorsal columns at the midthoracic level stimulated sensory axons to grow into the gray matter at the lesion site and up to 4 mm beyond the rostral end of the injury (Neumann and Woolf, 1999). Sciatic nerve crush has also enabled a small number of cut dorsal root axons to regenerate for at least 2 mm longitudinally within the dorsal horn, although not within the degenerated dorsal columns (Chong et al., 1999). Much of the conditioning lesion effect appears to be due to elevations in intraneuronal cAMP (Neumann et al., 2002; Qiu et al., 2002), with activation of downstream pathways involving polyamines (Cai et al., 2002). Injecting analogs of cAMP into lumbar DRG promotes sensory axon regeneration within the transected dorsal columns that is similar in extent to that seen after a conditioning lesion (Neumann et al., 2002; Qiu et al., 2002). Regeneration also appears to require the expression of at least two growth-associated proteins, GAP-43 and CAP-23 (Bomze et al., 2001). Transgenic mice that expressed both of these proteins, but not strains that expressed either alone, exhibited a 60-fold increase in the number of DRG neurons that sent axons into a peripheral nerve graft placed into the dorsal columns at the thoraco-cervical junction. These studies have emphasized the importance of enhancing the intrinsic neuronal response to injury. They have also led to treatments that might be applied to clinical injuries.

Most of the earlier approaches consist of strategies for insuring the long-term administration of supplemental neurotrophic factors to the cut dorsal roots or dorsal columns. The paucity of growth factors has been postulated to contribute to regenerative failure within the CNS since the time of Cajal. Increased exposure to neurotrophic factors may be at least partially responsible for the conditioning lesion effect. Exogenous growth factors have been shown to promote neurite outgrowth from adult sensory neurons in culture (Kimpinski and Mearow, 2001). Also, NGF infusion has been reported to stimulate sensory axons cut within the dorsal columns to regenerate across peripheral nerve grafts and into the rostral host dorsal columns (Oudega and Hagg, 1996).

Several studies have examined whether diffusible factors derived from fetal spinal cord transplants (Itoh et al., 1999), exogenously supplied growth factors (Iwaya et al., 1999), or growth factors in combination with transplants (Houle and Johnson, 1989), promote dorsal root regeneration into spinal cord previously injured by removal of the dorsal horn. Cut dorsal roots are known not to regenerate into adult spinal cord even when the DREZ is bypassed and they are directly apposed to the injured cord (Carlstedt, 1985b). They regenerated more robustly when they were sandwiched against the injured cord by a fetal spinal cord transplant or by a fibrin glue ball containing NT-3, BDNF, CNTF or GDNF. Under these conditions, both large and small caliber dorsal root axons regenerated into the adjacent host spinal cord, formed dense plexuses with arborizations in the ventral horn, and established synapses on spinal cord neurons. Regenerated dorsal roots remained within a few millimeters of the insertion site, however, and long-distance growth through gray or white matter was not observed. NT-3 administered by injection of genetically modified adenovirus into the ventral horn also stimulated cut dorsal roots to regenerate into the adjacent dorsal horn and intermediate gray matter (Zhang et al., 1998).

Remarkable examples of dorsal root regeneration have been obtained by chronic intrathecal administration of neurotrophic factors (Ramer et al., 2000). In this study, all classes of DRG neurons regenerated axons across the DREZ and into the central portion of the dorsal root when stimulated by neurotrophic factors for which they expressed the specific, high-affinity receptor. GDNF was the most effective of the factors studied. It stimulated elongation of all phenotypic classes of DRG axons. Regenerated axons grew into the dorsal horn and were able to elicit stimulus-driven postsynaptic activity from dorsal horn neurons. Most importantly, animals receiving GDNF or NGF, which stimulated regeneration of small-caliber nociceptive axons, exhibited behavioral recovery as measured by tests of nociceptive function. Small-caliber, primary afferents that express calcitonin gene-related peptide (CGRP) regenerated even more robustly into the dorsal horn when adenovirus, genetically modified to express NGF or fibroblast growth factor-2 (FGF2), was injected into the DREZ 16 days after dorsal rhizotomy (Romero et al., 2001). Regenerated axons were located in normal and ectopic target areas within the immediate zone of transgene expression and they contributed to the recovery of thermal sensation.

Future directions

Current strategies have improved primary afferent axon regeneration by making the terrain more permissive and enhancing the metabolic response to injury. Relatively few sensory axons extend into the gray or white matter close to the injury site in the dorsal root or dorsal column, and long-distance projections have been rare. Similar efforts to promote long-distance regeneration have been more successful in other systems. For example, blocking antibodies to the myelin-associated inhibitory molecule NI250/Nogo (Schnell and Schwab, 1990) and applying exogenous NT-3 (Grill et al., 1997) have promoted regeneration of corticospinal axons with partial recovery of locomotor function. BDNF produced similar results when administered to cut rubrospinal axons (Liu et al., 1999). Why primary sensory axons are more refractory to these interventions is still unknown. Several treatments that have stimulated regeneration of CNS axons have not yet been applied to DRG axons. Rather than utilizing individual physiologically important molecules, these strategies aim to combine the effects of a large number of molecules. David and colleagues, for example, reasoned that many molecules associated with myelin are likely to block regeneration and that only some of them have been identified. Therefore, they immunized mice against myelin homogenates and produced more extensive regeneration of corticospinal axons than hitherto possible when using other methods (Huang et al., 1999). This approach toward improving the permissivity of the terrain could be made even more inclusive by expanding the number of targeted molecules to include components of the glial scar. An analogous strategy for increasing the neuronal response to injury aims to combine the power of multiple neurotrophic factors by modulating intracellular downstream-signaling pathways common to more than one factor (Lehmann et al., 1999). Broadly targeted approaches to both the neuronal response and the environment could also be combined. Because recovery, not only after dorsal root avulsion, but also after spinal cord injury, will require more complete regeneration of injured sensory axons than hitherto possible, the search for more effective treatments will continue.

Acknowledgments

Research in the author's laboratory is supported by the Research Service of the Department of Veterans Affairs, NIH NS24707, The Eastern Paralyzed Veterans Association, and The International Spinal Research Trust.

Abbreviations

BDNF	brain-derived neurotrophic factor
CAP-23	a growth-associated protein
CHL1	close homolog of L1*
CNS	central nervous system
CNTF	ciliary neurotrophic factor
DREZ	dorsal root entry zone
DRG	dorsal root ganglion
GAP-43	another growth-associated protein
GDNF	glial-cell line-derived neurotrophic factor
L1	a growth-promoting, cell-adhesion molecule
MAG	myelin-associated glycoprotein
NGF	nerve growth factor
NT-3	neurotrophin-3
OMgp	oligodendrocyte-myelin glycoprotein
PNS	peripheral nervous system

*Chemical composition not yet delineated

References

Aldskogius, H. and Kozlova, E.N. (1998). Central neuron-glial and glial-glial interactions following axon injury. Prog. Neurobiol., 55: 1–26.

Andersen, L.B. and Schreyer, D.J. (1999). Constitutive expression of GAP-43 correlates with rapid, but not slow regrowth of injured dorsal root axons in the adult rat. Exp. Neurol., 155: 157–164.

Bomze, H.M., Bulsara, K.R., Iskandar, B.J., Caroni, P. and Skene, J.H.P. (2001). Spinal axon regeneration evoked by replacing two growth cone proteins in adult neurons. Nat. Neurosci., 4: 38–43.

Cai, D., Deng, K., Mellado, W., Lee, J., Ratan, R.R. and Filbin, M.T. (2002). Arginase I and polyamines act downstream from cyclic AMP in overcoming inhibition of axonal growth MAG and myelin in vitro. Neuron, 35: 711–719.

Carlstedt, T. (1985a). Regenerating axons form nerve terminals at astrocytes. Brain Res., 347: 188–191.

Carlstedt, T. (1985b). Dorsal root innervation of spinal cord neurons after dorsal root implantation into the spinal cord of adult rats. Neurosci. Lett., 55: 343–348.

Carlstedt, T., Dalsgaard, C.J. and Molander, C. (1987). Regrowth of lesioned dorsal root nerve fiber into the spinal cord of neonatal rats. Neurosci. Lett., 74: 14–18.

Chong, M.S., Woolf, C.J., Turmaine, M., Emson, P.C. and Anderson, P.N. (1996). Intrinsic versus extrinsic factors in determining the regeneration of the central processes of rat dorsal root ganglion neurons: the influence of a peripheral nerve graft. J. Comp. Neurol., 370: 97–104.

Chong, M.S., Woolf, C.J., Haque, N.S.K. and Anderson, P.N. (1999). Axonal regeneration from injured dorsal roots into the spinal cord of adult rats. J. Comp. Neurol., 410: 42–54.

Dent, L.J., McCasland, J.S. and Stelzner, D.J. (1996). Attempts to facilitate dorsal column axonal regeneration in a neonatal spinal environment. J. Comp. Neurol., 372: 435–456.

Fitch, M.T. and Silver, J. (1999). Beyond the glial scar: cellular and molecular mechanisms by which glial cells contribute to CNS regenerative failure. In: Tuszynski M.H. and Kordower J. (Eds.), CNS Regeneration. Academic Press, San Diego, pp. 55–88.

Founier, A.E. and Strittmatter, S.M. (2001). Repulsive factors and axon regeneration in the CNS. Curr. Opin. Neurobiol., 11: 89–94.

Franklin, R.J.M. and Barnett, S.C. (2000). Olfactory ensheathing cells and CNS regeneration: the sweet smell of success?. Neuron, 28: 15–18.

George, R. and Griffin, J.W. (1994). Delayed macrophage responses and myelin clearance during Wallerian degeneration in the central nervous system: the dorsal radiculotomy model. Exp. Neurol., 129: 225–236.

Grill, R., Murai, K., Blesch, A., Gage, F.H. and Tuszynski, M.H. (1997). Cellular delivery of neurotrophin-3 promotes corticospinal axonal growth and partial functional recovery after spinal cord injury. J. Neurosci., 17: 5560–5572.

Houle, J.D. and Johnson, J.E. (1989). Nerve growth factor (NGF)-treated nitrocellulose enhances and directs the regeneration of adult rat dorsal root axons through intraspinal neural tissue transplants. Neurosci. Lett., 103: 17–23.

Houle, J.D., Skinner, R.D., Garcia-Rill, E. and Turner, K. (1996). Synaptic evoked potentials from regenerating dorsal root axons within fetal spinal cord tissue transplants. Exp. Neurol., 139: 278–290.

Huang, D.W., McKerracher, L., Braun, P.E. and David, S. (1999). A therapeutic vaccine approach to stimulate axon regeneration in the adult mammalian spinal cord. Neuron, 24: 639–647.

Imaizumi, T., Lankford, K.L., Waxman, S.G., Greer, C.A. and Kocsis, J.D. (1998). Transplanted olfactory ensheathing cells remyelinate and enhance axonal conduction in the demyelinated dorsal columns of the rat spinal cord. J. Neurosci., 18: 6176–6185.

Itoh, Y. and Tessler, A. (1990a). Ultrastructural organization of regenerated adult dorsal root axons within transplants of fetal spinal cord. J. Comp. Neurol., 292: 396–411.

Itoh, Y. and Tessler, A. (1990b). Regeneration of adult dorsal root axons into transplants of fetal spinal cord and brain: a comparison of growth and synapse formation in appropriate and inappropriate targets. J. Comp. Neurol., 302: 272–293.

Itoh, Y., Waldeck, R.F., Tessler, A. and Pinter, M.J. (1996). Regenerated dorsal root fibers form functional synapses in embryonic spinal cord transplants. J. Neurophysiol., 76: 1236–1245.

Itoh, Y., Mizoi, K. and Tessler, A. (1999). Embryonic central nervous system transplants mediate adult dorsal root regeneration into host spinal cord. Neurosurgery, 45: 849–856.

Iwaya, K., Mizoi, K., Tessler, A. and Itoh, Y. (1999). Neurotrophic agents in fibrin glue mediate adult dorsal root regeneration into spinal cord. Neurosurgery, 44: 589–595.

Jenkins, R., McMahon, S.B., Bond, A.B. and Hunt, S.P. (1993). Expression of c-Jun as a response to dorsal root and peripheral nerve section in damaged and adjacent intact primary sensory neurons in the rat. Eur. J. Neurosci., 5: 751–759.

Kimpinski, K. and Mearow, K. (2001). Neurite growth promotion by nerve growth factor and insulin-like growth factor-1 in cultured adult sensory neurons: role of phosphoinositide 3-kinase and mitogen activated protein kinase. J. Neurosci. Res., 63: 486–499.

Kozlova, E.N., Stromberg, I., Bygdeman, M. and Aldskogius, H. (1994). Peripherally grafted human foetal dorsal root ganglion cells extend axons into the spinal cord of adult host rats by circumventing dorsal root entry zone astrocytes. Neuroreport, 5: 2389–2392.

Lehmann, M., Fournier, A., Selles-Navarro, I., Dergham, P., Sebok, A., Lecler, N., Tigyi, G. and McKerracher, L. (1999). Inactivation of Rho signaling pathway promotes CNS axon regeneration. J. Neurosci., 19: 7537–7547.

Liu, L., Persson, J.K.E., Svensson, M. and Aldskogius, H. (1998). Glial cell responses, complement, and clusterin in the central nervous system following dorsal root transection. Glia, 23: 221–238.

Liu, Y., Kim, D., Himes, B.T., Chow, S.Y., Schallert, T., Murray, M., Tessler, A. and Fischer, I. (1999). Transplants of fibroblasts genetically modified to express BDNF promote regeneration of adult rat rubrospinal axons and recovery of forelimb function. J. Neurosci., 19: 4370–4387.

Liu, L., Rudin, M. and Kozlova, E.N. (2000). Glial cell proliferation in the spinal cord after dorsal rhizotomy or sciatic nerve transection in the adult rat. Exp. Brain Res., 131: 64–73.

Messersmith, E.K., Leonardo, E.D., Shatz, C.J., Tessier-Lavigne, M., Goodman, C.S. and Kolodkin, A.L. (1995). Semaphorin III can function as a selective chemorepellent to pattern sensory projections in the spinal cord. Neuron, 14: 949–959.

Navarro, X., Valero, A., Gudino, G., Fores, J., Rodriguez, F.J., Verdu, E., Pascual, R., Cuadras, J. and Nieto-Sampedro, M. (1999). Ensheathing glia transplants promote dorsal root regeneration and spinal reflex restitution after multiple lumbar rhizotomy. Ann. Neurol., 45: 207–215.

Neumann, S. and Woolf, C.J. (1999). Regeneration of dorsal column fibers into and beyond the lesion site following adult spinal cord injury. Neuron, 23: 83–91.

Neumann, S., Bradke, F., Tessier-Lavigne, M. and Basbaum, A.I. (2002). Regeneration of sensory axons within the injured spinal cord induced by intraganglionic cAMP elevation. Neuron, 34: 885–893.

Oudega, M. and Hagg, T. (1996). Nerve growth factor promotes regeneration of sensory axons into adult rat spinal cord. Exp. Neurol., 140: 218–229.

Oudega, M., Rosano, C., Sadi, D., Wood, P.M., Schwab, M.E. and Hagg, T. (2000). Neutralizing antibodies against neurite growth inhibitor NI35/250 do not promote regeneration of sensory axons in the adult rat spinal cord. Neuroscience, 100: 873–883.

Pasterkamp, R.J., Anderson, P.N. and Verhaagen, J. (2001). Peripheral nerve injury fails to induce growth of lesioned ascending dorsal column axons into spinal cord scar tissue expressing the axon repellent Semaphorin 3A. Eur. J. Neurosci., 13: 457–471.

Pindzola, R.R., Doller, C. and Silver, J. (1993). Putative inhibitory extracellular matrix molecules at the dorsal root entry zone of the spinal cord during development and after root and sciatic nerve lesions. Dev. Biol., 156: 34–48.

Qiu, J., Cai, D., Dai, H., McAtee, M., Hoffman, P.N., Bregman, B.S. and Filbin, M.T. (2002). Spinal axon regeneration induced by elevation of cyclic AMP. Neuron, 34: 895–903.

Ramer, M.S., Priestley, J.V. and McMahon, S.B. (2000). Functional regeneration of sensory axons into the adult spinal cord. Nature, 403: 312–316.

Ramer, M.S., Duraisingam, I., Priestley, J.V. and McMahon, S.B. (2001). Two-tiered inhibition of axon regeneration at the dorsal root entry zone. J. Neurosci., 21: 2651–2660.

Ramon-Cueto, A. and Nieto-Sampedro, M. (1994). Regeneration into the spinal cord of transected dorsal root axons is promoted by ensheathing glia transplants. Exp. Neurol., 127: 232–244.

Ramon-Cueto, A., Cordero, M.I., Santos-Benito, F.F. and Avila, J. (2000). Functional recovery of paraplegic rats and motor axon regeneration in their spinal cords by olfactory ensheathing glia. Neuron, 25: 425–435.

Richardson, P.M. and Issa, V.M.K. (1984). Peripheral injury enhances central regeneration of primary sensory neurones. Nature, 309: 791–793.

Romero, M.I., Rangappa, N., Garry, M.G. and Smith, G.M. (2001). Functional regeneration of chronically injured sensory afferents into adult spinal cord after neurotrophin gene therapy. J. Neurosci., 21: 8408–8416.

Schnell, L. and Schwab, M.D. (1990). Axonal regeneration in the rat spinal cord produced by an antibody against myelin-associated neurite growth inhibitors. Nature, 343: 269–272.

Schreyer, D.J. and Skene, J.H.P. (1993). Injury-associated induction of GAP-43 expression displays axon branch specificity in rat dorsal root ganglion neurons. J. Neurobiol., 24: 959–970.

Sims, T.J. and Gilmore, S.A. (1994a). Regeneration of dorsal root axons into experimentally altered glial environments in the rat spinal cord. Exp. Brain Res., 99: 25–33.

Sims, T.J. and Gilmore, S.A. (1994b). Regrowth of dorsal root axons into a radiation-induced glial-deficient environment in the spinal cord. Brain Res., 634: 113–126.

Sugawara, T., Himes, B.T., Kowada, M., Murray, M., Tessler, A. and Battisti, W.P. (1999a). Putative inhibitory extracellular matrix molecules do not prevent dorsal root regeneration into fetal spinal cord transplants. Neurorehabil. Neural Repair, 13: 135–147.

Sugawara, T., Itoh, Y. and Mizoi, K. (1999b). Immunosuppressants promote adult dorsal root regeneration into the spinal cord. Neuroreport, 10: 3949–3953.

Tessler, A., Himes, B.T., Houle, J. and Reier, P.J. (1988). Regeneration of adult dorsal root axons into transplants of embryonic spinal cord. J. Comp. Neurol., 270: 537–548.

Wictorin, K., Brundin, P., Gustavii, B., Lindvall, O. and Bjorklund, A. (1990). Reformation of long axon pathways in adult rat central nervous system by human forebrain neuroblasts. Nature, 347: 556–558.

Wong, J. and Oblinger, M.M. (1990). A comparison of peripheral and central axotomy effects on neurofilament and tubulin gene expression in rat dorsal root ganglion neurons. J. Neurosci., 10: 2215–2222.

Zhang, Y., Dijkhuizen, P.A., Anderson, P.N., Leiberman, A.R. and Verhaagen, J. (1998). NT-3 delivered by an adenoviral vector induces injured dorsal root axons to regenerate into the spinal cord of adult rats. J. Neurosci. Res., 54: 554–562.

Zhang, Y., Roslan, R., Lang, D., Schachner, M., Lieberman, A.R. and Anderson, P.N. (2000). Expression of CHL1 and L1 by neurons and glia following sciatic nerve and dorsal root injury. Mol. Cell. Neurosci., 16: 71–86.

Zhang, Y., Tohyama, K., Winterbottom, J.K., Haque, N.S.K., Schachner, M., Lieberman, A.R. and Anderson, P.N. (2001). Correlation between putative inhibitory molecules at the dorsal root entry zone and failure of dorsal root axonal regeneration. Mol. Cell. Neurosci., 17: 444–459.

CHAPTER 15

Effects of an embryonic repair graft on recovery from spinal cord injury

Saburo Kawaguchi*, Tsutomu Iseda and Takeshi Nishio

Department of Integrative Brain Science, Kyoto University Graduate School of Medicine, Kyoto 606-8501, Japan

Abstract: It is widely believed that mammalian CNS axons have little regenerative capacity because their environment is non-permissive to regrowth. This viewpoint is based, in large part, on the fact that in virtually all previous studies on regeneration following spinal cord injury, regenerated axonal projections have been few in number, quite short, and considered to be mostly aberrant. As a result, motor recovery has been very limited in both experimental preparations and the human. In this chapter, we describe use of a neonatal, spinally transected animal model in which selected spinal cord segments were carefully replaced with equivalent tissue from embryonic tissue of the same species. We demonstrate that the new spinal environment is indeed permissive, and reconstruction is possible of neural connections, which are similar to the pre-injury, normal projections. Moreover, the distribution and number of regenerated axons are closely related to the extent of functional motor recovery. Our results suggest that contrary to doctrinaire thought, the mammalian CNS possesses a remarkable capacity for regrowth. For this to be efficacious, however, regenerating axons must contact the inherent, pre-injury guidance system, whose cues were used for establishing appropriate neural connections in the developing animal, and are retained in the adult. It is argued that by use of these guidance cues, regenerating axons that traverse the site of a spinal cord injury, can project on to locate their pre-injury pathways and targets, and thereby restore function.

Introduction

Innumerable attempts have been made to produce neural repair after a spinal cord injury. For example, grafts have been made into the injured site of (1) embryonic spinal cord (ESC) tissue (Howland et al., 1995; Miya et al., 1997), (2) peripheral nerve (PN) segments (Cheng et al., 1996), (3) transplanted cultured Schwann cells (Xu et al., 1997), (4) olfactory ensheathing cells (Li et al., 1997; Ramón-Cueto et al., 2000), (5) activated macrophages (Rapalino et al., 1998), (6) embryonic stem cells (McDonald et al., 1999; Liu et al., 2000), and (7) monoclonal antibodies that neutralize axon-growth-inhibiting proteins like Nogo-A (Thallmair et al., 1998; Brosamle et al., 2000) and other inhibitory molecules (Huang et al., 1999).

To this point, however, the functional efficacy of such strategies has been limited. The regenerated projections into the injured site have been few in number, quite short, and considered to be mostly aberrant. Consequently, the recovery of function has been poor (Murray and Tessler, Chapters 13 and 14 of this volume).

To achieve a marked recovery of function it seems necessary to reconstruct neural connections within the spinal cord that are similar to the previously normal ones in terms of their number, length and termination targets. Is this possible? We believe so because such regeneration has been achieved in the cerebellofugal projections of the kitten. In this model, regenerated projections were shown to course in the normal path, terminate in the normal target areas and elicit normal functional activity (Kawaguchi et al., 1986; see also McGeer et al., 1987).

*Corresponding author: Tel.: +81-75-753-4481; Fax: +81-75-753-4486; E-mail: kawag@leto.eonet.ne.jp

It appears likely that (1) path- and target-finding of axons during ontogeny is controlled by guidance cues (Silver et al., 1982; Serafini et al., 1996) and (2) regeneration recapitulates ontogeny (Kawaguchi et al., 1991). Thus, the growth of regenerating axons appears to be regulated by guidance cues. During ontogeny, an appropriately arranged organization of axonal guidance cues presumably exists along the presumptive pathway of a given projection system (Kawaguchi et al., 1991). Furthermore, it seems likely that such cues persist in a stable form after the projection has innervated its target (Kawaguchi et al., 1986). Therefore, it is reasonable to propose that following an injury in the central nervous system (CNS), a robust regrowth of axons and reconstruction of neural circuitry similar to the previously normal ones are possible if axons reentering the host spinal cord can meet an appropriate set of guidance cues caudal to the site of spinal injury. To test this hypothesis in the lesioned spinal cord of the neonatal rat, we grafted embryonically homologous spinal tissue to replace spinal cord segments and analyzed the extent of neuronal and functional recovery.

Methods

In neonatal rats, spinal cord segments were removed at the midthoracic level and ESC segments were grafted into the vacancy (Iwashita et al., 1994; Asada et al., 1998; Hase et al., 2002a,b). For such grafting, 1.5–2 segments (length, ~1.5–2.0 mm) of thoracic cord (T10–11) were resected by complete transection at two sites followed by aspiration, and a section of fetal ESC segments equal in length was prepared from dated pregnant rats at embryonic day 14. Particular care was taken to minimize traumatic injuries to the stumps of spinal cord in both the host and graft tissues. It was also extremely important to closely align the host stumps and graft tissue in their correct rostrocaudal and dorsoventral orientation. It must be recognized, however, that such surgery was very difficult and the host–graft contacts were inevitably more-or-less erratic.

Control operated animals were either left ungrafted or grafted with a segment of PN. Intact animals were used for comparison of locomotor performance during development.

Neural connections across the test (open or grafted) site were examined (1) morphologically by anterograde and retrograde tracing methods (Iwashita et al., 1994; Asada et al., 1998; Hase et al., 2002b), (2) electrophysiologically by recording the EMG (electromyography) of relevant muscles during postoperative walking on a treadmill (Hase et al., 2002a), and (3) behaviorally (Hase et al., 2002a,b) by use of the BBB (Basso, Beattie and Bresnahan) open-field locomotor scale (Basso et al., 1995). These observations were made for up to 6 weeks postsurgery.

For anterograde tracing, a total of 0.2 μl of 5% wheat-germ agglutinin-conjugated horseradish peroxidase (WGA–HRP) was injected into the cerebral sensorimotor cortex bilaterally by small (0.02 μl) multiple injections through a glass micropipette connected to a Hamilton syringe. For retrograde tracing, 0.6–0.8 μl of 2% Fast Blue was injected into the lumbar enlargement.

EMGs were recorded from m. triceps brachii in the forelimbs and m. quadriceps femoris in the hindlimbs during locomotion on a treadmill or in an open field. An EMG analysis was used to distinguish between different patterns of locomotion. These included the normal *alternate pattern* in which alternation of limbs occurred in the sequence: one hindlimb, followed by either the ipsi- or contralateral forelimb, then, the other hindlimb and finally, the remaining forelimb. A second, frequently observed gait was the *cruciate pattern* in which alternation of the forelimbs and hindlimbs occurs successively.

The BBB scoring method, which provides a rating of the quality of locomotion from 0 (complete paraplegia) to 21 (normal), was originally developed for assessment of functional recovery in a contusion injury model using the mature animal (Basso et al., 1995). Recent studies have shown, however, that the scale is applicable to neonatal models of spinal cord injury (Wang et al., 1998; Hase et al., 2002a).

Results

The tested groups of operated animals included nongrafted, PN-grafted and ESC-grafted groups. Figs. 1–3 show exemplary examples of their postoperative morphology and locomotor performance.

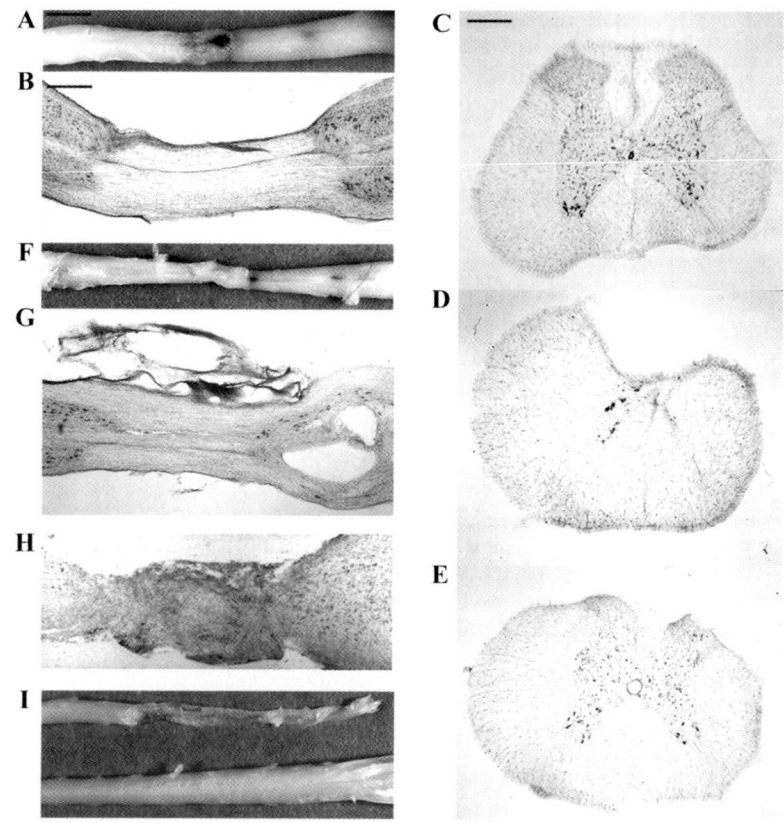

Fig. 1. Examples of the external appearance and cytoarchitecture of the spinal cord of the three test groups of spinal-injured animals. (A) Photograph of an exterior view of the spinal cord taken from a rat of the ESC group. Note that the graft site is at the center of the photograph. This animal had a high BBB rating (19/21). EMG patterns from the same animal are shown in Fig. 2A–B. (B) Photomicrograph of a horizontal section of the spinal cord shown in A. Note that no glial boundary was recognized at the host–graft interfaces. (C–E) Cross-sections taken from another rat of the ESC group with a high BBB rating (20/21). (C) Note that normal cytoarchitecture was preserved rostral to the graft. (D) In the graft, itself, note that there were few neurons and the gray matter was reduced and distorted, whereas the white matter was well developed. (E) Note that almost normal cytoarchitecture was again seen in the host spinal cord caudal to the graft, albeit the gray matter was slightly distorted and the central canal was enlarged. (F) Photograph of an exterior view of the spinal cord taken from a rat of the ESC group with a lower BBB rating (13/21). (G) Photomicrograph of a horizontal section of the spinal cord shown in F. Note that a large cavity and diminished gray matter were seen around the caudal stump of the host spinal cord. (H) Photograph of a sagittal view of the spinal cord taken from a rat of the PN group. Its BBB rating was 9/21. Note that a clear boundary consisting of a different cytoarchitecture was observed at the host–graft interfaces. (I) Photograph of an exterior view of the spinal cord taken from a rat of the Null (upper) and normal control (lower) group of the same age. The rat of the Null group had a BBB rating of 0/21 throughout the entire observation period (27 days). Note that all its lumbosacral segments were almost completely absorbed and eradicated, thereby indicating the occurrence of extensive secondary injury, which was presumably caused by a chain reaction. In A, B, F, G, H and I, rostral is to the left side of the photograph. Scale bar shown in the B panel: 5 mm (A, F, I), 1 mm (B, G, H) and 100 μm (C–E). (Reproduced from Hase et al., 2002a with permission from Mary Ann Liebert, Inc.)

Ungrafted group

In all of the ungrafted animals, the removed portion of the spinal cord remained empty. The space also became relatively larger than expected on the basis of the extent of the original removal. This indicated the occurrence of secondary injury. The upper photograph in Fig. 1I shows the severest such injury we observed. It resulted in the complete absorption of the spinal cord caudal to the operated site. In this group, no neurons were labeled in selected supraspinal nuclei after Fast Blue had been injected into the lumbar

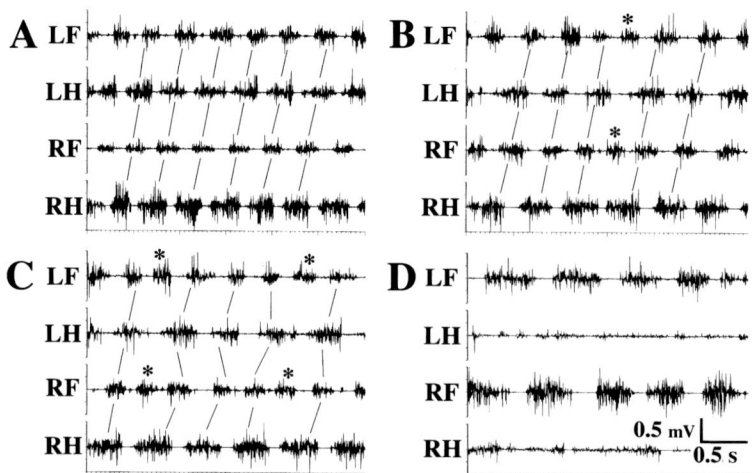

Fig. 2. Representative EMG patterns of four-limb muscles during the elaboration of walking by animals in two of our three test groups. (A) Recordings from a rat of the ESC group, which had a BBB score of 19/21 (spinal morphology shown in Fig. 1A–B). Note the parallel oblique lines, which show the phase relations between the EMG discharges in the four limbs. They demonstrate the animal's use of the normal alternate pattern of walking (see Methods), which is similar to that exhibited by intact rats of the same age. (B) These recordings were made in the same rat as in A. Note the occurrence of an *odd step* (at asterisks; see Results for further details). (C) Recordings from another rat of the ESC group, but with a lower BBB rating (13/21). Note that this animal exhibited a mixture of distorted alternate and cruciate pattern of walking, and a more frequent occurrence of odd steps. In this example, it bears emphasis that the assignment of lines and asterisks is somewhat arbitrary. (D) Recordings made in a rat of the PN group, with a BBB rating of 11/21. Note that its hindlimb EMG potentials were small in amplitude and long in duration. EMG recording sites: m. triceps brachii in the left (Lf) and right (Rf) forelimb; m. quadriceps femoris in the left (Lh) and right (Rh) hindlimb. (Reproduced from Hase et al., 2002a with permission from Mary Ann Liebert, Inc.)

enlargement. EMGs recorded from the hindlimbs of these animals were minute to nonexistent. Furthermore, the BBB rating of these animals' locomotor performance remained low (mean 3.3/21) up to end of the observation period (37 days postsurgery).

PN-grafted preparations

In this group, the graft was well united with the host spinal cord (Fig. 1H). Anterogradely labeled pyramidal tract fibers entered the PN segment but most failed to project beyond the caudal end of the graft. The few that did terminated just beyond the caudal graft–host interface. Many neurons in upper brain structures were labeled with Fast Blue after it had been injected into the graft–host interface. In sharp contrast, however, only a few neurons in the brainstem raphe nuclei were labeled when the tracer was injected into the lumbar enlargement (Asada et al., 1998). EMGs recorded from the hindlimbs were small in amplitude and long in duration compared with those of the intact control and ESC groups, and no hind–forelimb coordination was evident (Fig. 2D). By the end of the observation period, the BBB rating of these animals' locomotor performance was somewhat better (mean 9.8/21) than that of the ungrafted group.

ESC-grafted preparations

In ESC-grafted animals, the extent of anatomical regeneration varied from animal to animal. Presumably this variability was dependent, in part, on variations in the alignment and adherence of the host–graft interfaces (see the previous section).

In a subgroup of these animals ($n=6/20$), their postoperative elaboration of locomotion was near-normal, i.e., they had a BBB score of 18–20. In this subgroup, the grafted segments united firmly with the parenchyma of the host tissue to form a seamless and continuous spinal cord with no glial boundary at the host–graft interfaces (Fig. 1A–E). In these cases, pyramidal tract fibers not only crossed the graft, but, in addition, they extended well into the lumbar

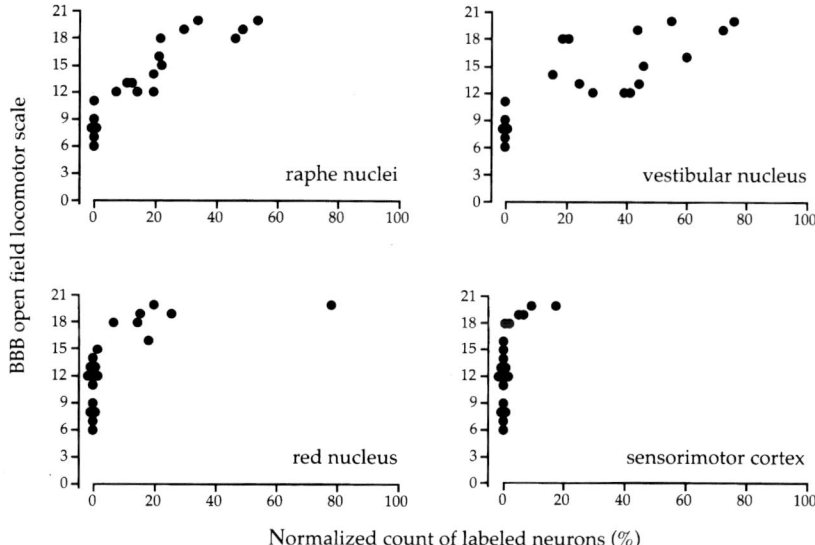

Fig. 3. Relationship between neural reconnections and locomotor performance. Each filled circle represents the results for a single animal. Ordinate, each animal's BBB score; abscissa, the same animal's number of neurons in selected supraspinal sites labeled retrogradely with Fast Blue. In rats with a BBB score of 6–11 ($n = 6/20$), no neurons were labeled in the tested supraspinal nuclei, whereas in rats with a BBB score of 12–14 (another 6/20), neurons were labeled in the raphe and vestibular nuclei, but not in the red nucleus and sensorimotor cortex. In another 2/20 rats, their BBB score 15–16, and neurons were labeled in 3/4 sites (i.e., not in the sensorimotor cortex). Finally, there were 6/20 rats with a BBB score of 18–20. They had labeled neurons in all four of the test sites. For each tested supraspinal site, a linear regression analysis of the X versus Y values was carried out on subsets of the data in which the percentage of labeled neurons was above zero. Significant positive correlations were shown for association of the BBB score with the number of neurons in the raphe nuclei ($P = 0.0002$) and the sensorimotor cortex ($P = 0.016$), but not in the red and vestibular nucleus. (Reproduced from Hase et al., 2002b with permission from Blackwell Science Ltd.)

segments (Iwashita et al., 1994; Asada et al., 1998), with some even reaching the coccygeal segments. Retrogradely labeled neurons were observed in the sensorimotor cortex, and the red, vestibular and raphe nuclei (Hase et al., 2002b). EMGs recorded from the limbs of this subgroup revealed their use of the *alternate pattern* of locomotion (Fig. 2A) in a manner that was very similar to that used by normal animals. These animals, however, showed occasionally odd steps, i.e., the step-cycle sequence involves a normal activation of forelimb muscles but no muscle activation in one or both hindlimbs. For example, the ESC-grafted rat whose EMG patterns are shown in Fig. 2A–B elaborated ~1 odd step per 23 cycles of normal hindlimb–forelimb coordination.

In the second subgroup of ESC-grafted animals ($n = 2/20$), their postoperative elaboration of locomotion was moderately effective (BBB score, 15 and 16). In this subgroup, retrogradely labeled neurons were observed in the red, vestibular and raphe nuclei but not in the cerebral cortex (Hase et al., 2002b).

In the third subgroup of ESC-grafted animals ($n = 6/20$), their postoperative elaboration of locomotion was less effective (BBB score, 12–14). In this subgroup, the grafted segments and the host tissue united with each other but the graft and neighboring segments became atrophic (Fig. 1F), and cavities were formed around the host/graft junction (Fig. 1G). Retrogradely labeled neurons were observed in the vestibular and raphe nuclei but not in the red nucleus and cerebral cortex (Hase et al., 2002b). In these animals, there was a marked deficiency in the extent and coherence of EMG activation among the four limbs. As a result, it was difficult to identify the pattern of locomotion used and whether individual step cycle sequences were normal or *odd*. For example, Fig. 2C shows the EMG patterns of a rat in this subgroup. The EMGs reveal an admixture of *alternate*, *cruciate* and odd step-cycle sequences.

In the fourth, final subgroup of the ESC-grafted animals ($n = 6/20$), their postoperative elaboration of locomotion was poor (BBB score, 6–11) and no retrogradely labeled neurons were observed in supraspinal nuclei (Hase et al., 2002b).

Figure 3 summarizes the association between the locomotor performance of the ESC-grafted animals and their distribution of labeled neurons in the tested supraspinal structures. Their locomotor performance was clearly related to both the number and distribution of labeled neurons in these supraspinal structures.

Comment

In this chapter, we have presented results, which included the use of a neonatal animal model in which selected spinal cord segments were replaced with equivalent neural tissue from the embryo. In such animals, it was shown that the distribution and number of regenerated projections across the graft were closely related to the extent of functional motor recovery (Asada et al., 1998; Hase et al., 2002a,b). To achieve a marked recovery of function it appears essential to reconstruct neural connections similar to normal ones in their number, degree of extension and sites of termination. Is this possible? We think so on the basis of our demonstration of such regeneration not only in neonates but also in slightly older animals (3 weeks of age; Inoue et al., 1998; Kikukawa et al., 1998) and even in the adult animal (Nishio et al., 2001). Thus, we opine that counter to the doctrinaire position, the ESC-grafted spinal environment is permissive to axonal regrowth, even to the extent of regenerating axons having the capacity to locate their preinjury pathways and targets caudal to the lesion, and thereby restore function.

Figure 4 provides evidence in support of the existence of an appropriately arranged guidance system caudal to the site of the spinal graft.

Figure 4A, B and D show a very successful case of ESC grafting. Note that pyramidal tract axons entered and passed through the graft, reentered the host spinal cord, reached the coccygeal segments (Fig. 4D), and terminated in their normal (preinjury)

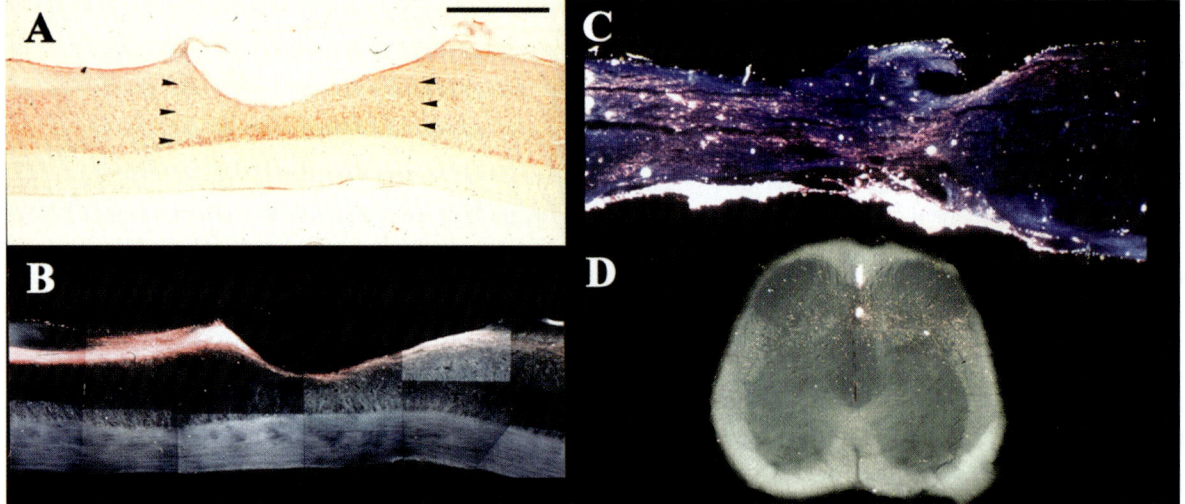

Fig. 4. Pathways and termination of regenerated pyramidal tract axons. (A) Photomicrograph of a sagittal section of the spinal cord taken from an ESC-grafted rat (36 days after grafting at 2 days postbirth). Arrowheads indicate the extent of the graft. (B) Dark-field photomicrograph of the section shown in A. The pinkish-colored portion is the pyramidal tract, which was labeled anterogradely with WGA–HRP following its injection into the sensorimotor cortex. (C) Dark-field photomicrograph of a sagittal section of the spinal cord taken from an ESC-grafted rat (21 days after grafting at 2 days postbirth). Note that pyramidal tract axons entered dispersedly into the graft but reentered into the host spinal cord in the more-or-less correct path. (D) Dark-field photomicrograph of a cross-section of the coccygeal segment taken from the ESC-grafted rat shown in A–B. A bundle of axons occupied the depth of the dorsal funiculus just as in the normal pyramidal tract and it terminated in the normal target areas (Rexed's laminae 3–5). Scale bar: 1 mm (A, B, C) and 300 μm (D). (Reproduced, in part, from Iwashita et al., 1994 with permission from Nature Publishing Group, and Asada et al., 1998 with permission from Elsevier Science.)

targets (Rexed's laminae 3–5). In the graft and immediately neighboring caudal spinal sites, however, the pyramidal axons did not course along their normal path in the deep part of the dorsal funiculus. Rather, they regenerated through the most superficial part of the cord dorsum. Nonetheless, they subsequently (more caudally) regained their appropriate (preinjury) location. This suggested that guidance cues in the funicular depth of the graft, and the immediately rostral and caudal host spinal tissue were ineffective and/or there were less cues in the superficial part of the graft's cord dorsum. Subsequently (more caudally), however, the guidance system was intact and efficacious.

Figure 4C shows for a case of less successful ESC grafting that pyramidal tract axons entered the graft dispersedly, as if to suggest that inappropriate guidance was available at or near the rostral host–graft interface. Nonetheless, once the axons had regenerated through the graft and reentered the host spinal cord, they converged and followed the normal course thereafter.

It is conceivable that in the earlier cases, the guidance system for pyramidal axons in the graft and the immediately surrounding host spinal tissue was disorganized by surgery, but maintained more caudally in the host spinal cord. Once the regrowing axons entered the correct path caudal to the graft, they were confined to it all the way to their targets. This suggests that the guidance system contains inhibitory mechanisms that suppress the growth of axons external to their correct, preinjury path.

Thus, we believe that in many analogous studies, the goal was to make the graft–intact spinal cord interface and its immediate surrounding areas more permissive to regeneration. Such attempts may well have perturbed the normal, prelesion guidance system for regenerating axons. If so, this would explain the relative lack of functionally effective regeneration in these studies.

Acknowledgments

This work was supported by: JSPS-RFTF96100203, Special Coordination Funds for Promoting Science and Technology, Science and Technology Agency, Japan; Health and Labor Sciences Research Grants for Research on Psychiatric and Neurological Diseases and Mental Health, Ministry of Health, Labor and Welfare, Japan; and an award from the Uehara Memorial Foundation.

Abbreviations

BBB*	Basso, Beattie and Bresnahan
ESC	embryonic spinal cord
PN	peripheral nerve
EMG	electromyography
Nogo-A**	a myelin protein produced by oligodendrocytes that inhibits axon growth
WGA–HRP	wheat germ agglutinin conjugated with horseradish peroxidase

*A descriptor used for the authors' locomotor rating scale
**This protein has been purified and its gene cloned

References

Asada, Y., Kawaguchi, S., Hayashi, H. and Nakamura, T. (1998). Neural repair of the injured spinal cord by grafting: comparison between peripheral nerve segments and embryonic homologous structures as a conduit of CNS axons. Neurosci. Res., 31: 241–249.

Basso, D.M., Beattie, M.S. and Bresnahan, J.C. (1995). A sensitive and reliable locomotor rating scale for open field testing in rats. J. Neurotrauma, 12: 1–21.

Brosamle, C., Huber, A.B., Fiedler, M., Skerra, A. and Schwab, M.E. (2000). Regeneration of lesioned corticospinal tract fibers in the adult rat induced by a recombinant, humanized IN-1 antibody fragment. J. Neurosci., 20: 8061–8068.

Cheng, H., Cao, Y. and Olson, L. (1996). Spinal cord repair in adult paraplegic rats: partial restoration of hindlimb function. Science, 273: 510–513.

Hase, T., Kawaguchi, S., Hayashi, H., Nishio, T., Asada, Y. and Nakamura, T. (2002a). Locomotor performance of the rat after neonatal repairing of spinal cord injuries: quantitative assessment and electromyographic study. J. Neurotrauma, 19: 267–277.

Hase, T., Kawaguchi, S., Hayashi, H., Nishio, T., Mizoguchi, A. and Nakamura, T. (2002b). Spinal cord repair in neonatal rats: a correlation between axonal regeneration and functional recovery. Eur. J. Neurosci., 15: 969–974.

Howland, D.R., Bregman, B.S., Tessler, A. and Goldberger, M.E. (1995). Transplants enhance locomotion in neonatal kittens whose spinal cords transected: a behavioral and anatomical study. Exp. Neurol., 135: 123–145.

Huang, D.W., McKerracher, L., Braun, P.E. and David, S. (1999). A therapeutic vaccine approach to stimulate axon regeneration in the adult mammalian spinal cord. Neuron, 24: 639–647.

Inoue, T., Kawaguchi, S. and Kurisu, K. (1998). Spontaneous regeneration of the pyramidal tract after transection in young rats. Neurosci. Lett., 247: 151–154.

Iwashita, Y., Kawaguchi, S. and Murata, M. (1994). Restoration of function by replacement of spinal cord segments in the rat. Nature, 367: 167–170.

Kawaguchi, S., Miyata, H. and Kato, N. (1986). Regeneration of the cerebellofugal projection after transection of the superior cerebellar peduncle in kittens: morphological and electrophysiological studies. J. Comp. Neurol., 245: 258–273.

Kawaguchi, S., Murata, M. and Kurimoto, Y. (1991). Ontogenesis of the cerebellofugal projection in the rats. Dev. Brain Res., 61: 285–289.

Kikukawa, S., Kawaguchi, S., Mizoguchi, A., Ide, C. and Koshinaga, M. (1998). Regeneration of dorsal column axons after spinal cord injury in young rats. Neurosci. Lett., 249: 135–138.

Li, Y., Field, P.M. and Raisman, G. (1997). Repair of adult rat corticospinal tract by transplants of olfactory ensheathing cells. Science, 277: 2000–2002.

Liu, S., Qu, Y., Stewart, T.J., Howard, M.J., Chakrabortty, S., Holekamp, T.F. and McDonald, J.W. (2000). Embryonic stem cells differentiate into oligodendrocytes and myelinate in culture and after spinal cord transplantation. Proc. Natl. Acad. Sci. USA, 97: 6126–6131.

McDonald, J.W., Liu, X.Z., Qu, Y., Liu, S., Mickey, S.K., Turetsky, D., Gottlieb, D.I. and Choi, D.W. (1999). Transplanted embryonic stem cells survive, differentiate and promote recovery in injured rat spinal cord. Nat. Med., 12: 1410–1412.

McGeer, P.L., Eccles, J.C. and McGeer, E.G. (1987). Molecular Neurobiology of the Mammalian Brain, 2nd ed., Plenum Press, New York, pp. 448–450.

Miya, D., Giszter, S., Mori, F., Adipudi, V., Tessler, A. and Murray, M. (1997). Fetal transplants alter the development of function after spinal cord transection in new born rats. J. Neurosci., 17: 4856–4872.

Nishio, T., Kawaguchi, S., Iseda, T., Hase, T., Kojima, K. and Kawasaki, T. (2001) Spinal cord repair in adult paraplegic rats by glial transplantation: recovery of walking with hind–forelimb coordination. IUPS 2001 Abstract, ID No. 2331.

Ramón-Cueto, A., Cordero, M.I., Santos-Benito, F.F. and Avila, J. (2000). Functional recovery of paraplegic rats and motor axon regeneration in their spinal cord by olfactory ensheathing glia. Neuron, 25: 425–435.

Rapalino, O., Lazarov-Spiegler, O., Agranov, E., Velan, G.J., Yoles, E., Fraidakis, M., Solomon, A., Gepstein, R., Katz, A., Belkin, M., Hadani, M., Schwartz, M. (1998). Implantation of stimulated homologous macrophages results in partial recovery of paraplegic rats. Nat. Med., 4: 814–821.

Serafini, T., Colamarino, S.A., Leonardo, E.D., Wang, H., Beddington, R., Skarnes, W.C. and Tessier-Lavigne, M. (1996). Netrin-1 is required for commissural axon guidance in the developing vertebrate nervous system. Cell, 87: 1001–1014.

Silver, J., Lorenz, S.E., Wahlsten, D. and Coughlin, J. (1982). Axonal guidance during development of the great commissures: descriptive and experimental studies in vivo, on the role of performed glial pathways. J. Comp. Neurol., 210: 10–29.

Thallmair, M., Metz, G.A., Z'graggen, W.J., Raineteau, O., Kartje, G.L. and Schwab, M.E. (1998). Neurite growth inhibitors restrict plasticity and functional recovery following corticospinal tract lesions. Nat. Neurosci., 1: 124–131.

Xu, X.M., Chen, A., Guenard, V., Kleitman, N. and Bunge, M.B. (1997). Bridging Schwann cell transplants promote axonal regeneration from both the rostral and caudal stumps of transected adult rat spinal cord. J. Neurocytol., 26: 1–16.

CHAPTER 16

Determinants of locomotor recovery after spinal injury in the cat

Serge Rossignol*, Laurent Bouyer, Cécile Langlet, Dorothy Barthélemy, Connie Chau, Nathalie Giroux, Edna Brustein, Judith Marcoux, Hugues Leblond and Tomás A. Reader[†]

Center for Research in Neurological Sciences, Department of Physiology, University of Montreal, Faculty of Medicine, Montreal, QC, H3C 3J7, Canada

Abstract: After a spinalization at the most caudal thoracic spinal segment, the cat can recover locomotion of the hindlimbs when they are placed on a moving treadmill. This chapter summarizes some of the determinants of such a dramatic recovery of motor function. Fundamental to this recovery is undoubtedly the genetically based spinal locomotor generator, which provides an essential rhythmicity to spinal motoneurons and hence the musculature. Other factors are also important, however. Sensory feedback is essential for the correct expression of spinal locomotion because spinal cats, devoid of cutaneous feedback from the hindfeet, are incapable of plantar foot placement. The neurochemical environment also adapts to spinalization, i.e., the loss of all modulation by descending monoaminergic pathways. Post-transection spinal rhythmicity then becomes more dependent on glutamatergic mechanisms. Finally, we argue that the mid-lumbar spinal segments evolve to play a crucial role in the elaboration of spinal locomotion as their inactivation abolishes spinal locomotion. In summary, the above findings suggest that the recovery of spinal locomotion is determined by a number of factors, each of which must now be more fully understood in the ever-continuing effort to improve the rehabilitation of spinal-cord-injured subjects.

Introduction

In most animal species, it is possible to elicit bilateral and coordinated walking movements of the hindlimbs after a complete chronic section of the spinal cord in the thoracic region (Delcomyn, 1980; Grillner, 1981; Rossignol, 1996, 2000; Rossignol et al., 1996, 2000). Cats spinalized at T13 can walk with their hindlimbs and make plantigrade contact to sustain the weight of their hindquarters. They can adapt their locomotion to the varying speed of the treadmill and, if an object perturbs the progression of one leg, the limb can generate a coordinated hyperflexion to bring the foot above and over the obstacle.

Among the most obvious defects of spinal locomotion are, of course, a lack of voluntary control, an absence of coordination of the fore- and the hindlimbs when all four paws are placed on the treadmill belt, and a quasi-total absence of balance control. Indeed, even when the animals can sustain the weight of the hindquarters, balance must be provided by guiding the hindquarters with the tail. Less obvious and more variable defects are a foot drag in the initial part of the swing phase and the presence of scissoring between the hindlimbs. Despite these defects, spinal locomotion in the cat is very impressive.

Our work has aimed at understanding how some physiological and neurochemical mechanisms evolve

*Corresponding author: Tel.: +514-343-6366; Fax: +514-343-6113; E-mail: serge.rossignol@umontreal.ca
[†]Dr. Tomás Reader passed away in May 2002.

in the spinal cord in the weeks and months following spinalization and lead to the expression of the locomotor pattern. A better understanding of the pathophysiology of the spinal cord can advance understanding of regulatory locomotor mechanisms that operate in the normal state. Our observations have also lead us to speculate about promising avenues of research which may eventually advance understanding and management of spinal cord injuries in the human.

The spinal locomotor generator

Spinal locomotion, which has been observed in several animal species, points to the existence of spinal circuitry capable of generating the locomotor pattern. This spinal locomotor generator must be innate because it can be expressed even if kittens are spinalized before they could possibly 'learn' to walk (Grillner, 1973; Forssberg et al., 1980a,b; Grillner and Wallén, Chapter 1 of this volume). In addition, a detailed locomotor pattern is observed in paralyzed spinal cats when injected with L-DOPA (Grillner and Zangger, 1979).

The locomotor pattern recorded on a treadmill during natural walking is not simply a flexion–extension pattern but, rather, one in which each muscle has more-or-less its own 'signature'. Figure 1, which shows the general experimental set-up for recording locomotion, also illustrates some raw electromyographic (EMG) recordings of this complex locomotor pattern in the intact cat. The knee flexor/hip extensor, semitendinosus (St), has a double burst of activity whereas the hip flexor/knee extensor, sartorius (Srt), is delayed relative to St and has a single burst in flexion. The ankle extensor, GL, typically starts abruptly whereas the activity in the knee extensor, VL (especially the contralateral VL in this example), usually builds up gradually.

How similar is the spinal locomotor pattern to that of the intact cat? Only a few days after a low thoracic spinalization, if the hindlimbs are placed on a treadmill while the forelimbs stand on a stationary platform, strong manual stimulation of the perineum or the base of the tail elicits small alternate rhythmic movements of the otherwise flaccid hindlimbs. These movements occur mainly at the hips, with the feet dragging on their dorsum and with very little knee and ankle flexion such that the hindlimbs move more-or-less rhythmically while being passively extended at the hip. The animals are unable to sustain the weight of their hindquarters, however. This inability to stand at this stage is probably due to disconnection of the vestibulospinal and reticulospinal pathways. Indeed, similar large, but incomplete, lesions of the ventral and ventrolateral quadrants on both sides of the cord lead to an inability to walk for several weeks with the hindlimbs, albeit quadrupedal locomotion does indeed recuperate subsequently (Brustein and Rossignol, 1998; Rossignol et al., 1999).

After this initial stage, there is considerable variation in the locomotor recovery following a complete spinalization (Murray, Tessler, and Kawaguchi et al., Chapters 13, 14 and 15 of this volume). In some cats, there is a fast progression within the first 10 days whereas other cats may need 2–3 weeks, or even more, of daily treadmill training to express a full pattern of spinal locomotion. When this plateau phase is attained, the hindlimbs can follow the treadmill belt up to speeds of 0.8 m/s without perineal stimulation. The paws make plantigrade contact and the force developed is adequate to sustain the weight of the hindquarters. There is often a pronounced adductor tonus so that a separator has to be placed between the hindlimbs to prevent them from impeding each other. Figure 2 compares the kinematics and average EMG activity in the same cat before and several days after spinalization and regular treadmill training. Although the spinal cat cannot voluntarily initiate locomotion, the movement of the treadmill is sufficient to induce locomotion although, initially, this is greatly facilitated by stimulation of the perineum.

The Fig. 2-type animal has a clear vestibular deficit since it cannot keep its balance during walking for more than a few steps. The cycle length is generally shorter after spinalization for the same walking speed and the feet tend to drag on the belt in the initial part of the swing phase. The angular excursion of the various hindlimb joints after spinalization is indeed very similar to that in the intact cat. The EMG pattern is also generally similar to that of the intact state, albeit with some noticeable

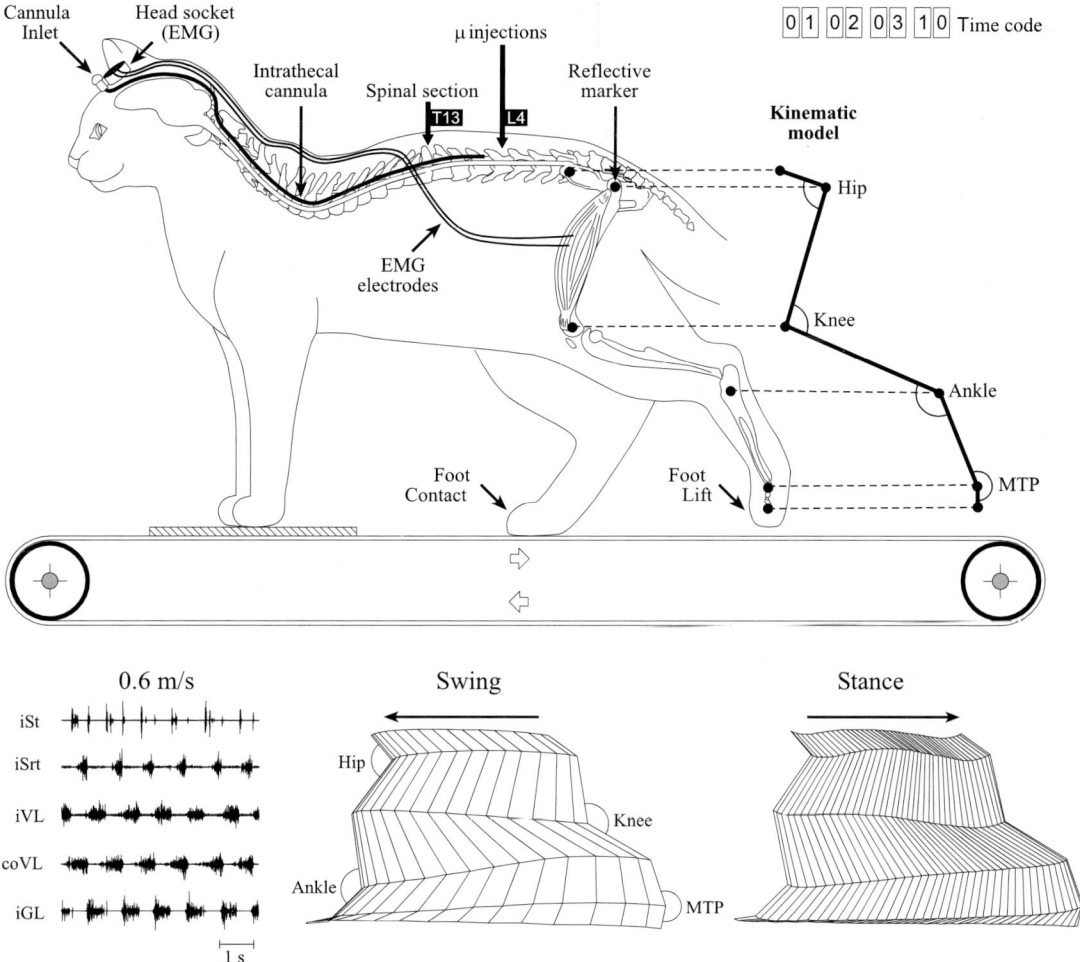

Fig. 1. General methodology for the study of locomotion in the spinal cat. The animal is placed with its forelimbs standing on a stationary platform while its hindlimbs walk on the treadmill (arrows within the belt indicate the direction of movement). Pairs of EMG wires are implanted into various muscles (only one pair is represented here) and an intrathecal cannula is inserted through the atlanto-occipital ligament down to ∼ L4. The multipin EMG connector, as well as the cannula inlet, is cemented to the skull. Reflective markers are placed at various points on the limb and the angle measurements taken in the indicated orientations. For each video field (16.7 ms between fields), the coordinates of the reflective markers are obtained and the hindlimb movement reconstructed as indicated in the kinematic model (MTP, metatarsophalangeal joint). From such data, the swing and stance phases of each cycle can be reconstructed as shown in the figure. Note that to prevent overlap of the stick figures, each one is displaced by an amount equal to the displacement of the foot along the horizontal axis. The foot-contact and foot-lift are also measured to determine cycle length and duration, and also to synchronize EMG events when needed. The digital time code at the upper right is used to synchronize video and EMG recordings. The spinal lesion is made at T13 and microinjections made at L3–L4 levels. The insert at the bottom left represents some characteristic EMGs taken from various muscles: St, semitendinosus; Srt, sartorius; VL, vastus lateralis and GL, gastrocnemius lateralis; i, ipsilateral to the camera or in the foreground of the figure; co, contralateral. (Reproduced, in part, from Rossignol et al., 2002 with permission from Elsevier Science.)

differences. The EMGs are more clonic after spinalization and the activation delay between St and Srt is usually absent. This may explain why spinal cat tend to drag their feet at the onset of the swing phase since it appears that the hip joint may start to flex at the same time as the knee whereas normally knee flexion first clears the foot before the hip flexes. This specific defect might be related to

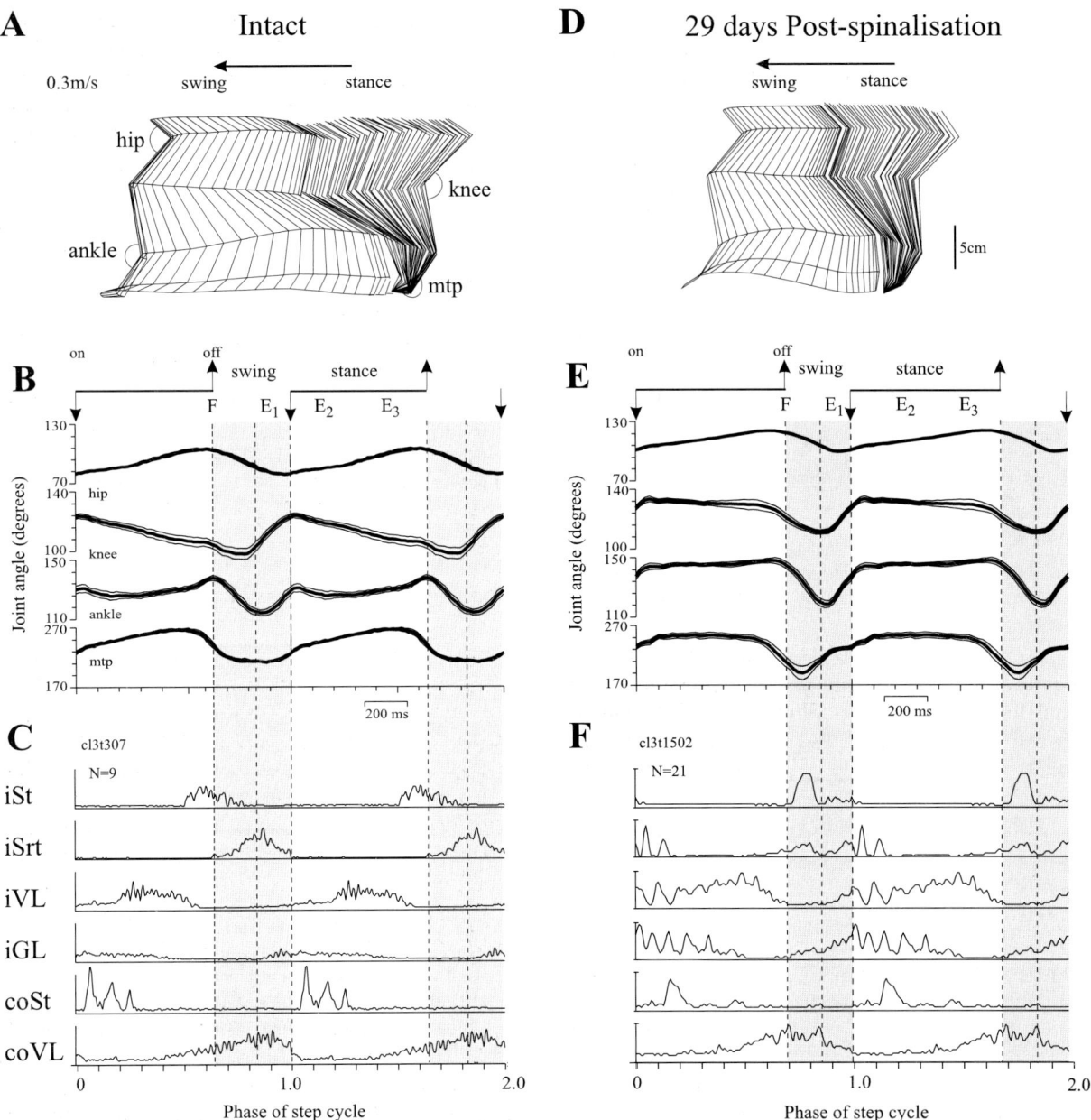

Fig. 2. Comparison between step cycles of the hindlimb (at 0.3 m/s) in an intact cat and in the same cat 29 days postspinalization. (A) Stick figures reconstructed from a video sequence of one normal step cycle. For the stance phase, the point on the paw is fixed whereas during the swing phase, the stick figures are displaced from each other by an amount equal to the displacement of the foot. The orientation of joint measurements is shown. The calibration is provided in D. (B) Angular excursion of the four joints averaged over 10 cycles. Flexion always corresponds to downward deflections of the angular traces. The vertical dotted lines separate various epochs of the step cycle (Philippson, 1905; F and E_1, swing phase; E_2 and E_3, stance phase). (C) Average of rectified EMG traces of nine cycles. The cycle is normalized to 1 and is repeated twice for clarity of illustration at turning points. The average is synchronized on foot contact. (D) The same format as in A. Note the period of foot drag in the initial part of the swing phase is represented by a line under the paw. Note also the shorter step cycle. (E) The same format as in B. (F) The same format as in C but for 21 cycles in a cat spinalized at T13, 29 days previously. Note that the EMGs are more clonic ('spiky') than in the control situation.

disconnection of the corticospinal and rubrospinal pathways because a cat with a lesion restricted to the dorsolateral quadrants may exhibit a quadrupedal locomotor pattern but also show a paw drag at the onset of the swing phase (Jiang and Drew, 1996; Rossignol et al., 1999).

In summary, a great deal of the basic detailed locomotor pattern is still expressed after spinalization. Specific defects observed can in large part be attributed to the loss of the normal control provided by the descending pathways (from the sensorimotor cortex, brainstem and supralesion propriospinal system) as revealed by partial lesion studies (for review, see Rossignol et al., 1999).

Locomotor training and spinal plasticity

Having lost all of the descending control pathways and mechanisms, what changes may occur within the spinal cord to enable the reexpression of the locomotor pattern?

Quite early work stressed the importance of locomotor training to facilitate the expression of the spinal locomotor pattern (Shurrager and Dykman, 1951; Shurrager, 1955). Later, it was shown that younger animals usually have a greater propensity to recover a fuller locomotor capacity (Bregman and Goldberger, 1983; Howland et al., 1995a,b). Interestingly, some other studies (Eidelberg et al., 1980) showed that adult spinal cat could not display a coordinated quadrupedal locomotion on a treadmill.

Except in very exceptional cases, all our adult spinal cats have been able to walk after spinalization at T13 (Barbeau and Rossignol, 1987; Bélanger et al., 1996; Chau et al., 1998a; Rossignol et al., 2000). We trained each animal for several weeks, however, with the training for several minutes/session on every day or every second day. In each session, the animal had the opportunity to support its bodyweight. This was achieved by holding the hindquarters over the moving treadmill belt, favoring foot contact by optimizing the manual positioning of the animal, and using concomitant perineal stimulation.

Since the alpha-2 noradrenergic agonist, clonidine, induces locomotion in spinal cat (Forssberg and Grillner, 1973; Barbeau et al., 1987; Chau et al., 1998b) and maintains its effects for 4–6 h, we used this time window to train cats on the treadmill daily after a single intrathecal injection of clonidine (Chau et al., 1998a). These experiments showed that daily injections of clonidine enabled an earlier and more regular locomotor training than would otherwise have been possible. It also resulted in a marked improvement of locomotor recovery on the treadmill such that by days 6–11 after spinalization, all cats could walk with a plantar foot contact and hindlimb weight support.

The previous findings suggest that the underlying mechanisms of plastic changes after spinalization may be modified by early, intensive locomotor training. Furthermore, spinal cats trained to stand or walk execute the trained task more effectively, thereby suggesting that specific plastic changes must occur within the spinal cord in association with the specific training task (de Leon et al., 1998).

The importance of sensory inputs

The importance of sensory inputs in the control of locomotion has been emphasized repeatedly (Andersson et al., 1981; Grillner, 1981; Rossignol et al., 1988; Pearson, 1993). After spinalization, timing and other cues that originate from descending pathways, including long propriospinal ones, are all unavailable, such that the spinal locomotor generator must now optimize its function using whatever information is available through sensory feedback. Although we have reported earlier that spinal cats generate appropriate reflexes when the dorsum of the paw is stimulated during the swing or stance phase, thereby leading to the appearance of phase-dependent reflex responses (Forssberg et al., 1975; Rossignol, 1996; Rossignol et al., 2000), only a few subsequent studies have dealt with the importance of specific sensory feedbacks for the correct expression of the locomotor pattern. In a pioneering work, Goldberger (1977) showed in the cat that after a unilateral lumbosacral rhizotomy which essentially removed all afferent feedback from one hindlimb, the animal could gradually recover locomotion. After adding an L1 hemisection on the operated side, however, the cat could no longer use the deafferented hindlimb. In contrast, the hemisected cat, without deafferentation, could

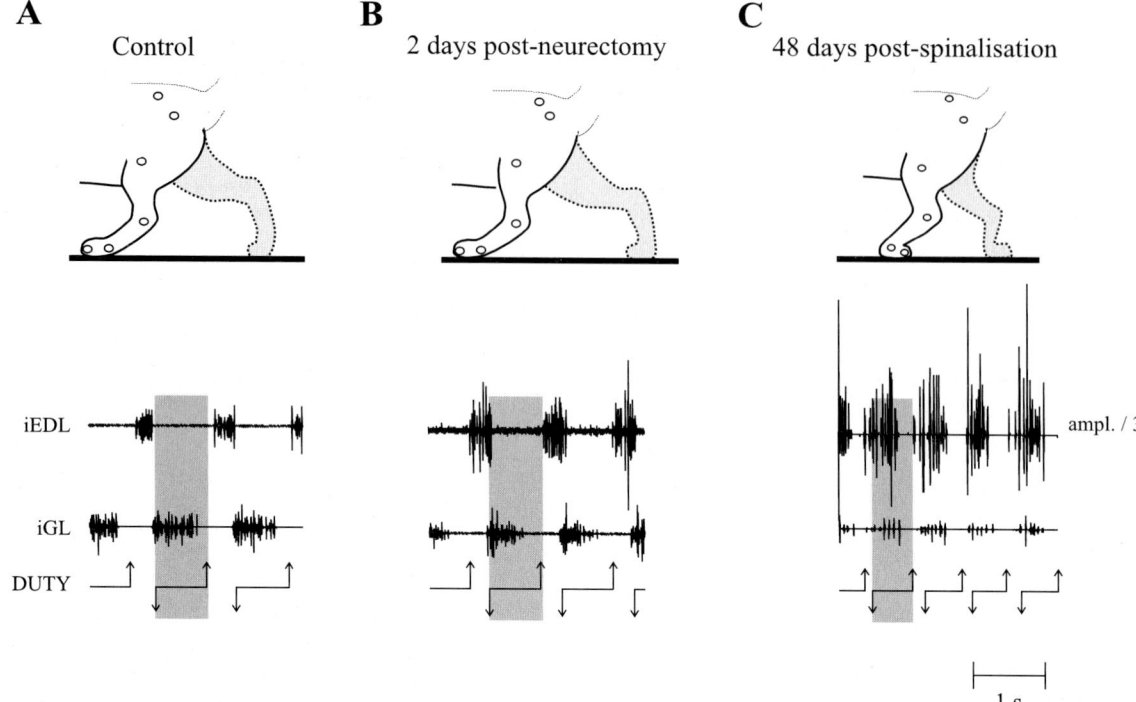

Fig. 3. Effect of cutaneous denervation of the hindpaw on the placement of the foot during locomotion. (A, top) Line drawing traced from a video image extracted from a control walking session at 0.5 m/s. (A, bottom) EMGs from the left-ankle flexor, extensor digitorum longus (iEDL) and left-ankle extensor, gastrocnemius lateralis (iGL) are synchronized with left-foot duty cycle (DUTY; down arrow, foot contact; up arrow, foot lift; solid line, stance phase). The gray box highlights the alternation in flexor/extensor activity. (B) Two days postneurectomy, foot placement (top) and alternating EMG activity (bottom) are still present, but flexor amplitude is increased. (C, top) Following a subsequent spinalization at T13, the same cat permanently loses plantar foot placement during the stance phase. (C, bottom) The pattern of EMG activity is also modified, with the flexor burst now extending into the stance phase. Because the EMG of EDL is very much increased, the gain has been divided by 3 for illustration purposes.

usually walk rather well with both hindlimbs. This Goldberger's study emphasized the importance of sensory inputs when an animal compensates for a central lesion.

Figure 3, taken from our recent work on the spinal cat (Bouyer and Rossignol, 1998, 2001), suggests that cutaneous inputs from the hindpaws are crucial for the correct placement of the foot on the treadmill. We showed in the cat that after sectioning all the cutaneous nerves at the ankle level, the animal recovers treadmill locomotion within ~24–48 h (Fig. 3B). It took several days for this preparation to recover the ability to walk on a horizontal ladder, however. At this latter time, if the spinal cord was cut at T13, the animal could no longer make plantar contact on the treadmill belt (Fig. 3C), even after several weeks of training. Rather, the paws were dragged throughout the whole swing phase and during the stance phase; the weight was supported on the dorsum of the paws. These results showed that cutaneous inputs are essential for the correct positioning of the foot in spinal locomotion.

In the same study, another cat preparation was first spinalized and allowed to recover the usual spinal locomotion over the next 40 days. Then, the five cutaneous nerves supplying the foot were cut one after the other on successive days, i.e., between 40 and 110 days postspinalization. After each nerve cut, there were some locomotor deficits (paw drag, misplacement of the foot at landing) but, following training, the spinal cat was capable of recovering adequate foot placement within a few days as long as some cutaneous input was still available from the paw. When the last cutaneous nerve was cut at day 110 postspinalization, the foot placement became abnormal as in the cat sample (Fig. 3) that was denervated before spinalization, and this deficit did not recover even after 70 days of further locomotor training. These observations suggest that spinal cats have some degree of spinal plasticity which enables them to recover a proper foot contact when sensory input is gradually removed. In other words, even in the spinal state, the cat can adapt to a major sensory deficit. There is a limit to this compensation, however. The spinal cat needs at least some cutaneous feedback from the foot for its appropriate placement, whereas the normal cat is capable, through other mechanisms, to achieve proper foot placement even in the absence of all cutaneous feedback from the foot.

The fate of spinal neurotransmitters and pharmacological effects

The preceding section suggests that the expression of the spinal locomotor generator evolves over time after a spinal lesion and that locomotor training, which undoubtedly provides important sensory feedback (particularly cutaneous feedback), augments the recovery process. Although it is indeed possible to relate the most obvious deficits to the severance of the loss of selected descending pathways, the consequences are far less clear of sectioning descending monoamine pathways, whose cell bodies are located in supraspinal structures and which normally provide noradrenaline (NA) and serotonin (5-HT) to the spinal cord, among other structures.

We have studied the fate of three receptors of neurotransmitters (alpha-1, alpha-2 NA, 5-HT1A) at various times after spinalization (Giroux et al., 1999). These receptors are upregulated for a few weeks after spinalization before returning to or below their control values. How important is the stimulation of these receptors for the expression of spontaneous spinal locomotion? As mentioned earlier, clonidine, an alpha-2 NA receptor, can induce locomotion within a few minutes after its application in a spinal cat. Since spontaneous spinal locomotion is re-expressed in absence of NA, which is now absent because the relevant descending pathways have been severed, it must be concluded that these NA receptors are not essential for the expression of spontaneous spinal locomotion. We have tested this idea more directly by injecting the alpha-2 blocker, yohimbine, and found no effect on spinal locomotion whereas, in the same cat before spinalization, yohimbine had a pronounced deleterious effect on locomotion (Giroux et al., 2001).

Thus, although alpha-2 agonists can trigger locomotion, other receptors must be activated when spontaneous spinal locomotion is expressed. We have therefore injected AP-5, an NMDA antagonist, which completely blocks locomotion in spinal cat whereas, in the same cats in the intact state, it only produces sag of the hindquarters. It is therefore likely that one of the major changes taking place after spinalization is a reorganization of the spinal locomotor generator such as to become more dependent on excitatory amino acids, and possibly other agents. Our preliminary results involving autoradiography of NMDA, AMPA and kainate receptors at various times after spinalization suggest that these receptors are maintained post-spinalization in an upregulated state. This is contrary to the NA and 5-HT receptors which, as mentioned earlier, eventually return to control or subcontrol values.

The importance of midlumbar segments

Recent studies suggest that the midlumbar segments are particularly important for the expression of spinal locomotion. Acute experiments performed in the decerebrated cat spinalized one week earlier have shown that localized injections of clonidine at L3–L4 can induce locomotion. On the other hand, yohimbine injected at these segments blocks locomotion (Marcoux and Rossignol, 2000). It is thus probable that the integrity of these midlumbar segments becomes a very important determinant in the organization of spinal locomotion. This is reminiscent of work on neonatal rats (Kiehn and Kjaerulff, 1998; Cazalets, 2000) in which the upper lumbar segments also appeared to play a critical role in the generation of locomotion. Similarly it is also

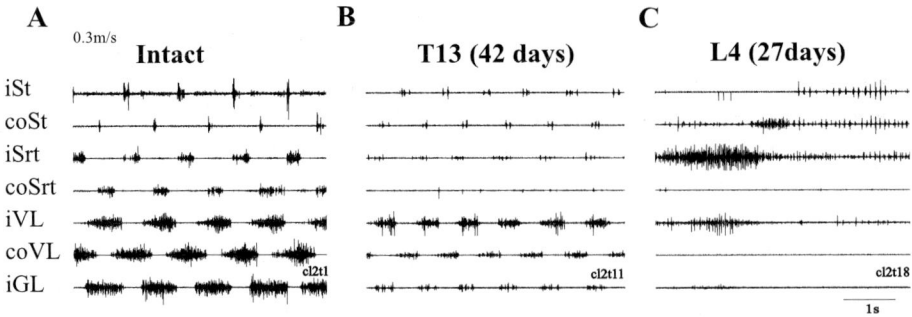

Fig. 4. Effect on locomotion of a first spinalization at T13, and then, a second one at L4. (A) Locomotion in the intact state. (B) Same cat, 42 days after spinalization at T13. Note the well-organized EMGs, which are reduced in amplitude, however. (C) Same cat, 27 days after a second spinalization at L4; note that in this case, EMG activity is present but disorganized.

compatible with results from experiments in which embryonic 5-HT cells were grafted caudal to a spinal lesion in adult rats. Locomotion was expressed only in cases wherein the 5-HT innervation reached the upper lumbar segments (Gimenez y Ribotta et al., 2000).

Currently, we are investigating further the importance of these midlumbar segments in the cat (Langlet and Rossignol, unpublished). Figure 4 illustrates three EMG sequences: first of a normal cat walking on a treadmill (Fig. 4A); and then, 42 days after a complete transection at T13 (Fig. 4B). Although the postlesion EMG amplitudes were smaller, the animal walked quite effectively on the treadmill. After a second spinal section at L4, however, and despite intensive attempts at training for 27 days (Fig. 4C), we were unable to reelicit locomotion even though the EMGs could still be activated by various other stimuli, thereby assuring that the motoneurones were still functional. For example, in response to placing a tape on the dorsum of the foot or after dipping the foot in water, there were surprisingly vigorous paw shakes, which involved several proximal and distal muscles, but still no locomotion. This observation on a chronic spinal preparation corroborates the finding that the acute spinal cat is also unable to walk on the treadmill after spinalization at L4 (Marcoux and Rossignol, 2000). We are now examining the effects of intraspinal electrical stimulation of these midlumbar spinal segments in the hope of activating the full pattern of locomotion from these segments (Barthelémy, Leblond and Rossignol, unpublished).

Conclusions

This brief review emphasizes two key points. (1) After spinalization in animals, the various locomotor deficits can be attributed to the severance of the loss of descending pathways which normally exert modulatory influences on the locomotor pattern. (2) The principal determinants for the expression of the spinal locomotor pattern are: the activation and training of a spinal locomotor pattern generator; the stimulation of sensory inputs; the changes occurring in the neurochemical balance of receptors of various neurotransmitters in the spinal cord; and, finally, the importance of midlumbar segments which appear to play a key role in the expression of spinal locomotion. It is hoped that the identification of some of the changes occurring in the cord after spinal lesions will lead to a better understanding of the recovery of function after spinal injury and thereby lead to better management and treatment of spinal-cord-injured humans (Barbeau and Rossignol, 1994; Colombo et al., 1998; Barbeau et al., 1999; Wernig et al., 1999; Wieler et al., 1999).

Acknowledgments

We would like to thank Drs. Trevor Drew and Becky Farley for reviewing a draft of this article, and Janyne Provencher, France Lebel, Claude Gagner, Philippe Drapeau, and Claude Gauthier for their technical contributions. The work was supported in part by the Canadian Institute for Health Research and the Spinal Cord Research Foundation.

Abbreviations

5-HT	5-hydroxytryptamine
5-HT	serotonine or 5-hydroxytryptamine
AMPA	alpha-amino-3-hydroxy-5-methyl-4-isoxazole propionate
AP-5 or APV	2-amino-5-phosphonovalerate
co	contralateral
E1	1st extension phase
E2	2nd extension phase
E3	3rd extension phase
EDL	extensor digitorum longus
F	flexion phase
GL	gastrocnemius lateralis
i	ipsilateral
L1	L3, L4; lumbar 1, lumbar 3, lumbar 4
MTP	metatarsophalangeal
NA	noradrenaline
NMDA	N-methyl-D-aspartate
SMPTE	Society of Motion Picture and Television Engineers
Srt	sartorius
St	semitendinosus
T13	thoracic 13
VL	vastus lateralis

References

Andersson, O., Forssberg, H., Grillner, S. and Wallen, P. (1981). Peripheral feedback mechanisms acting on the central pattern generators for locomotion in fish and cat. Can. J. Physiol. Pharmacol., 59: 713–726.

Barbeau, H. and Rossignol, S. (1987). Recovery of locomotion after chronic spinalization in the adult cat. Brain Res., 412: 84–95.

Barbeau, H. and Rossignol, S. (1994). Enhancement of locomotor recovery following spinal cord injury. Curr. Opin. Neurol., 7: 517–524.

Barbeau, H., Julien, C. and Rossignol, S. (1987). The effects of clonidine and yohimbine on locomotion and cutaneous reflexes in the adult chronic spinal cat. Brain Res., 437: 83–96.

Barbeau, H., McCrea, D.A., O'Donovan, M.J., Rossignol, S., Grill, W.M. and Lemay, M.A. (1999). Tapping into spinal circuits to restore motor function. Brain Res. Rev., 30: 27–51.

Bélanger, M., Drew, T., Provencher, J. and Rossignol, S. (1996). A comparison of treadmill locomotion in adult cats before and after spinal transection. J. Neurophysiol., 76: 471–491.

Bouyer, L. and Rossignol, S. (1998) The contribution of cutaneous inputs to locomotion in the intact and the spinal cat. In: Kiehn, O., Harris-Warrick, R.M., Jordan, L.M., Hultborn, H. and Kudo, N. (Eds.), Neuronal Mechanisms for Generating Locomotor Activity. Ann. N.Y. Acad. Sci., New York, pp. 508–512.

Bouyer, L. and Rossignol, S. (2001). Spinal cord plasticity associated with locomotor compensation to peripheral nerve lesions in the cat. In: Patterson M.M., Grau J.W., Wolpaw J.R., Willis W.D. and Edgerton V.R. (Eds.), Spinal Cord Plasticity: Alterations in Reflex Function. Kluwer Academic Publishers, Boston, pp. 207–224.

Bregman, B.S. and Goldberger, M.E. (1983). Infant lesion effect: I. Development of motor behavior following neonatal spinal cord damage in cats. Dev. Brain Res., 9: 103–117.

Brustein, E. and Rossignol, S. (1998). Recovery of locomotion after ventral and ventrolateral spinal lesions in the cat. I. Deficits and adaptive mechanisms. J. Neurophysiol., 80: 1245–1267.

Cazalets, J.-R. (2000). Organization of the spinal locomotor network in neonatal rat. In: Kalb R.G. and Strittmater S.M. (Eds.), Neurobiology of Spinal Cord Injury. Humana Press, Totowa, NJ, pp. 89–111.

Chau, C., Barbeau, H. and Rossignol, S. (1998a). Early locomotor training with clonidine in spinal cats. J. Neurophysiol., 79: 392–409.

Chau, C., Barbeau, H. and Rossignol, S. (1998b). Effects of intrathecal alpha1- and alpha2-noradrenergic agonists and norepinephrine on locomotion in chronic spinal cats. J. Neurophysiol., 79: 2941–2963.

Colombo, G., Wirz, M. and Dietz, V. (1998). Effect of locomotor training related to clinical and electrophysiological examinations in spinal cord injured humans. Ann. N.Y. Acad. Sci., 860: 536–538.

Delcomyn, F. (1980). Neural basis of rhythmic behavior in animals. Science, 210: 492–498.

de Leon, R.D., Hodgson, J.A., Roy, R.R. and Edgerton, V.R. (1998). Full weight-bearing hindlimb standing following stand training in the adult spinal cat. J. Neurophysiol., 80: 83–91.

Eidelberg, E., Story, J.L., Meyer, B.L. and Nystel, J. (1980). Stepping by chronic spinal cats. Exp. Brain Res., 40: 241–246.

Forssberg, H. and Grillner, S. (1973). The locomotion of the acute spinal cat injected with clonidine i.v. Brain Res., 50: 184–186.

Forssberg, H., Grillner, S. and Rossignol, S. (1975). Phase dependent reflex reversal during walking in chronic spinal cats. Brain Res., 85: 103–107.

Forssberg, H., Grillner, S. and Halbertsma, J. (1980a). The locomotion of the low spinal cat. I. Coordination within a hindlimb. Acta Physiol. Scand., 108: 269–281.

Forssberg, H., Grillner, S., Halbertsma, J. and Rossignol, S. (1980b). The locomotion of the low spinal cat: II. Interlimb coordination. Acta Physiol. Scand., 108: 283–295.

Gimenez y Ribotta, M., Provencher, J., Feraboli-Lohnherr, D., Rossignol, S., Privat, A. and Orsal, D. (2000). Activation of locomotion in adult chronic spinal rats is achieved by transplantation of embryonic raphe cells reinnervating a precise lumbar level. J. Neurosci., 20: 5144–5152.

Giroux, N., Rossignol, S. and Reader, T.A. (1999). Autoradiographic study of alpha1- and alpha2-noradrenergic and serotonin1A receptors in the spinal cord of normal and chronically transected cats. J. Comp. Neurol., 406: 402–414.

Giroux, N., Reader, T.A. and Rossignol, S. (2001). Comparison of the effect of intrathecal administration of clonidine and yohimbine on the locomotion of intact and spinal cats. J. Neurophysiol., 85: 2516–2536.

Goldberger, M.E. (1977). Locomotor recovery after unilateral hindlimb deafferentation in cats. Brain Res., 123: 59–74.

Grillner, S. (1973). Locomotion in the spinal cat. In: Stein R.B., Pearson K.G., Smith R.S. and Redford J.B. (Eds.), Control of Posture and Locomotion. Plenum Press, New York, pp. 515–535.

Grillner, S. (1981). Control of locomotion in bipeds, tetrapods, and fish. In: Brookhart J.M. and Mountcastle V.B. (Eds.), Handbook of Physiology. The Nervous System, Vol. II, American Physiological Society, Bethesda, pp. 1179–1236.

Grillner, S. and Zangger, P. (1979). On the central generation of locomotion in the low spinal cat. Exp. Brain Res., 34: 241–261.

Howland, D.R., Bregman, B.S. and Goldberger, M.E. (1995a). The development of quadrupedal locomotion in the kitten. Exp. Neurol., 135: 93–107.

Howland, D.R., Bregman, B.S., Tessler, A. and Goldberger, M.E. (1995b). Development of locomotor behavior in the spinal kitten. Exp. Neurol., 135: 108–122.

Jiang, W. and Drew, T. (1996). Effects of bilateral lesions of the dorsolateral funiculi and dorsal columns at the level of the low thoracic spinal cord on the control of locomotion in the adult cat: I. Treadmill walking. J. Neurophysiol., 76: 849–866.

Kiehn, O. and Kjaerulff, O. (1998). Distribution of central pattern generators for rhythmic motor outputs in the spinal cord of limbed vertebrates. Ann. N.Y. Acad. Sci., 860: 110–129.

Marcoux, J. and Rossignol, S. (2000). Initiating or blocking locomotion in spinal cats by applying noradrenergic drugs to restricted lumbar spinal segments. J. Neurosci., 20: 8577–8585.

Pearson, K.G. (1993). Common principles of motor control in vertebrates and invertebrates. Annu. Rev. Neurosci., 16: 265–297.

Philippson, M. (1905). L'autonomie et la centralisation dans le système nerveux des animaux. Trav. Lab. Physiol. Inst. Solvay (Bruxelles), 7: 1–208.

Rossignol, S. (1996). Neural control of stereotypic limb movements. In: Rowell L.B. and Sheperd J.T. (Eds.), Handbook of Physiology, Sec. 12. Exercise: Regulation and Integration of Multiple Systems. American Physiological Society, Oxford, pp. 173–216.

Rossignol, S. (2000). Locomotion and its recovery after spinal injury. Curr. Opin. Neurobiol., 10: 708–716.

Rossignol, S., Lund, J.P. and Drew, T. (1988). The role of sensory inputs in regulating patterns of rhythmical movements in higher vertebrates. A comparison between locomotion, respiration and mastication. In: Cohen A., Rossignol S. and Grillner S. (Eds.), Neural Control of Rhythmic Movements in Vertebrates. Wiley and Sons Co, New York, pp. 201–283.

Rossignol, S., Chau, C., Brustein, E., Bélanger, M., Barbeau, H. and Drew, T. (1996). Locomotor capacities after complete and partial lesions of the spinal cord. Acta Neurobiol. Exp., 56: 449–463.

Rossignol, S., Drew, T., Brustein, E. and Jiang, W. (1999). Locomotor performance and adaptation after partial or complete spinal cord lesions in the cat. In: Binder M.D. (Ed.), Peripheral and Spinal Mechanisms in the Neural Control of Movement. Elsevier, New York, pp. 349–365.

Rossignol, S., Bélanger, M., Chau, C., Giroux, N., Brustein, E., Bouyer, L., Grenier, C.-A., Drew, T., Barbeau, H., Reader, T. (2000). The spinal cat. In: Kalb R.G. and Strittmatter S.M. (Eds.), Neurobiology of Spinal Cord Injury. Humana Press, Totowa, NJ, pp. 57–87.

Rossignol, S., Chau, C., Giroux, N., Brustein, E., Bouyer, L., Marcoux, J., Langlet, C., Barthelémy, D., Provencher, J., Leblond, H., Barbeau, H., Reader, T.A. (2002). The cat model of spinal injury. Prog. Brain Res., 137: 151–168.

Shurrager, P.S. (1955). Walking in spinal kittens and puppies. In: Windle W.F. (Ed.), Regeneration in the Central Nervous System. C.C. Thomas, Springfield, pp. 208–218.

Shurrager, P.S. and Dykman, R.A. (1951). Walking spinal carnivores. J. Comp. Physiol. Psychol., 44: 252–262.

Wernig, A., Nanassy, A. and Muller, S. (1999). Laufband (treadmill) therapy in incomplete paraplegia and tetraplegia. J. Neurotrauma, 16: 719–726.

Wieler, M., Stein, R.B., Ladouceur, M., Whittaker, M., Smith, A.W., Naaman, S., Barbeau, H., Bugaresti, J. and Aimone, E. (1999). Multicenter evaluation of electrical stimulation systems for walking. Arch. Phys. Med. Rehabil., 80: 495–500.

SECTION V

Biomechanical and imaging approaches in movement neuroscience

CHAPTER 17

Trunk movements and EMG activity in the cat: level versus upslope walking

Naomi Wada[1,*] and Kenro Kanda[2]

[1]*Department of Veterinary Physiology, Yamaguchi University, Yamaguchi 753-8515, Japan*
[2]*Department of Central Nervous System, Tokyo Metropolitan Institute of Gerontology, Tokyo, 173-0015, Japan*

Abstract: This chapter addresses the neural control of spinal-column behavior during locomotion. Kinematic and EMG measurements were obtained from the adult cat during its level and upslope treadmill walking. Increasing the grade of upslope walking augmented horizontal movements of the spinal column and decreased its lateral movements. In conjunction, there were significant increases in the amplitude, duration, and pattern of EMG bursts in relevant spinal column musculature. During even steeper upslope walking, three EMG bursts were evident. They were phase-locked to the outward, downward and backward movements of the spinal column, respectively. Our results suggest that the component of the locomotor pattern generator that produces rhythmical spinal column movements must generate a wide variety of EMG bursts in spinal column muscles, as dependent in part on sensory input from the spinal column and its musculature.

Introduction

The spinal column of a terrestrial mammalian tetrapod like the cat or dog moves rhythmically during locomotion. Its lateral movement is substantial during symmetrical gaits, like walking and trotting, as is vertical movement during asymmetrical gaits, like galloping (for review, see Gambarian, 1974). At first glance, it might be thought that such locomotion-induced movements are attributable largely to the movement and position of the limbs, with the differences during symmetrical versus asymmetrical gait being due to the differences in the timing of limb contact and liftoff. The neural control of the trunk's epaxial and hypaxial muscles must also be considered, however. The epaxial ones include a medial (e.g.,

m. multifidus) and a lateral (m. longissimus, m. iliocostalis) group. The lateral ones have a large cross-sectional area at their lumbar level, where two bilateral EMG bursts occur during a single step cycle (Carlson, 1978). This activity might function to increase the stiffness of the trunk, and thereby damp its lateral movements (Carlson et al., 1979; English, 1980). Koehler et al. (1984) analyzed the rhythmic activity of motoneurons innervating m. longissimus during fictive locomotion, and suggested that such activity was quite different to that during natural locomotion. Thus, peripheral feedback, spinal pattern-generating activity and descending control signals all contribute to the overall control of locomotion (for review, see Orlovsky et al., 1999; Grillner and Wallén, Chapter 1 of this volume).

It must also be recognized that motoneuron activity during locomotion is task- and context-dependent. For example, Smith and collaborators

*Corresponding author: Tel.: +81-839-335885; Fax: +81-839-335885; E-mail: naomi@yamaguchi-u.ac.jp

(Carlson-Kuhta et al., 1998; Smith et al., 1998) studied differences in posture, hindlimb kinematics and EMG patterns in the cat during level versus inclined treadmill walking. They proposed that differences in all three were due to the difference in force output by the hindlimbs during these two tasks, as required to continuously move the body mass forward and upward. Postural changes during inclined walking, however, must also involve adjustments of the head, trunk, tail and forelimbs in addition to those of the hindlimbs. In particular, the integrated control of trunk movements and posture seem essential for stable walking to be performed under a variety of external conditions and contexts. To emphasize this point, we report here on our kinematic and EMG studies of the trunk, in addition to the fore- and hindlimbs, during level versus upslope walking on a treadmill (see also F. Mori et al., Chapter 19 of this volume).

Experimental features

Our experiments were performed on adult cats that had been trained to walk on a motorized treadmill. Under barbiturate anesthesia, 10–14 pairs of insulated, multistranded, stainless-steel wires were implanted bilaterally into m. longissimus between T5 and L5. The leadoff wires were extended subcutaneously to a multipin connector mounted on the skull. The treadmill's walkway was kept horizontal ($0°$), or inclined to different grades (5, 10, 15, 20, $25°$). The treadmill's speed was kept constant (45–60 m/min) during all grades of upslope walking. The movements of multiple points along the spinal column were recorded, as were lines connecting the midpoint between the eyes to the sacrum, forelimbs and hindlimbs, using two high-speed video cameras and a three-dimensional, motion-analysis system.

Limb kinematics during level versus upslope walking

Figure 1 shows that increasing the grade of upslope walking altered hindlimb but not forelimb movements. Figure 1A shows gait diagrams for a level ($0°$) versus upslope ($25°$) step. The paw-contact sequences were similar at all grades, and consistent with a 'lateral sequence', i.e., the paws contacted the ground in the order left hindlimb, left forelimb, right hindlimb and right forelimb. These gait diagrams also show that the single step cycle could be divided into eight phases (patterns) of limb contact with the treadmill belt. We focused on the most prevalent overall pattern (Fig. 1A), which was used by 7/9 cats. It included phases in which two (ipsilateral couplet), three (tripod) and four (quadruped) paws were simultaneously on the treadmill belt. Figure 1B shows that the average step-cycle duration and relative duration of the swing phase did not reduce significantly as the upslope grade was increased. The rate of swing-phase movement was significantly decreased, however, for the hindlimb, but not the forelimb. Figure 1C shows exemplary kinematic data for selected joints of the forelimb (shoulder, elbow, wrist) and hindlimb (hip, knee, ankle) during level versus upslope walking. Irrespective of the grade, the patterns of joint angular displacement were stereotyped and similar for all cats. The main feature of Fig. 1C is its demonstration that increasing the upslope grade prolonged the stance phase of the hindlimbs, and increased their range of angular movements. Such changes were not present in the forelimbs, however.

Spinal-column kinematics during level versus upslope walking

Figure 2A shows that the extent of spinal-column movement and its trajectory were dependent upon the point of measurement (e.g., T5 vs. L7) and the upslope grade ($0°$ vs. $25°$). This finding is expanded upon in Fig. 2B, which shows a close relationship between the point of spinal measurement (T5 vs. T10 vs. L1 vs. L5) and the four limbs' patterns of paw contact and liftoff. The times of paw contact and liftoff were usually found near the most lateral, caudal, cranial and lowest position of the spinal-column trajectories. These trajectories all differed from one another. These findings suggest that the movements of the spinal column during walking are strongly affected by the movement and position of the four limbs. Figure 2B also shows that increasing the upslope grade led to: (1) a decrease in the range

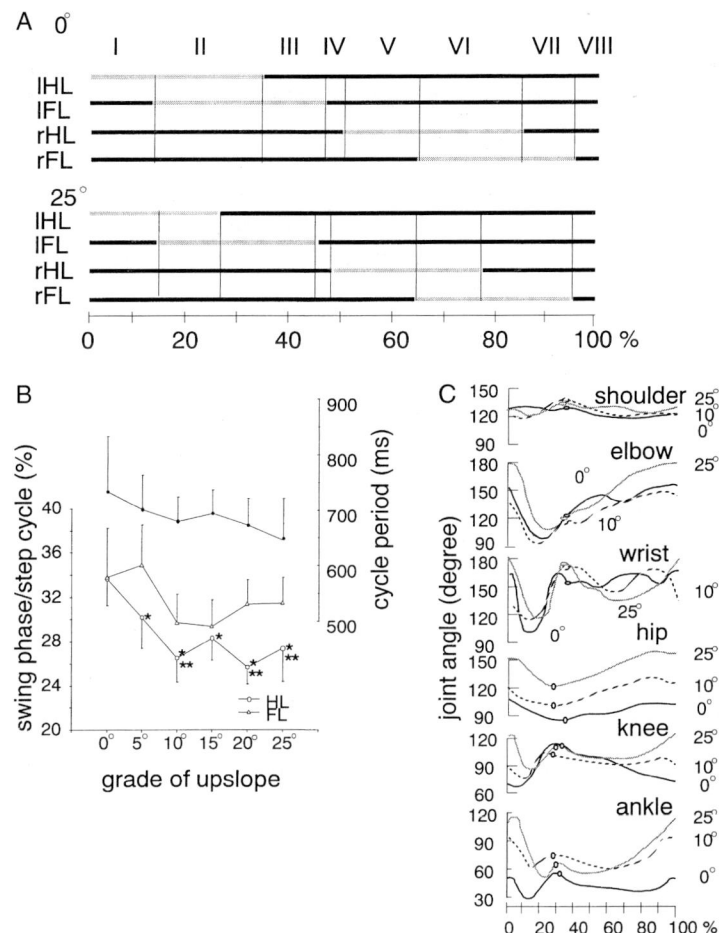

Fig. 1. Limb movements during level versus upslope walking. (A) Gait diagrams (technique of Hildebrand, 1959; averaged data from three cats) for a level (0°) versus upslope (25°) step. The relative time (0–100%) is shown for a single step cycle at both grades. The durations of paw contact (dark horizontal line) versus paw-off-belt (shaded horizontal line) are shown for the left (l) and right (r) hindlimbs (HLs), with all limbs referenced to the rHL cycle. The step cycle is divided into eight phases (I–VIII), each with a distinctive pattern of paw contacts, i.e., two (couplet), three (tripod) or four (quadruped) paws simultaneously on the belt. The shown footfall pattern was the most commonly used one, i.e., phase I, tripod; phase II, ipsilateral couplet; phase III, tripod; phase IV, quadruped; phase V, tripod; phase VI, ipsilateral couplet; phase VII, tripod; and phase VIII, quadruped. Note that changing the grade from 0° to 25° had the most effect on phases II, III, VI and VII. This meant that increasing the grade of upslope prolonged the stance phase of hindlimbs. (B) Changes in the duration of the step cycle (right-side ordinate; upper line with filled-circle data points) and the relative duration of its swing phase (left-side ordinate) in the left hindlimb (middle line with open-triangle data points) and left forelimb (lower line with open-circle data points) as a function of treadmill grade (0–25°). Vertical lines are for 50% of the SE of average values for two cats. The average step-cycle duration (range, 500–900 ms) and relative duration of the swing phase (range, ∼ 24–38%) reduced somewhat as the grade was increased, but these changes were not significant. Not shown is the rate (°/s) of swing-phase movement. This parameter reduced value in the hindlimbs, but not the forelimbs, as the upslope grade was increased. * and ** indicate a significant difference for this latter parameter at 0° and 5°, respectively. (C) Typical averaged forelimb and hindlimb kinematics (displacement vs. relative time) for a level (0°) versus upslope (10°, 25°; see right-side lettering) step at the shoulder/elbow/wrist and hip/knee/ankle joints. The solid lines show the average values (three cats) for 15 consecutive steps. These averages were quite similar to the records of the individual steps of individual animals. Graphs begin and end with paw liftoff. Note that at the onset of the swing phase, the three forelimb joints and the knee/ankle of hindlimb were in the flexion (F) phase of the step, in order to elevate the paw. Later in the swing phase, each joint extended (E1 phase; i.e., to lower the paw for contact). The hip joint was flexed during the swing phase and extended after contact. During E2, the elbow/wrist/shoulder and knee/ankle yielded, and this yield was usually sustained briefly. Note that as the upslope grade increased, the angular curves of the hindlimb joints increased in parallel, and their overall range enlarged. Finally, note that the range of all three hindlimb-joint movements increased as a function of treadmill grade. This did not occur in the forelimb joints.

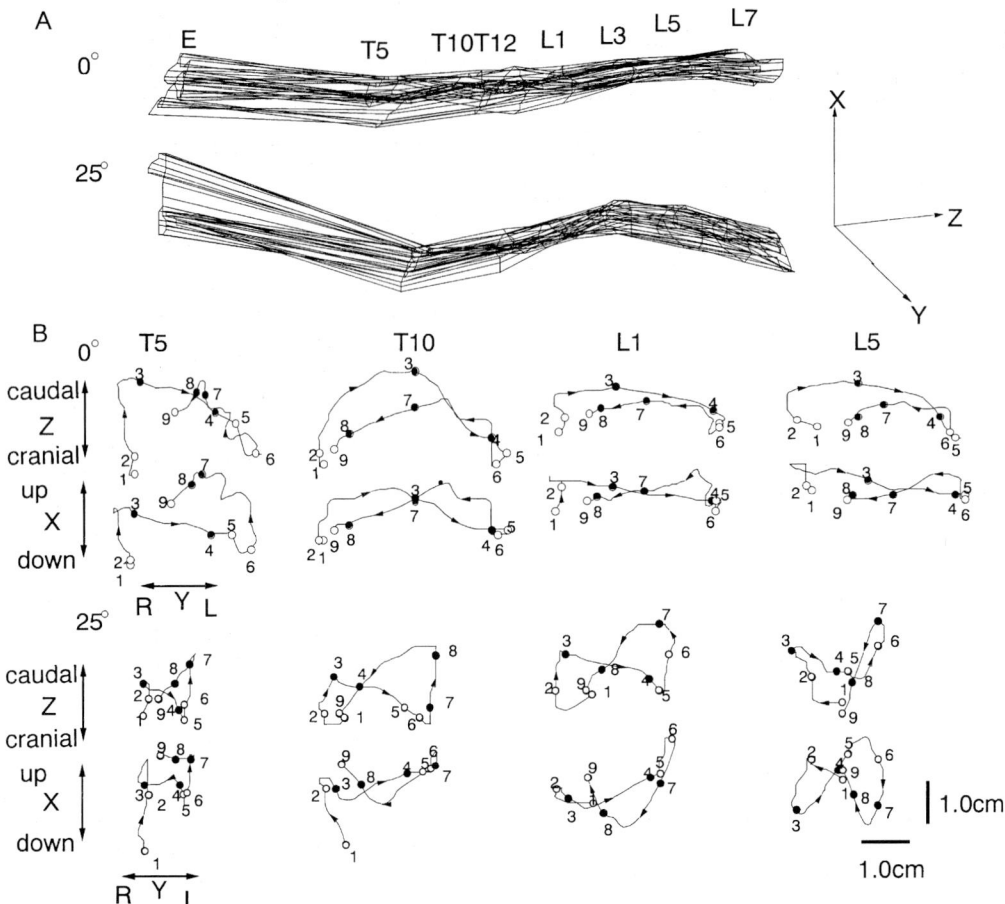

1: lHL liftoff, 2: lFL liftoff, 3: lHL contact, 4: lFL contact,
5: rHL liftoff, 6: rFL liftoff, 7: rHL contact, 8: rFL contact, 9: lHL liftoff.

Fig. 2. Spinal-column movements during level versus upslope walking. (A) Stick diagrams of the spinal column based on seven measurement points along its length (T5 to L7) relative to E, the midpoint between the eyes. The right-side schematic illustrates the vertical (X), horizontal (Y) and transverse (Z) axis. Note that at a 25° upgrade, flexion of the spinal column was much more pronounced than at the 0° grade. (B) Trajectories of spinal measurement points T5, T10, L1 and L5 during a single step at 0° versus 25°. Spinal-column movements to the right (R) and left (L) are shown on the Y axis (Y) in the horizontal (X–Y) and vertical (X–Z) planes. Paw contact and liftoff and of the individual limbs are designated by numbers and symbols. Note that as a single step cycle unfolds, there is a single cycle of movement, which is indicated at the bottom of the panel. Finally, note that increasing the upslope grade from 0° to 25° caused changes in the trajectories of a variety of movements in all three planes (i.e., the bottom 1–9 code).

of lateral movement, but an increase in the range of horizontal movement; and (2) changes in virtually all of the movement trajectories. It seems likely that a decreased lateral movement was caused by a shortened duration of the ipsilateral couplet support phase (II and VI), and that an increased horizontal movement was caused by increasing the angular movements of the hindlimb joints (see Fig. 1C). Furthermore, it appeared that an increase in the propulsive power of a hindlimb could alter the spinal-column movements. Finally, Fig. 2 gives credence to the idea that an upslope-grade-induced

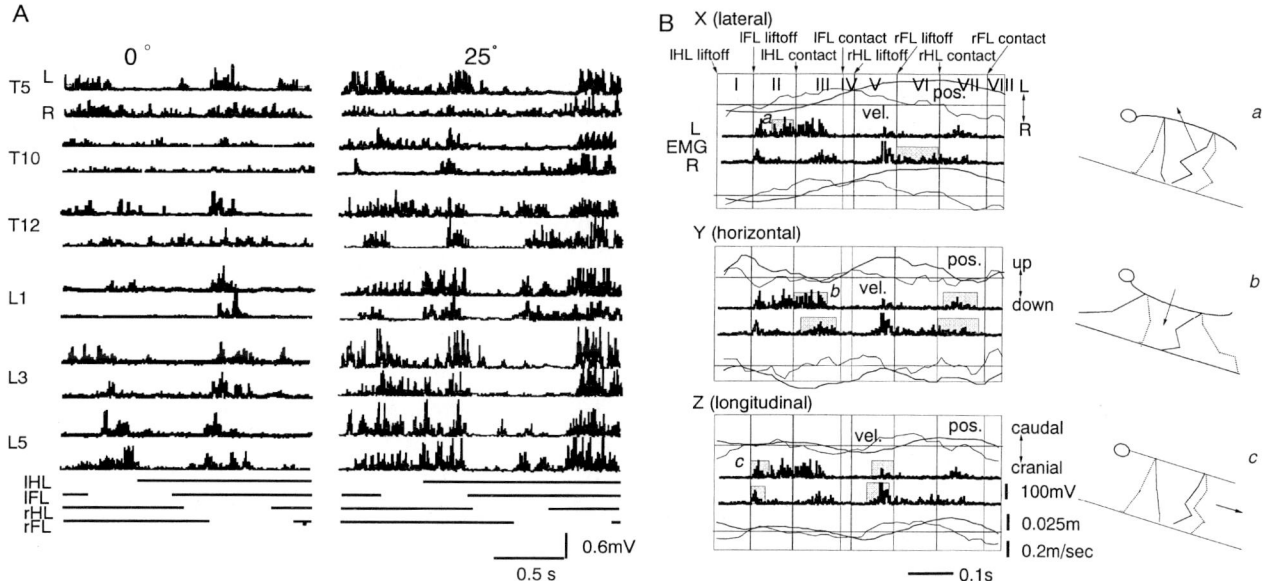

Fig. 3. EMG activity in m. longissimus during level versus upslope walking. (A) Shown are exemplary bursts of EMG activity in the left (L) and right (R) muscles of a single cat. The EMGs were recorded at various vertebral levels (T5 to L5) during a single step at 0° versus 25° grade. Note that the amplitude, duration and phase of these EMG bursts were grade-dependent. At the bottom of the panel, the footfall patterns (viz., Fig. 1A) are shown. In general, at all vertebral levels, there were two EMG bursts in phases II–III and VI–VII. These were of different amplitude and duration at these different levels, however. (B) Analogous, single-step EMG activity at the L5 level in another cat at a 25° upslope grade. Also shown are the position (pos.) and velocity (vel.) at the spinal-column point adjacent to the two sites of EMG recording. Note that these are presented for the X, Y and Z coordinates of Fig. 2. The sketches on the far-right side of the panel show the movements of the spinal column relative to the positions of the fore- and hindlimbs. EMG bursts were observed that were concomitant with lateral (*a*), downward (*b*) and backward (*c*) movements of the spinal column. The lateral (*a*) bursts occurred mainly in phases II and VI (viz., Fig. 1A) of level walking. During upslope walking, *b* and (shorter) *c* bursts followed by an *a* burst were prominent in phases III and VII. Note further that the *a* bursts usually started after the onset of a spinal movement from the right to the left side, and they continued until the midphase of the step. During such *a* bursts, the cat brought one hindlimb forward, and lateral bending of the spinal column was most prominent. The *b* bursts started before the lowest vertical position of the spinal column was reached, and they continued until the onset of an upward spinal-column movement. During these *b* bursts, the cat brought one forelimb forward and the contralateral hindlimb was in the late phase of the stance phase. At this stage, the spinal column was stretched, and its downward movement was most prominent. The *c* bursts started after the L5 point along the spinal column had reached its most cranial position. At this stage, one hindlimb was in the midswing phase while the other was in midstance.

prolongation of the stance phase of the hindlimbs led to an increase in the upward propulsion of the trunk, and thereby the movements of the spinal column.

The EMG of m. longissimus during level versus upslope walking

Figure 3A shows exemplary, bilateral EMG patterns in the m. longissimus of a single cat. The EMGs were recorded at different vertebral levels (T5 to L5) throughout a single step at a grade of 0° versus 25°. In general, the smallest and shortest EMG bursts were found at T12 and L1. Increasing the upslope grade caused an increase in the amplitude and duration of the EMG bursts. Figure 3B shows similarly exemplary data from another cat. It reveals associations between EMG patterns and the three-dimensional movements of the spinal column throughout a single step at a 25° grade. The movement velocity at each measured level of the spinal column was *decreased* by the EMG bursts, thereby revealing that the EMG activity of m.

longissimus functions to damp the locomotion-induced movements of the spinal column.

Neural control of trunk muscles during walking

Koehler et al. (1984) reported that an alternating EMG discharge of m. longissimus between one side and the other was observed during a single cycle of rhythmic activity of the hindlimb flexors. The discharge pattern during fictive locomotion in the spinalized cat was completely different from that in the intact, freely moving one. These findings suggest the singular importance of peripheral-afferent feedback and supraspinal control in the neural control of the trunk muscles during locomotion. Figure 3 supports this idea, and further suggests that neuronal activity of both peripheral and central origin can produce different kinds of m. longissimus EMG bursts that increase the stiffness of the spinal column. To date, we do not know if this muscle's three EMG bursts (Fig. 3) result from the activity of *separate* groups of motoneurons. We do have evidence that trunk–muscle motoneurons can be divided into different groups according to their synaptic input patterns from various muscle and species of cutaneous afferent supplying the trunk (unpublished data of Wada and colleagues) and hindlimb (Wada et al., 1999; Wada and Kanda, 2001). These findings suggest that individual motoneurons may play different roles in the control of the trunk, i.e., to increase its stiffness, and produce its vertical, lateral bending and other movements. Thus, there might well be different groups of motoneurons producing the different kinds of EMG bursts reported in the present chapter. Clearly, the present findings indicate that whatever neural mechanisms come into play in the control of trunk–muscle activity during locomotion, they are both task- and context-dependent.

Acknowledgments

We thank our colleague Junko Akatani for his assistance with the data collection and kinematic analysis. This research was supported in part by the Japanese Ministry of Education, Science and Culture.

Abbreviations

EMG	electromyogram/electromyographic
FL	forelimb
HL	hindlimb
L	left
L1–L5–L7	first, fifth and seventh lumbar vertebra, respectively
R	right
T5–T10	fifth and tenth thoracic vertebra, respectively

References

Carlson, H. (1978). Morphology and contraction properties of cat lumbar back muscles. Acta Physiol. Scand., 103: 180–197.

Carlson, H., Halbertsma, J. and Zomlefer, M. (1979). Control of trunk during walking in the cat. Acta Physiol. Scand., 105: 251–253.

Carlson-Kuhta, P., Trank, T.V. and Smith, J.L. (1998). Forms of forward quadrupedal locomotion. II. A comparison of posture, hindlimb kinematics, and motor patterns for upslope and level walking. J. Neurophysiol., 79: 1687–1701.

English, A.W. (1980). The functions of the lumbar spine during stepping in the cat. J. Morphol., 165: 55–66.

Gambarian, P.P. (1974). How Mammals Run, Anatomical Adaptation. New York, Toronto, John Wiley and Sons.

Hildebrand, M. (1959). Motions of the running cheetah and horse. J. Mammal., 40: 481–495.

Koehler, W.J., Schomburg, E.D. and Steffens, H. (1984). Phasic modulation of trunk muscle afferents during fictive spinal locomotion in cats. J. Physiol. (Lond.), 353: 187–197.

Orlovsky, G.N., Deliagina, T.G. and Grillner, S. (1999). Quadrupedal locomotion in mammals: Neuronal Control of Locomotion from Mollusc to Man, Part III. Oxford University Press, London, pp. 157–246.

Smith, J.L., Carlson-Kuhta, P. and Trank, T.V. (1998). Forms of forward quadrupedal locomotion. III. A comparison of posture, hindlimb kinematics, and motor patterns for downslope and level walking. J. Neurophysiol., 79: 1702–1716.

Wada, N. and Kanda, K. (2001). Neuronal pathways from group-I and -II muscle afferents innervating hindlimb muscles to motoneurons innervating trunk muscles in low-spinal cats. Exp. Brain Res., 136: 263–268.

Wada, N., Shikaki, N., Tokuriki, M. and Kanda, K. (1999). Neuronal pathways from low-threshold hindlimb cutaneous afferents to motoneurons innervating trunk muscles in low-spinal cats. Exp. Brain Res., 128: 543–549.

CHAPTER 18

Biomechanical constraints in hindlimb joints during the quadrupedal versus bipedal locomotion of *M. fuscata*

Katsumi Nakajima[1,2,*], Futoshi Mori[2], Chijiko Takasu[2], Masahiro Mori[2], Kiyoji Matsuyama[3] and Shigemi Mori[2]

[1]*Department of Physiology, Kinki University School of Medicine, Osaka-Sayama 589-8511, Japan;*
[2]*Department of Biological Control System, National Institute for Physiological Sciences, Okazaki 444-8585, Japan;*
[3]*Department of Physiology, Sapporo Medical University, Sapporo 060-8556, Japan*

Abstract: The operant-trained Japanese monkey, *Macaca fuscata*, can walk with both a quadrupedal (Qp) and a bipedal (Bp) gait on the surface of a treadmill belt, which moves at different speeds. The animal can also learn to transform its locomotor pattern from Qp to Bp, and vice versa, without a break in forward walking speed. This non-human primate model provides an intriguing opportunity to compare the kinematics of multiple body segments during Qp and Bp walking of the same subject. We found that *M. fuscata* selects a postural strategy and limb-kinematic parameters appropriate for the execution of both gaits. We propose that the basic locomotor rhythm-generating mechanisms of the brainstem and spinal cord, which are genetically endowed and relatively automatized, are used for the execution of both the Qp and Bp gait. The latter requires in addition, however, some higher-level circuitry which is shaped substantially by motor learning mechanisms.

Introduction

Bipedal (Bp) locomotion is an everyday activity for humans and selected advanced nonhuman primates. Despite its apparent simplicity, it is a complex CNS-controlled activity. It requires the integrated control of multiple motor segments, including the head, neck, trunk and limbs (Sherrington, 1906; Mori et al., 2001a; Nakajima et al., 2001). To advance understanding of its neural control mechanisms, a wide variety of experimental paradigms have been used (Orlovsky et al., 1999). For example, in human subjects, Herman et al. (1976) focused upon biomechanical information about movements of the lower limb. Their measures of angular displacement of the hip, knee and ankle revealed a precise temporal and spatial ordering, i.e., angular motion sequentially determined in both the swing (SW) and stance (ST) phase of the step cycle. Another approach was that of Nilson and Thorstensson (1989) who recorded three orthogonal ground reaction force components in the weight-bearing limbs during human Bp walking and running. They found complex interactions between the vertical and horizontal forces needed for propulsion and equilibrium, which changed with the speed and mode of progression. In yet a third approach, Fukuyama et al. (1997) used single photon emission computed tomography (SPECT) on normal subjects to identify several brain regions whose activity increased during walking (see also Shibaski et al., Chapter 20 of this volume).

For the Bp locomotion of nonhuman primates, most previous studies are those of anthropologists

*Corresponding author: Tel.: +81-72-366-0221; Fax: +81-72-366-0206; E-mail: nakajima@med.kindai.ac.jp

and biologists seeking to elucidate its kinematics, and the relationship between morphology and species-specific locomotor behavior (Hildebrand, 1967; Rose, 1973; Fleagle and Mittermeier, 1980; Rollinson and Martin, 1981; Ishida et al., 1985; Okada, 1985). Far fewer studies have been undertaken from a movement neuroscience perspective, however (in toto, see Lawrence and Kuypers, 1968a,b; Eidelberg et al., 1981; Vilensky et al., 1992; Mori et al., 1996, 2001a; Nakajima et al., 2001).

Our group's long-term goal is to elucidate CNS mechanisms controlling Bp locomotion in the nonhuman primate, wherein surgical intervention and associated electrophysiological recording are feasible. We hope to compare and extrapolate such-discovered mechanisms to those that might operate in the human. To this end, we have developed a nonhuman primate model, the developing and mature Japanese monkey, *Macaca fuscata*. We have trained this animal by operant conditioning to walk bipedally on the surface of a moving treadmill belt (Nakajima et al., 1997, 2001; Mori et al., 2000, 2001b; Tachibana et al., 2000; see also F. Mori et al., Chapter 19 of this volume).

An intriguing capability of *M. fuscata* is that it can transform its locomotor patterns from Qp to Bp and vice versa without a break in forward speed on the moving treadmill belt (Nakajima et al., 2001). This suggests that its CNS can rapidly select and combine integrated subsets of posture- and locomotor-related neural control mechanisms appropriate for the task at hand (Mori et al., 2001b). In this chapter, we explore the possibility that a comparison of the kinematics of hindlimb movements during the Bp versus Qp locomotion of the same animal may yield biomechanical information relevant to CNS control mechanisms, including some concerning motor learning. In the following sections, we first describe some critical biomechanical factors for the elaboration of Bp locomotion. Then, we postulate on how the CNS might control such factors.

Features of Qp versus Bp locomotion

The generic monkey's Qp walking is characterized by diagonal movements of the fore- and hindlimbs (diagonal sequence gait; Prost, 1965; Iwamoto and Tomita, 1966; Hildebrand, 1967; Mori et al., 1996).

Fig. 1A shows a monkey's elaboration of a single Qp step cycle.

From right to left in Fig. 1, note how the monkey extended the left (L) hindlimb with a maximum backward excursion at the termination of the ST phase. It then moved this limb forward to initiate the SW phase. Note that this limb's placement, at nearly maximum forward excursion, was about where the imaginary center of body mass would project (Preuschoft et al., 1988; Nakajima et al., 2001). The monkey then moved the L hindlimb backward to initiate the next ST phase. The final photograph shows when this limb began to come forward again at the termination of this ST phase. During the earlier single step cycle, the body's weight was supported sequentially by four, and then three limbs. Sometimes the latter could be two diagonal limbs. At a treadmill speed of 1.0 m/s, the body axis angle (i.e., between the ear-hip and the horizontal) was $\sim 13°$. The total excursion distance of the L hindlimb was ~ 40 cm. As speed increased progressively from 0.4 to 1.5 m/s, the body axis angle changed little (10 to 12–13°), whereas the limb excursions had relatively larger additions (e.g., L hindlimb from ~ 25 to ~ 45 cm).

Bipedal walking is characterized by double and/or single support phases of the L and R hindlimbs (Muybridge, 1955; Herman et al., 1976; Grillner et al., 1979). The propulsive force to carry the body mass forward is generated by the weight of the body mass 'sliding down an incline' with forward limb movements. Fig. 1B shows the same monkey's elaboration of a single Bp step cycle on the same day and at the same treadmill speed. Again from right to left, note how the animal first elaborated a double support phase by placing of the L and R hindlimbs backward and forward, respectively. The L hindlimb was at the terminal ST phase whereas the R hindlimb was at the early ST phase. The monkey then lifted its L hindlimb from the treadmill belt and moved it forward. Next, it again elaborated a double support phase. During the SW phase of the L hindlimb, the bodyweight was fully supported by the R hindlimb alone (single support phase). The monkey then initiated the next SW phase of the R hindlimb. The final photograph shows the instance at which the L hindlimb reached the terminal ST phase. A generality we found for Bp locomotion was that forward movements of the L and R hindlimbs were often

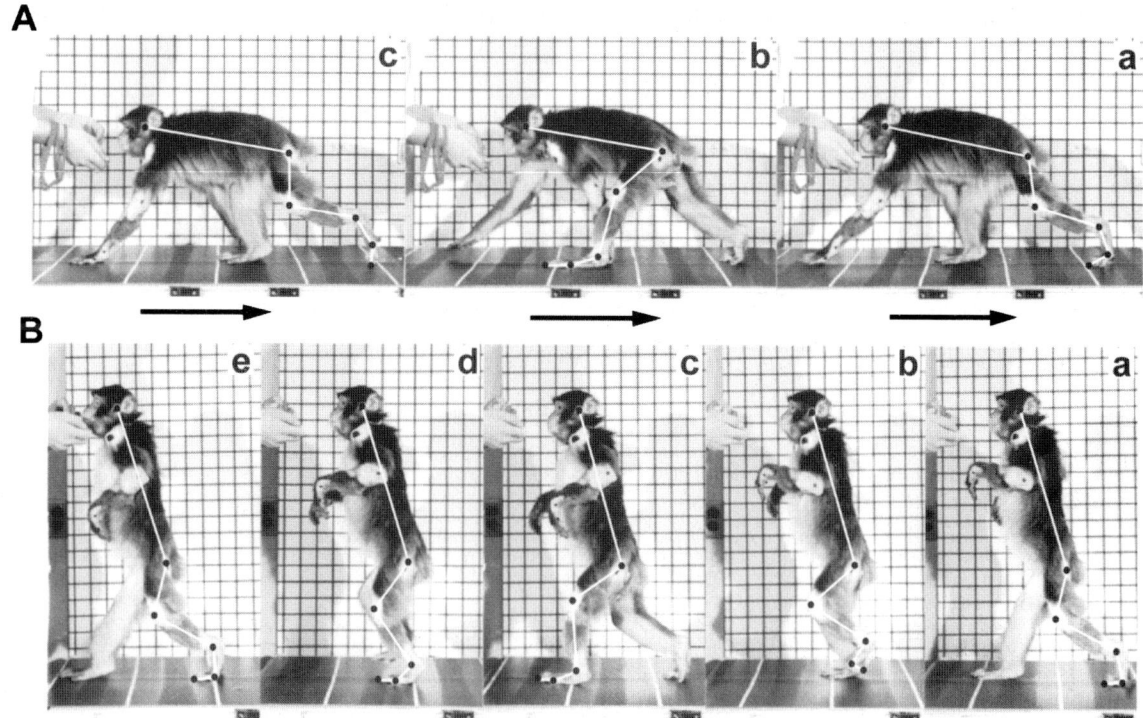

Fig. 1. Representative photographs of Qp versus Bp locomotion in the same monkey. The animal (male; bodyweight, 5.0 kg) was walking on the surface of a moving treadmill belt. The distance between adjacent parallel lines on this belt was 20 cm. Note further how a 5-cm^2 grid was ruled on the background to facilitate the measurements described below. Details of the operant conditioning methods were described previously (Mori et al., 1996). (A) Serial photographs (10 frames/s) of Qp locomotion during a single step cycle, with the monkey walking from right to left on a treadmill belt moving (1.0 m/s) in the opposite direction (arrows under upper panel). Note: (1) reward (pieces of fruit) presented by the experimenter in front of the monkey's mouth; (2) stick figures superimposed on the monkey in each photograph, comprising the lines connecting body surface markings at the ear, pivot points of the major hindlimb joints (hip, knee, ankle, metatarsophalangeal) and tip of the toe; and (3) interval (a) to (b) was 0.4 s and (b) to (c) was 0.5 s. (B) Similar photographs of Bp locomotion during a single step cycle. The (a) to (b) interval was 0.1 s, and all the others were 0.2 s. Note the dramatic qualitative difference in the configuration of the body's segments for Qp versus Bp locomotion.

accompanied by those of proximal parts of the R and L forelimbs, respectively (i.e., a diagonal sequence).

Fig. 1B shows that at a treadmill speed of 1.0 m/s, the body axis angle was ~75°, and the total excursion distance of the hindlimbs was shorter (~30 cm) than that during Qp locomotion (~40 cm). As speed increased from 0.4 to 1.5 m/s, the monkey inclined its upright posture and decreased its body axis angle from ~80° to ~70°. With an increase in the speed from 0.4 to 1.0 m/s, the total excursion distance of the hindlimbs increased from 20 to 30 cm. There was no further increase at higher speeds, however.

In its natural habitat, *M. fuscata* often stands and walks bipedally when it needs its forelimbs to gather food (Iwamoto, 1985). In such instances, the animal's posture and movement are characterized by flexed hip and knee joints, which result in a forwardly inclined, not-fully-upright posture (Okada, 1985). In contrast, our trained animal has a much more upright posture with extended hip and knee joints. To quantify how this training effect influences locomotion, we began to measure kinematic parameters, beginning with the duration of the SW and ST phases of the step cycle, and its cycle frequency.

Stepping parameters as a function of forward speed

During Qp and Bp locomotion in nonprimate quadrupeds, nonhuman primates, and the human, the SW and ST phases and step-cycle frequency are interactive parameters (Goslow et al., 1973; Herman et al., 1976; Grillner et al., 1979; Halbertsma, 1983; Nilsson et al., 1985; Mori et al., 1996). It is well established that as forward speed increases, the duration of the ST phase progressively shortens, whereas that of the SW phase remains relatively constant. It is the shortened ST phase that increases the step-cycle frequency. Figure 2 summarizes these relationships during the Qp versus Bp locomotion of two monkeys, while they walked at speeds from 0.4 to 1.5 m/s. At higher speeds, both monkeys converted to running, which gait is not addressed in this chapter.

Figure 2 shows that as forward speed increased from 0.4 to 1.5 m/s, the average duration of the ST phase for the two animals during Qp locomotion reduced from ∼0.9 to ∼0.4 s, whereas the SW phase remained at ∼0.3 s. The associated increase in step-cycle frequency was ∼0.9–1.5 Hz. During Bp locomotion, the corresponding changes were: ST phase, 0.7–0.3 s; SW phase, constant at ∼0.2 s; and, step-cycle frequency, ∼1.1–∼2 Hz.

The previous data suggest that at least at a coarse grain level, to increase forward speed, these two monkeys used the same overall CNS strategy for *both* Qp and Bp locomotion. At a slightly finer grain, note that Fig. 2 clearly shows that step-cycle frequency was *higher* for Bp versus Qp locomotion at all the tested speeds, and that this difference increased as speed increased. Note further that the duration of the ST and SW phases was *shorter* for Bp versus Qp locomotion at all the tested speeds, and that this difference decreased somewhat as speed increased. Not shown in Fig. 2 is data showing that the forward and backward excursion length of the hindlimbs was also *considerably shorter* for Bp versus Qp locomotion at all the tested speeds, and that this difference increased somewhat as speed increased. Taken together, these results show that *M. fuscata* increased the speed of its trained Bp locomotion by *an increase in the stepping frequency of the hindlimbs* whereas it increased the speed of its natural Qp locomotion by *an increase in the total excursion distance of the hindlimbs*.

Joint kinematics during Qp versus Bp locomotion

In five species of nonhuman primate (chimpanzee, gibbon, spider monkey, Japanese monkey and hamadyras baboon), Okada (1985) recorded angular displacement of hip, knee and ankle during Bp

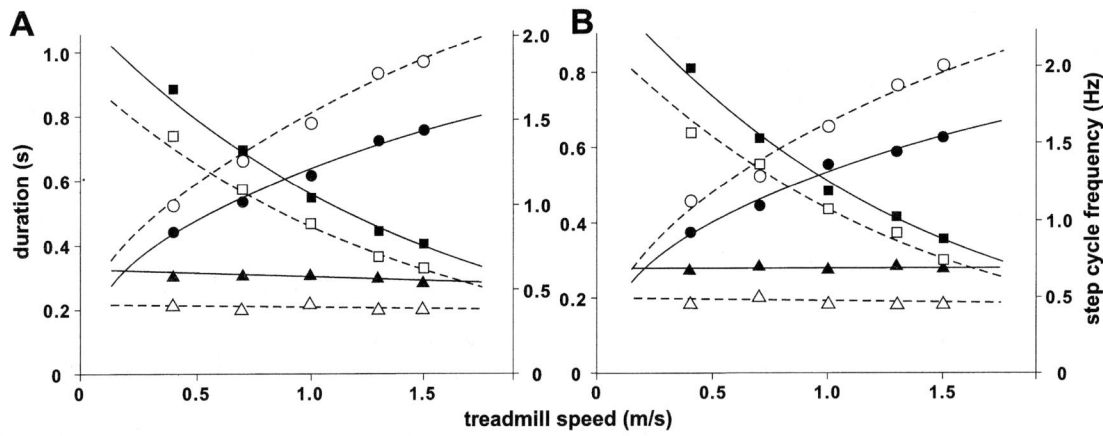

Fig. 2. Walking speed versus step-cycle parameters. Data from two monkeys (A and B) is presented. X axes show treadmill speed (m/s). Left-side Y axes show duration (s) of the SW (triangles) and ST (squares) phases of the step cycle. Right-side Y axes show step-cycle frequency (Hz), which parameter is shown by circles in the plots. Note that the data points are for the left hindlimb, and that they are mean values calculated from five successive step cycles. They are filled for Qp (solid lines) and open for Bp (broken lines) locomotion, which was executed on the same day. Exponential curves were fitted to plots of treadmill speed versus ST duration and step-cycle frequency, and straight lines to plots of speed versus SW duration. Further kinematic details have been provided previously (Mori et al., 1996). Note the quantitative differences in the patterns of interdependent associations between the three key kinematic parameters for Qp versus Bp locomotion.

walking and compared the measurements to analogous ones for the human. They found that the joint motion patterns were fundamentally different for the human versus nonhuman primate. In the human, the hip joint extends steadily from the beginning to the end of the ST phase, whereas knee flexion is delayed until the latter half of this phase. In the nonhuman primate, hip extension is delayed until the latter half of the ST phase, and the knee joint flexes steadily from the beginning to the end of this phase. This suggested to the authors that propulsive force is contributed largely by the hip joint during human Bp walking, whereas the knee joint has this function in the nonhuman primate. In contrast to Okada's interesting and valuable findings, we have found that when trained to walk bipedally, *M. fuscata* has hindlimb kinematics *quite similar* to those of the human.

Figure 3 shows representative angular changes in the hip, knee and ankle joints of a single monkey during its Qp versus Bp locomotion. For both gait patterns, each joint moved cyclically, as is well known. At any constant treadmill speed, step-cycle frequency was higher, and cycle duration correspondingly shorter for Bp versus Qp walking.

Hip joint

During Qp walking, Fig. 3A shows that the hip angle changes had a saw-tooth profile. For the L hip, this angle was ∼110° at its maximum extension (i.e., just before termination of the ST phase) and it then

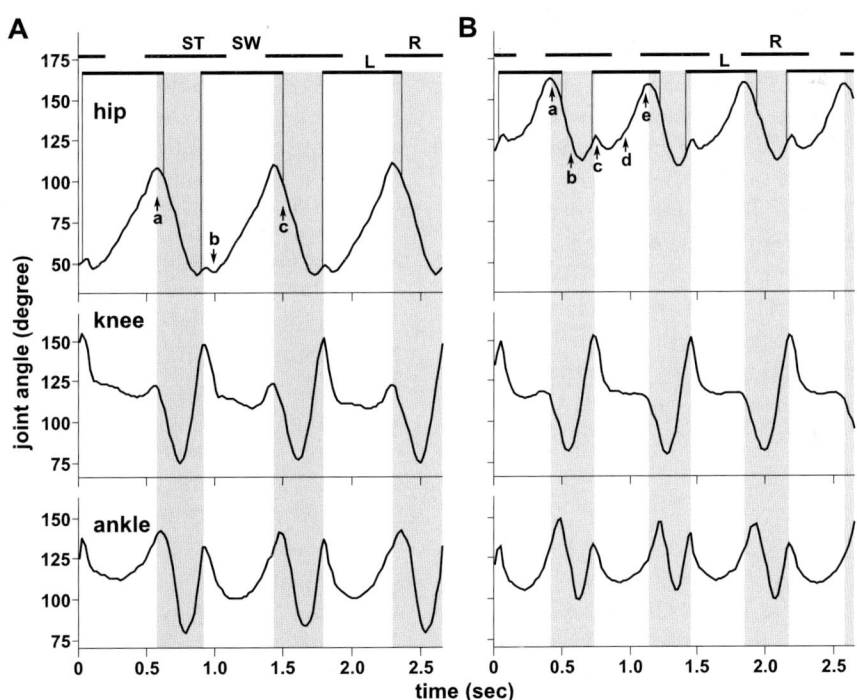

Fig. 3. Joint angles during Qp versus Bp locomotion. Each panel shows slightly more than three successive step cycles. Horizontal bars above the top two panels show the ST- and SW-phase intervals for the left (L) and right (R) hindlimbs. The vertically oriented shaded areas are to aid the reader. They show the time between the maximum extension reached by the hip to the subsequent maximum extension reached by the knee and ankle, i.e., the period when the L hindlimb was freed from the weight-bearing constraints. In turn, in the open area, the ground reaction force and the weight of body mass acted on the L hindlimb joints. (A) Representative plots of angular changes (sampling interval, 20 ms) of the L hip (top trace), knee (middle) and ankle (bottom) joints during Qp locomotion at 1.0 m/s. The a–c arrows in the top panel correspond in time to the a–c photographs in Fig. 1A. (B) Analogous plots during Bp locomotion on the same day, and at the same speed. The a–e arrows in the top panel correspond in time to when the a–e photographs in Fig. 1B were taken. Note that the periods of both Bp ST and SW phases were shorter than those for Qp locomotion.

decreased near-linearly throughout most of the subsequent SW phase to a maximum flexion angle of ~45°. Thus, the mobile (excursion) range of the hip joint was about ~65°. During Bp locomotion (Fig. 3B), the hip joint angle also had a saw-tooth profile but it was shifted to a higher range of angles, which reached ~165° at maximum extension and ~115° at maximum flexion. Another difference was that this joint's mobile range was smaller for Bp (~50°) versus Qp (~65°) walking.

Knee joint

Figure 3 shows that during both Qp and Bp walking, the knee joint underwent a flexion–extension excursion within the SW phase. In Qp locomotion, Fig. 3A shows that for this phase, this joint reached a maximum flexion angle of ~75° and then a maximum extension of ~150°, i.e., a mobile range of ~75°. During the subsequent ST phase, it underwent a more limited excursion: first flexing to midrange, then flexing a little further, and finally extending a little. Thus, the mobile range was much smaller (~30°). Figure 3B shows that during Bp locomotion, the knee joint's excursions were similar to those during Qp locomotion for the SW phase. During the subsequent ST phase, however, after the initial flexion to midrange, there was no further excursion in either direction. Thus, while the mobile range during the SW phase was essentially the same as that during Qp locomotion, its profile was somewhat different.

Ankle joint

Figure 3 shows that the profile of the joint angle movements resembled those of the knee joint, particularly in the SW phase. Figure 3A shows that for Qp locomotion, the maximum extension of the ankle (~140°) was at the end of the ST phase. The peak flexion at midswing was ~80°, i.e., a mobile range of ~60°. Figure 3B shows that for Bp locomotion, the ankle angle profile was quite similar to that of the knee joint, but with slightly higher angles and a lesser mobile range (~40°). For the ST phase, there were obvious qualitative differences between the knee and the ankle, however. Fig. 3A shows that for the ST phase of Qp walking, the ankle joint first flexed and then rebounded back to a slightly more extended angle in a curvilinear fashion, and with a mobile range of ~40°. Fig. 3B shows a relatively similar profile for the ST phase of Bp walking, but a slightly lesser mobile range (~30°).

Summary on joint angle kinematics

Figure 3 shows that the conversion from normal Qp to trained Bp locomotion brings on some kinematic reconfigurations of the hindlimbs, presumably related to biomechanical constraints. These included smaller mobile ranges of the hip and ankle joints, and an upward shift of angular waveform trajectories, particularly in the hip joint. Tellingly, the profile of angular change of the knee and ankle joints was quite similar for Qp and Bp locomotion, except for a slight change at the knee joint during the ST phase. It is attractive to speculate that to achieve the conversion from Qp to Bp locomotion, *M. fuscata* has to acquire a new hip joint motion, which generates propulsive force in a fashion quite similar to that of the human.

Postural considerations

The traditional Japanese folklore method of having *M fuscata* perform Bp walking is to reward the animal for grasping a light wooden rod positioned across the back of its shoulders. Attached to the rod is a string, which the trainer gently pulls upwards and forwards. Accordingly, the monkey walks bipedally (see Fig. 4 in Iwamoto, 1985). In such-trained monkeys, Hayama (1965) documented a compensatory curvature of the spine in the trained monkey. In his study, Bp training began when the animals were 2–3 years of age. The authors found a gradual acquisition of a lordosis of the lumber spine that became quite pronounced (see also Preuschoft et al., 1988). It resembled the development of lordosis in human children between the ages of 1 and 5 years. Hayama (1965) suggested that this morphological change in *M. fuscata* was an adaptation to the mechanical demands of an upright posture. In the postmortem examination of a such-trained monkey, a pronounced hypertrophy of girdle muscles was found (personal communication from S. Gotoh). This could contribute to the stabilization of the lower spinal column relative to the pelvis and femurs. We have preliminary X-ray data on the spinal column of our own-trained monkeys. It, too, reveals a pronounced lordosis of the lumbar spine as in the studies by Preuschoft et al. (1988). In addition, we, too, found an exercise-dependent hypertrophy of girdle and hindlimb muscles. Such developmental and plastic changes in the musculoskeletal system must have profound implications for the neural circuitry that brings on Bp locomotion in *M. fuscata*.

Summary and comment

For the elaboration of Bp locomotion, the normally Qp *M. fuscata* must acquire new human-like postural strategies to maintain equilibrium and generate propulsive force by the hindlimbs against gravity (Nakajima et al., 2001). By long-term locomotor learning, our monkeys have learned to walk bipedally (Nakajima et al., 1997, 2001; Mori et al., 2000, 2001b; Tachibana et al., 2000). To achieve this skill, it would seem that they must either modify preexisting neural networks, which allow Qp locomotion, and/or develop new neural networks that are added to the preexisting neural networks. We showed earlier that the basic relationships between the ST and SW phases of the step cycle and the latter's frequency are essentially the same for Qp and Bp walking. This suggests that *M. fuscata* employs Qp brainstem-spinal mechanisms for the elaboration of Bp locomotion, and in addition, higher CNS mechanisms for the selection of an appropriate postural strategy and limb-kinematic parameters to move the weight-bearing hindlimbs. In other words, the accommodation of the additional biomechanical demands of Bp locomotion require that the CNS of *M. fuscata* make use of a relatively higher set of neural networks than are required for this animal's genetically endowed and relatively automatized Qp locomotion. In keeping with this idea, recent SPECT measurements have shown that multiple sites in the CNS are activated during execution of Bp locomotion by both normal subjects and patients with Parkinson's disease (Fukuyama et al., 1997; Hanakawa et al., 1999; see also present Shibasaki chapter). Which combination of locomotor parameters (e.g., equilibrium, ground reaction force, stepping rhythm) is controlled by each of these brain sites is still not known, however. This quest can now be advanced by detailed invasive and noninvasive neurobiological studies of operant-trained *M. fuscata*.

Acknowledgments

This study was supported by grants to S.M. and K.N. from the Uehara Memorial Foundation; a Grant in Aid for Scientific Research (B12480247) to S.M. from the Ministry of Education, Science, Sports, Culture and Technology of Japan; a Grant in Aid for Comprehensive Research on Aging and Health to S.M. from the Ministry of Health and Labor of Japan; and, a Grant in Aid for Scientific Research on Priority Areas (C12210149) to K.N. from the Ministry of Education, Science, Sports, Culture and Technology of Japan.

Abbreviations

Bp	bipedal
CNS	central nervous system
L	left
Qp	quadrupedal
R	right
SPECT	single photon emission computed tomography
ST	stance phase of step cycle
SW	swing phase of step cycle

References

Eidelberg, E., Walden, J.G. and Nguyen, L.H. (1981). Locomotor control in macaque monkeys. Brain, 104: 647–663.

Fleagle, J.G. and Mittermeier, R.A. (1980). Locomotor behavior, body size, and comparative ecology of seven Surinam monkeys. Am. J. Phys. Anthropol., 52: 301–314.

Fukuyama, H., Ouchi, Y., Matsuzaki, S., Nagahama, Y., Yamauchi, H., Ogawa, M., Kimura, J. and Shibasaki, H. (1997) Brain functional activity during gait in normal subjects: a SPECT study. Neurosci. Lett., 228: 183–186.

Goslow, G.E., Jr., Reinking, R.M. and Stuart, D.G. (1973). The cat step cycle: hindlimb joint angles and muscle length during unrestrained locomotion. J. Morphol., 141: 1–42.

Grillner, S., Halbertsma, J., Nilsson, J. and Thorstensson, A. (1979). The adaptation to speed in human locomotion. Brain Res., 165: 177–182.

Halbertsma, J.M. (1983). The stride cycle of the cat: the modeling of locomotion by computerized analysis of automatic recordings. Acta Physiol. Scand. Suppl., 521: 1–75.

Hanakawa, T., Katsumi, Y., Fukuyama, H., Honda, M., Hayashi, T., Kimura, J. and Shibasaki, H. (1999). Mechanisms underlying gait disturbance in Parkinson's disease. A single photon emission computed tomography study. Brain, 122: 1272–1282.

Hayama, S. (1965). Morphological studies of *Macaca fuscata* II. The sequence of epiphyseal union by roentogenographic estimation. Primates, 6: 249–269.

Herman, R., Wirta, R., Bampton, S. and Finley, F.R. (1976). Human solutions for locomotion, I. Single limb analysis. In: Herman R.M., Grillner S., Stein P.S.G. and Stuart D.G. (Eds.), Neural Control of Locomotion, Advances in Behavioral Biology, Vol. 18, Plenum Press, New York, pp. 13–49.

Hildebrand, M. (1967). Symmetrical gaits of primates. Am. J. Phys. Anthrop., 26: 119–130.

Ishida, H., Kumakura, H. and Kondo, S. (1985). Primate bipedalism and quadrupedalism: comparative electromyography. In: Kondo S. (Ed.), Primate Morphophysiology, Locomotor Analyses and Human Bipedalism. University of Tokyo Press, Tokyo, pp. 59–79.

Iwamoto, M. (1985). Bipedalism of Japanese monkeys and carrying models of hominization. In: Kondo S. (Ed.), Primate Morphophysiology, Locomotor Analyses and Human Bipedalism. University of Tokyo Press, Tokyo, pp. 251–260.

Iwamoto, M. and Tomita, M. (1966). On the movement order of four limbs while walking and the body weight distribution to fore and hind limbs while standing on all fours in monkeys (in Japanese). J. Anthrop. Soc. Nippon, 74: 228–231.

Lawrence, D.G. and Kuypers, H.G.J.M. (1968a). The functional organization of the motor system in the monkey. I. The effects of bilateral pyramidal lesions. Brain, 91: 1–14.

Lawrence, D.G. and Kuypers, H.G.J.M. (1968b). The functional organization of the motor system in the monkey. II. The effects of lesion of the descending brain-stem pathways. Brain, 91: 15–36.

Mori, S., Miyashita, E., Nakajima, K. and Asanome, M. (1996). Quadrupedal locomotor movements in monkeys (*M. fuscata*) on a treadmill: kinematic analyses. Neuroreport, 7: 2277–2285.

Mori, F., Tachibana, A., Takasu, C., Nakajima, K. and Mori, S. (2000). When walking bipedally, the Japanese monkey, M. Fuscata, employs anticipatory and reactive mechanisms to external perturbations. Soc. Neurosci. Abstr., 26: 461.

Mori, S., Matsuyama, K., Mori, F. and Nakajima, K. (2001a). Supraspinal sites that induce locomotion in the vertebrate central nervous system. In: Ruzica E., Hallett M. and Jankovic J. (Eds.), Gait Disorders, Advances in Neurology, Vol. 87, Lippincott Williams & Wilkins, Philadelphia, pp. 25–40.

Mori, F., Tachibana, A., Takasu, C., Nakajima, K. and Mori, S. (2001b). Bipedal locomotion by the normally quadrupedal Japanese monkey, M. Fuscata: strategies for obstacle clearance and recovery from stumbling. Acta Physiol. Pharmacol. Bulg., 26: 147–150.

Muybridge, E. (1955). The Human Figure in Motion. Dover Publications, New York.

Nakajima, K., Matsuyama, K. and Mori, S. (1997). How do the Japanese monkey walking on the treadmill change their locomotion from quadrupedal to bipedal patterns?. Neurosci. Res. Suppl., 21: S192.

Nakajima, K., Mori, F., Takasu, C., Tachibana, A., Okumura, T., Mori, M. and Mori, S. (2001). Integration of upright posture and bipedal locomotion in non-human primates. In: Dengler R. and Kossev A.R. (Eds.), Sensorimotor Control. IOS Press, Amsterdam, pp. 95–102.

Nilson, J. and Thorstensson, A. (1989). Ground reaction forces at different speeds of human walking and running. Acta Physiol. Scand., 136: 217–227.

Nilsson, J., Thorstensson, A. and Halbertsma, J. (1985). Changes in leg movements and muscle activity with speed of locomotion and mode of progression in humans. Acta Physiol. Scand., 123: 457–475.

Okada, M. (1985). Primate bipedal walking: comparative kinematics. In: Kondo S. (Ed.), Primate Morphophysiology, Locomotor Analyses and Human Bipedalism. University of Tokyo Press, Tokyo, pp. 47–58.

Orlovsky, G.N., Deliagina, T.G. and Grillner, S. (1999). Neural Control of Locomotion from Mollusc to Man. Oxford University Press, New York.

Preuschoft, H., Hayama, S. and Gunther, M.M. (1988). Curvature of the lumbar spine as a consequence of mechanical necessities in Japanese macaques trained for bipedalism. Folia Primatol., 50: 42–58.

Prost, J.H. (1965). The methodology of gait analysis and gaits of monkeys. Am. J. Phys. Anthrop., 23: 215–240.

Rollinson, J. and Martin, R.D. (1981). Comparative aspect of primate locomotion with special reference. Symp. Zool. Soc. Lond., 48: 377–427.

Rose, M.D. (1973). Quadrupedalism in primates. Primates, 14: 337–357.

Sherrington, C.S. (1906). The Integrative Action of the Nervous System. Yale University Press, New Haven.

Tachibana, A., Mori, F., Nakajima, K., Takasu, C., Mori, M. and Mori, S. (2000). Developmental features of acquisition on an upright standing posture and bipedal locomotion by the Japanese monkey, *M. Fuscata*. Soc. Neurosci. Abstr., 26: 461.

Vilensky, J.A., Eidelberg, E., Moore, A.M. and Walden, J.G. (1992). Recovery of locomotion in monkeys with spinal cord lesions. J. Mot. Behav., 1: 284–292.

CHAPTER 19

Reactive and anticipatory control of posture and bipedal locomotion in a nonhuman primate

Futoshi Mori[1,*], Katsumi Nakajima[1], Atsumichi Tachibana[1], Chijiko Takasu[1], Masahiro Mori[1], Toru Tsujimoto[2], Hideo Tsukada[3] and Shigemi Mori[1]

[1]*Department of Biological Control System and* [2]*Center for Brain Experiment, National Institute for Physiological Sciences, Okazaki 444-8585, Japan*
[3]*PET Center, Central Research Laboratory, Hamamatsu Photonics KK, Hamakita 434-8601, Japan*

Abstract: Bipedal locomotion is a common daily activity. Despite its apparent simplicity, it is a complex set of movements that requires the integrated neural control of multiple body segments. We have recently shown that the juvenile Japanese monkey, *M. fuscata*, can be operant-trained to walk bipedally on moving treadmill. It can control the body axis and lower limb movements when confronted by a change in treadmill speed. *M. fuscata* can also walk bipedally on a slanted treadmill. Furthermore, it can learn to clear an obstacle attached to the treadmill's belt. When failing to clear the obstacle, the monkey stumbles but quickly corrects its posture and the associated movements of multiple motor segments to again resume smooth bipedal walking. These results give indication that in learning to walk bipedally, *M. fuscata* transforms relevant visual, vestibular, proprioceptive, and exteroceptive sensory inputs into commands that engage both anticipatory and reactive motor mechanisms. Both mechanisms are essential for meeting external demands imposed upon posture and locomotion.

Introduction

The elaboration of bipedal (Bp) locomotion by the human and selected nonhuman primates requires that the central nervous system (CNS) recruit and integrate lower-order automatic and higher-order volitional control mechanisms. Lower-order control is comprised of several posture- and locomotor-related subprograms stored in the brainstem and spinal cord (Mori, 1987; S. Mori et al., Chapter 33 of this volume). Conceptually, higher-order control involves at least three major subsystems: an *integration center* for postural and locomotor commands; an *execution center* for the initiation, sustenance, goal-directed modification and termination of locomotion (Mori, 1997; S. Mori et al., Chapter 33 of this volume); and, a central *program*, which integrates selected cognitive and emotive information and stores it as 'locomotor memory' (McFadyen and Bélanger, 1997). These centers appear to reside within the cerebellum, basal ganglia and cerebral cortex, and their interconnecting pathways (Brooks and Thach, 1981; Armstrong, 1986; Garcia-Rill and Skinner, 1987; Kably and Drew, 1998; Mori et al., 1999b). By integration of these lower- and higher-control mechanisms, Bp and quadrupedal (Qp) animals continuously adjust their posture and locomotion, and respond appropriately to environmental perturbations.

*Corresponding author: Tel.: +81-564-55-7772; Fax: +81-564-52-7913; E-mail: fmori@nips.ac.jp

DOI: 10.1016/S0079-6123(03)43019-7

Accommodation to the external environment requires both reactive and anticipatory control mechanisms, which are presumed to reside in lower and higher parts of the CNS, respectively (Wall et al., 1981; Mori, 1987; Patla et al., 1991; Lange et al., 1996; McFadyen and Bélanger, 1997). The use of these two CNS mechanisms provides the basis for successful Bp navigation on slanted, uneven and obstacle-impeded walking surfaces, all with a seamless integration of total posture and movement of individual motor segments. For this, the CNS must take into account the ongoing state of the spinal mechanisms and locomotor apparatus, e.g., the phase of the locomotor cycle (Arshavsky et al., 1986). In addition, CNS-controlled, visuomotor coordination is apparently one of the key factors for the successful execution of unaided locomotion (Georgopoulos and Grillner, 1986; Drew, 1991; Patla et al., 1991). Inadequate incorporation of proprioceptive, exteroceptive, visual and vestibular information into the motor commands from the *integration center* may be one of the causes of disintegration of a number of motor segments, thereby resulting in stumbling and other unstable forms of locomotion.

We have recently shown that the Japanese monkey, *Macaca fuscata*, is a valuable nonhuman primate Bp model for potentially advancing understanding of CNS mechanisms that contribute to the control of postural and locomotor behavior (Mori et al., 1996; Nakajima et al., 2001). This animal is normally Qp but it can be operant-trained to use Bp locomotion on the surface of a moving treadmill belt. In the latter mode, we have recently examined *M. fuscata*'s reactive and anticipatory motor behavior during various walking tasks. It should be mentioned here that the monkey's Bp walking is digitigrade and it differs from the plantigrade walking of humans, which is characterized by heel-strike and toe-off (Carlsöö, 1972). Nevertheless, both single and double support phase by the lower limbs were observed during execution of monkey's Bp walking. In the following sections, we first present a kinematic analysis of the animal's reactive capability during Bp locomotion on a slanted treadmill belt (Mori et al., 1999a). Next, we show its anticipatory and reactive capability when accommodating an obstacle on a horizontal treadmill belt (Mori et al., 2001a; see also Nakajima et al., Chapter 18 of this volume).

Reactive locomotor patterns during slanted walking

Bp locomotion in humans and nonhuman primates must simultaneously meet several requirements, including: (1) antigravity support, (2) stepping movements, (3) equilibrium and (4) propulsion (Martin, 1967; Mori, 1997). The force for the latter is provided by the weight of the body 'sliding down an incline' with forward movement. To accommodate a constantly changing environment, the neural mechanisms for the requirements must be coordinated. For example, uphill and downhill (slanted) Bp walking by humans require specific modifications in lower-limb kinematics and the supporting body posture (Martin, 1967; Brandell, 1977; Wall et al., 1981; Kawamura et al., 1991; Lange et al., 1996; Leroux et al., 1999). In uphill versus level walking, the legs need to generate a larger acceleration force in order to transfer the center of body mass forward. In parallel, postural adjustments are also required to maintain body equilibrium and facilitate generation of forward propulsive force. In downhill walking, the legs need to generate a net deceleration force to brake forward transfer of the center of body mass. Again, postural adjustments facilitate this braking. We have examined these responses during the skilled Bp locomotion of *M. fuscata* with a focus on the extent to which its movements resemble those of the human.

Prior to slanted walking, *M. fuscata* was operant-trained to use both a Qp and Bp gait while walking on a level treadmill belt at different speeds (for methods, see Mori et al., 1996; Nakajima et al., 2001). Once mastered, the animal was challenged with slanted tasks.

Figure 1 shows the exemplary walking patterns of *M. fuscata* on uphill ($+15°$) and downhill ($-15°$) grades at a fixed treadmill speed (1.3 m/s). Superimposed on the drawings are exemplary walking patterns of the same monkey on a level grade at the same treadmill speed. Lines have been drawn on the animal sketches to depict relevant kinematic and joint angles: ear–hip angle and the angles at the hip, knee, ankle and metatarsophalangeal (MTP) joints. The line between ear and hip represents the body's axis.

Figure 1A–B show the instantaneous postural shift when the monkey placed the foot of its left,

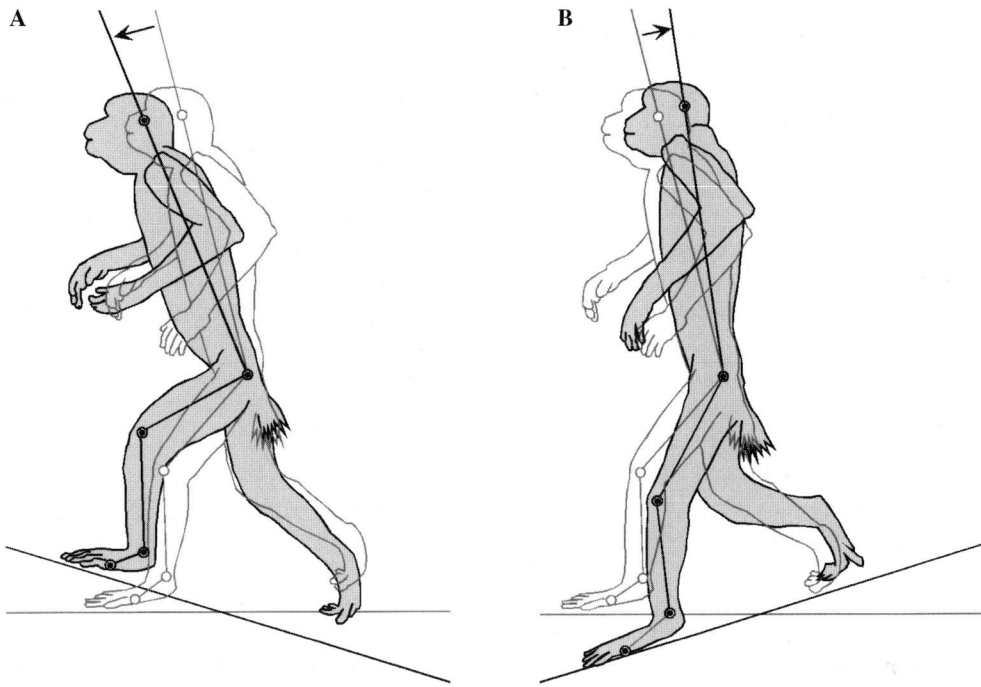

Fig. 1. Schematic drawings of *M. fuscata* during Bp level versus slanted walking at the same speed. Open drawings show body position and stride on a level (0°) treadmill belt. Filled drawings show the same animal during uphill (A, +15°) and downhill (B, −15°) walking. In all cases, the belt speed was 1.3 m/s. In this and Fig. 2, the treadmill belt was moving from left to right, and the monkey was walking from right to left. Drawings are superimposed with reference to the hip joint. Arrows indicate the changes in body axis from level to uphill (A), and level to downhill (B) walking.

forward limb on the moving treadmill belt (i.e., touchdown; onset of the stance phase of the left limb). Next, the monkey lifted the foot of its right, rearward limb up from the surface of the treadmill belt (i.e., takeoff; onset of the swing phase of the right limb). In uphill walking (Fig. 1A), the monkey inclined its body axis maximally during the stance phase of both limbs. The extent of forward body axis inclination was much larger than that observed during level walking. Throughout a single uphill versus level step cycle, the monkey exhibited: (1) a larger flexion of the hip joint, lesser extension of the knee joint and a larger ankle dorsiflexion during the mid-swing and early stance phase; and (2) a larger knee joint extension in the late stance phase. In downhill walking (Fig. 1B), the monkey declined its body axis maximally during the stance phase of both limbs. The extent of body axis declination was much larger than that observed during level walking.

Throughout a single downhill versus level step cycle, the monkey exhibited: (1) a larger extension of the knee joint and lesser flexion of the hip joint during the late swing and early stance phase; and (2) a lesser extension of the hip joint and larger flexion of the knee joint during the late stance and early swing phase. These coordinated hip–knee flexion–extension patterns were also observed in knee–ankle joints.

Figure 1 also shows that forward inclination and backward declination of the body axis increased proportionately with an increase in treadmill angle (upward or downward, respectively) and treadmill speed. An uphill increase in treadmill angle from 0° to 7° to 15° at a constant treadmill speed resulted in an increase in the maximum degree of forward inclination from 15° to 24° to 37° at a fixed treadmill speed (1.3 m/s), respectively. Similarly, a downhill increase in treadmill angle from 0° to −7° to −15° at the same treadmill speed resulted in a decrease in the maximum inclination degree of body axis from 15° to 8° to 2°, respectively. This relationship between changes in treadmill angle and maximum inclination/declination of the body axis was near linear across all three of the tested treadmill speeds (0.7, 1.0 and 1.3 m/s).

During level treadmill walking at a fixed speed (0.7 m/s), the duration of the stance and swing phase of the step was ~0.60 and ~0.25 s, respectively. Uphill and downhill walking involved more prolonged and shortened stance-phase duration, respectively. This relationship was maintained across the tested treadmill speeds. The duration of the swing phase of the step did not change significantly, however. We found a linear relationship between treadmill angle and stride length, i.e., a progressively longer stride for an increase in treadmill angle from −15° to +15°. We also found associations between stride length, treadmill angle and treadmill speed. During uphill walking, step-cycle frequency decreased as stride length and treadmill speed was increased, the reverse occurring during downhill walking. Healthy humans make quite similar adjustments to those described for slanted walking (Wall et al., 1981; Kawamura et al., 1991; Leroux et al., 1999). Taken together, these results suggest that the operant-trained Bp locomotion of *M. fuscata* is smooth and versatile under varying external conditions, and its functional coupling between the lower limbs and body posture is quite similar to that of humans. The results demonstrate that for Bp walking, the monkey has acquired optimal CNS parameters for the coordination of multiple motor segments during slanted walking.

Anticipatory and reactive adjustments during obstacle-encountered treadmill walking

During everyday human locomotion, the feet often collide with unexpected obstacles, thereby requiring compensatory postural and gait adjustments to prevent stumbling and falling, and reestablish smooth walking. Such adjustments have been studied experimentally in the human by having the subject walk on an obstacle-obstructed pathway (Eng et al., 1994) or a moving treadmill belt (Schillings et al., 1996). Similarly, Dietz et al. (1987) studied stumbling-corrective reactions in humans by accelerating and decelerating the speed of a treadmill belt. These studies all demonstrated that humans use both reactive and anticipatory postural adjustments to perturbations encountered while walking (McFadyen and Bélanger, 1997). Such adjustments occur in a proactive way during all phases of the step cycle, with visuomotor coordination a critical factor for anticipatory adjustments. In order to successfully judge distances and modify step length to attain a target or avoid an obstacle, the human needs but intermittent visual sampling of the immediate environment (Georgopoulos and Grillner, 1986; Drew, 1991; Patla et al., 1991). We have tested the extent to which *M. fuscata* can perform similarly during its Bp locomotion.

Figure 2 shows that we trained *M. fuscata* to step over an adjustable-height rectangular block (width: 25 cm; length: 5 cm; height: 2.4, 5.0 and 7.0 cm) that was placed on the left side of a treadmill belt. The obstacle was arranged to confront the left limb every 4–6 steps, as dependent on belt speed. In this situation, the monkey could be motivated to walk continuously for ~3–5 min, with such continuity termed a single trial. In a single trial, the monkey encountered the obstacle ~40–70 times. A 3–4-min rest was needed between trials to insure that the animal maintained a relatively constant level of reward-based motivation (Mori et al., 2001a). Successive trials involved use of different combinations of fixed treadmill speed (0.7–1.3 m/s) and obstacle height.

During initial trials, the monkey often stumbled when the toe of the trailing left foot stepped on the obstacle's top (horizontal) surface, or slipped up on the initially encountered (vertical) surface. After several sequential trials, the number of stumbles in a single trial was reduced, i.e., the monkey gradually learned to clear the obstacle at various walking speeds, by use of what in humans has been termed a 'hip–knee flexion strategy' (McFadyen and Winter, 1991). This involved the monkey increasing the extent of flexion of the trailing left limb's hip and knee joint while simultaneously, the leading right limb alone supported the center of body mass and maintained equilibrium. When the obstacle's height was raised, there was a corresponding increase in hip/knee joint flexion of the trailing limb so as to produce sufficient clearance space above the obstacle.

Figure 2A shows *M. fuscata*'s exemplary walking pattern during the control (no-obstacle) condition. Figures 2B–D show patterns observed at different times during the perturbation (obstacle-on) condition: B, with the in-coming obstacle 3–4 steps ahead and out-of-sight; and C, a successful clearance; and D, recovery after an unsuccessful clearance.

In the control (Fig. 2A) state, the center of body mass was supported by the leading (right) limb alone when the trailing left limb was at its mid-swing phase with moderate flexion at the hip and knee joints. When the in-coming obstacle was still out-of-sight (Fig. 2B), but probably anticipated, the hip and knee flexion extent of trailing limb was more pronounced. This effect was exaggerated even further during successful clearance of the obstacle (Fig. 2C), which also included a further dorsiflexion of the ankle. When the animal encountered the obstacle during the period of early- to mid-swing phase, it easily cleared the obstacle. With some presumably visual information, it could adjust the trajectory of left foot even during the late period of swing phase and successfully clear the obstacle. Figure 2 does not show that in this latter case, the trailing limb at the swing phase undertook a longer and more prolonged stride to maintain equilibrium. During this task, there was no significant change in the body's axis, and the head was maintained at a relatively constant position.

When the trailing foot failed to clear the obstacle (Fig. 2D), stumbling occurred routinely. Such stumbling usually involved either stepping on, or slipping up, the obstacle during the late swing phase of the trailing left limb's step. Figure 2D shows the pronounced postural perturbation associated with slipping up the obstacle. Immediately after foot–obstacle contact, the monkey slightly moved its body axis backward. This was followed immediately by a rapid and pronounced ($\sim 40°$) forward movement of the body axis. Subsequently, the animal: (1) shortened the swing period of the leading limb; (2) flexed the left and right lower limb joints to lower the center of body mass to the treadmill belt; (3) extended its left (and/or right) forelimb forward and/or downward; and finally (4) extended its lower limb joints to raise the lowered

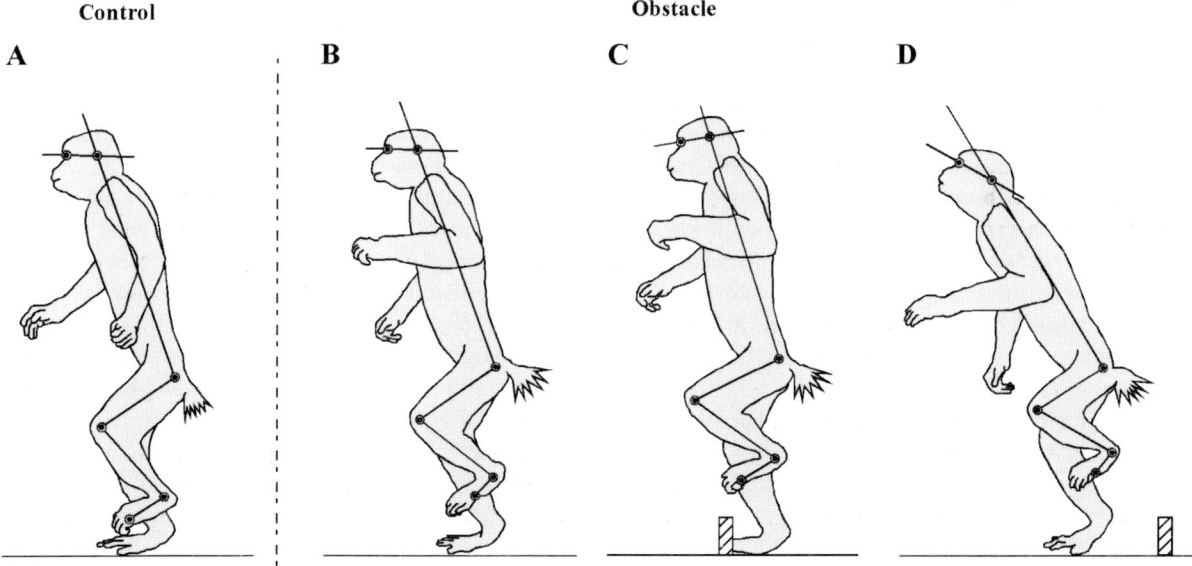

Fig. 2. Schematic drawings of the mid-swing phase of the left hindlimb during normal versus obstacle-encountered, level Bp walking. (A) Control (no-obstacle) walking at 1.0 m/s. (B–D) Walking at the same speed, but in the obstacle-encountered situation. (B) Obstacle 3–4 steps ahead of monkey and out-of-sight. (C) A successful clearance of the obstacle; and (D) the defensive posture that occurred after stumbling over the obstacle.

body mass to the preperturbed position. The first three compensatory reactions of multiple motor segments served to stabilize the perturbed posture. When the animal stepped on the obstacle, it immediately initiated a new trailing limb's swing phase from the top surface of the obstacle. In such instance, landing of the leading right foot on the treadmill belt always preceded the beginning of the trailing limb's swing phase. Alternatively, if the trailing limb's foot slipped up the obstacle, the animal initiated the new swing phase from the belt surface. To clear the obstacle, the monkey adjusted the trajectory of the trailing limb with the toes fully flexed. These reactive compensations for the perturbed posture finally made the animal possible to restore the position of the head, body and limb segments to preperturbed one, and to walk safely and smoothly.

Recovery of walking required the restoration of (1) the center of body mass, (2) the head position and (3) the body axis to its predisturbed position in space. Interestingly, the correction of the head position always occurred first, followed by that of the body axis and upper and lower limbs. Perhaps head position in space is a critical determinant of the nature and extent of reactive responses to postural perturbations encountered during walking.

From the preceding paragraphs, it can be seen that *M. fuscata* recruited both anticipatory and reactive control mechanisms to prepare for the incoming obstacle, even when it was still out-of-sight, and to restore the perturbed posture after stumbling over the obstacle. The latter mechanism was also used to adjust the trajectory of the trailing limb with the integration of visual information and motor output. We have also observed that the occurrence of stumbling in a single trial decreased as the monkey encountered successive obstacles. Such observations suggest that the monkey could store a novel 'obstacle clearance strategy' in its *'locomotor memory'* repertoire. This was presumably acquired by motor learning.

Comment and summary

Overall, the kinematic analyses of *M. fuscata*'s Bp locomotion indicate that this animal and the human have similar kinematics for the integration of posture and locomotion, particularly for the hip joint. In both, hip extension proceeds from touchdown (beginning of stance phase) to takeoff (beginning of swing phase) of the foot. When the Bp-walking monkey encountered an obstacle, it changed its foot trajectory to produce a larger-than-usual successful clearance over the obstacle (hip and knee flexion strategy). Moreover when the monkey stumbled over the obstacle, it quickly recovered from its perturbed posture and movement, and continued its smooth, well-coordinated Bp walking. These findings suggest that the monkey's CNS received and transformed salient visual, vestibular, proprioceptive and exteroceptive sensory information into output (command) motor signals appropriate for the integration of multiple body segments, this being essential for successful accommodation of the obstacle. Such a transformation process, including recall of *'locomotor memory'*, would require continual modification of the ongoing postural and locomotor control signals, which include anticipatory and reactive components. The studies described earlier show that the normally Qp *M. fuscata* is a valuable nonhuman primate model for studying the CNS control of not only Bp locomotion and accompanying posture, but also compensations to stumbling and falling, and eventually several other types of movement disturbance.

Shibasaki's Kyoto University group has recently used single photon emission computed tomography (SPECT) to study corridor (Fukuyama et al., 1997) and treadmill (Hanakawa et al., 1999) walking by normal adult subjects. Their results have shown a high bilateral activation in the primary sensorimotor-, supplementary motor- and visual cortex, as well as in the basal ganglia and cerebellum (see also Shibasaki et al., Chapter 20 of this volume). Presumably, the CNS of the Bp-walking monkey display analogous activation patterns. To test this idea, we have recently developed a noninvasive neuroimaging protocol. Positron emission tomography (PET; Watanabe et al., 1997) is being used to measure cerebral glucose metabolism after epochs of Bp locomotion. A preliminary study has already shown that multiple brain regions are activated (Mori et al., 2001b), just as they are in humans. Some of these brain regions may reflect plastic changes in neural circuitry involved in long-term locomotor learning. This PET project should provide the opportunity to study experimentally how each of the activated brain regions contributes to the control of Bp posture and locomotion, including reactive and anticipatory control mechanisms. It should also contribute to further understanding of CNS circuitry required for the initiation, maintenance and recall of Bp locomotion, and thereby give more rigor to terms like *integration center*, *execution center*, *central program* and *locomotor memory*. These terms are currently vague, but they are nonetheless useful in the quest to understand how the CNS can achieve a seamless integration of multiple motor segments for the execution of Bp locomotion.

Acknowledgments

The authors express sincere appreciation to Dr. Hirotaka Onoe, Tokyo Metropolitan Institute for Neuroscience, for his help with data processing and continuous encouragement, and Dr. Carol Boliek, University of Alberta, for her critical review and editing of the original version of this manuscript. We also appreciate the technical support of Hamamatsu Photonics KK personnel, including Messrs. Takeharu Kakiuchi, Masami Futatsubashi and Dai Fukumoto. This study was supported by a Grant-in Aid for Encouragement of Young Scientists to F.M. and a Grant-in Aid for General Scientific Research to S.M. from the Ministry of Education, Science, Sports, Culture and Technology of Japan; and, a Grant in Aid on Comprehensive Research on Aging and Health to S.M. from the Ministry of Health and Welfare of Japan.

Abbreviations

Bp	bipedal
CNS	central nervous system
M. fuscata	*Macaca fuscata*
MTP	metatarsophalangeal
PET	positron emission tomography
Qp	quadrupedal
SPECT	single photon emission computed tomography

References

Armstrong, D.M. (1986). Supraspinal contributions to the initiation and control of locomotion in the cat. Prog. Neurobiol., 26: 273–361.

Arshavsky, Y.I., Gelfand, I.M. and Orlovsky, G.N. (1986). Cerebellum and Rhythmical Movements. Springer-Verlag, Berlin.

Brandell, B.R. (1977). Functional roles of the calf and vastus muscles in locomotion. Am. J. Phys. Med., 56: 59–74.

Brooks, V.B. and Thach, W.T. (1981). Cerebellar control of posture and movement. In: Brooks V.B. (Ed.), Handbook of Physiology. American Physiological Society, Bethesda, pp. 877–946.

Carlsöö, S. (1972). How Man Moves. Heinemann, London.

Dietz, V., Quitern, J.Q. and Sillem, M. (1987). Stumbling reactions in man: significance of proprioceptive and pre-programmed mechanisms. J. Physiol. (Lond.), 386: 149–163.

Drew, T. (1991). Visuomotor coordination in locomotion. Curr. Opin. Neurobiol., 1: 652–657.

Eng, J.J., Winter, D.A. and Patla, A.E. (1994). Strategies for recovery from a trip in early and late swing during human walking. Exp. Brain Res., 102: 339–349.

Fukuyama, H., Ouchi, Y., Matsuzaki, S., Nagahama, Y., Yamauchi, H., Ogawa, M., Kimura, J. and Shibasaki, H. (1997). Brain functional activity during gait in normal subjects: a SPECT study. Neurosci. Lett., 228: 183–186.

Garcia-Rill, E. and Skinner, R.D. (1987). The mesencephalic locomotor region. II. Projections to reticulospinal neurons. Brain Res., 411: 13–20.

Georgopoulos, A.P. and Grillner, S. (1986). Visuomotor coordination in reaching and locomotion. Science, 245: 1209–1210.

Hanakawa, T., Katsumi, Y., Fukuyama, H., Honda, M., Hayashi, T., Kimura, J. and Shibasaki, H. (1999). Mechanisms underlying gait disturbance in Parkinson's disease: a single photon emission computed tomography study. Brain, 122: 1271–1282.

Kably, B. and Drew, T. (1998). Corticoreticular pathways in the cat. II. Discharge activity of neurons in area 4 during voluntary gait modifications. J. Neurophysiol., 80: 406–424.

Kawamura, K., Tokuhiro, A. and Takachi, H. (1991). Gait analysis of slope walking: a study on step length, stride width, time factors and deviation in the center of pressure. Acta Med. Okayama, 45: 179–184.

Lange, G.W., Hintermeister, R.A., Schlegel, T., Dillman, C.J. and Steadman, J.R. (1996). Electromyographic and kinematic analysis of graded walking and the implications for knee rehabilitation. J. Orthop. Sports Phys. Ther., 23: 294–301.

Leroux, A., Fung, J. and Barbeau, H. (1999). Adaptation of the walking pattern to uphill walking in normal and spinal-cord injured subjects. Exp. Brain Res., 126: 359–368.

Martin, J.P. (1967). The Basal Ganglia and Posture. Pitman Medical Publishing, London.

Massion, J. (1992). Movement, posture and equilibrium: interaction and coordination. Prog. Neurobiol., 38: 35–86.

McFadyen, B.J. and Bélanger, M. (1997). Neuromechanical concepts for the assessment of the control of human gait. In: Allard P., Cappozzo A., Lundberg A. and Vaughan C.L. (Eds.), Three-Dimensional Analysis of Human Locomotion. John Wiley and Sons, Chichester, pp. 49–66.

McFadyen, B.J. and Winter, D.A. (1991). Anticipatory locomotor adjustments during obstructed human walking. Neurosci. Res. Commun., 9: 37–44.

Mori, F., Nakajima, K., Gantchev, N., Matsuyama, K. and Mori, S. (1999). A new model for the study of the neurobiology of bipedal locomotion: the Japanese monkey, *M. fuscata*. In: Gantchev N. and Gantchev G.N. (Eds.), From Basic Motor Control to Functional Recovery. Academic Publishing House, Sofia, pp. 47–51.

Mori, F., Tachibana, A., Takasu, C., Nakajima, K. and Mori, S. (2001a). Bipedal locomotion by the normally quadrupedal Japanese monkey, *M. fuscata*: strategies for obstacle clearance and recovery from stumbling. Acta Physiol. Pharmacol. Bulg., 26: 147–150.

Mori, F., Tachibana, A., Nakajima, K., Takasu, C., Tsujimoto, T., Tsukada, H., Okumura, T., Mori, M. and Mori, S. (2001b). Bipedal locomotion in the Japanese monkey, *M. fuscata*: higher-order control mechanisms. Soc. Neurosci. Abstr., 27: 1077.

Mori, S. (1987). Integration of posture and locomotion in acute decerebrate cats and in awake, freely moving cat. Prog. Neurobiol., 28: 161–195.

Mori, S. (1997). Neurophysiology of locomotion: recent advances in the study of locomotion. In: Masdeu J.C., Sundarsky L. and Wolfson L. (Eds.), Gait Disorders of Aging. Lippincott-Raven, Philadelphia, pp. 55–78.

Mori, S., Miyashita, E., Nakajima, K. and Asanome, M. (1996). Quadrupedal locomotor movements in monkeys (*M. fuscata*) on a treadmill: kinematic analyses. Neuroreport, 7: 2277–2285.

Mori, S., Matsui, T., Kuze, B., Asanome, M., Nakajima, K. and Matsuyama, K. (1999). Stimulation of a restricted region in the midline cerebellar white matter evokes coordinated quadrupedal locomotion in the decerebrate cat. J. Neurophysiol., 82: 290–300.

Nakajima, K., Mori, F., Takasu, C., Tachibana, A., Okumura, T., Mori, M. and Mori, S. (2001). Integration of upright posture and bipedal locomotion in no-human primates. In: Dengler R. and Kossev A.R. (Eds.), Sensorimotor Control. IOS Press, Amsterdam, pp. 95–102.

Patla, A.E., Pretice, S.D., Robinson, C. and Neufeld, J. (1991). Visual control of locomotion: strategies for changing direction and for going over obstacles. J. Exp. Psychol. Hum. Percep. Perform., 17: 603–634.

Schillings, A.M., Van Wezel, B.M.H. and Duysens, J. (1996). Mechanically induced stumbling during human treadmill walking. J. Neurosci. Methods, 67: 11–17.

Wall, J.Z.C., Nottrodt, J.W. and Charteris, J. (1981). The effect of uphill and downhill walking on pelvic oscillations in the transverse plane. Ergonomics, 24: 807–816.

Watanabe, M., Okada, H., Shimizu, K., Omura, T., Yoshikawa, E., Kosugi, T., Mori, S. and Yamashita, T. (1997). A high resolution animal PET scanner using compact PS-PMT detectors. IEEE Trans. Nucl. Sci., 44: 1277–1282.

CHAPTER 20

Neural control mechanisms for normal versus Parkinsonian gait

Hiroshi Shibasaki[1,2,*], Hidenao Fukuyama[2] and Takashi Hanakawa[2]

[1]Department of Neurology and [2]Human Brain Research Center, Kyoto University Graduate School of Medicine, Kyoto 606-8507, Japan

Abstract: The non-invasive methods which can be applied to the study of the cerebral control mechanisms for human gait are limited because of technical constraints. In particular, the subject's head has to be fixed for most measurements. Despite this problem, SPECT-detected, rCBF activation studies have shown that the cerebral cortex participates in the control of the normal volitional walking of healthy human subjects. The active brain areas include the foot and trunk regions of the primary sensorimotor cortex, and the supplementary motor area (SMA), lateral premotor cortex, and cingulated gyrus. Selected subcortical structures also exhibit walking-related activity, including the dorsal brainstem and cerebellum. These same regions are active in patients with Parkinson's disease (PD), but significantly less so in the right SMA, left precuneus, and right cerebellar hemisphere. For the "kinesie paradoxale" of PD patients, the right lateral premotor area plays an important role in the visually induced gait improvement brought on by adding transverse lines to the walking path.

Introduction

Humans control their locomotion either volitionally or automatically, the latter with very little attention to the task. In fact, both are probably continually involved to varying degree in the locomotion of daily life. Therefore, it is reasonable to propose that while subcortical structures, such as the brainstem locomotor region and basal ganglia, may play important roles in automatic walking, the cerebral cortex may also become involved with changes in the level of effort and attentional demands.

Brain mechanisms for the control of human gait can be studied noninvasively through the use of only a limited number of neuroimaging techniques. Conventional ones, like positron emission tomography (PET) and functional magnetic resonance imaging (fMRI), have allowed us to study brain activation, as indicated by increased regional cerebral blood flow (rCBF) and/or increased cerebral metabolic rates of glucose (rCMRglu). These techniques cannot be applied to daily behaviors such as walking and voiding, however, because the subject's head must be fixed on the scanner bed throughout the scanning procedure. On the other hand, single photon emission computed tomography (SPECT) allows the use of radioactive tracers with a relatively long half-life to be injected well before the scanning procedure (Fukuyama et al., 1996). Thus, scanning to detect tissue radioactive concentration can indeed be performed after different types of behaviors have occurred. Fukuyama et al. (1997) was the first to apply SPECT technology during the walking of normal adult human subjects. 99mTc-HMPAO was injected intravenously while they walked along a corridor, and subsequently the initial scanning for the walking condition was undertaken. After the initial scanning, subjects received another injection of the same radioactive tracer and the second scanning for the resting (control) condition was undertaken. The difference between the two scans demonstrated

*Corresponding author: Tel.: +81-75-751-3601; Fax: +81-75-751-3202; E-mail: shib@kuhp.kyoto-u.ac.jp

which regions of the brain became active during overground walking (see also F. Mori et al., Chapter 19 of this volume). Hanakawa et al. (1999a,b) has since applied this technique to the study of mechanisms underlying the gait disorders characteristic of patients with Parkinson's disease (PD).

Although the latter type of neuroimaging provides accurate spatial information, it does not provide temporal information with respect to the time of gait initiation. Consequently, Vidailhet et al. (1993, 1995) simultaneously recorded forebrain EEGs and EMGs from the muscles involved in gait. The EEGs were back-averaged with respect to the time of gait initiation in order to obtain the Bereitschaftspotential (BP) associated with gait initiation. This strategy was complicated by the EEG record being inevitably contaminated with gait-induced movement artifacts. To obviate this problem, Yazawa et al. (1997) obtained contingent negative variation (CNV) recordings rather than the BP. Since the CNV is larger than the BP, the former can be back-averaged with a lesser number of sweeps than are needed for the BP average. As a result, they obtained clear-cut recordings of the cortical activity that preceded cued gait initiation. The studies using SPECT imaging and this CNV approach are focused upon in the next section.

Which brain structures are involved in the volitional control of human gait?

The SPECT technique developed by Fukuyama et al. (1997) for overground walking was subsequently applied to the constant-speed treadmill walking of normal adult subjects (Hanakawa et al., 1999a). Figure 1 shows that walking-related cortical activity was observed in the foot, leg and trunk regions of primary sensorimotor cortex, the premotor region (medial and lateral area 6), and the cingulate gyrus, superior parietal cortex, visual cortex, dorsal brainstem and cerebellum. The activation in all these regions incorporates, however, a summation of the

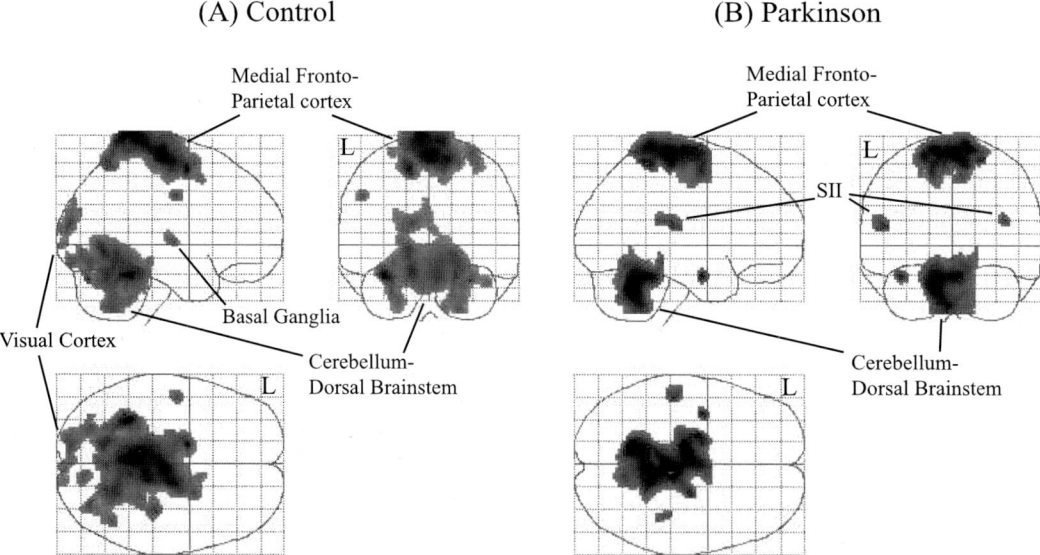

Fig. 1. Locomotion-induced brain activity in control versus age-matched PD subjects. Three orthogonal projections are shown for control (A) and PD (B) subjects. The data were obtained by SPECT in an rCBF study while the subjects walked for 5 min on a treadmill at a constant speed. At each voxel level in this and Fig. 2, the statistical threshold was set at $P < 0.001$. The regions showing significant activation are similar in the two groups. In the control versus PD group, however, activation of the medial frontal lobe extended more rostrally, and that of infratentorial structures more widely. Abbreviations: L, left side; SII, second somatosensory cortex. (Reproduced, in part, from Hanakawa et al., 1999a with permission from Oxford University Press.)

overall neural activity that occurred over a period of 5 min of continuous treadmill walking.

The Fig. 1 measurements presumably reflect multiple brain functions throughout the elaboration of walking, including those related to the planning as well as the execution of the gait, and CNS responses to sensory feedback from virtually all the affected modalities (e.g., vision, somatosensory, vestibular). The Fig. 1 results are quite similar to analogous results obtained from animal studies (Mori, 1997). From these data, it is reasonable to propose that the primary foot representation in the sensorimotor cortex, and the medial and lateral premotor areas, cingulate gyrus, dorsal brainstem and cerebellum are *all* actively engaged in the volitional control of human walking.

What CNS disturbances are evident during Parkinsonian gait?

When PD subjects perform treadmill walking, they exhibit a typical Parkinsonian gait, i.e., a short step length and an increased cadence (steps/min). We found that the SPECT-detected regions of brain activation were the same as those activated during the walking of normal subjects (Hanakawa et al., 1999a). Figure 2 shows, however, that the level of rCBF increase in PD versus normal subjects was significantly lower in the left supplementary motor region (SMA), right precuneus and left cerebellar hemisphere. On the other hand, PD subjects showed relatively more activity in the left temporal cortex, right insula, left cingulate cortex and cerebellar vermis.

Some of the Fig. 2 findings, particularly underactivity in SMA, may reflect an insufficient facilitatory input from the basal ganglia to the SMA. This has been proposed to be a fundamental pathophysiological mechanism in PD (DeLong, 1990; Playford et al., 1992; Georgiou et al., 1993; Samuel et al., 1997). We interpreted underactivity in the cerebellar hemisphere, in contrast to overactivity in the cerebellar vermis, as due to a loss of lateral gravity shift in Parkinsonian gait, based on the assumption that the patients with PD fail to shift the body's

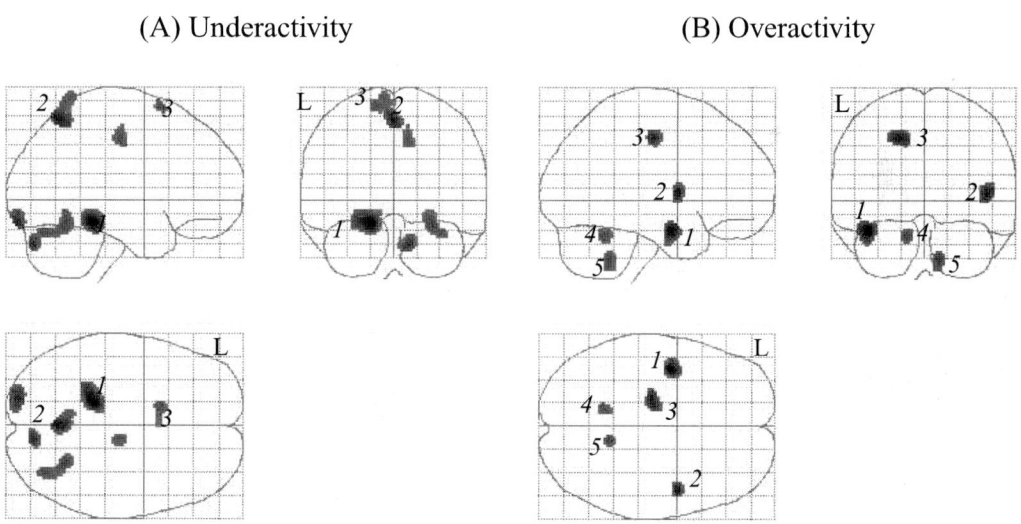

Fig. 2. Relative under- versus overactivity in locomotion-induced brain activity in PD versus control, age-matched subjects. (A) Relative underactivity in PD versus control subjects was found in the left cerebellar hemisphere (1), precuneus region (2), and left medial frontal region (3; corresponds to the pre-SMA). (B) Relative overactivity in PD versus control subjects was found in the left middle temporal region (1), right insula (2), left cingulate cortex (3) and cerebellar vermis (4–5). (Reproduced, in part, from Hanakawa et al., 1999a with permission from Oxford University Press.)

center of gravity onto one foot resulting in small shuffling steps (Hanakawa et al., 1999a).

Why do PD patients walk better upstairs than on a level surface?

It is well known that PD patients with a 'frozen' gait walk better upstairs than on a level surface. Similarly, they find it relatively easier to step across lines drawn transverse versus parallel to the walking direction ('kinesie paradoxale'; see Glickstein and Stein, 1991). To advance understanding of this vision-based phenomenon, we undertook a SPECT-rCBF activation study in which PD versus normal subjects walked on a constant-speed treadmill with two different types of visual stimuli (Hanakawa et al., 1999b). The cadence of the normal subjects was not affected by either stimulus. The cadence of the PD patients, however, was faster than normal subjects during parallel-line stimulus, and when confronted with the transverse stimuli, PD patients slowed their cadence. For both test groups, the locomotion-induced rCBF increase was greater for the transverse lines versus parallel lines in the posterior parietal cortex and cerebellar hemispheres.

Figure 3 shows that the only brain region where the difference in activation for transverse versus parallel lines was greater in the PD subjects was the right lateral premotor area. In contrast, during parallel-line stimulation, the rCBF increase in this region was smaller for the PD versus normal subjects. Thus, while walking with transverse-line stimulation, the rCBF of the right lateral premotor area increased in the PD subjects to the normal-subject level, thereby indicating that the activation of this particular region by visual stimuli must play an important role in the improvement of Parkinsonian gait. This region is known to receive abundant visual input from the parietal cortex, whereas the input from the basal ganglia is relatively sparse. In contrast, the more medial premotor area receives a strong facilitatory input from the basal ganglia, which is decreased in PD (DeLong, 1990; Playford et al., 1992; Georgiou et al., 1993; Samuel et al., 1997).

The above results also suggest that the right lateral premotor area, which receives abundant visual input and is less affected by PD pathophysiology,

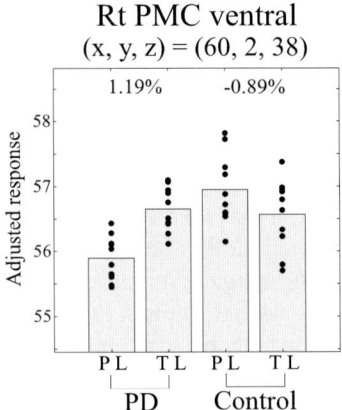

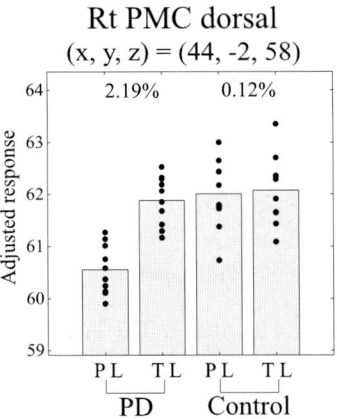

Fig. 3. Effects of visual stimulation on walking-related brain activity: PD versus age-matched control subjects. The Y axes (adjusted responses) show walking-induced rCBF increases (in %) for the ventral and dorsal regions of the right (Rt) lateral premotor cortex (PMC) in PD versus control subjects. The measurement sites' x, y and z stereotaxic coordinates are provided in mm. Note that while walking between two parallel lines (PL), the mean rCBF increase in both regions was less for PD versus control subjects (PMC ventral, 1.19% less; PMC dorsal, 2.19% less). The differences were much less, however (PMC ventral, 0.89% more; PMC dorsal, 0.12% less), while the subjects walked across transverse lines (TL). (Reproduced, in part, from Hanakawa et al., 1999b with permission from John Wiley and Sons.)

may compensate for the medial premotor area, whose excitability is reduced in PD subjects due to insufficient facilitatory input from the basal ganglia. Yet to be determined is why the right lateral premotor area, but not the left, may serve as the trigger for the gait improvements (i.e., slowed cadence, increased strides) in PD patients. Likewise, the question as to specifically why transverse-drawn lines but not parallel-drawn lines are effective for remedying paradoxical PD gait remains open for the moment.

What role does SMA play in the timing of gait initiation?

For normal versus PD subjects, Vidailhet et al. (1993) recorded the BP that preceded a stepping movement while standing versus making a simple foot movement (pseudostep) while sitting. In normal subjects, the BP was larger for a standing step versus a sitting pseudostep, but no such difference was observed in PD subjects. The investigators interpreted their data as indicating a PD-induced impairment of the preparation and assembly of the complex sequences of movement necessary to initiate walking.

In normal subjects, Yazawa et al. (1997) compared the scalp distribution of CNVs that preceded externally cued gait initiation in standing (in a simple reaction time paradigm with an interstimulus interval of 2 s) to that of CNVs preceding a simple foot dorsiflexion task in sitting. The late CNV in the gait initiation task started about 1 s before the imperative stimulus (S2), and was largest at the vertex midline (Cz electrode). Figure 4A shows that this late CNV was symmetrically distributed on the scalp, and declined steeply toward the parietal region. Figure 4B shows that in the foot dorsiflexion task, the late CNV was also maximal at Cz with asymmetric distribution (larger ipsilateral to the moved foot), and it was clearly seen also over the parietal region. The late CNV at Cz was significantly larger in the gait initiation versus foot dorsiflexion task. As for the foot dorsiflexion task, there was no difference in the amplitude or distribution of CNV between the standing versus sitting posture.

The above findings led the investigators to propose that the cerebral cortex has a different pattern of activation for the initiation of externally triggered gait versus making a simple foot movement. Furthermore, the late CNV in the former was significantly larger, but extended posteriorly to a lesser degree. Assuming that the foot area of the primary motor cortex is equally active in gait initiation and foot movement, the above finding was interpreted to suggest a greater contribution of the SMA to the CNV during gait versus the tested foot movement. Recall that in the late CNV, SMA is thought to generate mainly a radial component with respect to the scalp surface. Thus, the authors proposed that the bilateral SMA may play a more important role in gait initiation than in simple foot movement (Yazawa et al., 1997).

Summary and comment

SPECT-detected rCBF activation studies have shown that the cerebral cortex participates in the control of the normal volitional walking of healthy subjects. The active components include the foot and trunk regions of the primary sensorimotor cortex, and the SMA, lateral premotor cortex and cingulate gyrus. Subcortical structures also show walking-related activity, including the dorsal brainstem and cerebellum. These same regions are activated in PD patients, but significantly less so in the right SMA, left precuneus and right cerebellar hemisphere. For the 'kinesie paradoxale' of PD patients, the right lateral premotor area plays an important role in the visually induced gait improvement brought on by adding transverse lines to the walking path. Further advances in our understanding of the CNS control of human locomotion now seem feasible. For example, a particularly promising approach should be to combine the use of near infrared spectroscopy with the transient reversible functional disruption induced by transcranial magnetic stimulation. This should provide a new opening for study of the role of higher brain structures in locomotion and other forms of movement.

Acknowledgments

We would like to thank Dr. Becky Farley, the University of Arizona, for reviewing a draft of this article. The work was supported, in part, by research grants for Scientific Research on Priority Areas (C 12210012) and Scientific Research (B 13470134), Japan Ministry of Education, Culture, Sports, Science and Technology; and, a grant from the Research for the Future Program (JSPS-RFTF97L00201), Japan Society of the Promotion of Science.

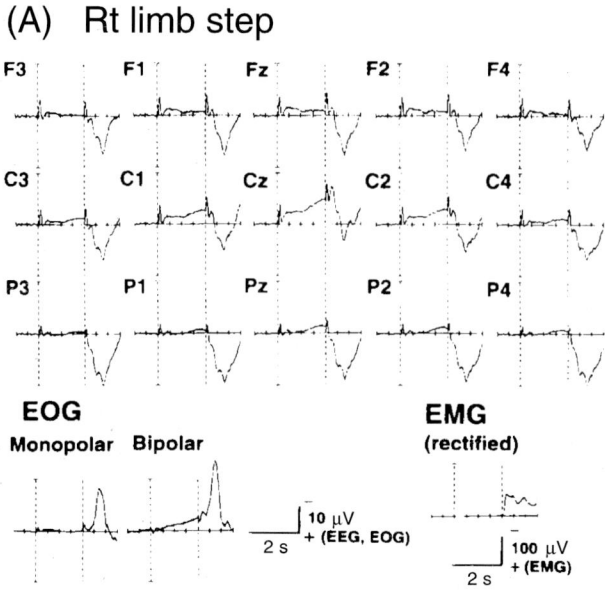

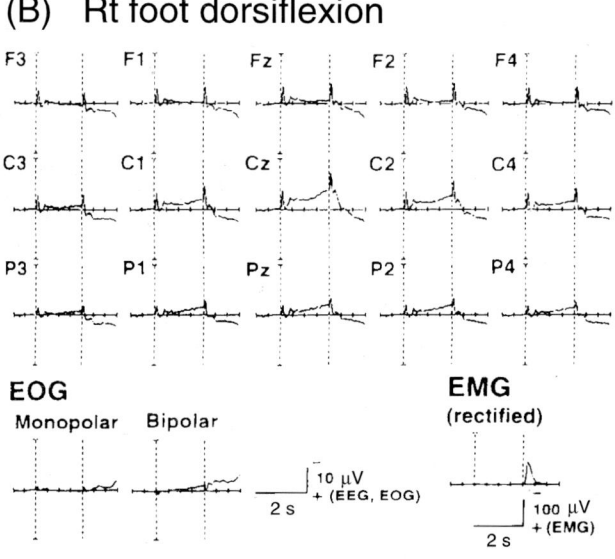

Fig. 4. The CNV associated with two leg movements in normal subjects. Grand average waveforms of the CNV are shown for 10 normal subjects. (A) A right-limb step was initiated from the standing position. (B) The right foot was dorsiflexed while in the sitting position. Both movements were the response task in a simple reaction–time paradigm. The responses were to the second stimulus (S2), which was presented as the imperative one 2 s after the warning stimulus (S1). In both tasks, the late CNV started ~1 s before S2 and it gradually enlarged until the onset of S2. The CNV was maximal at Cz in both conditions. For the A task, the distribution of the CNV was relatively symmetrical. For the B task, however, it was larger in the right versus left hemisphere (i.e., compare responses registered by electrode C2 vs. C1 and C4 vs. C3). For A versus B, note also the late CNV at Cz was larger, and it declined more steeply toward the parietal region. Abbreviations: C, central; F, frontal; P, parietal; z, midline. (Reproduced, in part, from Yazawa et al., 1997 with permission from Elsevier Publishing Company.)

Abbreviations

BP	Bereitschaftspotential
CNS	central nervous system
CNV	contingent negative variation
EEG	electroencephalogram
EMG	electromyography/electromyogram
fMRI	functional magnetic resonance imaging
L	left side
PD	Parkinson's disease
PET	positron emission tomography
rCBF	regional cerebral blood flow
rCMRglu	regional cerebral metabolic rate of glucose
SII	secondary somatosensory cortex
SMA	supplementary motor area
SPECT	single photon emission computed tomography

References and further reading

Andersen, P. and Lundberg, A. (1997). John C. Eccles (1903–1997). Trends Neurosci., 20: 324–325.

Arbas, E.A., Levine, R.B. and Strausfeld, N.J. (1997). Invertebrate nervous systems. In: Dantzler W.H. (Ed.), Comparative Physiology, Vol. II, Oxford University Press, New York, pp. 751–852 Sec. 13.

Barker D. (Ed.) (1962). Symposium on Muscle Receptors. Hong Kong University Press, Hong Kong.

DeLong, M.R. (1990). Primate models of movement disorders of basal ganglia origin. Trends Neurosci., 13: 281–285.

Fukuyama, H., Matsuzaki, S., Ouchi, Y., Yamauchi, H., Nagahama, Y., Kimura, J. and Shibasaki, H. (1996). Neural control of micturition in man examined with single photon emission computed tomography using 99m-Tc-HMPAO. Neuroreport, 7: 3009–3012.

Fukuyama, H., Ouchi, Y., Matsuzaki, S., Nagahama, Y., Yamauchi, H., Ogawa, M., Kimura, J. and Shibasaki, H. (1997). Brain functional activity during gait in normal subjects: a study. Neurosci. Lett., 228: 183–186.

Georgiou, N., Iansek, R., Bradshaw, J.L., Phillips, J.G., Mattinley, J.B. and Bradshaw, J.A. (1993). An evaluation of the role of internal cues in the pathogenesis of Parkinsonian hypokinesia. Brain, 116: 1575–1587.

Glickstein, M. and Stein, J. (1991). Paradoxical movement in PD. Trends Neurosci., 14: 480–482.

Hanakawa, T., Katsumi, Y., Fukuyama, H., Hayashi, T., Honda, M., Kimura, J. and Shibasaki, H. (1999a). Mechanisms underlying gait disturbance in PD. A single photon emission computed tomography study. Brain, 122: 1271–1282.

Hanakawa, T., Fukuyama, H., Katsumi, Y., Honda, M. and Shibasaki, H. (1999b). Enhanced lateral premotor activity during paradoxical gait in PD. Ann. Neurol., 45: 329–336.

Mori, S. (1997). Neurophysiology of locomotion: recent advances in the study of locomotion. In: Masdeu J.C., Sudarsky L. and Wolfson L. (Eds.), Gait Disorders of Aging; Falls and Therapeutic Strategies. Lippincott-Raven, Philadelphia, pp. 55–78.

Orlovsky, G.N., Deliagina, T.G. and Grillner, S. (Eds.) (1999) Quadrupedal locomotion in mammals. In: Neuronal Control of Locomotion—From Mollusc to Man, Part III. Oxford University Press, London, pp. 157–246.

Playford, E.D., Jenkins, I.H., Passingham, R.E., Nutt, J., Frackowiak, R.S.J. and Brooks, D.J. (1992). Impaired mesial frontal and putamen activation in PD: a positron emission tomography study. Ann. Neurol., 32: 151–161.

Samuel, M., Ceballos-Baumann, A.O., Turjanski, N., Boecker, H., Gorospe, A., Linazasoro, G., Homes, A.P., DeLong, M.R., Vitek, J.L., Thomas, D.G.T., Quinn, N.P., Obeso, J.A., Brooks, D.J. (1997). Pallidotomy in PD increases supplementary motor region and prefrontal activation during performance of volitional movements. An $H_2^{15}O$ PET study. Brain, 120: 1301–1313.

Vidailhet, M., Stocchi, F., Rothwell, J.C., Thompson, P.D., Day, B.L., Brooks, D.J. and Marsden, C.D. (1993). The Bereitschaftspotential preceding simple foot movement and initiation of gait in PD. Neurology, 43: 1784–1788.

Vidailhet, M., Atchinson, P.R., Stocchi, F., Thompson, P.D., Rothwell, J.C. and Marsden, C.D. (1995). The Bereitschaftspotential preceding stepping in patients with isolated gati ignition failure. Mov. Disord., 10: 18–21.

Yazawa, S., Shibasaki, H., Ikeda, A., Terada, K., Nagamine, T. and Honda, M. (1997). Cortical mechanism underlying externally cued gait initiation studied by contingent negative variation. Electroencephalogr. Clin. Neurophysiol., 105: 390–399.

CHAPTER 21

Multijoint movement control: the importance of interactive torques

Caroline J. Ketcham, Natalia V. Dounskaia and George E. Stelmach*

Motor Control Laboratory, Arizona State University, Tempe, AZ 85287-0404, USA

Abstract: The underlying mechanisms of the neural control of movement have long been explored, with a focus primarily on central control aspects and often overlooking the intrinsic mechanical properties of the motor system. To fully understand the control and regulation of movements, the biomechanical properties of the moving subject, specifically interactive torques, must be considered in the design, evaluation, and interpretation of empirical data. We first discuss the difficulty of extrapolating information from a wide variety of tasks due to their varying inherent task constraints. Examples are subsequently given where a biomechanical perspective provides a more informative interpretation of existing data. Finally, we focus on research examining the role of interactive torques with a discussion of how discoordinated movements may be explained by an inability to modulate interactive torques. Inclusion of biomechanical considerations in motor control research is a step toward incorporating multilevel methodologies and interpretations into the field, and providing a more comprehensive understanding of the neural control and regulation of movement.

Introduction

Researchers have long sought to understand the underlying mechanisms of movement control. Over the years, such control has been examined from various perspectives spanning several disciplines including psychology, bioengineering, neuroscience, physiology and kinesiology. Researchers in these areas continue to make small strides to insightful comprehension of how complex movement is controlled and regulated. It has long realized that the motor system involves both central and peripheral components, which interact in a complex manner and ultimately result in coordinated movement (Grillner and Wallén, and Cordo and Gurfinkel, this volume). Researchers have largely focused, however, on the central control aspects. Usually, they have overlooked the inherent intrinsic mechanical properties of a linked motor system. In this chapter, we raise the issue that this neglect of examining biomechanical properties has hindered progress by providing a limited view of how movement is controlled and regulated. Researchers need to recognize and appreciate the necessity of examining the constraints imposed on a given anatomical structure by dynamic movement. Such an approach provides a more detailed and comprehensive view of movement production.

In this chapter, we first discuss the difficulty of extrapolating information from a wide variety of tasks due to the inherent task constraints they possess. Subsequently, the importance of the biomechanical properties of the system is highlighted, specifically interactive torques. We then provide two examples of experiments in which a biomechanical perspective can provide a more insightful interpretation of the existing data.

While we recognize that there are many important mechanical factors to be considered, the scope of our

*Corresponding author: Tel.: +480-965-9847;
Fax: +480-965-8108; E-mail: stelmach@asu.edu

discussion is limited to the role of torques and the subsequent modulation of them by active control to maintain coordinated movements. Several experiments are detailed in which the task was manipulated to change the contribution of interactive torques, thereby providing insight into parameters presumably considered by the central nervous system (CNS). Furthermore, we address how the inability to effectively modulate interactive torques can explain some discoordinated movements observed in patients with neurological impairments.

Conceptual views of movement control

Over the several past decades, CNS processes have been studied from a range of viewpoints in an effort to reduce the computational complexity needed for movement control. These have included higher-order control perspectives, such as movement planning (Henry and Rogers, 1960; Schmidt, 1980; Heuer et al., 1995; Shea et al., 2001) and information processing (Fitts, 1954; Klapp, 1996; Valle et al., 2000). This research, while acknowledging the periphery, focused on central processes as the major mechanism for controlling movement. Other studies have examined lower levels of control, such as feedback in the system via segmental reflexes (Vallbo, 1974; Gordon and Ghez, 1991; Aminoff and Goodin, 2000), or muscle properties, including the mass-spring hypothesis (Bizzi et al., 1976; Rothwell, 1994; Winter et al., 1998) and/or the equilibrium-point hypothesis (Bizzi et al., 1976; Feldman, 1986; Schotland et al., 1989; Latash, 1993; Ghafouri and Feldman, 2001; Mah, 2001). These two hypotheses propose that movement is produced by a neural pattern that specifies a sequence of equilibrium points within a workspace (Schotland et al., 1989). The inverse dynamics perspective, namely sensorimotor transformation, has addressed control from a multilevel perspective. It proposes that to perform movement, the parameters of the task must be transformed into neural commands while simultaneously integrating the properties of the motor system to ultimately result in coordinated joint torque patterns (Abend et al., 1982; Atkeson, 1989; Lacquaniti, 1989; Kawato et al., 1990; Soechting and Flanders, 1991). Another perspective that has attempted to reduce the computational complexities for understanding control is the dynamical systems perspective (Kelso and Tuller, 1984; Kelso et al., 1991; Thelen et al., 1993; Perry, 1998; Schoner, 2002). Movements in this perspective arise from CNS interactions with the environment that are controlled via coordinative structures, which are not organized in a hierarchical framework. While many of the above studies have provided insight into motor control processes, most do not explicitly address, or even consider, the intrinsic biomechanical properties of the motor system, and how such properties can influence movement control and regulation.

Task considerations

During daily movements, much of the performance is determined by the task to be performed as well as the anatomical components of the movement that must be dealt with to efficiently execute the task. It has been shown that the functionality at the endpoint determines the path of the effector (Rosenbaum, 1991). For example, if one wants to pour juice into a glass that is turned upside-down in the cupboard, one reaches for the glass such that the arm pronates and the palm faces up when grasping the glass, and then as the arm lifts and brings it out of the cupboard it supinates so that juice can be poured into the glass. This is in contrast to if the glass is upright in the cupboard. The arm is unable to turn in a complete counterclockwise direction without injury due to the anatomical structure and the biomechanical properties of the limb. The role of the biomechanical constraints in this example is straightforward. The same properties apply in all tasks to some degree, however, and are considered by the motor system. Biomechanical properties can be equivalent to constraints when movement is compromised as a result of a specific anatomical or biomechanical property. Several studies have examined the limitations of movement by constraining the motor system in an effort to understand and detail the control and regulation of movement (Koshland et al., 1991; Sainburg et al., 1993; Cooke and Virji-Babul, 1995; Ghez and Sainburg, 1995; Gordon et al., 1995; Sainburg et al., 1995; Bastian et al., 1996; Dounskaia et al., 1998; Gribble and Ostry, 1999; Koshland et al.,

1999; Verschueren et al., 1999; Dounskaia et al., 2000a,b; Ketcham et al., 2000, 2001a,b; Dounskaia et al., 2002a; Galloway and Koshland, 2002; Seidler et al., 2002). Researchers constrain the system in defined ways, including, in particular: movement rate and/or direction; application of perturbations to assess how the system rebounds or deals with the perturbation; and movements which reach the anatomical limits of the system. Seldom have researchers operationalized control and regulation, however, by exploiting the system from a biomechanical perspective.

As a simplistic example, aiming movements in the sagittal plane differ greatly from movements executed in the frontal plane. Many times these are compared in the literature, however, to make inferences about the control of aiming movements (Fitts, 1954; Walker et al., 1997; Murata and Iwase, 2001; Ketcham et al., 2002). Movement in these two planes requires vastly different control mechanisms due to their different biomechanical constraints. Clearly, task configuration and movement plane must be considered from a biomechanical and anatomical perspective to achieve a more complete and detailed explanation of control.

Mechanical properties

The mechanical properties of the motor system have long been acknowledged. Bernstein (1967, 1984) was seminal in discussing the role of the biomechanics as a critical component in the control of movement. He emphasized that the biomechanical properties of the system are an inherent part of movement control, and when not considered, they may confound the results as well as interpretation of experimental data. Consideration of the inherent limitations of the motor system include, for example: the range of motion of joints, which change as a function of dynamic limb position due to the viscoelastic properties of the joint (Engin and Chen, 1987; Peindl and Engin, 1987; Yamaguchi et al., 1990; Pigeon et al., 1996); the type of joint involved in the movement (e.g., hinge vs. rotation joints); properties of motor units and their activation mechanisms (Enoka, 1995; Doherty, 2000); length and mass of individual limb segments (Yamaguchi et al., 1990; deLeva, 1996); muscle parameters, including membrane conduction velocity (Lemay and Crago, 1996; Pigeon et al., 1996; Chang et al., 1999); and activation dynamics (He et al., 1991) to specify a few.

All of the above have both structural and functional properties, are components of any and all movements, and must be considered to some degree by the CNS in its control of the motor system. The contribution of these properties varies at different points in the motion of a joint (Pigeon et al., 1996). Furthermore, multijoint movements in a linked system obey the physical laws of moving objects. This means that forces of movement at one joint act on all adjacent joints, thereby creating a need for the CNS to deal with interactive torques. Subsequently, the combination of multijoint dynamic positions has an even greater influence on the type of CNS control required. This occurs as a result of the anatomical and biomechanical structures of the limb, including, for example, biarticular muscles, whose properties change as a function of two-joint configurations.

A desired movement emerges from adjustments of neural signals to the specific mechanical properties of the involved limb segments. Joint interactions are the most pronounced mechanical factor influencing multiarticular movements. The limb segments generate these interactions during motion. Therefore, the movements about joints cannot be controlled independently from each other.

Examples of not considering interactive torques

Interpretations of data, which overlook the biomechanical properties, pose problems for the progress of research. Interpretations may not be wrong. Rather, they would be more robust if the mechanical properties of the system were considered. One example of a study that did not incorporate biomechanical properties into the interpretation of control was that of Kelso et al. (1991). It addressed the neural control of single- versus multijoint limb movement patterns. Subjects were asked to produce in-phase and anti-phase wrist/elbow motions. Previous literature had shown that the in-phase pattern is more stable than the anti-phase pattern. Subjects started in the anti-phase position and maintained the anti-phase movement as speed increased. They performed the task with the forearm both supine and prone for trials of 43-s duration. Of interest was whether subjects transitioned to the in-phase pattern of movement as speed increased, and at what speed this transition occurred. In-phase was defined as the situation in which the elbow and wrist flexed and extended in synchrony. Conversely, anti-phase was defined as when the elbow and wrist flexed and extended in a 180° phase shift such that as one flexed, the other extended. Phase transitions occurred when the forearm was in the supine- but not the prone position (Fig. 1). Spatially and biomechanically, the two movements were similar, however.

Based on the Fig. 1-type of result, Kelso et al. (1991) concluded that there are neurophysiological mechanisms that are dependent on the context of the task, specifically the direction of movement and proprioceptive feedback information. The inclusion of biomechanical properties, however, would have led the authors to a succinct and straightforward conclusion (Dounskaia et al., 1998). When the forearm is in a supinated position, flexion of the elbow results in flexion of the wrist as a result of interactive torques acting on the wrist due to movement of the elbow. The same will occur with extension of the elbow. In contrast, when the forearm is pronated, flexion of the elbow results in extension of the wrist, and vice versa, again as a result of the interactive torques. Thus, biomechanical constraints are a likely cause of the difference between in-phase and anti-phase movements in the Kelso et al.'s (1991) experiment.

Massaquoi and Hallett (1996) provide another example in which the biomechanical properties of the system were not considered. They tested cerebellar patients' ability to perform repetitive, straight-line movements. As their movement speed increased, the patients produced curved- rather than straight-line trajectories. This was interpreted as being due to a deficient rate of rotation at the shoulder and a slower rate of torque development, thereby producing ataxic movements. While this interpretation may hold up, another likely consideration is that the cerebellar patients were unable to efficiently and effectively counteract interactive torques, thereby leading to the elaboration of curved trajectories. If so, the problem may have resulted from an inefficient modulation at the elbow rather than a deficient rate of movement development at the shoulder. Similar biomechanical interpretations have been fruitful in other research on cerebellar patients (Bastian et al., 1996; see also Thach and Bastian, this volume).

Analysis of interactive torques

Several experiments have demonstrated kinematic invariance as speed, direction and amplitude of movement change in two-dimensional and three-dimensional motor tasks (Morasso, 1981; Soechting and Lacquaniti, 1981; Hollerbach and Flash, 1982; Hollerbach and Flash, 1983). Such invariance suggests that the CNS must compensate for, or deal with, interactive torques in a very refined and systematic way. This past decade, several researchers have assessed multijoint movements and considered interactive torques, the role they play, and how they are compensated for (Koshland et al., 1991; Sainburg

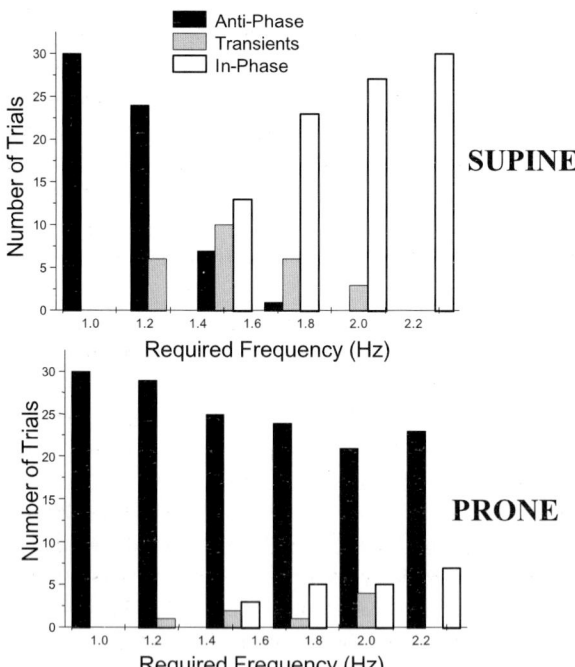

Fig. 1. Number of trials in behavioral movement patterns as movement speed increases. (Reproduced, in part, from Kelso et al., 1991 with permission from Springer-Verlag.)

et al., 1993; Cooke and Virji-Babul, 1995; Ghez and Sainburg, 1995; Sainburg et al., 1995; Bastian et al., 1996; Dounskaia et al., 1998; Gribble and Ostry, 1999; Koshland et al., 1999; Sainburg et al., 1999; Verschueren et al., 1999; Dounskaia et al., 2000a,b, 2002a,b; Galloway and Koshland, 2002).

An understanding of interactive torques acting on a limb allows researchers to discern different modes of control in both single- and multijoint movements. For example, Gribble and Ostry (1999) examined single-joint elbow flexion, single-joint shoulder flexion and multijoint movements involving both elbow and shoulder flexion. Their data suggest that in all cases, CNS signals to the muscles were adjusted in a predictive manner to compensate for expected (anticipated) interactive torques. In the shoulder-flexion condition, for example, activity in the elbow occurred prior to movement onset. Presumably, this stabilized the elbow and allowed completion of the single-joint movement. Thus, it would seem that the CNS must predict interactive torques in an upcoming movement, integrate this information, and then signal to musculature in all relevant joints the magnitude of active control required to deal with, or compensate for, the upcoming interactive torques. The Gribble and Ostry (1999) study provided an exemplary account of the refinement of the CNS control system. It also demonstrated that even the simplest movements require synchronous control of multiple components.

Further studies have demonstrated that compensatory muscle activity occurs before peripheral feedback can influence a movement (Hollerbach and Flash, 1982; Koshland and Hasan, 1994; Cooke and Virji-Babul, 1995; Koshland et al., 2000). Thus, the prediction of upcoming torques must be learned (Bernstein, 1984). This idea is well established in literature on motor development (McDonald et al., 1989; Schneider et al., 1989; Vereijken et al., 1992; Thelen et al., 1993), although the mechanism of learning is still a debatable question. One idea gaining favor is that interjoint dynamics and the prediction of upcoming torques are what are learned. Distinctions have been inferred between anticipatory and after-effect forces during a briefly perturbed movement (Shadmehr and Mussa-Ivaldi, 1994; Wolpert et al., 1995; Shadmehr and Moussavi, 2000). The authors have not suggested explicitly that the CNS predicts interactive torques. Rather, they propose that the CNS specifies the parameters of force and velocity to move in a specified direction and that this specification is learned. Such specification must, however, include a prediction of the upcoming interactive torques of the movement and implement this knowledge into the relevant CNS control mechanism(s). Furthermore, in many movements, changes in force fields lead primarily to changes in interactive torques and the modulation necessary to maintain the appropriate trajectory. Subjects have demonstrated the ability to learn these force fields and apply the appropriate 'internal model' (Wolpert et al., 1995) or 'motor primitive' (Thoroughman and Shadmehr, 2000). When given two perturbations in sequence, however, subjects are unable to switch between two learned internal models (Karniel and Mussa-Ivaldi, 2002).

The role of interactive torques has been extended even further in the literature. For example, Gribble and Ostry (1999) demonstrated that in a multijoint movement, phasic EMG activity at the focal/proximal joint preceded movement, and EMG bursts at the distal joint were related directly to the magnitude of anticipated interactive torques. The control of interactive torques has also been shown to exhibit a hierarchical feature in which a proximal limb segment creates torques that have a greater influence on a distal segment than vice versa (Dounskaia et al., 1998; Dounskaia et al., 2002a,b; Galloway and Koshland, 2002). Furthermore, movements of a distal segment tend to be corrective and performed to reach desired kinematic outcomes. Such hierarchical organization needs to be assessed in a variety of movements and effectors to determine if it is an underlying principle of CNS control.

Another approach to advance understanding of the CNS control implications of interactive torques is to examine what leads to a disruption (or breakdown) in the feedforward mechanism of control such that movements become distorted. Consider, for example, the role of interactive torques in cyclical movements (Sainburg et al., 1993; Ghez and Sainburg, 1995; Gordon et al., 1995; Sainburg et al., 1995; Dounskaia et al., 1998; Gribble and Ostry, 1999; Verschueren et al., 1999; Dounskaia et al., 2002a,b; Galloway and Koshland, 2002; Seidler et al., 2002). Many of these studies have shown that increasing movement speed can exceed the possibility of utilizing sensory feedback control. Specifically, it has been shown that as interactive torques become too large to compensate for effectively, movements distort in a stereotypical and predictable manner, namely toward a biomechanically stable pattern (Dounskaia et al., 1998; Dounskaia et al., 2000; Dounskaia et al., 2002a,b). Understanding breakdowns in movement coordination, and how they are influenced by interactive torques, provides insight into how the motor system is controlled by the CNS and, furthermore, how this control degrades in older adults and those with neurological impairments.

Role of interactive torques in multijoint movements

It is thought that the CNS utilizes interactive torques to execute multijoint drawing movements. For example, Galloway and Koshland (2002) assessed multijoint movements involving the shoulder, elbow and wrist. Subjects made point-to-point targeted movements in the horizontal plane with their arm in a mechanical apparatus that slid along the table. These movements were discrete, with some being curved, some straight and some instructed to be single-joint (i.e., elbow and shoulder flexion and extension). Kinematics and kinetics were measured. They found that dynamics differed depending on the joint they assessed. The shoulder muscle torque usually determined the joint acceleration, with interactive torques being minimal. In the more distal joints (wrist and elbow), net torque was a combination of muscle torque and interactive torque. The contribution and role changed systematically depending on target direction and joint amplitude. Thus, CNS control must have differed across joints.

In another study, Dounskaia et al. (2000a) assessed interjoint coordination during a handwriting task performed on a digitizing tablet. Subjects were instructed to draw a circle and straight lines using (1) only their wrist, (2) only their fingers, (3) an equivalent pattern using simultaneous flexion and extension of the wrist and fingers and (4) a nonequivalent pattern in which wrist flexed while the fingers extended and vice versa. Movements were performed at four speed levels. Across coordination patterns, variability and temporal characteristics of the movements changed. Overall, the equivalent pattern was more similar to a single-joint movement. In contrast, the nonequivalent pattern was executed slower than the other patterns, particularly when the subjects were instructed to move as fast as possible. Furthermore, in the circle-drawing condition, movements transitioned toward oval shapes, with decreases in the relative phase between wrist and fingers. Similar tendencies were shown in left-handed individuals, but at a different drawing slant (Fig. 2). Thus, the authors suggested that the slant of handwriting movements is adjusted to the biomechanical structure of the hand.

Dounskaia and colleagues (Ketcham et al., 2000a; Ketcham et al., 2000b; Ketcham et al., 2001; Dounskaia et al., 2002a,b) have extended this line of research by examination of repetitive circle, line and ellipse drawing in the horizontal plane, using shoulder and elbow motion. It was expected that preferred joint combinations would be easier to maintain as movement speed increased. The rationale was that influences of interactive torques on biomechanically disadvantaged joint combinations place a greater burden on muscle activation. Muscle activity must be appropriately timed and of optimal magnitude. Coordination patterns that utilize interactive torques to drive the movement should be maintained at higher speeds than those that require greater modulation.

Subjects were instructed to repetitively draw a shape template first with vision, and then without it. There were nine templates: a circle, four ovals and four lines oriented such that different joint combinations were emphasized. The trunk and wrist were immobilized such that a two-joint movement was maintained.

Forces acting on each joint are a combination of active and passive forces from all segments included in the linked effector. Thus, analyses of the role of these forces provide insight into the mechanisms of control. To determine the torque components acting at the shoulder and elbow, the arm was modeled as a linkage of two rigid segments (the upper and lower arm) constrained to move in the horizontal plane. Segment masses, locations of mass center and moments of inertia about transverse axes passing through the joint centers were estimated on the basis of the subject's weight and height.

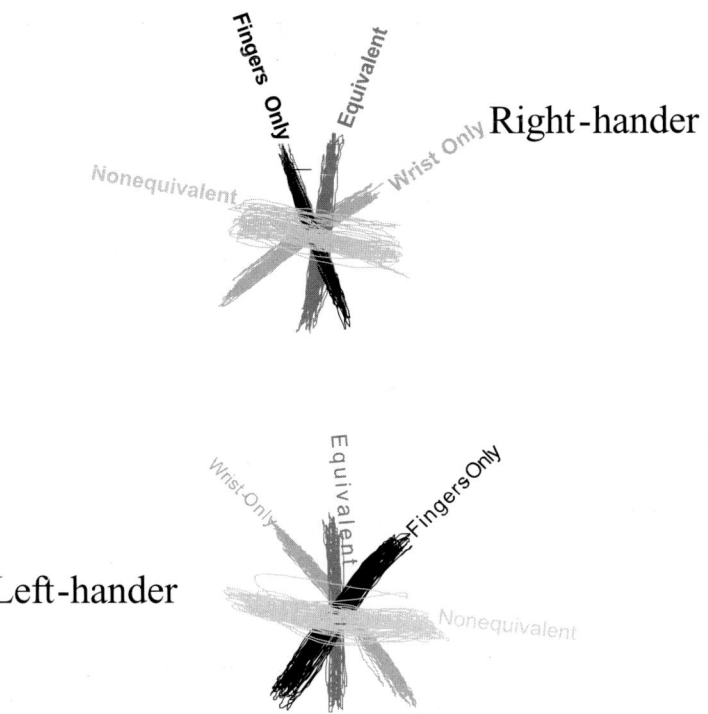

Fig. 2. Example trajectories for line drawing under four coordination patterns for individual right- versus left-handed subjects. (Reproduced, in part, from Dounskaia et al., 2000a with permission from Springer-Verlag.)

Two characteristics of muscle activity were quantified. First, EMG 'burstiness' was calculated for each muscle as a ratio between the maximal and minimal EMG values within each movement cycle, averaged across cycles of the trial. To compute the level of muscle alternation, each EMG value obtained from a flexor was divided by the corresponding extensor's value. To reveal the extent muscle activity directly caused accelerations/decelerations at the joints, phase offset was computed for each joint between the net torque and bursts of activity of the muscles spanning this joint.

Several informative outcomes have resulted from this research. First, Dounskaia et al. (2002a) examined how movement coordination changed across shapes specifically in line drawing at a 1.5-Hz cycling frequency. Joint amplitudes and the relative phase between joints varied widely across shapes. Kinetic analysis revealed that the role of muscle torque in movement was different across the two joints. In 8/9 of the movements, muscle torque accelerated and decelerated the shoulder and coped with interactive torques created by the elbow. Interactive torques created by the shoulder, however, played a substantial role in elbow acceleration and deceleration. Muscle torque adjusted this passive elbow movement to maintain movement trajectory. The EMG analysis was in agreement with the kinetic analysis. These data suggest that the CNS control strategy for this multijoint movement is to exploit and use the intersegmental dynamics of the two-joint system to produce the desired movements.

Similarly, Dounskaia and colleagues (Ketcham et al., 2001b; Dounskaia et al., 2002b) examined circle, ellipse and line drawing movements as cycling frequency was increased up to 2.5 Hz, and vision was occluded. Increases in speed and occlusion of vision lead to systematic deviations in endpoint trajectory, as measured by kinematic analyses. Furthermore, kinetic analysis revealed that each movement shape required a defined pattern of regulation of interactive torque with active torque. Deviations in endpoint were associated with changes in the control pattern. The increase in movement speed

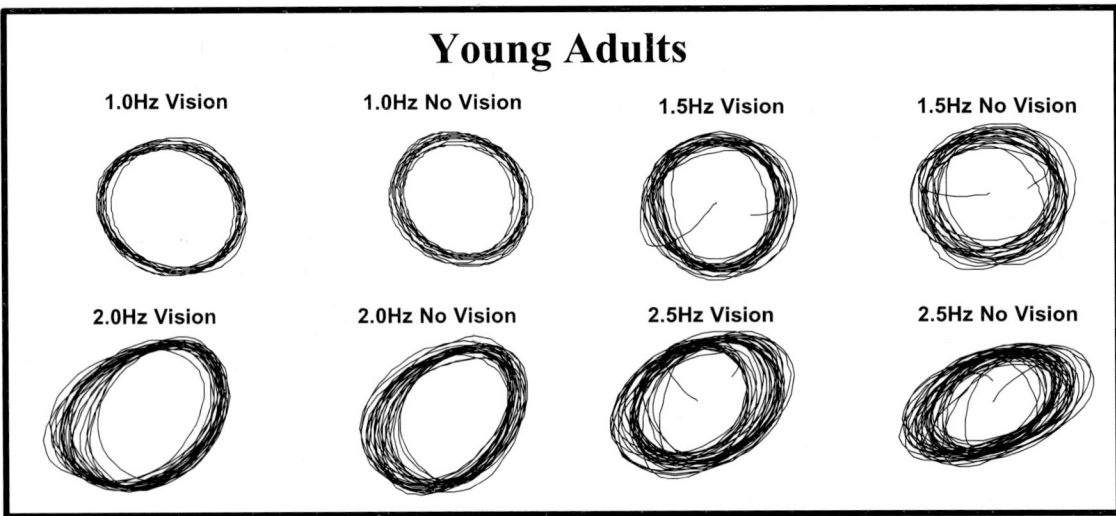
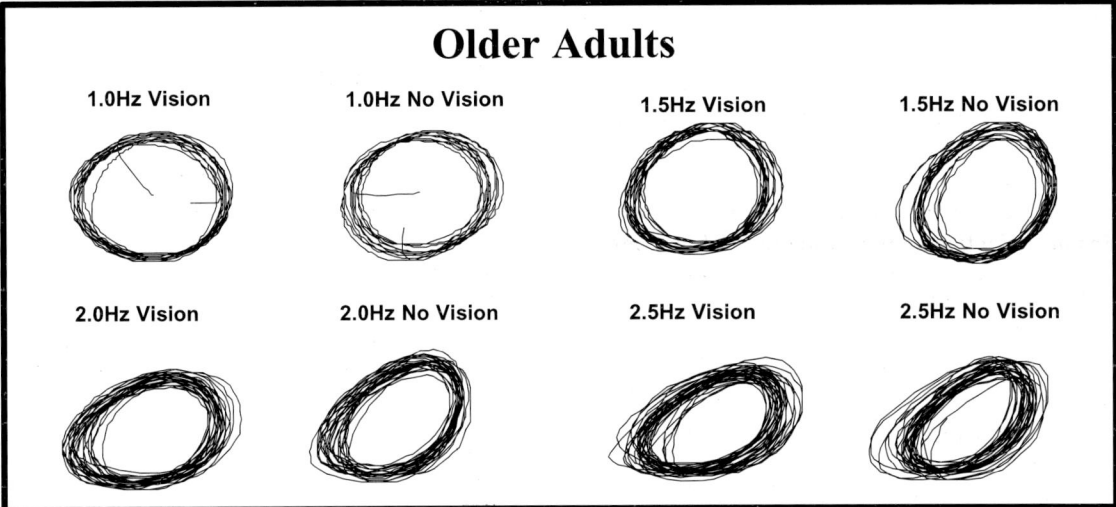
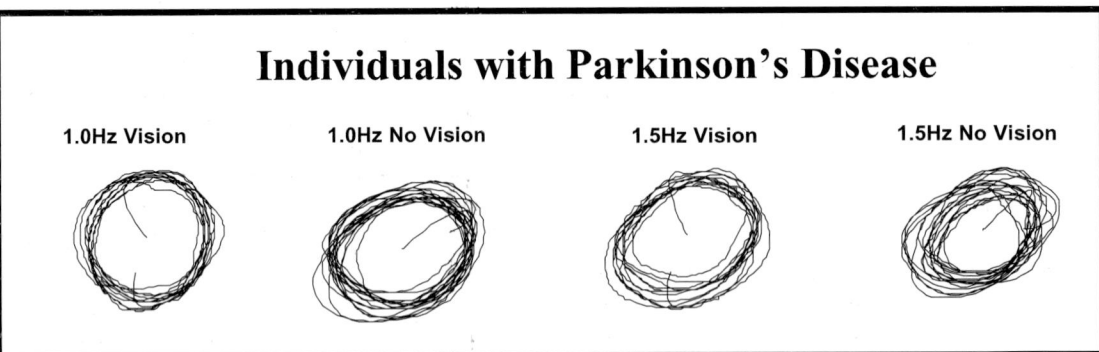

Fig. 3. Example trajectories for horizontal circle drawing by young, older and Parkinsonian adults at increasing cycling frequencies, both with and without vision.

and occlusion of vision constrained the system, which along with the biomechanical constraints, led to a varying complexity of joint coordination patterns. These depended on the role of interactive torques in the production of movements at each joint.

Interactive torques provide insight into discoordinated movements

We have also conducted studies like the earlier ones in older adults and Parkinsonian patients (Ketcham et al., 2000, 2001a,b). As speed increased (2.5 Hz for older adults, 1.5 Hz for Parkinson's patients) and when vision was occluded, endpoint trajectories became distorted, as seen previously in young adults (Ketcham et al., 2000, 2001b; Dounskaia et al., 2002b). Exemplary trajectories for circle drawing for all groups are shown in Fig. 3. Distortion occurred in older adults at 2.0 Hz and 1.5 Hz in Parkinsonian patients. This is in contrast to 2.5 Hz in young adults. Kinematic analysis revealed quantitative changes in diameter ratio for the circle and slope of the longest diameter for the four ellipses.

We also demonstrated that older adults and Parkinson patients lose the subtleties of fine control needed to maintain accurate movement trajectories as speed increased and when vision was occluded. This loss of fine modulation appeared to result from a reduced ability to accurately modulate interactive torques. Similar losses were found for young adults at higher movement speeds (Ketcham et al., 2000). This suggested that at these high speeds in both normal subjects and patients, the limitations of movement performance were reached. In young adults, this occurred at much higher speeds and was likely due to the biomechanical constraints of the system. These constraints changed as a function of aging and neurological disease.

Concluding remarks

The analysis of interactive torques in the control and regulation of movement has been fruitful. Several research groups have begun to utilize this method to ask questions of about how movement is controlled and regulated. This is by no means the only approach required to elucidate CNS control mechanisms. Rather, research should consider multilevel methodologies, with new interpretations likely to emerge. One major consideration is that the motor system is comprised of a multitude of components that must work in harmony with each other.

The symphonic properties of the motor system should lead us toward an understanding of the control of the interactions and connections of all of its components. This is an enormous task, which requires more comprehensive analyses than hitherto attempted. The field of movement neuroscience has certainly advanced understanding of how the CNS controls isolated parts of the system. The next task is to show how the CNS makes these parts work together, without undo emphasis or neglect of any single piece. To this end, we believe that the study of the CNS control of interactive torques will be at the forefront.

Acknowledgments

This work was supported by NIH grants awarded to George E. Stelmach: NIA-AG 147676; NINDS-NS 39352; and NINDS-NS 33173.

Abbreviations

CNS   central nervous system
EMG   electromyogram/electromyography

References

Abend, W., Bizzi, E. and Morasso, P. (1982). Human arm trajectory formation. Brain, 105: 331–348.

Aminoff, M.J. and Goodin, D.S. (2000). Studies of the human stretch reflex. Muscle Nerve, 9(Suppl.): S3–S6.

Atkeson, C.G. (1989). Learning arm kinematics and dynamics. Annu. Rev. Neurosci., 12: 157–183.

Bastian, A.J., Martin, T.A., Keating, J.G. and Thach, W.T. (1996). Cerebellar ataxia: abnormal control of interaction torques across multiple joints. J. Neurophysiol., 76: 492–509.

Bernstein, N.A. (1967). The Co-ordination and Regulation of Movements. Pergamon Press, Oxford.

Bernstein, N. (1984). Biodynamics of locomotion. In: Whiting H.T.A. (Ed.), Human Motor Actions: Bernstein Reassessed. North-Holland, Amsterdam, pp. 171–222.

Bizzi, E., Pollit, A. and Morasso, P. (1976). Mechanisms underlying achievement of final head position. J. Neurophysiol., 39: 435–444.

Chang, Y.W., Su, F.C., Wu, H.W. and An, K.N. (1999). Optimum length of muscle contraction. Clin. Biomech., 14: 537–542.

Cooke, J.D. and Virji-Babul, N. (1995). Reprogramming of muscle activation patterns at the wrist in compensation for elbow reaction torques during planar two-joint arm movements. Exp. Brain Res., 106(1): 177–180.

deLeva, P. (1996). Adjustments to Zatsiorsky-Seluyanov's segment inertia parameters. J. Biomech., 29: 1223–1230.

Doherty, T.J. (2000). Effects of short-term training on physiologic properties of human motor units. Can. J. Appl. Physiol., 25: 194–203.

Dounskaia, N.V., Swinnen, S.P., Walter, C.B., Spaepen, A.J. and Verschueren, S.M. (1998). Hierarchical control of different elbow-wrist coordination patterns. Exp. Brain Res., 121: 239–254.

Dounskaia, N., Van Gemmert, A.W. and Stelmach, G.E. (2000a). Interjoint coordination during handwriting-like movements. Exp. Brain Res., 135: 127–140.

Dounskaia, N.V., Swinnen, S.P. and Walter, C.B. (2000b). A principle of control of rapid multijoint movements: the leading joint hypothesis. In: Winters J.M. and Crago P.E. (Eds.), Biomechanics and Neural Control of Posture and Movement. Springer-Verlag, New York, pp. 390–404.

Dounskaia, N.V., Ketcham, C.J. and Stelmach, G.E. (2002a). Commonalities and differences in control of various drawing movements. Exp. Brain Res., 146: 11–25.

Dounskaia, N.V., Ketcham, C.J. and Stelmach, G.E. (2002b). Influence of biomechanical constraints on horizontal arm movements. Motor Control, 6: 366–387.

Engin, A.E. and Chen, S.M. (1987). Kinematic and passive resistive properties of human elbow complex. J. Biomech. Eng., 109: 318–323.

Enoka, R.M. (1995). Morphological features and activation patterns of motor units. J. Clin. Neurophysiol., 12: 538–559.

Feldman, A.G. (1986). Once more on the equilibrium-point hypothesis (model) for motor control. J. Mot. Behav., 18: 17–54.

Fitts, P.M. (1954). The information capacity of the human motor system in controlling the amplitude of movement. J. Exp. Psychol., 47: 381–391.

Galloway, J.C. and Koshland, G.F. (2002). General coordination of shoulder, elbow and wrist dynamics during multijoint arm movements. Exp. Brain Res., 142: 163–180.

Ghafouri, M. and Feldman, A.G. (2001). The timing of control signals underlying fast point-to-point arm movements. Exp. Brain Res., 137: 411–423.

Ghez, C. and Sainburg, R. (1995). Proprioceptive control of interjoint coordination. Can. J. Physiol. Pharmacol., 73: 273–284.

Gordon, J. and Ghez, C. (1991). Muscle receptors and spinal reflexes: the stretch reflex. In: Kandel E.R., Schwartz J.H. and Jessell T.M. (Eds.), Principles of Neural Science, 3rd ed., Elsevier, Amsterdam, pp. 564–580.

Gordon, J., Ghilardi, M.F. and Ghez, C. (1995). Impairments of reaching movements in patients without proprioception. I. Spatial errors. J. Neurophysiol., 73: 347–360.

Gribble, P.L. and Ostry, D.J. (1999). Compensation for interaction torques during single- and multijoint limb movement. J. Neurophysiol., 82: 2310–2326.

He, J., Levine, W.S. and Loeb, G.E. (1991). Feedback gains for correcting small perturbations to standing posture. IEEE Trans. Autom. Control, 36: 322–332.

Henry, F.M. and Rogers, D.E. (1960). Increased response latency for complicated movements and a 'memory drum' theory of neuromotor reaction. Res. Q., 31: 448–458.

Heuer, H., Schmidt, R.A. and Ghodsian, D. (1995). Generalized motor programs for rapid bimanual tasks: a two-level multiplicative-rate model. Biol. Cybern., 73: 343–356.

Hollerbach, J.M. and Flash, T. (1982). Dynamic interactions between limb segments during planar arm movement. Biol. Cybern., 44: 67–77.

Karniel, A. and Mussa-Ivaldi, F.A. (2002). Does the motor control system use multiple models and context switching to cope with a variable environment?. Exp. Brain Res., 143: 520–524.

Kawato, M., Maeda, Y., Uno, Y. and Suzuki, R. (1990). Trajectory formation of arm movement by cascade neural network model based on minimum torque-change criterion. Biol. Cybern., 62: 275–288.

Kelso, J.A.S. and Tuller, B. (1984). A dynamical basis for action system. In: Gazanniga M.S. (Ed.), Handbook of Cognitive Neuroscience. Plenum, New York, pp. 321–356.

Kelso, J.A.S., Buchman, J.J. and Wallace, S.A. (1991). Order parameters for the neural organization of single, multijoint limb movement patterns. Exp. Brain Res., 85: 432–444.

Ketcham, C.J., Dounskaia, N.V., Leis, B. and Stelmach, G.E. (2000). Interactive torques contribute to multijoint coordination impairments in Parkinsonian patients. Soc. Neurosci. Abstr., 26: 163.

Ketcham, C.J., Dounskaia, N. and Stelmach, G.E. (2001a). Older adults demonstrate trajectory distortions in multijoint coordination. Soc. Neurosci. Abstr., 27: 834.1.

Ketcham, C.J., Dounskaia, N.V. and Stelmach, G.E. (2001b). Biomechanical constraints imposed on multijoint coordination lead to trajectory distortion as a function of speed. Proceedings, 12th Annual Spring Brain Conference, Sedona, AZ, p. A18.

Ketcham, C.J., Seidler, R.D., Van Gemmert, A.W. and Stelmach, G.E. (2002). Age-related kinematic differences as influenced by task difficulty, target size, and movement amplitude. J. Gerontol. B Psychol. Sci. Soc. Sci., 57: P54–P64.

Klapp, S.T. (1996) Reaction time analysis of central motor control. In: Zelaznik, H.N. (Ed.), Advances in Motor Learning and Control. Human Kinetics, Champaign, IL, pp. 13–35.

Koshland, G.F. and Hasan, Z. (1994). Selection of muscles for initiation of planar, three-joint arm movements with different final orientations of the hand. Exp. Brain Res., 98: 157–162.

Koshland, G.F., Hoy, M.G., Smith, J.L. and Zernicke, R.F. (1991). Coupled and uncoupled limb oscillations during paw-shake response. Exp. Brain Res., 83: 587–597.

Koshland, G.F., Marasli, B. and Arabyan, A. (1999). Directional effects of changes in muscle torques on initial path during simulated reaching movements. Exp. Brain Res., 128: 353–368.

Koshland, G.F., Galloway, J.C. and Nevoret-Bell, C.J. (2000). Control of the wrist in three-joint arm movements to multiple directions in the horizontal plane. J. Neurophysiol., 83: 3188–3195.

Lacquaniti, F. (1989). Central representations of human limb movement as revealed by studies of drawing and handwriting. Trends Neurosci., 8: 287–291.

Latash, M.L. (1993). Control of Human Movement. Human Kinetics, Champaign, IL.

Lemay, M.A. and Crago, P.E. (1996). A dynamic model for simulating movements of the elbow, forearm, and wrist. J. Biomech., 29: 1219–1330.

Mah, C.D. (2001). Spatial and temporal modulation of joint stiffness during multijoint movement. Exp. Brain Res., 136: 492–506.

Massaquoi, S. and Hallett, M. (1996). Kinematics of initiating a two-joint arm movement in patients with cerebellar ataxia. Can. J. Neurol. Sci., 23: 3–14.

McDonald, P.V., Van Emmerik, R.E.A. and Newell, K.M. (1989). The effects of practice on limb kinematics in a throwing task. J. Mot. Behav., 21: 245–264.

Morasso, P. (1981). Spatial control of arm movements. Exp. Brain Res., 42: 223–227.

Morasso, P. (1983). Three dimensional arm trajectories. Biol. Cybern., 48: 187–194.

Murata, A. and Iwase, H. (2001). Extending Fitts' law to a three-dimensional pointing task. Hum. Mov. Sci., 20: 791–805.

Peindl, R.D. and Engin, A.E. (1987). On the biomechanics of human shoulder complex—II. Passive resistive properties beyond the shoulder complex sinus. J. Biomech., 20: 119–134.

Perry, S.B. (1998). Clinical implications of a dynamical systems theory. Neuroreport, 9: 4–10.

Pigeon, P., Yahia, L. and Feldman, A.G. (1996). Moment arms and lengths of human upper limb muscles as functions of joint angles. J. Biomech., 29: 1365–1370.

Rosenbaum, D.A. (1991). Human Motor Control. Academic Press, San Diego.

Rothwell, J. (1994). Control of Human Voluntary Movement, 2nd ed., Chapman and Hall, London.

Sainburg, R.L., Poizner, H. and Ghez, C. (1993). Loss of proprioception produces deficits in interjoint coordination. J. Neurophysiol., 70: 2136–2147.

Sainburg, R.L., Ghilardi, M.F., Poizner, H. and Ghez, C. (1995). Control of limb dynamics in normal subjects and patients without proprioception. J. Neurophysiol., 73: 820–835.

Sainburg, R.L., Ghez, C. and Kalakanis, D. (1999). Intersegmental dynamics are controlled by sequential anticipatory, error correction, and postural mechanisms. J. Neurophysiol., 81: 1045–1056.

Schmidt, R.A. (1980). Past and future issues in motor programming. Res. Q. Exerc. Sport, 51: 122–140.

Schneider, K., Zernicke, R.F., Schmidt, R.A. and Hart, T.J. (1989). Changes in limb dynamics during the practice of rapid arm movements. J. Biomech., 22: 805–817.

Schoner, G. (2002). Timing, clocks, and dynamical systems. Brain Cogn., 48: 31–51.

Schotland, J.L., Lee, W.A. and Rymer, W.Z. (1989). Wiping reflex and flexion withdrawal reflexes display different EMG patterns prior to movement onset in the spinalized frog. Exp. Brain Res., 78: 649.

Seidler, R.D., Alberts, J.L. and Stelmach, G.E. (2002). Changes in multi-joint performance with age. Motor Control, 6: 19–31.

Shadmehr, R. and Mussa-Ivaldi, F.A. (1994). Adaptive representation of dynamics during learning of a motor task. J. Neurosci., 14: 3208–3224.

Shadmehr, R. and Moussavi, Z.M. (2000). Spatial generalization from learning dynamics of reaching movements. J. Neurosci., 20: 7807–7815.

Shea, C.H., Lai, Q., Wright, D.L., Immink, M. and Black, C. (2001). Consistent and variable practice conditions: effects on relative and absolute timing. J. Mot. Behav., 33: 139–152.

Soechting, J.F. and Flanders, M. (1991). Deducing central algorithms of arm movement control from kinematics. In: Humphrey D.R. and Freund H.J. (Eds.), Motor Control: Concepts and Issues. Wiley, New York, pp. 293–306.

Soechting, J.F. and Lacquaniti, F. (1981). Invariant characteristics of a pointing movement in man. J. Neurosci., 1: 710–720.

Thelen, E., Corbetta, D., Kamm, K., Spencer, J.P., Schneider, K. and Zernicke, R.F. (1993). The transition to reaching: mapping intention and intrinsic dynamics. Child Dev., 64: 1058–1098.

Thoroughman, K.A. and Shadmehr, R. (2000). Learning of action through adaptive combination of motor primitives. Nature, 407: 742–747.

Vallbo, A.B. (1974). Human muscle spindle discharge during isometric voluntary contractions: amplitude relations between spindle frequency and torque. Acta Psychol. Scand., 90: 319–336.

Valle, M.S., Bosco, G. and Poppele, R. (2000). Information processing in the spinocerebellar system. Neuroreport, 11: 4075–4079.

Vereijken, B., Van Emmerik, R.E.A., Whiting, H.T.A. and Newell, K.M. (1992). Free(z)ing degrees of freedom in skill acquisition. J. Mot. Behav., 24: 133–142.

Verschueren, S.M., Swinnen, S.P., Cordo, P.J. and Dounskaia, N.V. (1999). Proprioceptive control of multijoint movement: unimanual circle drawing. Exp. Brain Res., 127: 171–181.

Walker, N., Philbin, D.A. and Fisk, A.D. (1997). Age-related differences in movement control: adjusting submovement structure to optimize performance. J. Gerontol. Psychol. Sci., 52B: P40–P52.

Winter, D.A., Patla, A.E., Prince, F., Ishac, M. and Gielo-Perczak, K. (1998). Stiffness control of balance in quiet standing. J. Neurophysiol., 80: 1211–1221.

Wolpert, D.M., Ghaharamani, Z. and Jordan, M.I. (1995). Are arm trajectories planned in kinematic or dynamic coordinates? An adaptation study. Exp. Brain Res., 103: 460–470.

Yamaguchi, G.T., Sawa, A.G.U., Moran, D.W., Fessler, M.J. and Winters, J.M. (1990). A survey of human musculotendon actuator parameters. In: Winters J.M. and Woo S.L.Y. (Eds.), Multiple Muscle Systems: Biomechanics and Movement Organization. Springer-Verlag, New York.

SECTION VI

Descending command issues

CHAPTER 22

How the mesencephalic locomotor region recruits hindbrain neurons

Igor Kagan[2] and Mark L. Shik[1],*

[1]Department of Zoology, Tel Aviv University, Tel Aviv 69978, Israel
[2]Department of Biomedical Engineering, Technion, Haifa 32000, Israel

Abstract: This chapter summarizes experiments which were designed to reveal how repetitive electrical stimulation of the mesencephalic locomotor region (MLR) recruits nearby hindbrain neurons into activity, such that locomotion can ensue in the tiger salamander, *A. tigrinum*. The MLR stimulus strength was subthreshold or near-threshold for locomotor movements to ensue. Such relatively weak stimulation of the MLR produced locomotor movements after a relatively long delay, which featured neuronal interactions in the hindbrain. MLR-evoked spike responses of single hindbrain neurons were recorded before locomotor movements began. This allowed consideration of the build-up of the hindbrain neuronal activity, which was subsequently impressed upon the spinal cord such as to evoke locomotor movements. Each train of MLR stimulus pulses evoked monosynaptic responses in but a small proportion of the hindbrain's neurons. Rather, oligosynaptic responses were routinely evoked, even in the "input" neurons that were activated monosynaptically. Consecutive stimulus volleys recruited a given neuron after a variable number of synaptic translations. It is argued that the hindbrain's input neurons excited a much larger number of other hindbrain neurons. By this means, an MLR-evoked, short-lived propagating wave of excitation (i.e., ~ 2–4 successive synaptic activations) can be spread throughout the hindbrain.

Introduction

Several divisions of the brain participate in the control of a movement. Take, for example, the induction of locomotion in various vertebrates, including caudate amphibians, by repetitive stimulation of the mesencephalic locomotor region (MLR; Shik, 1997; Grillner and Wallén, Takakusaki et al., and S. Mori et al., this volume). The excited neurons project to the hindbrain (Orlovsky, 1970). At this site, monosynaptically excited neurons initiate the processing that leads to activation of the appropriate amount and composition of the hindbrain's neuronal population, which, in turn, activates the relevant spinal locomotor networks. To achieve this task at the hindbrain level, how many successive synaptic excitations (translations) occur in relevant neurons after each input volley from the MLR?

In a previous study on the rough skin newt, *T. granulosa*, we recorded extracellularly the impulses of hindbrain neurons before MLR-induced locomotor movements began (Bar-Gad et al., 1999). The latency of time-locked synaptic responses was predominately at ~ 13, 18, 23 and 28 ms. These distinctive latencies appeared intermittently when the train of repetitive MLR stimulus pulses had an inter-stimulus (pulse) interval (IStI) of 100–200 ms. Each volley from the MLR evoked a propagation of activity in hindbrain neurons with a characteristic synaptic translation time of 5–6 ms. The activity was short-lived and ceased after three or four translations, at which time hindbrain neuronal activity terminated or became disengaged from the MLR stimulus. In that study, however, we mostly used trains of MLR stimulus pulses that were subthreshold

*Corresponding author: Tel.: +972-3-640-7390;
Fax: +972-3-6409403; E-mail: mrkshik@post.tau.ac.il

DOI: 10.1016/S0079-6123(03)43022-7

for evoking locomotor movements. Now, we are able to record the activity of the same hindbrain neuron in several successive periods of repetitive MLR stimulation, including the period of transition from rest to real locomotion.

In this chapter, we show how experiments on the tiger salamander, *Ambystoma tigrinum*, have provided an estimation of the processing of MLR stimulus volleys in the hindbrain, using stimulus trains that were both subthreshold and at the threshold for evoking locomotor movements. The information so gained provides insight into the nature of autonomous hindbrain processing, this being a key component of the multilevel control of locomotion.

Methods

Most of the methods have been described previously (Bar-Gad et al., 1999) and are provided below only in brief.

Locomotion was evoked in *A. tigrinum* by repetitive electrical stimulation of the MLR. The stimulus current and frequency were adjusted so that the locomotor movements began ~ 15 s after stimulus onset. It was assumed that the hindbrain neurons recruited by a near-locomotor-threshold train of stimulus pulses were mainly those that participated in the preparation's transition from rest to locomotion. The distance between the sites of MLR stimulation and the recorded hindbrain neurons was ~ 4–5 mm.

The stimulus train (trial) that was at the threshold for locomotor movements usually consisted of 5–12 µA pulses at an IStI of 80–200 ms. It was delivered for ~ 15 s, i.e., until locomotor movements began. The inter-trial pauses were for 2 min, and the total number of such trials per stimulus/recording session was 2–11. The neuron's extracellularly recorded impulses (i.e., spike discharges) were discriminated off-line. Each set of impulses and its corresponding train of stimulus pulses were then converted into point processes.

Measurement abbreviations and terminology

We define: L_n, the latency of the nth impulse, as the time between the impulse (neuronal spike) and the immediately preceding stimulus pulse; $IImI_n$, the inter-impulse interval between the nth and the $n+1$th impulse; IStI, inter-stimulus (pulse) interval of a train of stimuli delivered to MLR; T, the duration of an IStI; and k_n, the number of stimulus pulses $(0, 1, \ldots, m)$ between the impulses n and $n+1$. Note that $IImI_n = k_n T + (L_{n+1} - L_n)$. All time variables are in ms.

An IImI of a duration approximately equal to an integer multiple of T and containing k stimulus pulses is termed integer kT interval. Integer intervals were formed by the successive time-locked responses (e.g., intervals in Fig. 1 inset, *not marked* by asterisks: $k=1$). Delayed responses (see Results) could form noninteger $(k+1/2)T$ and $[(k+1)-1/2]T$ intervals (e.g., intervals in Fig. 1 inset, marked by asterisks: $k=1$, thus one-and-one half intervals were formed). Such noninteger intervals are abbreviated $k+$ and $(k+1)-$ (e.g., $1+$ and $2-$ in Fig. 1 inset). The $k+$ and the $(k+1)$ IImIs contain k and $k+1$ stimulus pulses, respectively.

Finally, the firing ratio is defined as the inter-stimulus interval divided by the average inter-impulse interval.

Results

In this chapter, we focus on the behavior of 16 hindbrain neurons, which we studied in nine experiments on four animals. In all, there were ~ 100 trials, 80 of which contained >20 consecutive impulse responses.

Basic variations in the impulse patterns

The following variations in the mode of firing were observed routinely. As shown in Fig. 1, they included: (1) an alternation of k values between successive impulses; (2) fluctuations of the IImI and the latency of time-locked impulses; and (3) the intermittent presence of delayed (relative to the preceding stimulus pulse) impulses.

The k value commonly alternated among two or three adjacent integers. For example, at a firing ratio of 0.3, k varied irregularly between 2, 3 and 4. At a firing ratio of 0.5, k was 1, 2 or 3. At a firing ratio of 0.8, k alternated between 1 and 2, or 0, 1 and 2 (Fig. 1). At an IStI of 100–200 ms, k normally

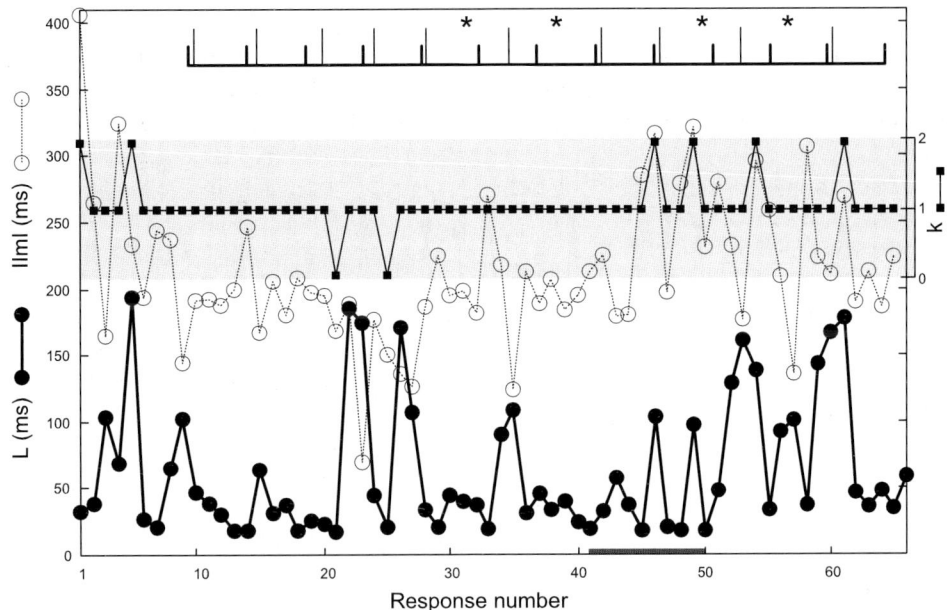

Fig. 1. Commonly observed variations in the firing of a hindbrain neuron during a train of stimulus pulses delivered to the MLR. This figure shows variations in the inter-impulse interval (IImI) and response latency (L) values (left-side ordinates) for 65 consecutive impulses (neuronal spike responses) of a hindbrain neuron to a train of MLR stimulus pulses (IStI = 200 ms). Also shown is the number of stimulus pulses (k) between successive impulse responses (right-side ordinate). Note that the delayed impulses prevented an increase of $k=2$ intervals to more than 300 ms during alternating k discharge. The *inset* at the top of the figure shows a part of the stimulus train (short thick vertical lines) and its associated impulse train (longer thin vertical lines). The thick horizontal line under the latency values (i.e., starting after the 40th spike response) shows where the inset epoch occurred within the trial. In this inset, note the intermittent presence of delayed impulses (i.e., in the 6th and 10th IStI). Four asterisks (*) denote the noninteger (1+) and (2−) inter-impulse intervals.

alternated around 1 (i.e., 1 to 2, or 1 to 0) during near-locomotor-threshold stimulus trains, but around 2 during subthreshold ones. At an IStI of 300–500 ms, the evoking stimulus trains remained commonly sub-locomotor-threshold and k alternated between 1, 2 or 3 in a stimulus-strength-dependent manner. The firing ratio commonly decreased when the MLR stimulus threshold for evoking locomotion gradually increased in the later phases of the stimulus/recording sessions.

Figure 1 shows that variations in the k and latency values were exhibited during epochs of both steady-state discharge and either decreasing or increasing discharge. The k trend generally resulted from a gradual change in the occurrence of k and $k+1$ intervals. The latency values fluctuated irregularly mainly among two ranges (see also Fig. 3C). Delayed impulses were encountered mostly either in the middle (e.g., 3rd and 9th impulses in Fig. 1) or at the end (e.g., 5th and 22nd impulses in Fig. 1) of an IStI. Therefore, they were usually recognized at an IStI > 120 ms. In general, when a hindbrain neuron responded to a sub-locomotor-threshold stimulus train, it exhibited a longer average IImI than when responding to a near-locomotor-threshold train.

Abrupt shifts in hindbrain neuronal discharge

Figure 2 shows that throughout a single trial, three types of abrupt shift in the firing of hindbrain neurons could occur either separately or in combination. They included: (1) a k shift in which a new k value could emerge, or one of the preexisting k values could disappear; (2) a shift in the modal or minimal latency value of time-locked impulses; and (3) the emergence of delayed impulses. These abrupt shifts

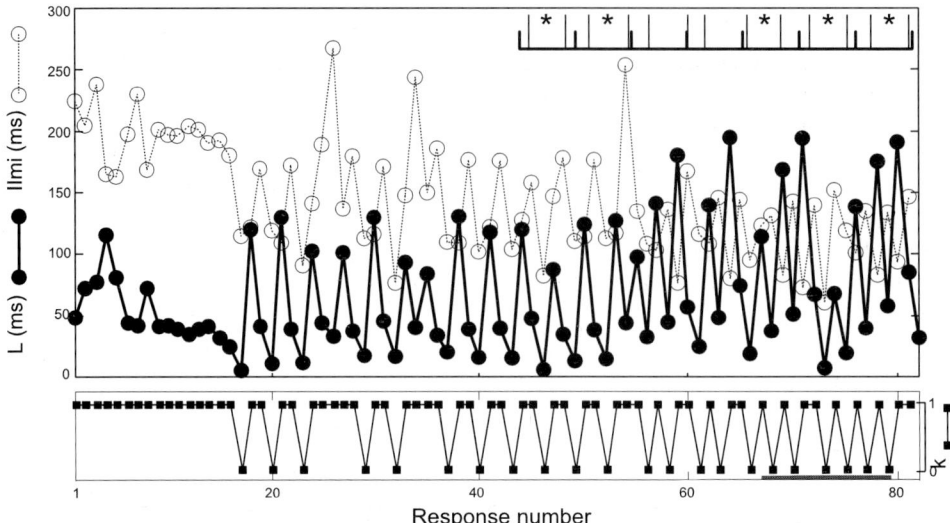

Fig. 2. An example of abrupt shifts in a hindbrain neuron's firing during MLR stimulation. This figure is organized like Fig. 1, except for the k values being presented at the bottom. For this sequence of impulses, the IStI was 200 ms. The figure shows that after ~15 initial inter-impulse intervals at $k=1$, there was a near-regular k alternation from 1 to 0. The *inset* shows an epoch during which five $k=0$ inter-impulse intervals were interspersed with $k=1$ intervals. Note the two abrupt shifts in regular patterns of the response latency (L: i.e., shifts at 17th and 54th impulse).

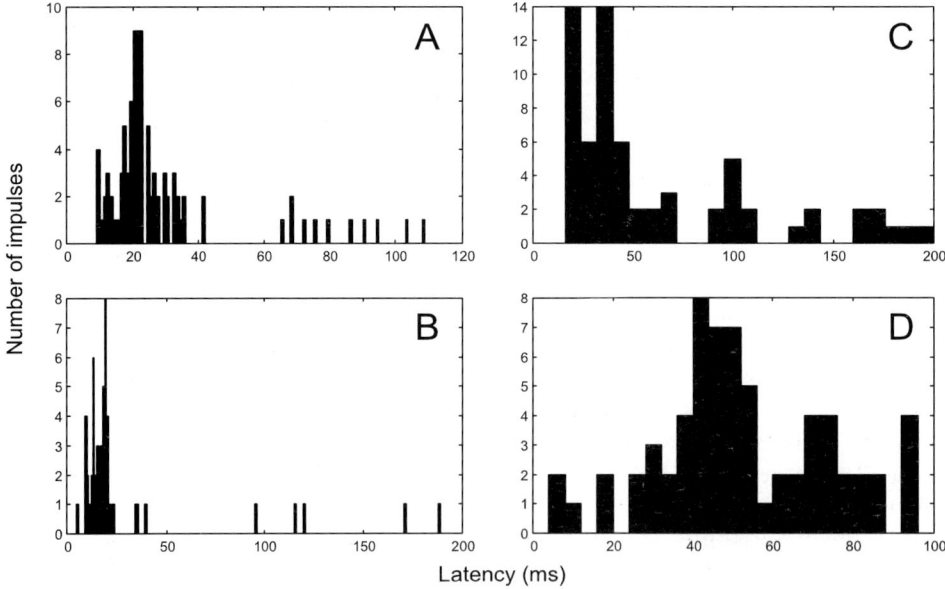

Fig. 3. Latency histograms for the impulses of four exemplary hindbrain neurons during MLR stimulation. (A) Responses (bin width, 1 ms) of the hindbrain neuron shown in Fig. 4 to MLR stimulation at an IStI of 120 ms (6 μA pulses). (B) Another neuron's responses (bin width, 1 ms) to MLR stimulation at an IStI of 200 ms (7 μA pulses). (C) Responses (bin width, 8 ms) of the hindbrain neuron shown in Fig. 1 to MLR stimulation at an IStI of 200 ms (9 μA pulses). (D) Responses (bin width, 4 ms) of a fourth neuron to MLR stimulation at an IStI of 100 ms (7 μA pulses). Two- or three-modal distribution of time-locked impulses (reflecting several synaptic translations of the propagating activity throughout the hindbrain) are seen in A–C but a broad unimodal distribution of nonlocked responses is seen in D. See text for discussion.

normally occurred during near-locomotor-threshold stimulus trains. They were also seen, however, in response to sub-locomotor-threshold stimulation.

A longer modal latency could be accompanied by an increase in the k value. For example, at an IStI of 500 ms, one hindbrain neuron exhibited a shift in its modal latency from 26 to 32 ms during $k = 1$ firing. This shift was abruptly followed by an alternation of k to values that ranged from 1 to 4. At an IStI of 200 ms, the same cell's modal latency increased from 12 to 35 ms, and the k value changed from 1–2 to ≥ 3. In yet another neuron, at IStIs of 120 ms there was a latency shift from 13–24 to 31–41 ms, and the abrupt appearance of delayed impulses.

Time-locked and delayed impulses of hindbrain neurons

The time-locked impulses mostly had a latency of 15–40 ms. Usually the range of latencies was 15–30 ms, with 1–3 modes (see examples in Figs. 1–3). A latency of 13–15 ms was characteristic of monosynaptic responses (Fig. 3A and B, first mode). About 7–9 ms of this value involved the conduction time from the point of stimulation (cf. Bar-Gad et al., 1999). Figure 3A shows that disynaptic responses (latency, 18–22 ms) predominated, however. The number of responses with a longer latency (oligosynaptic responses) could be either higher (Fig. 3B, third mode) or not (Fig. 3C, second mode).

Time-locked, mid-IStI and prestimulus impulses could usually be distinguished at an IStI ≥ 200 ms, but rarely at an IStI < 120 ms. At IStIs ≤ 200 ms, the latency of time-locked impulses of some neurons alternated between 15, 21, 28 and 35 ms. At IStIs of 500 ms, however, the time-locked impulses of these same neurons occurred at 28–55 ms. In other neurons, their time-locked impulses aggregated near 35 ms at an IStI of 200 ms, but around 21 and 35 ms at an IStI of 100 ms.

The successive time-locked impulses formed integer kT inter-impulse intervals. But the delayed mid-IStI impulses were commonly preceded and followed by noninteger IImIs (see Methods). The inset in Fig. 1 provides two examples of a delayed impulse that occurred in the middle of an IStI. Each of these mid-IStIs impulses terminated one integer-and-one half $[k + 1/2]T$ interval and began another one $[(k + 1) - 1/2]T$. These noninteger intervals are marked by asterisks in the Fig. 1 inset. These paired $(k+)$ and $[(k+1)-]$ intervals restrained fluctuations in the duration of IImIs during alternations of the k values. The IImIs within such a $[k+, (k+1)-]$ pair were usually of similar duration but some $k + 1$ IImIs could have either a longer or shorter duration than the k ones. The neuron in Fig. 1 produced both time-locked and several delayed impulses at an IStI of 500 ms, too.

Doubling of the discharge rhythm of hindbrain neurons

A doubling of the firing rhythm of hindbrain neurons occurred when both a time-locked and a delayed impulse were generated in the same IStI. For example, in Fig. 2 note that after some mid-IStI impulses emerged, the duration of the IImIs began to alternate with $0+$ and $1-$ IImIs, both with duration of about half the IStI. Then, prestimulus impulses also began to appear (see Fig. 2 inset). When this neuron was responding at an IStI of 500 ms, the delayed impulse was at a mid-IStI when the immediately previous impulse had a latency < 16 ms. If the latency of the time-locked impulse was > 16 ms, however, a delayed impulse appeared either near the end of the corresponding IStI, or it was not generated.

In another neuron (see Fig. 3B), time-locked impulses appeared mostly at 14, 19 and 24 ms, and the k value alternated around 2. Again, the impulses between the $0+$ and $1-$ IImIs occurred at the mid-IStI when a time-locked impulse with a latency of ~ 14 ms began a $0+$ IImI. Delayed impulses were prestimulus, however, when the latency of the preceding time-locked impulse was > 19 ms. Figure 4 shows that after this neuron displayed a period of time-locked impulses, two delayed (47th and 49th) impulses appeared. Shortly thereafter, the first $0+$ IImI occurred, and, finally, IImIs with a $k > 1$ value disappeared altogether. Each delayed impulse now contributed to a $0+$ IImI. All the $0+$ IImIs, and some of the $1-$ ones, had a duration that was half that of the IStI.

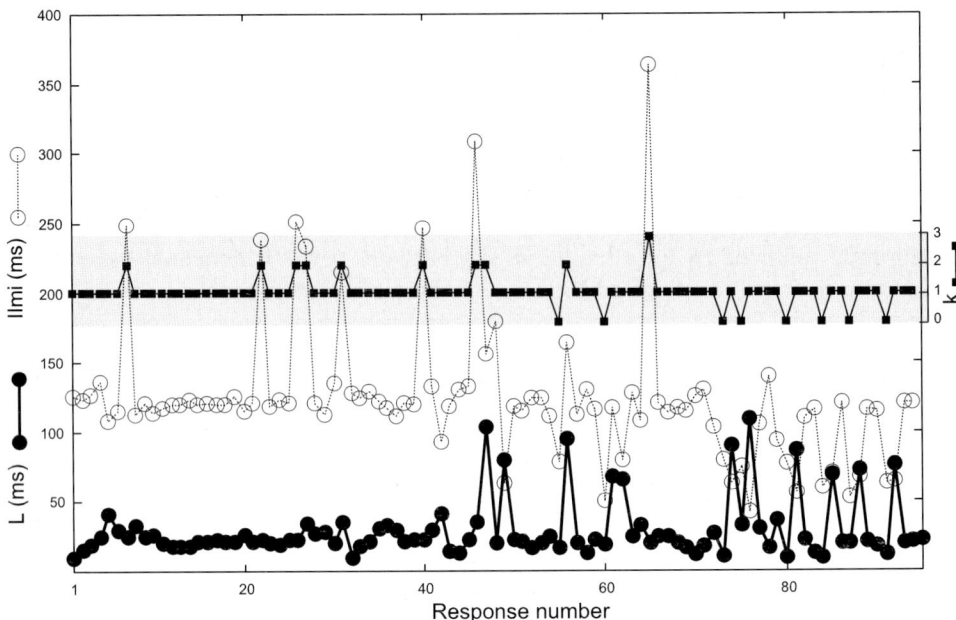

Fig. 4. An example of doubling the rhythm of a hindbrain neuron's discharge. This figure is organized as in Fig. 1 (but without an inset) and Fig. 5 below. The neuron's responses were to an IStI of 120 ms (6 µA stimulus pulses). Note that the duration of the IImIs first alternated between values of near-120 ms and near-240 ms (i.e., the latter was 100% greater), and later, between near-120 ms and near-60 ms (i.e., the latter was 50% less).

Nonlocked responses of hindbrain neurons

For the present purposes, a nonlocked response was defined as one in which its *average* latency had a value of approximately half the duration of the IStI. Nonlocked impulses were usually distributed uniformly throughout IStI. In contrast to time-locked and delayed responses, they did not form regular patterns. Some neurons generated integer intervals at an IStI of 500 ms but noninteger ones also occurred at an IStI of 200 ms. At an IStI of 100 ms, the latency distribution of the responses could have a maximum of ~45 ms (see Fig. 3D). In such cases, categorizing the type of response became equivocal.

For nonlocked impulses, the duration of their IImIs was broadly distributed, with a modal duration that was not related to the duration of the IStI. For example, one neuron's impulses were nonlocked at an IStI of 1000 ms. The modal duration of $k=1$ IImIs was 250 ms, while that for $k=0$ IImIs was 350 ms. The latency of this neuron's time-locked responses was scattered across a span of 42–62 ms, at an IStI of 500 ms. The neuron's IImIs varied from 380 to 650 ms, even during a period of constant $k=1$ discharge. At an IStI of 200 ms, however, this cell's k value shifted from 1 to an alternation between 1 and 2, and its impulses divided into time-locked and delayed ones (Fig. 5). In this neuron, as in several others, firing continued at a slow rate during a few seconds after the conclusion of the stimulation train. Evidently, this neuron was being driven by some distinct sources, which could affect its firing under certain conditions.

Background discharge in the absence of stimulation was exhibited by three of the total sample of 16 neurons, and all three had nonlocked responses to MLR stimulation. At an IStI of 1000 ms, the mean IImI of one of them decreased from 94 to 54 ms, and the mean's standard deviation reduced from 50 to 25 ms. During this cell's background discharge, the distribution of the duration of its IImIs had two modes at 70 and 140 ms. During the stimulus train, this cell's corresponding modes shifted to 40 and 80 ms. For another of these three neurons, the distribution of IImIs of the background discharge had three modes at 100, ~220 and

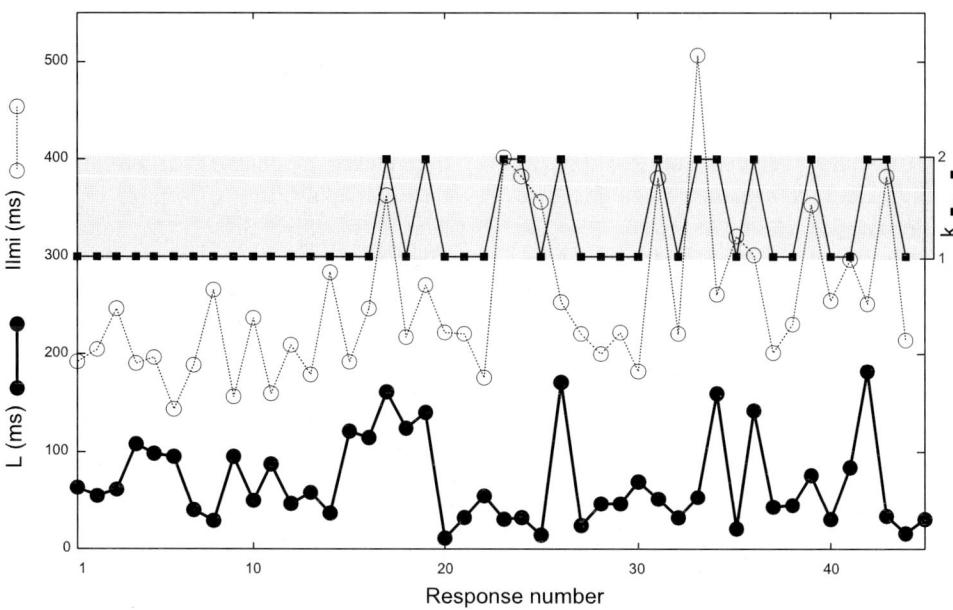

Fig. 5. An example of the progressively decreasing rate of a hindbrain neuron's firing during MLR stimulation. This figure is organized like Fig. 1 (but without an inset). The IStI was 200 ms (10 μA stimulus pulses). Delayed impulses ($L > 100$ ms) became clearly distinct from time-locked ones when the k value began to alternate between 1 and 2. This neuron continued to fire at a slow rate after the end of the stimulation train.

~320 ms. The modal duration was of 140 ms, however, at an IStI of 500 and 200 ms. The duration of the third neuron's IImIs during its background discharge had a mean value of 200 ms and one, two or three modes at ~130, 260 and 380 ms, in different recordings. These values were the same during sub-locomotor-threshold MLR stimulus trains. During near-locomotor-threshold stimulation, however, the duration of the mean IImI could decrease to 100 ms.

Discussion

Basic variations in hindbrain neuron responses to MLR stimulation

Some hindbrain neurons responded monosynaptically to at least a portion of the stimulus volleys from the MLR. Such cells presumably recruited other hindbrain neurons oligosynaptically. When near-threshold stimulus pulses were delivered, the synaptic translation time was 5–8 ms. A key point is that this delay time depended on the time course of the excitatory postsynaptic potential rather than the synaptic delay (Babalian and Shapovalov, 1984; Matsushima et al., 1989; Wu and Wang, 1995).

Propagation of activity among hindbrain neurons initially increased but ceased sharply after 3–4 synaptic translations (Fig. 3A–C). The short-lived hindbrain wave of excitation, when involving a given neuron, could ordinarily not reexcite this neuron after additional translations. It would seem that the excitatory wave *propagated* (i.e., excited *new* neurons) rather than reverberated (reexcited the same neurons). During a given trial, the pathway of the wave (i.e., from neuron to neuron in the hindbrain) and the latency for activation of a given neuron varied. These basic variations prevented prediction of whether a hindbrain neuron would or would not be excited by a given MLR volley. Again, it would seem that a substantial and persistent variability of the amplitude and duration of postsynaptic potentials could have contributed to these variations (cf. Parker and Grillner, 2000; Hatta et al., 2001).

Significance of near-locomotor-threshold MLR volleys

The hindbrain neurons activated by a near-locomotor-threshold MLR volley included a small subset of monosynaptically activated neurons. Even in these cells, the majority of responses were di- or oligosynaptic. Indeed, most hindbrain neurons were recruited after 2–4 translations. It would seem that the greater the excitation delay, the more the recruitment of hindbrain neurons must have depended on interactions among the latter, later-activated cells. We propose that the MLR facilitates such interactions by virtue of first monosynaptically activating the most appropriate 'input' hindbrain neurons. Subsequent translations would then result in recruitment of the appropriate amount and composition of hindbrain neurons to ensure that locomotion could then be brought about by the relevant spinal networks.

We further propose that the stronger the MLR's input volley to the hindbrain, the less the number of translations among the latter's neurons, and the more definitive the role of the MLR. Application of suprathreshold stimuli to MLR engaged more of the hindbrain neurons monosynaptically. Correspondingly, the hindbrain becomes less autonomous. In such instance, the predominant and shortest pathway from the MLR to the spinal cord would be the reticulospinal one (see Orlovsky, 1970; Sirota et al., 2000).

Delayed and nonlocked impulses

Delayed impulses could not be generated by the short-lived, MLR-evoked excitatory wave in the hindbrain. Rather, we propose that such impulses revealed the active contribution of hindbrain neurons in processing the MLR volley. The results suggest that it was a matter of chance as to whether a given IImI, which had been initiated by a time-locked impulse, would be terminated by a similar or a delayed impulse. Nonetheless, when both a time-locked and delayed impulse occurred within the same IImI, certain rules were evident. These included: (1) a short-latency impulse could start a noninteger IImI, whereas a long-latency one could not; (2) an IImI initiated by time-locked impulse could be designated random-like as an integer or noninteger one when its k value was > 2; and (3) if an IImI had a k value of < 1, its final (closing) impulse could be time-locked or delayed, depending on the latency of the interval's initial (opening) impulse. The relative degree to which these local rules applied to single hindbrain neurons contributed to their idiosyncratic responses to the MLR volleys.

Evolution of hindbrain neuronal behavior throughout a stimulus train delivered to MLR

In the MLR-activated hindbrain, changes in the firing pattern of single responding neurons, and the number of activated neurons throughout the time course of a single stimulus train, culminated when locomotor movements were about to ensue. This finding was supported by observations on the increase in the test neurons' firing ratio throughout the evolution of a stimulus train. Such frequency facilitation of the responses of neurons with an initially low firing ratio was also observed previously in hindbrain neurons of the cat (Selionov and Shik, 1990) and rough skin newt (Bar-Gad et al., 1999). Similarly, repetitive stimulation of the cerebral–buccal neuron in *Aplysia californica* was shown to enhance monosynaptic excitatory postsynaptic potentials in its target neurons (Sanchez and Kirk, 2000). Note further that the stimulation of mossy fibers decreases spike threshold in granule cells in the rat cerebellum (Armano et al., 2000). Moreover, reticulospinal cells can exhibit nonlinear amplification and generate high-frequency discharge in the lamprey (Di Prisco et al., 2000). The balance of actions of ATP and adenosine (Dale, 1998), or serotonin modulation of glutamate receptors (Li and Zhuo, 1998), might contribute to the gradual evolution of neuronal discharge too.

The latency between arrival of the input volley from the MLR at the hindbrain and the response of one of the latter's neurons could increase or decrease throughout a stimulus train. Different types of abrupt shifts might presumably influence distinct target neurons. Hindbrain neurons could experience facilitation, depression or both, throughout the time course of the same repetitive input from the MLR. All of these behaviors were evident in the hindbrain and their cumulative effect was a key component of the preparation's transition from rest to MLR-evoked locomotion.

Contribution of a single hindbrain neuron to the initiation of locomotion

The results showed that for an MLT stimulus train to reach the threshold for locomotion to ensue there was a trade-off between the strength of the train's stimulus pulses and their inter-pulse interval. This inverse rule was not obeyed by the individual activated-hindbrain neurons, however. Such a neuron could exhibit a doubling of its firing frequency in one near-locomotor-threshold MLR stimulus train, but not in another. Similarly, a given hindbrain neuron could generate two impulses in the IStIs of both near- and sub-locomotor-threshold trains. Such a neuron could exhibit either an augmented or decreased rate of firing during two *identical* stimulus trains, or during trials with the same *average* duration of IImIs. Similarly, a hindbrain neuron could exhibit a doubling of its firing rate while an adjacent one was generating nonlocked responses.

Concluding thoughts

In the authors' opinion, unstable neuronal discharge is a normal feature of neuronal behavior in the elaboration of movement. It reveals a way by which a single hindbrain neuron contributes to the processing of the input volley from the MLR. The experiments showed that when the MLR was stimulated at near-threshold strength for locomotion to ensue, the activation of hindbrain neurons occurred largely after 2–4 successive synaptic translations. During normal locomotion, it is likely that while interacting hindbrain neurons are developing their output to the spinal cord, they are responding to excitatory input from both the MLR and other higher command centers. Viewed in this light, the facultative nature of the MLR command to the locomotor hindbrain is functionally advantageous. To gain a better understanding of the origin and role of abrupt shifts in the firing of hindbrain neurons during MLR stimulation, it will next be desirable to record simultaneously from several neurons.

Acknowledgments

We are grateful to Dmitry Kagan for his help in converting the raw recordings into sets of point time series.

Abbreviations

IImI	inter-impulse (neuronal spike) interval
IStI	inter-stimulus (pulse) interval
k	number of stimulus pulses between consecutive impulse (neuronal spike) responses
L	latency
MLR	mesencephalic (midbrain) locomotor region
T	duration of an IStI

References

Armano, S., Rossi, P., Taglietti, V. and D'Angelo, E. (2000). Long-term potentiation of intrinsic excitability at the mossy fiber-granule cell synapse of rat cerebellum. J. Neurosci., 20: 5208–5216.

Babalian, A.L. and Shapovalov, A.I. (1984). Synaptic actions produced by individual ventrolateral fibres in frog lumbar motoneurones. Exp. Brain Res., 54: 551–563.

Bar-Gad, I., Kagan, I. and Shik, M.L. (1999). Behavior of hindbrain neurons during the transition from rest to evoked locomotion in a newt. In: Binder M.D. (Ed.), Peripheral and Spinal Mechanisms in the Neural Control of Movement, Prog. Brain Res., Vol. 123, Elsevier, New York, pp. 285–294.

Dale, N. (1998). Delayed production of adenosine underlies temporal modulation of swimming in frog embryo. J. Physiol. (Lond.), 511: 265–272.

Di Prisco, G.V., Pearlstein, E., Le Ray, D., Robitaille, R. and Dubuc, R. (2000). A cellular mechanism for the shift of a sensory input into a motor command. J. Neurosci., 20: 8169–8176.

Hatta, K., Ankri, N., Faber, D.S. and Korn, H. (2001). Slow inhibitory potentials in the teleost Mauthner cell. Neuroscience, 103: 561–579.

Li, P. and Zhuo, M. (1998). Silent glutamatergic synapses and nociception in mammalian spinal cord. Nature, 393: 695–698.

Matsushima, T., Satou, M. and Ueda, K. (1989). Medullary reticular neurons in Japanese toad: morphologies and excitatory inputs from the optic tectum. J. Comp. Physiol. A, 166: 7–22.

Orlovsky, G.N. (1970). Connections of the reticulospinal neurons with the 'locomotor region' of the brain stem. Biofizika, 15: 171–177.

Parker, D. and Grillner, S. (2000). The activity-dependent plasticity of segmental and intersegmental synaptic connections in the lamprey spinal cord. Eur. J. Neurosci., 12: 2135–2146.

Sanchez, J.A.D. and Kirk, M.D. (2000) Short-term synaptic enhancement modulates ingestion motor programs of Aplysia. J. Neurosci., 20, RC85: 1–7.

Selionov, V.A. and Shik, M.L. (1990). Two types of responses of medullary neurons to microstimulation of locomotor and inhibitory points of the brain stem. Neurofiziologia, 22: 257–266.

Shik, M.L. (1997). Locomotor patterns elicited by electrical stimulation of the brain stem in the mudpuppy. Motor Control, 1: 354–368.

Sirota, M.G., Di Prisco, G.V. and Dubuc, R. (2000). Stimulation of the mesencephalic locomotor region elicits controlled swimming in semi-intact lampreys. Eur. J. Neurosci., 12: 4081–4092.

Wu, G.-Y. and Wang, S.-R. (1995). Excitatory and inhibitory transmission from the optic tectum to nucleus isthmi and its vicinity in amphibians. Brain Behav. Evol., 46: 43–49.

CHAPTER 23

Role of basal ganglia–brainstem systems in the control of postural muscle tone and locomotion

Kaoru Takakusaki*, Junko Oohinata-Sugimoto,
Kazuya Saitoh and Tatsuya Habaguchi

Department of Physiology, Asahikawa Medical College, Asahikawa 078-8510, Japan

Abstract: This chapter argues that a basal ganglia–brainstem system throughout the mesopontine tegmentum contributes to an automatic control of movement that operates in conjunction with voluntary control processes. Activity of a muscle tone inhibitory system and the locomotion executing system can be steadily balanced by a net excitatory cortical input and a net inhibitory basal ganglia input to these systems. We further propose that dysfunction of the basal ganglia–brainstem system, together with that of the cortico-basal ganglia loop, underlies the pathogenesis of motor disturbances expressed in basal ganglia diseases.

Introduction

Insight into the organization of the motor disturbances in basal ganglia diseases is critical for understanding this structure's role in the control of movements. Voluntary movements are always associated with automatic control processes that are performed unconsciously (Grillner and Wallén, Chapter 1 of this volume). For example, initiation and termination of locomotion and avoiding obstacles during locomotion are volitional processes that require accurate control, part of which is achieved automatically (S. Mori et al., Chapter 33 of this volume). Similarly, the subject is largely unaware of the automatic control of rhythmic limb movements, postural muscle tone and the postural reflexes that accompany locomotion. The fact that each such aspect of locomotion is seriously impaired in Parkinsonian patients (Murray et al., 1978; Morris

*Corresponding author: Tel.: +81-166-68-2332;
Fax: +81-166-68-2339; E-mail: kusaki@asahikawa-med.ac.jp

et al., 1994) shows that the basal ganglia must play a crucial role in integrating the volitional and automatic aspects of the descending control of posture and movement.

During the past decade, the cortico-basal ganglia loop has come to be recognized as important for the volitional control of movement (Alexander and Crutcher, 1990; Delong, 1990). Less understood and discussed, however, is the role of the basal ganglia in the automatic aspects of regulating muscle tone and rhythmic limb movements during locomotion. There is recent emphasis on the importance for the control of locomotion of basal ganglia-outflow directly toward the brainstem, particularly to the pedunculopontine tegmental nucleus (PPTN; Garcia-Rill, 1991; Inglis and Winn, 1995; Reese et al., 1995; Pahapill and Lozano, 2000; Garcia-Rill et al. and Skinner et al., Chapters 27 and 28 of this volume). Because the PPTN and its adjacent areas include the midbrain locomotor region (MLR; Grillner, 1981) and a region inhibiting muscle tone (Kelland and Asdourian, 1989; Lai and Siegel, 1990; Takakusaki et al., 1997), it is critical to establish precisely how the basal ganglia–brainstem system controls muscle tone and locomotion.

In this chapter, we propose that automatic aspects of the descending control of movements are largely achieved via the GABAergic basal ganglia–brainstem system. We first describe basic neural structures and circuitry (architecture) in the brainstem and spinal cord that are involved in the control of postural muscle tone and locomotion. We then consider how the basal ganglia–brainstem system comes into play. Finally, the functional significance of this system is discussed in relation to the pathogenesis of motor disturbances in basal ganglia disorders.

Basic neural architecture

The neural architecture of muscle tone inhibitory systems is perceived somewhat differently among researchers. It is generally agreed, however, that cholinoceptive and cholinergic neurons in the pontine reticular formation excite reticulospinal neurons in the medullary inhibitory region. These provide a postsynaptic drive that inhibits motoneurons either directly or via an inhibitory interneuron (Chase and Morales, 1990). A similar action is exerted from the mesopontine tegmentum (Kelland and Asdourian, 1989; Lai and Siegel, 1990; Takakusaki et al., 1997). In this study, this latter region corresponds to the ventrolateral part of the PPTN. Its repetitive stimulation delivered bilaterally suppresses postural muscle tone in the decerebrate cat. The findings to this point suggest the architecture of the inhibitory system shown in Fig. 1A.

Figure 1A proposes that PPTN stimulation may activate cholinoceptive pontine neurons, which then excite medullary reticulospinal neurons and spinal interneurons to inhibit alpha-motoneurons of extensor and flexor muscles (Takakusaki et al., 1994).

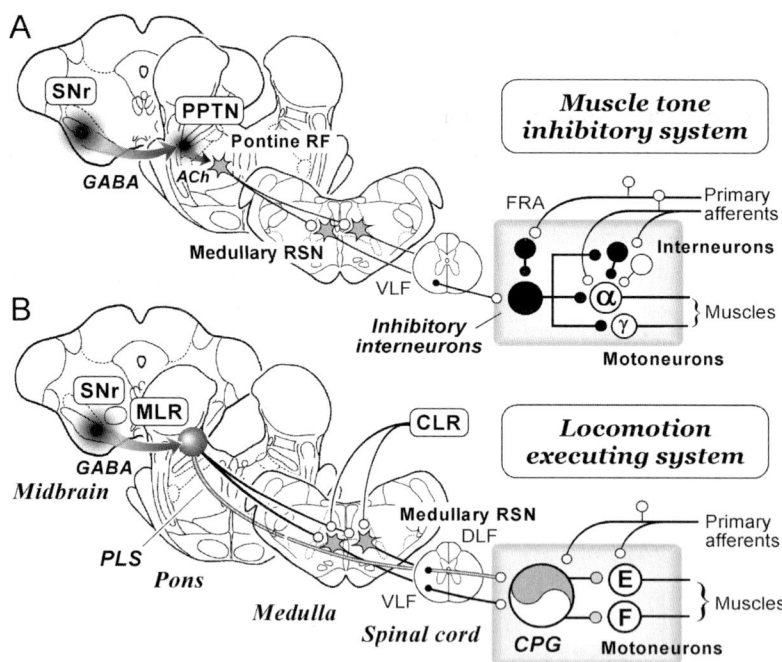

Fig. 1. Models of descending control systems for muscle tone and locomotion. Neural architecture of the muscle tone inhibitory system (A) and the locomotion executing system (B). See text for further explanation. Abbreviations: ACh, acetycholine; α, alpha-motoneuron; CLR, cerebellar locomotor region; CPG, central pattern generator; DLF, dorsolateral funiculus; E, extensor motoneurons; F, flexor motoneurons; FRA, flexion reflex afferents; GABA, gamma-aminobutyric acid; γ, gamma-motoneuron; PLS, pontomedullary locomotor strip; PPTN, pedunculopontine tegmental nucleus; RF, reticular formation; RSN, reticulospinal neuron; SNr, substantia nigra pars reticulata; VLF, ventrolateral funiculus.

Possibly suppressed in parallel are alpha-motoneurons and interneurons intercalated in segmental reflex pathways (Takakusaki et al., 2001). It has also been shown that stimulating the pontine and medullary inhibitory regions in decerebrate preparations reduces the activity of coerulospinal neurons, these comprising one of descending excitatory systems (Mileykovskiy et al., 2000). Thus, suppression of muscle tone is possibly induced by an activation of the inhibitory system together with an inactivation of descending excitatory systems. We suppose that this inhibitory system contributes to the regulation of muscle tone during movements when the animal is both intact and alert (Takakusaki et al., 2001). The inhibitory system is also thought to induce muscular atonia during the rapid-eye-movement (REM) phases of sleep (Chase and Morales, 1990).

It is well known that repetitive stimulation of the MLR evokes controlled locomotion in decerebrate preparations. There is considerable confusion, however, concerning whether the MLR is equivalent to the cholinergic area of the PPTN (Grillner et al., 1997). Most studies demonstrate that the MLR is located in the cuneiform nucleus (CNF), which is in the vicinity of the PPTN. Based on such work (Grillner, 1981; Mori, 1987; Rossignol, 1996), Fig. 1B shows the authors' current perception of the locomotion executing system.

Figure 1B proposes that at least two major pathways descend from the MLR. One is via the medial medullary reticulospinal tract and the other is via the pontomedullary locomotor strip (PLS), which projects through the dorsolateral funiculus. These pathways activate the locomotor central pattern generator (CPG), whose output alternatively generates EPSPs/action potentials and IPSPs to produce the locomotor rhythm. Recently, Mori et al. (1999) found that stimulating the mid-part of cerebellar white matter (the hook bundle of Russell) induced locomotion via the medial medullary reticulospinal tract in decerebrate cats. Cortical projections to the MLR have not yet been reported. They are conceivably mediated via polysynaptic connections through the subthalamic locomotor region (SLR; Rossignol, 1996).

It was reported that a patient with a lesion in the dorsolateral mesopontine tegmentum did not lose muscle tone during REM sleep ('REM without atonia'; Culebras and Moore, 1989). Also, a patient with a lesion in the dorsal part of mesopontine tegmentum could not stand and walk (Masdeu et al., 1994). These clinical case reports suggest that both a muscle tone inhibitory region and the MLR are realities in the mesopontine tegmentum of the human.

The GABAergic basal ganglia–brainstem systems

The mesopontine tegmentum including the PPTN receives a dense GABAergic projection from the basal ganglia, including, in particular, input from the substantia nigra pars reticulata (SNr; Inglis and Winn, 1995; Reese et al., 1995). To test the Fig. 1 proposal, we investigated the function of this system by activating or inactivating it during PPTN/MLR-controlled muscle tone and locomotion in the decerebrate cat, with the striatum, thalamus and cerebral cortex removed, but the SNr preserved.

Microinjection of GABAergic antagonists such as bicuculline or picrotoxin (1–5 mM, 0.1–0.25 µl) into the ventral part of the PPTN suppressed postural muscle tone (Habaguchi et al., 1998). On the other hand, injecting these substances into the CNF and its adjacent area elicited locomotion (cf. Garcia-Rill and Skinner, 1985). In addition, the effects of electrical stimulation of the PPTN/MLR were attenuated or diminished after injecting GABAergic agonists (GABA or muscimol; 2–10 mM; 0.2–0.5 µl) into each region (Takakusaki et al., 2003). These findings showed that GABAergic afferents to the PPTN and the MLR control muscle tone and locomotion, respectively.

We next examined whether SNr stimulation altered the level of muscle tone and locomotion via the GABAergic nigrotegmental projection (Takakusaki et al., 2003). Repetitive stimulation (100 Hz, 50 µA) of the SNr alone did not change muscular activity. However, stimulation of this structure at a lower strength (20–30 µA) attenuated *PPTN-induced* muscle tone suppression, and at a higher strength (50 µA), it finally blocked the PPTN-effect. In addition, SNr stimulation at a low strength (20–30 µA) reduced the number of *MLR-activated* step cycles, increased the duration of the stance phase, and disrupted the rhythmic alternation of limb movements. Stimulation of the SNr at a higher strength (50 μA) eventually stopped MLR-activated locomotion. Furthermore, the onset of the MLR-evoked locomotion was delayed by concomitant SNr stimulation of progressively increasing strength. Finally, injection of GABAergic antagonists into the PPTN/MLR further blocked the effects of SNr stimulation. Taken together, the results suggest that muscle tone and locomotion are subject to modulation by the GABAergic nigrotegmental projection, which not only maintains rhythmic limb movements but also controls the initiation of locomotion (Table 1).

We have shown that PPTN stimulation hyperpolarizes the membrane potentials of extensor and flexor motoneurons via a postsynaptic inhibitory mechanism (Takakusaki et al. 1997). We showed that SNr stimulation did not directly affect the excitability of motoneurons, but it clearly eliminated a concomitant PPTN-induced postsynaptic inhibition (Habaguchi et al., 1998). This suggests that SNr stimulation has an inhibitory effect on the muscle tone inhibitory system. Such a disinhibitory mechanism may de-inactivate descending excitatory systems and finally restore the excitability of motoneurons to increase the level of muscle tone.

In subthalamic cats, in which the SLR is preserved, stimulation of the SNr not only blocks MLR-induced locomotion but in addition, it inhibits spontaneous locomotion (Takakusaki et al., 2003). Since spontaneous locomotion in the subthalamic cat appears even after destroying most of the MLR, it would appear that the SLR can directly activate medullary reticulospinal neurons such as to evoke locomotion (Rossignol, 1996). SNr neurons also project to the medullary reticular formation (MRF; Schneider et al., 1985). It is thereby possible that SNr stimulation can inhibit the locomotion executing system at the level of both the mesopontine tegmentum and the MRF.

During fictive locomotion, we have observed that SNr stimulation does not suppress motoneuron discharge. Rather, it greatly reduces the amplitude and duration of the hyperpolarizing phases of the membrane oscillations of hindlimb extensor motoneurons (an unpublished observation). These findings show that SNr stimulation can inhibit locomotor CPG activity without decreasing the overall level of muscle tone. Iwakiri et al. (1995) demonstrated that a group of medullary reticulospinal neurons, which were involved in muscle tone suppression, discharged during MLR-induced locomotion. If the muscle tone inhibitory system contributes to the hyperpolarizing phase of motoneuron membrane oscillations during locomotion, SNr-induced locomotor suppression may be due to an inhibition of *both* the muscle tone inhibitory system and the locomotion executing system.

On the basis of these results, the authors' current opinion is that the basal ganglia control postural muscle tone and locomotion by a *combined* inhibition/disinhibition of *both* the muscle tone inhibitory system and the locomotion executing system.

Normal and abnormal operation of the basal ganglia–brainstem system

Motor cortical neurons do not exhibit altered discharge during steady-state locomotion but the discharge changes when the experimental preparation has to overcome obstacles (Drew et al., 1996; see also Chapter 23). Thus, cortical processing seems unnecessary during the automatic execution of locomotion. Rather, we propose that high-level processing may occur in the basal ganglia–brainstem in the absence of conscious awareness. In contrast, when a locomotory subject encounters obstacles, each foot must be placed with a high degree of accuracy. This accuracy requires a precise visuomotor coordination, in which the subject may have to modify the limb trajectory in each step, in order to achieve appropriate foot placements (Georgopoulos and Grillner,

Table 1. A comparison of Parkinsonian gait and SNr-affected locomotion in the decerebrate cat

Gait component	Parkinsonism	SNr stimulation
Gait velocity	Decrease	Decrease
Stride length	Decrease	Decrease
Cadence	Relative decrease	Decrease
Double support period	Increase	Increase
Onset of gait initiation	Delay	Delay

The features of Parkinsonian gait deficits are based on descriptions obtained from Murray et al. (1978), Morris et al. (1994) and Pahapill and Lozano (2000).

1989). In Fig. 2A, we propose that the cortico-basal ganglia loop can help serve this purpose.

On the basis of the results and viewpoints described, we propose in Fig. 2A a hypothetical model of how the basal ganglia achieve an integration of the volitional and automatic control of movement.

Figure 2A proposes that a cortico-basal ganglia loop is involved in the control of movements that requires volition, cognition and attention. In contrast, a basal ganglia–brainstem system seems required for the automatic regulation of muscle tone and rhythmic limb movements. The PPTN and the pontomedullary reticular formation receive corticofugal inputs mainly from the motor cortical areas (Matsuyama and Drew, 1997; Matsumura et al., 2000). These findings suggest that the muscle tone

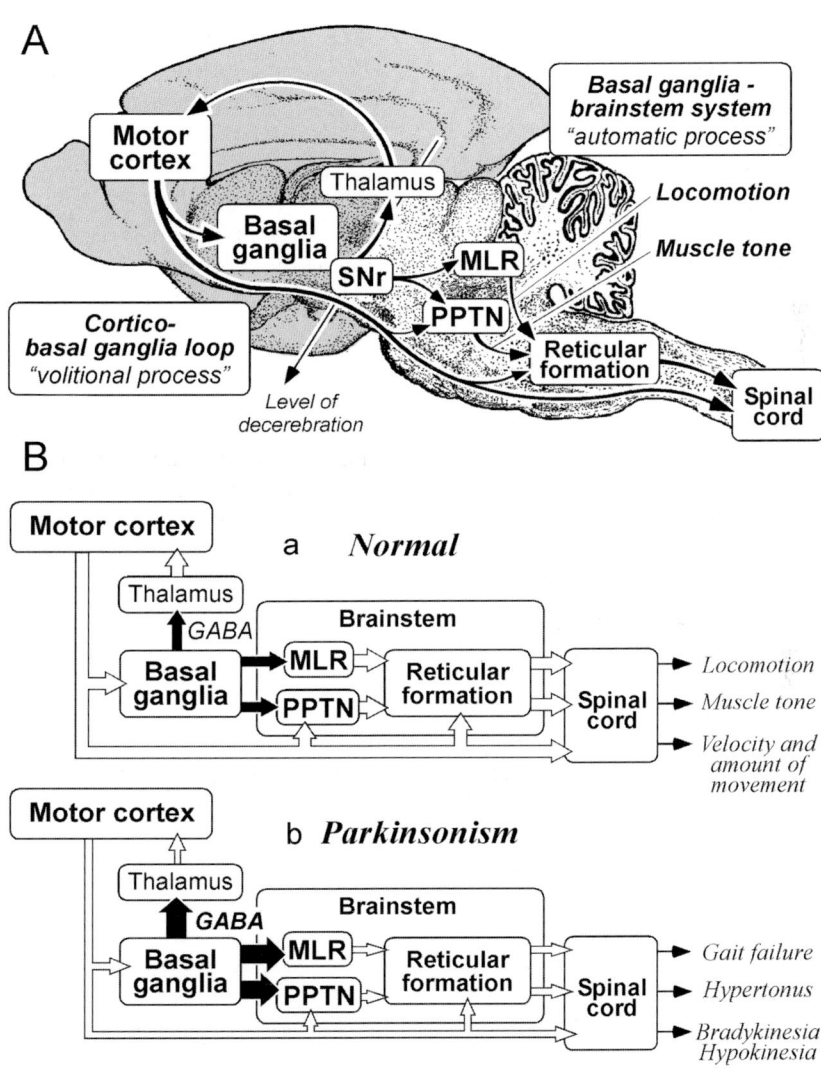

Fig. 2. Volitional and automatic control of normal and Parkinsonian movement. (A) Integration by the basal ganglia of volitional and automatic control. (B) Basal ganglia control of movements in normal (a) and Parkinsonian (b) states. White and black arrows indicate excitatory and inhibitory projections, respectively. Size of arrows indicates the strength of selected effects. See text for further explanation.

inhibitory system and the locomotion executing system are under the influence of both a *net* cortical excitation and basal ganglia inhibition.

If the Fig. 2A model is correct, how is dysfunction of the basal ganglia–brainstem system involved in basal ganglia motor disorders? Figure 2A addresses this issue for Parkinsonism. Motor cortical areas receiving basal ganglia use the corticospinal tract to control the velocity and amount of voluntary movement (Delong, 1990). In contrast, the basal ganglia–brainstem system controls muscle tone and locomotion via the PPNT/MLR. In the Parkinsonian state, GABAergic basal ganglia output is thought to be overactive (Fig. 2Bb; Alexander and Crutcher, 1990; Delong, 1990). Thus, an excessive GABAergic inhibition upon thalamocortical neurons may decrease the velocity and amount of movement (bradykinesia and hypokinesia, respectively). Such an increase in GABAergic inhibition together with a concomitant decrease in cortical excitation to the PPTN-influenced muscle tone inhibitory system may consequently increase the level of muscle tone (hypertonus). The effect of GABAergic inhibition on the MLR-implicated locomotion executing system may also disturb gait performance. Additionally, less activity of the premotor cortex may disturb the motor programming required for precise gait control (Pahapill and Lozano, 2000). Since gait disturbances in Parkinsonian patients resemble the locomotor pattern induced by SNr stimulation (Table 1), we propose that a dysfunction of the basal ganglia–brainstem system is the primary basis for Parkinson-induced gait impairments. In contrast, decreasing output from the basal ganglia may increase the velocity and amount of movements (hyperkinesia) and reduce the level of muscle tone (hypotonus). In summary, the Fig. 2Bb model provides a rational explanation of the pathogenesis of basal ganglia motor disorders such as both the hypokinetic–hypertonic syndrome (Parkinsonism) and hyperkinetic–hypotonic syndromes (Huntington's chorea, ballism, L-DOPA-induced dyskinesia).

Acknowledgments

We thank Professor Takashi Sakamoto, Department of Physiology, Asahikawa Medical College, for his reviewing this manuscript. This study was supported by the Japanese Grants-in-Aid for Scientific Research (C) and Priority Areas (A) and by a grant from Uehara Memorial Foundation to K.T.

Abbreviations

ACh	acetycholine
CLR	cerebellar locomotor region
CNF	cuneiform nucleus
CPG	central pattern generator
DLF	dorsolateral funiculus
E	extensor motoneurons
EPSPs	excitatory postsynaptic potentials
F	flexor motoneurons
FRA	flexion reflex afferents
GABA	gamma-aminobutyric acid
IPSP	inhibitory postsynaptic potential
L-DOPA	L-dihydroxyphenylalanine
MLR	midbrain locomotor region
MRF	medullary reticular formation
PLS	pontomedullary locomotor strip
PPTN	pedunculopontine tegmental nucleus
REM	rapid-eye-movement
RF	reticular formation
RSN	reticulospinal neuron
SNr	substantia nigra pars reticulata
SLR	subthalamic locomotor region
VLF	ventrolateral funiculus

References

Alexander, G.E. and Crutcher, M.D. (1990). Functional architecture of basal ganglia circuits: neural substrates of parallel processing. Trends Neurosci., 13: 267–271.

Chase, M.H. and Morales, F.R. (1990). The atonia and myoclonia of active (REM) sleep. Annu. Rev. Psychol., 41: 557–584.

Culebras, A. and Moore, J.T. (1989). Magnetic resonance findings in REM sleep behavior disorder. Neurology, 39: 1519–1523.

Delong, M.R (1990). Primate models of movement disorders of basal ganglia origin. Trends Neurosci., 13: 281–289.

Drew, T., Jiang, W., Kably, B. and Lavoie, S. (1996). Role of the motor cortex in the control of visually triggered gait modifications. Can. J. Physiol. Pharmacol., 74: 426–442.

Garcia-Rill, E. (1991). The pedunculopontine tegmental nucleus. Prog. Neurobiol., 36: 363–389.

Garcia-Rill, E. and Skinner, R.D. (1985). Chemical activation of the mesencephalic locomotor region. Brain Res., 330: 43–54.

Georgopoulos, A.P. and Grillner, S. (1989). Visuomotor coordination in reaching and locomotion. Science, 245: 1209–1210.

Grillner, S. (1981). Control of locomotion in bipeds, tetrapods and fish. In: Brooks V.B. (Ed.), Handbook of Physiology, The Nervous System, Vol. II, American Physiological Society, Bethesda, pp. 1179–1236.

Grillner, S., Georgopoulos, A.P. and Jordan, L.M. (1997). Selection and initiation of motor behavior. In: Stein P.S.G., Grillner S., Selverson A.I. and Stuart D.G. (Eds.), Neurons, Networks, and Motor Behavior. MIT Press, Cambridge, MA, pp. 3–19.

Habaguchi, T., Takakusaki, T., Suginoto, J., Saitoh, K. and Sakamoto, T. (1998). Nigral innervation of the pedunculopontine tegmental nucleus (PPN) neurons that control postural muscle tone in cats. Soc. Neurosci. Abstr., 24: 664.

Inglis, W.L. and Winn, P. (1995). The pedunculopontine tegmental nucleus: where the striatum meets the reticular formation. Prog. Neurobiol., 47: 1–29.

Iwakiri, H., Oka, T., Takakusaki, K. and Mori, S. (1995). Stimulus effects of the medial pontine reticular formation and the mesencephalic locomotor region upon medullary reticulsopinal neurons in acute decerebrate cats. Neurosci. Res., 23: 47–53.

Kelland, M.D. and Asdourian, D. (1989). Pedunculopontine tegmental nucleus-induced inhibition of muscle activity in the rat. Behav. Brain Res., 34: 213–234.

Lai, Y.Y. and Siegel, J.M. (1990). Muscle tone suppression and stepping produced by stimulation of midbrain and rostral pontine reticular formation. J. Neurosci., 10: 2727–2734.

Matsumura, M., Nambu, A., Yamaji, Y., Watanabe, K., Imai, H., Inase, M., Tokuno, H. and Takada, M. (2000). Organization of somatic motor inputs from the frontal lobe to the pedunculopontine tegmental nucleus in the macaque monkey. Neuroscience, 98: 97–110.

Matsuyama, K. and Drew, T. (1997). The organization of the projection from the pericruciate cortex to the pontomedullary brainstem of the cat: a study using the anterograde tracer. Phaseolus vulgaris leucoagglutinin. J. Comp. Neurol., 389: 617–641.

Masdeu, J.C., Alampur, U., Cavaliere, R. and Tavoulareas, G. (1994). Astasia and gait failure with damage of the pontomesencephalic locomotor region. Ann. Neurol., 35: 619–621.

Mileykovskiy, B.Y., Kiyashchenko, L.I., Kodama, T., Lai, Y.Y. and Siegel, J.M. (2000). Activation of pontine and medullary motor inhibitory regions reduces discharge in neurons located in the locus coeruleus and the anatomical equivalent of the midbrain locomotor region. J. Neurosci., 20: 8551–8558.

Mori, S. (1987). Integration of posture and locomotion in acute decerebrate cats and in awake, freely moving cats. Prog. Neurobiol., 28: 161–195.

Mori, S., Matsui, T., Kuze, B., Asanome, M., Nakajima, K. and Matsuyama, K. (1999). Stimulation of a restrict region in the midline cerebellar white matter evokes coordinated quadrupedal locomotion in the decerebrate cat. J. Neurophysiol., 82: 290–300.

Morris, M.E., Iansek, R., Matyas, T.A. and Summers, J.J. (1994). The pathogenesis of gait hypokinesia in Parkinson's disease. Brain, 117: 1169–1181.

Murray, M.P., Sepic, S.B., Gardner, G.M. and Downs, W.J. (1978). Walking patterns of men with Parkinsonism. Am. J. Phys. Med., 57: 278–294.

Pahapill, P.A. and Lozano, A.M. (2000). The pedunculopontine nucleus and Parkinson's disease. Brain, 123: 1767–1783.

Reese, N.B., Garcia-Rill, E. and Skinner, R.D. (1995). The pedunculopontine nucleus—auditory input, arousal and pathophysiology. Prog. Neurobiol., 42: 105–133.

Rossignol, S. (1996). Neural control of stereotypic limb movements. In: Rowell L.B. and Shepherd J.T. (Eds.), Handbook of Physiology, Sec. 12. Oxford University Press, New York, pp. 173–216.

Schneider, J.S., Manetto, C. and Lidsky, T.I. (1985). Substantia nigra projection to medullary reticular formation: relevance to oculomotor and related motor functions in the cat. Neurosci. Lett., 62: 1–6.

Takakusaki, K., Shimoda, N., Matsuyama, K. and Mori, S. (1994). Discharge properties of medullary reticulospinal neurons during postural changes induced by intrapontine injections of carbachol, atropine and serotonin, and their functional linkages to hindlimb motoneurons in cats. Exp. Brain Res., 99: 361–374.

Takakusaki, K., Hobaguchi, T., Nagaoka, T. and Sakamoto, T. (1997). Stimulus effects of the pedunculopontine tegmental nucleus (PPN) on hindlimb motoneurons in cats. Soc. Neurosci. Abstr., 23: 762.

Takakusaki, K., Kohyama, J., Matsuyama, K. and Mori, S. (2001). Medullary reticulospinal tract mediation of generalized motor inhibition in cats: parallel inhibitory mechanisms acting on motoneurons and on interneuronal transmission in reflex pathways. Neuroscience, 103: 511–527.

Takakuksaki, K., Habaguchi, T., Oohinata-Sugimoto, J., Saitoh, K. and Sakamoto, T. (2003). Basal ganglia efferents to the brainstem centers controlling postural muscle tone and locomotion: A new concept for understanding motor disorders in basal ganglia dysfunction. Neuroscience 119: 293–308.

CHAPTER 24

Locomotor role of the corticoreticular–reticulospinal–spinal interneuronal system

Kiyoji Matsuyama[1,*], Futoshi Mori[2], Katsumi Nakajima[2],
Trevor Drew[3], Mamoru Aoki[1] and Shigemi Mori[2]

[1]*Department of Physiology, Sapporo Medical University, School of Medicine, Sapporo 060-8556, Japan*
[2]*Department of Biological Control System, National Institute for Physiological Sciences, Okazaki 444-8585, Japan*
[3]*Department of Physiology, University of Montreal, Faculty of Medicine, Montreal, QC H3C 3J7, Canada*

Abstract: In vertebrates, the descending reticulospinal pathway is the primary means of conveying locomotor command signals from higher motor centers to spinal interneuronal circuits, the latter including the central pattern generators for locomotion. The pathway is morphologically heterogeneous, being composed of various types of in-parallel-descending axons, which terminate with different arborization patterns in the spinal cord. Such morphology suggests that this pathway and its target spinal interneurons comprise varying types of functional subunits, which have a wide variety of functional roles, as dictated by command signals from the higher motor centers. Corticoreticular fibers are one of the major output pathways from the motor cortex to the brainstem. They project widely and diffusely within the pontomedullary reticular formation. Such a diffuse projection pattern seems well suited to combining and integrating the function of the various types of reticulospinal neurons, which are widely scattered throughout the pontomedullary reticular formation. The corticoreticular–reticulospinal–spinal interneuronal connections appear to operate as a cohesive, yet flexible, control system for the elaboration of a wide variety of movements, including those that combine goal-directed locomotion with other motor actions.

Introduction

The neural control of locomotion in vertebrates involves continuous interactions between various kinds of neural subsystems which are widely distributed throughout the central nervous system (CNS) (Grillner, 1985; Mori et al., 1992; Grillner and Wallén, S. Mori et al., Chapters 1 and 33 of this volume). Of these subsystems, the corticoreticular (CR) pathway, the reticulospinal (RS) pathway and the spinal interneuronal (IN) system as a whole form a continuous, anatomical system that links the cerebral cortex to the spinal cord (SC) (Drew et al., Chapter 25 of this volume). The CR axons originate primarily from the premotor (area 6) and primary motor (area 4) regions of the sensorimotor cortex, and descend with the corticospinal (CS) axons through the internal capsule and cerebral peduncle (Brodal, 1981). These CR axons, some of which arise as collaterals of CS axons, terminate bilaterally in the medial pontomedullary reticular formation (PMRF), from which long-descending RS axons originate (Kuypers, 1981). The pontine and medullary RS axons descend ipsilaterally and bilaterally, respectively, throughout the full length of the SC, terminating in the internuncial layer of the ventral

*Corresponding author: Tel.: +81-11-611-2111 (ext. 2669);
Fax: +81-11-644-1020; E-mail: matsuk@sapmed.ac.jp

horn (VH), where several different types of INs are located (Brodal, 1981; Holstege and Kuypers, 1982; Peterson, 1984).

As reviewed by Armstrong (1986), the motor cortex participates in the regulation and modification of the basic locomotor pattern. The majority of motor cortical neurons, including corticofugal neurons such as CS and CR neurons, are modulated in phase with the rhythmic activities of different limb muscles during locomotion in cats walking on a treadmill, and they significantly increase their discharge frequency during voluntary gait modifications (Drew, 1993; Kably and Drew, 1998b). There is also evidence that when the CS axons are transected at the caudal medullary level, stimulation of the pyramids releases locomotion in mesencephalic cats, thus indicating that the corticobulbar axons, including CR axons, must be responsible for this effect (Shik et al., 1968). In addition, stimulation of the motor cortex excites RS neurons mono- and polysynaptically in anesthetized cats (Pilyavsky and Gokin, 1978).

Several RS unit-recording studies have shown that many RS neurons exhibit rhythmic discharges during locomotion evoked by stimulation of the mesencephalic locomotor region (MLR) in decerebrate cats (Orlovsky, 1970b), or during fictive locomotion in decerebrate, paralyzed cats (Perreault et al., 1993). In intact cats walking on a treadmill (Drew et al., 1986; Matsuyama and Drew, 2000a), the majority of RS neurons exhibit phasic modulation that is correlated to rhythmic muscle activity and/or rhythmic limb movements. Furthermore, when gait modifications, such as walking on an inclined plane and stepping over an obstacle are required, RS neurons exhibit changes in discharge activity, which coincide with the changes in muscle activity (Matsuyama and Drew, 2000b; Prentice and Drew, 2001). These changes are associated with the postural adjustment associated with the gait modifications. Therefore, the RS system as a whole may play a role in modulating interlimb coordination in each locomotor cycle and in producing coordinated postural responses during locomotion.

There is also clear evidence that many lumbar INs located in the intermediate region and VH of the SC show rhythmic discharges during MLR-evoked locomotion and fictive locomotion in decerebrate cats (Feldman and Orlovsky, 1975; Huang et al., 2000) and fictive scratching in spinal cats (Berkinblit et al., 1978). It has also been demonstrated that stimulation of the ventrolateral funiculus (VLF) at the upper cervical SC, which activates pathways including the RS tract, evokes fictive forelimb stepping and rhythmical discharges of cervical INs in decerebrate cats with the lower thoracic SC transected (Yamaguchi, 1986; Terakado and Yamaguchi, 1990). These results suggest that the RS pathway is, in large part, responsible for the activation of spinal IN circuits that produce sustained stepping movements of the limbs, i.e., the central pattern generator (CPG) for locomotion (Grillner, 1985). In this chapter, we will provide detailed information on: (1) the cortical areal specificity of the CR projection; (2) the morphological heterogeneity of the RS pathway; and (3) the morphology and physiology of lumbar commissural INs (CINs). We will then discuss the possible functional role of the CR–RS–spinal IN system in the control of locomotion.

Cortical areal specificity of the CR projection

Since the motor cortical areas are organized in a topographical fashion (Hassler and Muhs-Clement, 1964; Porter and Lemon, 1993), it is natural to think that the whole CR pathway is composed of multiple subcomponents with different cortical origins and functions. The organization of the CR pathway, however, is poorly understood, particularly with respect to the pattern of the projections from different cortical areas. Therefore, to advance understanding of the locomotor role of the CR pathway, we examined interareal differences of its projections within the brainstem (Matsuyama and Drew, 1997). In this study, we injected the anterograde neural tracer, phaseolus vulgaris-leucoagglutinin (PHA-L) into four physiologically identified subdivisions of the cat pericruciate cortex (fore- and hindlimb representations of area 4; areas 6aβ and 6aγ) (Matsuyama and Drew, 1997). Generally, as described previously (Keizer and Kuypers, 1984; Rho et al., 1997), PHA-L labeling from area 6 was more numerous and dense throughout the PMRF than was that from area 4. From the quantitative analyses of the density and distribution of PHA-L labeling in the brainstem, we could specify the characteristic pattern of CR projections from each of the four cortical areas (Matsuyama and Drew, 1997). The number of labeled pyramidal tract (PT) axons that originated from area 6 was much more numerous than that from area 4. The labeled axons from area 4 were of large diameter, however, and they projected caudally to the SC. In contrast, most of those from area 6 were of small diameter, and few descended to the SC. At the caudal pontine and medullary level, PT axons from area 6 decreased in number more rapidly than those from area 4, primarily because numerous fine PT axons from area 6 ran dorsally and directly entered the PMRF. Figure 1A shows that a large proportion of CR axons from area 6 are more likely to be direct CR axons, whereas those from area 4 arise primarily as collaterals from CS axons (see also Kably and Drew, 1998a).

Comparison of the distribution of terminal labeling from all four cortical areas further revealed the gross termination pattern of each CR projection in the PMRF. Generally, the number of terminal endings from area 6 distributed at the pontomedullary level was much greater than that from area 4. In addition, a high percentage of the terminal endings resulting from the area-6 injection at this level were restricted to the PMRF (area 6aβ, 53%; area 6aγ, 71%) whereas in the case of the area-4 injections, a much smaller percentage was distributed in this region (forelimb region of area 4, 32%; hindlimb region, 11%; Matsuyama and Drew, 1997). Conversely, there was a very large projection from area 4 to the pontine nuclei and other brainstem structures, and a correspondingly smaller projection from area 6. Within the PMRF, terminal endings from area 6, as well as the forelimb region of area 4, were distributed with an ipsilateral predominance at the rostral pons, but more bilaterally at the caudal pons and medulla. A smaller relative number of terminal endings from the hindlimb region of area 4 were distributed in the PMRF, and they were localized mainly in the medullary RF. These findings indicate that although CR projections from different cortical areas overlap extensively, each projection is organized in a specific fashion depending on the cortical area from which the CR axons originate.

In contrast to fastigoreticular terminals which have many close appositions to the somata of reticular neurons (Homma et al., 1995), CR terminals

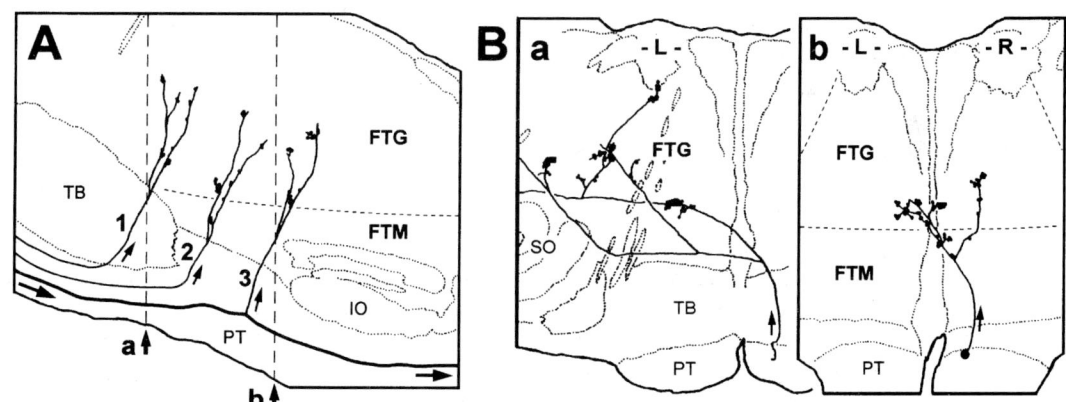

Fig. 1. Projections of CR axons. (A) Parasagittal view of the projections of two types of CR axon: 1 and 2 directly enter the PMRF, whereas 3 arise as an axon collateral from a CS axon. Vertical broken lines indicated by thick upward arrows (at a and b) correspond to the B levels of brainstem transverse section. (B) Transverse views of the arborization pattern of two other CR axons that originated in cortical areas 6aγ (a) and 4 (b; forelimb representation). These axons' projections were reconstructed from serial transverse sections at the level of the caudal pons (a) and rostral medulla (b). The a axon is a direct CR one, and that in b is a collateral from a CS axon indicated by a large dot. Arrows show the axons' trajectories, and small dots represent the location of their terminal swellings. Abbreviations: FTG, gigantocellular tegmental field; FTM, magnocellular tegmental field; IO, inferior olive; L, left side; PT, pyramidal tract; R, right side; SO, superior olive; TB, trapezoid body. (B panels reproduced, in part, from Matsuyama and Drew, 1997 with permission from Wiley-Liss.)

were rarely closely apposed to the somata in the PMRF (Matsuyama and Drew, 1997). This probably indicated that CR terminals contact primarily the dendritic processes of reticular neurons, including RS neurons. Reconstruction of single CR axons further revealed the morphological features of their termination in the PMRF (Matsuyama and Drew, 1997). As shown in Fig. 1A, most CR axons turned dorsally and then distributed terminal axons from the ventral to dorsal parts of the PMRF. Single CR axons usually projected extensively within the PMRF in the transverse plane (Fig. 1B), but with a relatively narrow rostrocaudal extent (see Fig. 1A), and their terminals are far less densely aggregated than those distributed in other brainstem nuclei such as the pontine nuclei (see also Matsuyama and Drew, 1997). Since RS neurons are scattered throughout the whole PMRF (Brodal, 1981; Holstege and Kuypers, 1982), this diffuse projection pattern of each CR axon seems to be well suited to recruit varying types of RS neurons with different functions and locations.

Morphological heterogeneity of RS pathway

The medial PMRF from which the major RS pathways originate can be subdivided into four major nuclei according to their cytoarchitecture and location in the brainstem (Brodal, 1981): the nuclei reticularis pontis oralis (NRPo) and caudalis (NRPc) in the pons, and the nuclei reticularis gigantocellularis (NRGc) and magnocellularis (NRMc) in the medulla. Furthermore, according to Berman (1968), these four reticular nuclei are organized into two groups; the gigantocellular (FTG) and magnocellular tegmental fields (FTM). The FTG corresponds to three of the reticular nuclei (NRPo, NRPc, NRGc), while the FTM corresponds to the NRMc. In sharp contrast to such clear anatomical subdivisions, the PMRF is organized in a loose topographical fashion in relation to limb–trunk motor outputs, with a large overlap in the representation, and no clear segregation as is normally seen in the motor cortex (Peterson, 1984; Drew and Rossignol, 1990).

Since the contribution of the PMRF to motor control depends highly on the projection pattern of the RS pathways, we have performed a series of PHA-L tracing studies to characterize the morphology of RS axons originating from each reticular nucleus in the cat (Matsuyama et al., 1988, 1997, 1999a,b).

The results of the PHA-L tracing studies demonstrated the following general characteristics of the RS pathways originating from the FTG and FTM. As described previously (Holstege and Kuypers, 1982), PHA-L-labeled RS axons from the pontine FTG (NRPo and NRPc) descend purely ipsilaterally, while those from the medullary FTG (NRGc) and the FTM descend almost evenly bilaterally. Axons that descend contralaterally cross the midline at the medullary level. There is a clear difference in the axonal organization between FTG-RS and FTM-RS pathways. The former RS pathway is composed mainly of thick axons (diameter 3–10 µm at the cervical SC) descending in the ventral (VF) and VLF funiculi, while the latter pathway is composed of numerous thin axons (diameter < 3 µm at the cervical SC), which descend in the VF, VLF and the dorsolateral funiculus (DLF; Matsuyama et al., 1988, 1997, 1999a). Furthermore, FTG-RS axons terminate mainly in spinal IN (laminae VII–VIII) and axial MN nuclei at all levels of the SC, while FTM-RS axons terminate widely in laminae VI–VIII and axial and limb MN nuclei.

We have traced the trajectories of individual FTG-RS axons in continuity at multiple segments, and characterized intricate details of the innervation pattern of these axons along the rostrocaudal extent (Matsuyama et al., 1997, 1999b). Generally, as shown in Fig. 2A, FTG-RS axons give off multiple axon collaterals along their descent with an intercollateral interval of ∼5 mm in both the cervical and lumbar enlargements (2–3 vs. 1–2 collaterals/segment at the cervical vs. lumbar SC).

The diameter of the stem FTG-RS axons decrease as they descend caudally (Matsuyama et al., 1997, 1999b). The axon collaterals terminate primarily in lamina VIII and the adjacent lamina VII (Fig. 2B). The innervation pattern of each axon along the rostrocaudal extent is different from one stem axon to another. This indicates that the FTG-RS pathway is morphologically a heterogeneous system composed of various types of axons with different innervation patterns. Physiologically, FTG-RS neurons are also subdivided into several groups according to their

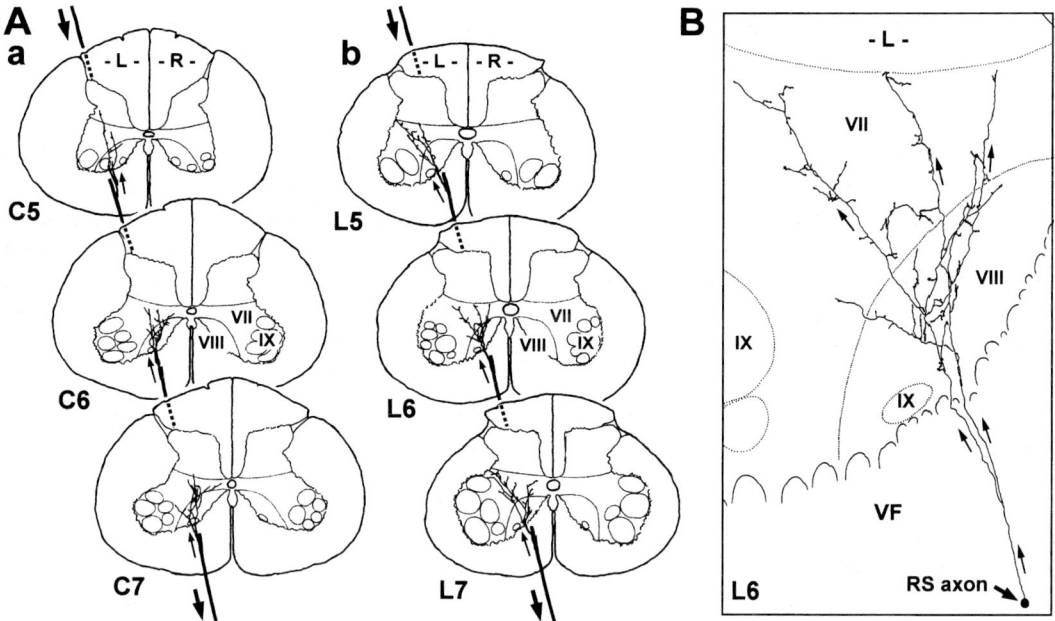

Fig. 2. Projections of RS axons. (A) Multisegmental innervation pattern of two single pontine RS axons in the cervical (a) and lumbar (b) enlargements. These axons originated in the left pontine FTG. The arborization patterns of three axon collaterals arising from the single stem axons were reconstructed from serial transverse sections (at C5–7 for a; L5–7 for b). (B) Fine arborization of a single axon collateral at the lumbar SC. This axon collateral is from the RS axon in Ab at the L6 level. Again, arrows indicate the trajectories of the axons and their successive collaterals. Abbreviations: VF, ventral funiculus; VII, VIII and IX, Rexed's laminae VII, VIII and IX. (Reproduced, in part, from Matsuyama et al., 1999b with permission from Wiley-Liss.)

discharge patterns during locomotion (e.g., EMG-related, locomotor-rhythm-related, unrelated, tonic, silent; Drew et al., 1986; Prentice and Drew 2001). Therefore, it is highly possible that the FTG-RS pathway is a heterogeneous, parallel-descending pathway composed of various types of neuronal elements with different functions and morphology. This heterogeneity might be related to the functional complexity of FTG-RS pathway. Since various types of RS neurons are intermingled within the FTG region, and since many individual RS axons branch profusely in the SC, stimulation of even a limited region of the FTG evokes a complex and mixed pattern of motor outputs (see also Drew and Rossignol, 1990).

Despite such a high degree of heterogeneity, there are three organizing principles related to the collateral termination pattern of individual FTG-RS axons. First, each axon collateral arborizes in a limited rostrocaudal region with a thickness of <1 mm. Second, axon collaterals arising from ~80% of FTG-RS axons innervate only the unilateral gray matter, ipsilateral to their stem axons, whereas those from the remaining axons (~20%) innervate the bilateral gray matter. Third, the termination areas of axon collaterals from a given FTG-RS axon are similar at each segmental level along the course of that axon, and different from one stem axon to another. Such commonality of the pattern of collateral termination along the rostrocaudal extent of parent axons is found throughout the cervical and lumbar SC. We therefore propose that a single RS neuron innervates a large number of INs at multiple segments throughout the full length of the SC, and of more functional importance, a single RS neuron may activate INs at similar locations at different SC levels.

Although preliminary, we have also analyzed the morphological features of FTM-RS axons (Matsuyama et al., 1999a). In contrast to FTG-RS axons, most FTM-RS axons have small diameters (<3 µm), and they give off axon collaterals at a low occurrence frequency (~1 collateral/segment at the cervical SC level). The terminal axons of FTM-RS axons are widely distributed in laminae VI–VIII and additionally in lamina IX, whereas those of FTG-RS axons are distributed primarily in laminae VIII–VII. This suggests that the RS pathways are morphologically divided into at least two subcomponents: a FTG-RS pathway, composed of thick axons with high collateralization and a FTM-RS pathway, composed of thin axons with low collateralization.

Several findings suggesting that both FTG-RS and FTM-RS pathways are highly implicated in the control of locomotion. Recall first that there are several specific areas identified as locomotion-inducing sites in decerebrate and thalamic cats: notably, the MLR in the midbrain (Shik et al., 1966), the subthalamic locomotor region (SLR) in the subthalamus (Orlovsky, 1969) and the cerebellar locomotor region (CLR) in the cerebellum (Mori et al., 1999). The most marked efferent projections from the former two regions are those to the PMRF in the brainstem (Bayev et al., 1988), and a large number of RS neurons receive excitatory inputs from these regions (Orlovsky, 1970a). In particular, the major efferents from the MLR project mainly to the FTM (Steeves and Jordan, 1984), and stimulation of this site activates FTM-RS neurons with a wide (10–100 m/s) range of axonal conduction velocity (Garcia-Rill and Skinner, 1987). In contrast, one of the major efferents from the CLR is the crossed fastigioreticular pathway that terminates in the PMRF (FTG > FTM; Homma et al., 1995). Stimulation of this site monosynaptically activates FTG-RS neurons with a relatively fast (50–100 m/s) conduction velocity (Mori et al., 2000). Therefore, we suggest that both FTG-RS and FTM-RS pathways are essential for the generation and regulation of locomotor movements. This suggestion is strongly supported by the finding that lesions of the ventral and ventrolateral SC at low thoracic levels in the cat, which probably severs descending pathways including both RS pathways, cause pronounced locomotor deficits accompanying postural instability. These include an irregularity of the hindlimb's step-cycle duration and a step-by-step inconsistency of interlimb coupling between homolateral fore- and hindlimbs (Brustein and Rossignol, 1998).

Morphology and physiology of lumbar commissural INs

A wide variety of spinal INs involved in motor functions are distributed in the spinal internuncial layer. These can be classified into several groups according to the characteristics of their morphology and physiology (for review, see Scheibel and Scheibel, 1969; Jankowska, 1992). Most of these INs receive peripheral afferent signals mono- or polysynaptically and they project to axial and/or limb MNs, thereby contributing to a wide variety of spinal reflex pathways (Jankowska, 1992). During stepping and scratching, a large number of spinal INs, which are located in the lumbar enlargement, exhibit rhythmic activity related to that of hindlimb MNs (Feldman and Orlovsky, 1975; Berkinblit et al., 1978). The majority of the rhythmically modulated INs are located in the medial to lateral parts of the intermediate gray matter, as well as in the VH, i.e., in laminae VI–VII. Among these are Ia INs, which mediate Ia reciprocal inhibition of antagonists, and burst rhythmically in phase with their parent muscle during real stepping. By inhibiting antagonistic MNs, they may contribute to the generation of alternating activity of extensor and flexor muscles in each limb (Feldman and Orlovsky, 1975).

The previous studies have provided important information about several of the characteristics of spinal INs, including their pattern of activity during locomotion, their location and their neural inputs. This information is essential for understanding the locomotor roles of spinal INs. Because the function of each IN is largely related to its structure and termination pattern, however, we think it is also important to further detail the morphology of spinal INs. For example, we have recently examined the fine morphology of RS-activated lumbar INs, and recorded their discharge characteristics during the generation of fictive locomotion (Matsuyama et al., 1999c). For this, we utilized an intra-axonal recording technique combined with an intra-axonal injection of neurobiotin (NB). The microelectrode was positioned in the VF at L4–7 segments in decerebrate, paralyzed cats. This technique allowed us to obtain both morphological and physiological characteristics for each IN.

In fully stained INs, the morphology revealed included their dendrites and collateral branching pattern. Their stem axons cross the midline, and their somata are located mainly in lamina VIII, and also in laminae VI–VII. The majority (~80%) discharge rhythmically during MLR-evoked fictive locomotion. They always exhibit one discharge peak in each locomotor cycle, and their discharge is phase-related to the discharge of left- or right-side hindlimb extensor nerve activity. Such INs are activated orthodromically by stimulation of the FTG (NRPc or NRGc). They are particularly responsive at short latency (~4 ms) to contralateral NRGc stimulation. Since ~40% of NRGc-RS axons cross the midline and descend in the contralateral VF and VLF (Matsuyama et al., 1988), we think that the activity of lamina VIII INs is strongly controlled by descending inputs via the crossed FTG-RS pathway. Furthermore, because a majority of lamina VIII INs which received excitatory RS inputs showed locomotor-related rhythmic activity, we also suggest that these neurons may be subcomponents of the locomotor CPG for each hindlimb.

The lamina VIII INs which we identified are divided into three types according to their axonal course. As shown in Fig. 3A, their stem axons cross the midline at the SC level where their own somata are located, and enter the contralateral VF. The axon then, either bifurcates rostrally or caudally, ascends or descends in the VF, giving off multiple axon collaterals (~5 collaterals/segment) for a few segments. These collaterals terminated primarily in lamina VIII and adjacent VII, and some additionally terminate in hindlimb MN nuclei (lamina IX; see Fig. 3B). Therefore, the lamina VIII INs which we identified can be considered as CINs. This coincides well with the previous findings reported by Scheibel and Scheibel (1969), which emphasized that a high population (~85%) of spinal INs located in lamina VIII are CINs.

The preliminary anterograde neural tracing study using the tracer, biotinylated dextran amine (Matsuyama and Aoki, 2001), has revealed some aspects of the gross projection pattern of lamina VIII CINs. Their axons cross the midline through the anterior commissure forming thick axon bundles, and they course caudally and/or rostrally in the contralateral VF. These axons commonly give off axon collaterals projecting primarily in laminae VIII–VII and additionally in lamina IX. The density of these axon projections is highest at the SC level where the tracer is injected, and they decrease rapidly with distance from the injection site. These findings suggest that lamina VIII CINs located on the left and right sides are tightly coupled to each other at the same SC level. Such mutual connection of CINs might be well suited for the intrasegmental integration and coordination of the locomotor rhythm generated on both sides of the SC, this being essential for the generation of reciprocal rhythmic movements around similar joints of the left and right limbs (Kjaerulff and Kiehn, 1997). Furthermore, as shown in Fig. 3C, a majority of lamina VIII CINs that we identified receives excitatory inputs via the crossed FTG-RS pathway. Thus, these CINs and the crossed RS pathway may form one of the essential neural systems responsible for the generation and regulation of the reciprocal locomotor rhythm.

Summary and comments

It has been well established that the RS pathway is an all-encompassing descending pathway, which mediates locomotor command signals to spinal IN circuits from various higher locomotor centers including the CLR, MLR and SLR (Grillner, 1985; Mori et al., 1992, 2000). The RS system has also been suggested to play an important role in a number of additional motor behaviors, including orienting and postural movements (Peterson, 1984; Drew et al., 1986; Mori et al., 1992). These motor behaviors commonly require the simultaneously coordination of activity in the head and limbs and between different limbs. A large part of this coordinating activity is thought to be 'hardwired' and mediated, in part, by the diffuse pattern of projection of individual RS axons (Kuypers, 1981; Peterson, 1984). The findings in the PHA-L study strongly support this idea, because each RS axon has a common propensity to innervate an exceptionally large number of laminae VIII–VII INs throughout the full length of the SC (Matsuyama et al., 1997, 1999b). Therefore, assuming that a spinal IN circuit located at each segment generates a local locomotor activity, we can imagine that the RS system as a whole may provide the means to achieve an intersegmental

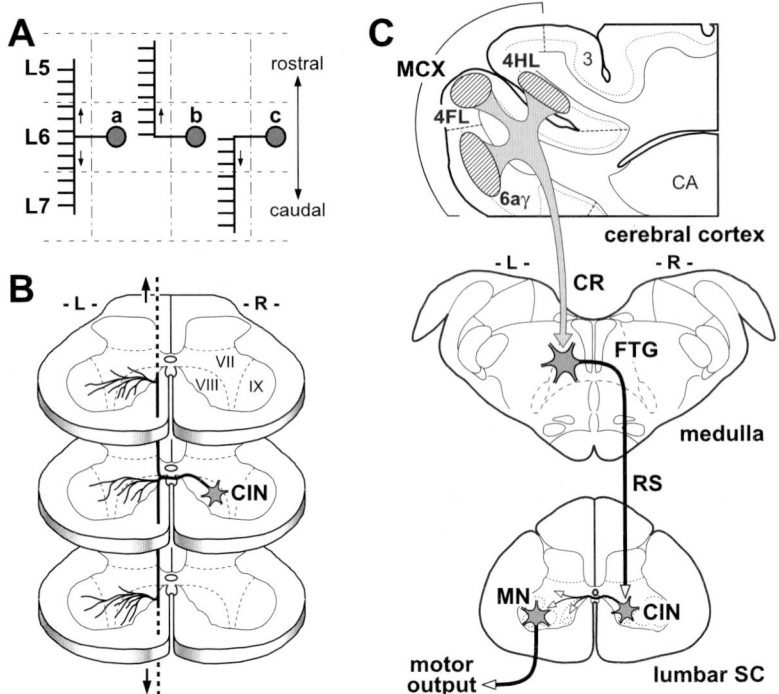

Fig. 3. Lumbar CINs and their descending control. (A) Axonal morphology of three types of lumbar lamina VIII commissural CINs. Filled circles, solid lines and horizontal short bars indicate the cells' somata, stem axons and axon collaterals, respectively. Vertical chain and horizontal broken lines indicate the midline of the SC and the border of SC segments, respectively. (B) Innervation pattern of an exemplary lamina VIII CIN. Note the cell's multiple axon collaterals at, or close to, the SC segment where its soma is located. The axon's collaterals terminate primarily in laminae VIII and the adjacent VII, with a few also in lamina IX. (C) Summary of the neural connections of the CR pathway, crossed FTG-RS pathway and lamina VIII CINs. CR axons from cortical areas 4 and 6 terminate in the PMRF where the RS pathway, including its crossed FTG-RS component, originates. Crossed FTG-RS axons descend in the contralateral VF throughout the full length of the SC, and their collateral terminations are primarily in laminae VIII and the adjacent VII. The lamina VIII CINs, which receive excitatory inputs from the crossed FTG-RS pathway, innervate primarily the contralateral laminae VIII–VII, with some terminations also in lamina IX. The top drawing is based on a parasagittal section of the right-side cortex. Broken lines indicate boundaries between cytoarchitectonically defined areas, and dotted lines indicate the position of layer V. The three-hatched areas in the top drawing correspond to the PHA-L injection sites in the fore- and hindlimb representation of area 4, and in area 6aγ (see text for details). Abbreviations: CA, caudate nucleus; 4FL and 4HL, fore- and hindlimb representations of area 4; MCX, motor cortex; 3, area 3.

integration of locomotor activities generated by numerous local spinal IN circuits, which are distributed along the rostrocaudal axis of the SC, and located on both sides of the SC. This arrangement may facilitate an integration and coordination of interlimb and/or limb–trunk locomotor rhythmic activity, something, which is necessary for the elaboration of automatic and synergistic locomotor movements.

Moreover, the PHA-L studies have demonstrated that the RS pathway is a heterogeneous, parallel-descending pathway composed of various types of axons with different arborization patterns in the SC (Matsuyama et al., 1997, 1999b). Such morphological heterogeneity suggests that this pathway and its target spinal INs provide varying types of functional subunits, which have a wide variety of functional roles under various movement command signals from the higher motor centers such as the motor cortex. Furthermore, the CR pathways can also be considered as parallel-descending pathways that are composed of various CR axons arising from different cortical areas with different functions. Thus, the CR–RS–spinal IN system might be considered as a multisynaptic, parallel-descending system, composed of a great variety of neural elements. Of these elements, CR neurons terminating in the medullary FTG, RS neurons with crossed axons and spinal lamina VIII CINs may constitute a functionally unified system responsible for the generation and regulation of reciprocal left side–right side locomotor movements (see Fig. 3C).

Finally, what is the primary function of the overall CR–RS–spinal IN system in the control of locomotion? Recall the classical finding that decorticate animals can stand and walk spontaneously but purposelessly, using movement patterns that are stereotypical like those of an automaton (Schaltenbrand and Cobb, 1931). This strongly suggests that the neocortex plays an important role in the goal-directed, diversified locomotion of intact animals. The PHA-L study (Matsuyama and Drew, 1997) showed that each CR neuron projects widely and diffusely within the PMRF. This pattern is well suited to combining and integrating the function of the various types of RS neurons, which are scattered widely throughout the PMRF. Therefore, we suggest that the functionally united CR–RS–spinal IN system serves to provide the optimally flexible neural substrate required for the elaboration and refinement of the wide variety of locomotor patterns used in goal-directed locomotion when both self-induced movements are subject to external perturbations.

Acknowledgments

We would like to thank Masahiro Mori and Chijiko Takasu for their excellent technical assistance. This study has been supported by Grants in Aid for Scientific Research from the Ministry of Education, Culture, Sports, Science and Technology of Japan (11170253, 12680804 and 13035039).

Abbreviations

CIN	commissural interneuron
CLR	cerebellar locomotor region
CNS	central nervous system
CPG	central pattern generator
CR	corticoreticular
CS	corticospinal
EMG	electromyogram
FTG	gigantocellular tegmental field
FTM	magnocellular tegmental field
IN	interneuron
MLR	mesencephalic locomotor region
MN	motoneuron
NB	neurobiotin^{TM}
NRGc	nucleus reticularis gigantocellularis
NRMc	nucleus reticularis magnocellularis
NRPc	nucleus reticularis pontis caudalis
NRPo	nucleus reticularis pontis oralis
PHA-L	phaseolus vulgaris-leucoagglutinin
PMRF	pontomedullary reticular formation
PT	pyramidal tract
RS	reticulospinal
SC	spinal cord
SLR	subthalamic locomotor region
VF	ventral funiculus
VH	ventral horn
VLF	ventrolateral funiculus
VI, VII, VIII, IX	Rexed's laminae VI, VII, VIII, IX

References

Armstrong, D.M. (1986). Supraspinal contributions to the initiation and control of locomotion in the cat. Prog. Neurobiol., 26: 273–361.

Bayev, K.V., Beresovskii, V.K., Kebkalo, T.G. and Savoskina, L.A. (1988). Afferent and efferent connections of brainstem locomotor regions: study by means of horseradish peroxidase transport technique. Neuroscience, 26: 871–891.

Berkinblit, M.B., Deliagina, T.G., Feldman, A.G., Gelfand, I.M. and Orlovsky, G.N. (1978). Generation of scratching. I. Activity of spinal interneurons during scratching. J. Neurophysiol., 41: 1040–1057.

Berman, A.L. (1968). The Brain Stem of the Cat: A Cytoarchitectonic Atlas with Stereotaxic Coordinates. University of Wisconsin Press, Madison.

Brodal, A. (1981). Neurological Anatomy in Relation to Clinical Medicine. 3rd edition. Oxford University Press, New York, pp. 180–293.

Brustein, E. and Rossignol, S. (1998). Recovery of locomotion after ventral and ventrolateral spinal lesion in the cat. I. Deficits and adaptive mechanisms. J. Neurophysiol., 80: 1245–1267.

Drew, T. (1993). Motor cortical activity during voluntary gait modifications in the cat. I. Cells related to the forelimbs. J. Neurophysiol., 70: 179–199.

Drew, T. and Rossignol, S. (1990). Functional organization within the medullary reticular formation of intact unanesthetized cat. I. Movements evoked by microstimulation. J. Neurophysiol., 64: 767–781.

Drew, T., Dubuc, R. and Rossignol, S. (1986). Discharge pattern of reticulospinal and other reticular neurons in chronic, unrestrained cats walking on a treadmill. J. Neurophysiol., 55: 375–401.

Feldman, A.G. and Orlovsky, G.N. (1975). Activity of interneurons mediating reciprocal Ia inhibition during locomotion. Brain Res., 84: 181–194.

Garcia-Rill, E. and Skinner, R.D. (1987). The mesencephalic locomotor region. II. Projections to reticulospinal neurons. Brain Res., 411: 13–20.

Grillner, S. (1985). Neurobiological bases of rhythmic motor acts in vertebrates. Science, 228: 143–149.

Hassler, R. and Muhs-Clement, K. (1964). Architektonischer Aufbau des sensomotorischen und parietalen Cortex der Katze. J. Hirnforsch., 6: 377–420.

Holstege, G. and Kuypers, H.G.J.M. (1982). The anatomy of brain stem pathways to the spinal cord in cat. A labeled amino acid tracing study. In: Kuypers H.G.J.M. and Martin G.F. (Eds.), Descending Pathway to the Spinal Cord, Prog. Brain Res., Vol. 57, Elsevier, New York, pp. 145–175.

Homma, Y., Nonaka, S., Matsuyama, K. and Mori, S. (1995). Fastigiofugal projection to the brainstem nuclei in the cat: an anterograde PHA-L tracing study. Neurosci. Res., 23: 89–102.

Huang, A., Noga, B.R., Carr, P.A., Fedirchuk, B. and Jordan, L.M. (2000). Spinal cholinergic neurons activated during locomotion: localization and electrophysiological characterization. J. Neurophysiol., 83: 3537–3547.

Jankowska, E. (1992). Interneuronal relay in spinal pathways from proprioceptors. Prog. Neurobiol., 38: 335–378.

Kably, B. and Drew, T. (1998a). Corticoreticular pathways in the cat. I. Projection patterns and collateralization. J. Neurophysiol., 80: 389–405.

Kably, B. and Drew, T. (1998b). Corticoreticular pathways in the cat. II. Discharge activity of neurons in area 4 during voluntary gait modifications. J. Neurophysiol., 80: 406–424.

Keizer, K. and Kuypers, H.G.J.M. (1984). Distribution of corticospinal neurons with collaterals to lower brainstem reticular formation in cat. Exp. Brain Res., 54: 107–120.

Kjaerulff, O. and Kiehn, O. (1997). Crossed rhythmic synaptic input to motoneurons during selective activation of the contralateral spinal locomotor network. J. Neurosci., 17: 9433–9447.

Kuypers, H.G.J.M. (1981) Anatomy of descending pathways. In: Brooks, V.B. (Ed.), Handbook of Physiology, Sec. 1, The Nervous System, Vol. II, Motor Control, Part 1. American Physiological Society, Bethesda, pp. 597–666.

Matsuyama, K. and Aoki, M. (2001). Intraspinal projection pattern of lumbar commissural neurons of cats. Neurosci. Res., Suppl. 25: S157.

Matsuyama, K. and Drew, T. (1997). Organization of the projections from the pericruciate cortex to the pontomedullary brainstem of the cat: a study using the anterograde tracer Phaseolus vulgaris-leucoagglutinin. J. Comp. Neurol., 389: 617–641.

Matsuyama, K. and Drew, T. (2000a). Vestibulospinal and reticulospinal neuronal activity during locomotion in the intact cat: I. Walking on a level surface. J. Neurophysiol., 84: 2237–2256.

Matsuyama, K. and Drew, T. (2000b). Vestibulospinal and reticulospinal neuronal activity during locomotion in the intact cat: II. Walking on an inclined plane. J. Neurophysiol., 84: 2257–2276.

Matsuyama, K., Ohta, Y. and Mori, S. (1988). Ascending and descending projections of the nucleus reticularis gigantocellularis in the cat demonstrated by the anterograde neural tracer, Phaseolus vulgaris leucoagglutinin (PHA-L). Brain Res., 460: 124–141.

Matsuyama, K., Takakusaki, K., Nakajima, K. and Mori, S. (1997). Multisegmental innervation of single pontine reticulospinal axons in the cervico-thoracic region of the cat: anterograde PHA-L tracing study. J. Comp. Neurol., 377: 234–250.

Matsuyama, K., Drew, T., Mori, F. and Mori, S. (1999a). The cortico-reticulo-spinal system: organization of the cortico-reticular projection and fine architecture of the reticulospinal pathway in the cat. In: Gantchev G.N., Mori S. and Massion J. (Eds.), Motor Control Today and Tomorrow. Academic Publishing House 'Prof. M. Drinov', Sofia, pp. 45–56.

Matsuyama, K., Mori, F., Kuze, B. and Mori, S. (1999b). Morphology of single pontine reticulospinal axons in the lumbar enlargement of the cat: a study using the anterograde tracer PHA-L. J. Comp. Neurol., 410: 413–430.

Matsuyama, K., Mori, F. and Mori, S. (1999c). Fine structure of a reticulospinal-interneuronal system for coordinating locomotor rhythm in cats. In: Gantchev N. and Gantchev G.N. (Eds.), From Basic Motor Control to Functional Recovery: Concepts, Theories and Models Present State and Perspective. Academic Publishing House 'Prof. M. Drinov', Sofia, pp. 14–18.

Mori, S., Matsuyama, K., Kohyama, J., Kobayashi, Y. and Takakusaki, K. (1992). Neuronal constituents of postural and locomotor control systems and their interactions in cats. Brain Dev., 14 (Suppl.): S109–S120.

Mori, S., Matsui, T., Kuze, B., Asanome, M., Nakajima, K. and Matsuyama, K. (1999). Stimulation of a restricted region in the midline cerebellar white matter evokes coordinated quadrupedal locomotion in the decerebrate cat. J. Neurophysiol., 82: 290–300.

Mori, S., Matsui, T., Mori, F., Nakajima, K. and Matsuyama, K. (2000). Instigation and control of treadmill locomotion in high decerebrate cats by stimulation of the hook bundle of Russell in the cerebellum. Can. J. Physiol. Pharmacol., 78: 945–957.

Orlovsky, G.N. (1969). Spontaneous and induced locomotion of the thalamic cat. Biophysics, 14: 1154–1162.

Orlovsky, G.N. (1970a). Connexions of the reticulo-spinal neurons with the locomotor regions of the brainstem. Biophysics, 15: 178–186.

Orlovsky, G.N. (1970b). Work of the reticulo-spinal neurons during locomotion. Biophysics, 15: 761–771.

Perreault, M.C., Drew, T. and Rossignol, S. (1993). Activity of medullary reticulospinal neurons during fictive locomotion. J. Neurophysiol., 69: 2232–2247.

Peterson, B.W. (1984). The reticulospinal system and its role in the control of movement. In: Barnes C.D. (Ed.), Brainstem Control of Spinal Cord Function. No. 6 as Research Topics in Physiology. Academic Press, London, pp. 27–86.

Pilyavsky, A.I. and Gokin, A.P. (1978). Investigation of the cortico-reticulo-spinal connections in cats. Neuroscience, 3: 99–103.

Porter, R. and Lemon, R. (1993). Corticospinal Function and Voluntary Movement. No. 45 as Monographs of the Physiological Society. Oxford Science Publication, Oxford.

Prentice, S.D. and Drew, T. (2001). Contribution of the reticulospinal system to the postural adjustments occurring during voluntary gait modifications. J. Neurophysiol., 85: 679–698.

Rho, M.J., Cabana, T. and Drew, T. (1997). Organization of the projections from the pericruciate cortex to the pontomedullary reticular formation of the cat: a quantitative retrograde tracing study. J. Comp. Neurol., 388: 228–249.

Schaltenbrand, G. and Cobb, S. (1931). Clinical and anatomical studies on two cats without neocortex. Brain, 53: 449–488.

Scheibel, M.E. and Scheibel, A.B. (1969). A structural analysis of spinal interneurons and Renshaw cells. No. 11 as UCLA Forum in Medical Sciences. In: Brazier M.A.B. (Ed.), The Interneuron. University of California Press, Los Angeles, pp. 159–208.

Shik, M.L., Severin, F.V. and Orlovsky, G.N. (1966). Control of walking and running by means of electrical stimulation of the midbrain. Biophysics, 11: 755–765.

Shik, M.L., Orlovsky, G.N. and Severin, F.V. (1968). Locomotion of the mesencephalic cat evoked by pyramidal stimulation. Biophysics, 13: 127–135.

Steeves, J.D. and Jordan, L.M. (1984). Autoradiographic demonstration of the projections from the mesencephalic locomotor region. Brain Res., 307: 263–276.

Terakado, Y. and Yamaguchi, T. (1990). Last-order interneurones controlling activity of elbow flexor motoneurons during forelimb fictive locomotion in the cat. Neurosci. Lett., 111: 292–296.

Yamaguchi, T. (1986). Descending pathways eliciting forelimb stepping in the lateral funiculus: experimental studies with stimulation and lesion of the cervical cord in decerebrate cats. Brain Res., 379: 125–136.

CHAPTER 25

Cortical and brainstem control of locomotion

Trevor Drew[1,*], Stephen Prentice[2] and Bénédicte Schepens[3]

[1]*Department of Physiology, University of Montreal, Faculty of Medicine, Montreal, QC H3C 3J7, Canada*
[2]*Department of Kinesiology, University of Waterloo, Waterloo, ON N2L 3G1, Canada*
[3]*Unité de Physiologie et Biomécanique de la Locomotion, Département d'Éducation Physique et de Réadaptation, Université Catholique de Louvain, 1348 Louvain-La-Neuve, Belgium*

Abstract: While a basic locomotor rhythm is centrally generated by spinal circuits, descending pathways are critical for ensuring appropriate anticipatory modifications of gait to accommodate uneven terrain. Neurons in the motor cortex command the changes in muscle activity required to modify limb trajectory when stepping over obstacles. Simultaneously, neurons in the brainstem reticular formation ensure that these modifications are superimposed on an appropriate base of postural support. Recent experiments suggest that the same neurons in the same structures also provide similar information during reaching movements. It is suggested that, during both locomotion and reaching movements, the final expression of descending signals is influenced by the state and excitability of the spinal circuits upon which they impinge.

Introduction

In the intact animal, goal-directed locomotion depends on the interaction between many different structures within the central nervous system (CNS). As reviewed in previous chapters (Grillner and Wallén, Rossignol et al., Chapters 1 and 16 of this volume), the spinal cord is capable of generating many aspects of the normal locomotor pattern. To adapt this basic pattern to the terrain over which the animal walks, however, the spinal cord depends critically on inputs from peripheral afferents (Pearson, this volume) and supraspinal structures. Following the original concepts of Kuypers and colleagues (Lawrence and Kuypers, 1968a,b; Kuypers, 1963), one may consider, in a very general sense, that supraspinal control is divided among two systems. One is involved in the fine control of locomotion (the lateral system, including the cortico- and rubrospinal tracts). The other provides the postural support on which the fine control is superimposed (the medial system, including the reticulo- and vestibulospinal tracts). In reality, these two systems overlap to some extent and they do not act separately, but rather in close cooperation.

In this chapter, we discuss the contribution of two such pathways, one from each system, to the control of locomotion. The focus is particularly on situations wherein there is a requirement for a voluntary modification of the basic locomotor rhythm. Two points are emphasized. First, all signals descending from supraspinal structures must be integrated at the spinal level with signals generating the basic locomotor rhythm and signals from peripheral afferents. This arrangement might constrain the efficacy of the descending signals but it provides the advantage that descending commands are *readily integrated* into the ongoing locomotor rhythm. The second key point is that the same CNS structures, and, in many cases, the same neurons, contribute not only to the control of locomotion, but also to more discrete voluntary movements, such as reaching. We propose that the logical extension of this idea is that

*Corresponding author: Tel.: +514-343-7061; Fax: +514-343-6113; E-mail: trevor.drew@umontreal.ca

the same neural circuits within the spinal cord are used for the control of *both* locomotion and reaching (see also Grillner and Wallén, Chapter 1 of this volume).

Cortical contribution to the control of locomotion

One's view of the function of any given CNS structure often depends critically upon the circumstances under which it is studied (see also Bloedel, this volume). In the case of the cortical contribution to the control of locomotion, it was long considered that the role was facultative rather than essential because cortical lesions have so little effect on the ability of most animals to walk (see citations in Armstrong and Drew, 1984a,b; Drew et al., 1996a). A classical examination of this issue by Liddell and Phillips (1944) showed, however, that cats with a bilateral pyramidotomy exhibit major locomotor deficits when challenged to walk either from rung to rung on a horizontal ladder or along the surface of a curved pipe. Such demonstrations have suggested that the major contribution of the motor cortex during locomotion in the cat is likely to occur when the animal has to modify its basic locomotor rhythm to accommodate an uneven locomotor terrain.

Technical advances in the last 20 years have provided the means to record neuronal activity from the motor cortex in the same type of challenging (skilled) locomotor tasks for which cortical integrity seems essential. All such studies have shown that the activity of a majority of cortical neurons, including those identified as projecting at least as far as the more caudal regions of the pyramidal tract [pyramidal tract neurons (PTNs)] is modified during tasks that require skillful changes in gait. These changes have included placing the paw accurately on the rungs of a horizontal ladder (Armstrong, 1988; Amos et al., 1990), and stepping over barriers on the ground (Beloozerova and Sirota, 1993) or attached to a moving treadmill belt (Drew, 1988, 1993; Drew et al., 1996a).

Most cortical neurons show an increase of their discharge activity during voluntary gait modifications. Usually, this increase is phase-locked to periods of modified activity in the physiological flexor muscles. In this respect, cortical neurons behave, in general, in the same manner as analogous neurons in nonhuman primates trained to make a reaching movement. During locomotion, however, voluntary modifications in gait have to be superimposed upon the basic locomotor rhythm. Nevertheless, the underlying problem of control is similar in the two situations. To produce the required limb trajectory in both goal-directed locomotion and arm reaching, descending command signals must ensure activation of the correct muscles in their appropriate spatiotemporal order, and for their requisite duration. Analysis of the temporal relationships of the neuronal discharge patterns of populations of motor cortical neurons during voluntary gait modifications suggests that much of the spatiotemporal organization of modified limb-muscle EMG activity is specified in the signals of individual PTNs. As illustrated in Fig. 1A, different PTNs discharge at discrete times during the gait cycle.

Although we view this PTN descending command as a relatively low-level signal that specifies much of the required change in muscle activity, it must be emphasized that expression of the descending command is unlikely to be either direct or simple. Experiments in which the motor cortex, or pyramidal tract, has been stimulated during locomotion (Orlovsky, 1972; Armstrong and Drew, 1985; Rho et al., 1999) suggest that the effects of a corticospinal volley are mediated by interneuronal (IN) pathways that are influenced by, or part of, the spinal central pattern generator (CPG) for locomotion (Grillner, 1981). As such, expression of the pattern will be influenced by the level of activity in these IN pathways. In addition, PTNs collateralize at the spinal level (Futami et al., 1979) and the discharge in any one group of PTNs is unlikely to be restricted to any one small group of synergistically acting muscles. Nevertheless, there is evidence that the synaptic efficacy of cortical projection neurons with motoneurons (MNs) may be greater with some of the muscles than with others (Bennett and Lemon, 1996). We therefore suggest that the final expression of the descending command is a result of its integration with the underlying pattern of activity in IN locomotor networks. The conceptual model of Fig. 1B illustrates one way for different populations of PTNs to interact with the spinal locomotor CPG. It shows that PTNs,

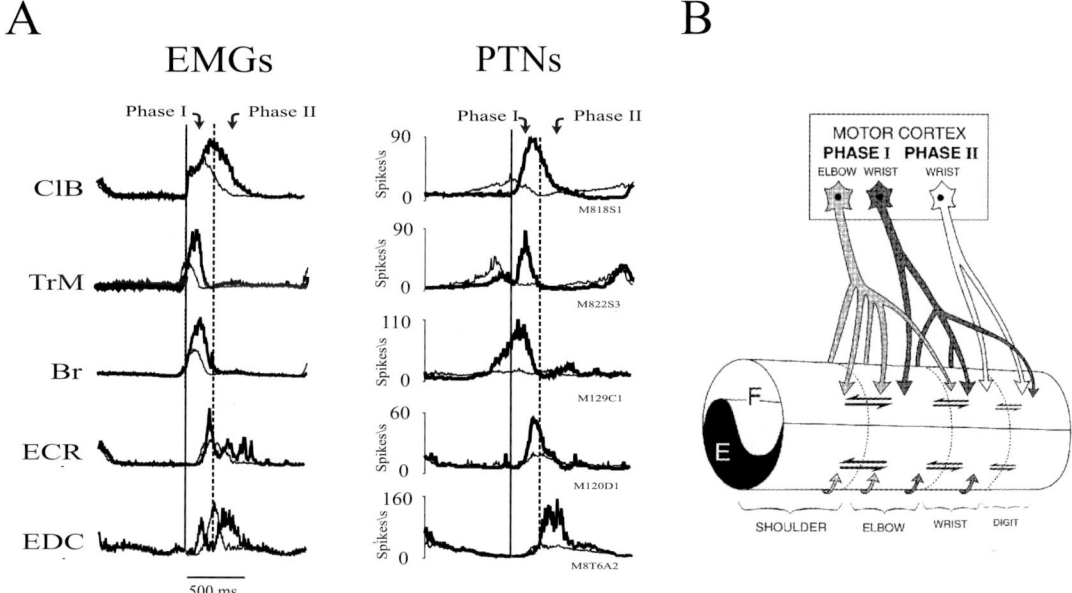

Fig. 1. Temporal relation between PTN and forelimb muscle activity during voluntary gait modifications. (A) Sequential activation patterns of selected forelimb muscles and PTNs during steps over an obstacle. Each trace is centered on the onset of EMG activity in the ClB muscle contralateral to the motor cortical recording site (vertical line). The duration of ClB activity approximately defines the swing phase of the forelimb. Thinner lines indicate activity patterns during unobstructed locomotion and thicker lines show activity during a step over an obstacle. The vertical dotted line divides the transport phase of gait modification (Phase I) from the subsequent period when the paw is oriented for repositioning on the treadmill belt (Phase II). Each period of EMG activity was recorded simultaneously with the PTN illustrated to the right. The time base supplied below the EMG traces provides an averaged value based on the data from each of the five recordings. It also applies to the PTN traces. (B) Conceptual model showing one way in which the descending signal from the motor cortex may integrate with the basic locomotor rhythm. The model is based on the 'Unit CPG' concept of Grillner (1982). Each group of PTNs is suggested to act preferentially on one of the modules. Interactions between modules, and the collateralization of the groups of PTNs aid in assuring a smooth, coordinated movement. Abbreviations: Br, brachialis; ClB, cleidobrachialis; ECR, extensor carpi radialis; EDC, extensor digitorum communis; TrM, teres major. (Reproduced, in part, from Lavoie and Drew, 2002 with permission from The American Physiological Society, and from Drew, 1991b with permission from Elsevier Science.)

whose discharge temporally covaries with the activity of elbow flexors such as brachialis, would have a preferential input to the spinal IN networks related to these muscles but would also influence IN networks controlling muscles around contiguous joints. Similarly, the model shows other PTNs, whose discharge covaries with activity of muscles acting around the wrist, such as extensor carpi radialis, projecting primarily to INs of the wrist module, and also influencing contiguous modules. Still other PTNs, such as those discharging later in swing phase of the step, when extensor digitorum communis is active, might have a more limited distribution of spinal IN effects. This type of arrangement would provide a flexible control system that allows the motor cortex to differentially modify the activity of muscles active at different parts of the swing phase, while still ensuring that the overall movement is coordinated and smoothly integrated into the basic locomotor rhythm. As such, changes in the discharge characteristics of different groups of PTNs could readily produce the modifications in intralimb coordination required to step over obstacles of different shape and size.

The integration of movement and posture

Modifications of gait are potentially destabilizing and they must be accompanied by postural modifications which provide the support on which voluntary movements are superimposed. Although there is an extensive literature on the biomechanical responses that accompany movement (for review, see Horak and Macpherson, 1996), there is little information on the neural mechanisms responsible for the production of such postural responses, or those responsible for the integration of movement and posture. Theoretical considerations of this problem (Massion, 1992; also Massion et al., Chapter 2 of this volume) suggest that the cortical command for movement may furnish one of the feedforward signals used for the production of the postural responses that anticipate and accompany voluntary movements, including gait modifications. There is considerable evidence to suggest that the pontomedullary reticular formation (PMRF) is involved in the production of postural responses (Mori, 1987; Mori et al., 1992; see also S. Mori et al., Chapter 33 of this volume). It is therefore likely that the corticoreticular projections that link the motor areas of the cortex with reticulospinal neurons (RSNs) in the PMRF are a component of pathways involved in integrating posture and movement.

To test this hypothesis, we examined the anatomy and physiology of corticoreticular connections (see also Matsuyma et al., Chapter 24 of this volume). First, in anatomical studies, we confirmed the results of Kuypers (1958) and Berrevoets and Kuypers (1975) by showing a large projection from the pericruciate cortex of the cat to the PMRF. We extended these results by demonstrating that a substantial proportion ($\sim 25\%$) of this projection was from the motor cortex (Area 4), including those parts which were electrophysiologically identified as representing the forelimb (Matsuyama and Drew, 1997; Rho et al., 1997). These results demonstrated an anatomical substrate for motor commands to effectuate movement by way of RSNs in the PMRF. In electrophysiological experiments, we then demonstrated that a large proportion of the cortical neurons that projected to the caudal level of the pyramidal tract (and thus likely on to the spinal cord), as well as to the PMRF, exhibited phasic increases in their discharge during gait modifications that occurred in the swing phase of the step (Kably and Drew, 1998a,b). This finding provides a mechanism that might facilitate the integration of posture and movement by ensuring that a copy of the descending command for movement is forwarded to one of the brainstem centers most likely to be involved in the control of posture. This signal could be used to adjust both the timing and magnitude of the postural responses that accompany movement.

Contributions of the reticulospinal system to the control of posture during locomotion

A large body of experimental data suggests that the PMRF makes an important contribution to the regulation of muscle tone during standing (Mori, 1987; Mori et al., 1992), reaching movements (Luccarini et al., 1990; Sakamoto et al., 1991) and locomotion (Orlovsky, 1970; Shimamura et al., 1982; Drew et al., 1986). In our own experiments (Prentice and Drew, 2001), we initially examined the discharge of RSNs within the PMRF during the same type of voluntary gait modifications that were used in the above-described studies on the motor cortex. A major finding of Prentice and Drew (2001) was that a majority of the tested RSNs showed increased discharge as the cat stepped over an obstacle. Many RSNs showed *multiple* periods of increased activity (Prentice and Drew, 2001). Figure 2 shows that in some cases, RSNs exhibited an increase in activity as *each limb in turn* stepped over the obstacle. This is in contrast to the previously studied projection neurons in the motor cortex which exhibited a *single* period of increased activity while the cat stepped over an obstacle (using the limb represented in the area of cortex under study).

The complex pattern of RSN activity shown in Fig. 2 was found in 27% of the RSNs that exhibited modified discharge associated with a gait alteration. In many cases, these neurons received convergent and bilateral input from several regions of the motor cortex (Fig. 2). This suggests that the presence of such multiple bursts of RSN activity are indicative of their pattern of convergent cortical input. In addition to this corticoreticular input, RSN discharge reflects input from the cerebellum (Mori et al., 2000) and the limbs (Drew et al., 1996b). Thus, RSNs receive information that could ensure that the resultant RSN-influenced postural modification is integrated appropriately with the current state and posture of the animal.

It is important to realize, however, that modified RSN discharge is unlikely to *specify* a modified postural response *per se*. Consider, for example, the differing postural requirements when the cat steps over the obstacle with the left versus right forelimb. As the left forelimb passes over the obstacle, the animal must modify postural support in the right forelimb. As the right forelimb passes over the obstacle, however, it supports itself on the left forelimb (Lavoie et al., 1995). It therefore seems unlikely that multiple bursts of RSN like that shown in Fig. 2 can specify any one particular postural pattern.

How then is this complex RSN discharge used when the animal produces a movement-appropriate postural pattern? One possibility is that this transformation is the result of a combination of passive and active mechanisms that occur at the spinal level. Many RSNs have axons that project bilaterally to both cervical and lumbar enlargements, where they branch profusely within the spinal gray matter (Peterson et al., 1975; Matsuyama et al., 1988; Matsuyama et al., 1997; Matsuyama et al., this volume). Moreover, results from microstimulation studies in a variety of preparations (Orlovsky, 1972; Drew and Rossignol, 1984; Drew, 1991a; Degtyarenko et al., 1993; Floeter

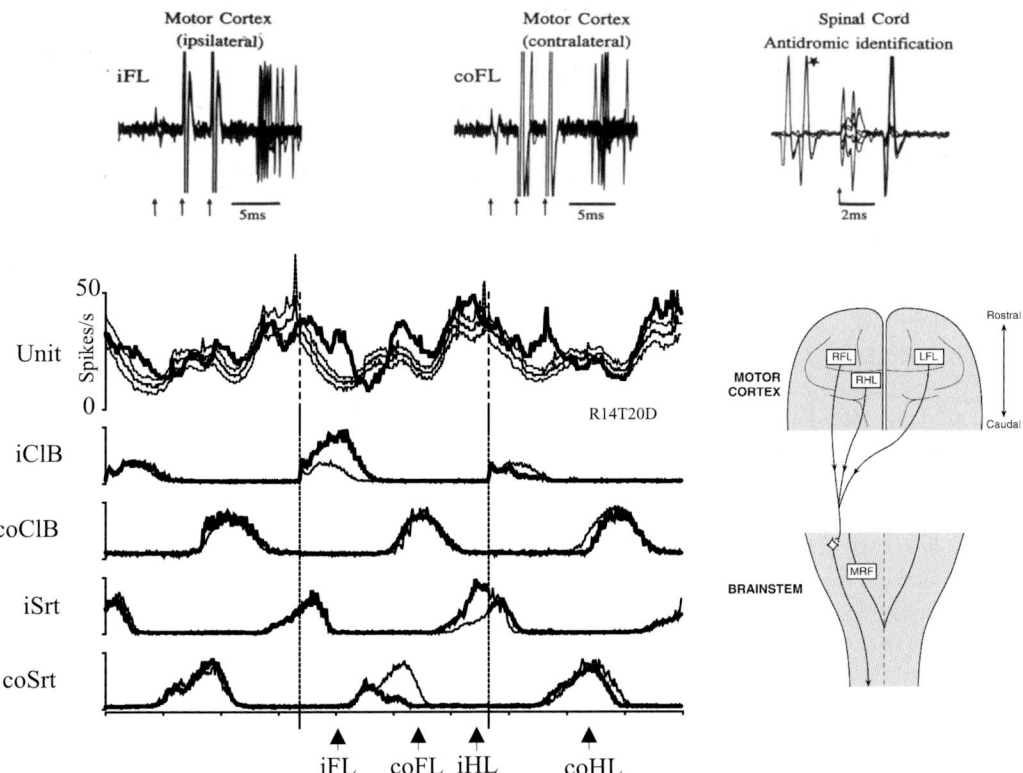

Fig. 2. Example of a RSN that increased its discharge at multiple times during forelimb steps over an obstacle. The cell was activated antidromically (latency, 2.5 ms) from the spinal cord and orthodromically (at 11 ms) from the forelimb representation of the motor cortex of both sides. This cell was also activated from the hindlimb representation of the ipsilateral motor cortex (not illustrated). Thick and thin lines in the average indicate activity during steps over the obstacle and control locomotion, respectively. For the unit recording, the central line indicates the average activity. The other two lines show the confidence level ($P < 0.01$) of the standard error of the mean. Muscles are identified as ipsilateral (i) or contralateral (co) according to their relationship to the recording site in the left PMRF. Arrows at the bottom of the figure indicate the approximate time that each limb passed over the obstacle. The schematic diagram on the bottom right of the figure illustrates the functional connections between the motor cortex and the PMRF that were identified for the illustrated neuron. Other abbreviations: FL, forelimb; HL, hindlimb; Srt, anterior head of sartorius.

et al., 1993; Perreault et al., 1994) suggest that RSNs transiently modify activity in both flexor and extensor muscles according to the phase of the step cycle in which the stimulation, or the natural descending volley, occurs. It is therefore possible that multiple periods of RSN activity are transmitted, in a phase-dependent manner, to those MNs that are in an excitable state at the time of a required gait modification. In such case, a descending RSN volley will tend to reinforce activity in muscles that are already active. For example, increased RSN discharge during passage of the left forelimb over an obstacle would tend to facilitate flexor activity in the left limb and extensor activity in the right limb. In contrast, as the right limb passes over the obstacle, increased activity in the same neuron would tend to facilitate activity in flexor muscles of the right limb and extensor muscles in the left limb. This possibility is illustrated schematically in Fig. 3 which proposes a mechanism by which RSNs discharging with multiple bursts of activity and having widely branching axons could produce appropriate modifications of postural activity. It is a sophisticated mechanism because it produces appropriate postural patterns, regardless of the speed and gait of the animal (which both change interlimb coordination), and regardless of the animal's overall posture (e.g., when walking on an incline).

We suggest that this phase-dependent distribution of the descending RSN signal provides a mechanism by which more general aspects of postural reorganization can be determined. Superimposed upon this general pattern, however, we suggest that there will be more specific descending signals that ensure that the general pattern is appropriately adapted to the specific demands of the task. Such adaptive modifications could be provided, for example, by other RSNs with more specific termination patterns within the spinal cord. This would allow for differential modifications of muscles in any *single* limb. In addition, we have suggested (Prentice and Drew, 2001) that the final expression of postural responses may also be contingent upon the activity in other pathways transmitting the descending command for the movement, such as the corticospinal pathway. In such a case, there would be modifications in the level of limb-muscle EMG activity only if the spinal IN networks receive *convergent*, and *augmented*, signals from both the corticospinal and the reticulospinal tracts.

Contributions of the reticulospinal system to the control of reaching

We have also examined the contribution of the reticulospinal system to the control of reaching (Schepens and Drew, 2000). Cats were trained to stand quietly on four, separate force platforms and to make a reaching movement of either the left or right forelimb in order to retrieve a food reward. We recorded the discharge of PMRF neurons, including identified RSNs, during both locomotion and these reaching movements. The discharge activity of many neurons was modulated during *both* behaviors. A major finding was that most reticular neurons exhibited qualitatively similar, modulated discharge during reaches of both the left and right forelimb. An example is shown in Fig. 4.

Figure 4 shows that during reaching movements of the left limb (ipsilateral to the recording site in the PMRF), the neuron exhibited a pronounced increase in its discharge rate, which preceded the onset of activity in this limb's flexor muscles. A slightly elevated firing rate was maintained throughout the time that the reaching limb was off the ground, with postural support distributed among the supporting limbs. During reach with the other (right) limb, the cell also showed a period of increased activity which was relatively, albeit not precisely, similar to that recorded during reaching movements with the left limb.

The pattern of activity illustrated in Fig. 4 is reminiscent of that we observed previously in a population of RSNs studied during locomotion. As during locomotion, postural requirements during reaching are different during left versus right reaches. Indeed, the analysis of ground reaction forces and limb-muscle EMG activity during the reaching task revealed that the postural modifications were completely reciprocal during left and right reaches. Thus, as during locomotion, the increased reticular neuron activity during the reach cannot specify the details of the postural pattern. Rather, it presumably provides a more general signal that specifies the timing and the magnitude of the requisite postural activity. Given the similarity of the constraints in the

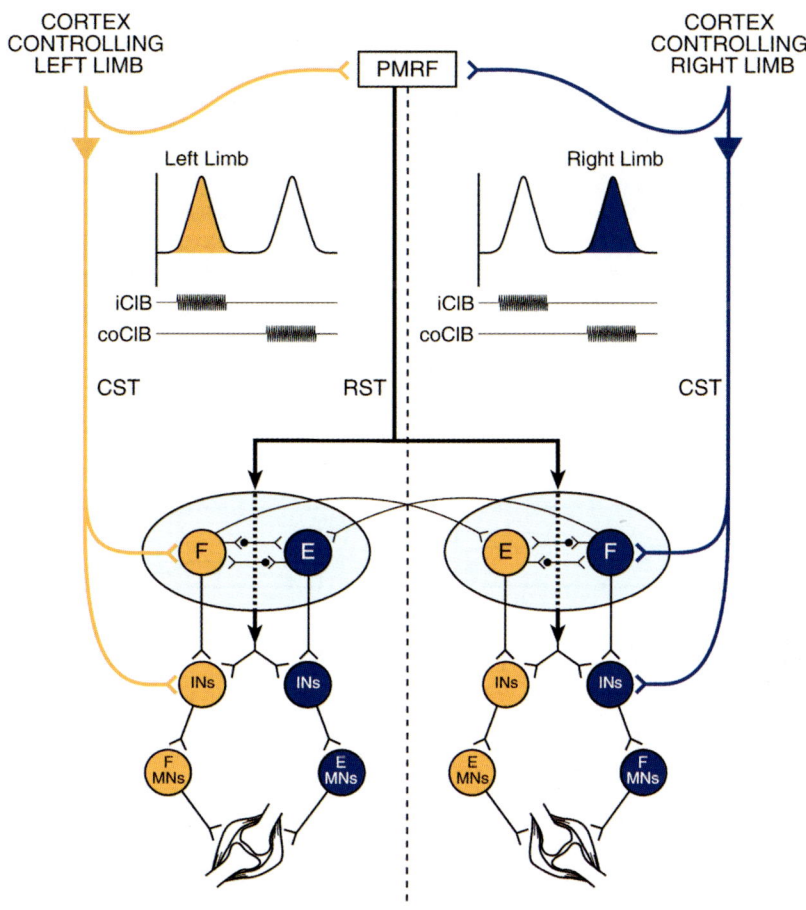

Fig. 3. Conceptual model showing how the PMRF may contribute to the control of posture during locomotion. This model assumes that a population of RSNs in the PMRF receives input from the forelimb representation of both the ipsilateral and contralateral motor cortex. This results in a double burst of activity in these cells. These RSNs are assumed to branch in the cervical spinal cord to contact IN pathways (Ins) controlling flexor (F) and extensor (E) muscles of each limb. These INs may be influenced by, or form part of, the locomotor CPG (represented by the gray oval). During the passage of the left forelimb over the obstacle, the first burst of activity (in orange) has access to spinal IN pathways (in orange) because of the intrinsic rhythmicity of the spinal cord, as well as convergent input from other descending pathways. These RSNs will therefore facilitate left-side flexor muscles and right-side extensors. During passage of the right forelimb (blue), the pathways open for transmission are modified such that the same population of RSNs now facilitates right-side flexors and left-side extensors. Other abbreviations: MNs, motoneurons; CST, corticospinal tract; RST, reticulospinal tract.

locomotor and reaching tasks, we propose that the same neural circuits are used in both activities. This suggestion is an extension of the original proposal of Georgopoulos and Grillner (1989). We further extend this proposal by suggesting that the spinal mechanisms that ensure reciprocity of postural responses within and between limbs during locomotion are involved in the production of the postural responses that occur during reaching. During locomotion, it is well established that intrinsic spinal mechanisms ensure an alternation in limb activity (Grillner and Wallén, Chapter 1 of this volume). During reaching, we now propose that the descending command for movement will activate these same circuits, but without also initiating locomotor rhythmicity.

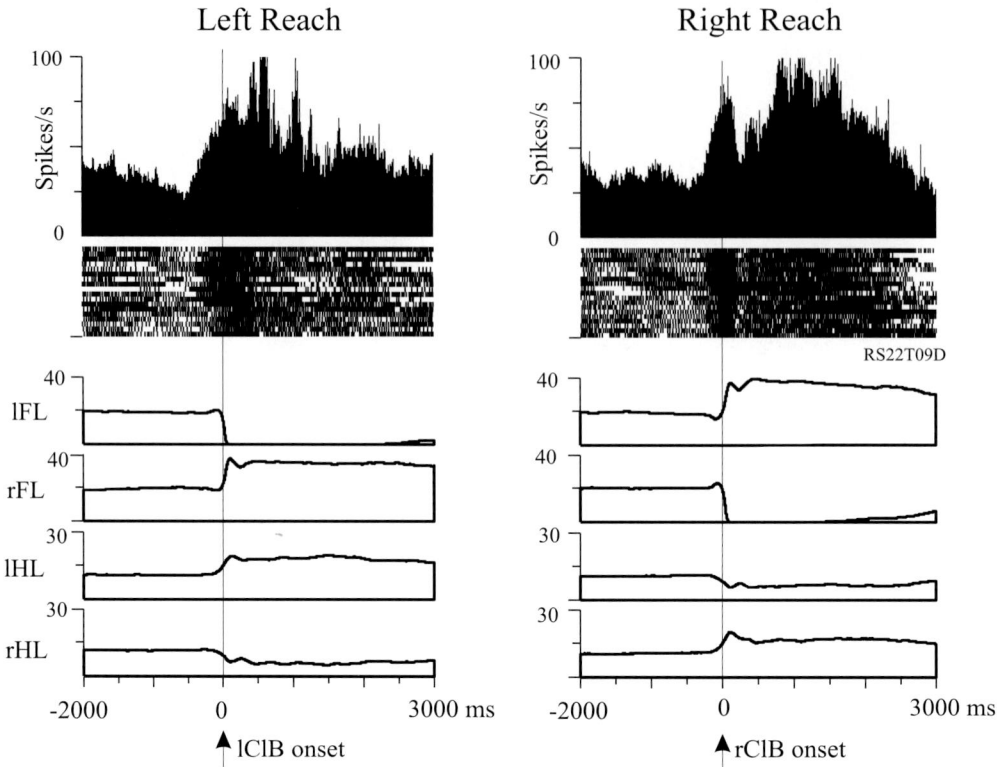

Fig. 4. Reaching-involved activity of an RSN. This cell exhibited increased activity during reaching movements of both the left (ipsilateral to the recording site) and right forelimb. This increased level of activity was maintained throughout the time that the reaching limb was off the ground. Each side of the figure shows, from top to bottom, the average activity of the RSN, a raster display of activity during each reach, and the vertical ground reaction forces (Newtons) measured under each of the four limbs of the standing cat. Note that although the discharge of the RSN was qualitatively similar for left- and right-limb reaches, the postural pattern of activity, upon which the movement was imposed, was reciprocal for both the fore- and hindlimbs. The data are synchronized to the onset of activity in ClB, which became active just before the paw was lifted from the support surface.

Conclusions

In this chapter, we have presented some concepts concerning the control of movement and posture during locomotion and reaching which have emerged from work in our and others' laboratories. Several points are noteworthy. First, the motor cortex specifies the changes in EMG activity required to produce appropriate modifications of limb trajectory during voluntary gait modifications. This descending signal is integrated with different subpopulations of spinal INs to ensure that gait modification is smoothly superimposed onto the basic locomotor rhythm. Second, the motor cortex also plays a role in regulating the activity of subcortical structures, which have their own specific functions in the control of posture and movement. Third, we have shown how corticoreticular projections might provide a signal to RSNs in the PMRF such that the latter cells can affect appropriate postural support to movement. Finally, we suggest that the corticospinal tract may also modulate the transmission of activity in spinal IN pathways that receive input from RSNs. This would provide a mechanism by which the descending command for a movement can also influence the postural responses that accompany that movement.

A major implication of these results is that the discharge of individual RSNs may not encode a signal that specifies the details of a required postural modification, but, rather, a command concerning the timing and magnitude of the modification.

We further suggest that the expression of an appropriate postural pattern is a function of the connectivity of RSNs within the spinal cord, the excitability of the IN spinal circuits upon which they impinge, and the presence or absence of a contingent signal from the corticospinal tract, or elsewhere. This overall organization provides sufficient flexibility of action to allow a relatively limited number of neurons to provide an infinite number of postural responses.

To conclude, the results speak to the issue of whether specific neural circuits are used for different behaviors, or whether common neuronal elements are activated for a large number of quite diverse behaviors. Individual RSNs discharge during (1) unperturbed locomotion, (2) voluntary gait modifications of locomotion, and (3) reaching movements of both the left and right forelimbs. This versatility suggests that neural circuits are shared at the level of both the brainstem and spinal cord. We further propose, as a general principle, that most neurons in structures involved in regulating motor activity during locomotion are similarly involved during reaching movements.

Acknowledgments

We would like to acknowledge the technical assistance of Natacha Da Silva, Marc Bourdeau, Philippe Drapeau and Jacques Bérichon. We thank M. Claude Gauthier for the illustration in Figure 3 and Drs. Elaine Chapman and Serge Rossignol for their comments on a draft of this chapter. (Supported by the Canadian Institutes of Health Research. Bénédicte Schepens was supported by the Human Frontier Science Program.)

Abbreviations

Br	brachialis
ClB	cleidobrachialis
CNS	central nervous system
CPG	central pattern generator
CST	corticospinal tract
E	extensor
ECR	extensor carpi radialis
EDC	extensor digitorum communis
EMG	electromyogram
F	flexor
FL	forelimb
HL	hindlimb
IN	interneuron
MN	motoneuron
PMRF	pontomedullary reticular formation
PTNs	pyramidal tract neurons
RSNs	reticulospinal neurons
RST	reticulospinal tract
Srt	sartorius
TrM	teres major

References

Amos, A., Armstrong, D.M. and Marple-Horvat, D.E. (1990). Changes in the discharge patterns of motor cortical neurones associated with volitional changes in stepping in the cat. Neurosci. Lett., 109: 107–112.

Armstrong, D.M. (1988). The supraspinal control of mammalian locomotion. J. Physiol. (Lond.), 405: 1–37.

Armstrong, D.M. and Drew, T. (1984a). Discharges of pyramidal tract and other motor cortical neurones during locomotion in the cat. J. Physiol. (Lond.), 346: 471–495.

Armstrong, D.M. and Drew, T. (1984b). Locomotor-related neuronal discharges in cat motor cortex compared with peripheral receptive fields and evoked movements. J. Physiol. (Lond.), 346: 497–517.

Armstrong, D.M. and Drew, T. (1985). Forelimb electromyographic responses to motor cortex stimulation during locomotion in the cat. J. Physiol. (Lond.), 367: 327–351.

Beloozerova, I.N. and Sirota, M.G. (1993). The role of the motor cortex in the control of accuracy of locomotor movements in the cat. J. Physiol. (Lond.), 461: 1–25.

Bennett, K.M.B. and Lemon, R.N. (1996). Corticomotoneuronal contribution to the fractionation of muscle activity during precision grip in the monkey. J. Neurophysiol., 75: 1826–1842.

Berrevoets, C.E. and Kuypers, H.G.J.M. (1975). Pericruciate cortical neurons projecting to brain stem reticular formation, dorsal column nuclei and spinal cord in the cat. Neurosci. Lett., 1: 257–262.

Degtyarenko, A.M., Zavadskaya, T.V. and Baev, K.V. (1993). Mechanisms of supraspinal correction of locomotor activity generator. Neuroscience, 52: 323–332.

Drew, T. (1988). Motor cortical cell discharge during voluntary gait modification. Brain Res., 457: 181–187.

Drew, T. (1991a). Functional organization within the medullary reticular formation of the intact unanaesthetized cat. III. Microstimulation during locomotion. J. Neurophysiol., 66: 919–938.

Drew, T. (1991b). Visuomotor coordination in locomotion. Curr. Opin. Neurobiol., 1: 652–657.

Drew, T. (1993). Motor cortical activity during voluntary gait modifications in the cat. I. Cells related to the forelimbs. J. Neurophysiol., 70: 179–199.

Drew, T. and Rossignol, S. (1984). Phase dependent responses evoked in limb muscles by stimulation of the medullary reticular formation during locomotion in thalamic cats. J. Neurophysiol., 52: 653–675.

Drew, T., Dubuc, R. and Rossignol, S. (1986). Discharge patterns of reticulospinal and other reticular neurons in chronic, unrestrained cats walking on a treadmill. J. Neurophysiol., 55: 375–401.

Drew, T., Jiang, W., Kably, B. and Lavoie, S. (1996a). Role of the motor cortex in the control of visually triggered gait modifications. Can. J. Physiol. Pharmacol., 74: 426–442.

Drew, T., Cabana, T. and Rossignol, S. (1996b). Responses of medullary reticulospinal neurones to stimulation of cutaneous limb nerves during locomotion in intact cats. Exp. Brain Res., 111: 153–168.

Floeter, M.K., Sholomenko, G.N., Gossard, J.-P. and Burke, R.E. (1993). Disynaptic excitation from the medial longitudinal fasciculus to lumbosacral motoneurons: modulation by repetitive activation, descending pathways, and locomotion. Exp. Brain Res., 92: 407–419.

Futami, T., Shinoda, Y. and Yokota, J. (1979). Spinal axon collaterals of corticospinal neurons identified by intracellular injection of horseradish peroxide. Brain Res., 164: 279–284.

Georgopoulos, A.P. and Grillner, S. (1989). Visuomotor coordination in reaching and locomotion. Science, 245: 1209–1210.

Grillner, S. (1981). Control of locomotion in bipeds, tetrapods and fish. In: Brooks V.B. (Ed.), Handbook of Physiology, Vol. II, American Physiological Society, Bethesda, pp. 1179–1236 Part 2.

Grillner, S. (1982). Possible analogies in the control of innate motor acts and the production of sound in speech. In: Grillner S. (Ed.), Speech Motor Control. Pergamon Press, Oxford, pp. 217–229.

Horak, F.B. and Macpherson, J. (1996). Postural orientation and equilibrium. In: Rowell L.B. and Shepherd J.T. (Eds.), Exercise: Regulation and Integration of Multiple Systems. American Physiological Society, New York, pp. 255–292.

Kably, B. and Drew, T. (1998a). The corticoreticular pathway in the cat: I. Projection patterns and collaterization. J. Neurophysiol., 80: 389–405.

Kably, B. and Drew, T. (1998b). The corticoreticular pathway in the cat: II. Discharge characteristics of neurons in area 4 during voluntary gait modifications. J. Neurophysiol., 80: 406–424.

Kuypers, H.G.J.M. (1958). Anatomical analysis of corticobulbar connexions to the pons and the lower brain stem in the cat. J. Anat., 92: 198–218.

Kuypers, H.G.J.M. (1963). The organization of the 'motor system'. Int. J. Neurol., 4: 78–91.

Lavoie, S. and Drew, T. (2002). Discharge characteristics of neurons in the red nucleus during voluntary gait modifications: a comparison with the motor cortex. J. Neurophysiol., 88: 1791–1814.

Lavoie, S., McFadyen, B. and Drew, T. (1995). A kinematic and kinetic analysis of locomotion during voluntary gait modification in the cat. Exp. Brain Res., 106: 39–56.

Lawrence, D.G. and Kuypers, H.G.J.M. (1968a). The functional organization of the motor system in the monkey. I. The effects of bilateral pyramidal lesions. Brain, 91: 1–15.

Lawrence, D.G. and Kuypers, H.G.J.M. (1968b). The functional organization of the motor system in the monkey. II. The effects of lesions of the descending brain-stem pathways. Brain, 91: 15–36.

Liddell, E.G.T. and Phillips, C.G. (1944). Pyramidal section in the cat. Brain, 67: 1–9.

Luccarini, P., Gahery, Y. and Pompeiano, O. (1990). Cholinoceptive pontine reticular structures modify the postural adjustments during the limb movements induced by cortical stimulation. Arch. Ital. Biol., 128: 19–45.

Massion, J. (1992). Movement, posture and equilibrium: interaction and coordination. Prog. Neurobiol., 38: 35–56.

Matsuyama, K. and Drew, T. (1997). The organization of the projections from the pericruciate cortex to the pontomedullary brainstem of the cat: a study using the anterograde tracer, Phaseolus vulgaris leucoagglutinin. J. Comp. Neurol., 389: 617–641.

Matsuyama, K., Takakusaki, K., Nakajima, K. and Mori, S. (1997). Multi-segmental innervation of single pontine reticulospinal axons in the cervico-thoracic region of the cat: anterograde PHA-L tracing study. J. Comp. Neurol. 377: 234–250.

Matsuyama, K., Ohta, Y. and Mori, S. (1998). Ascending and descending projections of the nucleus reticularis gigantocellularis in the cat demonstrated by the anterograde neural tracer, Phaseolus vulgaris leucoaggulation (PHA-L). Brain Res., 460: 124–141.

Mori, S. (1987). Integration of posture and locomotion in acute decerebrate cats and in awake, freely moving cats. Prog. Neurobiol., 28: 161–195.

Mori, S., Matsuyama, K., Kohyama, J., Kobayashi, Y. and Takakusaki, K. (1992). Neuronal constituents of postural and locomotor control systems and their interactions in cats. Brain Dev., Suppl. 14: S109–S120.

Mori, S., Matsui, T., Mori, F., Nakajima, K. and Matsuyama, K. (2000). Instigation and control of treadmill locomotion in high decerebrate cats by stimulation of the hook bundle of Russell in the cerebellum. Can. J. Physiol. Pharmacol., 78: 945–957.

Orlovsky, G.N. (1970). Work of the reticulo-spinal neurons during locomotion. Biophysics (USSR), 15: 761–771.

Orlovsky, G.N. (1972). The effect of different descending systems on flexor and extensor activity during locomotion. Brain Res., 40: 359–371.

Perreault, M.-C., Rossignol, S. and Drew, T. (1994). Microstimulation of the medullary reticular formation during fictive locomotion. J. Neurophysiol., 71: 229–245.

Peterson, B.W., Maunz, R.A., Pitts, N.G. and Mackel, R.G. (1975). Patterns of projection and branching of reticulospinal neurons. Exp. Brain Res., 23: 333–351.

Prentice, S.D. and Drew, T. (2001). Contributions of the reticulospinal system to the postural adjustments occurring during voluntary gait modifications. J. Neurophysiol., 85: 679–698.

Rho, M.-J., Cabana, T. and Drew, T. (1997). The organization of the projections from the pericruciate cortex to the pontomedullary reticular formation of the cat: a quantitative retrograde tracing study. J. Comp. Neurol., 388: 228–249.

Rho, M.-J., Lavoie, S. and Drew, T. (1999). Effects of red nucleus microstimulation on the locomotor pattern and timing in the intact cat: a comparison with the motor cortex. J. Neurophysiol., 81: 2297–2315.

Sakamoto, T., Gahery, Y. and Mori, S. (1991). Effects of bethanecol injection into pontine reticular formation upon postural changes accompanying a food retrieval task by a forelimb in a standing cat. In: Shimamura M., Grillner S. and Edgerton V.R. (Eds.), Neurobiological Basis of Human Locomotion. Japan Scientific Societies Press, Tokyo, pp. 45–50.

Schepens, B. and Drew, T. (2000). Reticulospinal neuronal activity during a reaching task in intact cats. Soc. Neurosci. Abstr., 26: 460.7.

Shimamura, M., Kogure, I. and Wada, S.I. (1982). Reticular neuron activities associated with locomotion in thalamic cats. Brain Res., 231: 51–62.

CHAPTER 26

Direct and indirect pathways for corticospinal control of upper limb motoneurons in the primate

Roger N. Lemon[1,]*, Peter A. Kirkwood[1], Marc A. Maier[2], Katsumi Nakajima[3] and Peter Nathan[4,†]

[1]*Sobell Department of Motor Neuroscience and Movement Disorders, Institute of Neurology, University College London, Queen Square, London WC1N 3BG, UK*
[2]*INSERM U483, Université Pierre et Marie Curie, Paris 75005, France*
[3]*Department of Physiology, Kinki University School of Medicine, Osaka-Sayama 589-8511, Japan*
[4]*The National Hospital for Neurology and Neurosurgery, Queen Square, London WC1N 3BG, UK*

Abstract: In the macaque monkey and in humans, the monosynaptic cortico-motoneuronal system is well developed. It allows the cortical motor areas to make an important direct contribution to the pattern of muscle activity during upper limb movements. There is, in addition, good anatomical evidence for descending corticospinal inputs being able to influence the premotoneuronal networks of the cervical spinal cord, and especially those operating at the segmental level of upper limb motoneurons. While oligosynaptic inhibition has been easy to demonstrate in the macaque, and may be a very important component of descending corticospinal control, it has proved much more difficult to detect signs of oligosynaptic excitation. In contrast, in the squirrel monkey, in which the cortico-motoneuronal system is far less developed, oligosynaptic excitation is prominent. There are important changes in the interplay between direct and indirect pathways in different primates, which may provide important clues on the nature of the corticospinal control of upper limb function.

Introduction

There is increasing interest in the human motor control literature in the mechanisms that translate cortical activity into purposeful movements. The huge increase in our knowledge of which cortical areas are active during particular types of movement, and which areas are involved in the planning and coordination of movement, has not been paralleled with a better understanding of what all this supraspinal activity means to the spinal machinery for movement. We do know that in humans and other primate species there has been a massive increase in the relative size of the cerebral cortex (Barton and Harvey, 2000). There has also been a proportional increase in the amount of neocortex giving rise to the corticospinal tract (CST) (Nudo and Masterton, 1990), together with the appearance of a particularly fast component in the latter. This component is relatively small in terms of its number of axons, but it appears to exert a pronounced influence over spinal mechanisms (Porter and Lemon, 1993). Furthermore, we know that cortical damage is generally more devastating in humans than in animal models (Passingham, 1993; Porter and Lemon, 1993; Creutzfeldt, 1995). There is obviously a great need to understand how the motor areas of the cerebral cortex interact with

*Corresponding author: Tel.: +020-7837-3611 (ext. 4184); Fax: +020-7419-7170; E-mail: rlemon@ion.ucl.ac.uk
†In memoriam (1914–2002).

spinal mechanisms (e.g., Drew et al., Rouiller et al., Weber and He, Chapters 25, 44 and 45 of this volume).

Premotoneuronal control

Fundamental to our understanding of the operation of the spinal machinery for movement is the concept that different populations of interneurons (INs) are intercalated between central and peripheral sources of control. Networks of INs are known to be able to generate many of the basic motor patterns, such as locomotion, scratching, etc. (Orlovsky et al., 1999). Recent evidence suggest that they may represent motor 'primitives' that can be assembled into the most complex of motor acts (Bizzi et al., 2000).

These INs provide an integrative network that allows for control at a premotoneuronal level, without involvement of the spinal MN, itself, the 'final common path' of Sherrington. All the information that may be important to achieve the precise timing and balance of activity in the MN pool can be integrated and filtered at a premotoneuronal level. This information includes that originating in a variety of descending motor pathways, ascending and descending propriospinal pathways, local segmental INs, as well as from peripheral inputs in primary sensory afferents. Several authorities have emphasized the potential benefits of this type of organization, which allows for a very considerable flexibility of control. There is also the possibility of a rapid updating of centrally driven movements by late changes in peripheral input (Baldissera et al., 1981; Georgopoulos and Grillner, 1989; Alstermark and Lundberg, 1992; Pierrot-Deseilligny, 1996; Jankowska, 2001; Prut et al., 2001).

Extensive nature of the premotoneuronal network

Interneuron and propriospinal networks are very extensive. Molenaar and Kuypers (1978) observed that large injections of HRP into the spinal cord at one level retrogradely labeled neurons over many different segments. Long propriospinal projections, for example linking the lumbar and cervical enlargements, arose mainly from the medial part of the spinal gray (Rexed's lamina VIII), while short propriospinal systems, providing linkage over a few local segments, arose mainly from the lateral parts of laminae V and VI, and lamina VII. More recent investigations, using transneuronal labeling techniques (Alstermark and Kummel, 1990a,b; Ugolini, 1992), have amply confirmed this, thereby showing that even a single upper limb muscle is controlled by an extensive group of local segmental INs (located in the same segments as the muscle's MNs) and PNs (propriospinal neurons) located in more rostral segments of the cervical spinal cord. Alstermark and Kummel (1990a) demonstrated that the labeling of the network can occur in an activity-dependent fashion. These experiments demonstrated that local segmental INs far outnumber those in the more rostral segments, and that PNs in the cervical cord are concentrated in the C3 and C4 segments, but are not restricted to these segments.

Corticospinal inputs to the premotoneuronal network

Descending corticospinal inputs may access this premotoneuronal network at multiple levels. In their classic study, Kuypers and Brinkman (1970) made very small lesions within the hand area of macaque primary motor cortex. They noted that although degeneration was most pronounced in the lower cervical segments, terminations could be found throughout the cervical cord, in both the dorsolateral and ventromedial regions of the intermediate zone. In a more recent study, we made a small, focal injection of the anterograde tracer WGA–HRP into the macaque M1 hand area (Maier et al., 2002). While the bulk of the corticospinal projections were focused in the contralateral C8-Th1 grey matter, all parts of the intermediate zone were labeled from C1 to C7. From C1 to C4, this labeling was particularly dense in the medial and lateral parts of laminae V and VI, but almost absent in their central parts. Within a given segment, there are corticospinal projections, albeit of different density, to most spinal laminae (Porter and Lemon, 1993, pp. 83–88; Dum and Strick 1996; Armand et al., 1997).

Thus, in attempting to understand how cortical commands are integrated into this IN network, we must first recognize that such integration can occur over many spinal segments. Because of the distributed nature of the IN network, descending activity involved in the control of a small group of synergistic muscles might act at many different spinal levels. Within a single segment, the projection can include a wide range of different targets distributed over many laminae.

Descending pathways with monosynaptic connections to MNs

Given the operational advantages of integrating descending commands into the premotoneuronal network, it is interesting that several descending pathways have been shown to produce monosynaptic actions on MNs, including fibers running in the rubro-, reticulo-, vestibulo- and corticospinal tracts. "... signals conveyed by the direct route bypass the integration with peripheral afferent information and intrinsic spinal activities before they reach the motoneurons" (Baldissera et al., 1981, p. 568). In other words, these monosynaptic connections allow a direct contribution to MN control. Interestingly, there is evidence that presynaptic inhibition of the terminals of these direct projections is absent. This further emphasizes that these direct connections have a *direct* impact on motor output, and suggests that their signals are transmitted to the 'final common path' without further modification (Nielsen and Petersen, 1994; Anissimova et al., 1998).

One important source of monosynaptic inputs comes from the cortico-motoneuronal (CM) component of the CST, which provides a direct pathway from cortex to muscle (Fig. 1). The discovery of the CM pathway in the early 1950s prompted three main lines of enquiry: first, *what could the function of such monosynaptic projections be?* Second, *how would their actions work alongside premotoneuronal control?* Finally, since it was soon discovered that the CM system was developed to different extents across species (Kuypers, 1981), *were there key differences in corticospinal function across species?*

The relative importance of cortico-motoneuronal effects

In their 1981 review, Baldissera, Illert and Hultborn pointed out that direct excitatory inputs to MNs

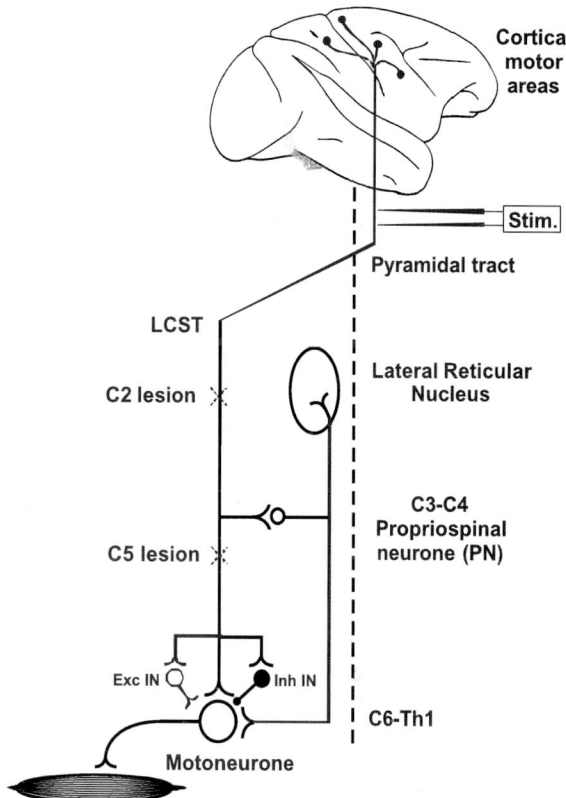

Fig. 1. Schematic diagram of pathways transmitting corticospinal actions to cervical MNs in the primate. Corticospinal projections originating in different cortical motor areas cross the midline (dashed line) and project to the cervical enlargement (segments C6-Th1) where they make direct monosynaptic connections with MNs and with segmental inhibitory interneurons (Inh IN) and excitatory INs (Exc IN). Corticospinal collaterals may also influence MNs via C3–C4 PNs. These PNs project monosynaptically onto MNs and also have ascending axon collateral to the LRN in the brainstem. The corticospinal system was activated via stimulating electrodes implanted in the medullary pyramidal tract (Stim.). Effects on MNs were assessed by recording the discharge of SMUs from a variety of upper limb muscles or from intracellular recordings from alpha-MNs. C3–C4 PN effects were identified by making a lesion in the dorsolateral funiculus at the C5 level to interrupt the fibers of the LCST, but leaving intact the more ventrally located PN axons. These effects can be further demonstrated to be of propriospinal, rather than brainstem, origin by showing that a subsequent C2 lesion to the LCST abolishes them.

from descending pathways were characterized by having small amplitude EPSPs. Activation of the entire monosynaptic input from a given pathway might produce compound EPSPs in MNs of ~ 2 mV.

For the CM system, the mean values of the compound EPSPs evoked in most upper limb MNs by stimulation of the fast-conducting component of the CST were of this order (Porter and Lemon, 1993). This would mean that such inputs could not, in and of themselves, bring a MN to discharge threshold. Indeed, stimulation of the corticospinal system in an awake animal or human subject may not produce overt contraction unless there is a background of voluntary activation of the muscle (Hess et al., 1987). Thus, it seems likely that these direct inputs work in parallel with the premotoneuronal control system. Possibly, they add the final spatio-temporal patterns of excitation needed for appropriate levels of MN recruitment and discharge.

In the case of the hand and finger muscles, however, CM effects on MNs are larger. For example, Porter and Lemon (1993) reported EPSPs of up to 7.5 mV. In the conscious macaque, it is possible to evoke clear twitches in relaxed hand muscles with single stimuli delivered to the pyramidal tract. Calculations made by Cheney et al. (1991) suggested that the total CM input to MNs supplying wrist extensor muscles could provide up to 60% of the facilitatory drive needed to maintain steady discharge of these MNs. In humans, transcranial electrical stimulation (TES) delivered through the scalp and activating corticospinal axons in the subcortical white matter evokes monosynaptic responses in single motor units (SMUs) in contralateral upper limb muscles. It is possible to estimate the size of the EPSPs underlying these responses as being at least 4–5 mV in amplitude for hand muscles (de Noordhout et al., 1999). This value is probably an underestimate, since it is technically difficult to measure unit responses at the intensities of TES needed to recruit the entire 'colony' of CM neurons projecting to the unit. In summary, while it is clear that CM actions must work in parallel with premotoneuronal control, their contribution can be considerable.

Possible functions mediated by CM connections with spinal MNs

The CM connection was first identified electrophysiologically by the Swedish group of Bernhard, Bohm and Petersen in 1953. In their seminal review, Bernhard and Bohm (1954) suggested that CM connections might be of particular importance for control of fine digit movement, but insisted that this might also involve more proximal muscles. "... The difference in the movement patterns between an animal with paws, e.g., the cat, and an animal with hands, e.g., the monkey, concerns not only the activity of the muscles serving the digits but, more or less, that of all the muscles of the extremities." (Bernhard and Bohm, 1954, p. 495).

Although much emphasis has been placed on a role of CM control of digit movement (Lemon, 1993; Bortoff and Strick, 1993), one should recall that Bernhard et al. (1953) showed that monosynaptic CM responses could be recorded from *both* upper and lower limb nerves. Thus, although CM projections to hand muscles may be larger than to other muscle groups, there is plentiful evidence that many upper and lower limb muscles receive CM projections (Jankowska et al., 1975; Rothwell et al., 1991; Palmer and Ashby, 1992; Nielsen et al., 1995).

Kuypers stressed a more general function for the CM projection. He suggested that it provided the capacity for 'fractionation' of movements and the control of small groups of muscles in a highly selective manner (Kuypers, 1978). This is reminiscent of the 'analytical' type of motor control described by Charles Sherrington, which he thought was necessary to selectively activate different muscles in space and time. Thus, CM control of foot muscles, for example when locomoting over an uneven, difficult terrain, might make use of the same basic operational control seen in the hand during skilled manipulation. The breaking down of relatively hardwired motor patterns can be regarded as an important feature of voluntary motor actions, and one that is probably essential to the acquisition of new motor skills (Wolpert et al., 2001).

Monosynaptic and oligosynaptic effects on spinal MNs from the CST in the macaque monkey

There are two main ways in which these effects have been detected. The first is to use the classical approach of stimulating the PT or motor cortex and searching for postsynaptic effects in intracellular recordings made in identified MNs. The main preparation that we have used for this is the chloralose-anesthetized macaque monkey, in which a high degree of excitability is maintained within the motor pathways. A second approach has been spike- or stimulus-triggered averages of muscle activity (SMUs or whole-muscle EMG) in the awake or lightly sedated monkey. The latter method is less invasive and it avoids the use of anesthetic agents, which may selectively depress or enhance transmission in particular pathways.

Monosynaptic excitation

Most upper limb MNs show a clear CM-EPSP response to stimulation of the PT (Porter and Lemon, 1993). This was the case for 76% of identified MNs reported by Maier et al. (1998). These EPSPs are characterized by very brief segmental delays (<1.1 ms) between arrival of the corticospinal volley at the spinal segment and onset of the EPSP. Many studies have confirmed that the fast-rising, postspike facilitation of upper-limb-muscle EMG activity is probably of CM origin (Porter and Lemon, 1993; Baker and Lemon, 1998). Olivier et al. (2001) recently reported responses of SMUs recorded from arm, wrist and hand muscles to PT stimulation in the awake or sedated macaque. All units showed constant, short-latency responses with a short duration consistent with CM excitation (mean duration of peak in PSTH 0.74 ± 0.25 ms).

Oligosynaptic excitation

In order to evoke oligosynaptic effects in spinal MNs, it is often necessary to use repetitive activation of the PT. This ensures that INs, which are intercalated between the PT and the target MN, are brought to discharge. Signs of oligosynaptic excitation from the cortex or pyramidal tract were largely lacking in early recordings from primate MNs (Phillips and Porter, 1964; Shapovalov, 1975; Fritz et al., 1985). Interactions between cortical stimulation and H-reflex inputs, however, did produce some longer-latency facilitation (Preston et al., 1967). Maier et al. (1998) focused attention on oligosynaptic effects and found them to be relatively uncommon: 19% of their sample showed 'late EPSPs', with segmental latencies beyond the monosynaptic range (>1.3 ms). These effects were evoked by repetitive but not single PT stimuli. Only 3% of their tested MNs showed effects within the disynaptic range. The origin of such EPSPs is unknown. They could arise by activation of segmental excitatory INs of the kind described by Perlmutter et al. (1998), or through propriospinal mechanisms (see below).

Olivier et al. (2001) examined the effect of PT stimulation on the discharge of rhythmically firing upper limb SMUs in monkeys that were either awake and performing a precision grip task, or lightly sedated with ketamine. These experiments were designed to enhance the possibility of finding oligosynaptic transmission to these SMUs. Weak PT stimuli were used, so that CM EPSPs generated from the PT were subthreshold for discharging the SMU on most trials. These EPSPs could then summate with any late EPSPs and might reach threshold. In addition, repetitive PT shocks were used to enhance oligosynaptic transmission. Despite this, none of the PSTHs of SMU activity showed any evidence of later peaks consistent with oligosynaptic action. In many, but by no means all, cases this may have been because of the powerful suppression which immediately followed the early excitation (see below).

Inhibition

There is no evidence for CM-evoked monosynaptic inhibition. Disynaptic inhibition is widespread. It is sometimes found alone (10% of MNs; Maier et al., 1998), but, more commonly, it follows immediately after the CM EPSP. Indeed, the segmental latencies of these IPSPs (1.3–1.9 ms) is such that they probably truncate the final phase of CM EPSPs, which typically have segmental latencies of 0.7–0.9 ms, and rise times of ~1 ms or more (Maier et al., 1997). Importantly, this means that the amplitudes of these CM EPSPs are underestimated. Disynaptic IPSPs can be strongly enhanced by repetitive PT stimulation.

These disynaptic inhibitory effects are much reduced by C5 CST lesions (see the next section), thereby suggesting that an important contribution comes from segmental inhibitory INs (cf. Jankowska et al., 1976; Perlmutter et al., 1998; Prut et al., 2001). This would be in keeping with the strong corticospinal projections to the regions of the intermediate zone in which these INs are located (Armand et al., 1997; Maier et al., 2002). The ubiquity of this type of inhibition testifies to the oliogosynaptic actions of the CST. It also speaks to the preservation of transmission through such pathways in the chloralose-anesthetized monkey. Short-latency suppression of EMG or SMU firing in spike-triggered averages from CM cells (Kasser and Cheney, 1985) and from cortical (Cheney et al., 1985; Lemon et al., 1987) and PT stimulation (Olivier et al., 2001) is also common, and is consistent with a disynaptic IPSP. This suppression is often profound and may cause complete silence in EMG/SMU activity.

We should not ignore the possibility that cortical neurons giving rise to CM actions also exert, via their extensive intraspinal collateral network, actions on INs. Single CM cells can exert both exatory and inhibitory actions (Kasser and Cheney, 1985; Bennett and Lemon, 1996).

Searching for C3–C4 propriospinal transmission to spinal MNs in the primate

The C3–C4 propriospinal system in the cat has been thoroughly investigated in a long series of outstanding papers by Lundberg, Illert, Alstermark, Isa and coworkers. In the cat, which lacks CM projections, this C3–C4 system represents an important pathway mediating cortical control of the forelimb, and particularly forelimb reaching movements. C3–C4 PNs transmit corticospinal excitation onward to forelimb MNs (Fig. 1). Descending monosynaptic connections to forelimb MNs travel in the ventral part of the lateral funiculus, and an ascending axon collateral travels up to the lateral reticular nucleus (LRN; Fig. 1). The latter is assumed to provide efference copy of PN activity to the cerebellum. In the cat, the C3–C4 system provides an integrative system for motor commands originating in the cortex, superior colliculus, red nucleus and reticular formation (Baldissera et al., 1981; Alstermark and Lundberg, 1992).

The approach developed in the cat for the study of this system has the elegant advantage of allowing study of premotoneuronal influences quite separate from the segmental apparatus. The C5 lesion, made in the dorsal half of the lateral funiculus, cuts the fibers of the lateral CST, and interrupts most of the corticospinal input to the lower cervical segments, in which the target MNs are located. The lesion leaves intact the corticospinal inputs to PNs located above the lesion, and their descending axons. This approach is particularly useful in the primate, where oligo-synaptic excitatory effects may be obscured both by monosynaptic (CM-evoked) EPSPs and, more importantly by the large and ubiquitous disynaptic inhibition. One disadvantage of this preparation is that it does not provide the opportunity to test for *segmental* oligosynaptic effects, which may, or may not, be important.

The completeness of the CST lesion is a critical issue. In the monkey, the fibers of the lateral corticospinal tract (LCST) extend more ventrally into the lateral funiculus than do those in the cat. The lesion, however, cannot be extended too far ventrally without fear of interrupting descending PN axons. Thus, even after large C5 lesions, stimulation of the pyramidal tract in the medulla evokes small amounts of surface positivity (Maier et al., 1998). The contribution of the uncrossed (and unlesioned!) CST remains unknown. It is classically considered to terminate on long-axoned PNs and to influence axial and girdle musculature (Kuypers, 1981). Its further study is clearly required, however.

Differences in corticospinal organization in two primate species

We have used the C5 lesion approach to study two species of monkey, the Old World macaque (*Macaca mulatta*) and the New World squirrel monkey (*Saimiri sciureus*; Maier et al., 1998; Nakajima et al., 2000). Our objective was to search for transmission of corticospinal excitation to upper limb MNs via C3–C4 PNs. Their interest was to investigate how such transmission operates alongside the CM system. Were well-defined differences in the influence of the CM system in the macaque versus squirrel monkey (Bortoff and Strick, 1993; Armand et al., 1997; Maier et al., 1997) paralleled by changes in C3–C4 transmission?

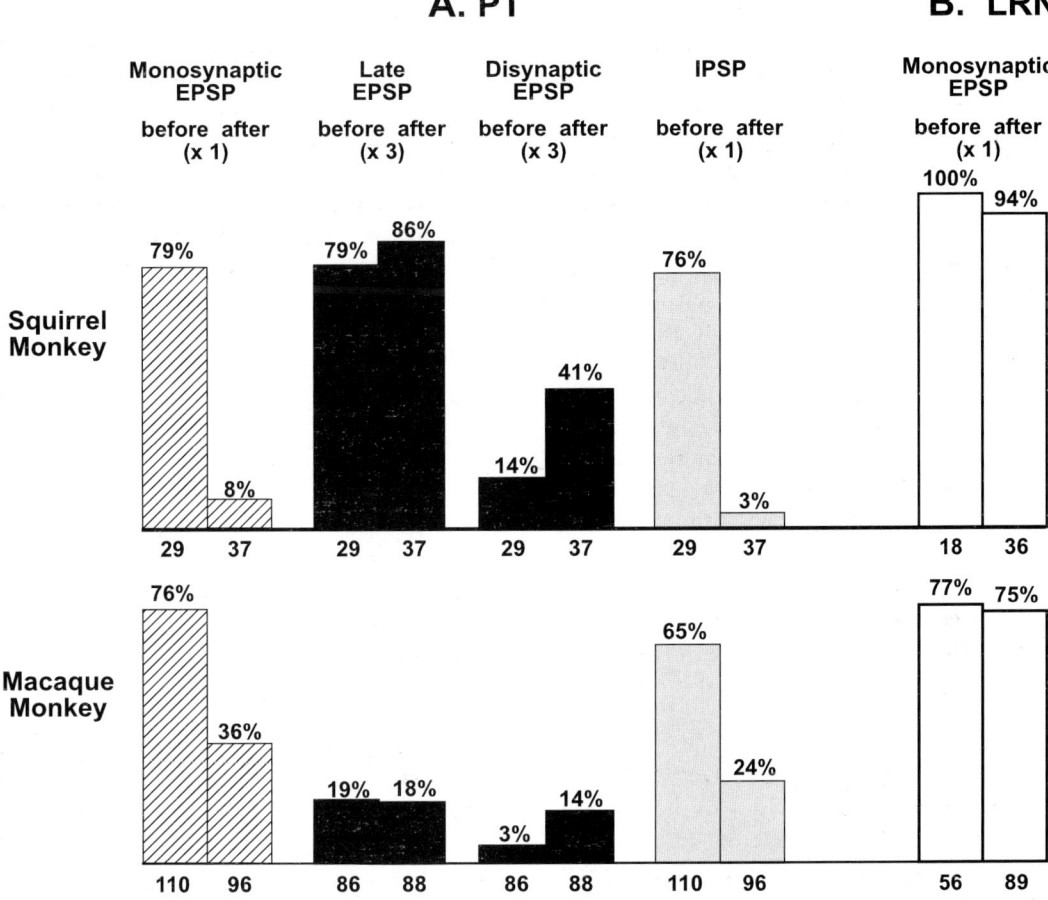

Fig. 2. Proportion of MN responses to PT (A) and LRN (B) stimulation in squirrel and macaque monkeys recorded before and after a C5 lesion. The proportion of MNs showing each type of response (New World squirrel monkey above; Old World macaque monkey below) is shown before (left column in each case) and after (right column) a C5 lesion interrupting the LCST. (A) Hatched columns: proportion showing monosynaptic CM EPSPs evoked by single PT stimuli. Filled columns: late, nonmonosynaptic EPSPs to repetitive (3 ×) PT stimulation; the proportion of MNs showing this type of response with a segmental delay within the disynaptic range is also shown. Grey columns: IPSPs evoked by single PT stimuli. (B) The percentage of MNs with EPSPs evoked by single stimuli delivered to the LRN before and after a C5 lesion. Stimulation strength, 200 μA. (Based on unpublished data from Maier et al., 1998; Nakajima et al., 2000.)

We obtained strikingly different results in the two studies (Fig. 2). In the macaque monkey, we did not find oligosynaptic EPSPs evoked by repetitive PT stimulation to be any more common after the C5 lesion (18% of tested MNs) than before it (19%). Only 14% of the MNs showed disynaptic excitation. Their frequency of occurrence was so low that we were unable to make any meaningful experiments with an additional C2 lesion. In the cat studies of Lundberg and colleagues, this additional lesion was made to ensure that oligosynaptic effects were indeed due to C3–C4 transmission, and did not involve, for example, a brainstem relay (Illert et al., 1977). Thus, while the results demonstrate that C3–C4 transmission does exist, one must conclude that the C5 lesion did not unmask widespread C3–C4 input to MNs. This contrasts with the cat, in which animal all forelimb MNs show such effects after a C5 lesion.

In contrast, in the squirrel monkey, C3–C4 transmission was a commonplace: 32/37 (86%) MNs tested after a C5 lesion showed robust late excitation with repetitive PT stimulation, and 41% of them showed disynaptic EPSPs (Nakajima et al., 2000; see Fig. 2). A C2 lesion abolished these effects, verifying that they were transmitted through C3–C4 PNs. In this same study, we confirmed that the mean amplitude of CM EPSPs in hand and forearm MNs was significantly smaller than in a comparable sample of macaque MNs (cf. Maier et al., 1997).

It is important to stress here that the results in the macaque do *not* suggest that a projection of C3–C4 PNs to upper limb MNs was absent, but rather that it was weak. In the cat, ascending collaterals from these neurons terminate in the LRN (Fig. 1). This projection is considered to be of strategic functional importance because it transmits to the cerebellum rapid feedback on the state of the PN excitability (Alstermark and Lundberg, 1992). Stimulation in the LRN activates this projection antidromically and it evokes monosynaptic EPSPs in many forelimb MNs (Alstermark and Sasaki, 1986). In our studies, we also searched for such effects, and found, in the macaque, short-latency EPSPs in 75% of tested MNs (Maier et al., 1998). These EPSPs from the LRN were small (mean amplitude, 1.2 mV) compared to those in the squirrel monkey (1.7 mV), and particularly so when compared with those in the cat (5.2 mV). This result supported the authors' view that transmission through this system is relatively weak. Even if there were relatively strong projections from the PT to C3–C4 PNs, the resulting disynaptic EPSPs might be small and infrequent (see Kirkwood et al., 2002).

Possible reasons for paucity of C3–C4 effects in the macaque

Before we assess the functional significance of these results, it is important to consider why Maier et al. (1998) had such difficulty in finding evidence of C3–C4 transmission from the PT to MNs in the macaque. A number of possible reasons for the result can be considered, but none of them provides good evidence against the fundamental conclusion that C3–C4 PN transmission in the macaque is relatively weak.

Anesthesia

We do not think this is a major factor, because the same anesthetic regime has been used in studies in all three species tested to date: cat, squirrel monkey and macaque monkey. Oligosynaptic effects were not found in the SMU experiments reported in Olivier et al. (2001), which were carried out without the depressive effects of general anesthesia.

Sample of MNs

A further possibility is that PN effects are only present in MNs supplying certain muscles (e.g., proximal vs. distal; extensor vs. flexor). There is good evidence in the cat that the C3–C4 PN system is particularly important for target reaching (Alstermark and Lundberg, 1992), and it has been shown that cat forelimb extensor MNs receive their largest EPSPs from LRN stimulation (Alstermark and Sasaki, 1986). Nevertheless, MNs of most forelimb muscles sampled by Alstermark and Sasaki showed a significant input from the LRN, thereby suggesting that if the C3–C4 system were organized along similar lines in the macaque to the cat, PN-mediated excitation from the PT should be widespread. The sample of MNs studied by Maier et al. (1998) and Olivier et al. (2001) included those supplying a variety of flexor and extensor muscles acting at the elbow, wrist and digits. We did not see any difference in the occurrence of PN-like effects in any particular muscle group. The inherent bias in intracellular recording made it likely that the intracellular sample of Maier et al. (1998) were dominated by large, fast-conducting MNs, while the SMU study (Olivier et al., 2001) was biased toward small, low-threshold MNs. Yet both studies indicated a paucity of PN effects in the macaque.

Lack of convergent facilitatory peripheral inputs to C3–C4 PNs

Work in the cat has indicated the importance of convergent peripheral inputs to the C3–C4 PNs. This has also been an essential feature of human studies, in which similar circuits have been assumed to exist (Alstermark and Lundberg, 1992; Pierrot-Deseilligny,

1996). In the anesthetized macaque, it may be that without such supportive peripheral inputs, there is insufficient excitation of these PNs to allow significant transmission of descending corticospinal excitation. Premotoneuronal control is particularly well suited, of course, for accomplishing this kind of precise interaction between descending and peripheral inputs. The latter does not appear to be essential for such transmission in the cat or squirrel monkey, because in these species, repetitive activation of PT inputs alone is sufficient to excite PNs. One could also argue that the experimental necessity in the human to provide a precisely timed peripheral input to PNs before descending transmission to MNs can be demonstrated (Pierrot-Deseilligny 1996; Nicolas et al., 2001) is, itself, a key characteristic of a rather weak system. This fact is in keeping with our own findings.

Feedforward and feedback inhibition of PNs

A quite different interpretation of the findings of Maier et al. (1998) came from the experiments of Alstermark et al. (1999). They confirmed the paucity of PN excitatory transmission in the macaque, but went on to show that after systemic administration of strychnine, repetitive PT stimulation now evoked pronounced oligosynaptic excitation in MNs. These authors suggested that under normal circumstances, such stimulation exerts a mixture of excitation and inhibition of PNs, with the latter mediated by inhibitory INs in the vicinity of the C3–C4 PNs. These neurons have been shown in the cat to mediate feedback inhibition from the periphery, and feedforward inhibition from descending pathways (Alstermark et al., 1984a,b). It was concluded that administration of strychnine blocked glycinergic transmission in these inhibitory pathways and revealed excitation. Thus, if stimulation of the whole PT exerted both inhibitory and excitatory effects on PNs, and inhibition dominated, little excitatory transmission to MNs would occur. Evidence for just such a suppressive action has since been produced in humans by Nicolas et al. (2001), using TMS.

The above are important observations, because they suggest that the C3–C4 system is under more inhibitory control in the macaque and human than in the cat. Why should this be? One interesting possibility is that while C3–C4 transmission might be depressed by gross stimulation of the PT, focused excitatory inputs, activated during specific motor acts, could enhance transmission. This raises new possibilities for experimental progress (see the next section). A more general point is that if there are fundamental differences in the balance of excitation/inhibition between species, then the caution advised by Maier et al. (1998) in interpreting results in man on the basis of a system fully worked out in the cat, is all the more appropriate.

Three further specific points about the results with strychnine experiments require emphasis. First, the *disynaptic* components of the published examples of the EPSPs evoked from the PT after strychnine were all of the order of 1 mV in amplitude (Alstermark et al., 1999). This supports the earlier conclusions about the rather weak connections between PNs and MNs in the macaque. Later oligosynaptic effects could be due, for example, to the action of other inhibitory inputs to MNs, and stress the rather gross and widespread actions of systemically administered strychnine. Second, there is as yet no direct evidence in the macaque for increased feedforward inhibition (Nakajima et al., 2000, Fig. 11) or that more intense PT stimulation suppresses PN transmission. Thus, in those MNs in which PN-mediated EPSPs were present, the EPSP amplitude *increased* with PT intensity (Fig. 3): there was no sign of these effects being suppressed by stronger PT shocks. In the SMU study, we did not find that later peaks were any more prevalent with weak PT shocks than with stronger ones (Fig. 4). Finally, if PT-evoked inhibition prevents monosynaptic excitation of PNs from the LCST, then it must be relatively fast and powerful to do so. Whether or not this is the case will need recordings from the PNs themselves.

Demonstration of functional CM and premotoneuronal control during natural motor acts

In assessing the importance of the different motor pathways transmitting excitation and inhibition to MNs, it is always necessary to demonstrate not only

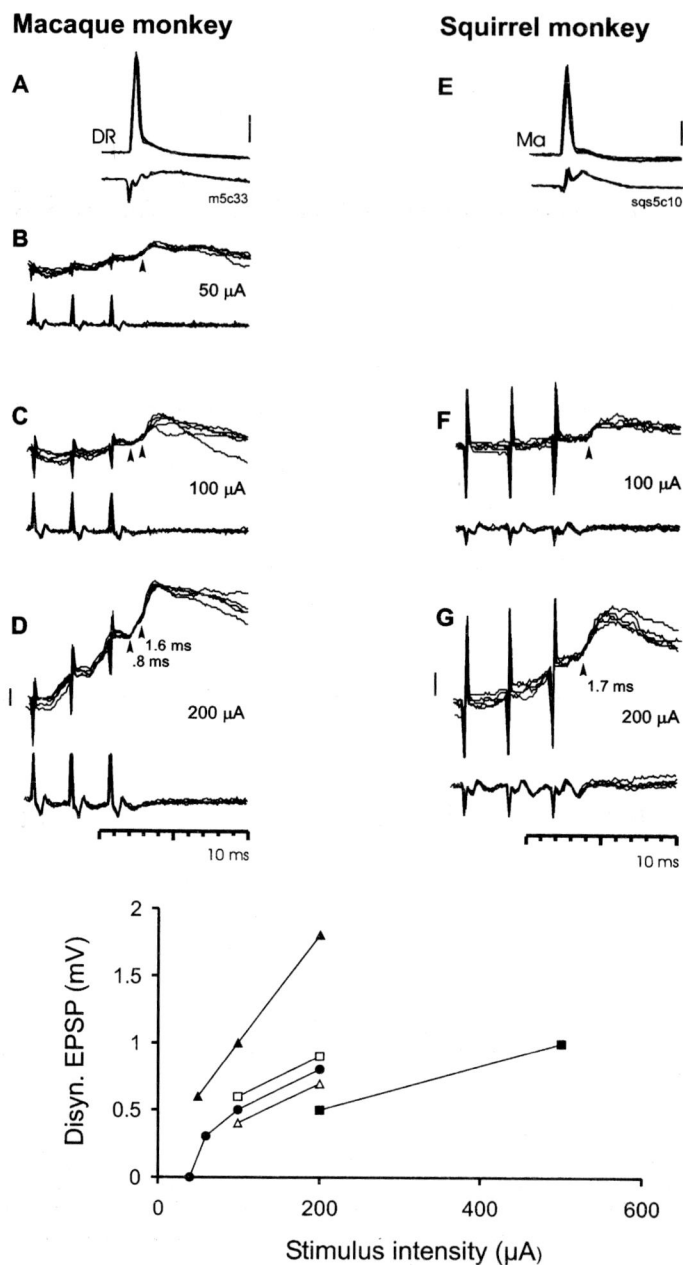

Fig. 3. Effects of increasing pyramidal tract stimulation intensity on C3–C4 propriospinal-mediated EPSPs in macaque and squirrel monkey MNs. (A–G) Examples from macaque (A–D) and squirrel monkey (E–G) of recordings made from MNs after a C5 lesion. In each panel, the upper record is the intracellular recording from a MN [antidromically identified from deep radial nerve (A) and median nerve at the axilla (E)], and the lower record is the surface recording from the same segment in which the MN was located. B–D and F–G show responses to repetitive (three-shock) stimulation to the PT at different intensities. In each MN, this evoked a late EPSP (arrowed) with a segmental latency in the disynaptic range (1.6 ms in B–D, 1.7 ms in F–G). Note that the amplitude of this disynaptic EPSP *increased* with increasing PT shock intensity. There was *no sign* of suppression from the PT. This is confirmed in the plots below which show intensity–amplitude curves for five macaque MNs. They all exhibited disynaptic EPSP responses to 3 × PT stimulation after a C5 lesion. (Based on unpublished observations of R.N.L., M.A.M., P.A.K. and K.N.)

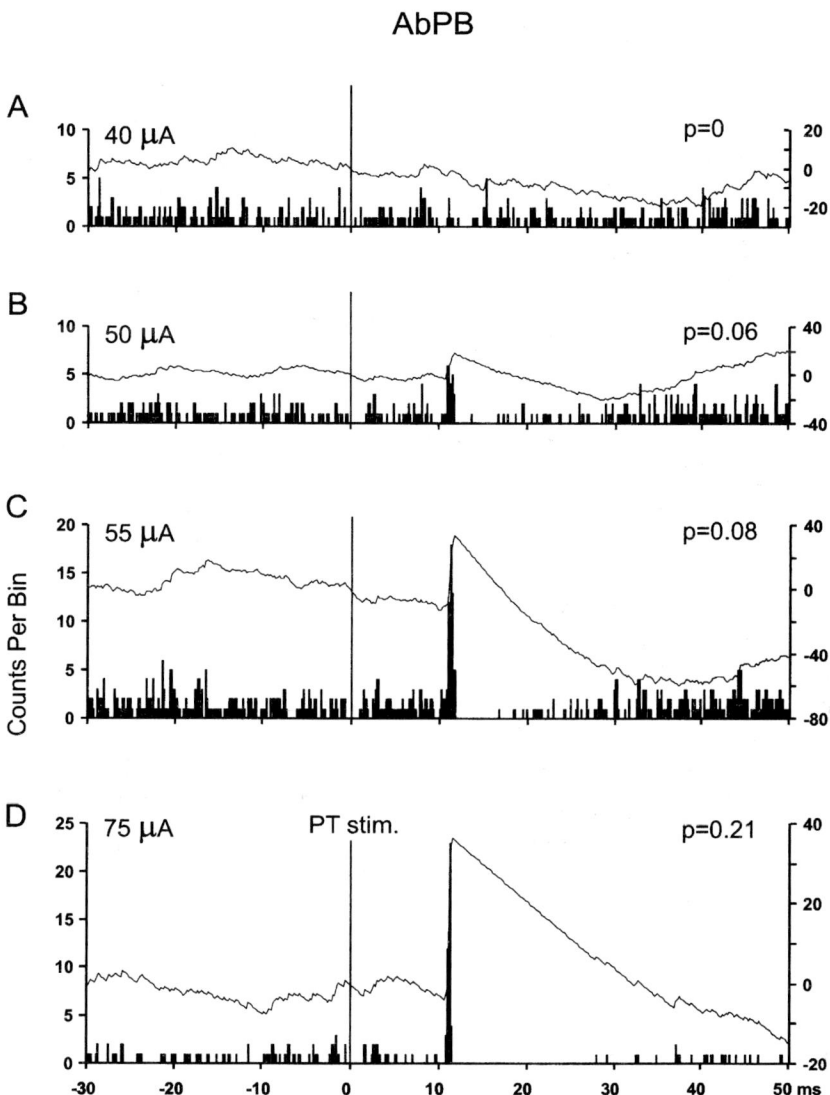

Fig. 4. Effect of increasing PT stimulation intensity on responses of a single thenar motor unit in the lightly sedated macaque monkey. Poststimulus histograms (PSTHs; left-hand scale) and cumulative probability histograms (CUSUMs) for a SMU recorded from the short thumb abductor (AbPB) in a sedated monkey (ketamine) during application of single PT stimuli at four different intensities. The motor unit was firing at a steady spontaneous rate throughout the delivery of the PT stimuli. There was no excitatory response at 40 μA (A), and higher intensities at 50 (B), 55 (C) and 75 μA (D) produced steadily higher response probabilities (P). Consistent with CM action, the response peak was very brief (< 1.0 ms) and constant. Given the low response probability of this CM-evoked peak, it is likely that any later, oligosynaptic excitatory input would have added to the underlying subthreshold CM EPSP and raised the MN to threshold. This was not observed, however, at either low or high PT intensities. The higher intensities produced pronounced postpeak suppression of unit firing (see text). Number of stimuli: A, 385; B, 386; C, 611 and D, 197. Binwidth: 0.2 ms.

the existence of the pathway, but also to show that it is active during functionally important motor acts. Critical components of the evidence for CM contributions during acts such as reach and grasp has come from experiments showing that (1) CM cells exhibit activity with an appropriate pattern and timing to facilitate the major muscle groups involved in these acts, (2) these cells exert facilitation of their target muscles throughout the motor act and (3) cells are recruited in a manner that reflected the pattern of facilitation across the different muscles, i.e., the 'muscle field' of the tested CM cell (Cheney and Fetz, 1980; Fetz and Cheney, 1980; Lemon et al., 1986; Bennett and Lemon, 1996; McKiernan et al., 1998; Baker et al., 2001).

Recent work from the Fetz laboratory (Perlmutter et al., 1998; Prut et al., 2001) has provided parallel evidence for a contribution of spinal segmental INs. These INs were recorded in awake, behaving monkeys from C6-Th1. They were identified by their postspike effects in spike-triggered averaging of EMGs recorded from upper limb muscles. Most of these INs had excitatory effects on their target muscles, this being consistent with a widespread premotoneuronal excitatory input at the segmental level. Some INs with inhibitory effects were also encountered, although technical reasons prevented a reliable assessment of the true extent of this population. In assessing the importance of premotoneuronal control from the cortex, it will be of critical importance to test the impact of PT and cortical inputs on these different INs. Given the prevalence of short latency inhibition from the PT (see the preceding section), we predict that many of the inhibitory INs will be activated from the PT. It will be much more exciting to discover what proportion of the segmental excitatory INs receives PT inputs. A further central issue that can be resolved by the Fetz method will be to discover whether C3–C4 propriospinal INs do actually show patterns of activity appropriate for control of upper limb movements. Evidence of this is still lacking.

Significance of species differences in the extent of C3–C4 propriospinal transmission

The most important conclusion to be drawn from this short review is that there are important species differences in the organization of corticospinal transmission to MNs. These differences can provide valuable insights into the mechanisms underpinning motor control. Maier et al. (1998) warned that "extrapolation of results from the cat to the primate should be made with considerable caution". We have suggested that there is a relative increase in the importance of the CM system for upper limb motor control in the macaque compared to the squirrel monkey and the cat. This change is paralleled by a relative decline in the significance of transmission through a C3–C4 system (Nakajima et al., 2000). If this hypothesis is correct, then in humans, with the best developed CM system for upper limb control, one might predict that transmission of corticospinal excitation through C3–C4 PNs is relatively unimportant.

Although there are many differences in motor behavior between the cat, monkey and human, we believe that changes in the relative influence of the CM and PN systems are mainly concerned with differences in dexterous manipulation. Cats have a limited capacity for independent digit movement (Boczek-Funcke et al., 1998), and even squirrel monkeys lack precision grip and do not use their digits for skilled manipulation (Costello and Fragaszy, 1988). These capacities are far more developed in the macaque and, of course, the human (Napier, 1961; Lawrence and Kuypers, 1968; Lawrence and Hopkins, 1976; Heffner and Masterton, 1983). The important role played by the PN system in the cat during reaching may have been increasingly taken over by CM control. Possibly, this occurred because precision grasp and manipulation become more dependent on the stability and accuracy of the reaching movement (Christel and Billard, 2002).

The possibility that PN transmission is under feedforward and feedback inhibition has been raised in both monkey and human studies (Alstermark et al., 1999; Nicolas et al., 2001). One suggestion is that during natural voluntary movements, inhibitory control might help to '... focus motor activity' (Nicolas et al., 2001, p. 918). This idea has yet to be tested. There are plentiful examples in the literature of oligosynaptic pathways that transmit excitation only during particular behaviors (Jankowska, 2001). They include, for example, disynaptic excitation of external intercostal MNs by muscle spindle afferents during inspiration (Kirkwood and Sears, 1982), and disynaptic excitation of flexor and extensor MNs during fictive scratching (Degtyarenko et al., 1998) and particular phases of locomotion (McCrea et al., 1995; McCrea, 1998; Quevedo et al., 2000). Thus, searching for motor acts in which it transmits net excitation to MNs may well reveal an important role for the C3–C4 system in primates.

Because of the indirect nature of experiments in human subjects, it is difficult to find conclusive evidence for or against the operation of the C3–C4 system during natural motor acts. There is evidence that unilateral damage of the anterolateral funiculus at the C5/C6 level in a human patient produced a collection of electrophysiological results consistent with the interruption of descending propriospinal axons (Marchand-Pauvert et al., 2001). Interestingly, this patient did not suffer any obvious motor deficits as a result of the loss of PN projection, possibly because of compensation by other systems, over a long postlesion period.

There is another remarkable case to consider: a patient (case 19 in Nathan et al., 1996), in whom the anterolateral cord was cut bilaterally, on one side at C5 and on the other at C6, for the relief of chronic pain. These lesions were confirmed by careful postmortem staining using the Marchi method (Fig. 5). If the descending axons of the PN system travel in the same region of the lateral funiculus as in the cat, as suggested by Marchand-Pauvert et al. (2001), then these surgical incisions should have interrupted them. Since "... upper limb function on both sides was completely normal within 24 h of surgery" (P. Nathan, personal communication), it is difficult to attribute an important function to these fibers in controlling the upper limb.

Conclusions

The comment made by Porter and Lemon (1993) that in primates "... the functional role of propriospinal pathways in natural, voluntary movement performance clearly needs further evaluation" (p. 137) still seems valid. This issue has clearly become "... enmeshed in controversy—the best of all criteria for recognizing fruitful scientific endeavors" (Gould, 2001, p. xv). It will be important to prevent our "all-too-human propensity for dichotomization" (Gould, 2001, p. xv), in which participating scientists tend to see the issue as a contest between two extreme views, neither of which is really tenable!

In the macaque monkey, there is a good anatomical basis for descending corticospinal inputs

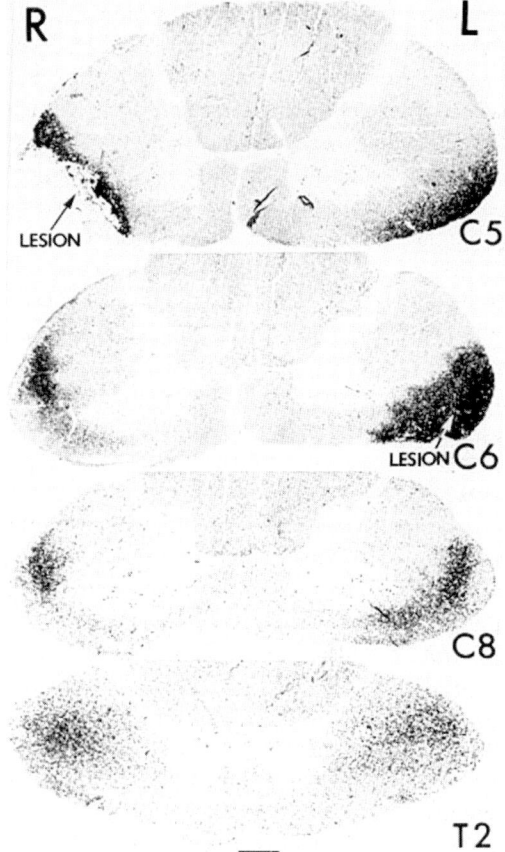

Fig. 5. Degeneration in the human spinal cord following cordotomy at the C5/C6 level. Marchi preparations of sections of the spinal cord from a patient who underwent anterolateral cordotomy for treatment of chronic pain ~123 days before death. They show the locations of axons degenerating as a result of the lesion. The incisions were made in the ventral parts of the lateral funiculus at the C5 level (on the right) and at C6 (on the left). The incisions spared most of the fibers of the CST, which are concentrated in the dorsolateral part of the lateral funiculus, but would have cut many propriospinal fibers in the ventral part. If those descending from C3 to C4 and transmitting corticospinal excitation to the upper limb travel in the same part of the lateral funiculus in human as in the cat, then the lesions shown would have interrupted these fibers (see text). Scale bar = 0.5 mm for C5, C6 and T2, and 1.0 mm for C8. (Based on unpublished data from case 19 in Nathan et al., 1996.)

to influence the premotoneuronal networks of the cervical spinal cord, and especially those operating at the segmental level of upper limb MNs. While oligo-synaptic inhibition has been easy to demonstrate, and may be a very important component of this descending corticospinal control, it has been proved much more difficult to detect signs of oligosynaptic excitation. This could reflect the methods used, which may be unsuitable for detecting such actions, or the experimental situations that have been studied to date. There is obviously scope for further research in both areas.

Finally, why have a direct CM system at all, if evolution shows that premotoneuronal control works so well? There has been considerable recent attention on the role of internal models for learning, planning and executing voluntary skilled movements (Wolpert et al., 2001). It is possible that these models have developed in parallel with the requirement for an increasingly feedforward, predictive type of voluntary control of the upper limb. Such control presumably involves the selection and recombination of innate spinal motor patterns (see Grillner and Wallén, Chapter 1 of this volume). The CM system may provide an executive pathway for such activity. According to this view, the internal model may carry out the role of interfacing descending commands with the spinal machinery for movement. Because the internal model provides an extremely accurate version of the desired movement, it is possible to rely largely on the use of commands that reach MNs directly, without the need for integration at the spinal premotoneuronal level. Thus, one might consider that these integrative mechanisms are now carried on at cortical and supraspinal levels, where they come under greater cognitive and sensory control.

Acknowledgments

We would like to thank all the colleagues who have worked with us on the studies cited in this review. The work was supported by The Wellcome Trust, MRC, International Spinal Research Trust and Brain Research Trust.

Abbreviations

C1, 2, etc.	spinal cervical segments
CM	cortico-motoneuronal
CST	corticospinal tract
EMG	electromyographic activity
EPSP	excitatory postsynaptic potential
HRP	horseradish peroxidase
IN	interneuron
IPSP	inhibitory postsynaptic potential
LCST	lateral corticospinal tract
LRN	lateral reticular nucleus
M1	primary motor cortex
MN	motoneuron
ms	millisecond
mV	millivolt
PN	propriospinal neurons
PSTH	poststimulus time histogram
PT	pyramidal tract
SMU	single motor unit
TES	transcranial electrical stimulation
TMS	transcranial magnetic stimulation
Th1	first thoracic segment
WGA–HRP	wheatgerm agglutinin–horseradish peroxidase

References

Alstermark, B. and Kummel, H. (1990a). Transneuronal transport of wheat germ agglutinin conjugated horseradish peroxidase into last order spinal interneurones projecting to acromio- and spinodeltoideus motoneurones in cat. 1. Location of labelled interneurons and influence of synaptic activity on the transneuronal transport. Exp. Brain Res., 80: 83–95.

Alstermark, B. and Kummel, H. (1990b). Transneuronal transport of wheat germ agglutinin conjugated horseradish peroxidase into last order spinal interneurones projecting to acromio- and spinodeltoideus motoneurones in cat. 2. Differential labelling of interneurons depending on movement type. Exp. Brain Res., 80: 96–103.

Alstermark, B. and Lundberg, A. (1992). The C3–C4 propriospinal system: target-reaching and food-taking. In: Jami L., Pierrrot-Deseilligny E. and Zytnicki D. (Eds.), Muscle Afferents and Spinal Control of Movement. Pergamon Press, Oxford, pp. 327–354.

Alstermark, B. and Sasaki, S. (1986). Integration in descending motor pathways controlling the forelimb in the cat 15. Comparison of the projection from excitatory C3-C4 propriospinal neurones to different species of forelimb motoneurones. Exp. Brain Res., 63: 543–556.

Alstermark, B., Lundberg, A. and Sasaki, S. (1984a). Integration in descending motor pathways controlling the forelimb in the cat. 10. Inhibitory pathways to forelimb motoneurones via C3-C4 propriospinal neurones. Exp. Brain Res., 56: 279–292.

Alstermark, B., Lundberg, A. and Sasaki, S. (1984b). Integration in descending motor pathways controlling the forelimb in the cat. 12. Interneurones which may mediate descending feed-forward inhibition and feed-back inhibition from the forelimb to C3-C4 propriospinal neurones. Exp. Brain Res., 56: 308–322.

Alstermark, B., Isa, T., Ohki, Y. and Saito, Y. (1999). Disynaptic pyramidal excitation in forelimb motoneurons mediated via C_3-C_4 propriospinal neurons in the *Macaca fuscata*. J. Neurophysiol., 82: 3580–3585.

Anissimova, N., Nielsen, J., Maier, M.A., Illert, M., Kummel, H. and Lemon, R.N. (1998). Presynaptic inhibition in the cat and monkey cervical spinal cord. Eur. J. Neurosci., 10: 63.14.

Armand, J., Olivier, E., Edgley, S.A. and Lemon, R.N. (1997). The postnatal development of corticospinal projections from motor cortex to the cervical enlargement in the macaque monkey. J. Neurosci., 17: 251–266.

Baker, S.N. and Lemon, R.N. (1998). Computer simulation of post-spike facilitation in spike-triggered averages of rectified EMG. J. Neurophysiol., 80: 1391–1406.

Baker, S.N., Jackson, A., Spinks, R. and Lemon, R.N. (2001). Synchronization in monkey motor cortex during a precision grip task I. Task-dependent modulation insingle-unit synchrony. J. Neurophysiol., 85: 869–885.

Baldissera, F., Hultborn, H. and Illert, M. (1981). Integration in spinal neuronal systems. In: Brookhart J.B. and Mountcastle V.B. (Eds.), Handbook of Physiology, The Nervous System, Vol. II, American Physiological Society, Bethesda, pp. 509–595.

Barton, R.A. and Harvey, P.H. (2000). Mosaic evolution of brain structure in mammals. Nature, 405: 1055–1058.

Bennett, K.M.B. and Lemon, R.N. (1996). Corticomotoneuronal contribution to the fractionation of muscle activity during precision grip in the monkey. J. Neurophysiol., 75: 1826–1842.

Bernhard, C.G. and Bohm, E. (1954). Monosynaptic corticospinal activation of FL motoneurones in monkeys. Acta Physiol. Scand., 31: 104–112.

Bernhard, C.G., Bohm, E. and Peters, N.I. (1953). Investigations on the organization of the corticospinal system in monkeys (Macaca mulatta). Acta Physiol. Scand., 29(Suppl. 106): 79–105.

Bizzi, E., Tresch, M.C., Saltiel, P. and d'Avella, A. (2000). New perspectives on spinal motor systems. Nat. Rev. Neurosci., 1: 101–108.

Boczek-Funcke, A., Kuhtz-Buschbeck, J.P., Raethjen, J., Paschmeyer, B. and Illert, M. (1998). Shaping of the cat paw for food taking and object manipulation: an X-ray analysis. Eur. J. Neurosci., 10: 3885–3897.

Bortoff, G.A. and Strick, P.L. (1993). Corticospinal terminations in two New-World primates: further evidence that corticomotoneuronal connections provide part of the neural substrate for manual dexterity. J. Neurosci., 13: 5105–5118.

Cheney, P.D. and Fetz, E.E. (1980). Functional classes of primate corticomotoneuronal cells and their relation to active force. J. Neurophysiol., 44: 773–791.

Cheney, P.D., Fetz, E.E. and Palmer, S.S. (1985). Patterns of facilitation and suppression of antagonist forelimb muscles from motor cortex sites in the awake monkey. J. Neurophysiol., 53: 805–820.

Cheney, P.D., Fetz, E.E. and Mewes, K. (1991). Neural mechanisms underlying corticospinal and rubrospinal control of limb movements. Prog. Brain Res., 87: 213–252.

Christel, M.I. and Billard, A. (2002). Comparison between macaques' and humans' kinematics of prehension: the role of morphological differences and control mechanisms. Behav. Brain Res., 131: 169–184.

Costello, M.B. and Fragaszy, D.M. (1988). Prehension in cebus and saimiri: 1. Grip type and hand preference. Am. J. Primatol., 15: 235–245.

Creutzfeldt, O.D. (1995). Cortex Cerebri: Performance, Structural and Functional Organisation of the Cortex. Oxford Science Publications, Oxford.

Degtyarenko, A.M., Simon, E.S., Norden-Krichmar, T. and Burke, R.E. (1998). Modulation of oligosynaptic cutaneous and muscle afferent reflex pathways during fictive locomotion and scratching in the cat. J. Neurophysiol., 79: 447–463.

de Noordhout, A.M., Rapisarda, G., Bogacz, D., Gerard, P., De Pasqua, V., Pennisi, G. and Delwaide, P.J. (1999). Corticomotoneuronal synaptic connections in normal man. An electrophysiological study. Brain, 122: 1327–1340.

Dum, R.P. and Strick, P.L. (1996). Spinal cord terminations of the medial wall motor areas in macaque monkeys. J. Neurosci., 16: 6513–6525.

Fetz, E.E. and Cheney, P.D. (1980). Postspike facilitation of forelimb muscle activity by primate corticomotoneuronal cells. J. Neurophysiol., 44: 751–772.

Fritz, N., Illert, M., Kolb, F.P., Lemon, R.N., Muir, R.B., van der Burg, J., Wiedemann, E. and Yamaguchi, T. (1985). The cortico-motoneuronal input to hand and forearm motoneurones in the anaesthetized monkey. J. Physiol. (Lond.), 366: 20P.

Georgopoulos, A.P. and Grillner, S. (1989). Visuomotor coordination in reaching and locomotion. Science, 245: 1209–1210.

Gould, S.J. (2001). Size matters and function counts. In: Falk D. and Gibson K.R. (Eds.), Evolutionary Anatomy of the Primate Cerebral Cortex. Cambridge University Press, Cambridge, pp. xiii–xvii.

Heffner, R.S. and Masterton, R.B. (1983). The role of the corticospinal tract in the evolution of human digital dexterity. Brain Behav. Evol., 23: 165–183.

Hess, C.W., Mills, K.R. and Murray, N.M.F. (1987). Responses in small hand muscles from magnetic stimulation of the human brain. J. Physiol. (Lond.), 388: 397–419.

Illert, M., Lundberg, A. and Tanaka, R. (1977). Integration in descending motor pathways controlling the forelimb in the cat. 3. Convergence on propriospinal neurons transmitting disynaptic excitation from the corticospinal tract and other descending tracts. Exp. Brain Res., 29: 323–346.

Jankowska, E. (2001). Spinal interneuronal systems: identification, multifunctional character and reconfigurations in mammals. J. Physiol. (Lond.), 533: 31–40.

Jankowska, E., Padel, Y. and Tanaka, R. (1975). Projections of pyramidal tract cells to alpha-motoneurones innervating hind-limb muscles in monkey. J. Physiol. (Lond.), 249: 637–667.

Jankowska, E., Padel, Y. and Tanaka, R. (1976). Disynaptic inhibition of spinal motoneurones from the motor cortex in the monkey. J. Physiol. (Lond.), 258: 467–487.

Kasser, R.J. and Cheney, P.D. (1985). Characteristics of corticomotoneuronal post-spike facilitation and reciprocal suppression of EMG activity in the monkey. J. Neurophysiol., 53: 959–978.

Kirkwood, P.A. and Sears, T.A. (1982). Excitatory postsynaptic potentials from single muscle spindle afferents in external intercostal motoneurones in the cat. J. Physiol. (Lond.), 322: 287–314.

Kirkwood, P.A., Maier, M.A. and Lemon, R.N. (2002) Interspecies comparisons for the C3-C4 propriospinal system: unresolved issues. In: Gandevia, S.C., Proske, U. and Stuart, D. (Eds.), Sensorimotor Control of Movement and Posture. Advances in Experimental Medicine and Biology. Kluwer/Plenum; New York, pp. 299–308.

Kuypers, H.G.J.M. (1978). The motor system and the capacity to execute highly fractionated distal extremity movements. Electroencephalogr. Clin. Neurophysiol., Suppl. 34: 429–431.

Kuypers, H.G.J.M. (1981). Anatomy of the descending pathways. In: Brookhart J.M. and Mountcastle V.B. (Eds.), Handbook of Physiology, The Nervous System, Vol. II, American Physiological Society, Bethesda, pp. 597–666.

Kuypers, H.G.J.M. and Brinkman, J. (1970). Precentral projections to different parts of the spinal intermediate zone in the rhesus monkey. Brain Res., 24: 29–48.

Lawrence, D.G. and Hopkins, D.A. (1976). The development of motor control in the rhesus monkey: evidence concerning the role of corticomotoneuronal connections. Brain, 99: 235–254.

Lawrence, D.G. and Kuypers, H.G.J.M. (1968). The functional organization of the motor system in the monkey. I. The effects of bilateral pyramidal lesions. Brain, 91: 1–14.

Lemon, R.N. (1993). Cortical control of the primate hand. The 1992 G.L. Brown Prize Lecture. Exp. Physiol., 78: 263–301.

Lemon, R.N., Mantel, G.W.H. and Muir, R.B. (1986). Corticospinal facilitation of hand muscles during voluntary movement in the conscious monkey. J. Physiol. (Lond.), 381: 497–527.

Lemon, R.N., Muir, R.B. and Mantel, G.W.H. (1987). The effects upon the activity of hand and forearm muscles of intracortical stimulation in the vicinity of corticomotor neurones in the conscious monkey. Exp. Brain Res., 66: 621–637.

Maier, M.A., Olivier, E., Baker, S.N., Kirkwood, P.A., Morris, T. and Lemon, R.N. (1997). Direct and indirect corticospinal control of arm and hand motoneurons in the squirrel monkey (*Saimiri sciureus*). J. Neurophysiol., 78: 721–733.

Maier, M.A., Illert, M., Kirkwood, P.A., Nielsen, J. and Lemon, R.N. (1998). Does a C3-C4 propriospinal system transmit corticospinal excitation in the primate? An investigation in the macaque monkey. J. Physiol. (Lond.), 511: 191–212.

Maier, M.A., Armand, J., Kirkwood, P.A., Yang, H.-W., Davis, J.N. and Lemon, R.N. (2002). Differences in the corticospinal projection from primary motor cortex and supplementary motor area to macaque upper limb motoneurons: an anatomical and electrophysiological study. Cereb. Cortex, 12: 281–296.

Marchand-Pauvert, V., Pradat-Diehl, P. and Pierrot-Deseilligny, E. (2001). Interruption of a relay of corticospinal excitation by a spinal lesion at C6-C7. Muscle Nerve, 24: 1554–1561.

McCrea, D.A. (1998). Neuronal basis of afferent-evoked enhancement of locomotor activity. Ann. N.Y. Acad. Sci., 860: 216–225.

McCrea, D.A., Shefchyk, S.J., Stephens, M.J. and Pearson, K.G. (1995). Disynaptic group I excitation of synergist ankle extensor motoneurones during fictive locomotion in the cat. J. Physiol. (Lond.), 487: 527–539.

McKiernan, B.J., Marcario, K., Karrer, J.H. and Cheney, P.D. (1998). Corticomotoneuronal postspike effects in shoulder, elbow, wrist, digit, and intrinsic hand muscles during a reach and prehension task. J. Neurophysiol., 80: 1961–1980.

Molenaar, I. and Kuypers, H.G.J.M. (1978). Cells of origin of propriospinal fiber and fibers ascending to supraspinal levels. An HRP study in cat and rhesus monkey. Brain Res., 152: 429–450.

Nakajima, K., Maier, M.A., Kirkwood, P.A. and Lemon, R.N. (2000). Striking differences in the transmission of corticospinal excitation to upper limb motoneurons in two primate species. J. Neurophysiol., 84: 698–709.

Napier, H.R. (1961). Prehensibility and opposability in the hands of primates. Symp. Zool. Soc. Lond., 5: 115–132.

Nathan, P.W., Smith, M. and Deacon, P. (1996). Vestibulospinal reticulospinal and descending propriospinal nerve fibres in man. Brain, 119: 1809–1833.

Nicolas, G., Marchand-Pauvert, V., Burke, D. and Pierrot-Deseilligny, E. (2001). Corticospinal excitation of presumed cervical propriospinal neurones and its reversal to inhibition in humans. J. Physiol. (Lond.), 533: 903–919.

Nielsen, J. and Petersen, N. (1994). Is presynaptic inhibition distributed to corticospinal fibres in man?. J. Physiol. (Lond.), 477: 47–58.

Nielsen, J., Petersen, N. and Ballegaard, M. (1995). Latency of effects evoked by electrical and magnetic brainstimulation in lower limb motoneurones in man. J. Physiol. (Lond.), 484: 791–802.

Nudo, R.J. and Masterton, R.B. (1990). Descending pathways to the spinal cord. IV: Some factors related to the amount of cortex devoted to the corticospinal tract. J. Comp. Neurol., 296: 584–597.

Olivier, E., Baker, S.N., Nakajima, K., Brochier, T. and Lemon, R.N. (2001). Investigation into non-monosynaptic corticospinal excitation of macaque upper limb single motor units. J. Neurophysiol., 86: 1573–1586.

Orlovsky, G.N., Deliagina, T.G. and Grillner, S. (1999). Neuronal Control of Locomotion: From Mollusc to Man. Oxford University Press, Oxford.

Palmer, E. and Ashby, P. (1992). Corticospinal projections to upper limb motoneurones in humans. J. Physiol. (Lond.), 448: 397–412.

Passingham, R.E. (1993). The frontal lobes and voluntary action: The Frontal Lobes and Voluntary Action. Clarendon Press, Oxford.

Perlmutter, S.I., Maier, M.A. and Fetz, E.E. (1998). Activity of spinal interneurons and their effects on forearm muscles during voluntary wrist movements in the monkey. J. Neurophysiol., 80: 2475–2494.

Phillips, C.G. and Porter, R. (1964). The pyramidal projection to motoneurones of some muscle groups of the baboon's forelimb. Prog. Brain Res., 12: 222–245.

Pierrot-Deseilligny, E. (1996). Transmission of the cortical command for human voluntary movement through cervical propriospinal premotoneurons. Prog. Neurobiol., 48: 489–517.

Porter, R. and Lemon, R.N. (1993). Corticospinal function and voluntary movement: Physiological Society Monograph. Oxford University Press, Oxford, pp. 428.

Preston, J.B., Shende, M.C. and Uemura, K. (1967). The motor cortex-pyramidal system: patterns of facilitation and inhibition on motoneurons innervating limb musculature of cat and baboon and their possible adaptive significance. In: Yahr M.D. and Purpura D.P. (Eds.), Neurophysiological Basis of Normal and Abnormal Motor Activities. Raven Press, New York, pp. 61–71.

Prut, Y., Perlmutter, S.I. and Fetz, E.E. (2001). Distributed processing in the motor system: spinal cord perspective. In: Nicolelis M.A.L. (Ed.), Advances in Neural Population Coding. Elsevier, Amsterdam, pp. 130267–130278.

Quevedo, J., Fedirchuk, S., Gosgnach, S. and McCrea, D.A. Group I disynaptic excitation of cat hindlimb flexor and bifunctional motoneurones during fictive locomotion. J. Physiol. (Lond.), 525: 549–564.

Rothwell, J.C., Thompson, P.D., Day, B.L., Boyd, S. and Marsden, C.D. (1991). Stimulation of the human motor cortex through the scalp. Exp. Physiol., 76: 159–200.

Shapovalov, A.I. (1975). Neuronal organization and synaptic mechanisms of supraspinal motor control in vertebrates. Rev. Physiol. Biochem. Pharmacol., 72: 1–54.

Ugolini, G. (1992). Transneural transfer of herpes simplex virus type 1 (HSV 1) from mixed limb nerves to the CNS. I. Sequence of transfer from sensory, motor, and sympathetic nerve fibres to the spinal cord. J. Comp. Neurol., 326: 527–548.

Wolpert, D.M., Ghahramani, Z. and Flanagan, J.R. (2001). Perspectives and problems in motor learning. Trends Cogn. Sci., 5: 487–494.

SECTION VII

Supraspinal sensorimotor interactions

CHAPTER 27

Arousal mechanisms related to posture and locomotion: 1. Descending modulation

Edgar Garcia-Rill*, Yutaka Homma and Robert D. Skinner

Department of Anatomy and Neurobiology, University of Arkansas Medical Sciences, Little Rock, AR 72205, USA

Abstract: Much of the controversy surrounding the induction of locomotion following stimulation of mesopontine sites, including the pedunculopontine nucleus (PPN), appears based on procedural differences, including stimulus onset, delay preceding stepping, and frequency of stimuli. The results reviewed in this chapter address these issues and provide novel information suggesting that descending projections from the PPN may exert a frequency-dependent effect. Stimulation at ~60 Hz (which induces prolonged tonic firing) may exercise a "push" towards locomotion (activation of pontine interneurons) as well as a "pull" away from decreased muscle tone (inhibiting giant pontine reticulospinal cells). Higher frequencies of stimulation (>100 Hz, which induces phasic burst-like activity) may "push" towards decreases in muscle tone, including the atonia of rapid eye movement sleep (activating giant pontine reticulospinal cells).

Introduction

Since the description of the mesencephalic locomotor region (MLR) (Shik et al., 1966; S. Mori et al., this volume), there has been considerable controversy regarding its anatomical localization. Since the MLR is still operationally defined as any site that produces controlled locomotion in decerebrate animals (Allen et al., 1996), anyone who has carried out such studies can find a number of different sites in the posterior midbrain that meet that criterion (reviewed in Reese et al., 1995). We concentrated on one site that could be identified immunocyto- and histochemically, the cholinergic pedunculopontine nucleus (PPN; see also Isa et al., Kobayashi et al., Skinner et al. and Takakusaki et al., Chapters 29, 41, 28 and 23 of this volume). It is important to realize, however, that the PPN is part of the reticular activating system (RAS), and, as such, must be activated appropriately in order to recruit locomotion following its stimulation. Much of the controversy surrounding the induction of locomotion following stimulation of mesopontine sites called the MLR appears based on procedural differences. The parameters usually involved are threefold: stimulus onset, delay-preceding stepping and frequency of stimuli. First, stimulation of more dorsal sites such as the cuneiform nucleus (and even the central nucleus of the inferior colliculus, especially in the rat) using sudden-onset pulses at threshold can be effective in inducing stepping. Such a paradigm applied to the PPN, however, will invariably induce a paroxysmal, startle-like response followed by uncontrolled escape or changes in muscle tone. Stimulation of the PPN instead requires slowly increasing (ramped) current levels. Recent studies found that low-amplitude stimulation applied immediately dorsal and at the dorsal edge of PPN-induced locomotion. When applied within PPN, it induced locomotion along with changes in muscle tone; when applied at the ventral edge and ventral to PPN, it induced changes in muscle tone (Takakusaki et al., 1997). These results are not unexpected given the sudden onset of stimulation employed. A second characteristic of MLR-induced stepping (regardless of stimulation site) is that locomotion is usually

*Corresponding author: Tel.: +501-686-5167;
Fax: +501-686-6382; E-mail: garciarilledgar@uams.edu

delayed, the first step ensuing only after 2 or more seconds of stimulation. This has prompted the suggestion that locomotion is 'recruited' rather than induced (Garcia-Rill, 1991; Reese et al., 1995), implying the presence of an intermediate process following stimulation, but in advance of stepping. A third important factor is frequency of stimulation. All of these mesopontine sites generally require a specific frequency range (40–60 Hz) before locomotion ensues, while the more posterior medioventral medulla (MED) locomotor site requires lower frequencies (5–20 Hz) to elicit stepping (Garcia-Rill and Skinner, 1987a). Spinal cord stimulation requires even lower frequencies (0.5–10 Hz) in neonates (Atsuta et al., 1990) as well as in adult animals (Iwahara et al., 1992). On the other hand, delivery of high-frequency trains of stimuli (≥ 100 Hz), especially instantaneously, tend to produce changes in muscle tone, which are, on occasion, accompanied by stepping (Lai and Siegel, 1990). We have been investigating the mechanisms involved in these peculiar requirements for inducing locomotion and provide some measure of understanding of these phenomena.

The reticular activating system (RAS)

The RAS includes a number of cell groups in the mesopontine tegmentum, namely the cholinergic PPN, the locus coeruleus and the raphe nuclei (Garcia-Rill, 1997). This region controls sleep–wake cycles and arousal as well as posture and locomotion. The RAS is a phylogenetically conserved system that modulates fight-or-flight responses. During waking, our ability to detect predator or prey is essential to survival. Under these circumstances, it is not surprising that the RAS is linked to control of the motor system in order to optimize attack or escape. During the vulnerable state of paradoxical sleep (which is controlled by the PPN), atonia (loss of postural muscle tone) keeps us from acting out our dreams. Cholinergic mesopontine neurons are known to fire in high-frequency bursts during paradoxical sleep, in relation to ponto-geniculo-occipital (PGO) waves concurrent with atonia. In fact, only our diaphragm and eye muscles appear to be acting out dream content. During waking, the RAS can modulate muscle tone and locomotion via reticulospinal systems. For example, in a standing individual, there is tonic activation of antigravity, mainly extensor, muscles. Before the first step can be taken, there must be flexion of the leg; therefore, there must be a release, or inhibition, from this postural, extensor, bias. The first sign of induced stepping from a standing position, therefore, will always be extensor inhibition and flexion. Interestingly, the startle response (SR), which is manifested as a rapid response to a supramaximal sensory stimulus, is basically a flexor response. This response shifts the standing individual from a standing position (extended) to a third baseman's 'ready' position (flexed). The SR is composed of a short latency activation of muscle activity (the 'ready' condition), which occurs too fast for RAS modulation, followed by a brief inhibition (the 'reset' state), then a long latency activation (the 'go' condition) (Miyazato et al., 1996). The brief, intermediate latency inhibition is thought to be part of the modulation of the SR by the RAS, and may represent a 'resetting' of motor programs that allow the subsequent selection of response strategies, and the triggering of attack or escape movements (Garcia-Rill, 1991; Reese et al., 1995; Garcia-Rill, 1997).

Descending modulation

There appear to be multiple descending pathways via which the cholinergic arm of the RAS, the PPN, modulates postural and locomotor control. In the oral pontine reticular formation, there is a paradoxical sleep-inducing region (pontine inhibitory area, PIA) which can be activated by cholinergic agonists to induce paradoxical sleep with atonia (Baghdoyan et al., 1984; Yamamoto et al., 1990). Lesions of this region produce an animal exhibiting paradoxical sleep without atonia (Sanford et al., 1994). Presumably, outputs from the PIA activate reticulospinal systems that lead to profound hyperpolarization of motoneurons, which is the mechanism responsible for the atonia (Chase et al., 1980). It is no wonder that electrical stimulation or injection of cholinergic agonists ventral (and medial) to PPN (in the vicinity of the PIA) will induce decreases in muscle tone (Takakusaki et al., 1993). On the other hand, cholinergic projections to the MED from the PPN appear to elicit increases in muscle tone and controlled locomotion (Garcia-Rill and Skinner, 1987a). Presumably, outputs from the MED activate reticulospinal systems that lead to the triggering of spinal pattern generators to induce stepping (Garcia-Rill and Skinner, 1987b). Electrical stimulation of the pontine reticular formation is known to induce decreased muscle tone at some sites, while producing stepping movements at other sites. This suggests the presence of a heterogeneous, distributed system of motor control. The required parameters of stimulation for eliciting these differing effects are important. Instantaneous, high-frequency trains (similar to high-frequency bursting activity in the range of RAS burst neurons) trigger pathways which lead to decreased muscle tone, while lower-frequency tonic stimulation appears to gradually 'recruit' locomotor movements (Garcia-Rill et al., 1996).

Current issues

New observations on the nature of descending effects of PPN efferents help us understand these requirements. Studies in the decerebrate cat showed that stimulation of the PPN induced prolonged responses (PRs) in caudal pontine reticular formation neurons (PnC) (Garcia-Rill et al., 2001). Using parameters of stimulation required for inducing locomotion (60 Hz, 0.5 ms pulses) but of a duration short enough not to induce fictive stepping (1 s), almost one-third of extracellularly recorded PnC neurons showed prolonged activation lasting on average over 12 s. The PRs were induced following stimulation using midfrequencies of stimulation (~ 60 Hz), but not using lower frequencies (10–30 Hz), and were of shorter duration or absent using high-frequency stimulation (90 Hz). Moreover, these stimulation frequency-dependent PRs were elicited in PnC neurons that did *not* project to the spinal cord (i.e., were nonreticulospinal; could be interneurons). Subsequent work in rat brainstem slices showed that two-thirds of intracellularly recorded PnC neurons had frequency-dependent PRs which lasted over 14 s on average following PPN stimulation (0.5 ms pulses at 60 Hz for 1 s). Figure 1 shows the effects of stimulus duration and stimulus frequency on PRs in rat PnC neurons. Briefly, increasing stimulus duration increased the duration of the PR, while stimulation at 60 Hz was more effective than stimulation at 10, 30 or 90 Hz. Thus, both stimulus duration and stimulus frequency are important parameters eliciting increased duration of the action-potential trains as well as the amplitude of the underlying depolarization (Homma et al., 2000). Figure 2A shows that PPN stimulation using 60 Hz trains induced the highest frequency attained in PnC neurons during PRs (~ 10 Hz firing rate) compared to 10, 30 or 90 Hz stimulation. Figure 2B shows that the duration of the PR was frequency-dependent, reaching a maximal duration at 60 Hz.

In general, these findings suggest that (1) PPN stimulation ~ 60 Hz produced prolonged activation of PnC neurons greatly outlasting the stimulation, and (2) 60 Hz stimulation induced a maximal firing rate in PnC neurons ~ 10 Hz. In addition, the depolarization induced by PPN stimulation was found to be due to muscarinic blockade of potassium channels, in keeping with increased input resistance observed during PRs (determined using $BaCl_2$ superfusion to block PPN stimulation-induced PRs) (not shown). These findings suggest that PPN stimulation-induced PRs may be due to increased excitability following closing of muscarinic receptor-modulated potassium channels, allowing PnC neurons to respond to a transient, frequency-dependent depolarization with long-lasting stable states. That is, PPN stimulation may recruit locomotion by gradually raising the excitability of PnC neurons (explaining why there is a delay before locomotion ensues following the onset of PPN stimulation), which then reach maximal firing rates ~ 10 Hz (explaining why stimulation frequencies ~ 60 Hz are required to induce stepping). Presumably, such maximal activation of PnC neurons is then relayed to downstream sites. This may be why stimulation frequencies in the 5–20 Hz range are required for inducing locomotion following stimulation of the MED, which does have reticulospinal projections which drive spinal pattern generators (Garcia-Rill and Skinner, 1987b). Interestingly, 10 Hz is the mean frequency of physiological tremor, the upper limit of individual movements, and is a signal thought to originate as a descending brainstem command (reviewed in Llinas, 2001). This command is thought to act as a cueing

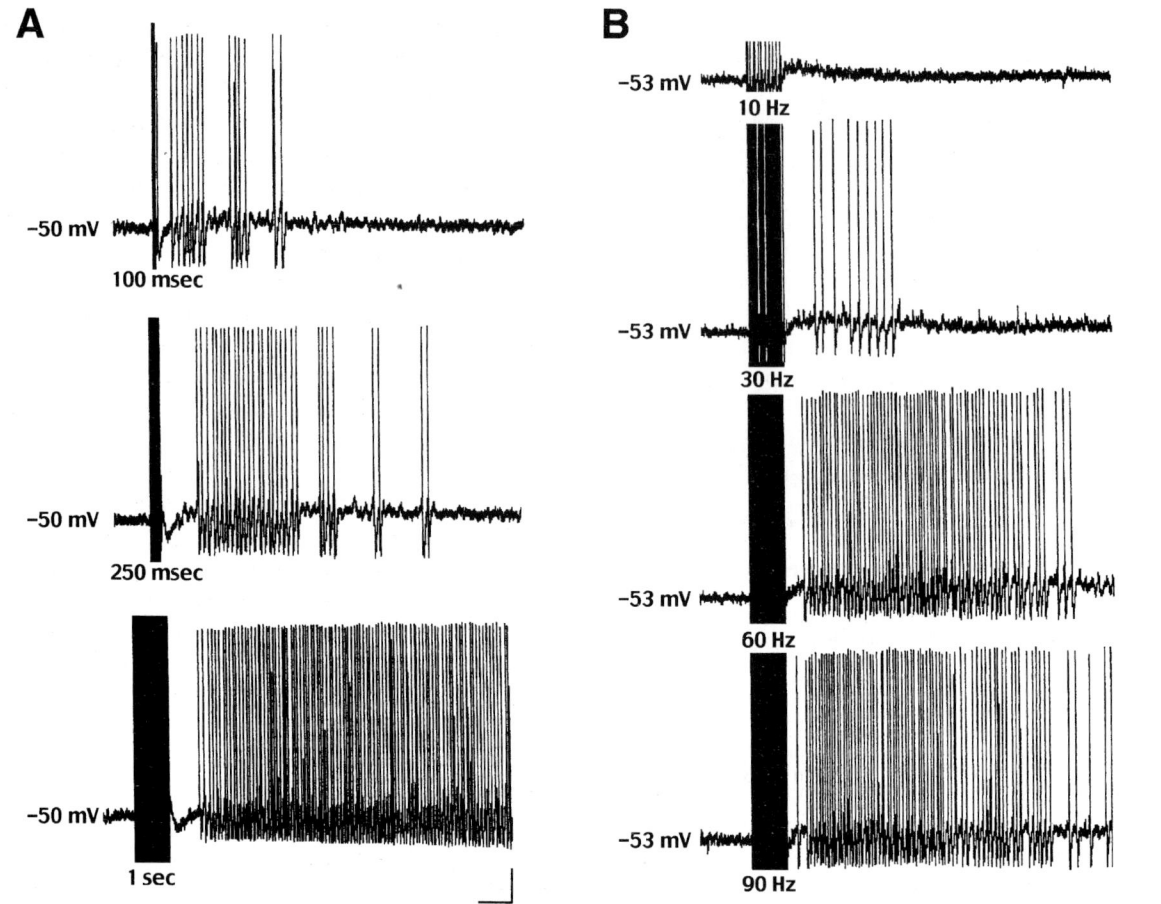

Fig. 1. Effects of stimulus duration and frequency on PRs. (A) Trains of different durations of PPN stimulation using 0.5 ms pulses at the same frequency and amplitude (60 Hz, 400 µA) showed increasing duration PRs (tonic action-potential activation along with an underlying depolarization of ~5 mV) as train duration was increased from 100 ms (top record) to 250 ms (middle record) to 1 s (bottom record). Note that the PR endured for many seconds following the 1-s train when the PnC neuron was held at −50 mV. (B) Trains of different frequencies of stimulation using 0.5 ms pulses at the same duration and amplitude (1 s, 400 µA) were applied to the PPN during recording of a PnC neuron held at −53 mV. Stimulation at 10 and 30 Hz was not as effective in inducing PRs, whereas 60 Hz stimulation elicited the longest duration PR superimposed on a long-lasting depolarization. In this PnC cell, 90 Hz stimulation induced a PR almost as long as when using 60 Hz stimulation (in other cells, 90 Hz stimulation was not as effective as 60 Hz stimulation). The amplitude of the membrane depolarization underlying the PR of action potentials increased from the prestimulation membrane potential with a stimulation frequency up to 60 Hz, then decreased at 90 Hz (not shown). Calibration bars: vertical, 10 mV; horizontal, 1 s.

function for synchronizing motoneurons, to provide inertia for overcoming friction and viscosity in muscles, and as a control system for binding inputs and outputs in time (Llinas, 2001).

Descending PPN projections to the PnC also may be involved in, for example, modulation of the SR (see reviews on the SR, Davis, 1984; Swerdlow et al., 1992; Koch, 1999). The pathway for the SR, which includes giant neurons in the PnC, has been delineated (Davis, 1984). These giant cells, however, make up only about 1% of PnC neurons (Koch, 1999). PPN lesions have been reported to reduce prepulse inhibition of the SR (Swerdlow et al., 1992), and auditory-responsive PnC neurons appear to be *inhibited* by cholinergic agonists (Koch, 1999), suggesting that the PPN modulates SR gating.

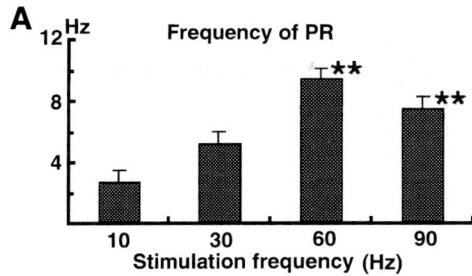

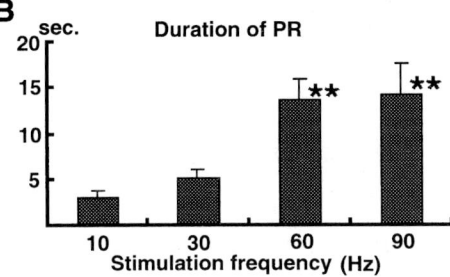

Fig. 2. Effects of stimulation frequency on the duration of the PR and on the firing rate of PnC cells during the PR. (A) Firing frequency of PnC neurons during the second second of the PR (firing frequency during the first second of the PR was not consistent across neurons) at different frequencies of stimulation. Stimulation using 60 Hz trains induced statistically significant higher firing frequencies than 10 or 30 Hz trains, whereas 90 Hz trains induced lower firing frequencies than 60 Hz trains, but still higher than 10 or 30 Hz trains. That is, maximal firing frequencies (~10 Hz in PnC neurons) during PRs were induced by PPN stimulation at 60 Hz. (B) In the same PnC cells, the mean ± SE duration of the PR was significantly longer following 60 and 90 Hz trains than 10 or 30 Hz trains of the same duration (1 s) and amplitude (400 μA). While the mean ± SE duration of the PR was significantly longer using 60 Hz trains than 10 or 30 Hz trains, there was no further increase in PR duration using 90 Hz trains, i.e., there may have been a plateau to this effect.

Studies on rat PnC neurons recorded in vitro showed the presence of (excitatory) PRs in most PnC neurons; a few PnC neurons were *inhibited*, however, following PPN stimulation and/or superfusion of the cholinergic agonist carbachol (Homma et al., 2000). This suggests that descending PPN projections activated by medium frequencies of stimulation may inhibit a few PnC cells (perhaps some SR-related neurons with spinal projections), while activating large numbers of PnC cells (possibly interneurons) to induce PRs, thereby eliciting a lasting change in state in these cells.

That is, *descending* PPN projections activated by stimulation ~60 Hz may exercise a 'push' toward locomotion (PRs in PnC neurons) as well as a 'pull' away from decreased muscle tone (inhibition of some PnC SR-related neurons), such as that observed in the SR (Koch, 1999) in order to promote stepping. In a similar fashion, *ascending* cholinergic projections may 'push' fast cortical rhythms by prolonged activation of thalamocortical relay neurons, while they may 'pull' away from slow cortical rhythms by inhibiting thalamic reticular neurons responsible for inducing slow waves, thereby promoting waking (see Chapter 21). A model of this hypothesis is provided in Fig. 3. It is possible that stimulation at higher frequencies (>90 Hz) may preferentially activate SR-related neurons, leading to decreases in muscle tone (especially extensor, antigravity muscle tone), perhaps mimicking the bursting during PGO waves. This may explain why high-frequency trains applied to the PPN might, via activation of the PIA and/or SR-related PnC neurons, lead to decreases in muscle tone (i.e., a stimulation frequency-dependent effect).

Future possibilities

The findings described earlier point the way to a number of exciting new studies. It will be important to determine if indeed descending PPN projections are segregated into excitatory inputs to interneurons and inhibitory inputs to SR-related neurons. The slice preparation, however, is limited in its inability to identify reticulospinal versus nonreticulospinal neurons or interneurons. Therefore, such studies must be carried out in the decerebrate fictive locomotion preparation. Moreover, electrophysiological and morphological analyses did not allow to determine which PnC neurons would develop into giant (SR-related) neurons, since both PR and non-PR neurons included those having low input resistance and large mean somatic area. Again, such studies require a decerebrate preparation. Moreover, it will be important to determine if PRs in PnC (presumably inter-) neurons are relayed to MED neurons, which in turn send reticulospinal projections to activate spinal pattern generators. Perhaps the most exciting project will be to determine the origin and mechanisms underlying the 10 Hz descending command which

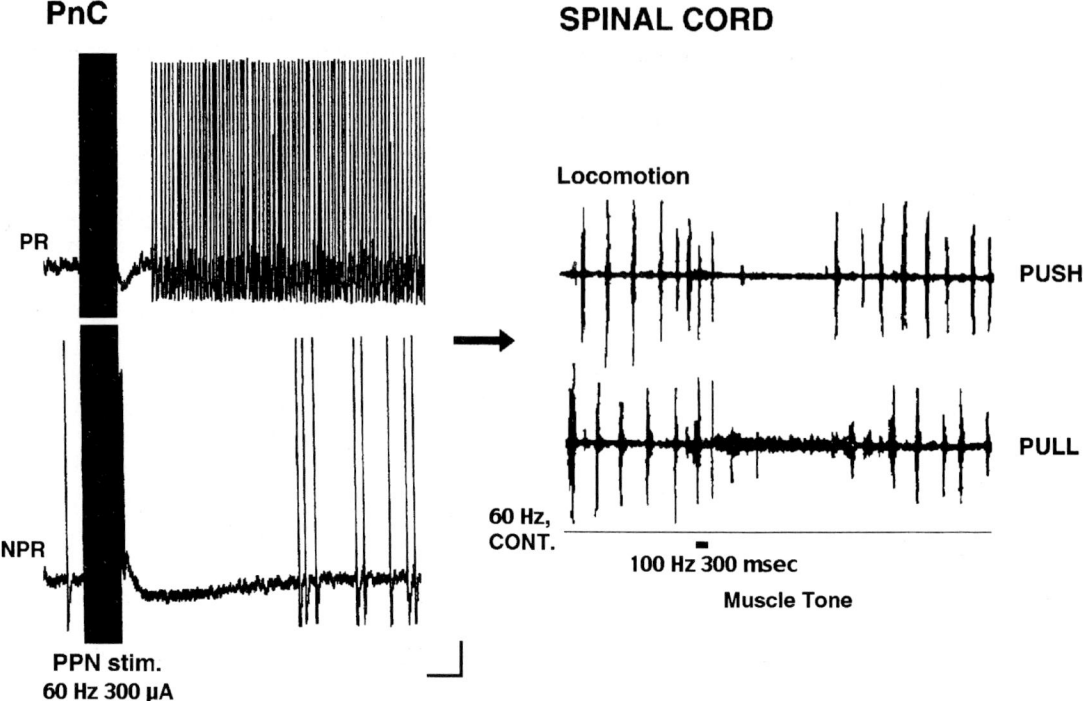

Fig. 3. Hypothetical role of descending PPN projections to PnC and their ultimate effects on the spinal cord. Left side: Example of the 65% of PnC neurons with PRs following PPN stimulation (top) and of one of the few PnC neurons hyperpolarized following PPN stimulation (bottom) using identical stimulation parameters (0.5 ms pulses, 60 Hz, 300 μA, 1 s trains). Right side: Electromyograms of antagonist muscles showing alternating activity (locomotion) in response to PPN stimulation in the decerebrate animal using 60 Hz pulses slowly ramped up to threshold and then applied continuously. The locomotion recruited was interrupted by a brief, high-frequency train (100 Hz, 300 ms) of stimuli to the PPN, leading to increased flexor muscle tone. Ultimately, these transient effects were overcome by the continued 60 Hz stimulation. The hypothetical PUSH-PULL mechanism proposed suggests that PPN stimulation at the appropriate frequencies (~60 Hz) for recruiting locomotion activates large numbers of PnC neurons (presumably interneurons which ultimately activate reticulospinal elements, perhaps via the MED) using cholinergic receptors to drive these cells into a new state (PUSH), while inhibiting other PnC neurons which normally inhibit muscle tone (PULL). Much additional evidence is needed to support this suggestion, although the findings described herein certainly lend some support to this hypothesis. Calibration bars apply to both sides. Left side, 10 mV and 1 s; right side, 500 μA and 1 s.

underlies physiological tremor and voluntary movement. Such information would have inestimable value in learning (1) how motoneurons induce and regulate movements, and (2) ways to compensate (prosthetically) for the loss of such descending commands.

Summary

In general, it is well accepted that the RAS, while controlling sleep/wake cycles and arousal, also modulates posture and locomotion, perhaps related to the execution of fight-or-flight responses. On the one hand, stimulation of various points in the posterior mesencephalon is known to induce controlled locomotion on a treadmill in the decerebrate cat and rat (Shik et al., 1966; Skinner and Garcia-Rill, 1984). Such locomotion-inducing sites include the PPN (reviewed in Garcia-Rill, 1991). On the other hand, sudden-onset stimulation of the PPN has been found to suppress muscle tone and induce stepping depending on stimulation site (dorsal edge, middle or ventral edge) (Lai and Siegel, 1990; Takakusaki et al., 1997). We proposed a firing frequency-dependent mechanism which can explain all of these observations by: (1) using amplitudes which are *slowly-ramped* up at medium frequencies of stimulation (~60 Hz; recall that reticulospinal pathways for locomotion are 'recruited' to produce alternation of flexor and extensor hindlimb nerves), and (2) using *sudden-onset* trains especially at high frequencies (>90 Hz). These observations demonstrate that locomotion can be inhibited and muscle tone suppressed (Garcia-Rill et al., 1996). Further research is required before more conclusive statements can be made regarding the frequency dependency of such complicated functions; this type of control system, however, would have great survival value when involved in fight versus flight responses.

Acknowledgments

This work was supported in part by USPHS grant NS20246.

Abbreviations

MED	medioventral medulla
MLR	mesencephalic locomotor region
PGO	ponto-geniculo-occipital
PIA	pontine inhibitory area
PnC	nucleus reticularis pontis caudalis
PPN	pedunculopontine nucleus*
PR	prolonged response
RAS	reticular activating system
SR	startle response

*Also called pedunculopontine tegmental nucleus (PPTN or PPTg)

References

Allen, L.F., Inglis, W.L. and Winn, P. (1996). Is the cuneiform nucleus a critical component of the mesencephalic locomotor region? An examination of the effects of excitotoxic lesions of the cuneiform nucleus on spontaneous and nucleus accumbens induced locomotion. Brain Res. Bull., 41: 201–210.

Atsuta, Y., Garcia-Rill, E. and Skinner, R.D. (1990). Characteristics of electrically induced locomotion in the rat in vitro brainstem-spinal cord preparation. J. Neurophysiol., 64: 727–735.

Baghdoyan, H.B., Rodrigo-Angulo, R.W., McCarley, R.W. and Hobson, J.A. (1984). Site-specific enhancement and suppression of desynchronized sleep signs following cholinergic stimulation of three brainstem regions. Brain Res., 306: 39–52.

Chase, M.H., Chandler, S.H. and Nakamura, Y. (1980). Intracellular determination of membrane potential of trigeminal motoneurons during sleep and wakefulness. J. Neurophysiol., 44: 349–358.

Davis, M. (1984). The mammalian startle response. In: Eaton R.C. (Ed.), Neural Mechanisms of Startle Behavior. Plenum Press, New York, pp. 287–342.

Garcia-Rill, E. (1991). The pedunculopontine nucleus. Prog. Neurobiol., 36: 363–389.

Garcia-Rill, E. (1997). Disorders of the reticular activating system. Med. Hypoth., 49: 379–387.

Garcia-Rill, E. and Skinner, R.D. (1987a). The mesencephalic locomotor region. I. Activation of a medullary projection site. Brain Res., 411: 1–12.

Garcia-Rill, E. and Skinner, R.D. (1987b). The mesencephalic locomotor region. II. Projections to reticulospinal neurons. Brain Res., 411: 13–20.

Garcia-Rill, E., Reese, N.B. and Skinner, R.D. (1996). Arousal and locomotion: from schizophrenia to narcolepsy. In: Holstege G., Bankler R. and Saper C. (Eds.), The Emotional Motor System, Prog. Brain Res, Vol. 107, Elsevier, Amsterdam, pp. 417–434.

Garcia-Rill, E., Skinner, R.D., Miyazato, H. and Homma, Y. (2001). Pedunculopontine stimulation induces prolonged activation of pontine reticular neurons. Neuroscience, 104: 455–465.

Homma, Y., Skinner, R.D. and Garcia-Rill, E. (2000). Effects of pedunculopontine nucleus stimulation on caudal pontine reticular formation neurons in vitro. Soc. Neurosci. Abstr., 26: 156.

Iwahara, T., Atsuta, Y., Garcia-Rill, E. and Skinner, R.D. (1992). Spinal cord stimulation-induced locomotion in the adult cat. Brain Res. Bull., 28: 99–105.

Koch, M. (1999). The neurobiology of startle. Prog. Neurobiol., 59: 107–128.

Lai, Y.Y. and Siegel, J.M. (1990). Muscle tone suppression and stepping produced by stimulation of midbrain and rostral pontine reticular formation. J. Neurosci., 10: 2727–2734.

Llinas, R. (2001). I of the Vortex: from Neurons to Self. MIT Press, Cambridge, MA, pp. 302.

Miyazato, H., Skinner, R.D., Reese, N.B. and Garcia-Rill, E. (1996). Midlatency auditory evoked potentials and the startle response in the rat. Neuroscience, 25: 289–300.

Reese, N.B., Garcia-Rill, E. and Skinner, R.D. (1995). The pedunculopontine nucleus-auditory input, arousal and pathophysiology. Prog. Neurobiol., 47: 105–133.

Sanford, L.D., Morrison, A.R., Graziella, L.M., Harris, J.S., Yoo, L. and Ross, R.J. (1994). Sleep patterning and behavior in cats with pontine lesions creating REM without atonia. J. Sleep Res., 3: 233–240.

Shik, M.L., Severin, F.V. and Orlovsky, G.N. (1966). Control of walking and running by means of electrical stimulation of the mid-brain. Biophysica, 11: 756–765.

Skinner, R.D. and Garcia-Rill, E. (1984). The mesencephalic locomotor region (MLR) in the rat. Brain Res., 323: 385–389.

Swerdlow, N.R., Caine, S.B., Braff, D.L. and Geyer, M.A. (1992). The neural substrates of sensorimotor gating of the startle reflex: a review of recent findings and their implications. J. Psychopharmacol., 6: 176–190.

Takakusaki, K., Matsuyama, K., Kobayashi, Y., Kohyama, J. and Mori, S. (1993). Pontine microinjection of carbachol and critical zone for inducing postural atonia in reflexively standing cats. Neurosci. Lett., 153: 185–188.

Takakusaki, K., Habauchi, T., Nagaoka, T. and Sakamoto, T. (1997). Stimulus effects of the pedunculopontine tegmental nucleus (PPTN) on hindlimb motoneurons in cats. Soc. Neurosci. Abstr., 23: 762.

Yamamoto, K., Mamelak, A.N., Quattrochi, J.J. and Hobson, J.A. (1990). A cholinoceptive desynchronized sleep induction zone in the anterolateral pontine tegmentum: locus of the sensitive region. Neuroscience, 39: 279–293.

CHAPTER 28

Arousal mechanisms related to posture and locomotion: 2. Ascending modulation

Robert D. Skinner*, Yutaka Homma and Edgar Garcia-Rill

Department of Anatomy and Neurobiology, University of Arkansas Medical Sciences, Little Rock, AR 72205, USA

Abstract: An intrinsic function of the reticular activating system (RAS) is its participation in fight vs. flight responses such that alerting stimuli simultaneously activate thalamocortical systems, as well as postural and locomotor systems, in order to enable an appropriate response. The P50 midlatency auditory-evoked potential appears to be an ascending manifestation of the cholinergic arm of the RAS in eliciting changes in arousal state. Abnormalities in the manifestation of the P50 potential are present in disorders which include: (1) dysregulation of sleep–wake cycles; (2) abnormalities in reflex/postural, especially, startle, responses; and (3) malfunctions in flight vs. flight responses. In general, the P50 potential appears to be upregulated (increased amplitude and/or decreased sensory gating) in disorders which are marked by upregulation of RAS outputs (hypervigilance), and downregulated in disorders characterized by decreased RAS outputs (hypovigilance). Many of the disorders discussed have a developmental etiology and a postpubertal age of onset.

Introduction

Early work showed that a specific group of nuclei in the pontomesencephalon appeared essential for the brainstem control of sleep–wake states (Moruzzi and Magoun, 1949). The main cell groups of the reticular activating system (RAS) are the cholinergic pedunculopontine nucleus (PPN), the noradrenergic locus coeruleus (LC) and the serotonergic raphe nuclei (RN). During waking, all the three cell groups are active. Cholinergic neurons become quiescent and then silent as slow wave sleep deepens, but then they increase their firing in relation to paradoxical (or rapid eye movement, REM) sleep, many exhibiting bursting in relation to ponto-geniculo-occipital (PGO) waves. Noradrenergic and serotonergic cell groups are active during waking and less so during slow wave sleep, falling virtually silent during REM sleep. Thus, in some states, these cell groups are coactive, while in others they are reciprocally active (reviewed in Garcia-Rill, 1991; Reese et al., 1995; Garcia-Rill, 1997). Ascending cholinergic, noradrenergic and serotonergic RAS outputs travel in parallel to the thalamus, hypothalamus, basal forebrain and other regions, while descending outputs travel to the cerebellum and pontine and medullary reticular formation. There are direct noradrenergic and serotonergic projections to the cortex and the spinal cord, while cholinergic afferents to these areas are relayed. This anatomical heterogeneity allows the RAS to activate a host of brain systems.

It is thought that waking is maintained by tonic activity in the RAS reinforced by sensory input. Basically, all of the primary sensory pathways conduct sensation-specific information via their respective lemniscal relays all the way to primary cortical areas. These are rapidly conducting pathways with fewer synapses and high synaptic security providing information for sensory discrimination, i.e., the "what is it?" Every sensory system activates the RAS in parallel, setting up a more slowly

*Corresponding author: Tel.: +501-686-5183; Fax: +501-686-6283; E-mail: skinnerrobertd@uams.edu

conducting, multisynaptic, low synaptic security, 'reticular' input. The less sensation-specific information is related to arousal, i.e., the "wake up, something happened!" It should be noted, however, that activation of the RAS does not indiscriminately elevate cortical activity. For example, a simple visual cortex cell, which responds to the orientation of a particular visual stimulus, will respond with increased firing if the RAS is stimulated in parallel, but the response of the cortical neuron will not lose its orientational selectivity. This potentiation of specific sensory information may be part of the attentional process. That is, the RAS may not only provide the necessary modulation for arousal, but also help form the basis for attentional processes.

An intrinsic function of the RAS is its participation in fight versus flight responses such that alerting stimuli *simultaneously* activate thalamocortical systems as well as postural and locomotor systems in order to enable an appropriate response. The previous chapter (Garcia-Rill et al., Chapter 27 of this volume) described the organization of descending modulation by the RAS, while this one summarizes the work in relation to the ascending modulation of arousal state by cholinergic elements of the RAS.

Ascending modulation

The P50 potential is a human midlatency auditory-evoked response evoked by a click stimulus and recorded at the vertex, which occurs at a latency of 40–70 ms. It also is referred to as the P1 potential because it is the first positive wave following the early brainstem and primary auditory responses. The latter are directly related to the auditory lemniscal pathway and the auditory cortex while the P50 potential is related to the nonlemniscal ascending pathways. The P50 potential has three main characteristics: (1) *sleep-state dependence*—it is present during waking and REM or rapid eye movement (REM) sleep, but is absent during deep slow wave sleep (Erwin and Buchwald, 1986a), i.e., is present when the cortex is activated and the EEG shows high frequencies; (2) *blocked by muscarinic cholinergic antagonists*, such as scopolamine, i.e., may be mediated, at least in part, by cholinergic neurons (Buchwald et al., 1991); and (3) *rapid habituation* at stimulation rates greater than 2 Hz, i.e., is not manifested by a primary afferent pathway, but perhaps by multisynaptic, low security synaptic elements of the RAS (Erwin and Buchwald, 1986a,b). The P50 potential decreases and disappears with progressively deeper stages of sleep and then reappears during REM sleep at full amplitude (Kevanishvili and von Specht, 1979; Erwin and Buchwald, 1986a). None of the earlier latency auditory-evoked potentials (brainstem or primary auditory cortex potentials) possess this characteristic. This suggests that the P50 potential is functionally related to states of arousal. In addition, it most likely is generated, at least in part, by cholinergic mesopontine cell groups, especially the PPN, since its cells are preferentially active during waking and REM sleep, but inactive during slow wave sleep (Garcia-Rill, 1991, 1997; Reese et al., 1995; see also Isa et al., Kobayashi et al., and Takakusaki et al., Chapters 29, 41 and 23 of this volume). Animal studies have established that the P13 potential is the rodent equivalent of the human P50 potential, having the same three characteristics as the P50 potential described previously (Miyazato et al., 1995). The fact that the vertex-recorded P13 potential is modulated by injections of neuroactive agents into the PPN lends further evidence that this potential has a cholinergic mesopontine origin (Miyazato et al., 1999).

The P50 potential thus appears to be an ascending manifestation of the cholinergic arm of the RAS in eliciting changes in arousal state. The relationship between the P50 potential (ascending arousal) and the descending effects on postural and locomotor state is shown in Fig. 1.

Briefly, the P50 potential occurs at a latency during which there is a decrement in the EMG, for example, of the blink reflex in the human (Fig. 1A). Similarly, the P13 potential in the rat occurs at the time of the intermediate inhibitory phase of EMG activity in neck EMG (Fig. 1B). The brief, intermediate latency inhibition is thought to be part of the modulation of the startle/blink response by the RAS, and may represent a 'resetting' of motor programs which allow the subsequent selection of response strategies, the triggering of attack or escape movements (Garcia-Rill, 1997; Garcia-Rill, Chapter 27 of this volume). It is likely, then, that abnormalities in the manifestation of the P50

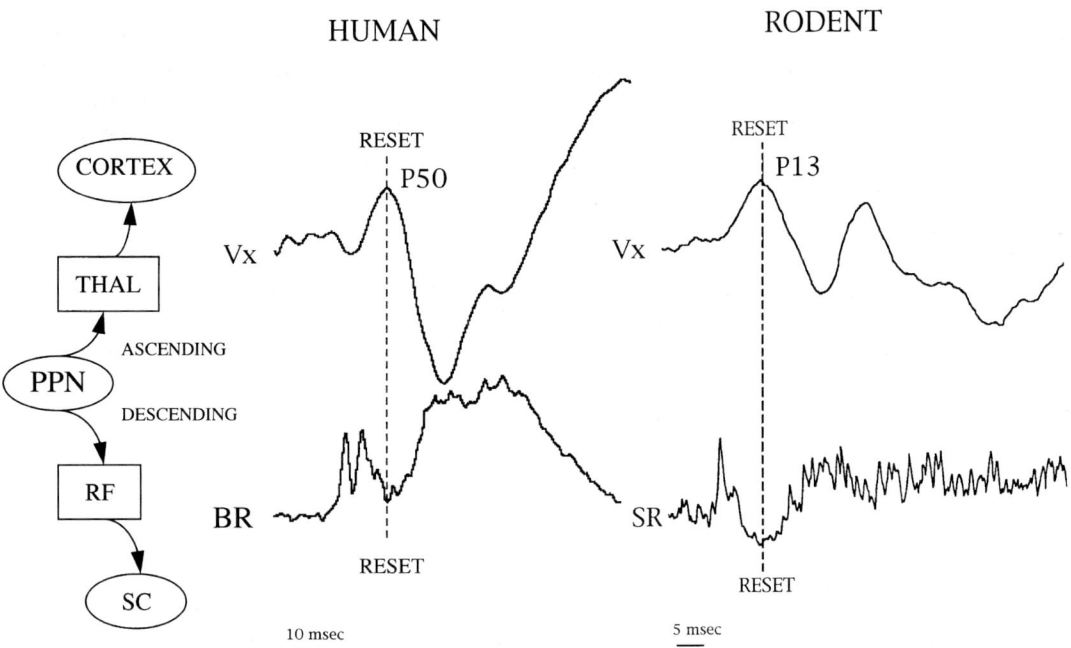

Fig. 1. Ascending (P50 and P13 potentials) and descending (blink and startle responses) manifestations of PPN projections. Left: The P50 auditory-evoked potential in relation to the human blink response and muscle 'reset' in preparation for movement. In the top recording, the peak of the averaged P50 potential (vertical dashed line) occurred concurrently with the 'reset' phase of the averaged eye blink response (BR) in the bottom recording in response to an auditory stimulus. Note the early phase of the BR, which began at about 25 ms, and the longer second phase of BR activity. These two phases were interrupted by a brief inhibitory period ('reset') coincident with the peak of the P50 potential. Right: The P13 auditory-evoked potential in relation to the rodent response and muscle 'reset' in preparation for movement. In the top recording, the peak of the averaged P13 potential (vertical dashed line) occurred concurrently with the 'reset' phase of the averaged startle response (SR) (neck EMG) in the bottom recording in response to an auditory stimulus. Note the early phase of the SR, which began at about 5 ms, and the longer second phase of SR activity. These two phases were interrupted by a brief inhibitory period ('reset') coincident with the peak of the P13 potential. The pathways outlined on the left indicate the role of the PPN in activating the cerebral cortex and modulating the SR. An auditory stimulus of sufficient loudness and novelty activates both *ascending* and *descending* projections of the PPN. Ascending projections activate thalamocortical fibers, which then activate the cortex. The P50 and P13 potentials are manifestations of this process in the respective species. Descending PPN projections modulate the SR in somatic muscles via the brainstem reticular formation (RF) and reticular projections to brainstem and spinal cord motoneurons. Other abbreviations: SC, spinal cord; THAL, thalamus.

potential should be present in disorders which include: (1) dysregulation of sleep–wake cycles; (2) abnormalities in reflex/postural, especially, startle, responses; and (3) malfunctions in fight versus flight responses.

Disorders of the RAS

In various psychiatric disorders, the P50 potential exhibits characteristic abnormalities, especially changes in sensory gating. A measure of sensory gating can be derived from the use of paired stimuli to test the level of habituation in the system being studied. Seminal studies in schizophrenic patients revealed that they do not inhibit the response to the second click stimulus under conditions in which normal subjects show a much reduced response, i.e., there was *decreased sensory gating* (Adler et al., 1982). Schizophrenic patients also show sleep-related symptoms, such as hypervigilance, reduced slow wave sleep, reduced REM sleep latency, exaggerated startle response and hallucinations (reviewed in Reese et al., 1995; Garcia-Rill, 1997). Postmortem studies on the brains of schizophrenic patients showed that in a population of nine institutionalized

(>30 years as inpatients), essentially intractable patients, there was an increase in PPN cell number (Karson et al., 1991). Recent studies have failed to replicate this finding in a small number of brains which were mainly from young outpatients, who were at least partially tractable (German et al., 1999; Manaye et al., 1999). In addition, the methodology in these studies was seriously flawed in: (1) failing to count a major portion of the cholinergic neuronal population by truncating the brainstem half way through the LC (this greatly underestimates the total number of mesopontine cholinergic neurons by eliminating most laterodorsal tegmental, LDT, cells); (2) failing to establish critical criteria for neurons in the PPN versus the LDT (increases only in PPN neurons were observed previously, so that shifting neurons to LDT would underestimate PPN cell counts); (3) counting neurons in every 10th or 20th section instead of every section or at least every 3rd section (making cell counts considerably less certain); (4) analyzing only one brain in the age range previously studied (another five were added later); and (5) not accounting for the vagaries of immunocytochemically (rather than histochemically) labeled neurons. Nevertheless, as previously suggested, increases in PPN cell number may only be present in *intractable* patients, perhaps those suffering from a much more severe form of the disorder stemming from developmental dysregulation. Younger, more tractable patients would *not* be expected to have cell increases or as severe a form of the disease (Garcia-Rill, 1997).

Recently, using a paired stimulus paradigm, decreased sensory gating of the P50 potential was described in subjects with posttraumatic stress disorder (PTSD), another disturbance marked by abnormalities of arousal and excitability. No difference was found between female rape victims with PTSD and male combat veterans with PTSD (Skinner et al., 1999). PTSD also is characterized by such sleep–wake state-related abnormalities as increased REM sleep drive, hyperarousal, hallucinations and exaggerated startle response (reviewed in Reese et al., 1995; Garcia-Rill, 1997). A similar decrease in sensory gating of the P50 potential was observed in patients in the late stages of Parkinson's disease (PD) (Teo et al., 1997), suggesting the presence of a sensory-gating deficit in this disorder.

The well-known symptoms of PD include resting tremor, rigidity, postural and gait abnormalities, bradykinesia and freezing episodes. In addition, a majority of untreated PD patients show sleep–wake disturbances, including 'light, fragmented sleep', reductions in slow wave sleep and frequent nocturnal arousals (reviewed in Teo et al., 1998). Interestingly, in PD patients who underwent bilateral pallidotomy for relief of symptoms of the disease, the decrease in sensory gating of the P50 potential was normalized (Teo et al., 1998). These studies were the first to suggest that mesopontine cholinergic neurons are upregulated in PD and that the normalization of P50 potential sensory gating following pallidotomy may be due to a reinhibition of the PPN (Teo et al., 1997, 1998). Studies in Huntington's disease (HD), another neurodegenerative disorder, suggest that sensory gating of this potential also is decreased (Uc et al., 2001). HD also is characterized by sleep dysregulation (Wiegand et al., 1991) and exaggerated startle response (Swerdlow et al., 1995). In addition, studies in depressed patients showed that decreased sensory gating is present (Clothier et al., 1998). Moreover, in a normal adolescent group (12–19 years), a small but significant decrease in sensory gating of the P50 potential, compared to three normal older age groups (24–39, 40–55 and 56–78 years), was found postpubertally (Rasco et al., 2000).

In other pathological states, it is the *amplitude* of the P50 potential that is reduced. It is reduced in Alzheimer's disease (AD) (Buchwald et al., 1989), autism (Buchwald et al., 1992) and narcolepsy (Boop et al., 1994), the latter being characterized by daytime somnolence, cataplexy, sleep paralysis and hypnagogic hallucinations. In general terms, the P50 potential appears to be *upregulated* (increased amplitude and/or decreased sensory gating) in disorders which are marked by upregulation of RAS outputs (hypervigilance), and *downregulated* in disorders characterized by decreased RAS outputs (hypovigilance). Many of the disorders studied have a developmental etiology and a postpubertal age of onset.

Current issues

Developmentally, REM sleep in man decreases from about 8 h/day in the newborn to about 1 h/day in the adult, and this decrease occurs mostly from birth until the end of puberty (Roffwarg et al., 1966). In the rat, the decrease in REM sleep occurs between 10 and 30 days of age, declining from over 75% of total sleep time to about 15% of sleep time (Jouvet-Mounier et al., 1970), a period which is also marked by significant hypertrophy in PPN neurons (Skinner et al., 1989). It has been suggested that the direction of these changes indicates that there is a REM sleep inhibitory process (RIP) that develops during the first 2 weeks of life in the rat (Vogel et al., 2000). This RIP may or may not be equivalent to the decrease seen in the human from birth through puberty. This hypothesis predicts that: (1) one or more inhibitory process becomes progressively stronger during this period (the PPN receives inhibitory serotonergic, noradrenergic, cholinergic and gabaergic inputs, all likely candidates); and (2) stimulation or blockade of this process will decrease or increase, respectively, the manifestations of REM sleep (Vogel et al., 2000). These studies point to a *switch in the polarity* of serotonergic inputs to PPN, suggesting a progressive increase in inhibition, during this developmental window (Homma et al., 2001). Figure 2 shows the distribution of responses of PPN neurons to the serotonergic agonist 5-carboxyamido-tryptamine (5-CT) between 12 and 21 days.

PPN neurons were recorded intracellularly from rat brainstem slices across this period and identified according to the presence of three types of PPN neurons based on their intrinsic membrane properties: type I (LTS current, noncholinergic), type II (A current, mostly cholinergic) and type III (A + LTS current, partly cholinergic), as previously described by various authors (Kang and Kitai, 1990; Leonard and Llinas, 1990; Kamondi et al., 1992). We found that the percent of type I (LTS) neurons increased from ∼10% at 12 days to ∼20% by 21 days, while the percent of type III (A + LTS) neurons decreased from ∼40% at 12–15 days to 0% at 17 days. This suggests that PPN neurons may differentiate into type I (LTS, bursting) neurons and type II (A, slow-firing) neurons across this critical stage in development. We tested the effects of the 5-HT$_1$ receptor agonist 5-CT and found that 12–16 day PPN neurons of all three types were depolarized (13%), unaffected (24%) or hyperpolarized (63%) by 5-CT superfusion, while 17–21 day PPN neurons were depolarized

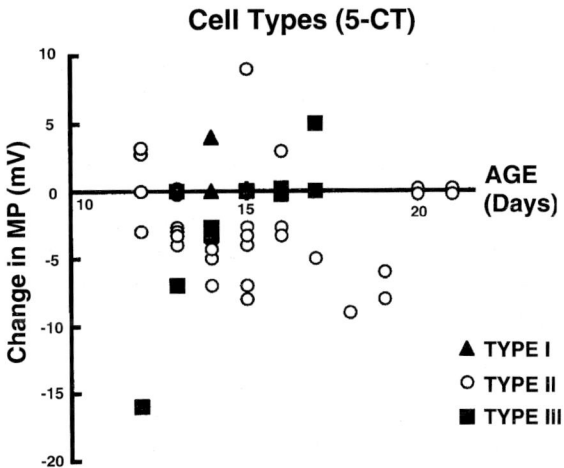

Fig. 2. Responses of PPN neurons in 12–21 day rat brainstem slices superfused with the serotonergic 5-HT$_1$ agonist 5-CT. Graph showing the extent of depolarization or hyperpolarization following superfusion of 5-CT (5 μM) in different PPN neurons as a function of age. All of the neurons recorded were verified histologically within the NADPH diaphorase-positive region of the PPN. PPN neurons were classified as type I, II or III depending on intrinsic current types. In general, the 5-HT$_1$-receptor agonist 5-CT both depolarized and hyperpolarized type II (mostly cholinergic) PPN neurons between 12 and 16 days of age, but hyperpolarized the same type cells between 17 and 21 days. Too few type I neurons have been studied, but 5-CT does not appear to hyperpolarize those cells studied to date. Type III neurons were hyperpolarized and one depolarized, but most were unaffected, as was a significant number of type II neurons. These findings suggest that 5-CT is both excitatory and inhibitory at ages during the beginning of the REM sleep drive decrease, but are purely inhibitory toward the later stages of this phenomenon. There is also the suggestion that the hyperpolarization induced in type II PPN neurons increases in amplitude (more negative membrane responses) with age. These findings support the hypothesis that a failure of the normal decrease in REM sleep drive across this age (a phenomenon terminating across puberty in the human but at around 30 days, prepubertally, in the rat) could lead to conditions characterized by permanent increases in REM sleep drive. Such an event would produce a condition marked by hypervigilance, sleep disturbances and exaggerated startle response.

(11%), unaffected (44%) or hyperpolarized (45%). When only type II, mostly cholinergic neurons were considered, at 12–16 days, 17% were depolarized and 70% were hyperpolarized, whereas at 17–21 days, none were depolarized while most were hyperpolarized. Moreover, the amplitude of the hyperpolarization appeared to increase with age. These preliminary findings suggest that serotonergic responses, especially in type II, mostly cholinergic PPN neurons, *switched from both excitatory and inhibitory before, to purely (and more strongly) inhibitory after, 17 days.* These findings indicate that a reorganization of PPN intrinsic membrane and serotonergic agonist responses occurs across this stage in development. The changes observed are in keeping with the proposed presence of a RIP during development.

A number of devastating disorders are marked by a *postpubertal increase in REM sleep drive*, and this induces a virtually permanent dysregulation in those afflicted. In schizophrenia, anxiety disorders and bipolar depression, increased REM sleep drive (increased REM duration, decreased REM latency, hypervigilance, etc., usually coupled with decreases in slow wave sleep) is a major, incapacitating symptom (Garcia-Rill, 1997). We know that the increased drive observed, for instance, in depression, can be normalized (perhaps in part by increasing inhibition to PPN neurons) following administration of serotonin reuptake inhibitors. *We hypothesize that some of the disorders with a postpubertal age of onset described previously may ensue because of a developmental failure of the normal decrease in REM sleep drive, essentially a blockade of the RIP.* That is, the RIP would normally tend to reduce REM sleep drive, but if it does not occur (i.e., if, for example, there is no switch in the polarity of serotonergic, noradrenergic or other inhibitory inputs to PPN neurons), a life-long hypervigilant state with exaggerated fight versus flight responses is a distinct possibility, such as is present in schizophrenia, bipolar disorder, etc.

Future possibilities

The P50 potential has a number of potential uses in exploring the functions of the ascending modulation of the RAS. Using a paired stimulus paradigm, the amplitude of the response following the first stimulus is a measure of the initial activation of the system, while the ratio of the amplitude of the second response in relation to the first is a measure of sensory gating or distractability. The P50 potential has excellent potential for studies on the effects of, for example, sleep deprivation and fatigue on distractability and attention. Because the P50 potential is a measure of ascending modulation by the RAS, it a more relevant marker of attentional mechanisms underlying cognitive performance than, for example, the SR, which is, after all, a measure of descending RAS modulation. Sensory gating of the P50 potential also can be achieved in one-fourth the time as SR sensory gating, due to the marked habituation of the SR and the need to significantly slow the frequency of stimulation. What does need to be determined is the degree to which decreased sensory gating of the P50 potential will influence cognitive performance. It can be assumed that processes/interventions requiring basic attentional mechanisms will affect P50 potential measures to a greater extent than those that do not.

Now that the rodent equivalent of the P50 potential has been identified and validated (Miyazato et al., 1995, 1999), the P13 potential can be used to explore animal models of a host of neurological and psychiatric disorders. The ability to manipulate developmental factors allows the determination of long-term effects of early insults on subsequent arousal, attentional activity and distractability. For example, developmental insults, such as mother deprivation, pain or early drug exposure, can be explored for their effects on P13 potential amplitude and sensory gating. It can be expected that, given that this potential reaches full expression only at puberty (Teneud et al., 2000), early insults may lead to dysregulation of P13 potential expression, but become manifested only after puberty (Teneud et al., 2001). Of course, given appropriate pharmacological intervention, such disturbances might be corrected or reduced.

Another line of potential future studies is the use of brainstem slices to determine factors that change across the critical period in development during the decrease in REM sleep drive. Given that serotonergic 5-HT$_1$ agonists may undergo a major reorganization across this developmental window, perhaps some of the other inhibitory inputs to PPN neurons also undergo changes. This may include noradrenergic, cholinergic and gabaergic inputs to PPN cells. Furthermore, it would be of interest to determine if a differential developmental dysregulation in each of these systems may lead to a different neuropsychiatric disorder. These are indeed exciting times in the study of a system which was first described over 50 years ago, and is now known to be involved in *all* of our sleeping and waking hours.

Acknowledgments

Supported in part by USPHS grant NS 20246.

Abbreviations

5-CT	5-carboxyamido-tryptamine
5-HT	5-hydroxytryptamine/serotonin
AD	Alzheimer's disease
EMG	electromyogram
HD	Huntington's disease
LC	locus coeruleus
LDT	laterodorsal tegmental nucleus
LTS	low threshold spike
PD	Parkinson's disease
PGO	ponto-geniculo-occipital
PPN	pedunculopontine nucleus*
PTSD	posttraumatic stress disorder
RAS	reticular activating system
REM	rapid eye movement
RIP	REM inhibitory process
RN	raphe nucleus

*Also called pedunculopontine tegmental nucleus (PPTN or PPTg)

References

Adler, L.E., Pachtman, E., Franks, R.D., Pecevich, M., Waldo, M.C. and Freedman, R. (1982). Neurophysiological evidence for a defect in neuronal mechanisms involved in sensory gating in schizophrenia. Biol. Psychiatry, 17: 639–654.

Boop, F.A., Garcia-Rill, E., Dykman, R. and Skinner, R.D. (1994). The P1: insights into attention and arousal. J. Pediatr. Neurosurg., 20: 57–62.

Buchwald, J.S., Erwin, R., Van Lancker, D. and Cummings, J.L. (1989). Midlatency auditory evoked responses: differential abnormality of P1 in Alzheimer's disease. Electroencephalogr. Clin. Neurophysiol., 74: 378–384.

Buchwald, J.S., Rubinstein, E.H., Schwafel, J. and Strandburg, R.J. (1991). Midlatency auditory evoked responses: differential effects of a cholinergic agonist and antagonist. Electroencephalogr. Clin. Neurophysiol., 80: 303–309.

Buchwald, J.S., Erwin, R., Van Lacker, D., Guthrie, D., Schwafel, J. and Tanguay, P. (1992). Midlatency auditory evoked responses: P1 abnormalities in adult autistic subjects. Electroencephalogr. Clin. Neurophysiol., 84: 164–171.

Clothier, J., Scruggs, J.T., Gamble, L.J., Skinner, R.D. and Garcia-Rill, E. (1998). The P1/P50 midlatency auditory evoked potential in depression. Sleep Res., 27A: 224.

Erwin, R.J. and Buchwald, J.S. (1986a). Midlatency auditory evoked responses: differential effects of sleep in the human. Electroencephalogr. Clin. Neurophysiol., 65: 383–392.

Erwin, R.J. and Buchwald, J.S. (1986b). Midlatency auditory evoked responses: differential recovery cycle characteristics. Electroencephalogr. Clin. Neurophysiol., 64: 417–423.

Garcia-Rill, E. (1991). The pedunculopontine nucleus. Prog. Neurobiol., 36: 363–389.

Garcia-Rill, E. (1997). Disorders of the reticular activating system. Med. Hypoth., 49: 379–387.

German, D.C., Manaye, K.F., Wu, D., Hersh, L.B. and Zweig, R.M. (1999). Mesopontine cholinergic and non-cholinergic neurons in schizophrenia. Neuroscience, 94: 33–38.

Homma, Y., Skinner, R.D. and Garcia-Rill, E. (2001). Pedunculopontine nucleus function during the developmental decrease in REM sleep. Soc. Neurosci. Abstr., 27: 596.9.

Jouvet-Mounier, D., Astic, L. and Lacote, D. (1970). Ontogenesis of the states of sleep in rat, cat, and guinea pig during the first postnatal month. Dev. Psychobiol., 2: 216–239.

Kamondi, A., Williams, J., Hutcheon, B. and Reiner, P. (1992). Membrane properties of mesopontine cholinergic neurons studied with the whole-cell patch-clamp technique: implications for behavioral state control. J. Neurophysiol., 68: 1359–1372.

Kang, Y. and Kitai, S.T. (1990). Electrophysiological properties of pedunculopontine neurons and their postsynaptic responses following stimulation of substantia nigra reticulata. Brain Res., 535: 79–95.

Karson, D.H., Garcia-Rill, E., Biedermann, J.A., Mrak, R., Husain, M. and Skinner, R.D. (1991). The brain stem reticular formation in schizophrenia. Psychiatr. Res., 40: 31–48.

Kevanishvili, Z. and von Specht, H. (1979). Human auditory evoked potentials during natural and drug-induced sleep. Electroencephalogr. Clin. Neurophysiol., 47: 280–288.

Leonard, C.S. and Llinas, R.R. (1990). Electrophysiology of mammalian pedunculopontine and laterodorsal tegmental neurons in vitro: implications for the control of REM sleep. In: Steriade M. and Biesold D. (Eds.), Brain Cholinergic Systems. Oxford Science, Oxford, pp. 205–223.

Manaye, K.F., Zweig, R., Wu, D., Hersh, L.B., Lacalle, S., Saper, C.B. and German, D.C. (1999). Quantification of cholinergic and select non-cholinergic mesopontine neuronal populations in the human brain. Neuroscience, 89: 759–770.

Miyazato, H., Skinner, R.D., Reese, N.B., Boop, F.A. and Garcia-Rill, E. (1995). A middle latency auditory evoked potential in the rat. Brain Res. Bull., 37: 265–273.

Miyazato, H., Skinner, R.D. and Garcia-Rill, E. (1999). Neurochemical modulation of the P13 midlatency auditory evoked potential in the rat. Neuroscience, 92: 911–920.

Moruzzi, G. and Magoun, H.W. (1949). Brain stem reticular formation and activation of the EEG. Electroencephalogr. Clin. Neurophysiol., 1: 455–473.

Rasco, L.M., Skinner, R.D. and Garcia-Rill, E. (2000). Effect of age on sensory gating of the sleep state-dependent P1/P50 midlatency auditory evoked potential. Sleep Res. Online, 3: 97–105.

Reese, N.B., Garcia-Rill, E. and Skinner, R.D. (1995). The pedunculopontine nucleus-auditory input, arousal and pathophysiology. Prog. Neurobiol., 47: 105–133.

Roffwarg, H.P., Muzio, J.N. and Dement, W.C. (1966). Ontogenetic development of the human sleep-dream cycle. Science, 152: 604–619.

Skinner, R.D., Conrad, N., Henderson, V., Gilmore, S. and Garcia-Rill, E. (1989). Development of NADPH diaphorase positive pedunculopontine neurons. Exp. Neurol., 104: 15–21.

Skinner, R.D., Rasco, L., Fitzgerald, J., Karson, C.N., Matthew, M., Williams, D.K. and Garcia-Rill, E. (1999). Reduced sensory gating of the P1 potential in rape victims and combat veterans with posttraumatic stress disorder. Depress. Anxiety, 9: 122–130.

Swerdlow, N.R., Paulsen, J., Braff, D.L., Butters, N., Geyer, M.A. and Swenson, M.R. (1995). Impaired prepulse inhibition of acoustic and tactile startle response in patients with Huntington's disease. J. Neurol. Neurosurg. Psychiatry, 58: 192–200.

Teneud, L., Homma, Y., Skinner, R.D. and Garcia-Rill, E. (2000). Development of the rodent midlatency auditory evoked P13 potential. Soc. Neurosci. Abstr., 26: 694.

Teneud, L., Homma, Y., Skinner, R.D. and Garcia-Rill, E. (2001). The rodent midlatency auditory evoked P13 potential following maternal deprivation. Soc. Neurosci. Abstr., 27: 976.1.

Teo, C., Rasco, L., Al-Mefty, K., Skinner, R.D. and Garcia-Rill, E. (1997). Decreased habituation of midlatency auditory evoked responses in Parkinson's disease. Mov. Disord., 12: 655–664.

Teo, C., Rasco, L., Skinner, R.D. and Garcia-Rill, E. (1998). Effect of pallidotomy on the disinhibition of the P1 midlatency auditory evoked potential in parkinsonism. Sleep Res. Online, 1: 62–70.

Uc, E., Skinner, R.D. and Garcia-Rill, E. (2001). Reticular activation system is dysregulated in Huntington's disease: a P50 auditory evoked potential study. Neurology, 56: A23.

Vogel, G.W., Feng, P. and Kinney, G.G. (2000). Ontogeny of REM sleep in rats: possible implications for endogenous depression. Physiol. Behav., 68: 453–461.

Wiegand, M., Moller, A.A., Lauer, C.J., Stoltz, S., Schreiber, W., Dose, M. and Krieg, J.C. (1991). Nocturnal sleep in Huntington's disease. J. Neurol., 238: 203–208.

CHAPTER 29

Switching between cortical and subcortical sensorimotor pathways

Tadashi Isa* and Yasushi Kobayashi

Department of Integrative Physiology, National Institute for Physiological Sciences, Okazaki 444-8585, Japan

Abstract: It is well known that the reaction times of visually guided saccades exhibit a bimodal distribution. Those with extremely short reaction times are termed 'express saccades'. In their case, visual input appears to be transformed into motor output via a 'short-loop', brainstem-mediated pathway. In contrast, those with longer reaction times are called 'regular saccades'. The latter are presumably executed via a cortically mediated, 'long-loop' sensorimotor pathway. The 'gate' that switches signal flow between the short and long loop is thought to be located in between the superficial and deeper layers of the superior colliculus (SC). Nonlinear signal amplification mechanisms, which operate in local circuits of the deeper SC layers may underlie this gating function, with switching of the gate regulated in a context-dependent manner by inputs from the cerebral cortex and basal ganglia.

Introduction

The sensorimotor loops in the mammalian CNS are multilayered. Sensory afferents can directly elicit motor responses via spinal- or brainstem-mediated reflex loops (Sherrington, 1906). In addition, a sensory signal may be transformed relatively quickly into a motor command via a 'transcortical reflex' pathway (Evarts and Fromm, 1981). Furthermore, when there is need for sensory input to be integrated into more complex cognitive processes such as judgment, inference and choice, its transformation into a motor command is by way of 'long loop' hierarchically arranged higher-order structures, including the association cortex, basal ganglia, etc. The focus of this chapter is on how signal processing between different sensorimotor loops is switched according to the behavioral context. We believe that the visually guided saccadic eye movement is an ideal model to address this issue (see also Chapter 41 of this volume).

It is generally accepted that the complexity of a sensorimotor transformation process is reflected in its reaction time (Posner, 1980): the more complex the process, the longer its reaction time. It is well known that in well-trained subjects, the reaction times of visually guided saccades (saccadic reaction times, SRTs) exhibit a bimodal or multimodal distribution. The earliest peak at 80–120 ms includes 'express saccades' with extremely short reaction times. Their precise range differs across reported studies. Later peaks are observed at 120–300 ms for 'regular saccades' (Fischer and Boch, 1983).

Importantly, SRTs are context-dependent. Fischer and Boch (1983) were the first to show that introduction of a several hundred-millisecond time gap between the fixation offset and target onset (gap paradigm) markedly shortened the SRTs (the 'gap effect'). Subsequently, it was reported that the

*Corresponding author: Tel.: +81-564-55-7859; Fax: +81-564-55-7790; ; E-mail: tisa@nips.ac.jp

predictability of target location, target size and daily practice affected the frequency of occurrence of express saccades (Fischer et al., 1984). Thus, the occurrence of express saccades depends on the attentional state of the subject (Fischer and Weber, 1993). To explain the multimodal distribution of SRTs, Fischer introduced a 'three-loop' model (Fischer and Weber, 1993). It is comprised of three hierarchically arranged sensorimotor loops of different transmission time, which exert their influence by affecting successively lower-order loops. The authors proposed that express saccades occur when the incoming sensory signal is processed by the loop with the shortest transmission time. In addition, they proposed that switching between the loops is modulated by the effect of attention. According to the three-loop model, the shortest (fastest) pathway is mediated by the primary visual cortex (V1) and the superficial layer of the superior colliculus (SC). The model proposes that slower saccades are mediated by other cortical routes that bypass the SC, e.g., the frontal eye field (FEF) and parietal cortex. A SC route for express saccades has been supported by Schiller et al. (1987), who showed that SC ablation obviated express saccades. There are many reservations, however, about the three-loop model (Schall and Hanes, 1993; Van Gisbergen and Minken, 1993), and recent physiological data do not support its details in their entirety (Dorris et al., 1997). Nonetheless, the basic thrust of the 'three-loop model', that signal flow is switched between loops depending on the attentional state, has set the stage for advancing understanding of the role of attention in the neural control of movement.

In this chapter, we use recently acquired electrophysiological data to propose that: (1) express saccades can be induced via a brainstem-mediated loop; and (2) the 'gate' that switches signal flow between brainstem-mediated and transcortically mediated loops is located between the superficial and a deeper layer of the SC.

Activity of SC neurons during express versus regular saccades

Figure 1 shows the visuomotor pathways thought to be involved in eliciting visually guided saccades. Figure 2 exemplifies the activity of SC neurons in the superficial layer (SGS; a 'visual' cell; i.e., one that

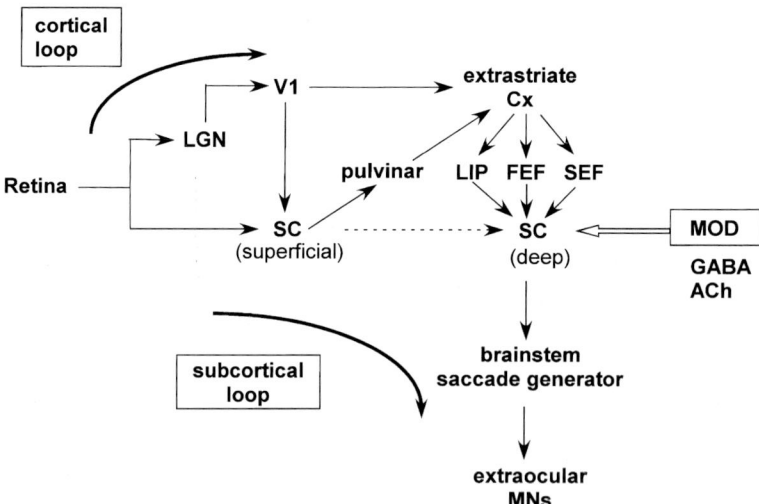

Fig. 1. Schematic diagram of the visuomotor pathways that regulate execution of visually guided saccades. Transcortical and subcortical (brainstem-mediated) loops. MOD, modulating factors (e.g., GABAergic and cholinergic systems), which switch signal flow between the cortical and subcortical loops at the level of the deeper layer of the SC. Other abbreviations: Cx, cortex; FEF, frontal eye field; LGN, lateral geniculate nucleus; LIP, lateral intraparietal sulcus; MNs, motoneurons; SEF, supplementary eye field; V1, primary visual cortex.

responds to visual stimuli presented in their receptive field) and in the intermediate layer (SGI; a 'visuomotor' cell; one that exhibits both visual activity and presaccadic activity) of a macaque monkey. Cell activity is shown during regular saccades in no-gap trials versus express saccades in gap trials.

The SGS neuron exhibited visual responses at a latency of 40–50 ms. Their magnitude was virtually the same during the regular versus express saccades, thereby suggesting that the enhancement of neuronal activity at the SGS level is not the direct cause of express saccades. In contrast, the visuomotor cell in SGI exhibited phasic visual responses. In the no-gap paradigm, this cell's responses at a latency of 40–50 ms, which was virtually the same as the visual response of the SGS neuron, did not appear to trigger saccades. The strong bursting responses, however, 50–200 ms after the initial visual response, did indeed appear to initiate the saccades (Fig. 2A). These longer SRTs fall in the range of regular saccades (120–300 ms). On the other hand, during express saccades in the gap paradigm, this neuron exhibited a gradual increase in activity during the gap period. This was followed by strong bursting activity at the same latency as this cell's visual responses during regular saccades. This bursting activity occurs almost simultaneously as the visual response in the SGS neuron (Fig. 2B). Such SGI cell behavior during the gap paradigm supports the previous work of Dorris et al. (1997). The gradual increase in activity during the gap period may be caused by both excitatory input from outside the SC and release from inhibitory inputs from fixation neurons located in either the rostral pole of the SC (Dorris and Munoz, 1995) or the substantia nigra pars reticulata (Hikosaka and Wurtz, 1983).

The Fig. 2B observation suggests that if an SGI neuron's activity is already sufficiently high at the time of target presentation, then its threshold for a bursting response is lowered. Its visual activity can then become pronounced enough to produce express saccades. It can be proposed that excitation from SGS neurons leads to the visual response of SGI neurons and directly triggers their bursting response immediately prior to express saccades. This hypothesis suggests that express saccades can be induced via a brainstem-mediated sensorimotor loop. For the moment, however, mechanisms leading to the visual response of SGI neurons are unclear. They may arise from the SGS, or the cerebral cortex, or both. Thus, to test this hypothesis, two of its key aspects should first be clarified: (1) the existence of a direct excitatory connection from the optic tract to the SGI-mediated via SGS neurons (i.e., an 'interlaminar connection of the SC'); and (2) the neuronal mechanism(s) underlying an all-or-none-like property in the induction of the bursting response of SGI neurons that triggers saccades.

The authors' views on these two issues are presented in the next section.

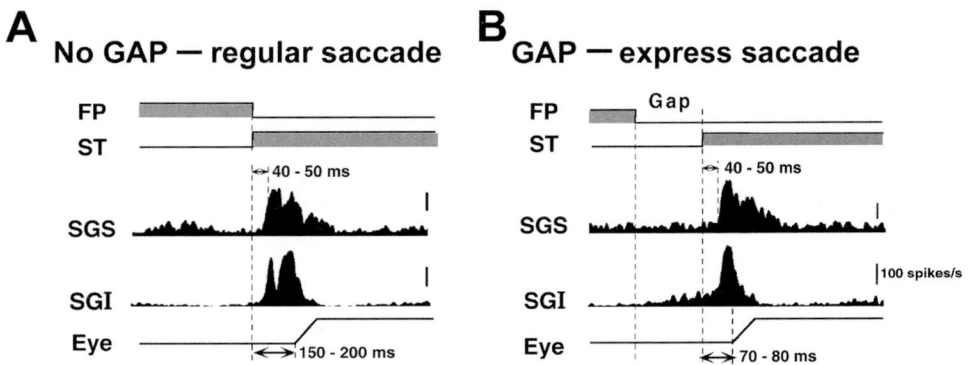

Fig. 2. Activity of SC neurons during saccades in the monkey. Spike density functions of exemplary neurons are shown. One was located in the superficial layer (SGS) of the SC, and one in the intermediate layer (SGI). (A) Responses during regular saccades in the no-gap paradigm. (B) Responses during express saccades in the gap paradigm. FP, presentation of fixation point; ST, presentation of saccade target; Eye, eye position.

Is the express saccade brainstem-mediated and does it require an interlaminar connection in the SC?

Existence of an interlaminar connection in the SC has been an open issue for over two decades. First, the connection seemed likely on the basis of demonstration by Maeda et al. (1979) of disynaptic EPSPs in SGI neurons following stimulation of the optic disk. Shortly thereafter, however, Edwards (1980) found no anatomical evidence of such a connection. Also negative, was the observation of Mays and Sparks (1980) that in select instances, such as double-step saccades, vigorous activity of SGS neurons does not always accompany bursting responses of neurons in the underlying layers of the monkey SC. More recently, however, anatomical studies have demonstrated the interlaminar connection (Mooney et al., 1988; Behan and Appel, 1992; Mooney et al., 1992; Hall and Lee, 1993). Furthermore, Hall and colleagues (Lee et al., 1997) and the authors' group (Isa et al., 1998a) have shown quite recently that electrical stimulation of the SGS induces excitatory postsynaptic currents/potentials in SGI neurons. These works were undertaken using the whole-cell, patch-clamp recording technique in brainstem slices obtained from the tree shrew and rat, respectively. In addition, the authors' group showed that release from GABAergic inhibition (by application of bicuculline) was essential for SGS-induced bursting responses in SGI neurons. In summary, the current state-of-the-play is that an anatomical interlaminar connection is clearly evident in the mammalian SC. Signal flow through this pathway, however, is subject to flexible modulation by GABAergic inhibition.

The above findings do not necessarily mean that express saccades are purely brainstem-mediated. It is still possible that cortical areas are also involved, especially the primary visual cortex and the FEF. This should be tested for, particularly in primates, in which role of cortical areas in control of saccades has been intensively studied. For example, FEF neurons have been shown to exhibit a 'gap effect' (Dias and Bruce, 1994; Everling and Munoz, 2000), thereby suggesting that they may be involved in producing express saccades. We have shown recently, however, that while local microinjection of nicotine into the monkey SC increases the frequency of express saccades, a stronger nicotine dose does not further shorten the SRTs (Aizawa et al., 1999). This is a strong evidence for the saccade signal passing through the *shortest possible pathway* after the local injection of nicotine into the SC. Even so, the evidence does not exclude the possibility that cortical areas are indeed involved in induction of the express saccades.

The neural basis of 'gating' and its modulation

When their activity level is sufficiently high, SGI neurons exhibit strong, visually induced, motor-linked bursts of discharge (Dorris et al., 1997). The induction of these bursting responses appears to be all-or-none-like. Such a 'threshold' effect may function as a 'gate' that switches on and off the signal flow between brainstem-mediated and cortically mediated sensorimotor loops. We have studied the neural possibility for such a gate, using brainstem slice preparations. We first investigated whether SGI neurons exhibit nonlinear intrinsic properties like the plateau potential (Hounsgaard et al., 1984). This possibility was excluded, because, in most cases, the f–I (spike frequency–stimulus current strength) relation was linear-like (Saito and Isa, 1999). On the other hand, in response to the type of synaptic activation described previously, SGI neurons displayed SGS-induced bursting responses only when the circuit was released from GABAergic inhibition by the application of bicuculline (Isa et al., 1998a). Such a bursting response often exhibited an all-or-none-like property. This was dependent on the activation of D-APV-sensitive, NMDA-type glutamate receptors (Saito and Isa, 2003).

Cholinergic inputs to the SGI, which were first demonstrated by Graybiel (1978), also lower the threshold of the SGI neurons' bursting response to excitatory inputs. We showed this by demonstrating that activation of nicotinic acetylcholine receptors (nAChRs) depolarizes the output neurons of the SGI (Isa et al., 1998b). Such depolarization reduces the 'Mg^{2+} block' of NMDA receptors, and thereby markedly enhances the EPSP level in SGI neurons, which leads, in turn to these cells' lower burst-response threshold to SGS stimulation. In summary, continuous cholinergic inputs to the SGI activate its neurons' nAChRs, and thereby lower the threshold for signal flow through the SGS–SGI pathway. This mechanism appears to be the most likely basis for the observation that injection of nicotine into the monkey SC increases the frequency of express saccades.

Discussion

The long-standing debate on the existence of an interlaminar connection in the SC has now been resolved to large extent. Studies using brainstem slice preparations from tree shrews (Lee et al., 1997) and rodents have clearly demonstrated the anatomical connection. Functional signal flow through the pathway appears to be dynamically modulated, however, at the level of the SGI. Importantly, this signal flow is behavioral context-dependent. The detailed analysis (Isa et al., 1998a) revealed that at least two factors (GABAergic inhibition and cholinergic facilitation) are implicated in modulating the gating threshold of this interlaminar connection. We concede, of course, that the current extrapolation of rodent slice results to primate behavior must now be verified by undertaking analogous slice experiments in the primate. Recall, however, that the fundamental cytoarchitecture of the SC is similar across mammalian species (Sprague, 1975). This certainly gives credibility and generality to the rodent slice results, including the previous interpretation of the results on the effects of nicotine microinjection into the monkey SC (Aizawa et al., 1999).

The SGI's GABAergic inhibition and cholinergic input

Anatomical and eletrophysiological studies suggest that one possible source of the GABAergic inhibition of SGI cells is local inhibitory interneurons excited by projections from fixation cells in the rostral pole of the SC (Munoz and Wurtz, 1993; Meredith and Ramoa, 1998). These fixation neurons may be under control of cortical areas such as the FEF, supplementary eye field and the substantia nigra pars reticulata (Aizawa et al., 1995). In addition, it is well known that GABAergic neurons in the substantia nigra pars reticulata directly inhibit the output SGI cells (May and Hall, 1984). This latter pathway has not been studied in slice preparations, however.

Cholinergic inputs to SGI cells originate in the pedunculopontine and dorsolateral tegmental nuclei of the midbrain (Hall et al., 1989; Yamamoto and Isa, unpublished observations). These latter neurons receive excitatory input from the frontal cortex (Matsumura et al., 2000) and inhibitory input from basal ganglia structures, including substantia nigra pars reticulata (Grofova and Zhou, 1998).

Sensorimotor-gating hypothesis

The previous results are the basis for the hypothesis that circuits originating in the cerebral cortex and basal ganglia play a key role in regulating sensorimotor gating between brainstem-mediated and transcortical loops. This hypothesis, which accommodates behavioral context-dependent effects on such gating, is based on the observations on visually guided saccades, which feature reaction times with clear-cut bi- and multimodal distributions. It remains open as to whether this oculomotor-based hypothesis can be generalized to other motor-control systems, such as those for forearm and digit movements. Analogous testing seems now feasible and necessary in these other systems.

Acknowledgments

We thank Chika Sasaki and Michi Seo for technical assistance. The studies were supported by grants from the Ministry of Education, Culture, Sports, Science and Technology of Japan, CREST of Japan Science and Technology Agency, Naito Memorial Foundation, Uehara Memorial Foundation, Mitsubishi Foundation and Daiko Foundation to T.I.

Abbreviations

CNS	central nervous system
D-APV	2-D-amino-5-phosphonovelerate
EPSP	excitatory postsynaptic potential
FEF	frontal eye field
GABA	gamma aminobutyric acid
NAChR	nicotinic acetylcholine receptor

NMDA N-methyl-D-aspartic acid
SC   superior colliculus
SGI   stratum griseum intermediale (intermediate gray layer of superior colliculus)
SGS   stratum griseum superficiale (superficial gray layer of superior colliculus)
SRTs saccadic reaction times
V1   primary visual cortex

References

Aizawa, H., Gerfen, C.R. and Wurtz, R.H. (1995) Afferent connection to fixation zone of monkey superior colliculus from frontal cortex and basal ganglia. Soc. Neurosci. Abstr., 21: 468.9.

Aizawa, H., Kobayashi, Y., Yamamoto, M. and Isa, T. (1999) Injection of nicotine into the superior colliculus facilitates occurrence of express saccades in monkeys. J. Neurophysiol., 82: 1642–1646.

Behan, M. and Appel, P.P. (1992). Intrinsic circuitry in the cat superior colliculus: projection from the superficial layers. J. Comp. Neurol., 315: 230–243.

Dias, E.C. and Bruce, C.J. (1994). Physiological correlate of fixation disengagement in the primate's frontal eye field. J. Neurophysiol., 72: 2532–2537.

Dorris, M.C. and Munoz, D.P. (1995). A neural correlate for the gap effect on saccadic reaction times in monkey. J. Neurophysiol., 73: 2558–2562.

Dorris, M.C., Paré, M. and Munoz, D.P. (1997). Neuronal activity in monkey superior colliculus related to the initiation of saccadic eye movements. J. Neurosci., 17: 8566–8579.

Edwards, S.B. (1980). The deep cell layers of the superior colliculus: their reticular characteristics and structural organization. In: Hobson J.A. and Brazier M.A.B. (Eds.), The Reticular Formation Revisited. Raven Press, New York, pp. 193–209.

Evarts, E.V. and Fromm, C. (1981). Transcortical reflexes and servo control of movement. Can. J. Physiol. Pharmacol., 59: 757–775.

Everling, S. and Munoz, D.P. (2000). Neuronal correlates for preparatory set associated with pro-saccades and anti-saccades in the primate frontal eye field. J. Neurosci., 20: 387–400.

Fischer, B. and Boch, R. (1983). Saccadic eye movements after extremely short reaction times in the monkey. Brain Res., 260: 21–26.

Fischer, B. and Weber, H. (1993). Express saccade and visual attention. Behav. Brain Sci., 16: 553–610.

Fischer, B., Boch, R. and Ramsperger, E. (1984). Express-saccades of the monkey: effects of daily training on probability of occurrence and reaction time. Exp. Brain Res., 55: 232–242.

Graybiel, A.M. (1978). A stereometric pattern of distribution of acetylcholinesterase in the deep layers of the superior colliculus. Nature, 272: 539–541.

Grofova, I. and Zhou, M. (1998). Nigral innervation of cholinergic and glutamatergic cells in the rat mesopontine tegmentum: light and electron microscopic anterograde tracing and immunohistochemical studies. J. Comp. Neurol., 395: 359–379.

Hall, W.C. and Lee, P. (1993). Interlaminar connections of the superior colliculus in the tree shrew. I. The superficial layer. J. Comp. Neurol., 332: 213–223.

Hall, W.C., Fitzpatrick, D., Klatt, L.L. and Raczkowski, D. (1989). Cholinergic innervation of the superior colliculus in the cat. J. Comp. Neurol., 287: 495–514.

Hikosaka, O. and Wurtz, R.H. (1983). Visual and oculomotor functions of monkey substantia nigra pars reticulata. II. Visual responses related to fixation of gaze. J. Neurophysiol., 49: 1254–1267.

Hounsgaard, J., Hultborn, H., Jespersen, B. and Kiehn, O. (1984). Intrinsic membrane properties causing a bistable behaviour of alpha-motoneurones. Exp. Brain Res., 55: 391–394.

Isa, T., Endo, T. and Saito, Y. (1998a). The visuomotor pathway in the local circuit of the rat superior colliculus. J. Neurosci., 18: 8496–8504.

Isa, T., Endo, T. and Saito, Y. (1998b). Nicotinic facilitation of signal transmission in the local circuits of the rat superior colliculus. Soc. Neurosci. Abstr., 24: 60.13.

Lee, P.H., Helms, M.C., Augustine, G.J. and Hall, W.C. (1997). Role of intrinsic synaptic circuitry in collicular sensorimotor integration. Proc. Natl. Acad. Sci. USA, 94: 13299–13304.

Maeda, M., Shibazaki, T. and Yoshida, K. (1979). Labyrinthine and visual inputs to the superior colliculus neurons. Prog. Brain Res., 50: 735–743.

Matsumura, M., Nambu, A., Yamaji, Y., Watanabe, K., Imai, H., Inase, M., Tokuno, H. and Takada, M. (2000). Organization of somatic motor inputs from the frontal lobe to the pedunculopontine tegmental nucleus in the macaque monkey. Neuroscience, 98: 97–110.

May, P.J. and Hall, W.C. (1984). Relationships between the nigrotectal pathway and the cells of origin of the predorsal bundle. J. Comp. Neurol., 226: 357–376.

Mays, L.E. and Sparks, D.L. (1980). Dissociation of visual and saccade-related responses in superior colliculus neurons. J. Neurophysiol., 43: 207–232.

Meredith, M.A. and Ramoa, A.S. (1998). Intrinsic circuitry of the superior colliculus: pharmacophysiological identification of horizontally oriented inhibitory interneurons. J. Neurophysiol., 79: 1597–1602.

Mooney, R.D., Nikoletseas, M.M., Hess, P.R., Allen, Z., Lewin, A.C. and Rhoades, R.W. (1988). The projection from the superficial to the deep layers of the superior colliculus: an intracellular horseradish peroxidase injection study in the hamster. J. Neurosci., 8: 1384–1399.

Mooney, R.D., Huang, X. and Rhoades, R.W. (1992). Functional influence of interlaminar connections in the hamster's superior colliculus. J. Neurosci., 12: 2417–2432.

Munoz, D.P. and Wurtz, R.H. (1993). Fixation cells in monkey superior colliculus. I. Characteristics of cell discharge. J. Neurophysiol., 70: 559–575.

Posner, M.I. (1980). Orienting of attention. Q. J. Exp. Psychol., 32: 3–25.

Saito, Y. and Isa, T. (1999). Electrophysiological and morphological properties of neurons in the rat superior colliculus. I. Neurons in the intermediate layer. J. Neurophysiol., 82: 754–767.

Saito, Y. and Isa, T. (2003). Local excitatory network and NMDA receptor activation generate a synchronous and bursting command from the superior colliculus. J. Neurosci., 23: 5854–5864.

Schall, J.D. and Hanes, D.P. (1993). Saccade latency in context: regulation of gaze behavior by supplementary eye field. Behav. Brain Sci., 16: 588–589 (Commentary).

Schiller, P.H., Sandell, J.H. and Maunsell, J.H.R. (1987). The effect of frontal eye field and superior colliculus lesions on saccadic latencies in the rhesus monkey. J. Neurophysiol., 57: 1033–1049.

Sherrington, C. (1906). The Integrative Action of the Nervous System. New Haven, Yale University Press.

Sprague, J.M. (1975) Mammalian tectum: intrinsic organization, afferent inputs, and integrative mechanisms. In: Ingle, D. and Sprague, J.M. (Eds.), Sensorimotor Function of the Midbrain Tectum, Neurosci. Res. Program Bull., Vol. 13, No. 2. MIT Press, Cambridge, MA, pp. 204–214.

Van Gisbergen, J.A.M. and Minken, A.W.H. (1993). Toward an alternative scheme for the generation of express saccades. Behav. Brain Sci., 16: 591–592 (Commentary).

SECTION VIII

Cerebellar interactions and control mechanisms

CHAPTER 30

Cerebellar activation of cortical motor regions: comparisons across mammals

Tetsuro Yamamoto*, Yoshihiro Nishimura, Toru Matsuura,
Hiroshi Shibuya, Min Lin and Toshihiro Asahara

Department of Physiology, Faculty of Medicine, Mie University, Tsu 514-8507, Japan

Abstract: This chapter discusses the nature of mammalian cerebello-thalamo-cortical projections. These play an important role in motor control, but with species differences evident in the innervation patterns of thalamo-cortical (T-C) fibers relaying cerebellar inputs. A phylogenetic comparison of the mode of cerebellar activation of cortical motor regions reveals that cerebellar inputs are relayed by the deep and superficial T-C projections in the cat, but predominately by the latter in rat and monkey. Another difference across mammals is the nature of cerebellar activation of cortical neurons. Fast-conducting pyramidal tract neurons in the cat routinely receive fast-rising EPSPs from the deep cerebellar nuclei (CN). This tendency is not observed in rat and monkey, however. These findings suggest that the responsiveness of cortical output neurons also shows species differences, these having bearing on the development of species-specific motor skills. Common to all three species, however, are fast-rising and large CN-EPSP responses of layer III pyramidal neurons to CN input. It is argued that layer III pyramidal neurons modulate cerebellar input to layer V pyramidal neurons, which latter cells provide command signals from the motor cortex to the lower centers.

Introduction

The cerebello-thalamo-cortical projection plays an important role in motor control, with the cerebellar nuclear (CN) inputs being relayed into the motor and association cortical regions via the thalamic ventral tier nuclei (Jones, 1985). This projection system has been studied electrophysiologically in the rat, cat and monkey (Sasaki et al., 1972, 1976, 1979; Yamamoto et al., 1979, 1983; Kawaguchi et al., 1983). It is intriguing phylogenetically that not only do the cortical receiving regions for cerebello-thalamic inputs differ across species, but so too do the thalamo-cortical (T-C) electrophysiological responses to cerebellar inputs. For example, Fig. 1 shows that in the rat, the surface-recorded CN-evoked response

is usually a negative-going potential followed (sometimes) by a small positive one (Yamamoto et al., 1979). In the cat, however, CN-evoked responses consist of surface positive–negative potentials in the motor cortex (MCx), whereas a pure surface-negative potential is observed in the lateral and middle suprasylvian gyri of the parietal cortex (PCx) and the anterior part of the ectosylvian gyrus of the insula (ICx; Sasaki et al., 1972; Noda and Oka, 1985). In the monkey, a surface-negative potential is observed largely in the lateral part (face and hand regions) of the MCx, whereas a surface-positive potential is induced in its medial part (hindlimb region) and in the PCx, as well (Sasaki et al., 1976, 1979; Yamamoto et al., 1983).

Strikingly, as shown in Fig. 1, the species' differences in CN-evoked responses are in sharp contrast to sensory-evoked potentials. Auditory-, somatosensory- and visual-evoked potentials are relatively similar across species, being generally

*Corresponding author: Tel.: +81-59-231-5005;
Fax: +81-59-231-5005;
E-mail: yamamoto@doc.medic.mie-u.ac.jp

comprised of surface positive–negative potentials (Creutzfeldt and Kuhnt, 1973; Libet, 1973; Keidel, 1976; Poggio and Mountcastle, 1980; Miyata et al., 1982; Kato et al., 1983). The uniqueness of the CN-evoked responses may reflect the evolutionary development of skillfulness in hand–finger manipulation and a capacity for a broader array of facial expressions among more- versus less-advanced species.

Field potentials in the cerebral cortex are presumably due to synaptic potentials (mainly EPSPs) induced by T-C inputs. Accordingly, laminar field potential analysis should indicate the cortical location of thalamic EPSPs. For example, a species difference in the profile of CN-evoked field potentials should mean that the laminar distribution of T-C terminals relaying CN information to the MCx also varies across species. To test this idea, we have examined comparative aspects of the mode of the MCx neuron responsiveness to CN inputs (see also Rouiller and Olivier, Chapter 44 of this volume).

Comparison of CN-evoked, surface-recorded potentials across mammals

The findings discussed in this section are summarized in the Fig. 1 sketches for rat versus cat versus monkey of schematic drawings of surface-recorded potentials,

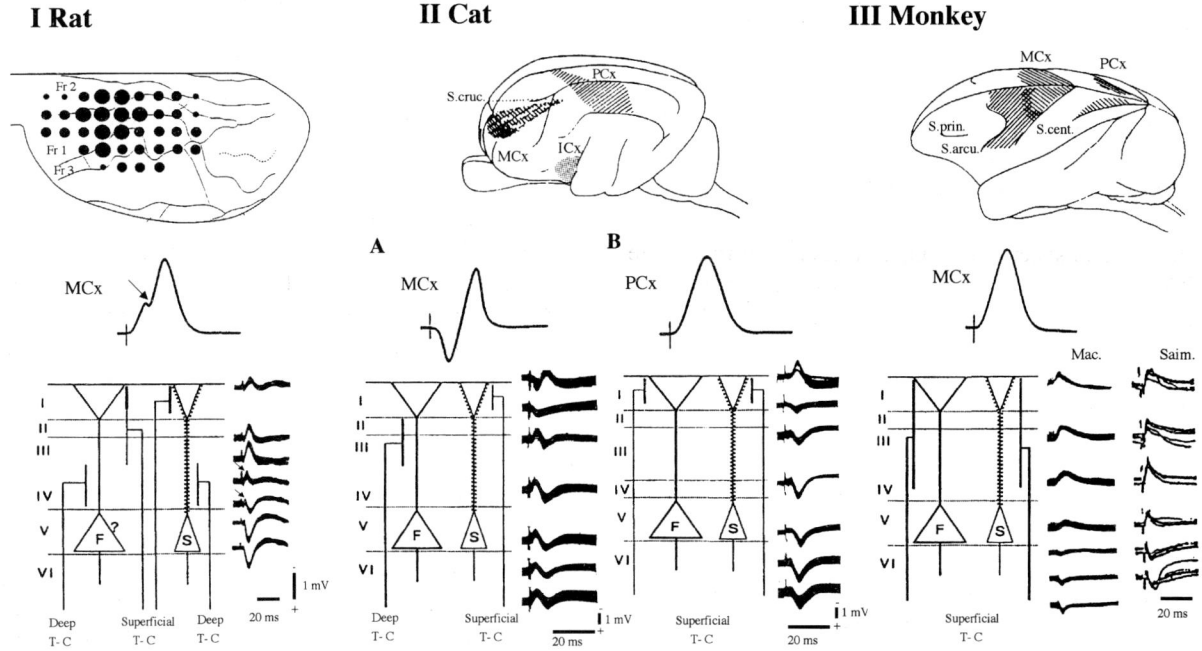

Fig. 1. Different modes of CN-activation across species and cortical regions. Schema of CN-responsive cortical regions in the rat (I), cat (II) and monkey (III). The sites of CN stimulation were all in the lateral nucleus. Upper row sketches shaded regions and dots indicate cortical regions responsive to CN input. Middle row traces, schematic profiles of cortical CN-evoked responses (negative upward). Lower row, left-side sketches, schematic drawings of cortical layers with fast (F)- and slow (S)-conducting PTNs, and the pattern of their deep and superficial innervation by thalamic afferents relaying CN inputs to the MCx. Lower row, right-side column, raw CN-evoked responses shown at their near-actual depth. In I, an arrow in the middle row trace indicates a positive notch on the rat's surface-negative potential. This is the equivalent of a negative potential (small arrows) in the depth recordings of the lower row, right-side column. In the rat, fast PTNs have not been truly identified and are shown with a question mark (?). Abbreviations: Fr 1, frontal area 1; Fr 2, frontal area 2; Fr 3, frontal area 3 (these three areas appropriate for the rat MCx; Zilles and Wree, 1995); ICx, insular cortex; Mac., *Macaca fuscata*; MCx, motor cortex; PCx, parietal cortex; S. arcu., arcuate sulcus; S. cent., central sulcus; S. cruc., cruciate sulcus; S. prin., principal sulcus; Saim., *Saimiri sciureus*. (Reproduced, in part, from Sasaki et al., 1972, 1979 with permission from Springer-Verlag, and from Yamamoto et al., 1979, 1983 with permission from Elsevier Science.)

the associated schemas of T-C terminal distributions and the raw depth profiled of the CN-evoked potentials.

Rat

In the rat, CN-induced responses in the MCx are composed of a dominant surface-negative potential with a small positive notch, which latter inflection is due to the deep T-C response (defined in the next section). The reversal level of surface-negative potentials is at a cortical depth of ~ 800 μm, but the small positive potentials reverse at 250 μm (see arrows in Fig. 1I, raw potential profiles).

Cerebellar nuclear-induced responses in the MCx must influence descending motor command signals, which, after all, are an integration of cortical designs for movement and posture. Sasaki et al. (1976) have suggested that the predominance of surface-negative potentials in the lateral part of the MCx might reflect the evolution of motor skills concerning hand–finger and facial movements in primates. Interestingly, Sasaki et al.'s (1976) suggestion might also be applicable to the rat, in which species the shape of the forelimb is quite different from that of the cat and primate. Contrary to general knowledge, the rat can indeed manipulate the fingers of its forelimbs, when small objects such as unshelled sunflower seeds are presented as food (Castro, 1972; Kalil and Schneider, 1975).

Cat

Electrophysiological studies of CN-evoked potentials have been performed extensively in the cat. Such studies clearly demonstrate that CN stimulation induces surface positive–negative potentials in the MCx, which reverse their polarity at a depth of ~ 250–300 μm (Sasaki et al., 1972). In the cat PCx and ICx, CN-induced potentials have a simple surface-negative profile, which reverses polarity at ~ 200–250 μm (Sasaki et al., 1972; Noda and Oka, 1985). EPSPs on apical dendrites of pyramidal neurons can induce electrical dipoles in the cortex. As a result, EPSPs in the superficial part of the apical dendrites induce surface-negative potentials, while those close to the somata produce surface-positive potentials. Accordingly, Sasaki et al. (1972) reasoned that the overall response in the MCx is actually composed of two T-C responses, while that in PCx consists of a single one. He named them the *superficial* and *deep* T-C responses, by virtue of the laminar location of the associated EPSPs induced by T-C inputs. These suggestions were subsequently confirmed by morphological studies on the distribution of degenerated and anterogradely labeled terminals (Strick and Sterling, 1974; Kawaguchi et al., 1983; Yamamoto et al., 1992a).

Monkey

In the monkey, the CN-evoked response is a surface-negative potential in the lateral part of MCx versus a surface-positive potential in both the medial part of the MCx and the PCx (Sasaki et al., 1976, 1979; Yamamoto et al., 1983). The surface-negative potentials reverse their polarity at a cortical depth of ~ 1000 μm. This reversal level is much deeper than that of the superficial T-C response in the cat, whereas the surface-positive potentials reverse at nearly the same depth.

The above variations in CN-evoked responses are due to the three species' different distribution pattern of T-C terminals. Those concerned for the superficial T-C response in the rat and monkey seem to be much denser and broader than in the cat. These interpretations are confirmed by morphological studies of the laminar distribution of labeled T-C terminals in the rat (Yamamoto et al., 1990a) and monkey (Nakano et al., 1992).

In the next section, we address whether the above species-dependent differences in field-potential responses and distribution of T-C terminals may affect the mode of CN activation of individual neurons in the MCx.

Comparison of CN activation of MCx neurons across mammals

We have investigated the intracellularly recorded EPSPs of MCx neurons to CN stimulation and compared their latency and duration across species. Figure 2 shows the cortical depth of recording versus latency of these CN-EPSPs for rat versus cat.

Rat

In the rat, the laminar distribution of CN-EPSP latencies (Fig. 2I) was like that of the cat MCx (Fig. 2IIA), even though the shape of the surface-recorded CN-evoked responses differed.

The laminar distribution of rat CN-EPSP latencies shown in Fig. 2I is possibly in keeping with the finding that the CN-evoked surface potential consists of a small (albeit distinct) surface-positive potential followed by a large negative potential. This is consistent with morphological findings on the distribution of PHA-L-labeled terminals (Yamamoto et al., 1990a). So far, aspiny or sparsely spiny layer V pyramidal neurons, which are termed fast pyramidal tract neurons (PTNs), seem not to exist in rats, however. Rather, almost all previous reports have shown rat layer V pyramidal neurons to have spiny apical dendrites (Donoghue and Kitai, 1981; Chagnac-Amitai et al., 1990). Therefore, the association between CN responsiveness and the morphology of layer V PTNs in the rat is still an open issue.

Cat

In the cat, it is well known that the layer V pyramidal neurons consist of two distinct types—fast versus slow PTNs—which presumably correspond to Betz versus other type of PTNs in the human MCx. These two PTN types respond to CN stimulation with different EPSP profiles: a rapid risetime at a shorter latency for fast PTNs versus a slower risetime at longer latency for slow PTNs (Yoshida et al., 1966; Deschenes et al., 1982; Noda and Yamamoto, 1984). In addition, fast PTNs have smooth or sparsely spiny apical dendrites with larger somata, whereas slow PTNs have dense, spiny apical dendrites (Deschenes et al., 1979).

We also recorded CN-EPSPs from neurons outside layer V. Those in layer III had a fast PTN-type response, whereas those in layer II had a slow PTN-type response (Noda and Yamamoto, 1984). We proposed that both neuron types receive monosynaptic inputs from the thalamus, and that the long latencies and slow PTN-type response in layer II pyramidal neurons were partly due to their reception of ultrafine T-C fibers relaying CN inputs.

We made analogous tests in the PCx, where CN-evoked surface potentials have a simple surface-negative profile. It was shown that latency differences are detected neither between fast versus slow PTNs, nor between cells in different cortical layers (Yamamoto and Oka, 1993). This finding might result from variations in thalamic innervation

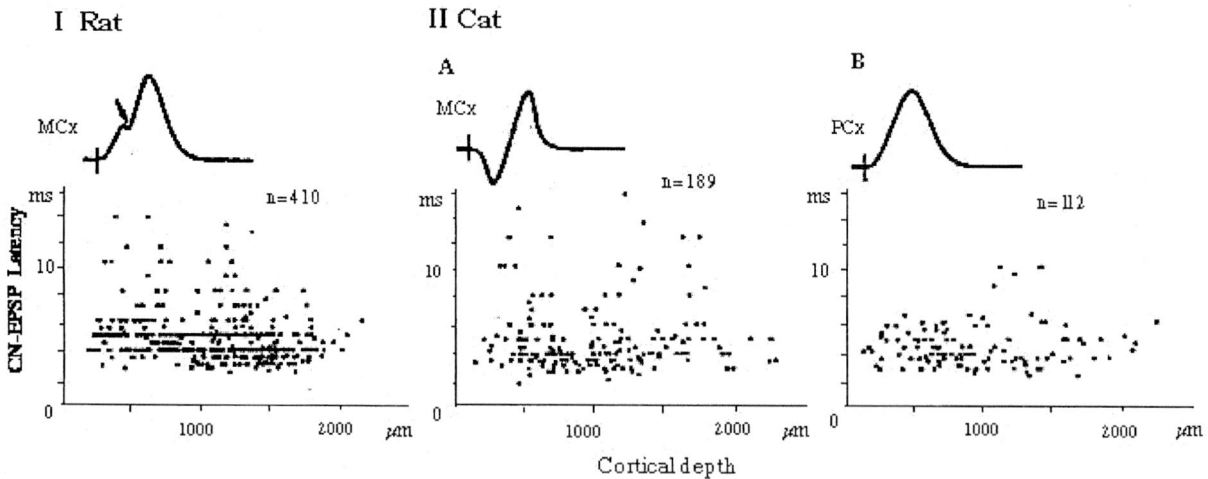

Fig. 2. Laminar distribution of CN-EPSP latencies in the MCx and PCx of the rat versus cat. Dot displays of recording depth versus CN-EPSP latency are shown below schematic traces of the CN-evoked responses. Note that the latency distribution is similar for the MCx of the rat (I) and cat (IIA), but somewhat different for the cat PCx (IIB). (Data from Noda and Yamamoto, 1984; Yamamoto and Oka, 1993.)

patterns for the MCx versus PCx. For example, thalamic terminals are distributed in layer I as well as layers III–V of the MCx, whereas they mainly project to layer I of the PCx (Strick and Sterling, 1974; Kawaguchi et al., 1983; Yamamoto et al., 1992a).

Monkey

The next step was to compare the CN-responsiveness of fast versus slow PTNs in a cortical region displaying a purely surface-negative potential response to CN input. We chose the lateral (hand–face) region of the monkey MCx (Sasaki et al., 1976, 1979; Hamada et al., 1981). Figure 3 shows intracellular recordings from its layer III versus V pyramidal neurons, which were identified either by intracellular staining or antidromic activation from the pyramidal tract.

As shown in Fig. 3, CN stimulation induces large EPSPs with a fast risetime in layer III pyramidal neurons. In contrast, CN-EPSPs in layer V pyramidal neurons were not so conspicuous even in fast PTNs, and they were often curtailed by IPSPs.

In Fig. 4, we have plotted the recording depth for these CN-EPSPs against their latency distribution (A), amplitude (B) and risetime (C). Fig. 4A shows that the latency distribution was: (1) quite similar to analogous EPSPs recorded in the cat PCx (Fig. 2IIB); and (2) not influenced by the depth of the recording. In contrast, the recording depth had a marked effect on EPSP amplitude and risetime, being larger and steeper, respectively for more superficial versus deeper located cells.

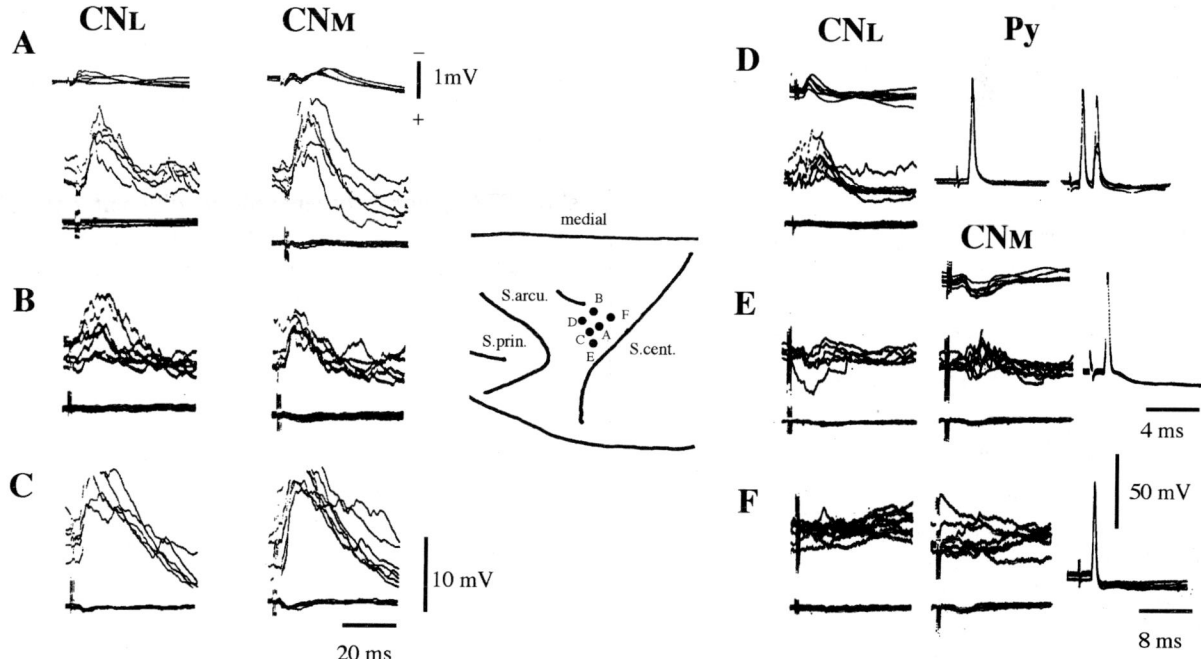

Fig. 3. Exemplary CN-EPSPs recorded in layer III and V pyramidal neurons of the hand region of the monkey MCx. (A–C) CN-EPSP responses of three layer III pyramidal neurons to stimulation of the lateral (L) and medial (M) cerebellar nucleus. The uppermost tracings are MCx-surface-recorded potentials evoked by CN stimulation. Subsequent pairs of microelectrode recordings are the intracellular (upper) and immediately extracellular (lower) versions of the EPSP responses. Latencies of the three CN-EPSPs were 3.5, 2.5 and 2.8 ms, respectively. Note that these EPSPs are relatively large and of rapid risetime. (D&E) Analogous CN-EPSP responses of two layer V pyramidal neurons (latency, 2 and 5 ms, respectively). Note that these EPSPs are smaller, and often followed by IPSPs. (F) No observable CN-EPSP in a third layer V pyramidal neuron. Also shown in D–F are the antidromic action potentials of the same three pyramidal cells evoked by pyramidal tract (Py) stimulation (antidromic latencies, 0.8, 1.0 and 2.2 ms, respectively). Calibrations: all surface recordings, voltage as in A (time bar in C); all PSP responses, as in C; action potentials, all 50 mV and 4 ms in D–E versus 8 ms in E. The middle MCx sketch shows the six recording sites for the A–F cells. Abbreviations are as in Fig. 1.

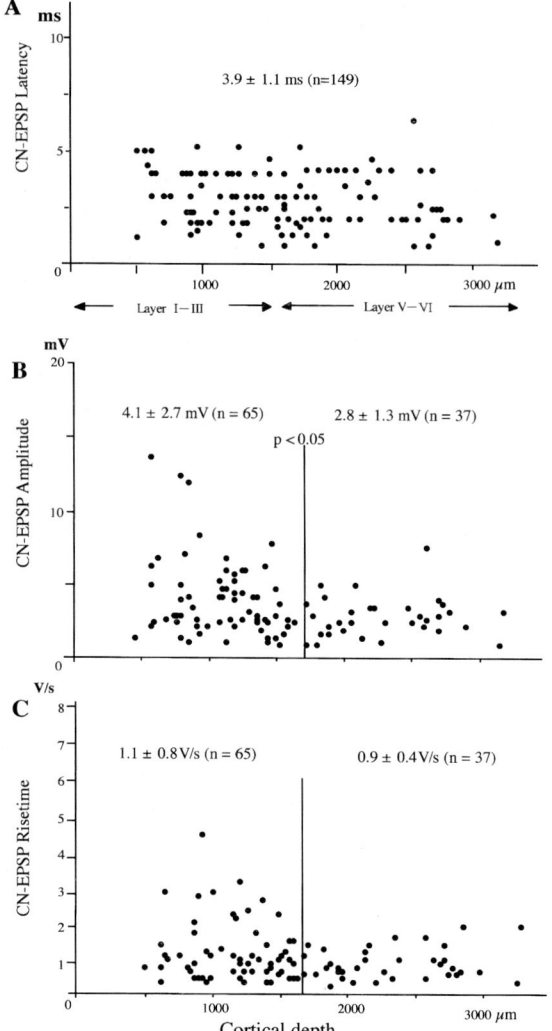

Fig. 4. Latency distribution, amplitude and risetime of CN-EPSPs in monkey MCx neurons. (A) Recording depth versus EPSP latency for 149 MCx neurons. Note that as in the cat (Fig. 2B), the latency value was independent of the recording depth. (B–C) Similar plots for EPSP amplitude (B) and risetime (C). For these two parameters, the cells are divided into two groups (vertical lines) at depths less versus greater than 1600 μm. Note that neurons in the more superficial versus deeper cortical layers received significantly larger EPSPs ($P < 0.05$), which were of more-or-less similar risetime.

Functional perspectives

Even with progressive encephalization from rat and cat to monkey, the role of the cerebello-thalamo-cortical projection in motor control, as well as its corresponding position in the hierarchical control of movement, must be relatively similar for these three species. How can we then explain the different mode of CN activation of MCx neurons across three 'phylogenetically close' mammalian species?

Using laminar field-potential analysis, Sasaki et al. (1976) were the first to note the different laminar profiles for cat versus monkey MCx, i.e., surface positive–negative potentials for the cat versus dominant surface-negative potentials in the monkey. Classically, surface-negative field potentials are thought to be induced by EPSPs in the superficial apical dendrites of pyramidal neurons, because at this location they can work as an active current sink, with a passive source near the soma. Recent data has suggested, however, that regenerative voltage-gated depolarizing channels exist in the cortical neurons' dendrites and spines (Turner et al., 1989). Thus, EPSPs in distal dendrites may be much more effective in exciting somata than was previously considered (Rall, 1962; Wilson, 1984). Magee and Cook (2000) demonstrated that dendritic EPSP amplitude increases with distance from the soma in the rat hippocampal CA1 pyramidal neurons using dual whole-cell patch-clamp recording. They showed that amplification of dendritic EPSP seems to be mainly due to a progressive increase in synaptic conductance and dendritic membrane properties play a minor role. Whatever mechanisms may concern in elevating the dendritic EPSP, the same mechanisms may also be involved in the pyramidal neurons of the cat and monkey. However, the data indicate that the superficial T-C inputs may not bring strong effects on layers V–VI pyramidal neurons as in the case of the cat PCx. Therefore, it seems to be likely that EPSPs near the somata are much more effective in triggering action potentials than those in the distal dendrites, with EPSPs in distal dendrites playing an important role in controlling overall neuronal excitability.

When we compare the effect of CN stimulation on pyramidal MCx neurons across mammals, it can be generalized that layer III neurons usually respond with much larger EPSPs than are seen in other cortical layers. This is commonly observed in other cortical regions, as well, e.g., the PCx and primary sensory cortex in the cat (Martin and Whitteridge, 1984; Yamamoto et al., 1990b; Yamamoto and Oka, 1993). In contrast, layer V pyramidal neurons, which

Rat

In the rat, in contrast to the cat and monkey, layer V pyramidal neurons display no clear morphological differences in the spine density of their apical dendrites, whereas some morphological differences have been reported for apical dendritic trajectory, as dependent on the neurons' intrinsic discharge patterns (Donoghue and Kitai, 1981; Chagnac-Amitai et al., 1990). In this study, we also detected some differences in such dendritic trajectory, but they were not dependent on the neurons' firing pattern (Yamamoto et al., 1992b). Concerning the mode of cerebellar activation, it was shown in Fig. 2I that the depth–latency relation is similar to that in the cat MCx (Fig. 2IIA). CN-EPSPs in layer V pyramidal neurons seemed not to be shortest, however. Among the identified neurons, layer III pyramidal neurons received more distinct CN-EPSPs versus those observed in other pyramidal neurons.

Cat

Two types of PTNs, fast and slow, have been identified in the cat MCx, PCx and SCx. It is only in the MCx, however, that these two neuron types respond differently to CN stimulation (Yoshida et al., 1966; Deschenes et al., 1982; Yamamoto et al., 1990b; Yamamoto and Oka, 1993). This finding may mean that the functional differentiation of cortical neurons depends on both their thalamic innervation pattern and their hierarchical position in motor-command signaling, i.e., do they modulate the motor command or are they part of the command, itself? For example, we could not detect latency differences for the thalamic-EPSPs of fast versus slow PTNs of the SCx, even though their latencies in other pyramidal neurons are shorter in the superficial versus deep layers (Yamamoto et al., 1990b). In both SCx and MCx, the T-C terminals are distributed mainly in layers III–V and sparsely in layer I. It is far more likely however, that SCx neurons function as modulators of MCx neurons, rather than contributing to the final motor command, as does a population of the latter neurons.

Monkey

In the monkey MCx, two types of PTNs have been identified and morphological distinction has been clearly shown by intracellular staining (Hamada et al., 1981). Cerebellar nuclear-evoked responses in MCx result in a routinely surface-negative potential, and we have never observed a differential activation of fast versus slow PTNs. Therefore, it appears that the differential activation of MCx layer V pyramidal neurons is unique to the cat. Concerning layer III pyramidal neurons, as shown in Fig. 4B–C, the amplitude and risetime of CN-EPSPs are remarkably larger than those in layers IV–VI pyramidal ones, even though the latency difference is not so conspicuous.

Concluding thoughts

The variation of CN-evoked, surface-recorded field potentials in the mammalian cerebral cortex is strikingly in contrast to the potentials evoked by primary sensory input. It is attractive to speculate that surface-negative evoked potentials in the cerebral cortex are a reflection of more flexibility in the control of excitability. We have routinely observed, however, that irrespective of the T-C projections and responses under consideration, layer III pyramidal neurons are activated more strongly than pyramidal neurons in other cortical layers, i.e., they have a less-latent and larger EPSPs with a shorter risetime. The same result obtains for cortical interneurons (Yamamoto et al., 1988). Note also that layer III pyramidal neurons are excitatory projection neurons within the cortex. They, with interneurons, are optimally positioned to modulate and integrate afferent input to other cortical neurons. On the other hand, layer V pyramidal neurons, which are the final output neurons of the MCx and other cortical regions, receive relatively weak excitation from any single afferent input. This arrangement may augment their capacity to act as a site of convergence of multiple inputs from subcortical as well as intracortical neurons, including layer III pyramidal neurons and interneurons. By this means, layer V pyramidal

are the final common output cells in the MCx, show species' specialties across mammals.

neurons might then have a finer capacity to grade their excitatory output to other cortical regions and lower centers. It should be possible to add rigor to this idea by further work along the lines presented above, which emphasizes a combination of electrophysiological and morphological approaches.

Acknowledgments

We thank Ms. Kaori Kawamura for her technical help. This research was supported by Grants-in-Aid from the Ministry of Education, Science and Culture of Japan (#06680808, #11680806), and by Awards from the Okasan-Katoh Foundation and the Epilepsy Research Foundation of Japan.

Abbreviations

CA1	field 1 of Ammon's horn
CN	cerebellar nuclear/nuclei
CN_L	lateral cerebellar nucleus
CN_M	medial cerebellar nucleus
EPSP	excitatory postsynaptic potential
F	fast
Fr1	frontal area 1 of Zilles
Fr2	frontal area 2 of Zilles
Fr3	frontal area 3 of Zilles
ICx	insular cortex
Mac	Macaca fuscata
MCx	motor cortex
PCx	parietal cortex
PTN	pyramidal tract neuron
Py	pyramidal tract
S	slow
S. arcu.	arcuate sulcus
S. cent.	central sulcus
S. cru.	cruciate sulcus
S. prin.	principal sulcus
Saim	Saimiri sciureus
SCx	somatosensory cortex
T-C	thalamo-cortical

References

Castro, A.J. (1972). Motor performance in rats. The effects of pyramidal tract section. Brain Res., 44: 313–323.

Chagnac-Amitai, Y., Luhmann, H.J. and Prince, D.A. (1990). Burst generating and regular spiking layer 5 pyramidal neurons of rat neocortex have different morphological features. J. Comp. Neurol., 296: 598–613.

Creutzfeldt, O.D. and Kuhnt, U. (1973). Electrophysiology and topographical distribution of visual evoked potentials in animals. In: Jung R. (Ed.), Visual Centers in the Brain, Handbook of Sensory Physiology, Vol. VII/3, Springer-Verlag, Berlin, pp. 595–646.

Deschenes, M., Labelle, A. and Landry, P. (1979). Morphological characterization of slow and fast pyramidal tract cells in the cat. Brain Res., 178: 251–274.

Deschenes, M., Landry, P. and Clercq, M. (1982). A reanalysis of the ventrolateral input in slow and fast pyramidal tract neurons of cat motor cortex. Neuroscience, 7: 2149–2157.

Donoghue, J.P. and Kitai, S.T. (1981). A collateral pathway to the neostriatum from corticofugal neurons of the rat sensory-motor cortex: an intracellular HRP study. J. Comp. Neurol., 201: 1–13.

Hamada, I., Sakai, M. and Kubota, K. (1981). Morphological differences between fast and slow pyramidal tract neurons in the monkey motor cortex as revealed by intracellular injection of horseradish peroxidase by pressure. Neurosci. Lett., 22: 233–238.

Jones, E.G. (1985). Thalamus. Plenum Press, New York.

Kalil, K. and Schneider, G.E. (1975). Motor performance following unilateral pyramidal tract lesions in the hamster. Brain Res., 100: 170–174.

Kato, N., Kawaguchi, S., Yamamoto, T., Samejima, A. and Miyata, H. (1983). Postnatal development of the geniculo-cortical projection in the cat: electrophysiological and morphological studies. Exp. Brain Res., 51: 65–72.

Kawaguchi, S., Samejima, A. and Yamamoto, T. (1983). Postnatal development of the cerebello-cerebral projection in kittens. J. Physiol. (Lond.), 343: 215–232.

Keidel, W.D. (1976). The physiological background of the electric response audiometry. In: Keidel W.D. and Neff W.D. (Eds.), Auditory System, Clinical and Special Topics, Handbook of Sensory Physiology, Vol. V/3, Springer-Verlag, Berlin, pp. 105–231.

Libet, B. (1973). Electrical stimulation of cortex in human subjects, and conscious sensory aspects. In: Iggo A. (Ed.), Somatosensory System, Handbook of Sensory Physiology, Vol. II, Springer-Verlag, Berlin, pp. 743–790.

Magee, J.C. and Cook, E.P. (2000). Somatic EPSPs amplitude is independent of synapse location in hippocampal pyramidal neurons. Nat. Neurosci., 3: 895–903.

Martin, K.A.C. and Whitteridge, D. (1984). Form, function and intracortical projections of spiny neurons in the striate visual cortex of the cat. J. Physiol. (Lond.), 353: 463–504.

Miyata, H., Kawaguchi, S., Samejima, A. and Yamamoto, T. (1982). Postnatal development of evoked responses in the auditory cortex of the cat. Jpn. J. Physiol., 32: 421–429.

Nakano, K., Tokushige, A., Kohno, M., Hasegawa, Y., Kayahara, T. and Sasaki, K. (1992). An autoradiographic study of cortical projections from motor thalamic nuclei in the macaque monkey. Neurosci. Res., 13: 119–137.

Noda, T. and Oka, H. (1985). Fastigial inputs to insular cortex in the cat: field potential analysis. Neurosci. Lett., 53: 331–336.

Noda, T. and Yamamoto, T. (1984). Response properties and morphological identification of neurons in the cat motor cortex. Brain Res., 268: 197–206.

Poggio, G.F. and Mountcastle, V.B. (1980). Functional organization of thalamus and cortex. In: Mountcastle V.B. (Ed.), Medical Physiology, 14th ed., Vol. 1, Mosby, St. Louis, pp. 227–253.

Rall, W. (1962). Electrophysiology of a dendritic cerebellar neuron model. Biophys. J., 2: 145–167.

Sasaki, K., Kawaguchi, S., Matsuda, Y. and Mizuno, N. (1972). Electrophysiological studies on cerebello-cerebral projections in the cat. Exp. Brain Res., 16: 75–88.

Sasaki, K., Kawaguchi, S., Oka, H., Sakai, M. and Mizuno, N. (1976). Electrophysiological studies on the cerebellocerebral projections in monkeys. Exp. Brain Res., 24: 495–507.

Sasaki, K., Jinnai, K., Gemba, H., Hashimoto, S. and Mizuno, N. (1979). Projection of the cerebellar dentate nucleus onto the frontal association cortex in monkeys. Exp. Brain Res., 37: 193–198.

Strick, P.L. and Sterling, P. (1974). Synaptic termination of afferents from the ventrolateral nucleus of the thalamus in the cat motor cortex. A light and electron microscope study. J. Comp. Neurol., 153: 77–106.

Turner, R.W., Meyers, D.E.R. and Barker, J.L. (1989). Localization of tetrodotoxin-sensitive field potentials of CA1 pyramidal cells in the rat hippocampus. J. Neurophysiol., 62: 1375–1387.

Wilson, C.J. (1984). Passive cable properties of dendritic spines and spiny neurons. J. Neurosci., 4: 281–297.

Yamamoto, T. and Oka, H. (1993). The mode of cerebellar activation of pyramidal neurons in the cat parietal cortex (areas 5 and 7): an intracellular HRP study. Neurosci. Res., 18: 129–142.

Yamamoto, T., Kawaguchi, S. and Samejima, A. (1979). Electrophysiological studies on the cerebello-cerebral projection in the rat. Exp. Neurol., 63: 545–558.

Yamamoto, T., Wagner, A., Hassler, R. and Sasaki, K. (1983). Studies on the cerebellocerebral and thalamocortical projection in squirrel monkeys (Saimiri sciureus). Exp. Neurol., 79: 27–37.

Yamamoto, T., Samejima, A. and Oka, H. (1988). Short latency activation of local circuit neurons in the cat somatosensory cortex. Brain Res., 461: 199–203.

Yamamoto, T., Kishimoto, Y., Yoshikawa, H. and Oka, H. (1990a). Cortical laminar distribution of rat thalamic ventrolateral fibers demonstrated by the PHA-L anterograde labeling method. Neurosci. Res., 9: 138–154.

Yamamoto, T., Samejima, A. and Oka, H. (1990b). The mode of synaptic activation of pyramidal neurons in the cat primary somatosensory cortex: an intracellular HRP study. Exp. Brain Res., 80: 12–22.

Yamamoto, T., Kishimoto, Y., Takagi, H., Yoshikawa, H. and Oka, H. (1992a). Morphological features of cat thalamoparietal projection fibers investigated by PHA-L immunohistochemistry and electron microscopy. Neurosci. Res., 14: 124–132.

Yamamoto, T., Shibuya, H., Kishimoto, Y., Takagi, H., Yoshikawa, H. and Oka, H. (1992b). Electrophysiological and morphological studies on the rat cerebellocerebral projection: intracellular staining and electron microscopic examination. Jpn. J. Physiol., (Suppl.) 42: S208.

Yoshida, M., Yajima, K. and Uno, M. (1966). Different activation of the two types of the pyramidal tract neurons through the cerebellothalamocortical pathway. Experientia, 22: 331–332.

Zilles, K. and Wree, A. (1995). Cortex: areal and laminar structure. In: Paxinos G. (Ed.), Forebrain and Midbrain, The Rat Nervous System, 2nd ed., Academic Press, New York, pp. 375–415.

CHAPTER 31

Task-dependent role of the cerebellum in motor learning

James R. Bloedel*

Departments of Health and Human Performance and Biomedical Sciences, Iowa State University, Ames, IA 50013, USA

Abstract: This chapter reviews several findings from our laboratory supporting the hypothesis that the cerebellum's role in motor learning is task-dependent. Namely, its contribution is dependent on the specific task being learned. Several studies are reviewed to demonstrate that the effect of temporary or permanent cerebellar lesions on a specific process such as storage varies depending on the behavior. Furthermore, this task-dependency is reflected also in the modulation of Purkinje cells and nuclear neurons recorded during the learning process. The behavioral correlates of this modulation are very paradigm specific. These observations support the above hypothesis and emphasize the importance of paradigm selection in designing experiments focused on elucidating the cerebellum's role in learning a specific motor behavior.

Introduction

The contribution of any one central nervous system (CNS) structure to the learning of a motor task in vertebrates has been a consistent issue of debate and discussion for the past century. This discussion began to intensify as a consequence of the theoretical study of Marr (1969) and the experimental findings related to two different reflex systems, the vestibular ocular reflex (VOR) and the eyeblink reflex. This early history has been reviewed many times (Thach et al., 1992; Thompson and Krupa, 1994; Bloedel and Bracha, 1995) and need not be repeated here. Only a few points will be reiterated.

First, the previous studies implicated the cerebellum in the learning of these behaviors, a structure that was viewed previously as a component of the CNS primarily, if not exclusively, involved in motor coordination. Second, and more pertinent to the issues that will be discussed here, the underlying objective that drove virtually all of these initial studies was to determine the storage site for the engram established during the learning of these behaviors. This focus on storage was developed without carefully considering the parcellation of the learning process into successive steps, each related to different mechanisms. These steps include acquisition, consolidation and short-term memory, as well as long-term memory or retention. Third, the inferences being drawn were based primarily on data from ablation experiments or studies in which the function of a cerebellar region was disrupted temporarily by microinjections of a local anesthetic or a pharmacological agent that suppressed synaptic transmission through a specific nuclear region. Fourth, the implications of the experiments examining the modification of reflex systems were extended to postulates regarding the cerebellum's role in the learning of volitional movements. The same role of the cerebellum in storage was proposed across many types of movements being studied at that time.

This review will examine the controversies resulting from these three perspectives and propose an alternative perspective that is now reasonably well supported in the literature. We argue that the

---

*Corresponding author: Tel.: +515-294-1785; Fax: +515-294-6100; E-mail: jbloedel@iastate.edu

cerebellum's role in motor learning is task- and even condition-dependent. Its role may vary depending upon the task being learned and the conditions under which the task is specified (see also Bracha, Chapter 32 of this volume).

Evolution of the controversy

Although the cerebellum was implicated in adaptive processes as a result of early ablation studies indicating that the VOR could not be modified following cerebellar ablations (Ito, 1984), the issue of engram storage was stimulated by the intriguing model of Marr (1969). It provided a cerebellar-specific mechanism for establishing a long-term change in synaptic efficacy within the cerebellar cortical circuitry. This was followed by experiments in which recordings of climbing fiber activity during VOR adaptation was believed to be consistent with Marr's hypothesis (for review, see Ito, 1984). Although the work related to VOR adaptation continued, the cerebellum's role in engram storage during the learning of motor tasks received strong support in the early 1980s. Data showed that ablating a discrete region of the cerebellar nuclei blocked the expression of a previously acquired associative behavior, the classical conditioning of the eyeblink reflex in the rabbit (McCormick et al., 1982).

The controversies following the above reports (Bloedel and Bracha, 1995; Bracha and Bloedel, 1996) did not challenge their fundamental finding. Rather, we focused on the contention that their observations demonstrated that the cerebellum is necessary and sufficient for the acquisition and retention of the classically conditioned eyeblink reflex. This argument was based in part on their finding that neither ablations nor muscimol-induced functional blocks changed the amplitude of the unconditioned reflex, even though the conditioned reflex was suppressed completely. These observations could *not* be duplicated in the laboratory of Welsh and Harvey (1989), or in authors' laboratory (Bracha et al., 1994). In contrast to previous findings, these latter studies showed a clear, consistent change in the unconditioned reflex following the ablative procedures. Although this controversy still exists, the issue of memory storage in the cerebellum is no longer as dependent as it was initially on the Welsh and Harvey (1989) and Bracha et al.'s (1994) findings. Nevertheless, these two reports illustrate the basis for the initial controversy and they indicate that disrupting cerebellar nuclear function may modify motor performance independent of any effects on retention (for additional discussion, see Bloedel and Bracha, 1995).

The argument that the cerebellum is necessary and sufficient for the classical conditioning of the eyeblink reflex was challenged also by experiments showing that plasticity supporting the conditioning of this behavior can be located *outside* the cerebellum independent of any plasticity that may occur *within* this structure. The initial finding supporting this view was the demonstration that the classically conditioned eyeblink could be acquired in the decerebrate, decerebellate rabbit (Kelly et al., 1990). Contrary to the contention of others (Nordholm et al., 1991), the authors' findings were based on evoked, nonspontaneous eyeblink reflexes that were truly associative in nature. For further discussion regarding distributed sites of plasticity in the learning of this reflex, see Bracha and Bloedel, 1996.

Even though other sites can support plasticity related to the acquisition of this reflex, there is now strong evidence that storage of engrams does indeed occur in the cerebellar nuclei. Bracha et al. (1998) used a unique experimental design to examine the effects of blocking protein synthesis, a process likely involved in establishing the plastic changes underlying long-term memory. In this study, rabbits were trained in a delay paradigm only during the infusion of anisomycin, a protein synthesis inhibitor, to determine whether the classically conditioned eyeblink reflex could be acquired under conditions in which protein synthesis was suppressed. The key findings and the outline of the experiments are shown in Fig. 1. First, the animals were conditioned using a standard delay paradigm for a three-day period during the infusion of anisomycin (D1–D3).

As seen in the Fig. 1 plots, the anisomycin-injected animals did much poorer than controls during this three-day period of training. On the fourth day (RE), a retention test was performed by assessing the responses in tone-only trials in the absence of any infusion. This test substantiated that the injected animals failed to acquire the conditioned reflex. Over

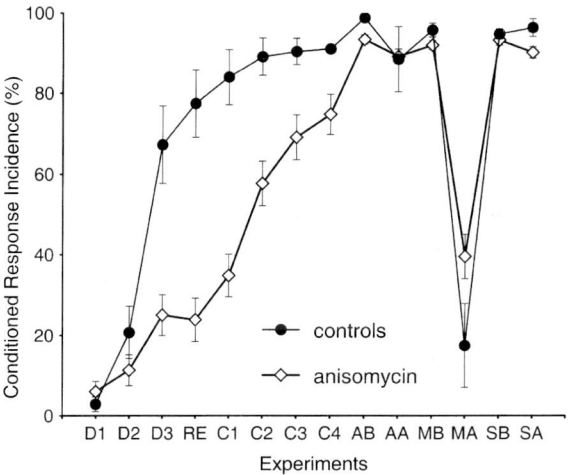

Fig. 1. Effect of the infusion of anisomycin during training required for acquisition of the classically conditioned eyeblink reflex. The plots show the group means ± S.E.M. The abbreviations along the abscissa are given in the text. (Reproduced from Bracha et al., 1998 with permission from Elsevier Science.)

the four subsequent days (C1–C4), all animals were conditioned using the delay paradigm in the absence of any injection. Notice that the injected animals were now able to acquire the reflex, thereby indicating that anisomycin did not damage the cerebellar nuclei. Next, after one day of assessing baseline responding (MB), muscimol was injected to determine whether the injections had been made in the nuclear region touted as the site for engram storage (MA). The strong reduction in CR responding indicated that the injections were made in the targeted region. Finally, the effects of saline injection were evaluated (SB and SA) to show that the effect on day MA was due to the injection of muscimol and not a general effect produced by the injection. These findings provided strong evidence that the acquisition of the classically conditioned eyeblink reflex was dependent on protein synthesis in the cerebellar nucleus and consequently may be the site of long-term memory storage. Subsequently, Thompson and colleagues (Krupa and Thompson, 1995, 1997) showed that blocking the critical cerebellar output projection during training does not disrupt acquisition and subsequent retention of the conditioned reflex, although blocking transmission in the critical cerebellar nuclear region during training definitely does. These data also suggest that it is unlikely that performance deficits alone can account for the results of the early ablation experiments. In summary, there is now strong evidence favoring the occurrence of plastic changes in the cerebellum during the classical conditioning of the eyeblink reflex in rabbits. The existence of distributed sites of plasticity outside the cerebellum associated with the learning of this behavior, however, is supported also by the literature (Bloedel and Bracha, 1995; Bracha and Bloedel, 1996).

Initial support for the concept of task-dependency

As mentioned in the previous section, the excitement created by initial studies examining the effects of temporary or permanent ablative procedures on the classically conditioned nictitating membrane reflex in the rabbit resulted in the generalization that these findings had direct implications for the learning of other types of motor behaviors, including volitional movements. In addition, data obtained from cerebellar patients were interpreted as supporting this notion. For example, one of these studies compared the capacity of cerebellar patients and normal subjects to perform a mirror-tracing task (Sanes et al., 1990). Error measurements were used to assess the extent of learning over successive trials. The data showing that the cerebellar patients exhibited greater errors than the control group after the task was practiced were interpreted as indicating that the patients had a learning deficit in this type of volitional task.

In the author's view, most of the initial studies of this question failed to account adequately for the performance deficit characterizing cerebellar patients. Consequently, an experiment was designed that compared the *rate* at which improvement occurred between groups rather than the *differences in the absolute magnitude* of the errors following practice (Timmann et al., 1996). The experiments employed a tracing task in which subjects were asked to practice tracing an irregularly shaped template and then to draw it from memory after mentally rotating the figure 90 degrees. As shown in Fig. 2, cerebellar patients clearly increased the quality of their performance as evidenced by the decrease in the error measurements over the practice trials.

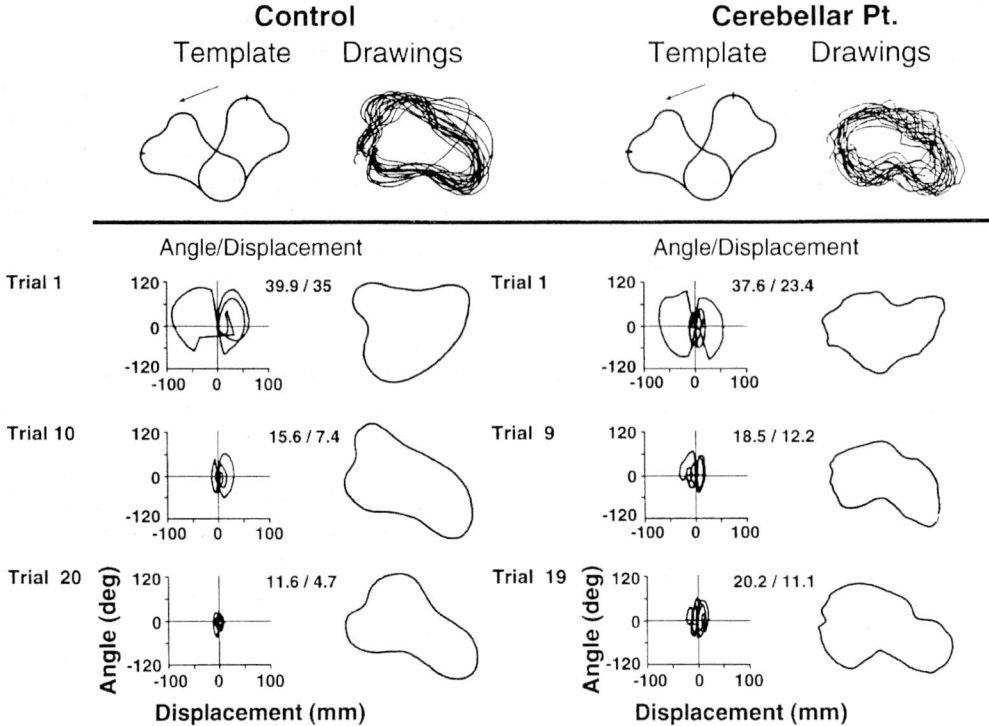

Fig. 2. Comparison of the capacity of the control subject and a cerebellar patient to acquire the template-drawing task described in the text. This patient had a right cerebellar infarct associated with moderate upper limb ataxia. The results from three different trials are shown for each subject. In this task, the template had to be drawn after mentally rotating it to the left following 20 practice trials with the template in the original position. The plot under 'Drawings' shows the drawings for all 20 trials superimposed on one another. The angle-versus-displacement plots show the combined changes in these types of errors across the three trials. These errors were measured and plotted continuously as the template was traced. Numbers in the upper right quadrant are the mean values for the two error measurements in each trial. It is apparent from both the plots of the error measurements, as well as the traced figures shown on the right for each trial, that both the control subject and the cerebellar patient improved their performance. (Reproduced from Timmann et al., 1996 with permission from Elsevier Science.)

Furthermore, a comparison of the plots of error magnitude as a function of trial number for both the patients and controls revealed that the patients' rate of improvement was not significantly different from that of normal subjects. These findings support the argument that the lesions in these patients did not render them incapable of improving their performance on this type of volitional task. In contrast to the inferences derived from the experiments involving the eyeblink reflex reviewed previously, patients with extensive cerebellar pathology could learn this type of motor behavior.

To pursue further the role of the cerebellum in the learning of volitional movements, experiments were performed employing cats that were trained to perform a template task. In all of these studies, the cat learned a sequence of movements required to move a manipulandum through a template consisting of 2–3 straight grooves to a target zone. In the first of these experiments (Milak et al., 1997), the cerebellar interposed and dentate nuclei were inactivated using muscimol after the cat had learned the sequence required to master the template. Each animal's performance was assessed after injection to determine if this procedure blocked retention as it did in the related studies of the conditioned eyeblink reflex. Although the expected ataxia impaired the performance of the movements, the animals clearly retained the capacity to negotiate the sequence of grooves required to move the manipulandum through the template. Thus, unlike the eyeblink experiments, muscimol inactivation of the cerebellar nuclei did

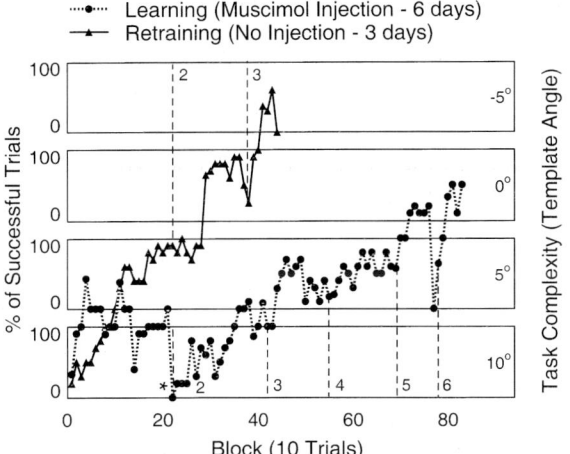

Fig. 3. Effects of muscimol on motor learning. Learning curves that characterize the capacity of a cat to move a manipulandum through inverted 'L' templates at successively smaller angles are shown. Simultaneously, the dentate and interposed nuclei were infused with muscimol. In a set of subsequent trials, the task was repeated in the absence of muscimol. Degrees are shown relative to 90 degrees (i.e., 0 degrees = 90 degrees). Each overall learning curve is constructed by stacking the individual learning curves obtained as the animal practiced the task at each angle. Dashed vertical lines and associated numbers from each curve indicate the days over which the learning occurred. Retraining in the absence of muscimol was performed after the last day of training during muscimol infusion. (Reproduced from Wang et al., 1998 with permission from The American Physiological Society.)

not block retention. This indicated a clear difference between classical conditioning of the eyeblink reflex and certain volitional movements regarding the essential role of the cerebellum in the retention of learned motor behaviors. This introduced the concept that the cerebellum's contribution to learning a movement may vary systematically depending on the behavior. Stated differently, the role of this structure in learning different behaviors may be task-dependent.

Related experiments evaluated the capacity of cats to learn the movement sequence required to master the template task after the infusion of muscimol (Wang et al., 1998). In this experiment, the angle between two grooves in the template was decreased progressively as the animal practiced the task (Fig. 3). After criterion performance was reached at one angle, the angle was reduced, and the cat continued to practice until criterion performance was again attained. This was repeated until the animal failed to reach criterion at a given angle after extensive practice. The plot in Fig. 3 for the learning/muscimol injection condition shows that this cat was capable of learning the task through an angle of 90 degrees over a six-day period of training during muscimol infusion into the interposed and dentate nuclei. Again, this failure to block acquisition contrasts considerably to the observations of Steinmetz et al. (1992). The Wang et al.'s (1998) study showed clearly that, in the intact rabbit, the classically conditioned eyeblink reflex cannot be reacquired during any procedure that renders dysfunctional the critical cerebellar nuclear region.

Together the previous experiments provide strong evidence that the role played by the cerebellum in the learning of specific volitional tasks is quite different than in the acquisition and retention of the classically conditioned eyeblink reflex. Conceptually, this implied to us that the cerebellum's role in motor learning is task-dependent, i.e., the specific memory process in which the cerebellum is involved varies as a function of the specific task being learned (Bloedel et al., 1996).

Role of the cerebellum in task acquisition

The study of Wang et al. (1998) reviewed in the previous section indicated that the cerebellum was not necessary for the acquisition of the particular task employed in those experiments. Other observations in the same experiment, however, emphasize that a structure can be *important* to a process without being *essential* for it to occur. Following the learning of the task during muscimol infusion, the capacity of the animal to perform the task without any infusion was tested. Interestingly, despite the absence of ataxia, the cat could not perform the task at the same level of difficulty acquired during the practice sessions under muscimol infusion (Fig. 3). The angle had to be increased substantially before criterion performance could be achieved. In fact, a new learning curve was generated as the animal again practiced the task with the cerebellum functionally intact. Although other interpretations are tenable, we inferred from these observations that the system preferentially relearns the behavior in order to optimize the performance of the task through the inclusion of the cerebellum into the circuits involved in controlling the required motor sequence. Preliminary studies (Bloedel et al., 1996) indicate that the cerebellum is critical for at least one feature of the performance of these movements, i.e., the capacity to perform the behavior in a very stereotypic manner from trial to trial. This capacity to select a preferred performance strategy and use it over consecutive trials may be cerebellar-dependent.

Recording experiments from this laboratory also support the cerebellum's participation in task acquisition. In these studies (Milak et al., 1995), a microwire recording technique was used to record from multiple cerebellar nuclear neurons during the learning of the template task described previously. An analysis was developed to quantify neuronal responses over the time course of the learning. As shown in Fig. 4, the period of learning was divided into stages based on the extent to which the task was performed to criterion across successive trials. In this way, it was possible to identify multiple stages in the acquisition of the strategy required to perform the task. The amplitude of the task-related modulation was quantified based on the response frequency relative to background frequency and then plotted as a function of the stages of learning.

The data from this experiment showed that the amplitude of the responses of the majority of recorded neurons increased dramatically at the stage of learning at which an effective strategy was just being identified for moving the manipulandum through the template. This is illustrated for one set of cells in Fig. 4. Learning occurred independent of whether a given cell also had a large response to the error condition at the time

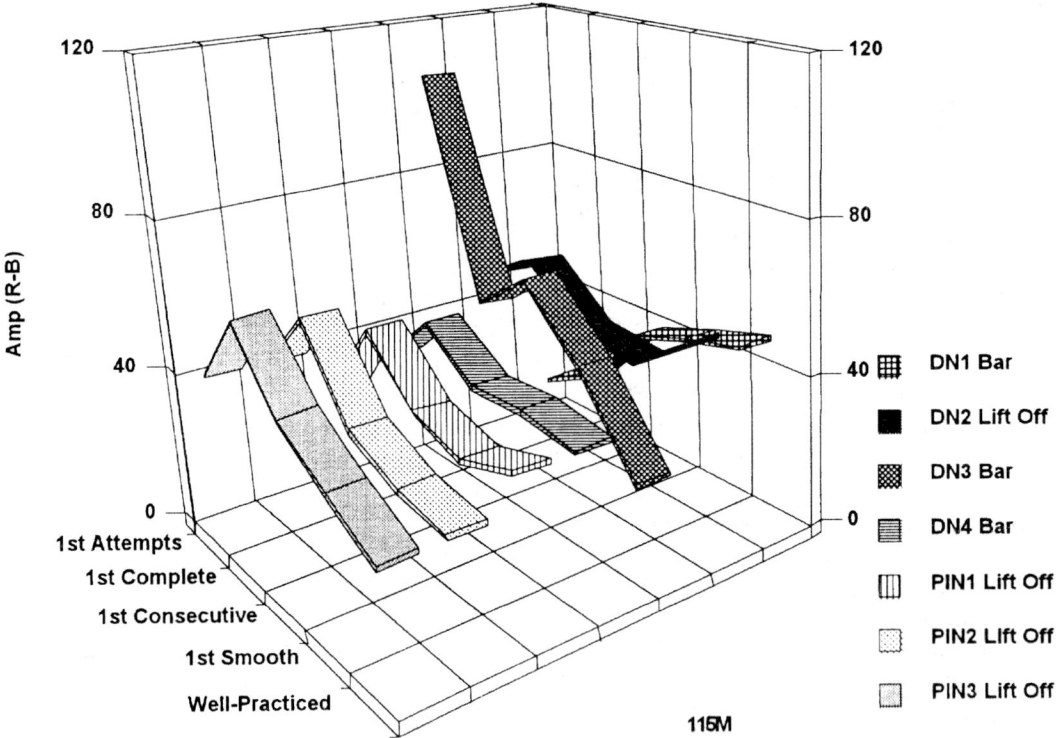

Fig. 4. Event-related modulation of the discharge of cerebellar nuclei neurons. Ribbon plots show changes in event-related modulation of four dentate neurons (DN) and three posterior interposed neurons (PIN) recorded throughout the period during which a cat was learning to move a manipulandum through an L-shaped template. The stages of learning are indicated along the indicated axis. Trials were combined into blocks to generate the histograms used to compute response amplitude at each learning stage. Smooth movements were defined as those in which there was only one peak in the velocity profile for each limb of the template. The amplitude of the responses is expressed as the mean discharge frequency within the response window minus background discharge rate (R−B). (Reproduced from Milak et al., 1995 with permission from Springer-Verlag.)

practice was first initiated (shown for cell DN3 in Fig. 4). This increased responsiveness could not be correlated with any kinematic change in the movement being performed. In fact, as the movement speed and percentage of trials performed at criterion both increased, the amplitude of the modulation decreased progressively as the movement became over-practiced. These data are consistent with the MRI findings of Friston et al. (1992), which showed that activation of the cerebellar nuclei increased during the intermediate stages of learning and then decreased as the task became over-practiced.

The contribution of the cerebellum to task acquisition may not be restricted to volitional behaviors. Interestingly, the cerebellum may play an important role in the acquisition of the classically conditioned eyeblink reflex, in addition to its likely role in the retention of this behavior. In order to provide support for this notion, Bracha et al. (1997) first showed that the visual threat response is a classically conditioned reflex that occurs normally during development. Next, they demonstrated that this reflex was present in all of the tested cerebellar patients despite the fact that it was acquired before the occurrence of pathology. They then demonstrated that these same patients were unable to acquire the classically conditioned eyeblink response in the laboratory when subjected to a classical tone-electrical stimulus delay paradigm. These findings indicate that the lesions did not affect the critical storage site for a conditioned eyeblink reflex acquired prior to the occurrence of the pathology. The same lesions, however, prevented the acquisition of the conditioned eyeblink reflex in the laboratory, and clearly after the pathology had been acquired. These findings are most consistent with the interpretation that the cerebellum plays an important role in the acquisition of this reflex. It is also possible that this effect is species-dependent, since cerebellar lesions clearly affected the retention of previously acquired eyeblink reflexes in the rabbit.

Task- and condition-dependence in the learning of volitional movements

A simple extrapolation of the above findings suggests that the role played by the cerebellum in learning motor tasks can be differentiated on the basis of whether or not the movement being learned is volitional or reflexive in nature. Subsequent studies, however, showed clearly that this inference is an oversimplification. Importantly, the extent to which the cerebellum plays a critical role in the acquisition of a task can be condition-dependent as well as task-dependent. In studies from the laboratory that have been published only in preliminary form (Shimansky et al., 1995), the capacity of cats to adapt to an elastic perturbation of a reaching movement was assessed under two conditions. In the first, the perturbations were applied to the wrist in successive trials, a condition in which the occurrence of a perturbation in the next trial was clearly predictable. In the second, the perturbations were applied in randomly distributed trials so that the occurrence of the perturbation could not be predicted prior to its application.

The capacity to acquire and retain both the predictable and unpredictable versions of the task was tested in intact cats and in cats with the interposed and dentate nuclei inactivated with muscimol. Cats with the cerebellar nuclei inactivated were able to acquire and retain the task using the predictable paradigm, but they could do neither in the unpredictable paradigm. Learning in the unpredictable paradigm was clearly cerebellar-dependent. These findings illustrate that the extent to which the acquisition of an adaptive task requires the cerebellum can be related to the specific condition under which the movement is to be performed. In this example, the predictable nature of the perturbation determines whether or not the cerebellum is essential for the learning to occur.

The acquisition of other unique volitional tasks is also cerebellar-dependent. One such task is the learning of irregular shapes using only kinesthetic cues. This was demonstrated in a set of experiments comparing the capacity of cerebellar patients and normal controls to learn the shape of irregularly shaped objects without using any visual cues (Shimansky et al., 1997). Each subject was required to learn the shape of these objects by moving a stylus through a groove while blindfolded. Blindfolded subjects were then asked to recognize the original template from a group of five test templates by 'tracing' them individually using a stylus. Only the control subjects were capable of learning under this condition. Figure 5 illustrates another interesting finding from this study. After performing the trials required for the learning component of the experiment, each subject was asked to draw the original template from memory, first blindfolded (A and C) and then with full vision (B and D). The drawings of the normal subject in three successive trials indicated that this individual was capable of generating a reasonable likeness of the original template under both conditions (A and B). The cerebellar patient, however, was not nearly as successful in this task (C and D). Interestingly, a portion of the template was duplicated quite well under both conditions. The remainder of the figure was completely wrong, however, and this errant representation of a part of the template was produced consistently across all trials and in both conditions. We interpreted these observations as suggesting that the cerebellum is required for generating an accurate representation of object shapes when only kinesthetic cues are available during acquisition. This experiment also demonstrates another type of volitional task whose acquisition is cerebellar-dependent.

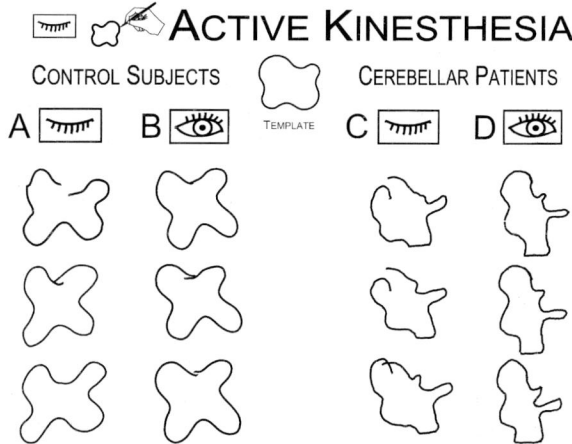

Fig. 5. Comparison of the capacity of control versus cerebellar patients to learn the shapes of irregular templates in the 'active kinesthesia condition'. Subjects attempted to learn each shape by tracing the original template (shown between B and C) over five successive sets of trials while blindfolded. Both patients and controls were then asked to draw the shape of the original template, first blindfolded (A and C) and then with eyes open (B and D). Traces under each condition for each subject were taken from a succession of trials from top to bottom of each column. (Reproduced from Shimansky et al., 1997 with permission from Cold Spring Harbor Laboratory Press.)

Conclusions

The data reviewed in the preceding sections support the notion that the cerebellum's role in motor learning is task- and even condition-dependent. Very importantly, this task-dependency is not based merely on large differences in behavior such as the difference between volitional and reflexive movements. Rather, the conditions under which a movement is to be learned are critical for determining whether either acquisition or retention is cerebellar-dependent. For example, as reviewed earlier, the cerebellum is required to adapt to unpredictable but consistent perturbations of a reaching movement (Shimansky et al., 1995). This condition is very different than that characterizing another volitional task in which subjects were required to use kinesthetic cues to learn the shapes of objects (Shimansky et al., 1997). Interestingly, both tasks require the on-line processing of kinesthetic cues for the acquisition and performance of the learned behavior. If so, this class of conditions may define one type of task whose learning is cerebellar-dependent. The studies reviewed earlier, however, clearly indicate that the cerebellum is required for the learning of behaviors such as eyeblink conditioning in which this condition is irrelevant. Thus, there are clearly several other conditions that are pertinent for engaging the cerebellum in the learning of a specific motor behavior. In fact, the concepts that define the extent to which the learning of a task is cerebellar-dependent are just beginning to emerge.

Historically, the initial studies addressing task-dependency have focused on determining whether or not the learning of a certain behavior is cerebellar-dependent rather than examining the *mechanisms* underlying this structure's contribution to learning a specific movement. Nevertheless, these experiments provided the first step. Future studies making the most pertinent advances will focus on *mechanisms*, particularly those features of information processing for which the cerebellum is critical to the learning of the task. Clearly designed experiments will be required to identify which behavioral conditions are critical for the cerebellum's involvement, and subsequent studies will be necessary to examine the cerebellar-dependent changes in neuronal interactions, synaptic modifications and neurochemical changes that accompany the learning process. As suggested earlier, the diversity of mechanisms as well as the tasks likely to involve the cerebellum is considerable. These tasks undoubtedly include several that are 'nonmotor' or cognitive in nature (Schmahmann, 1997). They include the task discussed earlier in which the cerebellum was required for establishing the internal representation of objects based on certain types of sensory cues (Shimansky et al., 1997). The many observations supporting this view also emphasize that the search for a singular, inclusive role for the cerebellum in motor learning should be abandoned in favor of concepts that support the diversity of roles this structure plays in learning motor tasks.

The diversity of interactions in which the cerebellum participates during the learning of different behaviors is particularly apparent from recent studies in the laboratory. They employed multiple single-unit recording techniques to examine the modification of information processing in the cerebellum during the learning process. Although full reports on these experiments are not yet available, the initial observations are very illuminating. Recordings have been made during the acquisition of three of the tasks discussed earlier: the classical conditioning of the eyeblink reflex (Bloedel et al., 1998), the reach perturbation task (Shimansky et al., 1999) and a task requiring a cat to 'solve' a template consisting of 2–3 connected straight grooves (Milak et al., 1995). In each of these experiments, changes in event-related modulation were correlated with critical features of the paradigm while the behavior was being acquired. During the initial stages of the classical conditioning of the eyeblink reflex (Bloedel et al., 1998), response components of Purkinje cells and nuclear neurons were well timed to the interval between the conditioned and unconditioned stimuli. This interval is the *critical temporal feature* of the paradigm. The hypothesis is that this modulation transiently encodes the interstimulus interval in both the cerebellar cortex and its nuclei during early stages of the conditioning process.

The modulation of cerebellar neurons was quite different during adaptation to unexpected perturbations of the wrist applied during a reaching task (Shimansky et al., 1999). In this paradigm, the learning-related responses of these cells were correlated with the absence of an expected perturbation as the animal transitioned from the predictable to the unpredictable paradigm. Thus, the activity was correlated with the absence of an event, not a temporal feature of the task. Finally, during the acquisition of the template task (Milak et al., 1995), the depth of modulation of existing responses increased just as the pattern of movements required to negotiate the template were being expressed in a relatively consistent manner. (For a further description of these experiments, see the previous discussion of Fig. 4.) Interestingly, in all of the recent studies, modulation of cerebellar neurons was related to *selected critical features of the paradigm* rather than to a single feature or class of features applicable across all behaviors.

We also contend that the concept of task-dependency is applicable considerably beyond the cerebellum and its contribution to motor learning. Neuroscience is replete with efforts to ascribe a single function to a variety of other components of the CNS. This approach has precipitated great debates just as it has in the field of motor learning. In most areas, however, the actual progress has been slow. We propose that the understanding of CNS function will be advanced more rapidly and with fewer distractions if task-dependency is considered in the design and conduct of integrative experiments. No one paradigm, no matter how well conceived, will be adequate for defining comprehensively the function of a component of the CNS. Unfortunately, a positive finding often serves as the basis for supporting a singular function for a structure to the exclusion of all others. To address this problem, it is clearly advantageous to study a single CNS structure using *multiple paradigms*. Furthermore, interpreting data acquired from diverse paradigms on the basis of *task-dependency* is likely to lead to significant advances in the understanding of CNS interactions. It will certainly focus discussions on the most meaningful concepts, particularly in the area of motor learning.

Acknowledgments

This work was supported by NIH grants NS 21958 and NS 36752.

Abbreviations

B	background rate
C1–C4	conditioning days
CNS	central nervous system
CR	conditioned reflex
D1–D3	experimental days
D1–D4	four dentate neurons
MA	muscimol action
MB	baseline for muscimol injections
MRI	magnetic resonance imaging
PIN1–PIN3	three posterior interposed neurons
R	response rate
RE	retention evaluation
RE	retention evaluation
SA	saline action
SB	baseline for saline injections
SEM	standard error of the mean
VOR	vestibular ocular reflex

References

Bloedel, J.R. and Bracha, V. (1995). On the cerebellum, cutaneomuscular reflexes, movement control and the elusive engrams of memory. Behav. Brain Res., 68: 1–44.

Bloedel, J.R., Bracha, V., Shimansky, Y. and Milak, M.S. (1996). The role of the cerebellum in the acquisition of complex, volitional forelimb movements. In: Bloedel J.R., Ebner T.J. and Wise S.P. (Eds.), The Acquisition of Motor Behavior in Vertebrates. MIT Press, Cambridge, pp. 319–342.

Bloedel, J.R., Bracha, V. and Wunderlich, D.A. (1998). Evolution of activity in small populations of neurons in the cerebellar cortex and nuclei during acquisition of the conditioned eyeblink reflex in the rabbit. Soc. Neurosci. Abstr., 24: 1522.

Bracha, V. and Bloedel, J.R. (1996). The multiple-pathway model of circuits subserving the classical conditioning of withdrawal reflexes. In: Bloedel J.R., Ebner T.J. and Wise S.P. (Eds.), The Acquisition of Motor Behavior in Vertebrates. MIT Press, Cambridge, pp. 175–204.

Bracha, V., Webster, M.L., Winters, N.K., Irwin, K.B. and Bloedel, J.R. (1994). Effects of muscimol inactivation of the cerebellar interposed-dentate nuclear complex on the performance of the nictitating membrane response in the rabbit. Exp. Brain Res., 100: 453–468.

Bracha, V., Zhao, L., Wunderlich, D.A., Morrissy, S.J. and Bloedel, J.R. (1997). Patients with cerebellar lesions cannot acquire but are able to retain conditioned eyeblink reflexes. Brain, 120: 1401–1413.

Bracha, V., Irwin, K.B., Webster, M.L., Wunderlich, D.A., Stachowiak, M.K. and Bloedel, J.R. (1998). Microinjections of anisomycin into the intermediate cerebellum during learning affect the acquisition of classically conditioned responses in the rabbit. Brain Res., 788: 169–178.

Friston, K.J., Frith, C.D., Passingham, R.E., Liddle, P.F. and Frackowiak, R.S.J. (1992). Motor practice and neurophysiological adaptation in the cerebellum: a positron tomography study. Proc. R. Soc. Lond. B. Biol. Sci., 248: 223–228.

Ito, M. (1984). The Cerebellum and Neural Control. Raven Press, New York.

Kelly, T.M., Zuo, C.C. and Bloedel, J.R. (1990). Classical conditioning of the eyeblink reflex in the decerebrate-decerebellate rabbit. Behav. Brain Res., 38: 7–18.

Krupa, D.J. and Thompson, R.F. (1995). Inactivation of the superior cerebellar peduncle blocks expression but not acquisition of the rabbit's classically conditioned eye-blink response. Proc. Natl. Acad. Sci. USA, 92: 5097–5101.

Krupa, D.J. and Thompson, R.F. (1997). Reversible inactivation of the cerebellar interpositus nucleus completely prevents acquisition of the classically conditioned eye-blink response. Learn. Mem., 3: 545–556.

Marr, D. (1969). A theory of cerebellar cortex. J. Physiol. (Lond.), 202: 437–470.

McCormick, D.A., Clark, G.A., Lavond, D.G. and Thompson, R.F. (1982). Initial localization of the memory trace for a basic form of learning. Proc. Natl. Acad. Sci. USA, 79: 2731–2735.

Milak, M.S., Bracha, V. and Bloedel, J.R. (1995). Relationship of simultaneously recorded cerebellar nuclear neuron discharge to the acquisition of a complex, operantly conditioned forelimb movement in cats. Exp. Brain Res., 105: 325–330.

Milak, M.S., Shimansky, Y., Bracha, V. and Bloedel, J.R. (1997). Effects of inactivating individual cerebellar nuclei on the performance and retention of an operantly conditioned forelimb movement. J. Neurophysiol., 78: 939–959.

Nordholm, A.F., Lavond, D.G. and Thompson, R.F. (1991). Are eyeblink responses to tone in the decerebrate, decerebellate rabbit conditioned responses? Behav. Brain Res., 44: 27–34.

Sanes, J.N., Dimitrov, B. and Hallet, M. (1990). Motor learning in patients with cerebellar dysfunction. Brain, 113: 103–120.

Schmahmann J.D. (Ed.), (1997). Cerebellum and Cognition. Academic Press, San Diego.

Shimansky, Y., Wang, J.J., Bracha, V. and Bloedel, J.R. (1995). Cerebellar inactivation abolishes the capability of cats to compensate for unexpected but not expected perturbations of a reach movement. Soc. Neurosci. Abstr., 21: 914.

Shimansky, Y., Saling, M., Wunderlich, D.A., Bracha, V., Stelmach, G.E. and Bloedel, J.R. (1997). Impaired capacity of cerebellar patients to perceive and learn two-dimensional shapes based on kinesthetic cues. Learn. Mem., 4: 36–48.

Shimansky, Y., Bauer, R., Bracha, V. and Bloedel, J.R. (1999). Modulation of simultaneously recorded cerebellar cortical and nuclear cells during adaptation to expected and unexpected perturbations of reaching movements in cats. Soc. Neurosci. Abstr., 25: 1560.

Steinmetz, J.E., Logue, S.F. and Steinmetz, S.S. (1992). Rabbit classically conditioned eyelid responses do not reappear after interpositus nucleus lesion and extensive post-lesion training. Behav. Brain Res., 15: 103–115.

Thach, W.T., Goodkin, H.P. and Keating, J.G. (1992). The cerebellum and the adaptive coordination of movement. Annu. Rev. Neurosci., 15: 403–442.

Thompson, R.F. and Krupa, D.J. (1994). Organization of memory traces in the mammalian brain. Annu. Rev. Neurosci., 17: 519–549.

Timmann, D., Shimansky, Y., Larson, P.S., Wunderlich, D.A., Stelmach, G.E. and Bloedel, J.R. (1996). Visuomotor learning in cerebellar patients. Behav. Brain Res., 81: 99–113.

Wang, J.J., Shimansky, Y., Bracha, V. and Bloedel, J.R. (1998). Effects of cerebellar nuclear inactivation on the learning of a complex forelimb movement in cats. J. Neurophysiol., 79: 2447–2459.

Welsh, J.P. and Harvey, J.A. (1989). Cerebellar lesions and the nictitating membrane reflex: performance deficits of the conditioned and unconditioned response. J. Neurosci., 9: 299–311.

CHAPTER 32

Role of the cerebellum in eyeblink conditioning

Vlastislav Bracha*

Department of Biomedical Sciences, Iowa State University College of Veterinary Medicine, Ames, IA 50011, USA

Abstract: The mammalian cerebellum is thought to participate in motor control and motor learning. The specific cerebellar contribution to these processes is not clear, however. Advances in understanding cerebellar function have been relatively slow, because, at least in most cases, the cerebellum appears to play only an ancillary role in the behaviors studied to date. A remarkable exception is classical conditioning of eyeblink responses in the rabbit. In this model, an intact cerebellum is critical for both the acquisition and expression of conditioned responses. Recent experiments suggest that the cerebellar role in classical conditioning might be similar in all mammals, including the human. Moreover, anticipatory defensive reflexes in other effector systems show a similar dependence on the intermediate cerebellum. Further developments in our understanding of cerebellar function will depend on examination of a wider array of cerebellar-involved neural networks. There is also need for the development of new experimental approaches to associative learning in both the nonhuman primate and the human.

Introduction

The ultimate challenge to living organisms is their adaptation to an ever-changing environment. This adaptation is achieved either across generations on a genomic level, or on an individual level. The latter involves acquiring and retaining new behavioral solutions through learning and memory. According to current views, learning and memory are not monolithic processes subserved by memory-dedicated parts of the brain. On the contrary, several specialized and relatively separate neural circuits seem to be responsible for the storage of distinct types of information. Based on the location of relevant neural circuits, as well as on the type of stored information, multiple memory systems have been identified (Squire and Zola, 1996; Eichenbaum, 2001).

The cerebellum is thought to participate in the fine control of movements and motor learning. This notion is based on studies examining the performance of human and animal subjects with experimental or pathological cerebellar lesions in a variety of motor control and learning tasks (Holmes, 1939; Dow and Moruzzi, 1958; Thach et al., 1992). One of the puzzling features of cerebellar involvement in motor control and learning is that for most of the studied behaviors, lesions of the cerebellum do not abolish the behavior or form of learning. Rather, they produce only various degrees of partial degradation of the behaviors (for an overview, see Bloedel, 1992). A remarkable exception to this rule is classical conditioning of eyeblink responses in the rabbit (Fig. 1). Participation of the rabbit brain above the level of the midbrain is not required for eyeblink conditioning to occur and be retained (Oakley and Russell, 1977).[1] Small lesions of the

*Corresponding author: Tel.: +515-294-6278;
Fax: +515-294-2315; E-mail: vbracha@iastate.edu

---

[1] This independence of eyeblink conditioning from control by supra-midbrain structures is valid for the *delay-conditioning paradigm*, in which the conditioned stimulus (CS; elicits the learned response) and the unconditioned stimulus (US; elicits the inborn unconditioned reflex) partially overlap in time and coterminate. When the CS and US do not overlap, and the end of the CS is separated from the onset of the US by a short period of time, it is called the *trace-conditioning paradigm*. As with delay conditioning, trace conditioning depends on the integrity of the intermediate cerebellum. In contrast to delay conditioning, however, trace conditioning requires the participation of higher levels of the brain, most notably the hippocampus. This review is solely focused on data obtained when using delay-conditioning paradigms.

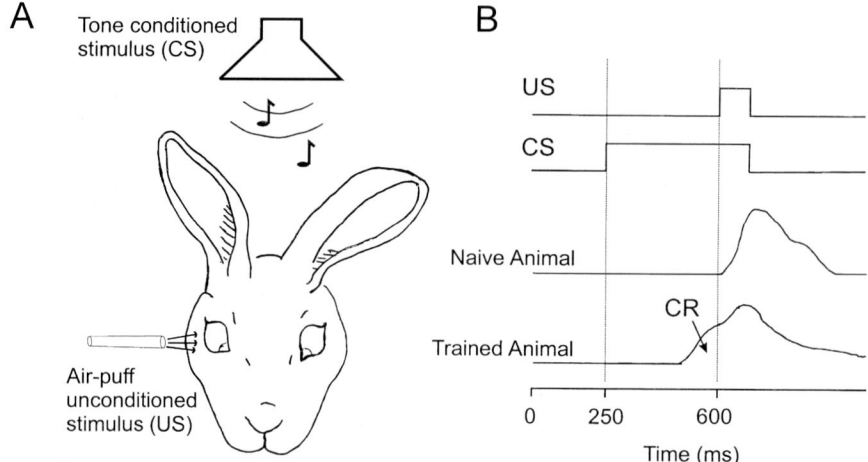

Fig. 1. Classical conditioning of the eyeblink response in the standard delay paradigm. Rabbits are presented with the combination of an air-puff unconditioned stimulus (US) and a tone conditioned stimulus (CS) (A and the top two traces in B). Recordings of eyelid movements (B, the third trace from the top) show that naïve animals exhibit only reflexive eyeblinks elicited by the US (this inborn reflex is also called the unconditioned response—UR). During repetitive presentations of the paired CS + US, animals learn to respond to a previously ineffective tone. This learned response is designated the conditioned response (CR, the last trace in B). In rabbits, both the CR and UR are predominantly unilateral and expressed on the side of the US application. Lesions of the intermediate cerebellum and associated brainstem structures block both the acquisition and expression of CRs.

cerebellar interposed nuclei, however, completely and permanently abolish the capacity to acquire new conditioned eyeblinks and prevent the expression of conditioned eyeblinks learned *before* the lesion (McCormick and Thompson, 1984). This surprising *total dependence* on the integrity of the intermediate cerebellum has spawned a highly dynamic line of research focused on the analysis of cerebellar contributions to eyeblink conditioning.

This chapter provides a short overview of current understanding of the cerebellar role in withdrawal reflex conditioning in experimental animals and discusses the applicability of knowledge derived from animal models to the neural substrates of eyeblink conditioning in the human.

Cerebellar involvement in eyeblink conditioning in the rabbit

Systematic lesion studies performed in several laboratories have successfully delineated components of circuits that are required for the acquisition and retention of classically conditioned eyeblink responses in the rabbit. Early experiments demonstrated that brain tissue above the level of midbrain is not necessary for conditioning to occur and for conditioned responses to be expressed (Oakley and Russell, 1977; Mauk and Thompson, 1987; Kelly et al., 1990). On the other hand, acquisition and retention of the eyeblink conditioned response (CR) are permanently abolished following lesions of the lateral pontine nuclei, inferior olive, cerebellar anterior-interposed nucleus, the red nucleus and the inferior, middle and superior cerebellar peduncles (for review, see Bracha and Bloedel, 1996; Thompson et al., 2000). Studies examining the effects of lesioning the cerebellar cortex (components HV and HVI) reported either a complete disruption of eyeblink conditioning (Yeo et al., 1985; Garcia et al., 1999) or its weakened but not abolished state (Lavond and Steinmetz, 1989; Harvey et al., 1993). All of the above structures exhibit electrophysiological correlates of the conditioning stimuli and CRs (for review, see Bracha and Bloedel, 1996). Based on these facts, and the known neuroanatomical connectivity, the circuit subserving classical conditioning of the eyeblink response in the rabbit can be depicted as an intermediate cerebellum-related network superimposed on the circuit controlling the inborn unconditioned eyeblink reflex (Fig. 2). It is assumed that the CS (typically a tone) enters the circuit predominantly through the pontine nuclei, although other points of entry (e.g., the trigeminal nuclei; or direct pathways to the cerebellum) have not been ruled out. The information about the US (typically a tactile stimulus applied to the surface of the eye or the periocular area) enters through the trigeminal nucleus, from which it propagates through both the inferior olive and pontine nuclei to the cerebellum. These inputs are processed by a network which eventually produces the properly timed activation of the motor nuclei. The relevant motor nuclei in the rabbit include the facial nucleus, the accessory abducens nucleus (both groups controlling eyelid closure) and motoneurons supplying the lavator palpebrae muscle (which elevates the upper eyelid).

Much of the attention in this field is directed toward discovering which parts of the network are modified during learning. Such research is frequently labeled as a 'search for the engram' or 'search for traces of memory'. For a detailed discussion of the issue of eyeblink-conditioning-related plasticity, see Bracha and Bloedel (1996), Thompson and Kim (1996) and Steinmetz (2000). At present, the likely places of plastic changes that contribute to eyeblink conditioning include the intermediate cerebellar cortex, anterior-interposed nucleus, pontine nuclei and the trigeminal complex. Also possibly contributing are not-yet-identified interneurons both in the polysynaptic components of unconditioned eyeblink circuits and the auditory pathways.

Most of the search for eyeblink-conditioning-related plasticity has focused on the intermediate cerebellum. The converging evidence from studies manipulating the network during CR acquisition [e.g., studies using temporary inactivation, protein synthesis inhibition (Bracha et al., 1998) and the inhibition of protein kinase (Chen and Steinmetz, 2000)] suggests that the interposed nuclei, and perhaps also the overlying portions of the cerebellar cortex, are places at which some type of plastic change is required for learning to proceed normally and the learned response to be generated and properly timed. It should be emphasized, however, that other parts of the network have not yet been

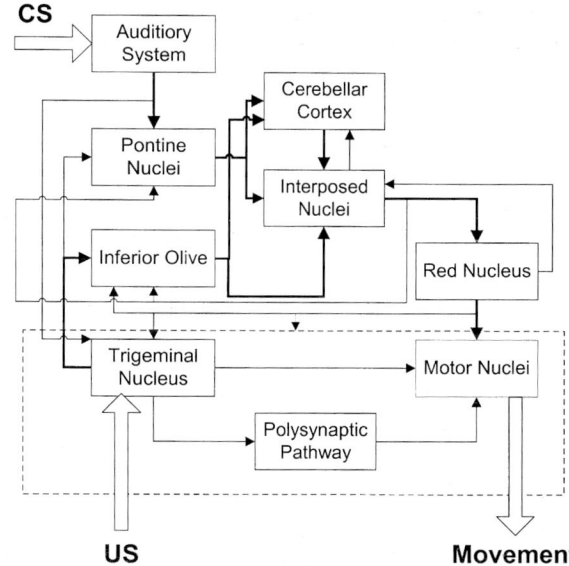

Fig. 2. Simplified schematic of a network that is required for the acquisition and expression of classically conditioned eyeblinks in the rabbit. The basic component of this network is the circuit controlling the unconditioned eyeblink (dashed box). Superimposed over the UR circuit is the intermediate cerebellum-centered network. Most of the early studies were designed on the premise that only part of the existing connectivity (bold arrows) within the network is relevant for eyeblink conditioning (Thompson, 1986). Major challenges for future investigations are: (1) analysis of the physiological role of multiple feedback loops, (2) further elucidating how the interposed nucleus and the red nucleus interface with eyeblink circuits to produce the coordinated action of several motoneuronal groups, and (3) showing how this basic network is integrated with the rest of the brain when it participates in more complex tasks, such as the eyeblink conditioning in the trace paradigm.

studied extensively, and their plastic contributions to this form of learning are still poorly understood.

The circuit depicted in Fig. 2 is among the *most completely described* neural networks involved in associative learning in the mammalian brain. As such, it offers tremendous potential for understanding the mechanisms of learning and memory. The following sections of this chapter address two fundamental questions: (1) are similar circuits involved in the classical conditioning of responses in other effector systems and other vertebrate species; and (2) is this network also involved in the classical conditioning of eyeblink responses in the human?

Phylogeny and the multifunctionality of intermediate cerebellum-related eyeblink conditioning networks

The phylogenetic emergence and universality of intermediate cerebellum-related networks have not yet been studied systematically. The unconditioned eyeblink reflex first appeared in terrestrial vertebrates that possess external eyelids. Knowledge is still quite limited about the capacity of this response to be classically conditioned in different vertebrate species. In the frog, for example, it seems that eyeblink circuits do not support eyeblink conditioning (F.P. Kolb, personal communication). In the turtle, eyeblink conditioning has not been examined at the behavioral level. Research using an in vitro preparation of the turtle brain, however, has demonstrated a functional analog of trigeminal reflex conditioning (Keifer et al., 1995). Interestingly, this turtle preparation does not require the cerebellum and the red nucleus for 'eyeblink' conditioning to occur (Anderson and Keifer, 1997). It is not clear, however, whether an intact turtle would require cerebellar participation in a similar way to that shown in the rabbit. Based on the available data, it appears that the circuits involved in eyeblink conditioning in the turtle differ from their mammalian counterparts in their dependency on cerebellum-related circuits. In bird species, substrates involved in eyeblink conditioning have not yet been studied.

In mammals other than rabbits, lesion studies have demonstrated that the interposed nucleus is required for eyeblink conditioning in rodents (mouse, rat; Skelton, 1988; Chen et al., 1999). The expression of classically conditioned eyeblinks is also affected in the cat following inactivation of the interposed nuclei (Kolb et al., 1994). These data indicate that the dependence of eyeblink conditioning on intermediate cerebellum-related neuronal networks is shared among different mammalian species. Presently, it is not clear whether these circuits are unique to mammals or whether they perform a similar function in other vertebrates.

Are intermediate cerebellum-related circuits involved selectively in eyeblink conditioning, or do they also participate in the conditioning of responses in other effector systems? Functionally, the eyeblink reflex represents a withdrawal reflex during which the surface of the body is protected from the potentially damaging effect of aversive stimuli. A functional analog of the eyeblink reflex in another effector system is the limb withdrawal reflex. Cerebellar involvement in limb withdrawal conditioning has been systematically examined in the cat. Permanent lesions of the superior cerebellar peduncle (Voneida, 2000), as well as temporary inactivation of the interposed nuclei (Kolb et al., 1997), disrupt the expression of classically conditioned fore- and hindlimb withdrawal reflexes. These recent data seem to contradict classic reports of studies performed in dogs, which claimed that permanent lesions of the cerebellum do not prevent classical conditioning of limb withdrawal (Popov, 1929; Gambaryan, 1960; Fanardjian, 1961). It is not clear whether the conflicting data from the dog and cat studies are due to species differences, or whether they can be explained by methodical differences between the experimental paradigms. Nevertheless, the available data demonstrate that the intermediate cerebellum-related network is multifunctional because it is involved in the acquisition and/or expression of classically conditioned responses in multiple withdrawal response systems in different mammalian species.

Cerebellar involvement in the rabbit is not limited to the classical conditioning of withdrawal responses. Eyeblink conditioning in normal subjects is accompanied by autonomic response components, most notably by the learned slowing of heart rate (i.e., classically conditioned bradycardia). While lesions of the intermediate cerebellum (which is crucial for eyeblink conditioning) do not affect conditioned bradycardia acquisition and retention, the acquisition rate of this response is severely affected by bilateral lesions of the posterior cerebellar vermis (Supple and Leaton, 1990; Supple and Kapp, 1993). This finding indicates that different parts of the cerebellum are involved in the control of skeleto-muscular and autonomic components of behavior learned during classical conditioning of responses to aversive stimuli.

Cerebellar involvement in eyeblink conditioning in humans

Intermediate cerebellum-related networks have become some of the best-described and understood circuits involved in a relatively simple form of associative learning/memory in mammals. The wealth of data and insights gained from animal models raise an important question: how does this knowledge relate to the substrates of the same type of learning in the human? The answer to this question is not straightforward, because the number of experimental approaches appropriate for human subjects is limited when compared to those conducted on animals. In essence, only two methods have been widely used on human subjects. The first utilizes imaging methods to reveal parts of the brain that are active during classical conditioning. The second approach uses patients with pathology restricted to specific parts of the brain, and then tests the subject's capacity to acquire and/or retain classically conditioned eyeblinks.

Imaging studies on the human

Four PET studies have examined brain regions involved in eyeblink conditioning (Molchan et al., 1994; Logan and Grafton, 1995; Blaxton et al., 1996; Schreurs et al., 1997). A further one has described the pattern of brain activation during classical conditioning of the limb flexion reflex (Timmann et al., 1996; for a recent review, see also McIntosh and Schreurs, 2000). All of these experiments documented that in normal subjects, the cerebellum exhibits activity changes during classical conditioning, but not during pseudoconditioning. The resolution of the data, however, does not permit a distinction to be drawn between relative activity changes in the cerebellar nuclei versus cortex. For the same reason, patterns of increased/decreased activation in the brainstem were not described. A noticeable feature of these studies is that they described a number of other involved structures (e.g., auditory cortex, basal ganglia, frontal cortex,) normally not implicated as essential for delay eyeblink conditioning in the rabbit. The functional significance of activity changes in the human forebrain is not clear. Elucidation of this question awaits careful studies involving patients with lesions restricted to clearly described regions of the brain. It could be that some or even most of the areas exhibiting the activity changes are involved in nonspecific cognitive and/or sensory components of the task. This point is best illustrated by the fact that although the basal ganglia exhibit significant changes of activity during conditioning, patients with basal ganglia pathologies can acquire conditioned eyeblinks at a rate comparable to that of normal controls (Woodruff-Pak and Papka, 1996; Sommer et al., 1999).

Patients with cerebellar pathologies exhibit CR acquisition deficits

In accordance with predictions based on animal models, patients with lesions and other types of cerebellar pathologies are deficient in their capacity to acquire classically conditioned eyeblinks (Lye et al., 1988; Solomon et al., 1989; Daum et al., 1993; Topka et al., 1993; for review, see Schugens et al., 2000). The learning deficit is profound because even 4 days of training (at 100 trials/day) does not produce a significant acquisition of CRs (Bracha et al., 1997). In contrast, normal subjects typically learn CRs in a *single* training session. Investigators using unilateral unconditioned stimuli reported that subjects with unilateral lesions exhibit unilateral deficits on the side ipsilateral to the damage. In general, these data confirm the basic prediction derived from the rabbit-conditioning model. In contrast to the previously cited studies on human patients, however, we found that patients with unilateral cerebellar lesions exhibit bilateral CR acquisition deficits when the unconditioned stimulus is a glabela tap which elicits symmetrical bilateral eyeblinks (Bracha et al., 1997). It is not clear whether the unexpected bilateral effect of unilateral lesions (in rabbits the deficit is strictly unilateral following a unilateral lesion) represents a lesion site- and size-dependent peculiarity of human eyeblink conditioning circuits, or whether it is due to 'tighter' bilateral eyeblink control in humans. Because of limited documentation on the extent of the cerebellar lesions in the examined population of patients, and because current radiological methods do not provide a resolution comparable to the histological analysis of the animal brain, understanding the topography of critical cerebellar components in humans is rudimentary. The differential contribution of separate cerebellar cortical areas, as well as the role of individual cerebellar nuclei, is simply unknown.

Similar to the reports describing deficits in the expression of the classically conditioned limb withdrawal in the cat (Kolb et al., 1997), Timmann et al. (2000) reported that patients with pathologies in the anterior and superior portions of the cerebellum have an impaired capacity to acquire the classically conditioned leg flexion response. These authors also documented that cerebellar patients are deficient in other types of conditioned tasks: the long-term habituation of the eyelid component of the acoustic startle response (Maschke et al., 2000), and fear-conditioned bradycardia (Maschke et al., 2002). Similar deficits in classically conditioned bradycardia responses, and in the long-term habituation of the startle response following bilateral lesions of the cerebellar vermis, were previously reported in the rat (Supple and Leaton, 1990).

In summary, data from studies of patients with cerebellar pathologies indicate that similar to other mammals, the cerebellum in humans is significantly involved in the acquisition of several types of CRs. This begs the question: does the human cerebellum play a similarly important role in CR retention?

Retention of classically conditioned eyeblinks in patients with cerebellar lesions

If human circuits controlling CR acquisition are similar to the networks described in the rabbit, it could be expected that lesions in relevant parts of the cerebellum affecting CR acquisition should prevent the expression of responses learned before the lesion. This fundamental prediction is difficult to test in human subjects for obvious reasons, i.e., the acquisition of cerebellar pathologies by human subjects is not under the experimenter's control. Therefore, one cannot train subjects just before the lesion and then, after the pathology develops, test for CR retention. This difficulty can be overcome only if one would be able to identify CRs that are acquired naturally and early during ontogenesis as a consequence of normal daily experiences. If such naturally conditioned eyeblinks could be found, and if a patient with cerebellar pathology acquired these responses before the pathology developed, one could test this subject for the retention of this learned response as well as for the acquisition of new responses.

To address the issue of conditioned eyeblink retention, we identified cerebellar patients unable to acquire new CRs, and in these subjects, we examined if they retain conditioned eyeblinks learned before onset of the cerebellar pathology. As the first candidate for a naturally conditioned eyeblink, we selected the visual threat eyeblink response (VTER). It can be readily elicited by having objects rapidly approach the subject's face. To test this response, we developed a new paradigm in which the VTER is elicited in each trial by a small ball that swings toward and eventually taps the subject's forehead (Bracha et al., 1997). We surmised that visually triggered eyeblinks are naturally acquired CRs for which the visual information of the approaching ball serves as the CS, while the cutaneous sensations elicited by the ball's impact on the forehead operate as the trigeminal US. Three facts support this assumption: (1) during ontogeny, the VTER develops significantly later than unconditioned trigeminal eyeblinks; (2) in contrast to the trigeminal UR, the VTER extinguishes (i.e., when not reinforced by the impact of the ball); and (3) the VTER exhibits the blocking effect (i.e., subjects fail to acquire CRs to a new CS when this new CS is presented simultaneously with the VTER CS). The second and third observations are characteristic properties of classically conditioned eyeblinks.

Surprisingly, patients who could not acquire new classically conditioned eyeblinks were able to produce VTERs (Bracha et al., 1997). This ability of CR acquisition-deficient cerebellar patients to generate previously learned responses was further confirmed in an additional study demonstrating that these patients can also perform the kinesthetic threat eyeblink response, which is another type of naturally conditioned eyeblink (Bracha et al., 2000). These data suggest that some of the cerebellar components of the neural circuits controlling CR acquisition in humans are not required for the long-term storage of corresponding memory traces.

There are two possible interpretations of the above-described acquisition data. It is possible that in humans, the cerebellum is required only during CR acquisition and that long-term plastic changes subserving the storage of well-learned responses are located at extracerebellar sites. Perhaps a more likely explanation is that the process of CR retention involves healthy parts of the cerebellum being spared in the subjects included in the authors' studies. For example, most of the subjects had lesions predominantly in the cerebellar cortex. It is possible, just as other authors have proposed for the rabbit model, that the cerebellar cortex is essential in the initial stages of learning and also for the modification of existing CRs (Raymond et al., 1996). This notion is supported by the finding that cerebellar patients have an impaired ability to extinguish naturally learned CRs (Bracha et al., 2000).

Future directions

The general thrust of available data indicates that there are significant parallels in the cerebellum's involvement in eyeblink conditioning in humans and other mammals. This fact emphasizes the need for further development of mouse, rat and rabbit models because they seem to address general mammalian mechanisms. As such, knowledge obtained from these models will most likely be applicable to the physiology of learning and memory in the human. The continued development of traditional animal models will depend on a more detailed understanding of the function of the involved circuits from the perspective of network operation, rather than conceptualizing the circuits as unidirectional pathways that merely relay the information to dedicated sites of plasticity (Bracha and Bloedel, 1996). Understanding the neuronal network's function will require approaches using high-resolution electrophysiological techniques combined with sophisticated methods of network manipulation using, in addition, neuropharmacological and molecular biology tools. Addressing the question of sites of plasticity in these networks will require the further sophistication of the tools that manipulate local mechanisms of plasticity without affecting on-line communication in the involved neural networks (Bracha et al., 1996).

Besides the development of traditional animal models, progress in understanding human substrates of this form of memory will require a better understanding of the phylogenetic underpinning of the involved mechanisms. On the other hand, a parallel development of technologies that could be used in human experiments is needed. Because of the methodical limits of human research, insights into human mechanisms of classical conditioning would greatly benefit from an accelerated development of a primate model of eyeblink conditioning (Clark and Zola, 1998). Human studies will be significantly advanced by the further sophistication of paradigms that examine CR retention in subjects with lesions of the cerebellum. These include (1) examining other components of intermediate cerebellum-related circuits (e.g., pontine nuclei, red nucleus), (2) increasing the resolution of current brain imaging methods, (3) constructing detailed maps of the parts of the brain activated during CR acquisition and retention, and (4) interfacing these maps with detailed maps of pathologies causing learning/retention deficits.

Acknowledgments

This research was supported by NIH Grants 2 R01 NS36210 and 5 R01 NS21958.

Abbreviations

CR	conditioned response
CS	conditioned stimulus
HV	fifth hemispheric lobulus (Larsell)
HVI	sixth hemispheric lobulus (Larsell)
PET	positron emission tomography
US	unconditioned stimulus
UR	unconditioned response
VTER	visual threat eyeblink response

References

Anderson, C.W. and Keifer, J. (1997). The cerebellum and red nucleus are not required for in vitro classical conditioning of the turtle abducens nerve response. J. Neurosci., 17: 9736–9745.

Blaxton, T.A., Zeffiro, T.A., Gabrieli, J.D.E., Bookheimer, S.Y., Carrillo, M.C., Theodore, W.H. and Disterhoft, J.F. (1996). Functional mapping of human learning: a positron emission tomography activation study of eyeblink conditioning. J. Neurosci., 16: 4032–4040.

Bloedel, J.R. (1992). Functional heterogeneity with structural homogeneity: How does the cerebellum operate? Behav. Brain Sci., 15: 666–678.

Bracha, V. and Bloedel, J.R. (1996). The multiple pathway model of circuits subserving the classical conditioning of withdrawal reflexes. In: Bloedel J.R., Ebner T.J. and Wise S.P. (Eds.), Acquisition of Motor Behavior in Vertebrates. MIT Press, Cambridge, pp. 175–204.

Bracha, V., Zhao, L., Wunderlich, D.A., Morrissy, S.J. and Bloedel, J.R. (1997). Patients with cerebellar lesions cannot acquire but are able to retain conditioned eyeblink reflexes. Brain, 120: 1401–1413.

Bracha, V., Irwin, K.B., Webster, M.L., Wunderlich, D.A., Stachowiak, M.K. and Bloedel, J.R. (1998). Microinjections of anisomycin into the intermediate cerebellum during learning affect the acquisition of classically conditioned responses in the rabbit. Brain Res., 788: 169–178.

Bracha, V., Zhao, L., Irwin, K.B. and Bloedel, J.R. (2000). The human cerebellum and associative learning: dissociation between the acquisition, retention and extinction of conditioned eyeblinks. Brain Res., 860: 87–94.

Chen, G. and Steinmetz, J.E. (2000). Microinfusion of protein kinase inhibitor H7 into the cerebellum impairs the acquisition but not the retention of classical eyeblink conditioning in rabbits. Brain Res., 856: 193–201.

Chen, L., Bao, S. and Thompson, R.F. (1999). Bilateral lesions of the interpositus nucleus completely prevent eyeblink conditioning in Purkinje cell-degeneration mutant mice. Behav. Neurosci., 113: 204–210.

Clark, R.E. and Zola, S. (1998). Trace eyeblink classical conditioning in the monkey: a nonsurgical method and behavioral analysis. Behav. Neurosci., 112: 1062–1068.

Daum, I., Schugens, M.M., Ackermann, H., Lutzenberger, W., Dichgans, J. and Birbaumer, N. (1993). Classical conditioning after cerebellar lesions in humans. Behav. Neurosci., 107: 748–756.

Dow, R.S. and Moruzzi, G. (1958). The Physiology and Pathology of the Cerebellum. University of Minnesota Press, Minneapolis.

Eichenbaum, H. (2001). From Conditioning to Conscious Recollection—Memory Systems of the Brain. Oxford University Press, New York.

Fanardjian, V.V. (1961). Influence of the cerebellum ablation on motor conditioned reflexes in dogs. Zh. Vyssh. Nerv. Deyat., 11: 920–926.

Gambaryan, L.S. (1960). Conditioned avoidance reflexes induced in dogs with cerebellar lesions. Physiol. Bohemoslov., 9: 261–266.

Garcia, K.S., Steele, P.M. and Mauk, M.D. (1999). Cerebellar cortex lesions prevent acquisition of conditioned eyelid responses. J. Neurosci., 19: 10940–10947.

Harvey, J.A., Welsh, J.P., Yeo, C.H. and Romano, A.G. (1993). Recoverable and nonrecoverable deficits in conditioned responses after cerebellar cortical lesions. J. Neurosci., 13: 1624–1635.

Holmes, G. (1939). The cerebellum of man. Brain, 62: 1–30.

Keifer, J., Armstrong, K.E. and Houk, J.C. (1995). In vitro classical conditioning of abducens nerve discharge in turtles. J. Neurosci., 15: 5036–5048.

Kelly, T.M., Zuo, C.C. and Bloedel, J.R. (1990). Classical conditioning of the eyeblink reflex in the decerebrate-decerebellate rabbit. Behav. Brain Res., 38: 7–18.

Kolb, F.P., Irwin, K.B., Winters, N.K., Bloedel, J.R. and Bracha, V. (1994). Involvement of the cat cerebellar interposed nucleus in the control of conditioned and unconditioned withdrawal reflexes. Soc. Neurosci. Abstr., 20: 1746.

Kolb, F.P., Irwin, K.B., Bloedel, J.R. and Bracha, V. (1997). Conditioned and unconditioned forelimb reflex systems in the cat: involvement of the intermediate cerebellum. Exp. Brain Res., 114: 255–270.

Lavond, D.G. and Steinmetz, J.E. (1989). Acquisition of classical conditioning without cerebellar cortex. Behav. Brain Res., 33: 113–164.

Logan, C.G. and Grafton, S.T. (1995). Functional anatomy of human eyeblink conditioning determined with regional glucose metabolism and positron emission tomography. Proc. Natl. Acad. Sci., 92: 7500–7504.

Lye, R.H., O'Boyle, D.J., Ramsden, R.T. and Schady, W. (1988). Effects of the unilateral cerebellar lesion on the acquisition of the eye-blink conditioning in man. J. Physiol. (Lond.), 403: 58.

Maschke, M., Drepper, J., Kindsvater, K., Kolb, F.P., Diener, H.C. and Timmann, D. (2000). Involvement of the human medial cerebellum in long-term habituation of the acoustic startle response. Exp. Brain Res., 133: 359–367.

Maschke, M., Schugens, M., Kindsvater, K., Drepper, J., Kolb, F.P., Diener, H.C., Daum, I. and Timmann, D. (2002). Fear conditioned changes of heart rate in patients with medial cerebellar lesions. J. Neurol. Neurosurg. Psychiatry, 72: 116–118.

Mauk, M.D. and Thompson, R.F. (1997). Retention of classically conditioned eyelid responses following acute decebration. Brain res., 403: 89–95.

McCormick, D.A. and Thompson, R.F. (1984). Cerebellum: essential involvement in the classically conditioned eyelid response. Science, 223: 296–299.

McIntosh, A.R. and Schreurs, B.G. (2000). Functional networks underlying human eyeblink conditioning. In: Woodruff-Pak D.S. and Steinmetz J.E. (Eds.), Eyeblink Classical Conditioning: Applications in Humans. Kluwer Academic Publishers, Boston, pp. 51–70.

Molchan, S.E., Sunderland, T., McIntosh, A.R., Herscovitch, P. and Schreurs, B.G. (1994). A functional anatomical study of associative learning in humans. Proc. Natl. Acad. Sci., 91: 8122–8126.

Oakley, D.A. and Russell, I.S. (1977). Subcortical storage of Pavlovian conditioning in the rabbit. Physiol. Behav., 18: 931–937.

Popov, N.F. (1929) The role of the cerebellum in the formation of conditioned motor reflexes. In: Fursikov, D.S., Gurevich, M.O. and Zalmanzon, A.N. (Eds.), Vishaya N'ervnaya D'eyatel'nost': Sbornik Trudov Instituta. Komunisticheskaya Akademiya, Institut Vyshey N'ervnoy D'eyatel'nosti, pp. 140–148.

Raymond, J.L., Lisberger, S.G. and Mauk, M.D. (1996). The cerebellum: A neuronal learning machine?. Science, 272: 1126–1131.

Schreurs, B.G., McIntosh, A.R., Bahro, M., Herscovitch, P., Sunderland, T. and Molchan, S.E. (1997). Lateralization and behavioral correlation of changes in regional cerebral blood flow with classical conditioning of the human eyeblink response. J. Neurophysiol., 77: 2153–2163.

Schugens, M., Topka, H. and Daum, I. (2000). Eyeblink conditioning in neurological patients with motor impairments. In: Woodruff-Pak D.S. and Steinmetz J.E. (Eds.), Eyeblink Classical Conditioning: Applications in Humans. Kluwer Academic Publishers, Boston, pp. 191–204.

Skelton, R.W. (1988). Bilateral cerebellar lesions disrupt conditioned eyelid responses in unrestrained rats. Behav. Neurosci., 102: 586–590.

Solomon, P.R., Stowe, G.T. and Pendlebury, W.W. (1989). Disrupted eyelid conditioning in a patient with damage to cerebellar afferents. Behav. Neurosci., 103: 898–902.

Sommer, M., Grafman, J., Clark, K. and Hallett, M. (1999). Learning in Parkinson's disease: eyeblink conditioning, declarative learning, and procedural learning. J. Neurol. Neurosurg. Psychiatry, 67: 27–34.

Squire, L.R. and Zola, S.M. (1996). Structure and function of declarative and nondeclarative memory systems. Proc. Natl. Acad. Sci., 93: 13515–13522.

Steinmetz, J.E. (2000). Brain substrates of classical eyeblink conditioning: a highly localized but also distributed system. Behav. Brain Res., 110: 13–24.

Supple, W.F., Jr. and Kapp, B.S. (1993). The anterior cerebellar vermis: essential involvement in classically conditioned bradycardia in the rabbit. J. Neurosci., 13: 3705–3711.

Supple, W.F., Jr. and Leaton, R.N. (1990). Cerebellar vermis: essential for classically conditioned bradycardia in the rat. Brain Res., 509: 17–23.

Thach, W.T., Goodkin, H.P. and Keating, J.G. (1992). The cerebellum and the adaptive coordination of movement. Annu. Rev. Neurosci., 15: 403–442.

Thompson, R.F. (1986). The neurobiology of learning and memory. Science, 233: 941–947.

Thompson, R.F. and Kim, J.J. (1996). Memory systems in the brain and localization of a memory. Proc. Natl. Acad. Sci., 93: 13438–13444.

Thompson, R.F., Swain, R., Clark, R. and Shinkman, P. (2000). Intracerebellar conditioning: Brogden and Gantt revisited. Behav. Brain Res., 110: 3–11.

Timmann, D., Kolb, F.P., Baier, C., Rijntjes, M., Müller, S.P., Diener, H.C. and Weiller, C. (1996). Cerebellar activation during classical conditioning of the human flexion reflex: a PET study. Neuroreport, 7: 2056–2060.

Timmann, D., Baier, P.C., Diener, H.C. and Kolb, F.P. (2000). Classically conditioned withdrawal reflex in cerebellar patients. 1. Impaired conditioned responses. Exp. Brain Res., 130: 453–470.

Topka, H., Walls-Solé, J., Massaquoi, S.G. and Hallett, M. (1993). Deficit in classical conditioning in patients with cerebellar degeneration. Brain, 116: 961–969.

Voneida, T.J. (2000). The effect of brachium conjunctivum transection on a conditioned limb response in the cat. Behav. Brain Res., 109: 167–175.

Woodruff-Pak, D.S. and Papka, M. (1996). Huntington's disease and eyeblink classical conditioning: normal learning but abnormal timing. J. Int. Neuropsychol. Soc., 2: 323–334.

Yeo, C.H., Hardiman, M.J. and Glickstein, M. (1985). Classical conditioning of the nictitating membrane response of the rabbit. II. Lesions of the cerebellar cortex. Exp. Brain Res., 60: 99–113.

CHAPTER 33

Integration of multiple motor segments for the elaboration of locomotion: role of the fastigial nucleus of the cerebellum

Shigemi Mori*, Katsumi Nakajima, Futoshi Mori and Kiyoji Matsuyama

Department of Biological Control System, National Institute for Physiological Sciences, Okazaki 444-8585, Japan

Abstract: This chapter provides a conceptual overview of the role and operation of higher structures of the central nervous system (CNS) in the control of posture and locomotion in the mammal, including the nonhuman primate and the human. Both quadrupedal and bipedal locomotion require the integrated neural control of multiple body segments against gravity. During development, and in selected instances in the adult, motor learning is required, particularly for merging anticipatory and reactive CNS processes, the latter being necessary after tripping and stumbling. We have recently found that the fastigial nucleus (FN) of the cerebellum in the cat plays a particularly important role in the control of locomotion, by virtue of its critical position in uniting the cerebro-cerebellar and the spino-cerebellar loops of neural activity that participate in the integrated control of multiple body segments. Further understanding of the CNS structures that achieve this integration has come from our recent study of an intact nonhuman primate, the Japanese monkey, *Macaca fuscata*, as it learns to elaborate bipedal locomotion rather than its normal quadrupedal fashion. Based on findings from these two animal species, we now present a model of the overall integrated control of posture and locomotion that features the combined operation of parallel and distributed neural circuitry throughout the CNS.

Introduction

This chapter presents an integration of current knowledge on the control of posture and locomotion by the central nervous system (CNS), with an emphasis on the potential role of the fastigial nucleus (FN) of the cerebellum. Heuristic constructs are proposed in the context of Sherrington's (1906) conceptualization of the association between posture and movement. Some hypotheses are presented on the elaboration and coordination of posture and locomotion, and there is emphasis on several unresolved issues.

In Sherrington's classic 1906 monograph he described animal posture as: "... *a posture of the animal as a whole—a total posture—is as much a complex built up of postures of portions of the animal—segmental postures (Bonnier)—as is the total movements of the animal—its locomotion—compounded of segmental movements*" (p. 341). He further stated that: "... *the posture of the head segments in many animals is dependent on the musculature not of the head segments themselves but of a long series of segments behind the head*" (p. 342) and "... *the vertebrate appendages called limbs are plurisegmental, but the individual segments constituting the limb form in respect of the limb a functional group of such solidarity that their reactions in the limb are at any time unitary*" (p. 343). In this monograph, Sherrington also proposed that the underlying mechanisms for the integration of multiple motor segments involve two distinct processes: *short reflexes*, which produce *local reactions* that serve to

*Corresponding author: Tel.: +81-564-55-7771; Fax: +81-564-52-7913; E-mail: mrsgm@nips.ac.jp

integrate a limited series of adjoining segments; and, *long reflexes*, which produce *nonlocal reactions* that serve to integrate a long or whole series of body segments. He further opined that the proprioceptive organs, such as the labyrinthine receptors, contribute to the integration of a short series of segments, whereas *distance-receptors* integrate the whole series of segments. The cerebellum was considered as the principal ganglion of proprioceptive organs.

By the time of Sherrington's 1906 monograph, a number of completed studies had demonstrated that cerebellar functions were diverse and multifaceted. The postulated functions then included: (a) maintenance of mechanical equilibrium, (b) coordination of volitional movements, (c) coordination of the timing of bodily actions, (d) being an organ for *muscular sense* and (e) being an organ which uses unconscious processes to exert a continual reinforcing action on the activity of all other CNS centers (Lewandowski, 1903; Luciani, 1904). Almost eight decades later, Ito (1984) critically discussed involvement of the cerebellum in a variety of body functions and proposed that there is synaptic plasticity in the cerebellar cortex such as to facilitate motor learning. Evidence from more recent studies suggests that the cerebellum contributes to the control of other behaviors, as well, including (a) tracking the movements of objects (Paulin, 1993), (b) cognition (Leiner et al., 1993), (c) further timing controls (Irvy, 1996), (d) enhancement of sensory data acquisition (Bower, 1997) and (e) dynamic motor learning as part of an adaptive programmable generator (Houk et al., 1996). Thach (1996) also emphasized motor learning with his suggestion that the cerebellum is critical for the context-linkage required for the learning of novel movements and patterns of muscle activation (see also, Thach and Bastian, Chapter 34 of this volume).

For the past 40 years, the authors' group has been engaged in the study of the role of the spinal cord, brainstem and cerebellum in the control of posture and locomotion. We have recently found that the FN of the cat cerebellum plays a particularly important role (Asanome et al., 1998; Mori et al., 1998, 1999, 2000, 2001). Further understanding has come from this work with a nonhuman primate, the Japanese monkey, *Macaca fuscata* (Mori et al., 1996). This animal can be operant-trained to walk on the surface of a moving treadmill belt and convert from its normal quadrupedal (Qp) to bipedal (Bp) locomotion (F. Mori et al., 2001a; Nakajima et al., 2001a; and this volume). Based on findings from the studies on these two animal species, we present a hypothesis on how the cerebellum, especially the FN, contributes to the integration of dynamic segmental postures. In keeping with Sherrington's (1906) views, we also argue that these evolving postures are, in reality, locomotion.

Characteristics requisite of a locomotion integration center

Locomotion and its associated postures are controlled continually by volitional or automatic means, or both (Grillner and Wallén, Chapter 1 of this volume). It follows that any neural substrate with the capacity to be a locomotion integration center needs to fuse together a number of locomotor and posture-related CNS subsystems into one system. Such a center must meet five criteria, as hypothesized and shown conceptually in Fig. 1.

First, descending command signals for locomotion must take into account multimodal afferent inputs. Changes in body configuration are first registered by both the labyrinthine receptors and other proprioceptive receptors embedded in the neck, trunk and limbs. Changes in the external world are received by *distance* receptors, such as the eyes and ears (Sherrington, 1906). By reception of multimodal afferent inputs, the integration center can compute the body's moment-to-moment configuration relative to the immediate and distant environment. Figure 1 emphasizes the need for both reactive and anticipatory control mechanisms (McFadyen and Belanger, 1997; F. Mori et al., Chapter 19 of this volume), with both based on the continuous reception and processing of exteroceptive and interoceptive information. Any disturbance in sensory data acquisition and central sensory-motor transformation results in an impairment of posture and locomotion, from the subtle to the extreme.

Second, the previous computation needs reference to the internal (reference) system (Fig. 1). The daily execution of locomotion contributes to the build up of postural and locomotor memory which are critical components of this internal system. Its other

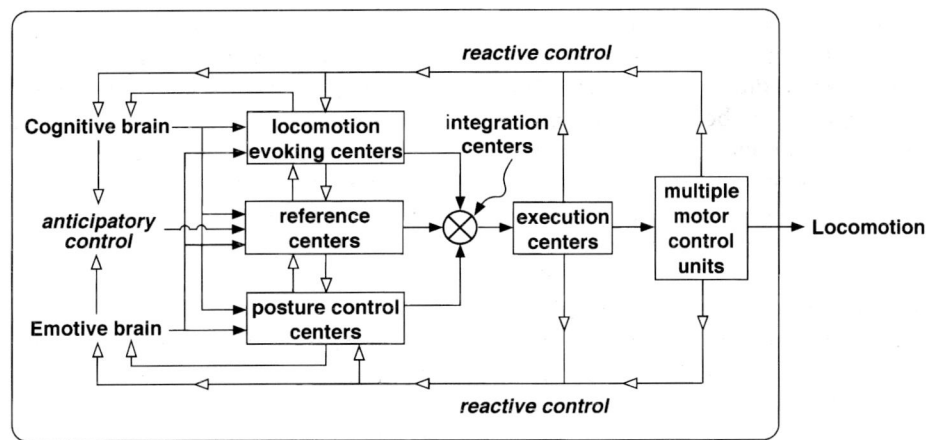

Fig. 1. A conceptualization of the overall integrated control of posture and locomotion including *anticipatory* and *reactive control*. Open and closed arrowheads represent the ascending and descending flow of signals, respectively. From rostro-anterior to caudo-posterior, the structures and their proposed processes include: *cognitive processing*; *emotive processing*; *locomotion-evoking centers*, such as the subthalamic (SLR) and mesencephalic (MLR) locomotor regions where repetitive electrical stimulation produces well-coordinated locomotion; *posture control centers*, such as the dorsal (DTF) and ventral tegmental filed (VTF) of pons where repetitive electrical stimulation produces suppression or augmentation of postural muscle tone; *internal model*, including a geocentric reference for posture and locomotor memory; *integration centers*, where the multirequirements of integrated posture and locomotion are partly achieved (including the cerebellum's FN and its CLR output); *execution centers*, which receive integration-center output and transmit locomotor-driving signals by way of multiple and parallel motor pathways; and *multiple motor control units*, including central pattern generators (CPG) and motor segments.

component includes a postural body scheme, i.e., a reference frame of body configuration. Under normal gravity conditions, the vertical geocentric reference value of posture is provided by labyrynthine, visual and proprioceptive inputs, which also serve as error-detecting signals with respect to that reference value (Massion, 1992). The reference system, which is probably stored in interconnecting networks at a high level of the CNS (Ito, 1984; Bloedel and Bracha, 1995; Thach, 1996; McFadyen and Belanger, 1997), serves to minimize impairments of posture and locomotion.

Third, we hypothesize that descending commands from the cognitive and emotive portions of the higher CNS, and activity of both locomotion-inducing sites and posture-control sites are constantly compared with that of the reference system (Fig. 1), with their collective output sent to the integration center. Such a system incorporates both anticipatory and reactive control processes. Note that the authors' model incorporates that component of reactive control that is responsive to corollary discharges of descending output signals, at all levels from the locomotor centers to the spinal cord. Such central feedback combined with peripheral feedback at the cerebral cortical level enables the animal conscious perception of its kinesthetic aspects of volitional and automatic adjustments of ongoing locomotion (Gandevia, 1996).

Fourth, it follows from the above that an integration center must participate in a comparator function: comparing *top-down* locomotor command signals with *bottom-up* feedback signals revealing the current state of locomotion. Top-down signals may originate in either cognitive or emotive brain structures (Mogenson et al., 1980) and affect the animal's exploratory/appetitive and/or defensive locomotor behaviors (Sinnamon, 1993; Jordan, 1999). For example, the nucleus accumbens, an older part of the striatum, receives input signals from limbic structures, including the hippocampus. It is a major site for initiation of motivation-driven locomotion (Mogenson et al., 1980; Holsetge, 1998). Bottom-up signals contribute to reactive control of posture and locomotion (Arshavsky et al., 1986).

Fifth, the integration center's efferent output is distributed by way of executing centers (see Fig. 3). The latter's function is to ensure that motor signals are sent to a number of different muscle-control systems such that multiple body segments are activated in a coordinated manner. These output signals guarantee that appropriate and timely forces are applied to relevant joints (Mori et al., 2000), the result being a smooth execution of locomotion, with correctly phased limb movements and degrees of postural muscle tone (Mori, 1987, 1989a).

The executing centers of Fig. 1 project to both common and functionally different descending pathways, such as reticulospinal (RS) and vestibulospinal (VS) pathways. These pathways mediate the descending command signals that activate a number of spinal MN columns, each of which innervates a specific set of muscles that operates on a specific motor segment. The descending command signals comprise those needed to activate spinal pattern generators that bring about the step cycle of each single limb (Rossignol, 1996), commissural INs that coordinate left- and right-limb motor segments (Matsuyama et al., Chapter 24 of this volume), and propriospinal INs that integrate fore- and hindlimb motor segments (Orlovsky et al., 1999). Some of the output signals also modulate brainstem and spinal circuitry involved in the elaboration of postures that support the execution of locomotion (Mori et al., 2000).

Fastigial nucleus as an integration center for the control of multiple motor segments

The cerebellar vermis integrates visual, vestibular, proprioceptive and exteroceptive afferent information originating from a wide variety of sources (Armstrong, 1986; Arshavsky et al., 1986). The vermis projects to the FN wherein both ascending and descending fibers take origin. The FN has inputs from the vestibular complex, lateral reticular nucleus and the spinocerebellar pathways (Brodal 1981). Ascending FN projections include: fastigiothalamic fibers (Anguat and Bowsher, 1970) which terminate mainly contralaterally in the ventroanterior–ventrolateral (VA–VL); and, ventromedial (VM) and central medial (CMN) thalamic nuclei (Asanuma et al., 1983a,b; Okumura et al., 2001). Thalamocortical fibers arising from VA–VL terminate in cortical areas 4 and 6 and some premotor areas (Rispal-Padel and Massion, 1970), from which originate corticothalamic (Rinvik, 1968) and corticoreticular projections (Wiesendanger, 1986), and projections to precerebellar nuclei (Matsuyama and Drew, 1997a). Originating from these latter nuclei are massive ponto-cerebellar mossy fibers, which project to the cerebellar cortex and vermis. Some tectal efferents also project into the visual area of the cerebellar vermis by way of the tecto-ponto-cerebellar pathway (Brodal, 1981). A multisynaptic cerebro-cerebellar loop is thereby formed between the cerebrum and cerebellum.

Note also that corticoreticular fibers and the collaterals of corticospinal fibers terminate in the medial pontomedullary reticular formation (mMRF; Matsuyama and Drew, 1997a; Matsuyama et al., 1999a). Single RS and VS axons give off axon collaterals as they descend from the cervical to lumbar segments (Matsuyama et al., 1997b, 1999b; Kuze et al., 1999). Reticulospinal axon collateral ramifies repeatedly and distributes terminal fibers to INs of the cervical, thoracic and lumbar segments. Collaterals of VS axons, in contrast to the rather uniform ramification patterns of RS axons across spinal segments, show a segment-specific dense ramification pattern, especially at the lumbosacral segments (Kuze et al., 1999). The RS and VS pathways are the most relevant motor pathways comprising the ventromedial motor system of Kuypers (1981). They are involved in the control of posture and locomotion in both cats (Shik and Orlovsky, 1976; Arshavsky et al., 1986; Mori et al.), and nonhuman primates (Lawrence and Kuypers, 1968a,b).

Descending fastigiofugal fibers projecting contralaterally include fastigioreticular and fastigiovestibular fibers (Walberg et al., 1962a,b; Homma et al., 1995), fastigiospinal fibers (Fukushima et al., 1977). Other fastigiofugal fibers terminate in the superior colliculus (Anguat and Bowsher, 1970) where tecto-reticulospinal fibers originate (Sasaki et al., 1996). Fibers projecting ipsilaterally include fastigioreticular and fastigiovestibular ones (Fig. 2). The majority of fastigiospinal fibers terminate in the interneurons

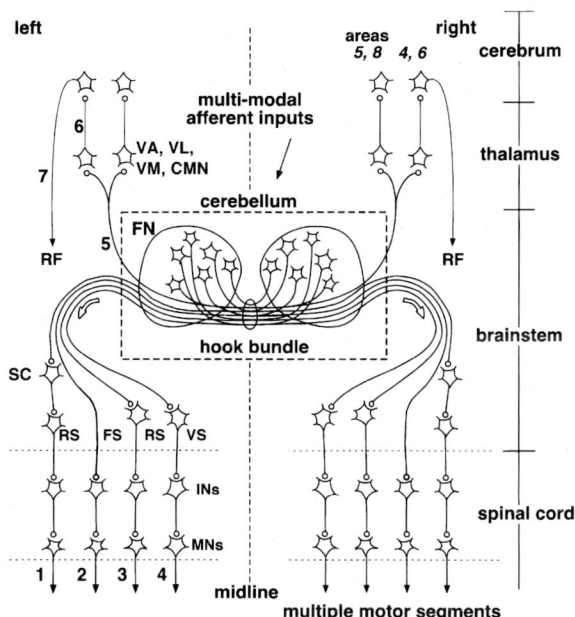

Fig. 2. CNS connections of the CLR. Crossing projections are shown of ascending and descending fastigiofugal pathways, including: 1, fastigio-tecto-reticulospinal (RS); 2, fastigiospinal (FS); 3, fastigio-reticulospinal (RS); 4, fastigio-vestibulospinal (VS); 5, fastigio-thalamic; 6, thalamo-cortical and 7, cortico-reticular. Abbreviations of thalamic nuclei include: VA, ventroanterior; VM, ventromedial; VL, ventrolateral and CMN, central medial. Other abbreviations are: FN, fastigial nucleus; INs, interneurons; MNs, motoneurons and RF, reticular formation. Note that the CLR is at the midline of the cerebellum's hook bundle of Russell.

(INs) of the upper cervical segments, and contributes to the control of neck extensor muscles. The tecto-RS pathway contributes to the coordinated control of head, neck and body movements (Sasaki et al., 1996). Spinocerebellar fibers and collaterals of spinocerebellar mossy fiber afferents project to the cerebellar vermis and the FN, respectively (Armstrong, 1986; Arshavsky et al., 1986). The neural connections between the cerebellum, FN, brainstem and spinal cord form a multisynaptic spinocerebellar loop, and provide a substrate for the control of posture and locomotion. The thoroughly studied mesencephalic locomotor region (MLR) recruits such a spinocerebellar loop to evoke locomotion (Orlovsky and Shik, 1976; Shik and Orlovsky, 1976; Arshavsky et al., 1986).

We have recently identified a cerebellar locomotor region (CLR) rostral to the most anterior portion of the FN in the midline of the cerebellar white matter (Mori et al., 1999, 2000). In a high-decerebrate cat, train-pulse microstimulation of the CLR evokes 'controlled' locomotion on the surface of a moving treadmill belt. The optimal CLR stimulus site corresponds to the hook bundle of Russell (Rasmussen, 1933), through which the crossed fastigiofugal fibers pass (Walberg et al., 1962a,b; Matsushita and Iwahori, 1971; Homma et al., 1995). Interestingly, FN stimulation, itself, fails to evoke locomotion, possibly because of the contaminated stimulus effect of Purkinje cell axons (Mori et al., 1999). These axons descend from the cerebellar vermis, pass through the FN and terminate in the vestibular complex, where long-descending VS pathways originate (Kuypers, 1981). CLR stimulation activates monosynaptically cells of origin of medullary RS and VS pathways, and with a much higher degree of synaptic security than achieved by MLR stimulation (Mori et al., 2000). In the intact, unrestrained cat, CLR stimulation evokes locomotion with accompanying arousal reactions (unpublished observations), which are similar to those evoked by stimulating the MLR and the subthalamic locomotor region (SLR; Mori et al., 1989b). These reactions include raised-head and looking-around behavior. Consistent with such responses, a CLR lesion impairs both posture and locomotion (Nakajima et al., 2001b).

It has long been known that higher CNS structures are required for goal-directed locomotion. For example, while the decorticate cat has normal muscle tone and the ability to stand and walk, its locomotion becomes inadequate in the face of environmental perturbations (e.g., barrier and ladder tests). Indeed, ablation of the motor cortex (MCx) is sufficient to render a cat incapable of performing space-linked stepping (Shaltenbrandt and Cobbs, 1930). Similarly, the discharge of MCx units is greater during walking along a horizontal ladder versus a level treadmill (Armstrong, 1986) or during barrier-encountered versus normal walking (Bellozerova and Sirota, 1988). The latter authors also found that a thalamic VL lesion renders a cat unable to handle obstacle-encountered locomotion even though its level and uphill walking are normal. Similarly, they found that the rhythmical modulation of MCx units during locomotion is decreased considerably after this thalamic lesion. Drew (1991) has suggested that the MCx participates in locomotor control when visual information is required for the control of leg trajectory and foot placement. Striade (1995) recently suggested that in addition to its role in controlling the postural axial and proximal musculature, the fastigio–VM–cortical system (Kyuhou and Kawaguchi, 1985) is part of the generalized arousal activating system.

We propose that the FN is positioned optimally to function as one of Fig. 1 model's integration centers by virtue of its critical position in uniting the cerebro-cerebellar and the spino-cerebellar loops. It is intriguing that Sadikot et al. (1992) found that fastigiothalamic fibers terminate in the CMN, wherein thalamostriatal projections originate (Fig. 2). These authors suggested that the net effect of CMN discharge to the striatum is to disinhibit the neurons of pallidal and nigral targets, such as the thalamus, the subthalamic nucleus, the superior colliculus and the brainstem tegmentum, thus modulating the activity of basal ganglia neural circuits. All of these target nuclei are involved in the control of posture and locomotion (Garcia-Rill, 1991).

A nonhuman primate model for studying reactive and anticipatory control mechanisms

We are currently analyzing the unrestrained normal Qp and operant-trained Bp locomotor behavior of a nonhuman primate, the Japanese monkey, *M. fuscata* (Mori et al., 1996; Nakajima et al., 2001a; F. Mori et al., Chapter 19 of this volume). Japanese monkeys are originally quadrupedal, but with long-term locomotor learning, they acquire a novel strategy of walking bipedally. Figure 3 shows how we first train *M. fuscata* to elaborate Qp locomotion on a treadmill. Food rewards are given intermittently by the experimenter who stands in front of the treadmill. During the animal's Qp locomotion, the experimenter quickly raises the food-containing hand. This resets the reward height to where Bp locomotion is required, with the monkey using a forelimb to grasp the food. During the transitional period from Qp to Bp locomotion, the monkey coordinates sequentially independent movements of multiple motor segments such as the eyes, head, neck, fore- and hindlimbs, in order to satisfy the dual purpose of freeing the forelimbs from the constraints of weight-bearing and adopting Bp locomotion.

During Qp locomotion, the monkey targets its eyes to the position of the reward. With upward

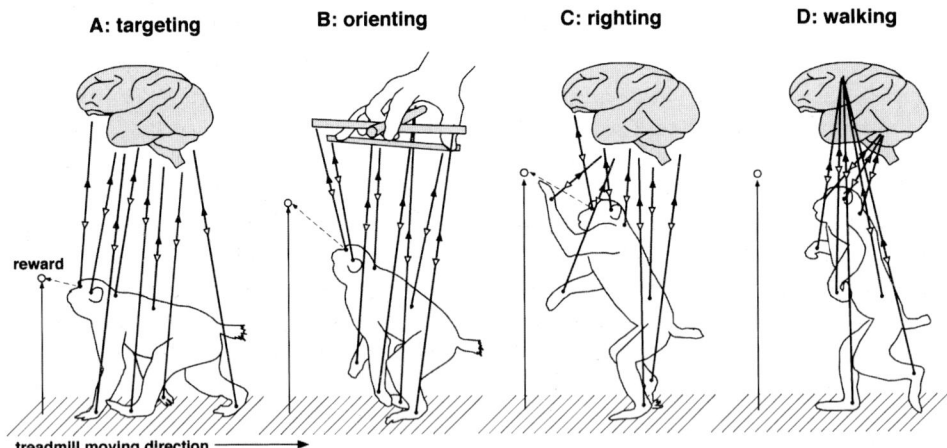

Fig. 3. The conversion from Qp to Bp locomotion. Each sketch was drawn from serial photographs taken at 10 frames/s (frame interval between each panel, ~200 ms). Downward and upward arrowheads on each line represent descending and ascending signals to and from each motor segment. Note the changes in the height of the reward. See the text for further details. (Reproduced from Nakajima et al., 2001a with permission from IOS Press.)

movement of the reward, the monkey orients to the new reward position by dorsiflexion of its head. Next, the monkey frees the weight-supporting forelimbs and rights its posture by a hip maneuver that involves a larger upward excursion of hip-joint angle. This enables the monkey to shift the weight of its center of body mass to the hindlimbs, and thereby walk bipedally. During the sequence of postural and locomotor changes, the monkey initiates reaching and grasping movements, extending either the freed left or right forelimb forward and upward to attain the reward. Kinematics of eye–head positions, body axis and major joint angles of the hindlimbs have revealed the significance of a *hip maneuver strategy* for the monkey's conversion from stable Qp to similarly stable, operant-learned Bp locomotion (Nakajima et al., 2001a). After acquiring the latter, *M. fuscata* also learns to recruit both reactive and anticipatory control mechanisms (Mori et al., 2001a and this volume).

A conceptualization of the integrated CNS control of posture and locomotion

The *M. fuscata* studies have prompted the Fig. 3 representations of the distributed and parallel CNS centers/pathways for the integrated control of posture and locomotion. The monkey is conceptualized as a puppet with multiple motor segments that must be coordinated in time and space to accomplish a required motor action. This coordination is provided by the CNS of monkey, which is represented in Fig. 3B as a puppet master, whose delicate use of five fingers of one hand can achieve the requisite controls. The puppet master utilizes multisensory afferent signals, including vision, to continuously refine the descending commands such that the puppet can elaborate Qp or Bp locomotion and their interconversions. The lines in Fig. 3 are analogous to efferent and afferent pathways which connect the CNS to multiple motor segments. These lines thereby emphasize the concept that both distributed and parallel centers, pathways and control mechanisms are needed for the seamless integration of posture and locomotion.

During the monkey's conversion from Qp to Bp locomotion, Fig. 3 shows that targeting, orienting and righting are executed in a smooth sequence. Targeting requires the coordinated activity of head and neck motor segments. Orienting requires the coordination of the head, neck, trunk and forelimb. Righting requires the coordination of the head, neck, trunk and hindlimb. All of these processes require multiple command signals descending *in parallel* from higher brain structures to the brainstem and the spinal cord and corresponding multiple ascending signals. Presumably, the command signals descend *in parallel* from a number of interconnected CNS regions, including the sensorimotor cortex, supplementary motor cortex, basal ganglia and cerebellum (Wiesendanger 1986; Massion, 1992). Depending on the external and internal requirements for the execution of locomotion, the weighting function of each CNS site may vary. The descending parallel command signals must play the dual role of (a) activating individual motor segments to provide segment-specific movement, and (b) integrating movement across multiple body segments. Interestingly, Sherrington recognized these needs almost a century ago (1906), but without the further understanding that came from the past four decades work on locomotion-inducing sites and spinal pattern generators (Grillner and Wallen, 1985; Mori, 1987; Rossignol, 1996; Orlovsky et al., 1999).

We have already demonstrated in cats that the vermis-FN complex is one of the main control centers of multiple body segments (Mori et al., 2001; Nakajima et al., 2001b). As schematically illustrated in Fig. 2, the vermis-FN complex receives multiple afferent inputs, whereas on the output side, five contralateral (four descending, one ascending) and three ipsilateral (two descending, one ascending) efferent pathways originate from the FN in parallel. The four descending pathways projecting contralaterally from the FN are comprised of fastigiospinal, fastigio-tecto-reticular, fastigioreticular and fastigio-vestibular fibers. The two pathways projecting ipsilaterally contain fastigioreticular and fastigio-vestibular fibers. These two pathways originate from both the left and right FN in mirror image. It is therefore likely that some long-descending RS and VS cells, which are located in either the left or right side, receive converging signals arising from both the ipsilateral and contralateral FN, thus increasing redundancy for secure activation of RS and VS cells.

The remaining RS and VS cells may receive signals arising mainly from the contralateral or ipsilateral FN. Again, using the puppet analogy, the puppet master can utilize the vermis-FN afferent and efferent complex to spatially and temporally coordinate segmental postural and multijoint movements.

The puppet model is supported by the pioneer studies of Botterell and Fulton (1938) in nonhuman primates. They showed that midline cerebellar lesions, with or without fastigial nuclei impairment, give rise to unique difficulties in stance and gait. The animals cannot stand up against gravity. If the lesions are unilateral, the animals fall to the side of the lesion, as does the cat (Sprague and Chambers, 1953). This suggests that the midline representation is of the whole body's segments, but only for circumscribed functions. Thach et al. (1992) has also found recently that fastigial inactivation in rhesus monkey by microinjection of muscimol (an analog of GABAa) impedes appropriate sitting, standing and walking, with frequent falls to the side of the lesion.

Concluding thoughts

Bipedal terrestrial locomotion is a routine daily activity for humans and advanced nonhuman primates. Although seemingly simple, this motor skill actually requires motor learning and the integrated control of multiple body segments against gravity. This integration includes the control of movement, an upright posture and balance. Also coming into play are anticipatory and reactive CNS processes, with the latter necessary after tripping and stumbling (F. Mori et al., 2001a and this volume). To advance understanding of these CNS mechanisms by use of interventional neuroscience techniques, we have trained the normally Qp Japanese monkey, *M. fuscata*, to use a freely moving Bp walking gait on the surface of a moving treadmill belt. Already, this animal model has provided us with kinematic data on the logic of Sherrington's (1906) conception of locomotion as series of evolving dynamic postures. In addition, data from our pilot, noninvasive imaging study have shown that multiple brain regions are activated in parallel during *M. fuscata*'s locomotion (F. Mori et al., 2001b; see also Shibasaki et al., Chapter 20 of this volume). These activated regions included the motor-, supplementary motor- and visual cortices and the cerebellum, with some intriguing differences in their activity patterns noted for Qp versus Bp movements. We plan to continue such investigations on *M. fuscata* in the hope that this approach will provide definitive information about the role and operation of higher CNS structures in the integrated control of posture and locomotion.

Acknowledgments

We express our sincere appreciation to Dr. Carol A. Boliek, University of Alberta, for her critical review and editing of the original version of this manuscript. This study was supported by grants to S.M. from the Uehara Memorial Foundation, the Ministry of Education, Science, Sports and Culture, and Technology of Japan (Grant in Aid for Scientific Research B12480247), and the Ministry of Health and Labor of Japan (Grant in Aid on Comprehensive Research on Aging and Health).

Abbreviations

Bp	bipedal
CNS	central nervous system
CLR	cerebellar locomotor region
CMN	central medial thalamic nucleus
FN	fastigial nucleus
MCx	motor cortex
MLR	mesencephalic locomotor region
mMRF	medial medullary reticular formation
Qp	quadrupedal
RS	reticulospinal
SLR	subthalamic locomotor region
VA	ventroanterior thalamic nucleus
VL	ventrolateral thalamic nucleus
VM	ventromedial thalamic nucleus
VS	vestibulospinal

References

Anguat, P. and Bowsher, D. (1970). Ascending projections of the medial cerebellar (fastigial) nucleus: an experimental study in the cat. Brain Res., 4: 49–68.

Armstrong, D.M. (1986). Supraspinal contribution to the initiation and control of locomotion in the cat. Prog. Neurobiol., 26: 273–361.

Arshavsky, Yu.I., Gelfand, I.M. and Orlovsky, G.N. (1986). Cerebellum and Rhythmical Movements. Springer-Verlag, Berlin.

Asanome, M., Matsuyama, K. and Mori, S. (1998). Augmentation of postural muscle tone induced by the stimulation of the decussating fibers in the midline of the cerebellar white matter in the acute decerebrate cat. Neurosci. Res., 30: 257–269.

Asanuma, C., Thach, W.T. and Jones, F.G. (1983a). Distribution of cerebellar terminations and their relation to other afferent terminations in the ventral thalamic region of the monkey. Brain Res. Rev., 5: 237–265.

Asanuma, C., Thach, W.T. and Jones, F.G. (1983b). Brainstem and spinal projections of the deep cerebellar nuclei in the monkey, with observations on the brainstem projections of the dorsal column nuclei. Brain Res. Rev., 5: 299–322.

Bellozerova, I.N. and Sirota, M.G. (1988). Role of motor cortex in control of locomotion. In: Gurfinkel V.S., Joffe M.E. and Massion J. (Eds.), Stance and Motion, Facts and Concepts. Plenum Press, New York, pp. 163–176.

Bloedel, J.R. and Bracha, V. (1995). On the cerebellum, cutaneomuscular reflexes movement control and the elusive engrams of memory. Behav. Brain Res., 68: 1–44.

Botterell, E.H. and Fulton, J.F. (1938). Functional localization in the cerebellum of primates. II. Lesions of midline structures (vermis) and deep nuclei. J. Comp. Neurol., 69: 47–62.

Bower, J.M. (1997). Is the cerebellum sensory for motor's task, or motor for sensory's task: the view from the whiskers of a rat? Prog. Brain Res., 114: 463–496.

Brodal, A. (1981). Neurological Anatomy in Relation to Clinical Medicine. Oxford University Press, London.

Drew, T. (1991). Visuomotor coordination in locomotion. Curr. Opin. Neurobiol., 1: 652–657.

Fukushima, K., Peterson, B.W., Uchino, Y., Coulter, J.D. and Wilson, V.J. (1977). Direct fastigiospinal fibers in the cat. Brain Res., 126: 309–328.

Gandevia, S.C. (1996). Kinesthesia: roles for afferent signals and motor commands. In: Rowell L.B. and Shepherd J.T. (Eds.), Handbook of Physiology, Sec. 12, Exercise: Regulation and Integration of Multiple Systems. Oxford University Press, New York, pp. 128–172.

Garcia-Rill, E. (1991). The pedunculopontine nucleus. Prog. Neurobiol., 36: 363–389.

Grillner, S. and Wallen, P. (1985). Central pattern generators for locomotion with special reference to vertebrates. Annu. Rev. Neurosci., 8: 233–261.

Holsetge, G. (1998). The emotional motor system in relation to the supraspinal control of micturition and mating behavior. Behav. Brain Res., 92: 103–109.

Homma, Y., Nonaka, S., Matsuyama, K. and Mori, S. (1995). Fastigiofugal projection to the brainstem nuclei in the cat: an anterograde PHA-L tracing study. Neurosci. Res., 23: 89–102.

Houk, J.C., Buckingham, J.T. and Barto, A.G. (1996). Models of the cerebellum and motor learning. Behav. Brain Sci., 19: 363–383.

Irvy, R.B. (1996). The representation of temporal information in perception and motor control. Curr. Opin. Neurobiol., 6: 651–657.

Ito, M. (1984). The Cerebellum and Neural Control. Raven Press, New York.

Jordan, L. (1999). Initiation of locomotion in mammals. In: Kiehn O., Harris-Warrick R.M., Jordan L.M., Hultborn H. and Kudo N. (Eds.), Ann. N.Y. Acad. of Sci., Vol. 860, The New York Academy of Science, New York, pp. 83–93.

Kuypers, H.G.J.M. (1981). Anatomy of descending pathways. In: Brookhart J.M. and Mountcastle V.B. (Eds.), Handbook of Physiology, The Nervous System, Vol. II, American Physiological Society, Bethesda, pp. 597–666.

Kuze, B., Matsuyama, K., Matsui, T., Miyata, H. and Mori, S. (1999). Segment-specific branching patterns of single vestibulospinal tract axons arising from the lateral vestibular nucleus in the cat: a PHA-L tracing study. J. Comp. Neurol., 414: 80–96.

Kyuhou, S. and Kawaguchi, S. (1985). Cerebellocerebral projection from the fastigial nucleus onto the frontal cortex in the cat. Brain Res., 347: 385–389.

Lawrence, D.G. and Kuypers, H.G.J.M. (1968a). The functional organization of the motor system in the monkey. I. The effects of bilateral pyramidal tract lesion. Brain, 91: 1–14.

Lawrence, D.G. and Kuypers, H.G.J.M. (1968b). The functional organization of the motor system in the monkey. II. The effects of lesions of the descending brain-stem pathways. Brain, 91: 15–36.

Leiner, H.C., Leiner, A.L. and Dow, R.S. (1993). Cognitive and language functions of the human cerebellum. Trends Neurosci., 16: 444–454.

Lewandowski, M. (1903). Ueber die Verrichtungen des Kleinhirns. Arch. F. Anat. u. Physiol. Physiol. Abt., 27: 129–191.

Luciani, L. (1904). Das Kleinhirn. Ergebn. Physiol., 3: 259–338.

Massion, J. (1992). Movement, posture and equilibrium: interaction and coordination. Prog. Neurobiol., 38: 35–56.

Matsushita, M. and Iwahori, N. (1971). Structural organization of the fastigial nucleus. I. Dendrites and axons. Brain Res., 25: 597–610.

Matsuyama, K. and Drew, T. (1997a). Organization of the projections from the pericruciate cortex to the pontomedullary brainstem of the cat: a study using the anterograde tracer *Phaseolus vulgaris-leucoagglutinin*. J. Comp. Neurol., 389: 617–641.

Matsuyama, K., Takakusaki, K., Nakajima, K. and Mori, S. (1997b). Multi-segmental innervation of single pontine reticulospinal axons in the cervico-thoracic region of the cat: anterograde PHA-L tracing study. J. Comp. Neurol., 377: 234–250.

Matsuyama, K., Drew, T., Mori, F. and Mori, S. (1999a). The cortico-reticulo-spinal system: organization of the corticoreticular projection and fine architecture of the reticulospinal pathway in the cat. In: Gantchev G.N., Mori S. and Massion J. (Eds.), Motor Control, Today and Tomorrow. Academic Publishing House of 'Prof. M. Drinov', Sofia, pp. 45–56.

Matsuyama, K., Mori, F., Kuze, B. and Mori, S. (1999b). Morphology of single pontine reticulospinal axons in the lumbar enlargement of the cat: a study using the anterograde tracer PHA-L. J. Comp. Neurol., 410: 413–430.

McFadyen, B.J. and Belanger, M. (1997). Neuromechanical concepts for the assessment of the control of human gait. In: Allard P., Cappozzo A., Lundberg A. and Vaughan C.L. (Eds.), Three-Dimensional Analysis of Human Locomotion. John Wiley & Sons, Chichester, pp. 49–66.

Mogenson, G.J., Jones, D.L. and Yim, C.Y. (1980). From motivation to action. Functional interface between the limbic system and the motor system. Prog. Neurobiol., 14: 69–97.

Mori, F., Tachibana, A., Takasu, C., Nakajima, K. and Mori, S. (2001a) Bipedal locomotion by the normally quadrupedal Japanese monkey, *M. fuscata*: strategies for obstacle clearance and recovery from stumbling. Acta Physiol. Pharmacol. Bulgarica, 26: 147–150.

Mori, F., Tachibana, A., Nakajima, K., Takasu, C., Tsujimoto, T., Tsukada, H. and Mori, S. (2001b). Bipedal locomotion in the Japanese monkey, *M. fuscata*: higher-order control mechanisms. Soc. Neurosci. Abstr., 27: 1077.

Mori, S. (1987). Integration of posture and locomotion in acute decerebrate cats and in awake, freely moving cats. Prog. Neurobiol., 28: 161–195.

Mori, S. (1989a). Contribution of postural muscle tone to full expression of posture and locomotor movements: multifaceted analyses of its setting mechanisms in the cat. Jpn. J. Physiol., 39: 785–809.

Mori, S., Matsuyama, K., Takakusaki, K. and Kanaya, T. (1988). The behaviour of lateral vestibular neurons during walk, trot and gallop in acute precollicular decerebrate cats. Prog. Brain Res., 76: 211–220.

Mori, S., Sakamoto, T., Ohta, Y., Takakusaki, K. and Matsuyama, K. (1989a). Site-specific postural and locomotor changes evoked in awake, freely moving intact cats by stimulating the brainstem. Brain Res., 505: 66–74.

Mori, S., Miyashita, E., Nakajima, K. and Asanome, M. (1996). Quadrupedal locomotor movements in monkeys (*M. fuscata*) on a treadmill: kinematic analyses. Neuroreport, 7: 2277–2285.

Mori, S., Matsui, T., Kuze, B., Asanome, M., Nakajima, K. and Matsuyama, K. (1998). Cerebellar induced locomotion: reticulospinal control of spinal rhythm generating mechanisms in cats: Ann. N.Y. Acad. Sci., Vol. 860, The New York Academy of Science, New York, pp. 94–105.

Mori, S., Matsui, T., Kuze, B., Asanome, M., Nakajima, K. and Matsuyama, K. (1999). Stimulation of a restricted region in the midline cerebellar white matter evokes coordinated quadrupedal locomotion in the decerebrate cat. J. Neurophysiol., 82: 290–300.

Mori, S., Matsui, T., Mori, F., Nakajima, K. and Matsuyama, K. (2000). Instigation and control of treadmill locomotion in high decerebrate cats by stimulation of the hook bundle of Russell in the cerebellum. Can. J. Physiol. Pharmacol., 78: 945–957.

Mori, S., Matsuyama, K., Mori, F. and Nakajima, K. (2001). Supraspinal sites that induce locomotion in the vertebrate central nervous system. In: Ruzicka E., Hallet M. and Jankovic J. (Eds.), Gait Disorders. Lippincott Williams & Wilkins, Philadelphia, pp. 25–40.

Nakajima, K., Mori, F., Takasu, C., Tachibana, A., Okumura, T., Mori, M. and Mori, S. (2001a). Integration of upright posture and bipedal locomotion in non-human primates. In: Dengler R. and Kossev A.R. (Eds.), Sensorimotor Control. IOS Press, Amsterdam, pp. 95–102.

Nakajima, K., Mori, F., Okumura, T., Tachibana, A., Mori, M., Takasu, C. and Mori, S. (2001b). What is the functional role of the cat cerebellar locomotor region (CLR) in postural and locomotor control?. Soc. Neurosci. Abstr., 27: 773.

Okumura, T., Czarkowska-Bauch, J., Nakajima, K. and Mori, S. (2001). Fastigiothalamic projection in the cat: an anterograde BDA tracing study. Soc. Neurosci. Abstr., 27: 773.

Orlovsky, G.N. and Shik, M.L. (1976). Control of locomotion: a neurophysiological analysis of the cat locomotor system. Int. Rev. Physiol. Neurophysiol. II., 10: 281–317.

Orlovsky, G.N., Deligiana, T.G. and Grillner, S. (1999). Neuronal Control of Locomotion from Mollusc to Man. Oxford University Press, Oxford.

Paulin, M.G. (1993). The role of the cerebellum in motor control and perception. Brain Behav. Evol., 41: 39–50.

Rasmussen, A.T. (1933). Origin and course of the fasciculus uncinatus (Russell) in the cat, with observations on other fiber tracts arising from the cerebellar nuclei. J. Comp. Neurol., 57: 165–197.

Rinvik, E. (1968). The corticothalamic projection from the pericruciate and coronal gyri in the cat. An experimental study with silver-impregnation methods. Brain Res., 10: 79–119.

Rispal-Padel, L. and Massion, J. (1970). Relations between the ventrolateral nucleus and the motor cortex in the cat. Exp. Brain Res., 10: 331–339.

Rossignol, S. (1996). Neural control of stereotypic limb movements. In: Rowell L.B. and Shepherd J.T. (Eds.), Handbook of Physiology, Sec. 12, Exercise: Regulation and Integration of Multiple Systems. Oxford University Press, New York, pp. 173–216.

Sadikot, A.F., Parent, A., Smith, Y. and Bolam, J.P. (1992). Efferent connections of the centromedian and parafascicular thalamic nuclei in the squirrel monkey: a light and electron microscopic study of the thalamostriatal projection in relation to striatal heterogeneity. J. Comp. Neurol., 320: 228–242.

Sasaki, S., Naito, K. and Oka, M. (1996) Firing characteristics of neurones in the superior colliculus and the pontomedullary reticular formation during orienting in unrestrained cats. In: Norita, M., Bando, T. and Stein, B. (Eds.), Extrageniculostriate Mechanisms Underlying Visually Guided Orientation Behavior, Prog. Brain Res., vol. 112. pp. 99–116. Elsevier, Amesterdam.

Shaltenbrandt, G. and Cobbs, S. (1930). Clinical and anatomical studies on two cats without neocortex. Brain, 53: 449–488.

Sherrington, C.S. (1906). The Integrative Action of the Nervous System. Yale University Press, New Haven.

Shik, M.L. and Orlovsky, G.N. (1976). Neurophysiology of locomotor automatism. Physiol. Rev., 56: 465–501.

Sinnamon, H.M. (1993). Preoptic and hypothalamic neurons and the initiation of locomotion in the anesthetized rat. Prog. Neurobiol., 41: 323–344.

Sprague, J.M. and Chambers, W.W. (1953). Regulation of posture in intact and decerebrate cat. I. Cerebellum, reticular formation and vestibular nuclei. J. Neurophysiol., 16: 451–463.

Striade, M. (1995). Two channels in the cerebellothalamocortical system. J. Comp. Neurol., 354: 57–70.

Thach, W.T. (1996). On the specific role of the cerebellum in motor learning and cognition: clues from PET activation and lesion studies in man. Behav. Brain Sci., 19: 411–431.

Thach, W.T., Goodkin, H.P. and Keating, J.G. (1992). The cerebellum and the adaptive coordination of movement. Annu. Rev. Neurosci., 15: 403–442.

Walberg, F., Pompeiano, O., Wrestrum, L.E. and Hauglie-Hanssen, E. (1962a). Fastigioreticular fibers in the cat. An experimental study with silver methods. J. Comp. Neurol., 119: 187–199.

Walberg, F., Pompeiano, O., Brodal, A. and Jansen, J. (1962b). The fastigiovestibular projection in the cat. An experimental study with silver impregnation methods. J. Comp. Neurol., 118: 49–76.

Wiesendanger, M. (1986). Recent developments in studies of the supplementary motor area of primates. Rev. Physiol. Biochem. Pharmacol., 103: 1–59.

CHAPTER 34

Role of the cerebellum in the control and adaptation of gait in health and disease

W. Thomas Thach[1],* and Amy J. Bastian[2]

[1]*Departments of Anatomy and Neurobiology, Program in Physical Therapy, Neurology and Neurological Surgery, and the Irene Walter Johnson Rehabilitation Research Institute, Washington University School of Medicine, St. Louis, MO 63110, USA*
[2]*Departments of Neurology, Neuroscience and Physical Medicine and Rehabilitation, Johns Hopkins Medical School and the Kennedy Krieger Institute, Baltimore, MD 21205, USA*

Abstract: In humans, inability to stand and walk is the most limiting of motor disabilities. In humans, upright stance and gait is the most sensitive indicator of cerebellar disease. From animal and human studies, much has been learned about how the cerebellum coordinates normal movement, and how it may play roles in normal motor adaptation and learning. Much of this work suggests that different parts of the cerebellum control stance and gait in different ways, and differently located lesions cause different deficits. What is not known is whether the cerebellum can compensate for stance and gait disorders caused by lesions in other parts of the nervous system, or whether one part of the cerebellum can compensate for deficits caused by lesion of another part. These issues have become increasingly important in rehabilitation research and practice.

Introduction

The cerebellum has long been known to control stance and gait. Cerebellar lesions in humans and animals cause incoordination of the many muscles that are involved, and may result in irregular gait and falling. Electrical stimulation of the cerebellum of animals may initiate gait (S. Mori et al.; present Chapter 33). Single unit recording and inactivation studies in animals show that the control is adaptive. An understanding of the adaptive capacity of the cerebellum in stance and gait control may help rehabilitate patients with disease of the cerebellum and other parts of the nervous system.

Human signs of cerebellar disease in stance and gait

In humans, cerebellar lesions affect stance and gait, and the deficit is recognized by motor abnormalities and also what the subject does to compensate for them. Thus, steps are irregular in time and placement, falls are frequent, the stance is overly wide-based, and the steps are short, with the foot barely clearing the ground. If the subject is forced to stand on a narrow base, there is an increase in postural sway, and a tendency to fall, with the eyes open or shut. If the subject is forced to walk on a narrow base, or tandem (heel-to-toe), there is a tendency to side step, in order to establish a widened base, and thereby prevent a fall. These are signs not only of disease, but also of the healthy remaining nervous system adapting to provide a compensatory response.

Few studies have described the general walking deficits of people with cerebellar damage (Hallett and Massaquoi, 1993; Bastian et al., 1998; Palliyath et al., 1998; Earhart and Bastian, 2001). One common finding across these studies is that the joints of the leg tend to move at abnormal rates relative to one another, and often with extreme variability. For example, subjects with cerebellar atrophy may have delays in the relative movement of the knee and ankle

*Corresponding author: Tel.: +314-362-3538;
Fax: +314-362-4370; E-mail: thachw@thalamus.wustl.edu

throughout the gait cycle (Hallett and Massaquoi, 1993). Palliyath et al. (1998) noted a similar deficit, with cerebellar subjects showing delays in the timing of peak knee flexion during the swing phase of walking. Earhart and Bastian (2001) have studied people with mild-to-moderate gait ataxia as they stepped onto 'wedges' of variable inclination during walking. They found that cerebellar subjects had difficulty making appropriate adjustments of the hip, knee and ankle joints to accommodate steeper wedges. Cerebellar subjects often reduced the number of moving joints by fixing the ankle joint in a dorsiflexed position on the steeper wedges.

Collectively, the above studies suggest that a general function of the cerebellum in walking is to control the relative movement of the leg joints during walking. But what are *specific* cerebellar deficits in stance and gait? Studies of focal lesions in humans, and especially in animals, help focus the question but as yet, such studies do not give the complete answer.

Is control of gait localized within the cerebellum?

Effect of focal cerebellar lesions on stance and gait in humans

Vestibulocerebellum

Dow and Moruzzi (1958) commented on the similarity of signs in the monkey, ape and human after damage of the flocculonodular lobe cortex, the 'vestibulocerebellum'. These include a head tilt (i.e., the occiput moving to the side of the lesion), a bilateral horizontal nystagmus (greatest on the side of the lesion), circling away from, and falling toward the side of the lesion.

Posterior vermis

Damage of the posterior vermis also causes ataxia of gait without ataxia of limbs. Bastian et al. (1998) studied five children (ages, 6–15 years), 1–24 months after a surgical midline section and resection of the midline inferior aspect of the vermis (lobules V–IX) to remove ventricle IV tumors. All the children had a deficit in tandem gait (Bastian et al., 1998). Postoperatively, all five patients had deficits involving eye movements, speech, limb movements, balance and gait that improved over several weeks. At the time of kinematic study, neurological exam revealed no dysarthria, intention tremor, dysmetria or truncal titubation, and a normal heel–knee–shin test. Kinematic analysis revealed only slight abnormalities in each of the study's tasks with the exception of tandem walking, which was markedly impaired. Patients were unable to complete successive steps without falling to either side, or stopping to regain their balance. Rather, they tended to remain in the stance phase for extended periods of time, as compared to control subjects. Remarkably, despite the inability to take successive steps during tandem walking, the patients were able to hop on either foot for at least five repetitions. They could reach, pinch, kick and walk and hop normally.

Anterior lobe

Victor et al. (1959) showed that alcoholic cerebellar atrophy is initially localized to anterior vermal and intermediate portions of the anterior lobe. The symptoms included ataxia of gait, an inability to hop on one leg, and ataxia during the heel–knee–shin test. Deficits in arm-reaching movements and speech were notably delayed until the atrophy had spread posteriorly. Victor et al. (1959) emphasized the congruence of the site of the atrophy to the leg area of Snider and Eldred's (1952) mossy fiber input map and the impairment of the legs, which occurred first and was the most severe deficit.

Different areas of the cerebellum control gait differently

Dow and Moruzzi (1958) noted that in humans, stance and gait are impaired by lesions of all three primary cerebellar subdivisions; the vestibulo-, spino- and cerebro-cerebellum. Their and others' results may be due, in part, to the fact that human pathology rarely respects anatomical let alone phylogenetic boundaries. Mauritz et al., 1979 studied stance on a pressure plate, and found that patients with vestibulocerebellar lesions exhibited a pronounced and irregular postural sway, with a tendency to fall in all directions. In contrast, anterior lobe spinocerebellar lesions produced a to-and-fro sway. Lateral hemisphere lesions produced the least deficit in stance.

All cerebellar infarcts can cause ataxia of stance and gait (Barth et al., 1993), though the specific mechanisms and clinical course of the ataxia may differ. We speculate that anterior inferior cerebellar artery (AICA) territory infarcts may result in ataxic gait and falls to the side of the lesion because of the involvement of the middle cerebellar peduncle, rather than involvement of a particular cerebellar territory, per se. The posterior inferior cerebellar artery (PICA) stroke syndrome commonly includes damage to the posterior–lateral and medial cerebellar cortex, inferior cerebellar peduncle and all the climbing fibers on that side, the dorsal- and cuneo-spino-cerebellar tracts, and the crossed descending 'hook bundle' of Russell from the fastigial nucleus. It is possible that PICA strokes can impair gait via damage to posterior–midline cerebellar cortex; alternatively, it may be via brainstem damage and not necessarily from the central cerebellar territory per se. The superior cerebellar artery (SCA) territory stroke syndrome commonly includes an ataxia of stance and gait. Again, it is not clear whether this is because of involvement of (1) the dentate nucleus, which is supplied solely by the SCA, (2) the anterior lobe cortex, or (3) the middle and superior cerebellar peduncles, the latter of which carries output from all three deep nuclei.

Note further that the above signs of acute focal cerebellar disease, regardless of their cause, may not lead to a permanent disability of stance and gait. Compensation can and often does occur. In contrast, generalized cerebellar disease, regardless of the cause, does indeed lead to permanent disability of stance and gait. This suggests that in focal cerebellar disease, the surviving portions may play an important role in the restoration of useful stance and gait. This, in turn, raises the question of what particular regions contribute to control of stance and gait, and how they might contribute to compensation. We next consider how this issue has been addressed by studies on the stance and gait effects of focal experimental cerebellar lesions in animals.

Localization of effects of focal cerebellar lesions on stance and gait in animals

In the monkey, Botterell and Fulton (1938a,b) made lesions in the vermis and lateral cerebellar hemispheres. The midline lesions, with or without involvement of the fastigial nuclei, gave rise to a unique set of stance and gait impairments. The animal could not stand up against gravity: if the lesions were unilateral, it fell to the side of the lesion. Previously, neurologists had considered cerebellar ataxias of gait to arise from an abnormality of the function of proximal musculature, as if the lesion affected only the truncal portion of a map of somatic motor representation in the cerebellum. Botterell and Fulton (1938a,b) rejected this explanation, and pointed out that abnormalities existed in movements of the elbow, hand and fingers, and the knee, foot and toes, in addition to those at the shoulder, trunk, and hip. Moreover, there were few deficits in arm-reaching and fending-off movements, which involve both proximal and distal musculature. Therefore, the essential abnormality was of *most* of the body musculature, but when used only in *selected* tasks; the midline representation was therefore of the whole body but only for circumscribed functions. This indeed was a pioneering concept in cerebellar localization. The idea was that, as in the cerebrum, there might be focal mapping of *functions* (involving many body parts) in addition to the focal mapping *of body parts*. Despite the seminal work of Botterell and Fulton, subsequent studies in the human have largely interpreted the stance deficit as being due to an abnormality in the control of the proximal trunk muscles (Brown, 1949).

Lesions of the lateral hemispheres (with or without dentate involvement) also involve many muscles of the body, but for an entirely different set of tasks (Botterell and Fulton, 1938b). The animal was shown to overshoot the mark when reaching for an object, and was 'clumsy' when manipulating small objects. When running headlong down a hallway, it often bumped into a wall, and was unable to stop in time. Vision was thought to be normal. The normal small hops that the animal makes in response to a perceived threat or aggression care were exaggerated into high leaps into the air. Again, it would seem that the lesion impaired cerebellar control mechanisms of the whole body, but only for a circumscribed and separate set of behaviors.

The above observations were confirmed and extended by Sprague and Chambers (1953) in cats with electrolytic lesions placed stereotaxically within selected nuclei. Following fastigial lesions, the animal fell toward the side of the lesion. Anterior fastigial lesions caused the ipsilateral hindlimb to flex and the opposite one to extend. Posterior fastigial lesions had more affect on the forelimbs, albeit the attitude (flexion vs. extension) was variable. Arm reaching and climbing was normal, however. These authors argued that the lesion destroyed not only the cells of the nuclei, but also the fibers that cross posteriorly (the hook bundle of Russell).

Shortly thereafter, (Chambers and Sprague, 1955a,b) provided evidence that stance and gait were relatively normal following lateral hemisphere lesions in the cat. Reaching resulted in overshoot, and pawing movements were 'clumsy'. Both in this study and the previous one of Botterell and Fulton (1938a,b) it was noted that tremor in arm-reaching movements was far more likely to occur following hemispheric versus midline cerebellar lesions. Chambers and Sprague also assumed and emphasized a differential representation of motor functions in lateral versus midline portions of the cerebellum.

Kane et al. (1988), Thach et al. (1992a,b) studied the effect of nuclear inactivation across deep cerebellar nuclei on the performance of trained single-joint wrist movements following an oscilloscopic display in the Rhesus monkey. In the same animal during the same sessions, the performance of several untrained multijointed tasks was studied, including sitting, standing, walking, reaching out for bits of food, and picking small bits of food out of deep narrow food wells with a precision pinch of the fingers. Figure 1 shows that each individual nuclear inactivation produced its own specific type of incapacitating motor impairment.

Fastigial inactivation impaired or prevented sitting, standing and walking, because of falls to the side of the lesion. This was interpreted as a deficit in equilibrium. Monkeys did not exhibit arm-reaching deficits, and were content to reach for food from a lying position until the effects of the inactivation wore off.

Interpositus inactivation resulted in an extensor bias with a failure of the foot to clear the ground. During the swing phase, the dorsum of the paw and digits were dragged along the floor. At the end of swing phase, the heel of the palm descended to the ground, and the digits, still touching the ground on the dorsal aspect, were then passively folded in flexion. This deficit was reminiscent of the impairment of contact placing of the foot described previously by Amassian and colleagues. By contrast, inactivation of the overlying cerebellar cortex, that projects to and tonically inhibits nucleus interpositus, causes locomotor movements to be hyperflexed. Udo et al. (1980) in awake and decerebrate locomoting cats transiently cooled then permanently ablated intermediate zone lobule V cerebellar cortex. This area projects to the forelimb area of n. interpositus. They observed hyperflexed movements in the ipsilateral forelimb. The duration of the swing phase was prolonged, and that of the stance phase shortened. They concluded that the intermediate zone lobule V cortex contributes to control of flexor/extensor activities and the proper timing of touchdown and lift-off of the forelimb. Another remarkable deficit of interposed lesions consisted of a severe action tremor at 3–5 Hz during arm reaching (Thach et al., 1992a,b).

Dentate inactivation had no apparent effect on sitting or stance. During walking, the leg and paw exhibited an increased height above ground during the swing phase of the step. This was most noted in the forelimbs, and reminiscent of the high hops described previously by Botterell and Fulton (1938a,b). The most severe dentate deficits were in the 'voluntary' movements of arm reaching and finger pinching. In arm reaching, there was an overshoot of the target by several centimeters. In finger pinching, there was an increased use of single-digit strategies when attempting to pinch and pick up food morsels.

The above tests thus showed differences across the three nuclei with respect to the control of stance and gait. Each of the nuclei appeared to contribute to stance and gait, but in quite different ways. It should also be noted that we have observed that the deficits in multijointed movements from inactivation in any zone are qualitatively more severe than deficits in simple, single-jointed wrist movements. This finding is consistent with the view that the principal role of the cerebellum is to coordinate multijointed compound movements (Thach et al., 1992b).

Mapping across the deep cerebellar nuclei

Figure 2 shows the mapping of deficits produced by focal nuclear inactivation by injection of muscimol in the first of three Rhesus monkeys. In the one monkey (shown), 30 injections stepped at 1 mm intervals from medial nucleus fastigius (NF) across nucleus interpositus (NI) and nucleus dentate (ND) to the white matter lateral to dentate over a range of 10 mm, producing deficits which were either pure or mixes of two adjacent (not three) nuclear syndromes.

In a second monkey, 43 injections were given in 1 mm steps across a grid 8 mm anterior–posterior by 8 mm medial–lateral, spanning interpositus, dentate, and adjacent white matter. In a third monkey, 23 injections in 1-mm steps across a grid 6-mm anterior–posterior by 8-mm medial–lateral centered on interpositus and dentate and adjacent white matter. The localization of reach and grasp to the dentate differs from a localization and coordination scheme proposed by Mason et al. (1998), which was also based on muscimol injections. They suggest that the coordination of fingers (in grasp) and shoulder (in reach) is localized to the central and posterior portions of nucleus interpositus (and not dentate), respectively. This interpretation was based on eight muscimol injections in one monkey (ranging approximately over 9 mm, anterior–posterior; 3 mm, medial–lateral) and seven injections in a second monkey (ranging approximately over 7 mm, anterior–posterior; 2 mm, medial–lateral). Their injection pattern appeared to reveal the reported separate sites of grasp (central) and reach (posterior) in the anterior–posterior dimension, though we have not seen this separation. We note that their injection pattern could not have distinguished between interpositus and dentate contributions, because of

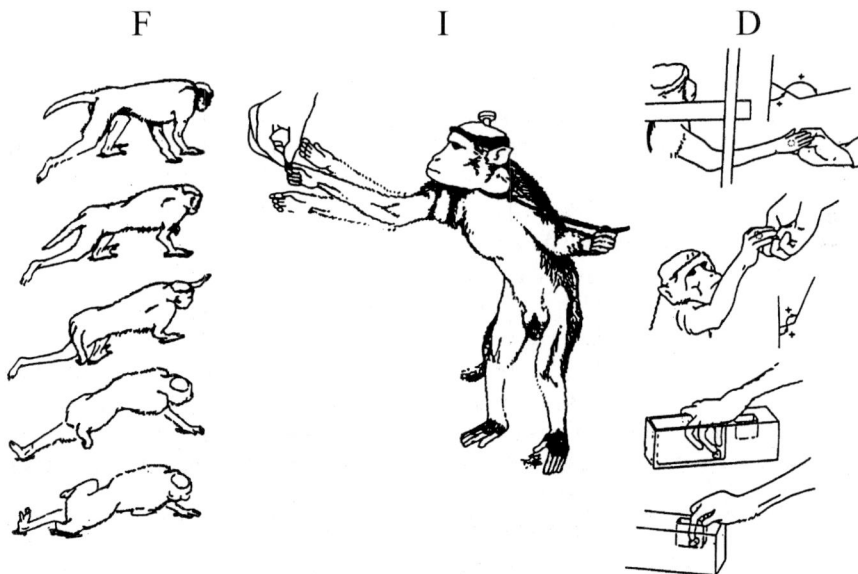

Fig. 1. Behavioral deficits in a Rhesus macaque after muscimol inactivation of deep cerebellar nuclei. Shown are sketches of impairments following injections into n. fastigius (F) interpositus (I) and dentatus (D). Note that inactivation of n. fastigius causes falls to the side of the injection. In contast, inactivation of n. interpositus produces an action tremor during reaching. In further contrast, inactivation of n. dentatus causes both uncoordinated reaching and pinching. (Reproduced from Thach et al., 1992b with permission from the Annual Review of Neuroscience, Volume 15 © 1992 by Annual Reviews http://www.AnnualReviews.org.)

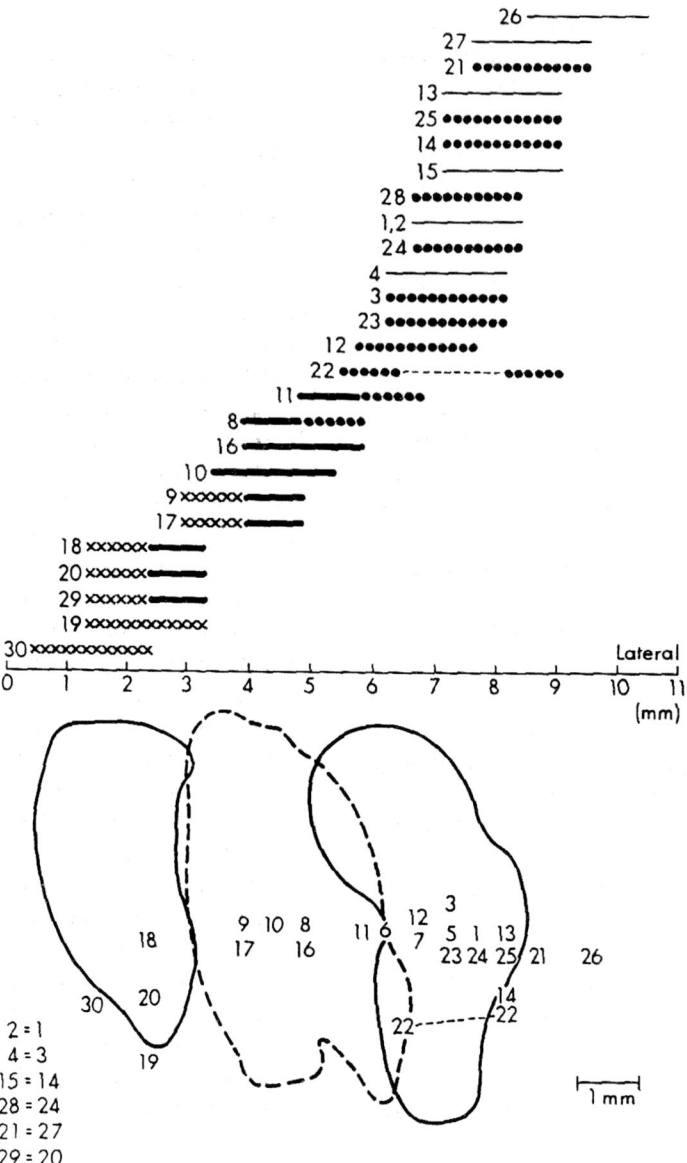

Fig. 2. Evidence for three distinct inactivation syndromes following muscimol injection into the three deep cerebellar nuclei. The upper part of the figure shows the coded inactivation syndrome that followed muscimol injection into the numbered cerebellar sites. The lower part shows, from left to right, the outlines of the deep nuclei from the medial edge of the fastigial nucleus (continuous line), to the interposed nucleus (dashed line) and, finally, the lateral edge of the dentate nucleus (continuous line). Injection sites are mapped relative to each nucleus. Muscimol injections into the center of each nucleus gave a 'pure' deficit of either: imbalance and falls (fastigial nucleus; 'xxx'); reaching tremor (interpositus nucleus; '—'); or, reach overshoot and uncoordinated pinch (dentate nucleus; '•••'). Some injections gave no observed deficit ('—'). These were often a second injection repeated at the one site, or one of two injections at closely adjacent sites. Injection site #22 was uncertain (i.e., whether on the medial versus lateral edge of the dentate nucleus). The subsequent deficit, however, was a 'pure' reach overshoot and uncoordinated pinch. Therefore, it was coded as a dentate deficit, but with interpolated dashed line to reflect the uncertainty of the injection site ('•••——•••'). Injections within 1 mm of the nuclear boundaries gave a combination of *two* deficits, i.e., those characteristic of the nuclei on either side of the boundary. Thus, injections within 1 mm of the fastigial-interpositus boundary gave deficits of imbalance and falls combined with reach tremor ('xxx—'), and injections within 1 mm of the interpositus-dentate boundary gave deficits of reach tremor combined with reach overshoot and uncoordinated pinch ('—•••').

the narrowness of their grid in the medial–lateral dimension and its centering on the interpositus/dentate border, which would result in muscimol spread to both nuclei.

Neural recording studies in animals

Single unit recording studies in the decerebrate cat during its controlled (by midbrain stimulation) walking and scratching movements have shown neural discharge in the fastigial nucleus phase-related with the movements. Little or no such correlation was found in the interpositus and dentate nuclei, however (Antziferova et al., 1980; Arshavsky et al., 1980; Orlovsky et al., 1999). These observations are consistent with a fastigial role specialized for the control of stance and gait.

It has been theorized that a key motor role of the interpositus is to make comparisons between the command for movement and the feedback from the movement, to sense errors, and to correct the latter quickly during the course of the movement (Eccles, 1967; Evarts and Thach, 1969; Allen and Tsukahara, 1974). In visually triggered movements, timing studies have shown that interpositus neurons begin to fire after discharge in dentate neurons and the motor cortex (Thach, 1978). In movements triggered by a somatosensory perturbation of the part to be moved (wrist), however, it was shown that the interpositus discharge comes first (Thach, 1978). A crucial observation was made by Strick (1983). He found that when a movement was made to oppose a perturbation, interpositus discharge led that of dentate neurons. When the movement was made in the direction of the perturbation, however, the discharge order was reversed.

The above results suggest that the discharge of interpositus neurons leads that of dentate neurons when the motor reaction is triggered by a somatosensory stimulus (e.g., returning to hold a position despite a perturbing displacement). In contrast, the discharge order is reversed when the motor task is initially counter-instinctive and has to be learned (e.g., moving in the direction the subject is being pushed).

Heretofore, the interposed nucleus (i.e., nuclei globose and emboliform in the human) has had no well-agreed-upon syndrome attached to its ablation. There are previous reports, however, of such an ablation producing an impairment in the execution of selected tasks; e.g., contact placing (Amassian and Rudell, 1978) and the nictitating membrane response (McCormick and Thompson, 1984; Yeo et al., 1984). These findings, together with the above-described observations on discharge patterns compatible with long-loop stretch reflex operation (Thach, 1978; Strick, 1983) suggest that this nucleus, too, exhibits neural discharge that is task-dependent.

The dentate nucleus has been thought to translate percepts and mental concepts into action plans for movement (Eccles, 1967; Evarts and Thach, 1969; Allen and Tsukahara, 1974; Brooks and Thach, 1981). In keeping with the idea, dentate neural discharge was reported to precede movement onset performed by trained monkeys, and even to lead the discharge of motor cortex neurons (Thach, 1975, 1978; Lamarre et al., 1983). Correspondingly, dentate inactivation was found to prolong visuo-motor reaction time. A reaction time delay of ∼100–200 ms was first reported by Holmes (1917) after human cerebellar injury, and it has been confirmed and localized to the dentate nucleus in a number of subsequent animal studies (Thach, 1975; Meyer-Lohman et al., 1977; Trouche and Beaubaton, 1980; Spidalieri et al., 1983). Furthermore, dentate inactivation was found to delay the onset of discharge of motor cortex neurons, and also the onset of movement (Meyer-Lohman et al., 1975; Spidalieri et al., 1983). This was interpreted to support the idea that the dentate nucleus participates in the initiating of volitional movement via the motor

---

(Contd.) No injection gave deficits of the combined fastigial-dentate syndromes ('xxx●●●'), or, of a *combination of all three* nuclear syndromes ('xxx—●●●'). Experiments in three monkeys gave consistent results. Judging from the spacing of injections and the resulting deficits, a diffusion distance of 1 mm was estimated from the site of muscimol injection. On the basis of these results, and those using anatomical tracing (Asanuma et al., 1983a,b) and unit recording (Thach et al., 1993), we have proposed that the body map is replicated within each nucleus, with the leg anterior, hand and arm central, and head and eyes posterior, and with each map subserving a separate control function: the fastigial nucleus for upright stance against gravity; the interpositus nucleus for agonist–antagonist muscle coordination at individual joints; and, the dentate nucleus for synergistic muscle coordination subserving external, goal-directed movements. (Reproduced from Thach et al., 1992a with permission from Springer-Verlag.)

cortex (Eccles, 1967; Evarts and Thach, 1969; Allen and Tsukahara, 1974; Brooks and Thach, 1981). The relatively greater deficit in many muscled, multijointed movements as compared to movements involving fewer joints and muscles is more consistent with the view that a major role of the cerebellum is to coordinate complex movements (Bastian et al., 1996; Goodkin and Thach, 2003a,b).

There is also evidence that dentate nucleus activity is important in the adaptive modification of walking to meet environmental demands. Rhythmic firing in the dentate nuclei of the cat occurs during walking across the rungs of a ladder (Marple-Horvat and Criado, 1999). Firing in both the dentate neurons and the complex spike discharge in Purkinje neurons were shown to be time-locked to movement of an individual step (Anderson and Armstrong, 1987; Armstrong and Marplehorvat, 1996). Armstrong and Marplehorvat (1996) have suggested that the cerebellum sends signals, containing information about the cat's distance from the obstacle, to relevant motor areas (e.g., thee motor cortex) in order to prime these areas for appropriate action.

Multiple somatomotor representations within the deep cerebellar nuclei: relation to control of locomotion

The acts of walking and scratching are by their very nature coordinated multijoint tasks, and what the cerebellum adds to their control is not yet agreed upon (cf., however, S. Mori et al.; Chapter 33). It is thought by many that the central patterns generators for these movements are located in the brainstem and spinal cord (Arshavsky et al., 1972a,b; Bloedel and Courville, 1981; Grillner and Wallén, Chapter 1). What the cerebellum adds is therefore presumably superimposed upon these fundamental motor synergies (but see Mori, Chapter 33).

Fastigial neurons have inputs from the vestibular complex, lateral reticular nucleus, and (indirectly) the spinocerebellar pathways (Jansen and Brodal, 1940; Brodal, 1981). The fastigial nucleus projects to (1) lateral and descending vestibular nuclei, (2) n. reticularis tegmenti pontis and n. prepositus hypoglossi, (3) the reticular gray of the midbrain, (4) the inferior olive (MAO), and (4) directly to contralateral motor neurons in the spinal cord (Asanuma et al., 1983d). It also has relatively weak connections to the 'cerebellar thalamus' (defined in the following section), along with stronger ones with the interpositus- and especially the dentate nucleus.

Interpositus neurons receive fast feedback from movement sensors and spinal motor programs via various spinocerebellar pathways, and descending input from the motor cortex via the pons (Bloedel and Courville, 1981). The interpositus nucleus projects to (1) the 'cerebellar' thalamus; (2) the magnocellular red nucleus (vanishingly small in humans, however); (3) the reticular nucleus of the pontine tegmentum; (4) the inferior olive (DAO); and (5) the intermediate grey of the spinal cord (see also Mori et al., 1999; S. Mori et al., Chapter 33 of this volume).

Neurons in the dentate nucleus and lateral cerebellar hemisphere receive descending input from frontal and parietal association cortex via the pons (Allen et al., 1978; Brodal, 1978). In turn, the dentate nucleus projects to (1) the cerebellar thalamus (see next section), (2) the parvocellular red nucleus, (3) the reticular nucleus of the pontine tegmentum, and (4) the inferior olive (PO).

The 'cerebellar' thalamus

The output of the cerebellum and its deep nuclei project to a target area within the thalamus, which is sufficiently free of other inputs that it can be termed the 'cerebellar thalamus' (Stanton, 1980; Kalil, 1982; Asanuma et al., 1983a; Asanuma et al., 1983d; Schell and Strick, 1983; Orioli and Strick, 1989). The target area includes several portions of the ventrolateral nucleus (VL) of the thalamus (VLc, VLps, VLo), the anterior portion of the ventroposterolateral nucleus (VPLo), and area X (the periarcuate area). The basal ganglia input to the thalamus is more anterior in VLo and the ventral anterior nucleus (VA), whereas the lemniscal input is more posterior in VPL. This cerebellar thalamus projects to area 4 (particular from VLc, VLps, VPLo, and parts of VLo) and to lateral area 6, the periarcuate area (X). Both the dentate and the interposed nuclei project in a completely overlapping fashion to the whole width (coronal) of the contralateral thalamic receiving area. The fastigial projection is sparse, bilateral, and restricted. It appears not to project to area X and VLo. The above connections show that at the cerebellar outflow to thalamic targets is multiple and that a body map is repeated across each nucleus.

From the known somatotopic mapping of the cerebellar thalamus onto the motor cortex (Strick, 1976a,b), and from the topographic projection of each deep cerebellar nucleus onto their common thalamic targets, somatotopic mapping may be inferred within each of the cerebellar nuclei. In the cerebellar thalamus, the head is medial, the tail lateral, the trunk dorsal, and the extremities ventral. In the cerebellar nuclei the head would then be caudal, the tail rostral, the trunk lateral, and extremities medial (Asanuma et al., 1983c). Such topography has been supported by neural recording and inactivation studies during movement (Thach et al., 1992a,b, 1993) It is also consistent with the electroanatomical input mapping studies in macaques of Allen et al., 1977, 1978). Finally, microstimulation of the deep cerebellar nuclei has shown a similar topography in baboons (Rispal-Padel et al., 1982) and rats (Cicirata et al., 1989, 1992).

Functional considerations

It is not yet clear whether all routes out of the cerebellum are equally responsible for gait control. Certainly, the direct projections from the fastigial nucleus to the vestibular nuclei and spinal cord would seem to be directly involved. A similar case can be made for the dentate nucleus projections to the reticular nuclei. The interposito-rubro-spinal pathway appears to be less important or even absent in the human. Finally, it is unknown to what extent cerebellar pathways to the thalamus and cerebral cortex are implicated in the control of gait (cf., however, S Mori et al., Chapter 33).

We have cited above evidence that the dentate nucleus plays a role in initiating reaching movements of the forelimb. Mori et al. (1999, 2000) (Chapter 33) have provided evidence that the fastigial nuclei and the hook bundle of Russell participate in the control of gait. The high-decerebrate cat was supported by a rubber hammock, with the feet contacting the moving surface of a treadmill. Electrical microstimulation of

their so-called cerebellar locomotor region (CLR) in the midline white matter provoked well-coordinated, bilaterally symmetrical, fore- and hindlimb movements. The gait cycle time and pattern were controlled by appropriate changes in stimulus intensity and treadmill speed. In the same animals, controlled locomotion was evoked by stimulating the mesencephalic locomotor region (MLR). With simultaneous subthreshold stimulation of both the CLR and MLR, locomotor movements were evoked that were identical to those during suprathreshold stimulation of each site alone. With simultaneous stimulation at suprathreshold strength at both sites, locomotion became more vigorous with a shortened cycle time. After lesions at either the CLR or MLR (unilateral or bilateral), locomotion was still evoked at the prior stimulus strength by stimulating the remaining site. These results showed that the fastigial nucleus is one of the supraspinal locomotion inducing sites and that it can independently and simultaneously trigger brainstem and spinal locomotor subprograms formerly believed to be the exclusive domain of brain stem regions, including the MLR and the subthalamic locomotor region (Grillner and Wallén; Chapter 1).

Does the cerebellum play a role in adapting and learning stance and gait?

Many studies have shown that the cerebellum plays a crucial role in motor learning (Ito, 1984; Thompson, 1990; Thach et al., 1992a,b). Very few studies, however, have considered the cerebellum's role in the adaptation of stance and gait.

Recent studies have addressed learning to couple a novel postural response to a novel perturbation of stance. Horak (1990) and Horak and Diener (1993) studied postural responses to a perturbation of stance. Subjects were instructed to stand on a platform, which was driven backward by a motor. One of the perturbation parameters that was varied was the distance of backward displacement: four distances at a given velocity. There were 190 trials of each, such that the subject had the opportunity to experience and adapt to these particular conditions. Only the last three trials were analyzed in a block of trials, after the opportunity for adaptation had occurred.

Control subjects adapted a plantar flexor specific torque to each of the four different distances (fixed velocity) with proportionate differences in EMG amplitude. Subjects with cerebellar cortical atrophy were distinctly different from controls. They showed no matching of specific torque to each displacement distance. Instead, they had excessive torques for all distances. This led to hyperactive responses with overshoot of the proper endpoint position, and without any stimulus-response scaling. Since the magnitude of the displacement (at fixed velocity) could *not* have been inferred from stimulus parameters at onset, the magnitude of the response had to be learned through trial and error. The adaptation was one of association, and not of scale. Cerebellar patients showed excessive activity of both antagonist and agonist muscles, which led to excessive responses and hypermetria. This was interpreted as a compensatory co-contraction as if to restrain and reduce the response that would otherwise have been even greater. The authors concluded that "... the anterior lobe of the cerebellum appears to play a critical role in modifying the magnitude of automatic postural responses to anticipated displacement conditions based on prior experience."

Two studies have examined the ability of humans with cerebellar damage to adapt walking. One study addressed whether cerebellar subjects could learn to predict and respond to perturbations during treadmill locomotion (Rand et al., 1998). This study showed that such subjects could adapt to transient changes in treadmill speed, but did so in a variable and inconsistent manner, without scaling the gait pattern specifically to the speed. The authors suggested that the cerebellum is important for specification of optimal strategies to be used during compensatory or learned movements. A second study addressed whether the cerebellum is important for an adaptive phenomenon induced by stepping in place on a rotating disk (Earhart et al., 2002). After performing this maneuver, healthy subjects inadvertently turn in circles when asked to step in-place on a stationary surface with eyes closed. This 'podokinetic after-rotation' (PKAR) is evidence that the nervous system has stored a new motor pattern induced by practice on the rotating disk; the PKAR decays quickly after practice off of the disk. Earhart et al. (2002) found that five of eight cerebellar subjects tested demonstrated a reduced PKAR magnitude, when compared to controls. Interestingly, the impaired PKAR magnitude was not associated with an impaired time constant of decay. The authors suggest that the cerebellum is important for regulation of the amplitude of the PKAR, though not the time course.

Yanagihara and colleagues (1992) have studied adaptive interlimb coordination during perturbed locomotion in the chronic decerebrate cat. Perturbations were applied to the stance phases of the left forelimb using a then-novel developed treadmill, which consisted of three compartments, one each for the left forelimb, the left hindlimb and both right limbs. During the perturbed locomotion, the treadmill belt for the left forelimb was driven at about twice the speed for the other limbs. During the first ~19–50 perturbed steps, the step cycles of both forelimbs showed marked fluctuations. Thereafter, the preparation achieved stable locomotion by using slightly shortened step cycle durations, and also by adjusting the duration of bisupport phases in the left and right forelimbs.

In the Yanagihara et al.'s (1993) study, the adult cat ($n=5$) was decerebrated at the precollicular level and ~4–5 days later, mounted on a treadmill. Spike potentials from Purkinje cells were recorded in the lateral vermis of lobule V in the cerebellar cortex. During stable locomotion, inferior olive climbing fiber input to Purkinje cells was only slightly modulated, i.e., there was a slight increase in spike-frequency during the swing phase of the ipsilateral forelimb. When the left contralateral forelimb alone was suddenly subjected to a faster belt velocity, the spike-frequency of the climbing fiber discharges was significantly increased during the late swing phase of the ipsilateral forelimb. These observations are in accordance with the general notion that inferior olive cell discharge represents error signals in the control of movement. This, in turn, is consistent with the theory that the inferior olive monitors motor behavior, i.e., if it recognizes an error, its neurons discharge and thereby 'teach' the Purkinje cells to fire, such as to correct the erroneous behavior (Marr, 1969; Albus, 1971; Ito, 1972).

Is such inferior olivary discharge actually involved in the adaptation of gait? Studies on the conscious behaving monkey have shown an increase in the discharge of inferior olivary climbing fibers during adaptations of motor behavior (Gilbert and Thach, 1977). Experimentally, stimulation of climbing fibers causes long-term depression (LTD) of Purkinje cell responses to their normally excitatory input from parallel fibers. It has been claimed that LTD is a key synaptic mechanism underlying the cerebellum's role in motor adaptation and learning (Ito, 1984, 1989).

Nitrous oxide (NO) is an intermediary in the induction of LTD of the excitatory synapses of parallel fibers onto Purkinje cells. Accordingly, Yanagihara and Kondo (1996) sought to establish a link between behavioral adaptation, LTD, and NO. They asked whether interference with NO might show that a specific perturbation of the LTD mechanism, and thereby result in an inability to adapt gait. To be specific for adaptation, their experimental manipulations had to be such as to not impair the baseline ability to ambulate. They injected either inhibitors of NO synthetase or NO scavengers into sites in the cerebellar cortex where inferior olive climbing fibers had been active in relation to adaptation. They found that *both* substances had the ability to block such adaptation without impairing normal gait. Thus, inactivating the NO system with small, circumscribed regions of the cerebellar cortex prevented the adaptation.

What is the message for rehabilitation?

The experimentally and clinically observed fact that stance and gait improve after focal cerebellar injury and not after diffuse atrophy suggests that the remaining intact parts of the cerebellum are helping to compensate for their deficits. The different parts of the cerebellum have different pathways projecting to the central pattern generators for stance and gait. It is conceivable that each part of the cerebellum can compensate for the loss of another part. For example, inactivation of the different deep cerebellar nuclei leads to a different set of deficits in stance and gait, thereby suggesting that each nucleus participates in the control of a different aspect of this integrated motor activity. Studies of activity (neuronal recording in the cat and monkey) and imaging studies in the cerebellum also suggest regional differences in how the cerebellum controls posture and movement (Shibasaki et al.; Chapter 20). The more medial parts of the cerebellum are specialized for antigravity stance. The intermediate part is specialized for modulating sensory reflex contributions to placing reactions, stepping, and holding still a certain position of a joint or limb. The lateral cerebellum is specialized for movements under exteroceptive guidance (e.g., moving to visual targets), movement planning, and mental imagination of movement. All of these categories of movement have been shown to adapt with trial and error practice. During practice, they pass from a state of conscious, effortful, imperfect control to one that is unconscious (automatic), easier, and closer to perfect (Lang and Bastian, 2002; Grillner and Wallén, Chapter 1). These differences have major implications for rehabilitation: they suggest that training strategies may be feasible, wherein the remaining cerebellar portions 'learn' to adjust the remaining connections so as best to compensate for the impaired functions. For example, following medial cerebellar lesions, exercises might be developed that focus on visuo-somesthetic controlled gait via intact lateral cerebellar structures. Similarly, one can envisage vestibular-visual exercises for damage to the intermediate cerebellum, and vestibular-somesthetic ones for subjects with lateral cerebellar lesions.

Is gait rehabilitation feasible for patients with diffuse cerebellar disease?

In this situation, a critically important structure for the learning and adjustment of coordinated multijoint movements is impaired or lost. One may have to resort to training strategies that altogether bypass the function of the cerebellum. For the upper extremity, a useful strategy might be to train patients to try to brace the limb and to move one joint at a time, ad seriatim. It is possible that such movements are more exclusively within the sphere of control of the motor cortex and spinal cord. If the nonmoving joints are not supported, however, there is poor fixation and the single-joint movements are not accurate (Bastian et al., 2000). It remains unknown, however, if an analogous strategy can be developed for lower extremity cerebellar ataxia, together with prosthetic support of nonmoving joints.

Is rehabilitation training necessary?

A commonality across studies of motor learning is that performance errors are reduced trial-by-trial through practice, until the improved performance is automatized. How much practice is required? We speculate that this can vary from several trials to many thousands of trials over months and even years. More study is required to determine the extent to which the cerebellum is essential in motor recovery from brain damage or even in normal motor leaning. Currently, it seems a likely contributor to both, and training is important in both these situations. Research and rehabilitative practice will likely continue to focus on understanding the role of the cerebellum in normal motor learning and recovery from brain damage.

Concluding thoughts

An inability to walk is perhaps the chief cause of movement disability in humans, and falling in the elderly and in patients with neurologic diseases is a leading cause of death. It is therefore disturbing that the efficacy of rehabilitation in these individuals has been questioned, and health care funding for rehabilitation has diminished. The cerebellum is now known to control, help initiate, and adjust stance and gait in humans and lower animals by mechanisms that are becoming increasingly well understood. This better understanding of cerebellar mechanisms responsible for the control of stance and gait and their adaptation promises a new emphasis on rehabilitation care that is *specific*, *effective* and *efficient*.

Acknowledgments

This work was supported by NIH grants NS 12777 (to W.T.T.) and HD 40289 (to A.J.B.)

Abbreviations

AICA	anterior inferior cerebellar artery
DAO	dorsal accessory olivary nucleus
MAO	medial accessory olivary nucleus
PICA	posterior inferior cerebellar artery
SCA	superior cerebellar artery
VL	ventrolateral thalamic nucleus
VLo	ventrolateral thalamic nucleus, oralis
VLc	ventrolateral thalamic nucleus, caudalis
VLps	ventrolateral thalamic nucleus, postrema
VPLo	ventroposterolateral thalamic nucleus, oralis

References

Albus, J.S. (1971). A theory of cerebellar function. Math. Biosci., 10: 25–61.

Allen, G.I. and Tsukahara, N. (1974). Cerebrocerebellar communication systems. Physiol. Rev., 54: 957–1006.

Allen, G.I., Gilbert, P.F.C., Marini, R., Schultz, W. and Yin, T.C.T. (1977). Integration of cerebral and peripheral inputs by interpositus neurons in monkey. Exp. Brain Res., 27: 81–99.

Allen, G.I., Gilbert, P.F.C. and Yin, T.C.T. (1978). Convergence of cerebral inputs onto dentate neurons in monkey. Exp. Brain Res., 32: 151–170.

Amassian, V.E. and Rudell, A. (1978). When does the cerebellum become important in coordinating placing movements? J. Physiol. (Lond.), 276: 35–36.

Antziferova, L.I., Arshavsky, Y.I., Orlovsky, G.N. and Pavlova, G.A. (1980). Activity of neurons of cerebellar nuclei during fictitious scratch reflex in the cat. I. Fastigial nucleus. Brain Res., 200: 239–248.

Armstrong, D.M. and Marplehorvat, D.E. (1996). Role of the cerebellum and motor cortex in the regulation of visually controlled locomotion. Can. J. Physiol. Pharmacol., 74: 443–455.

Arshavsky, Y.I., Berkinblit, M.B., Fuxson, O.I., Gel'fand, I.M. and Orlovsky, G.N. (1972a). Recordings of neurones of the dorsal spinocerebellar tract during evoked locomotion. Brain Res., 43: 272–275.

Arshavsky, Y.I., Berkinblit, M.B., Fuxson, O.I., Gel'fand, I.M. and Orlovsky, G.N. (1972b). Origin of modulation in neurones of the ventral spinocerebellar tract during locomotion. Brain Res., 43: 276–279.

Arshavsky, Yu.I., Orlovsky, G.N., Pavlova, G.A. and Perret, C. (1980). Activity of neurons of cerebellar nuclei during fictitious scratch reflex in the cat. II. The interpositus and lateral nuclei. Brain Res., 200: 249–258.

Asanuma, C., Thach, W.T. and Jones, E.G. (1983a). Distribution of cerebellar terminations and their relation to other afferent terminations in the ventral lateral thalamic region of the monkey. Brain Res. Rev., 5: 237–265.

Asanuma, C., Thach, W.T. and Jones, E.G. (1983b). Anatomical evidence for segregated focal groupings of efferent cells and their terminal ramifications in the cerebellothalamic pathway of the monkey. Brain Res. Rev., 5: 267–269.

Barth, A., Bogousslavsky, J. and Regli, F. (1993). The clinical and topographic spectrum of cerebellar infarcts:

a clinical-magnetic resonance imaging correlation study. Ann. Neurol., 33: 451–456.

Bastian, A.J., Mink, J.W., Kaufman, B.A. and Thach, W.T. (1998). Posterior vermal split syndrome. Ann. Neurol., 44: 601–610.

Bastian, A.J., Zackowski, K.M. and Thach, W.T. (2000). Cerebellar ataxia: torque deficiency or torque mismatch between joints?. J. Neurophysiol., 83: 3019–3030.

Bloedel, J.R. and Courville, J. (1981). Cerebellar afferent systems. In: Brooks V.B. (Ed.), Handbook of Physiology, The Nervous System, Sec. 1, Vol. II, American Physiological Society, Bethesda, pp. 735–830.

Botterell, E.H. and Fulton, J.F. (1938a). Functional localization in the cerebellum of primates. II. Lesions of midline structures (vermis) and deep nuclei. J. Comp. Neurol., 69: 47–62.

Botterell, E.H. and Fulton, J.F. (1938b). Functional localization in the cerebellum of primates. III. Lesions of hemispheres (neocerebellum). J. Comp. Neurol., 69: 63–87.

Brodal, P. (1978). The corticopontine projection in the rhesus monkey. Origin and principles of organization. Brain, 101: 251–283.

Brodal, A. (1981). Neurological Anatomy in Relation to Clinical Medicine. Oxford University Press, New York.

Brooks, V.B. and Thach, W.T. (1981). Cerebellar control of posture and movement. In: Brooks V.B. (Ed.), Handbook of Physiology, The Nervous System, Sec. 19, Vol. II, American Physiological Society, Bethesda, pp. 877–946.

Brown, J.R. (1949). Localizing cerebellar syndromes. J. Am. Med. Assoc., 141: 518–521.

Chambers, W.W. and Sprague, J.M. (1955a). Functional localization in the cerebellum. I. Organization in longitudinal cortico-nuclear zones and their control of posture, both extrapyramidal and pyramidal. J. Comp. Neurol., 103: 105–130.

Chambers, W.W. and Sprague, J.M. (1955b). Functional localization in the cerebellum. II. Somatotopic organization in cortex and nuclei. Arch. Neurol. Psychol., 74: 653–680.

Cicirata, F., Angaut, P., Panto, M.R. and Serapide, M.F. (1989). Neocerebellar control of the motor activity: experimental analysis in the rat. Comparative aspects. Brain Res. Rev., 14: 117–141.

Cicirata, F., Angaut, P., Serapide, M.F., Panto, M.R. and Nicotra, G. (1992). Multiple representation in the nucleus lateralis of the cerebellum: An electrophysiologic study in the rat. Exp. Brain Res., 89: 352–362.

Cooper, S.E., Martin, J.H. and Ghez, C. (2000). Effects of inactivation of the anterior interpositus nucleus on the kinematic and dynamic control of multijoint movement. J. Neurophysiol., 84: 1988–2000.

Dow, R.S. and Moruzzi, G. (1958). The Physiology and Pathology of the Cerebellum. University of Minnesota Press, Minneapolis.

Earhart, G.M. and Bastian, A.J. (2001). Cerebellar gait ataxia: selection and coordination of human locomotor forms. J. Neurophysiol., 85: 759–769.

Earhart, G.M., Fletcher, W.A., Horak, F.B., Block, E.W., Weber, K.D., Suchowersky, O. and Melvill Jones, G. (2002). Does the cerebellum play a role in podokinetic adaptation?. Exp. Brain Res., 146: 538–542.

Eccles, J.C. (1967). Circuits in the cerebellar control of movement. Proc. Natl. Acad. Sci. USA, 58: 336–343.

Evarts, E.V. and Thach, W.T. (1969). Motor mechanisms of the CNS: cerebrocerebellar inter-relations. Annu. Rev. Physiol., 31: 451–498.

Gilbert, P.F.C. (1974). A theory of memory that explains the function and structure of the cerebellum. Brain Res., 70: 1–18.

Gilbert, P.F.C. and Thach, W.T. (1977). Purkinje cell activity during motor learning. Brain Res., 128: 309–328.

Hallett, M. and Massaquoi, S. (1993). Physiologic studies of dysmetria in patients with cerebellar deficits. Can. J. Neurol. Sci., 20 Suppl. 3: S83–S92.

Holmes, G. (1917). The symptoms of acute cerebellar injuries due to gunshot injuries. Brain, 40: 461–535.

Horak, F.B. (1990). Comparison of cerebellar and vestibular loss on scaling of postural responses. In: Brandt T., Paulus W., Bles W., Dietrerich M., Krafczyk S. and Straube A. (Eds.), Disorders of Posture and Gait. Georg Thieme Verlag, Stuttgart, pp. 370–373.

Horak, F.B. and Diener, H.C. (1993). Cerebellar control of postural scaling and central set. J. Neurophysiol., 72: 479–493.

Ito, M. (1972). Neural design of the cerebellar control system. Brain Res., 40: 80–82.

Ito, M. (1984). The Cerebellum and Neural Control. Appleton-Century-Crofts, New York.

Ito, M. (1989). Long-term depression. Annu. Rev. Neurosci., 12: 85–102.

Kalil, K. (1982). Projections of the cerebellar and dorsal column nuclei upon the thalamus of the Rhesus monkey. J. Comp. Neurol., 195: 25–50.

Kane, S.A., Mink, J.W. and Thach, W.T. (1988). Fastigial, interposed, and dentate cerebellar nuclei: somatotopic organization and the movements differentially controlled by each. Soc. Neurosci. Abstr., 14: 954.

Lamarre, Y., Spidalieri, G. and Chapman, C.E. (1983). A comparison of neuronal discharge recorded in the sensorimotor cortex, parietal cortex, and dentate nucleus of the monkey during arm movements triggered by light, sound or somesthetic stimuli. Exp. Brain Res., Suppl. 7: 140–156.

Lang, C.E. and Bastian, A.J. (2002). Cerebellar damage impairs automaticity of a recently practiced movement. J. Neurophysiol., 87: 1336–1347.

Marple-Horvat, D.E. and Criado, J.M. (1999). Rhythmic neuronal activity in the lateral cerebellum of the cat during visually guided stepping. J. Physiol. (Lond.), 518: 595–603.

Marr, D. (1969). A theory of cerebellar cortex. J. Physiol. (Lond.), 202: 437–470.

Martin, T.A., Keating, J.G., Goodkin, H.P., Bastian, A.J. and Thach, W.T. (1996). Throwing while looking through prisms I. Focal olivocerebellar lesions impair adaptation. Brain, 119: 1183–1198.

Mason, C.R., Miller, L.E., Baker, J.F. and Houk, J.C. (1998). Organization of reaching and grasping movements in the primate cerebellar nuclei as revealed by focal muscimol inactivations. J. Neurophysiol., 79: 537–554.

Mauritz, K.H., Dichgans, J. and Hufschmidt, A. (1979). Quantitative analysis of stance in late cortical cerebellar atrophy of the anterior lobe and other forms of cerebellar ataxia. Brain, 102: 461–482.

McCormick, D.A. and Thompson, R.F. (1984). Cerebellum: essential involvement in the classically conditioned eyelid response. Science, 223: 296–299.

Meyer-Lohman, J., Hore, J. and Brooks, V.B. (1977). Cerebellar participation in generation of prompt arm movements. J. Neurophysiol., 40: 1038–1050.

Mori, S., Matsui, T., Kuze, B., Asanome, M., Nakajima, K. and Matsuyama, K. (1999). Stimulation of a restricted region in the midline cerebellar white matter evokes coordinated quadrupedal locomotion in the decerebrate cat. J. Neurophysiol., 82: 290–300.

Mori, S., Matsui, T., Mori, F., Nakajima, K. and Matsuyama, K. (2000). Instigation and control of treadmill locomotion in high decerebrate cats by stimulation of the hook bundle of Russell in the cerebellum. Can. J. Physiol. Pharmacol., 78: 945–957.

Orioli, P.J. and Strick, P.L. (1989). Cerebellar connections with the motor cortex and the arcuate premotor area: an analysis employing retrograde transneuronal transport of WGA–HRP. J. Comp. Neurol., 288: 612–626.

Orlovsky, G.N., Deliagin, T.G. and Grillner, S. (Eds.) (1999) Role of the cerebellum in locomotor coordination. Neural Control of Locomotion from Mollusc to Man. Oxford University Press, New York, pp. 215–236.

Palliyath, S., Hallett, M., Thomas, S.L. and Lebiedowska, M.K. (1998). Gait in patients with cerebellar ataxia. Mov. Disord., 13: 958–964.

Rispal-Padel, L., Cicirata, F. and Pons, C. (1982). Cerebellar nuclear topography of simple and synergistic movements in the alert baboon (Papio papio). Exp. Brain Res., 47: 365–380.

Schell, G.R. and Strick, P.L. (1983). The origin of thalamic inputs to the arcuate premotor and supplementary motor areas. J. Neurosci., 4: 539–560.

Snider, R.S. and Eldred, E. (1952). Cerebro-cerebellar relationships in the monkey. J. Neurophysiol., 5: 27–40.

Spidalieri, H.J., Busby, L. and Lamarre, Y. (1983). Fast ballistic arm movements triggered by visual, auditory, and somesthetic stimuli in the monkey. II. Effects of unilateral dentate lesion on discharge of precentral cortical neurons and reaction. J. Neurophysiol., 50: 1359–1379.

Sprague, J.M. and Chambers, W.W. (1953). Regulation of posture in intact and decerebrate cat. I. Cerebellum, reticular formation, vestibular nuclei. J. Neurophysiol., 6: 451–463.

Strick, P.L. (1983). The influence of motor preparation on the response of cerebellar neurons to limb displacements. J. Neurosci., 3: 2007–2020.

Thach, W.T. (1975). Timing of activity in the cerebellar dentate nucleus and cerebral motor cortex during prompt volitional movement. Brain Res., 69: 168–172.

Thach, W.T. (1978). Correlation of neural discharge with pattern and force of muscular activity, joint position and direction of intended next movement in motor cortex and cerebellum. J. Neurophysiol., 41: 654–676.

Thach, W.T., Kane, S.A. and Goodkin, H.P. (1998). Dentate/interposed cerebellar nuclear sites coordinating reach and grasp. Soc. Neurosci. Abstr., 24: 1406.

Thach, W.T., Kane, S.A., Mink, J.W. and Goodkin, H.P. (1992). Cerebellar output: multiple maps and modes of control in movement coordination. In: Llinas R. and Sotelo C. (Eds.), The Cerebellum Revisited. Springer-Verlag, New York, pp. 283–300.

Thach, W.T., Goodkin, H.G. and Keating, J.G. (1992). Cerebellum and the adaptive coordination of movement. Annu. Rev. Neurosci., 15: 403–442.

Thach, W.T., Perry, J.G., Kane, S.A. and Goodkin, H.P. (1993). Cerebellar nuclei: rapid alternating movement, somatotopy, and the control of muscle synergy. Rev. Neurol. (Paris), 11: 607–628.

Trouche, E. and Beaubaton, D. (1980). Initiation of a goal-directed movement in the monkey. Exp. Brain Res., 40: 311–321.

Udo, M., Matsukawa, K., Kamei, H. and Oda, Y. (1980). Cerebellar control of locomotion: effects of cooling cerebellar cortex in high decerebrate and awake walking cats. J. Neurophysiol., 44: 119–134.

Victor, M., Adams, R.D. and Mancall, E.L. (1959). A restricted form of cerebellar cortical degeneration occurring in alcoholic patients. Arch. Neurol., 1: 579–688.

Yanagihara, D. and Udo, M. (1994). Climbing fiber responses in cerebellar vermal Purkinje cells during perturbed locomotion in decerebrate cats. Neurosci. Res., 19: 245–248.

Yanagihara, D. and Kondo, I. (1996). Nitric oxide plays a key role in adaptive control of locomotion in cat. Proc. Natl. Acad. Sci. USA, 93: 13292–13297.

Yanagihara, D., Udo, M., Kondo, I. and Yoshida, T. (1993). A new learning paradigm: adaptive changes in interlimb coordination during perturbed locomotion in decerebrate cats. Neurosci. Res., 18: 241–244.

Yeo, C.H., Hardiman, M.J. and Glickstein, M. (1984). Discrete lesions of the cerebellar cortex abolish classically conditioned nictitating membrane response of the rabbit. Behav. Brain Res., 13: 261–266.

SECTION IX

Eye–head–neck coordination

CHAPTER 35

Current approaches and future directions to understanding control of head movement

Barry W. Peterson*

Department of Physiology, The Feinberg Medical School, Northwestern University, Chicago, IL 60611, USA

Abstract: This chapter reviews four key issues that must be addressed to advance our knowledge of control of head movement by the central nervous system (CNS). (1) Researchers must consider how the CNS utilizes the multiple muscle patterns that can produce the same head movement in carrying out tasks in an optimal way. (2) More attention must be paid to the dynamics of neck muscle activation that are required to implement head movements and show they are produced by CNS circuits. (3) Research is required to determine how the multiple pathways that impinge upon neck motor centers are utilized in a variety of tasks including eye–head gaze shifts, smooth head tracking, head stabilization and manipulating objects with the head. These pathways include corticospinal, vestibulospinal, reticulospinal (three subdivisions), fastigiospinal, tectospinal and interstitiospinal tracts. (4) Further analysis is needed to understand how vestibular signals are modulated during each of the above-mentioned tasks. This ambitious agenda is justified by the fact that the head–neck motor system is an ideal model for understanding issues of complex motor control.

Introduction

This chapter provides a review of the current state of work on head-movement control. What follows is a conceptual look at this system, as illustrated by diagrams rather than data from the author's and others' laboratories.

An interesting property of the head-movement control system is that it shares features of both the eye and arm-movement control systems. Like the eye, the head is used to control the direction of gaze. It thus requires mechanisms to keep its orientation stable in the face of external perturbations. As in the eye, evolution has solved this problem by providing strong, short-latency vestibular reflexes, the vestibulocollic reflexes (VCRs). Muscle-spindle-driven reflexes, which serve to adjust compliance of the arm, are also present in the neck (cervicocollic reflexes) but their monosynaptic portion is relatively

*Corresponding author: Tel.: +312-503-6216;
Fax: +312-503-5101; E-mail: b-peterson2@northwestern.edu

weaker than in the arm. The head also participates in gaze shifts that align the optic axes with an object to be viewed. Their stereotypic nature makes such gaze shifts easier to understand than arm movements whose goals and trajectories are exceedingly diverse.

The presence of both eye- and arm-related control systems is reflected in the wide variety of pathways impinging on regions of the cervical spinal cord that control the neck musculature. These include vestibulospinal, reticulospinal, corticospinal, tectospinal and fastigiospinal pathways. As described below, motor commands traversing these pathways are selectively deployed to accomplish various head-movement tasks.

This review examines four aspects of head motor control that shed light on the special properties of this neuromuscular system. The first two are the kinematics and dynamics of the head–neck neuromuscular system. The author then examines supranuclear commands for voluntary head movements carried by the various descending pathways that impinge upon neck motor nuclei. The fourth section examines neural control strategies for generating eye–head gaze shifts. A final section will discuss future directions for research on head-movement control.

Kinematics of the overcomplete head–neck musculoskeletal system

Various lines of evidence suggest that the head–neck system is overcomplete in that it has more than the minimum number of muscles required to control the motions of its seven joints (Pellionisz, 1988). This means that there will not be a single unique pattern of muscle activation associated with every head movement. Rather, a particular movement can be executed by many different muscle-activation patterns.

Figure 1 illustrates the contrast between kinematics of the oculomotor and head-movement systems. In the oculomotor system, three pairs of muscles control the three rotational degrees of freedom of the eye (yaw, pitch and roll). This corresponds to a situation with three unknowns and three equations. As a result, one can calculate a single unique set of activation strengths of each muscle pair that are required to produce each motion (Robinson, 1982).

Matrix A in Fig. 1 represents the eye movements that each muscle pair will generate. For instance, the middle row indicates that activation of the superior/inferior rectus pair (superior rectus activation coupled with suppression of inferior rectus) will produce negligible horizontal rotation, strong upward rotation (−0.91) and significant torsional rotation (0.42). Because the matrix is invertible, one can also compute the pattern of muscle activation needed to produce a particular motion as represented in the inverse matrix, B. For instance, the middle row indicates that a downward pitch motion requires activation of inferior rectus and inactivation of superior rectus with a net strength of −0.81 and also activation of the oblique muscle pair with a strength of 0.43 to counter the roll motion that the vertical recti would otherwise produce. There is also a very weak inactivation of LRMR.

The matrix in Fig. 1C portrays a simplified head/neck system with three muscles on both the right and left sides and two joints. Because the muscles do not align into push/pull pairs as in the eye, each must be

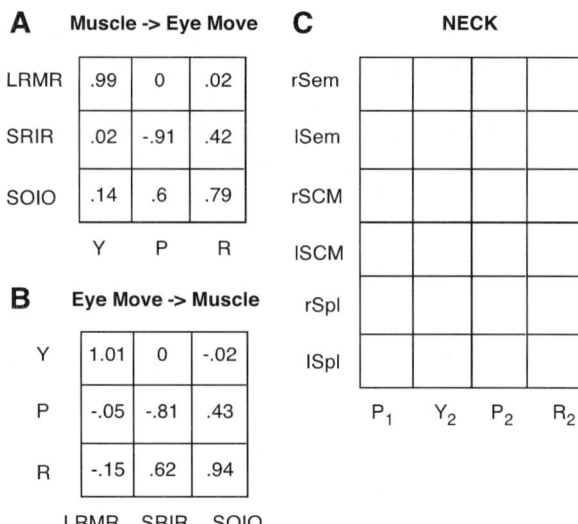

Fig. 1. Kinematic transformation matrices for eye and neck motor plants. In A and C, rows indicate muscle inputs, and columns show movement outputs. Values in A, which indicate the movement produced by each muscle pair, are from Robinson (1982). B is the inverse of A. Here, rows indicate movements, and columns show muscles must be activated to produce them. Movements produced by each muscle in the neck are still in doubt and thus the matrix elements in C have been left blank. Muscle abbreviations: LRMR, lateral and medial rectus; l and r SCM, left and right sternocleidomastoid; l and r Spi, left and right splenius; l and r Sem, left and right semispinalis; SOIO, superior and inferior oblique; SRIR, superior and inferior rectus. Other abbreviations: P, vertical or pitch eye movement; P_1, P_2 head pitch about joint 1 and 2, respectively; R, torsional or roll eye movement; R_2, Y_2, roll and pitch of head about joint 2.

considered as acting separately, resulting in six unknown activation strengths. Two types of joint are included. One can move only in the pitch direction, as does the joint between C1 and the skull. The other can move in all three directions as in the case of lower cervical joints. This results in four equations—too few to completely specify the activation of muscles required to produce a particular motion. Rather, there are many different solutions that can all produce the same resultant motion. We can fill in the matrix elements specifying the motions each muscle would produce but the 4 × 6 matrix is not invertible. Thus, we cannot specify a pattern of muscle activation that is necessary to produce a given motion.

Experimental examination of muscle-activation patterns associated with head movements has provided examples where multiple activity patterns were associated with a particular movement and others where a single pattern was observed. A single pattern across time and individual subjects was observed in highly constrained tasks such as stabilization of the head by the VCR during body rotation (Keshner et al., 1992; Banovetz et al., 1995; Uchino, this volume) and generation of isometric forces with the head (Vasavada et al., 2002). In contrast, multiple patterns were observed during tracking of a target with the head (Keshner and Peterson, 1988; Choi et al., 2000) and when subjects had to stabilize the head against loads applied with a weight-pulley system (Keshner et al., 1989). This suggests that the redundancy inherent in the overcomplete head/neck musculoskeletal system allows the central nervous system (CNS) to choose different patterns of muscle activation to execute the same movement in ways that optimize some aspect of the movement (see also Sugiuchi et al., this volume). Obvious distinctions are between rapid, predictive and fine, visually driven tracking and between force generation with low and high stiffness. The observed behavior suggests that different central neural pathways are used to drive movements that have such differing properties. This is an important question to be addressed in studies recording central neural activity associated with head movement. Can we indeed show that different premotor pathways are active when a movement is made utilizing different muscle synergies? Also, can we define the way in which each pattern is optimal and relate it to the task being performed when the pattern appears? Answers to these questions will greatly advance our understanding of motor systems.

Dynamics of head–neck system and its neural controllers

Dynamics refers to the way in which amplitude and timing (phase) of neural or movement signals vary with frequency of the head movement or of the stimulus that is eliciting it. Mechanics of the eye are dominated by viscosity of orbital tissues and spring-like forces of the extraocular muscles. Correspondingly, extraocular motoneuron (MN) activity has a burst component related to desired eye velocity and a tonic component related to eye position. Mechanics of the head/neck system are dominated by the inertia of the head and secondarily by viscoelastic properties of the muscles. Correspondingly, neck muscle activity should be triphasic with pulses of activity in agonist and antagonist muscles to accelerate and brake the head at movement onset and termination with possible tonic activity related to head deviation. Such signals were in fact reported in early studies (Zangemeister et al., 1982). The dynamics of MN and premotoneuron activity, however, has received relatively little attention in recent studies of eye–head gaze shifts, which have emphasized instead the use of common error signals to drive both the eye and head. This leaves unanswered questions regarding where in the final output pathway the appropriate dynamical signals are generated?

In contrast, studies of head stabilization have paid a great deal of attention to head/neck dynamics. Early recordings in the decerebrate cat revealed the expected second-order behavior with head position signals at low frequencies and head acceleration signals at high frequencies (Bilotto et al., 1982). Angular accelerations of the head activate regularly and irregularly firing afferents from the vestibular semicircular canals. It is the latter that preferentially drive vestibulocollic pathways (Highstein et al., 1987). Their dynamics include a pole term that converts angular acceleration to angular velocity at lower frequencies and a zero term that converts back to acceleration at higher frequencies above 1 Hz (Fernandez and Goldberg, 1971). These terms form the first part of the VCR transfer function in Fig. 2.

Wilson et al. (1979) used electrical polarization to bypass this first afferent-related term allowing them to observe the further processing of VCR signals in the CNS. As shown in the transfer function, this consists of a second pole-zero pair. The second pole converts the velocity signal produced by the first to an angular position signal. Again above 1 Hz, the second zero reverses this effect, yielding an angular acceleration signal at high frequencies. Thus, semicircular canal inputs are processed exactly as required to produce a second-order

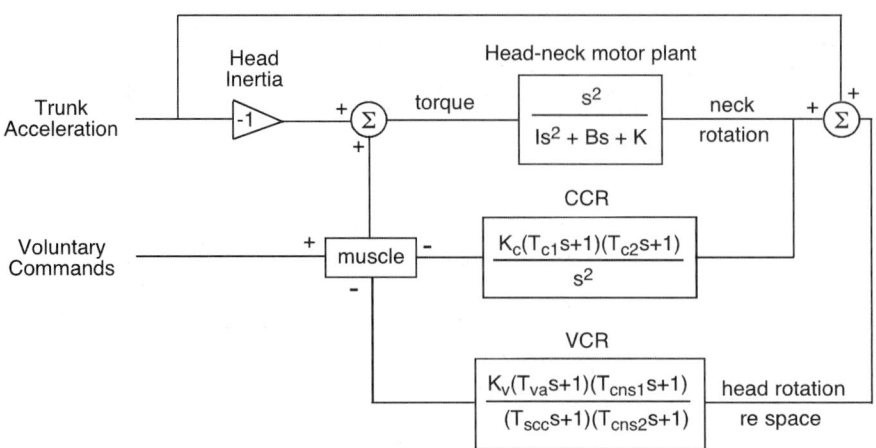

Fig. 2. Dynamics of horizontal head stabilization about a single joint. The symbol s is the LaPlace variable. I, B and K are inertia, viscosity and stiffness, respectively, of neck. Kc and Kv are constants weighting the action of the CCR and VCR, respectively, on neck musculature. See text for discussion of time constants (T).

position/acceleration control signal on neck MNs. Peterson et al. (1985) also observed the dynamics of responses to rotation of the body about an earth-fixed head, finding again a second-order signal. This behavior, which is represented by the cervicocollic reflex (CCR) transfer function in Fig. 2, likely reflects the properties of neck muscle spindles.

Measurement of the transfer functions of system components opened the way for dynamical modeling of the head stabilization system in cats (Goldberg and Peterson, 1986) and humans (Peng et al., 1996a,b). These models have proven remarkably effective in replicating and explaining the complex dynamic behavior of the head during horizontal rotations of the body (Keshner and Peterson, 1995; Keshner et al., 1995). Several conclusions can be reached from these models: (1) the VCR is not very effective in stabilizing horizontal head position in space at frequencies below 0.5 Hz (gain ~0.1); (2) the VCR and CCR are vital for preventing head oscillations in the 1–3 Hz range; and (3) a static vertical VCR appears to be important for stabilizing the head in an upright posture against the force of gravity. It is likely that the VCR also plays a role in stabilizing the head during linear motions of the body (Lacour and Borel, 1989; Borel and Lacour, 1992; Borel et al., 1994). This is a topic for future investigation.

Supranuclear commands for voluntary head movements

Commands for generating head movements reach neck MNs from a variety of sources that are shown in Fig. 3.

Several classes of vestibulospinal neurons project to neck motor pools. One group includes the bifurcating position-vestibular-pause (PVP) neurons that project to both extraocular and neck MNs (Isu et al., 1988). Their name denotes the fact that they carry an eye position signal and a vestibular semicircular canal signal and pause during saccadic gaze shifts. Perlmutter et al. (1998a) established that the canal input to these neurons is from a single canal. Responses of identified bifurcating neurons to linear head motions have not been studied but it is likely that they also receive input from otolith afferents sensitive to linear acceleration as observed in unidentified PVP neurons (Angelaki et al., 2001). The likely signal carried by these neurons is a command for stabilizing eye and head movements aligned with the coordinate frame of the three semicircular canals. This would then be distributed with the proper weights to extraocular and neck MNs to generate the required motions.

There are also vestibulospinal neurons that project only to the neck. These typically receive convergent

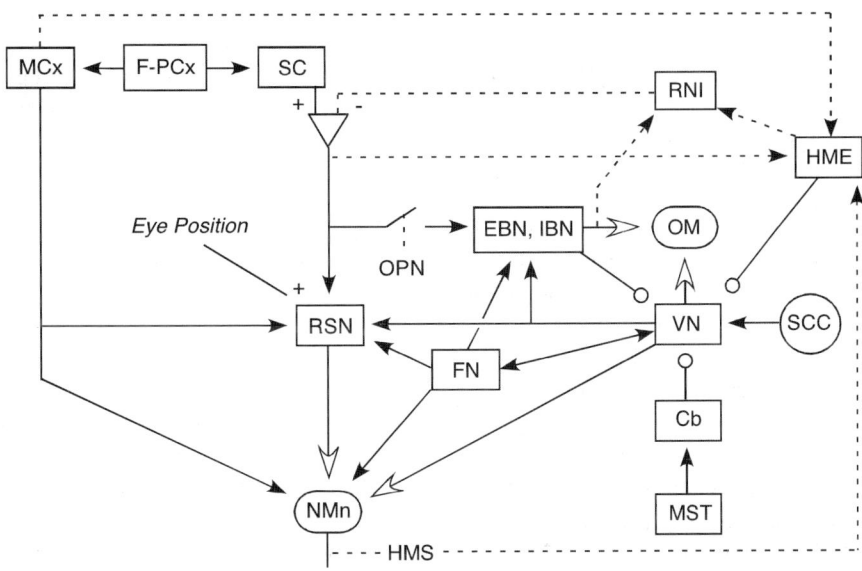

Fig. 3. Scheme showing processes involved in generating and controlling eye–head gaze shifts. Commands for saccadic gaze shifts originate in association areas of frontal and parietal cortex (F-PCx) and activate output systems originating in motor cortex (MCx) and SC. SC signals reach neck motoneurons (NMn) primarily via reticulospinal neurons (RSN) and oculomotor nuclei (OM) via excitatory and inhibitory burst neurons (EBN, IBN). Both head and eyes are stabilized in space via pathways originating in semicircular canals (SCC) and relaying in the vestibular nuclei (VN). The VN are also the relay for smooth pursuit commands relayed via area MST and cerebellar vermis and flocculus (Cb). The cerebellum also regulates eye and head movements via its fastigial nucleus (FN). Also included are three mechanisms that regulate gaze shifts: the eye saccade gate implemented by omnipause neurons (OPN); the resettable neural integrator (RNI) that turns off saccades when the correct goal is reached; and, a head-movement estimator (HME) that cancels the portion of VN activation due to commanded head movements. HMS are head-movement sensors—possibly muscle spindles. Solid arrows indicate excitatory connections; open circles, inhibitory connections; open arrows, mixed actions.

input from multiple canals that produce a unique preferred rotation direction that is likely aligned with that of the specific motor pools to which they project (Perlmutter et al., 1998a). It is not known how such vestibulocollic neurons respond to linear motions. Observations by Boyle et al. (1996) suggest that this population may include neurons that receive input from the cerebellar flocculus and exhibit an eye–head velocity discharge pattern. They could provide one pathway for mediating visually guided smooth head movements.

Finally, neck MNs also receive input from collateral branches of vestibulospinal neurons projecting to lower levels of the spinal cord (Perlmutter et al., 1998b). These long vestibulospinal neurons exhibit a wide variety of preferred angular rotation directions including directions that activate semicircular canals on the opposite side. Interestingly, it is such Type II neurons that appear to receive the strongest direct input from otolith afferents sensitive to linear accelerations (Angelaki et al., 2001). The signals carried by these neurons are likely to mediate postural-stabilizing responses.

Another major source of input to neck MNs is the reticulospinal system, which originates from several groups of neurons in the medial mesencephalic, pontine and medullary brainstem (Robinson et al., 1994). The best understood neurons are those that are located in the paramedian pontine reticular formation and send their axons to the spinal cord via the medial reticulospinal tract (mRST). The paramedian pontine reticular region is the final premotor relay in the network that controls horizontal gaze shifts. In addition to mRST neurons, it also contains excitatory burst neurons (EBNs) that drive horizontal eye movements via projections to abducens MNs and internuclear neurons (Strassman et al., 1986a,b). Both mRST and EBN neurons carry a signal related to gaze velocity. They receive strong excitatory inputs from the superior colliculus (SC), a structure that specifies the desired metrics of a gaze shift by a spatial code wherein discharge of neurons in a specific location on a retinotopic map leads to a saccade appropriate to foveate a stimulus falling on the corresponding portion of the retina. Transformation from the spatial code in the SC to the gaze velocity code in the reticular formation is thought to involve weighted projections from SC to the reticular formation and regulation by cerebellar circuits that project to the reticular formation via the cerebellar fastigial nucleus (Optican and Quaia, 2002). (Further aspects of this system are considered below). Neurons projecting in the mRST also receive input from frontal cortex and the vestibular system (Peterson et al., 1974; Peterson and Abzug, 1975), raising the possibility that they may participate in voluntary control of the proximal musculature and postural reflexes.

Less well understood are the projections to neck MNs from neurons in the gigantocellular reticular formation of the medial medulla. These descend primarily in the ipsilateral and contralateral lateral reticulospinal tracts (lRSTs) and have both excitatory and inhibitory actions on neck MNs (Peterson et al., 1978). The latter pay a key role in the atonia associated with rapid eye-movement sleep (Terzuolo et al., 1965). Like mRST neurons, these neurons receive inputs from frontal cortex, SC, fastigial nucleus and vestibular and somesthetic afferents (Peterson and Felpel, 1971; Eccles et al., 1975; Peterson et al., 1980). This raises the possibility that they serve a wide variety of functions related to voluntary head movements, orienting and postural stabilization. Kitama and colleagues (Kitama et al., 1995a,b), however, have reported that these functions may be separated. Neurons that play a key role in orienting movements produced by the tecto-reticulospinal system (Grantyn and Berthoz, 1985, 1987; Grantyn et al., 1987, 1992a,b; Grantyn et al., this volume; Sasaki, this volume) appear to receive relatively little vestibular input whereas other neurons in the pontine and medullary reticular formation respond vigorously to horizontal rotations and also receive input from the SC. These likely include the reticulospinal neurons examined by Bolton et al. (1992), which responded especially strongly to inputs from otolith organs. This raises the possibility that the reticulospinal system may contain separate channels for mediating visually driven orienting responses, and slow head stabilization and tracking movements. This suggestion requires further examination in animals with free heads.

A third reticulospinal projection arises from the region of central mesencephalic tegmentum corresponding to nucleus subcuneiformis (Robinson et al., 1994). Some neurons in this region fire prior to saccades while others discharge later, at the point when head movement is underway (Waitzman et al., 1996, 2002). Inactivation of the region results in hypermetric contralateral saccades (Waitzman et al., 2000), suggesting that it is involved in the regulation of gaze-shift amplitude. An intriguing possibility is that this region plays a role in apportioning the SC gaze command signal between eye and neck motor plants.

Corticospinal projections to the neck arise from several cortical areas including primary motor cortex, the dorsal and ventral premotor areas and the supplementary and cingulate motor areas (He et al., 1993, 1995). In primates, some of these projections appear to terminate on neck MNs (Dum and Strick, 1996). In the cat, however, the most direct excitatory cortico-neck MN pathway was disynaptic with a relay in the pontomedullary reticular formation. Stimulation of corticospinal fibers below a transected pyramid gave rise only to disynaptic inhibition followed by longer latency excitation (Alstermark et al., 1985, 1992). It is likely that cortico-reticulospinal connections are stronger than direct corticospinal connections in primates, as well. They would allow cortical access to orienting responses involving synergistic activation of large sets of axial MNs and to postural adjustments that form a vital part of many voluntary movements. The latter may also involve cortico-vestibulospinal pathways. The special role of the direct pathways remains to be explored experimentally.

Another descending pathway arises from the interstitial nucleus of Cajal, which functions as a controller for vertical gaze position (Fukushima, 1987; see also Fukushima et al., Chapter 37 of this volume). There are also projections from the SC and cerebellar fastigial nucleus that appear to terminate on interneurons within the neck segments of the spinal cord (Robinson et al., 1994). These may be the head-movement analog of projections of these structures to EBNs that drive eye saccades.

Neuromuscular strategies underlying eye–head gaze shifts

The most extensive analysis of head movements concerns their role in eye–head gaze shifts. Figure 4 summarizes key aspects of this behavior.

The lower traces in Fig. 4A indicate the movements of the eyes (in the head) and of the head in space during an 80° gaze shift. Gaze is the sum of the other two signals and reflects the point in space toward which the optic axes of the eyes are directed. In this example, the eyes and head begin moving at the same time (gaze shift onset indicated by the first dotted line) but this can vary (Zangemeister et al., 1981, 1982; Zangemeister and Stark, 1982). For small movements, the head is often delayed whereas for large movements or highly predictable movements, the head may begin to move before the eyes (Ron et al., 1993). In this case, the VOR drives the eyes in the opposite direction, thereby holding gaze stable until gaze onset.

The gaze shift ends 0.3 s after it began. At this time, the head is still moving in the direction of the gaze shift while the eyes are moving in the opposite direction under the influence of the VOR. In fact, eye velocity typically reverses direction before the end of the gaze shift, thereby reflecting an opposing interaction between gaze drive and the partially suppressed VOR (Tabak et al., 1996). It was originally thought that the VOR was normally active throughout a gaze shift. This would require the saccadic system to generate a much larger drive in order to overcome the vestibuloocular eye movements that would oppose the gaze shift (Bizzi et al., 1976). There is now wide agreement, however, that the VOR is suppressed during a gaze shift (Tomlinson and Bahra, 1986a; Ron et al., 1993; Tabak et al., 1996; Roy and Cullen, 1998). Careful measurements by Tabak et al. (1996) indicate that the degree of suppression in the human increases with amplitude of the gaze shift but is never complete. The drawing at the top of Fig. 4A is typical, showing a suppression of 50%. As discussed further below, there is also evidence that VCRs are suppressed during active, voluntary head movements (McCrea et al., 1999; Roy and Cullen, 2001).

Figure 4B illustrates another aspect of eye–head behavior. Head-movement components of gaze shifts depend on eye position in the head (Guitton et al., 1984; Tomlinson and Bahra, 1986b; Guitton and Volle, 1987; Freedman and Sparks, 1997; Goossens

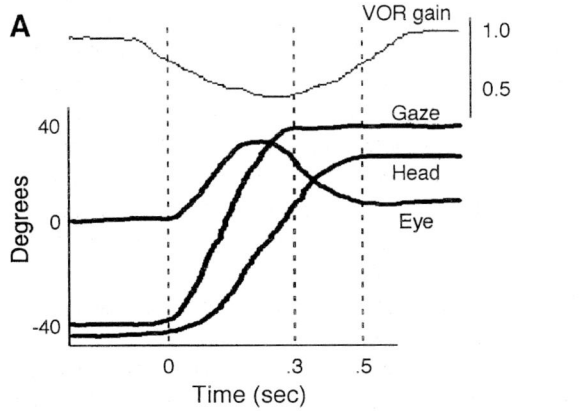
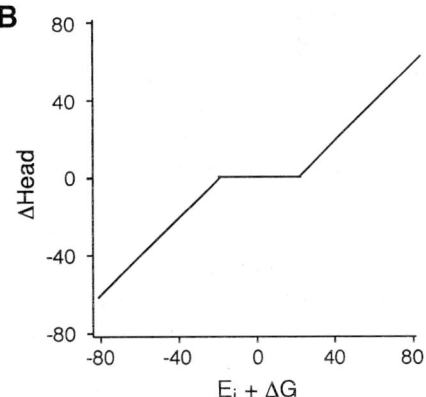

Fig. 4. Eye–head behavior during gaze saccades. (A) Traces at bottom show trajectories of head in space, eye in head and of their sum (gaze) during an 80° gaze saccade. Top trace plots VOR gain as measured by Tabak et al. (1996). Vertical lines indicate gaze shift onset, gaze shift end and head-movement end. (B) Plot of head-movement contribution to gaze shifts as a function of the predicted eye eccentricity at the end of the gaze shift if the head did not move (the sum of initial eye position in head and the change in gaze), as analyzed by Stahl (1999). Axes are in degrees.

and Van Opstal, 1997; Stahl, 1999; Ceylan et al., 2000). If a gaze shift can be accomplished by eye movements alone, leaving the eyes near central position, the head-movement component will be minimal. The abscissa on the Fig. 4B graph, which plots the sum of initial eye position and the gaze shift, reflects this behavior. If the initial eye position (E_i) is already in the direction of the gaze shift, $E_i + \Delta G$ will be large, and a larger head movement will be planned than if initial eye position is in the direction opposite to the gaze shift. The range of allowed eye position varies between individuals—those with large allowed ranges are eye movers while those with small ranges are head movers. The range can also be adjusted to match cognitive demands of the task. For instance, when subjects view the world through pinhole glasses, the range is greatly reduced so that gaze shifts are largely carried out by moving the head. This allows the target to remain visible through the pinhole (Ceylan et al., 2000). The simple form of apportioning gaze shifts can be accomplished by adding an eye position offset signal to the main reticulospinal pathway carrying gaze commands to the neck as shown in Fig. 3, and as observed experimentally (Grantyn and Berthoz, 1987). It is not clear, however, how the range is adjusted in accordance with task demands.

Figure 3 summarizes current knowledge of neural circuitry involved in generating eye–head gaze shifts. A major source of commands for such shifts is the intermediate-to-deep layers of the SC, which receives inputs from many sources, including gaze centers in the frontal and parietal cortex (F-PCx). SC discharge is in the form of a location code where firing of a particular neuron commands movements to a specific point on a retinotopic map. Firing intensity varies somewhat with the motivation of the animal and the corresponding velocity of the resulting gaze shift, but this variation is a secondary effect. Output of the colliculus is thus a *gaze displacement* command. This must be converted to a *gaze velocity* command that declines to zero as gaze reaches the target. Figure 3 shows conceptually how this can be done. A resettable neural integrator (RNI) keeps track of how far gaze has shifted since the saccade began and its output is subtracted from the SC signal to yield the desired gaze velocity command. There is evidence that this process may involve shifting of the locus of activity on the colliculus during the gaze shift toward regions that code progressively smaller gaze shifts (Munoz and Guitton, 1985; Munoz et al., 1991; Munoz and Wurtz, 1995) but the current models assign a key role in the process to the cerebellar vermis (Optican and Quaia, 2002) and its output pathway via the fastigial nucleus (FN).

The eye-movement component of the gaze shift passes through a gate representing the inhibitory action of omnipause neurons (OPN) before reaching EBNs and inhibitory burst neurons (IBNs). The presence of this gate explains why activity sometimes reaches head-movement circuits prior to activation of EBNs and IBNs. Activity of these bursters then drives oculomotor neurons (OM). It also charges the RNI and the eye-position integrator that maintains eccentric eye positions between gaze shifts (not shown in Fig. 3). A final action arises from the projection of IBNs to the vestibular nuclei where they are thought to inhibit PVP-type vestibuloocular neurons. This would account for the fact that the pause exhibited by these neurons is precisely aligned with the time of the saccadic gaze shift.

The head-movement component of the SC-controlled gaze shift reaches neck MNs primarily via reticulospinal neurons (RSNs). As explained above, there are separate populations of RSNs responsible for horizontal and vertical head movements located in the mesencephalic, pontine and medullary tegmentum. The box shown in Fig. 3 is a composite of these populations. It receives SC-, motor cortical- and vestibular input, which may be segregated in different populations of RSNs (Kitama et al., 1995a). This box also receives an eye-position signal, which has been observed in medial RSNs of the pons (Grantyn and Berthoz, 1987; Grantyn et al., 1992a). As explained above, this signal varies the activation of RSNs and hence the head-movement component of gaze shifts depending on the position of the eyes in the head.

A prominent input to EBNs, IBNs and RSNs arises from the fastigial nucleus (May et al., 1990; Robinson et al., 1994). Knowledge of the complex signals carried by fastigial neurons, and the way in which they are distributed to different premotoneurons, is currently quite limited. There is a similar lack of information on the role of pathways from motor and premotor cortical areas of the frontal lobe

(He et al., 1993, 1995; Dum and Strick, 1996) in producing head movements. Anatomical and electrophysiological observations indicate that these pathways have powerful actions on RSNs and neck MNs (Peterson et al., 1974; Alstermark et al., 1985, 1992) but the signals they carry during behavior have not been observed. There is also considerable controversy regarding whether the frontal eye field, a region traditionally thought to be the primary premotor cortical center for eye movements, also acts upon the neck musculature. Given its strong excitatory action on the SC, it would seem likely that such an action should exist but some groups have observed it (Tu and Keating, 2000) while others (Chen and Sparks, 2001) have not.

The vestibular nuclei are an essential component in organizing and driving stabilizing eye and head movements that compensate for externally imposed motions of the head. They do this both via powerful excitatory and inhibitory projections directly to extraocular and neck MNs and via projections to reticular neurons (EBN, IBN, RSN). Also shown is the direct inhibitory input from Purkinje cells of the flocculus and nodulus. Little is known about the latter but the former play a key role in relaying signals that allow animals to follow small moving targets (Lisberger et al., 1994). Smooth pursuit eye movements are mediated by flocculus-receiving neurons in the vestibular nuclei that exhibit eye–head, velocity type behavior (Zhang et al., 1995a,b). Some of these neurons appear to project to the neck, as well (Boyle et al., 1996), and may be one of the pathways that contributes to smooth pursuit head movements. There are also projections to and from the fastigial nucleus whose roles are poorly understood. One such role may be in organizing postural adjustments during such behaviors as locomotion.

As in the case of the VOR, there is evidence that the VCR is suppressed during gaze shifts. Part of this involves the pause in discharge of PVP neurons that was discussed. Many of these neurons bifurcate to project to both eye and neck motor pools (Isu et al., 1988). Therefore the portion of the VCR due to their action will be absent during the time of the gaze shift (but not the entire duration of the head movement). Several groups have also observed cancellation of the portion of vestibular input related to a voluntary head movement in other types of vestibular neurons (Boyle et al., 1996; McCrea et al., 1999; Roy and Cullen, 2001). These neurons do not cease discharging as do PVPs. Furthermore, they continue to be responsive to externally produced head movements. There is thus selective cancellation of just that portion of the vestibular signal produced by a voluntary head movement. This cancellation extends throughout the duration of the head movement, thus continuing beyond the end of the gaze shift. In Fig. 3, this cancellation is mediated by the box labeled 'head-movement estimator' (HME). This mechanism processes efference copy signals of motor commands from the SC and motor cortex channels, and head-movement feedback signals (HMS) to arrive at an estimate of the vestibular signal that would be produced by a voluntary head movement. This signal would then inhibit vestibular neurons. The cancellation appears to be less complete during slow tracking head movements (K. Cullen, personal communication), perhaps because the HME does not have access to signals arriving via the MST-cerebellar pathway. (MST is the abbreviation for the cortical visual area which senses image motion.)

Future directions

Compared to the wealth of knowledge available on oculomotor control, our knowledge of head-movement control is at a very early stage. Significant progress can be made by studying the head-movement system using paradigms developed for oculomotor studies. We need to know how neurons in each of the sites that project to the neck behave during several head-movement behaviors. Ideally, neural activity would be compared across eye–head gaze shifts, smooth head tracking, head stabilization and reaching for food with the mouth. Such studies will be much more valuable if the neurons investigated are identified (using electrical stimulation) as projecting to the neck, and perhaps also according to areas of the CNS, which project to them. This will be a major undertaking since it involves recordings from several cortical areas, the SC, three reticular regions [central mesencephalic reticular formation (CMRF), paramedian pontine reticular formation (PPRF), gigantocellularis], the medial, lateral and inferior vestibular nuclei and the cerebellar fastigial nucleus. It would be valuable to record from several such sites simultaneously in order to see how their activity shifts with the animal's behavior (see also Kobayashi et al., this volume).

Another approach that can be adopted from our oculomotor colleagues is the use of neural systems models like those used by Galiana and Guitton (1992). These are indispensable tools in understanding the function of neural networks like that shown in Fig. 3. We will also need models that deal with the complex kinematics and dynamics of the head–neck system. Studies reviewed above describe the start that has been made in this direction. Hopefully, we will eventually arrive at a neural network model that incorporates a realistic neck motor plant and emulates what is known about its neural control.

Construction of the latter type of model will depend upon an experimental approach that is not commonly employed in oculomotor studies. This involves correlation of the activity of neck muscles and specific classes of central neurons. Two methods are available. Spike-triggered averaging reveals direct actions of the test neuron on different neck motor pools. Correlation of neural and EMG activity reveals the relationship between them and how it changes across categories of behavior. Since the required analyses can be done off-line after data are obtained, they can provide valuable additional information about each neuron studied.

This is an ambitious agenda but the author believes that the effort will be worthwhile. As pointed out in Section 1, the head–neck system is in many ways an ideal model system for studying neural control of structures that have motor plants more complex than the very simple oculomotor plant. The presence of stereotyped behaviors such as gaze shifts and head stabilization has many advantages over the more complex behaviors that characterize arm movements. But, if necessary, one can also elicit more complex behaviors with the head that explore features such as load compensation during voluntary movements or anticipatory postural adjustments. Observations during free locomotion as pioneered by Mori and colleagues would also be very interesting (see S. Mori et al., Chapter 33 of this volume). Comprehensive understanding of motor behavior certainly requires knowledge of the control of axial muscle systems. The neck is an ideal place to start!

Acknowledgments

The author's work on head movements is supported by grant number NS 41288 from the National Institute of Neurological Disorders and Stroke.

Abbreviations

CCR	cervicocollic reflex
CMRF	central mesencephalic reticular formation
CNS	central nervous system
EBN	excitatory burst neuron
F-PCx	frontal and parietal cortex
HME	head-movement estimator
HMS	head-movement feedback signals
IBN	inhibitory burst neuron
lRST	mRST, lateral and medial reticulospinal tracts
MN	motoneuron
MST	cortical visual area sensing image motion
PPRF	paramedian pontine reticular formation
PVP	position-vestibular-pause (neuron)
RNI	resettable neural integrator
RSN	reticulospinal neuron
SC	superior colliculus
VCR	vestibulocollic reflex
VOR	vestibuloocular reflex

References

Alstermark, B., Pinter, M.J. and Sasaki, S. (1985). Pyramidal effects in dorsal neck motoneurones of the cat. J. Physiol. (Lond.), 363: 287–302.

Alstermark, B., Pinter, M.J. and Sasaki, S. (1992). Descending pathways mediating disynaptic excitation of dorsal neck motoneurones in the cat: brain stem relay. Neurosci. Res., 15: 42–57.

Angelaki, D.E., Green, A.M. and Dickman, J.D. (2001). Differential sensorimotor processing of vestibulo-ocular signals during rotation and translation. J. Neurosci., 21: 3968–3985.

Banovetz, J.M., Peterson, B.W. and Baker, J.F. (1995). Spatial coordination by descending vestibular signals. 1. Reflex excitation of neck muscles in alert and decerebrate cats. Exp. Brain Res., 105: 345–362.

Bilotto, G., Goldberg, J., Peterson, B.W. and Wilson, V.J. (1982). Dynamic properties of vestibular reflexes in the decerebrate cat. Exp. Brain Res., 47: 343–352.

Bizzi, E., Polit, A. and Morasso, P. (1976). Mechanisms underlying achievement of final head position. J. Neurophysiol., 39: 435–444.

Bolton, P.S., Goto, T., Schor, R.H., Wilson, V.J., Yamagata, Y. and Yates, B.J. (1992). Response of pontomedullary reticulospinal neurons to vestibular stimuli in vertical planes. Role in vertical vestibulospinal reflexes of the decerebrate cat. J. Neurophysiol., 67: 639–647.

Borel, L. and Lacour, M. (1992). Functional coupling of the stabilizing eye and head reflexes during horizontal and vertical linear motion in the cat. Exp. Brain Res., 91: 191–206.

Borel, L., Le Goff, B., Charade, O. and Berthoz, A. (1994). Gaze strategies during linear motion in head-free humans. J. Neurophysiol., 72: 2451–2466.

Boyle, R., Belton, T. and McCrea, R.A. (1996). Responses of identified vestibulospinal neurons to voluntary eye and head movements in the squirrel monkey. Ann. N.Y. Acad. Sci., 781: 244–263.

Ceylan, M., Henriques, D.Y., Tweed, D.B. and Crawford, J.D. (2000). Task-dependent constraints in motor control: pinhole goggles make the head move like an eye. J. Neurosci., 20: 2719–2730.

Chen, L.L. and Sparks, D.L. (2001). Initial head position modulates the direction and amplitude of gaze shifts evoked by microstimulation in the frontal eye field of head-unrestrained monkeys. Soc. Neurosci. Abstr., 27: 784.5.

Choi, H., Keshner, E. and Peterson, B.W. (2000) Primate neck muscle activation patterns during tracking in two postures. Soc. Neurosci. Abstr.

Dum, R.P. and Strick, P.L. (1996). Spinal cord terminations of the medial wall motor areas in macaque monkeys. J. Neurosci., 16: 6513–6525.

Eccles, J.C., Nicoll, R.A., Schwarz, W.F., Taborikova, H. and Willey, T.J. (1975). Reticulospinal neurons with and without monosynaptic inputs from cerebellar nuclei. J. Neurophysiol., 38: 513–530.

Fernandez, C. and Goldberg, J.M. (1971). Physiology of peripheral neurons innervating semicircular canals of the squirrel monkey. II. Response to sinusoidal stimulation and dynamics of peripheral vestibular system. J. Neurophysiol., 34: 661–675.

Freedman, E.G. and Sparks, D.L. (1997). Eye-head coordination during head-unrestrained gaze shifts in rhesus monkeys. J. Neurophysiol., 77: 2328–2348.

Fukushima, K. (1987). The interstitial nucleus of Cajal and its role in the control of movements of head and eyes. Prog. Neurobiol., 29: 107–192.

Galiana, H.L. and Guitton, D. (1992). Central organization and modeling of eye-head coordination during orienting gaze shifts. Ann. N.Y. Acad. Sci., 656: 452–471.

Goldberg, J. and Peterson, B.W. (1986). Reflex and mechanical contributions to head stabilization in alert cats. J. Neurophysiol., 56: 857–875.

Goossens, H.H. and Van Opstal, A.J. (1997). Human eye-head coordination in two dimensions under different sensorimotor conditions. Exp. Brain Res., 114: 542–560.

Grantyn, A. and Berthoz, A. (1985). Burst activity of identified tecto-reticulo-spinal neurons in the alert cat. Exp. Brain Res., 57: 417–421.

Grantyn, A. and Berthoz, A. (1987). Reticulo-spinal neurons participating in the control of synergic eye and head movements during orienting in the cat. I. Behavioral properties. Exp. Brain Res., 66: 339–354.

Grantyn, A., Ong-Meang Jacques, V. and Berthoz, A. (1987). Reticulo-spinal neurons participating in the control of synergic eye and head movements during orienting in the cat. II. Morphological properties as revealed by intra-axonal injections of horseradish peroxidase. Exp. Brain Res., 66: 355–377.

Grantyn, A., Berthoz, A., Hardy, O. and Gourdon, A. (1992a). Contributions of reticulospinal neurons to the dynamic control of head movements: presumed neck bursters. In: Berthoz A., Graf W. and Vidal P.P. (Eds.), The Head-Neck Sensory-Motor System. Oxford University Press, New York, pp. 318–329.

Grantyn, A., Hardy, O., Olivier, E. and Gourdon, A. (1992b). Relationships between task-related discharge patterns and axonal morphology of brainstem projection neurons involved in orienting eye and head movements. In: Shimazu H. and Shinoda Y. (Eds.), Vestibular and Brainstem Control of Eye and Head and Body Movements. Japan Scientific Society Press, Tokyo, pp. 255–273.

Guitton, D. and Volle, M. (1987). Gaze control in humans: eye-head coordination during orienting movements to targets within and beyond the oculomotor range. J. Neurophysiol., 58: 427–459.

Guitton, D., Douglas, R.M. and Volle, M. (1984). Eye-head coordination in cats. J. Neurophysiol., 52: 1030–1050.

He, S.Q., Dum, R.P. and Strick, P.L. (1993). Topographic organization of corticospinal projections from the frontal lobe: motor areas on the lateral surface of the hemisphere. J. Neurosci., 13: 952–980.

He, S.Q., Dum, R.P. and Strick, P.L. (1995). Topographic organization of corticospinal projections from the frontal lobe: motor areas on the medial surface of the hemisphere. J. Neurosci., 15: 3284–3306.

Highstein, S.M., Goldberg, J.M., Moschovakis, A.K. and Fernandez, C. (1987). Inputs from regularly and irregularly discharging vestibular nerve afferents to secondary neurons in the vestibular nuclei of the squirrel monkey. II. Correlation with output pathways of secondary neurons. J. Neurophysiol., 58: 719–738.

Isu, N., Uchino, Y., Nakashima, H., Satoh, S., Ichikawa, T. and Watanabe, S. (1988). Axonal trajectories of posterior canal-activated secondary vestibular neurons and their coactivation of extraocular and neck flexor motoneurons in the cat. Exp. Brain Res., 70: 181–191.

Keshner, E.A. and Peterson, B.W. (1988). Motor control strategies underlying head stabilization and voluntary head movements in humans and cats. Prog. Brain Res., 76: 329–339.

Keshner, F.A. and Peterson, B.W. (1995). Mechanisms controlling human head stabilization. I. Head-neck dynamics during random rotations in the horizontal plane. J. Neurophysiol., 73: 2293–2301.

Keshner, E.A., Campbell, D., Katz, R.T. and Peterson, B.W. (1989). Neck muscle activation patterns in humans during isometric head stabilization. Exp. Brain Res., 75: 335–344.

Keshner, E.A., Baker, J.F., Banovetz, J. and Peterson, B.W. (1992). Patterns of neck muscle activation in cats during reflex and voluntary head movements. Exp. Brain Res., 88: 361–374.

Keshner, E.A., Cromwell, R.L. and Peterson, B.W. (1995). Mechanisms controlling human head stabilization. II. Head-neck characteristics during random rotations in the vertical plane. J. Neurophysiol., 73: 2302–2312.

Kitama, T., Grantyn, A. and Berthoz, A. (1995a). Orienting-related eye-neck neurons of the medial ponto-bulbar reticular formation do not participate in horizontal canal-dependent vestibular reflexes of alert cats. Brain Res. Bull., 38: 337–347.

Kitama, T., Ohki, Y., Shimazu, H., Tanaka, M. and Yoshida, K. (1995b). Site of interaction between saccade signals and vestibular signals induced by head rotation in the alert cat: functional properties and afferent organization of burster-driving neurons. J. Neurophysiol., 74: 273–287.

Lacour, M. and Borel, L. (1989). Functional coupling of the stabilizing gaze reflexes during vertical linear motion in the alert cat. Prog. Brain Res., 80: 385–394 discussion 373–375.

Lisberger, S.G., Pavelko, T.A. and Broussard, D.M. (1994). Responses during eye movements of brain stem neurons that receive monosynaptic inhibition from the flocculus and ventral paraflocculus in monkeys. J. Neurophysiol., 72: 909–927.

May, P.J., Hartwich-Young, R., Nelson, J., Sparks, D.L. and Porter, J.D. (1990). Cerebellotectal pathways in the macaque: implications for collicular generation of saccades. Neuroscience, 36: 305–324.

McCrea, R.A., Gdowski, G.T., Boyle, R. and Belton, T. (1999). Firing behavior of vestibular neurons during active and passive head movements: vestibulo-spinal and other non-eye-movement related neurons. J. Neurophysiol., 82: 416–428.

Munoz, D.P. and Guitton, D. (1985). Tectospinal neurons in the cat have discharges coding gaze position error. Brain Res., 341: 184–188.

Munoz, D.P. and Wurtz, R.H. (1995). Saccade-related activity in monkey superior colliculus. II. Spread of activity during saccades. J. Neurophysiol., 73: 2334–2348.

Munoz, D.P., Pelisson, D. and Guitton, D. (1991). Movement of neural activity on the superior colliculus motor map during gaze shifts. Science, 251: 1358–1360.

Optican, L.M. and Quaia, C. (2002). Distributed model of collicular and cerebellar function during saccades. Ann. N.Y. Acad. Sci., 956: 164–177.

Pellionisz, A. (1988). Tensorial aspects of the multidimensional massively parallel sensorimotor function of neuronal networks. Prog. Brain Res., 76: 341–354.

Peng, G.C., Hain, T.C. and Peterson, B.W. (1996a). A dynamical model for reflex activated head movements in the horizontal plane. Biol. Cybern., 75: 309–319.

Peng, G.C., Hain, T.C. and Peterson, B.W. (1996b) How is the head held up? Modeling mechanisms for head stability in the sagittal plane. Proc. 18th Annu. Int. Conf. IEEE Eng. Med. Biol. Soc., Amsterdam.

Perlmutter, S.I., Iwamoto, Y., Baker, J.F. and Peterson, B.W. (1998a). Interdependence of spatial properties and projection patterns of medial vestibulospinal tract neurons in the cat. J. Neurophysiol., 79: 270–284.

Perlmutter, S.I., Iwamoto, Y., Barke, L.F., Baker, J.F. and Peterson, B.W. (1998b). Relation between axon morphology in C1 spinal cord and spatial properties of medial vestibulospinal tract neurons in the cat. J. Neurophysiol., 79: 285–303.

Peterson, B.W. and Abzug, C. (1975). Properties of projections from vestibular nuclei to medial reticular formation in the cat. J. Neurophysiol., 38: 1421–1435.

Peterson, B.W. and Felpel, L.P. (1971). Excitation and inhibition of reticulospinal neurons by vestibular, cortical and cutaneous stimulation. Brain Res., 27: 373–376.

Peterson, B.W., Anderson, M.E. and Filion, M. (1974). Responses of ponto-medullary reticular neurons to cortical, tectal and cutaneous stimuli. Exp. Brain Res., 21: 19–44.

Peterson, B.W., Pitts, N.G., Fukushima, K. and Mackel, R. (1978). Reticulospinal excitation and inhibition of neck motoneurons. Exp. Brain Res., 32: 471–489.

Peterson, B.W., Fukushima, K., Hirai, N., Schor, R.H. and Wilson, V.J. (1980). Responses of vestibulospinal and reticulospinal neurons to sinusoidal vestibular stimulation. J. Neurophysiol., 43: 1236–1250.

Peterson, B.W., Goldberg, J., Bilotto, G. and Fuller, J.H. (1985). Cervicocollic reflex: its dynamic properties and interaction with vestibular reflexes. J. Neurophysiol., 54: 90–109.

Robinson, D.A. (1982). The use of matrices in analyzing the three-dimensional behavior of the vestibulo-ocular reflex. Biol. Cybern., 46: 53–66.

Robinson, F.R., Phillips, J.O. and Fuchs, A.F. (1994). Coordination of gaze shifts in primates: brainstem inputs to neck and extraocular motoneuron pools. J. Comp. Neurol., 346: 43–62.

Ron, S., Berthoz, A. and Gur, S. (1993). Saccade-vestibulo-ocular reflex co-operation and eye-head uncoupling during orientation to flashed target. J. Physiol. (Lond.), 464: 595–611.

Roy, J.E. and Cullen, K.E. (1998). A neural correlate for vestibulo-ocular reflex suppression during voluntary eye-head gaze shifts. Nat. Neurosci., 1: 404–410.

Roy, J.E. and Cullen, K.E. (2001). Selective processing of vestibular reafference during self-generated head motion. J. Neurosci., 21: 2131–2142.

Stahl, J.S. (1999). Amplitude of human head movements associated with horizontal saccades. Exp. Brain Res., 126: 41–54.

Strassman, A., Highstein, S.M. and McCrea, R.A. (1986a). Anatomy and physiology of saccadic burst neurons in the alert squirrel monkey. I. Excitatory burst neurons. J. Comp. Neurol., 249: 337–357.

Strassman, A., Highstein, S.M. and McCrea, R.A. (1986b). Anatomy and physiology of saccadic burst neurons in the alert squirrel monkey. II. Inhibitory burst neurons. J. Comp. Neurol., 249: 358–380.

Tabak, S., Smeets, J.B. and Collewijn, H. (1996). Modulation of the human vestibuloocular reflex during saccades: probing by high-frequency oscillation and torque pulses of the head. J. Neurophysiol., 76: 3249–3263.

Terzuolo, C.A., Llinas, R. and Green, K.T. (1965). Mechanisms of supraspinal actions upon spinal cord activities: distribution of reticular and segmental inputs in cat's alpha-motoneurons. Arch. Ital. Biol., 103: 635–651.

Tomlinson, R.D. and Bahra, P.S. (1986a). Combined eye-head gaze shifts in the primate. I. Metrics. J. Neurophysiol., 56: 1542–1557.

Tomlinson, R.D. and Bahra, P.S. (1986b). Combined eye-head gaze shifts in the primate. II. Interactions between saccades and the vestibuloocular reflex. J. Neurophysiol., 56: 1558–1570.

Tu, T.A. and Keating, E.G. (2000). Electrical stimulation of the frontal eye field in a monkey produces combined eye and head movements. J. Neurophysiol., 84: 1103–1106.

Vasavada, A.N., Peterson, B.W. and Delp, S.L. (2002). Three-dimensional spatial tuning of neck muscle activation in humans. Exp. Brain Res., 147: 437–448.

Waitzman, D.M., Silakov, V.L. and Cohen, B. (1996). Central mesencephalic reticular formation (cMRF) neurons discharging before and during eye movements. J. Neurophysiol., 75: 1546–1572.

Waitzman, D.M., Silakov, V.L., DePalma-Bowles, S. and Ayers, A.S. (2000). Effects of reversible inactivation of the primate mesencephalic reticular formation. I. Hypermetric goal-directed saccades. J. Neurophysiol., 83: 2260–2284.

Waitzman, D.M., Pathmanathan, J., Presnell, R., Ayers, A. and DePalma, S. (2002). Contribution of the superior colliculus and the mesencephalic reticular formation to gaze control. Ann. N.Y. Acad. Sci., 956: 111–129.

Wilson, V.J., Peterson, B.W., Fukushima, K., Hirai, N. and Uchino, Y. (1979). Analysis of vestibulocollic reflexes by sinusoidal polarization of vestibular afferent fibers. J. Neurophysiol., 42: 331–346.

Zangemeister, W.H. and Stark, L. (1982). Types of gaze movement: variable interactions of eye and head movements. Exp. Neurol., 77: 563–577.

Zangemeister, W.H., Jones, A. and Stark, L. (1981). Dynamics of head movement trajectories: main sequence relationship. Exp. Neurol., 71: 76–91.

Zangemeister, W.H., Stark, L., Meienberg, O. and Waite, T. (1982). Neural control of head rotation: electromyographic evidence. J. Neurol. Sci., 55: 1–14.

Zhang, Y., Partsalis, A.M. and Highstein, S.M. (1995a). Properties of superior vestibular nucleus flocculus target neurons in the squirrel monkey. I. General properties in comparison with flocculus projecting neurons. J. Neurophysiol., 73: 2261–2278.

Zhang, Y., Partsalis, A.M. and Highstein, S.M. (1995b). Properties of superior vestibular nucleus flocculus target neurons in the squirrel monkey. II. Signal components revealed by reversible flocculus inactivation. J. Neurophysiol., 73: 2279–2292.

CHAPTER 36

The neural control of orienting: role of multiple-branching reticulospinal neurons

Shigeto Sasaki*, Kazuya Yoshimura and Kimisato Naito

Department of Neurophysiology, Tokyo Metropolitan Institute for Neuroscience, Tokyo 183-8526, Japan

Abstract: This chapter emphasizes the functional significance of the multiple-branching patterns of descending axons implicated in the control of movement. The example provided concerns orienting head movements, which are controlled by pathways from the superior colliculus (SC). Such control is mediated via cervical reticulospinal neurons (C-RSNs), which take origin in the nucleus reticularis pontis caudalis and nucleus reticularis gigantocellularis, and give off multiple collaterals along the full length of their axonal trajectory. Their projection is not only to lamina IX neck motor nuclei in upper cervical segments, but also to laminae VII–VIII in lower cervical segments. Thus, SC commands for head orienting are transmitted to both neck motoneurons and lower cervical spinal circuitry, which latter network controls appropriate postural adjustments by the coordinated control of motoneurons supplying the four limbs.

Introduction

Descending axons conveying motor command signals from supraspinal structures to spinal segmental centers give off multiple axon collaterals throughout their trajectory. These have been both visualized by intra-axonal staining and inferred from electrophysiological analysis of the terminal fields of antidromic spikes (Shinoda et al., 1986; Grantyn and Berthoz, 1987; Grantyn et al., 1987; Iwamoto et al., 1990; Isa and Sasaki, 1992a; Kakei et al., 1994; Matuyama et al., 1997; Sasaki and Iwamoto, 1999). Far less attention has been directed, however, to the *precise* functional significance of this extensive axonal branching.

In this chapter, we focus on such functional implications in the multiple branching within a descending tract comprising reticulospinal neurons (RSNs). These neurons are optimal for this purpose because their basic circuitry has been extensively studied in relation to orienting head movements.

*Corresponding author: Tel.: +81-0423-25-3881 (ext. 4216); Fax: +81-0423-21-8678; E-mail: shigeto@tmin.ac.jp

Pathways controlling the orienting of head movements

Figure 1A shows the principal pathways from the superior colliculus (SC) to neck motoneurons participating in the control of orienting head movements. The SC is a primary center for orienting (Hess, 1956; Sprague and Meikel, 1965). We have shown that there are two separate RSN pathways from the SC to neck motoneurons (Alstermark et al., 1985, 1992a,b,c; Isa and Sasaki, 1988, 1992a,b; Iwamoto and Sasaki, 1990; Iwamoto et al., 1990). One is by way of the nucleus reticularis pontis caudalis (NRPC) and the nucleus reticularis gigantocellularis (NRG). Figure 1A shows that by far the majority of RSNs projecting to neck motoneurons (termed cervical RSNs; C-RSNs) via this pathway continue their projection to the lower cervical (brachial segmental level) cord. They tend not to extend to the lumbar cord, however. This pathway is mainly involved in controlling head movements in the horizontal direction. It activates motoneurons that supply neck muscles such as the splenius, biventer cervicis and biventer complexus muscles.

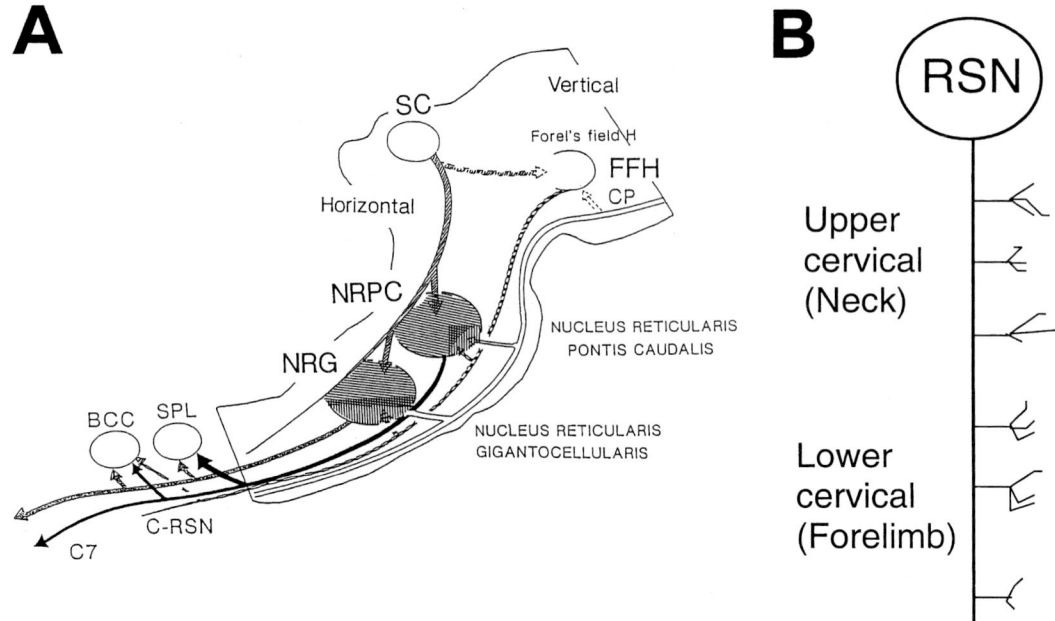

Fig. 1. Overview of the anatomical substrate for this chapter. (A) Schematic diagram of circuitry controlling orienting head movements. Critical circuitry is from the superior colliculus to cervical nuclei of neck motoneurons in the cat. Terms and abbreviations from rostral to caudal: Vertical, source of control of vertical head movements; FFH, H field of Forel; CP, cerebral peduncle; SC, superior colliculus; Horizontal, source of control of horizontal head movements; SPL, motor nucleus supplying splenius muscle; BCC, motor nuclei supplying biventer cervicus and biventer complex muscles; C-RSN, cervical reticulospinal neuron; C7, cervical segment 7. (B) Schematic of an exemplary C-RSN axon giving off multiple branches in the upper and lower cervical cord. (Reproduced from Sasaki et al., 1996 with permission from Elsevier Science.)

The other pathway is by way of the H field of Forel (FFH). As shown in Fig. 1A, the FFH neurons either project monosynaptically to neck motoneurons, or disynaptically via the NRG (Holstage and Cowie, 1989; Isa and Sasaki, 1992a,b). This pathway is mainly involved in controlling head movements in the vertical direction. It activates motoneurons that supply the biventer cervicis and biventer complexus muscles, but not the splenius muscle.

Figure 1B shows that the C-RSN axons that extend into the lower cervical cord give off further collaterals. Examples of these are shown in Fig. 2, with intra-axonal staining showing that they project mainly to neck motor nuclei (lamina IX), and far less into adjacent laminae VIII and VII (Iwamoto et al., 1988; Kakei et al., 1994; Sasaki and Iwamoto, 1999).

The descending projections of C-RSNs to the cervical brachial segment have also been studied by injecting HRP at the C6–7 segments (Fig. 3A; Sasaki, 1997). Figure 3B shows superimposed traces of six axons identified as C-RSNs. Their stem axons descend in the ventral funiculus close to the gray matter, where they give off many collaterals to laminae V–VIII. The axons terminate largely in VI–VIII, with very few terminals in the motor nuclei of lamina IX (Fig. 3C).

The results showed that the *same* orienting commands are transmitted to *both* neck and forelimb motoneurons, with the latter mainly via spinal interneurons.

Relation between orienting and forelimb kinetics

To test for the potential orienting function of C-RSN projections to forelimb motoneurons, we first analyzed forelimb kinetics during orienting movements. We trained cats to fixate a light spot back-projected on a tangent screen placed about 24 cm ahead of them and to follow the light spot as it

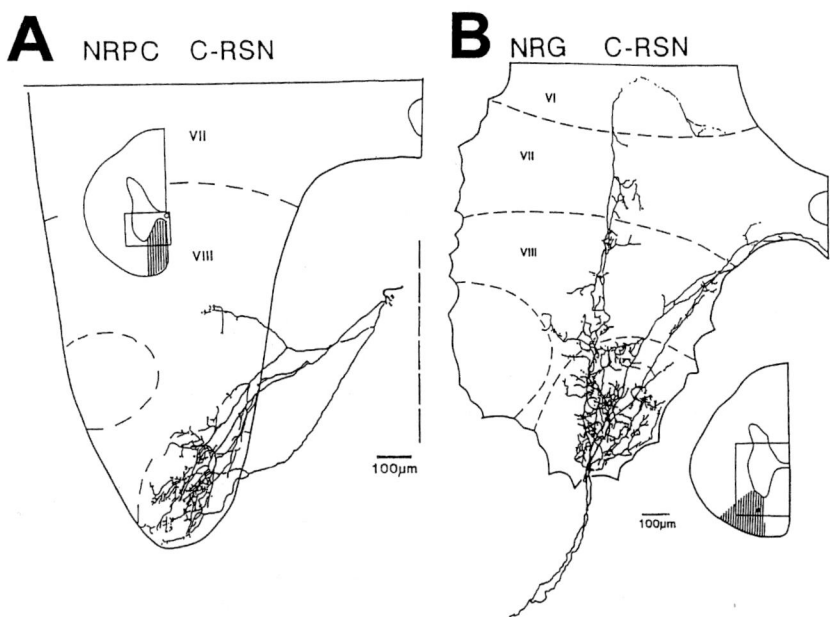

Fig. 2. Arborization of two RSN axons. Axonal trajectory at the upper (C2–3) level of the cervical spinal cord of one C-RSN from the NRPC (A) and one from the NRG (B). The axons of these neurons were identified by antidromic activation from the C7 (but not L2) segment, and by monosynaptic activation from the contralateral tectum and medullary pyramid. The cells' axons were then injected with HRP and visualized subsequently using the DAB method. Note Rexed's laminae VI–VIII. Rectangles in the insets show the displayed areas at lower magnification. (Reproduced from Sasaki et al., 1996 with permission from Elsevier Science.)

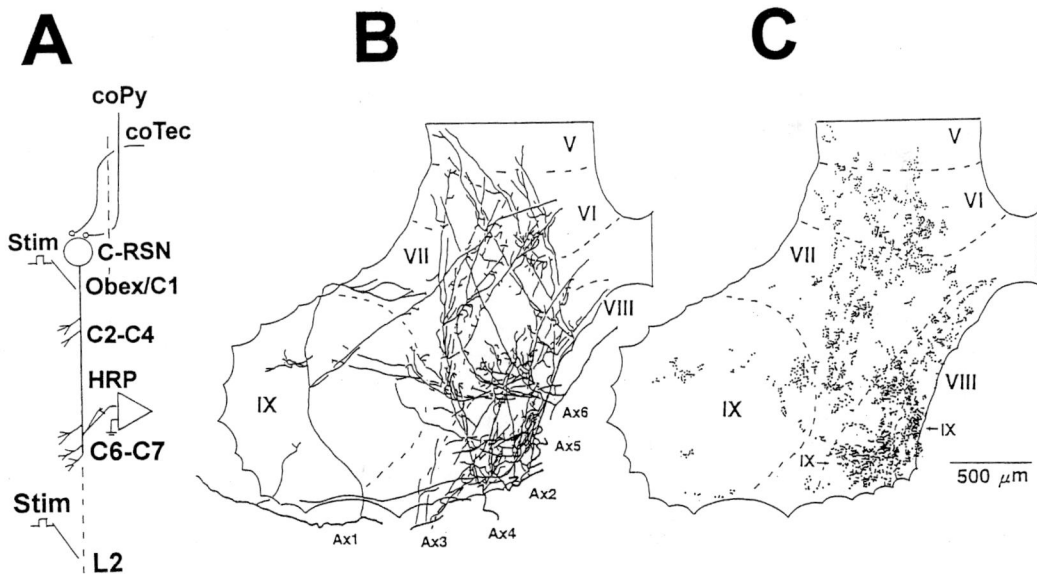

Fig. 3. Axonal trajectory of C-RSNs in the lower cervical cord (C7–C8). (A) Technique for identifying C-RSNs. Their impaled axons were shown by direct activation from the C1 (but not L2) segment and by monosynaptic activation from the contralateral pyramidal and tectum. (B). Superimposed traces of six axons. (C). Distribution of terminal boutons of the same six axons. (Reproduced from Sasaki, 1997 with permission from Elsevier Science.)

jumped from a fixation point to a target position. Figure 4A and C shows that for this task, the cat first moves the head slowly toward the target with a contraversive eye movement caused by the vestibulo-ocular reflex. This keeps the gaze in a fixed position in space. The cat then makes a saccade to the target, followed by a quick head movement. The gaze shift completes within 100 ms after the onset of the saccade. The head movement continues even after completion of the gaze shift. As the head keeps turning, the vestibulo-ocular reflex moves the eye toward the center of the orbit. A frame-by-frame analysis of such orienting movements reveals that the trunk, shoulder and a proximal part of the forelimbs move almost synchronously with the head (viz. Fig. 4A). The limb movements appear to facilitate the postural adjustments associated with the head movement.

The near-simultaneous onset of head and limb movements was further verified by recording the ground reaction forces exerted by the four limbs during orienting (Fig. 4B–C). The load in the right forelimb (orienting side) begins to decrease as the head moves. This load reduction is initially slight but it then increases rapidly, being synchronized with the saccade. The load then remains low for a time, before increasing rapidly, in phase with the shift of body axis toward the target (cf. Fig. 4A). The profile of the load change in the contralateral (left) forelimb is the inverse, near-mirror-image of the right forelimb. Likewise, the hindlimb load-changes (not shown in Fig. 4) mirrors those of the forelimbs, the right (orienting)-side one increasing and the left (contralateral)-side one decreasing.

This synchronization of limb load changes with head movement is observed in the majority of trials. It could be caused by C-RSNs transmitting the motor command to both neck and forelimb motoneurons. This possibility was tested by stimulating the medial half of NRPC and NRG where a majority of C-RSNs projecting to neck motoneurons originate (Iwamoto et al., 1990, cf. Fig. 1). Electrical stimulation of the reticular formation moves the head (in the horizontal direction), trunk and limbs in virtually the same

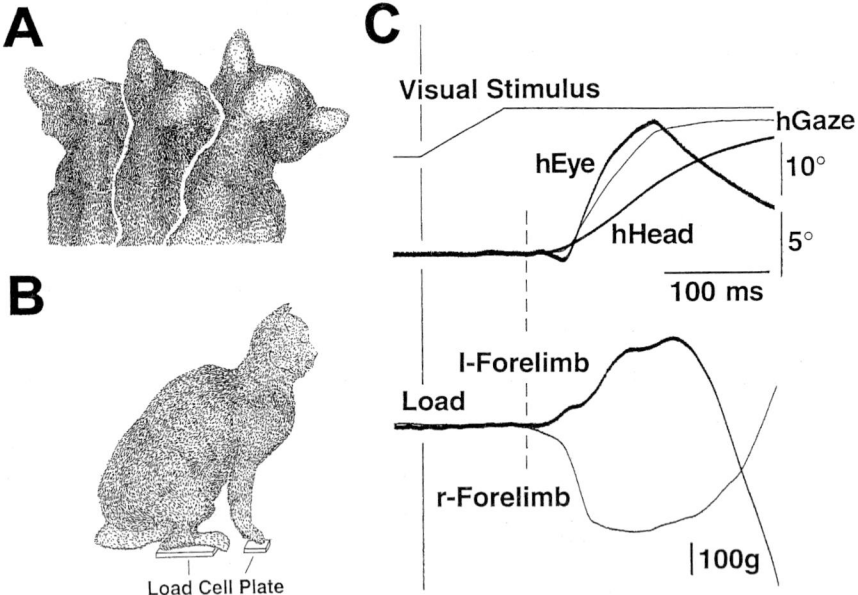

Fig. 4. Phase relations between orienting and load changes in the forelimbs. (A) Drawing of the head and upper trunk's movements during orienting. (B) Method of measuring ground reaction forces of the limbs. Cats were trained to stand on platforms, under which load cells were placed. (C) Simultaneous recording of orienting movements in the horizontal (h) plane and load change of the left (l) and right (r) forelimbs. Vertical dashed line indicates the onset of the head movement. The orienting movements are those of the gaze in space, eye in the orbit (i.e., gaze–head) and head in space. The head movement and gaze shift were measured by a magnetic search coil technique.

temporal pattern as in visually guided orienting. Hindlimb loads change almost synchronously with those of forelimbs (not indicated). We assume that this is produced by simultaneous activation of RSNs projecting to the lumbar segments (L-RSNs), where they are intermingled with C-RSNs (Iwamoto et al., 1990).

The above morphological, ground-reaction, and electrical stimulation results all strongly suggest that the same C-RSNs are involved in both orienting head movements and its requisite postural adjustments.

Firing pattern of C-RSNs during orienting

Previously, RSN neurons have been classified as phasic neurons versus phasic-sustained neurons on the basis of their firing patterns during head movement (Grantyn and Berthoz, 1985, 1987; Grantyn et al., 1987, 1992; Isa and Naito, 1995; Sasaki et al., 1996). Figure 5A shows that this classification is also appropriate for the presently studied C-RSNs. The phasic type fire in brief bursts associated with saccades at a lead time < 10 ms. This firing pattern closely resembles that of selected premotor neurons projecting to ocular motoneurons, i.e., ones termed medium lead burst neurons (Keller, 1974).

Figure 5B shows that in contrast, the phasic-sustained type of C-RSN discharges briefly in response to the visual stimulus at a latency of ∼40 ms. Thereafter, it ceases firing for a brief period, and then recommences discharge in a burst of firing just before the saccade. Subsequently, it displays a low rate of firing, which continues until the end of the head movement. The overall frequency profile of phasic-sustained discharge closely resembles the profile of the head movement, itself.

The Fig. 5 data also suggest that the phasic-sustained type of neuron commands the initial small change in forelimb load, whereas the subsequent large change in load is controlled by both phasic and phasic-sustained neurons.

Summary and comment

There is strong evidence to suggest that axon collaterals of C-RSNs within the lower cervical spinal cord participate in the control of the postural adjustments required during orienting head movements. We presume that this control involves activating spinal interneuronal circuitry that can

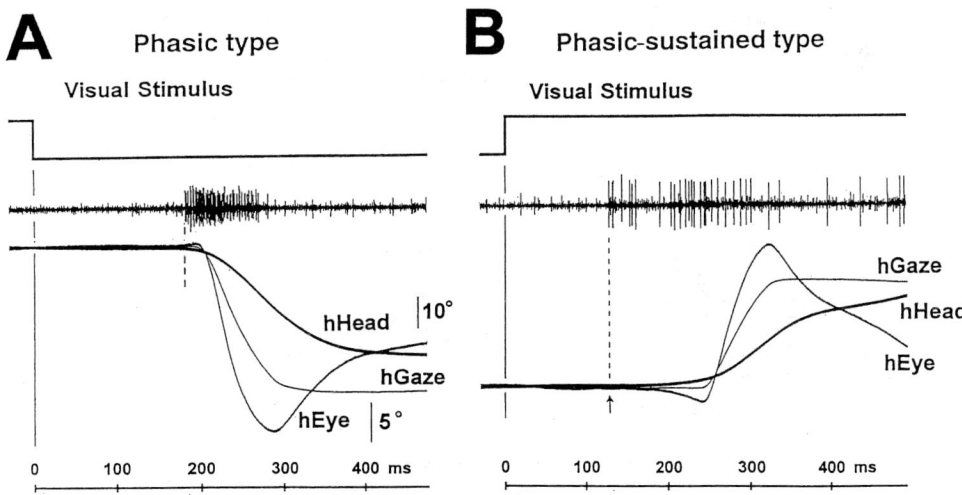

Fig. 5. Behavior of exemplary RSNs during orienting. Their discharge is shown relative to head, gaze and eye movements in the horizontal (h) direction. Note difference in the firing pattern of the two major subtypes of RSNs: phasic (A) versus phasic-sustained (B). Calibration bar: 5° for eye movements; 10° for head and gaze movements.

produce coordinated movements among the four limbs, i.e., circuitry largely in laminae in VI–VIII (Baldissera et al., 1981). This region is also known to receive inputs from vestibular and primary sensory afferents. Thus, the orienting postural adjustments appear to require a spinal interneuronal integration of several descending command signals, including those carried by C-RSNs, and perhaps also local spinal pattern-generating limb coordination activity, and sensory feedback.

Orienting head movements seem to be achieved by a coordinated activation of many neck muscles acting on the cervical vertebrae, head and limb girdles. Appropriate for this purpose are the wide and multiple axon collaterals of C-RSNs at the upper level of the cervical spinal cord.

In conclusion, it is important to delineate the functional role of the multiple collaterals of CNS axons along the full length of their trajectory (cf., present chapters of Matsuyama and Grantyn). In this article, we have provided a clear-cut example of the functional value of such branching, by describing how C-RSNs can participate in the control of the coordination of head orienting and limb posture.

Acknowledgments

This study was supported by grants from the Ministry of Education, Science, Sports and Culture of Japan.

Abbreviations

C-RSN	cervical reticulospinal neuron
FFH	H field of Forel (Forel's field H)
NRG	nucleus reticularis gigantocellularis
NRPC	nucleus reticularis pontis caudalis
RSN	reticulospinal neuron
SC	superior colliculus

References

Alstermark, B., Pinter, M.A. and Sasaki, S. (1985). Pyramidal effects in dorsal neck motoneurones of the cat. J. Physiol. (Lond.), 363: 287–302.

Alstermark, B., Pinter, M. and Sasaki, S. (1992a). Tectal and tegmental excitation in dorsal neck motoneurones of the cat. J. Physiol. (Lond.), 454: 517–532.

Alstermark, B., Pinter, M. and Sasaki, S. (1992b). Descending pathways mediating excitation of dorsal neck motoneurones in the cat: facilitatory interactions. Neurosci. Res., 15: 32–41.

Alstermark, B., Pinter, M.A. and Sasaki, S. (1992c). Descending pathways mediating excitation of dorsal neck motoneurones in the cat: brain stem relay. Neurosci. Res., 15: 42–57.

Baldissera, F., Hultborn, H. and Illert, M. (1981). Integration in spinal interneuronal systems. In: Brooks V.E. (Ed.), Handbook of Physiology, Sec. I, The Nervous System, Vol. II, Motor Control, American Physiological Society, Bethesda, pp. 509–595.

Grantyn, A. and Berthoz, A. (1985). Burst activity of identified reticulospinal neurons in the alert cat. Exp. Brain Res., 57: 417–421.

Grantyn, A. and Berthoz, A. (1987). Reticulospinal neurons participating in the control of synergic eye and head movements during orienting in the cat. I. Behavioral properties. Exp. Brain Res., 66: 339–354.

Grantyn, A., Ong-Meang Jacques, V. and Berthoz, A. (1987). Reticulospinal neurons participating in the control of synergic eye and head movements during orienting in the cat. II. Morphological properties as revealed by intra-axonal injections of horseradish peroxidase. Exp. Brain Res., 66: 355–377.

Grantyn, A., Hardy, O., Olivier, E. and Gourdon, A. (1992). Relationships between task-related discharge patterns and axonal morphology of brainstem projection neurons involved in orienting eye and head movements. In: Shimazu H. and Shinoda Y. (Eds.), Vestibular and Brainstem Control of Eye, Head and Body Movements. Japan Scientific Society Press and S. Karger, Tokyo, Basel, pp. 255–273.

Hess, W.R. (1956). Hypothalamus and Thalamus. Documentary Pictures. Georg Thieme, Stuttgart.

Holstage, G. and Cowie, R.J. (1989). Projection from the rostral mesencephalic reticular formation to the spinal cord. An HRP and autoradiographical tracing study in the cat. Exp. Brain Res., 75: 265–279.

Isa, T. and Naito, K. (1995). Activity of neurons in the medial pontomedullary reticular formation during orienting movements in alert head-free cats. J. Neurophysiol., 74: 73–95.

Isa, T. and Sasaki, S. (1988). Effects of lesion of paramedian pontomedullary reticular formation by kainic acid injection on the visually triggered horizontal orienting movements in the cat. Neurosci. Lett., 87: 233–239.

Isa, T. and Sasaki, S. (1992a). Descending projection of Forel's field H neurones to the brain stem and the upper cervical spinal cord in the cat. Exp. Brain Res., 88: 563–579.

Isa, T. and Sasaki, S. (1992b). Mono- and disynaptic pathways from Forel's field H to dorsal neck motoneurones in the cat. Exp. Brain Res., 88: 580–593.

Iwamoto, Y. and Sasaki, S. (1990). Monosynaptic excitatory connexions of reticulospinal neurones in the nucleus reticularis pontis caudalis with dorsal neck motoneurones in the cat. Exp. Brain Res., 80: 227–289.

Iwamoto, Y., Sasaki, S. and Suzuki, I. (1988). Descending cortical and tectal control of dorsal neck motoneurones via reticulospinal neurons in the cat. Prog. Brain Res., 76: 97–108.

Iwamoto, Y., Sasaki, S. and Suzuki, I. (1990). Input-output organization of reticulospinal neurones, with special reference to connexion with neck motoneurones in the cat. Exp. Brain Res., 80: 260–276.

Kakei, S., Muto, N. and Shinoda, Y. (1994). Innervation of multiple neck motor nuclei by single reticulospinal tract axons receiving tectal input in the upper cervical cord. Neurosci. Lett., 172: 85–88.

Keller, E.L. (1974). Participation of the medial reticular formation in eye movement generation in monkey. J. Neurophysiol., 37: 316–332.

Matuyama, K., Takakusaki, K., Nakajima, K. and Mori, S. (1997). Multisegmental innervation of single pontine reticulospinal axons in the cervico-thoratic region of the cat anterograde PHA-L tracing study. J. Comp. Neurol., 377: 234–250.

Sasaki, S. (1997). Axonal branching and termination of cervical reticulospinal neurons in the cat brachial segments. Neurosci. Lett., 228: 83–86.

Sasaki, S. and Iwamoto, Y. (1999). Axonal trajectories of the nucleus reticularis gigantocellularis neurons in the C2–C3 segments in cats. Neurosci. Lett., 264: 137–140.

Sasaki, S., Naito, K. and Oka, M. (1996). Firing characteristics of neurones in the superior colliculus and the pontomedullary reticular formation during orienting in unrestrained cats. Prog. Brain Res., 112: 99–115.

Shinoda, Y., Yamaguchi, T. and Futami, T. (1986). Multiple axon collaterals of single corticospinal axons in the cat spinal cord. J. Neurophysiol., 55: 425–448.

Sprague, J.M. and Meikel, T.H. (1965). The role of the superior colliculus in visually guided orienting in the cat. Exp. Neurol., 11: 114–146.

CHAPTER 37

Role of the frontal eye fields in smooth-gaze tracking

Kikuro Fukushima*, Takanobu Yamanobe, Yasuhiro Shinmei
Junko Fukushima and Sergei Kurkin

Department of Physiology, Hokkaido University School of Medicine, Sapporo 060-8638, Japan

Abstract: Visual and vestibular senses are essential for appropriate motor behavior in three-dimensional (3D) space. Discovery of relevant specific subdivisions in sensory and motor pathways in recent decades has considerably advanced our understanding of the overall neural control of movement. Such subdivisions must eventually be further delineated into functional neural circuits for purposeful motor acts. Two critical questions are where in the brain do such circuits operate, and by what means. In this chapter, these issues are addressed for smooth tracking eye-movement systems in the simian. These results show that contrary to current understanding, synthesis of the functionally similar eye-movement systems, smooth-pursuit and vergence, takes place in the frontal cortex. This processing, which is of higher order than previously supposed, enables primates to track and manipulate objects moving in 3D space with the utmost of efficiency.

Introduction

With the development of a high-acuity fovea, frontal-eyed primates acquire the ability to coordinate binocular eye movements in order to accurately track small objects moving in space. The smooth-pursuit system moves both eyes in the same direction to track target movement in the frontal plane, whereas the vergence system moves left and right eyes in the opposite direction to track a target moving toward or away from the observer. Understanding how these two systems are linked in the brain is a critical issue in advancing understanding of both the neural control of movement and how target movement in space is represented centrally. This is also important in advancing understanding of how primates maintain their stable balance during both standing and walking while simultaneously tracking and manipulating objects which are moving in three-dimensional (3D) space.

Smooth-gaze tracking system: smooth-pursuit and vergence eye movements

Current findings indicate that the smooth-pursuit and vergence systems are separate not only in their cortical and brainstem pathways (for review, see Mays et al., 1986; Gamlin, 1999; Leigh and Zee, 1999; Gamlin and Yoon, 2000), but also in their retinal inputs. Smooth-pursuit is driven primarily by visual signals related to the velocity of image-slip on the retina, whereas vergence tracking is driven by the disparity of images on the two retinas (i.e., position-related signals; Rashbus and Westheimer, 1961; Mays and Gamlin, 1995).

Despite the above differences, smooth-pursuit and vergence systems have some functional similarities. For example, they do not operate independently but, rather, they interact with the vestibular system in

---

*Corresponding author: Tel.: +81-11-706-5038;
Fax: +81-11-706-5041; E-mail: kikuro@med.hokudai.ac.jp

order to maintain the precision of eye movements in space (i.e., gaze; for review, see Robinson, 1981). This interaction requires calculation of gaze-velocity in order to match the eye-velocity in space to the actual target-velocity. Moreover, for the long delays involved in processing visual information, both systems must predict the direction and speed of target movement. Although prediction in the smooth-pursuit system is well established (Becker and Fuchs, 1985; Kettner et al., 1996), its underlying mechanisms are far from known (Suh et al., 2000).

To explain established pursuit characteristics, Robinson (1982) extended Young's internal positive feedback model (Yasui and Young, 1975). Figure 1 shows a simplified version of Robinson's model, with the smooth-pursuit system using an internal estimate of target-velocity (see also Robinson et al., 1986).

Gaze-velocity commands are generated from the estimate of target-velocity-in-space (Fig. 1A). In a condition in which head pursuit occurs minimally, this estimate is calculated using retinal image slip-velocity, efference copy of eye-velocity, and vestibular-velocity information. The latter two signals are used to calculate gaze-velocity. To drive ocular motoneurons during smooth-gaze tracking in the frontal plane, two stages of further signal conversion are necessary: (1) omnidirectional gaze-velocity signals must be sorted into roughly horizontal and vertical components, and (2) such gaze-velocity components must be converted into eye-velocity signals, and through the neural velocity-to-position-integrators (Robinson, 1981), which are thought to be common for all conjugate eye movements, these signals are sent to ocular motoneurons. The Robinson model (Fig. 1A) explains many

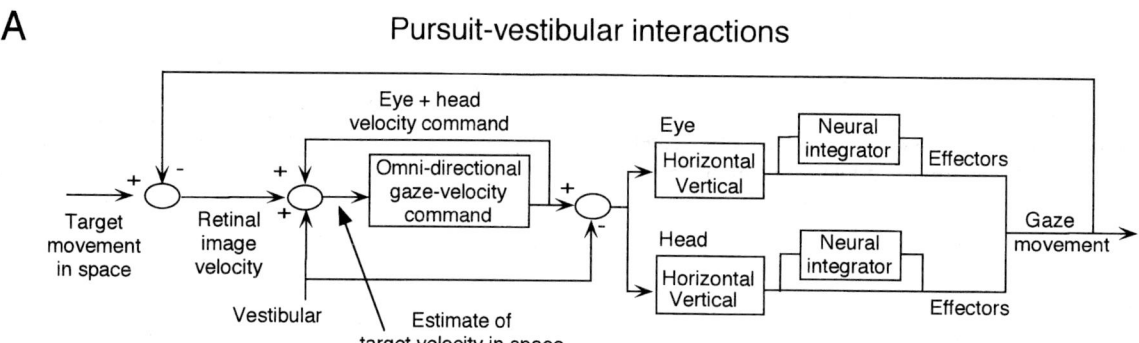

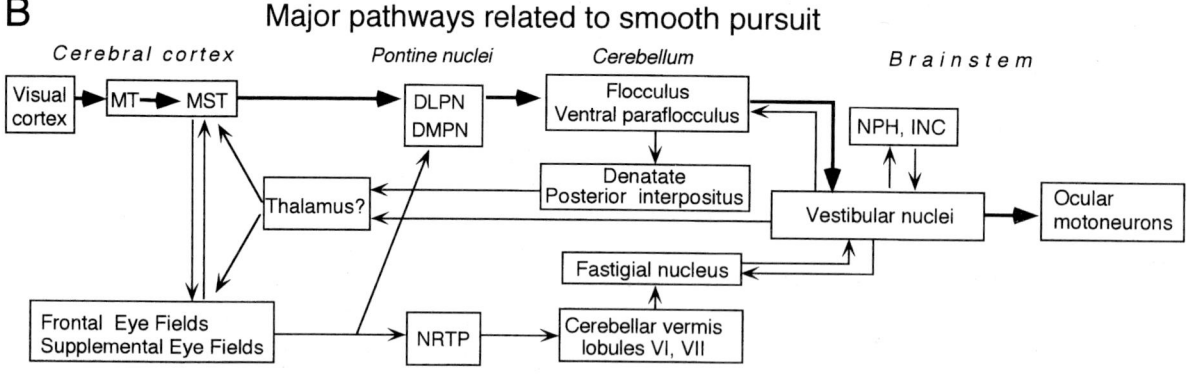

Fig. 1. An overview of the smooth pursuit system. (A) A simplified signal-flow model of pursuit-vestibular interactions (Robinson, 1982). (B) Major pathways related to smooth pursuit. Direct pathways are shown with thick lines (Kawano 1999). Unboxed brain areas for individual processing functions are shown in italics. 'Thalamus?' means that smooth-pursuit-related neurons have not yet been recorded from in the thalamus. For further explanation, see the text.

characteristics of smooth-pursuit eye movements since an internal estimate of target-velocity can also be used to predict target movement (cf. Barnes, 1993). The fundamental question of whether such an estimate of target-velocity is actually operational in the central nervous system (CNS) has not been clearly answered, however (cf. Robinson et al., 1986).

Major pathways for smooth-pursuit are shown in Fig. 1B. The middle temporal (MT) and middle superior temporal visual areas (MST) are essential for initiation and maintenance of smooth-pursuit. From these sites, there appear to be at least two descending pathways (for review, see Keller and Heinen, 1991; Leigh and Zee, 1999). The direct pathways send signals to extraocular motoneurons probably through the dorsolateral pontine nucleus (DLPN), the cerebellar flocculary lobe (flocculus and ventral paraflocculus) and the vestibular nuclei. This pathway has been identified as that for ocular following, i.e., a short-latency tracking response to movement of large field visual patterns (for review, see Kawano, 1999). Smooth-pursuit eye movements are thought to use the same pathway, at least in part (Kawano, 1999). The ventral paraflocculus is reported to project not only to the vestibular nuclei but also to the posterior interpositus and dentate nuclei (Nagao et al., 1997). Such projections may send possible feedback signals to the cerebral cortex (Fig. 1B). The brainstem circuits include pathways for the neural integrators (Fig. 1A; Robinson, 1981) that help control horizontal and vertical eye movements [nucleus prepositus hypoglossi (NPH) and the interstitial nucleus of Cajal (INC), respectively; see Fig. 1B; Fukushima et al., 1992; Fukushima and Kaneko, 1995].

Visual signals for target-velocity in the MST are also sent to the frontal eye (FEF) and supplemental eye fields. These areas are known to have reciprocal connections with the MST (Fig. 1B). The FEF projects to the DLPN, the dorsomedial pontine nuclei (DMPN), and the nucleus reticularis tegmenti pontis (NRTP), and the NRTP projects to the cerebellar dorsal vermis and fastigial nuclei. Reciprocal connections between the fastigial nucleus and vestibular nuclei are also well known.

The importance of gaze-velocity signals for generating smooth tracking of a moving target was first suggested by study of the cerebellar flocculary lobe, which contains many horizontal gaze-velocity Purkinje cells (Miles and Fuller, 1975; Lisberger and Fuchs, 1978; Miles et al., 1980; Stone and Lisberger, 1990). Since the preferred activation direction of these cells is roughly either horizontal or vertical (Miles et al., 1980; Stone and Lisberger, 1990; Shidara and Kawano, 1993; Krauzlis and Lisberger, 1996; Fukushima et al., 1999a), and since gaze-velocity cells are not among the majority of vertical Purkinje cells but rather have activity related more closely to eye movement (Fukushima et al., 1996, 1999a; see also Belton and McCrea, 2000 for horizontal flocculary Purkinje cells), the first sorting (i.e., omnidirectional gaze-velocity to vertical and horizontal gaze-velocity components) must take place before (rostral to) the flocculary lobe's Purkinje cells. This suggests that the conversion of retinotopic (i.e., omnidirectional) visual into gaze-velocity signals is generated up-stream to the flocculary lobe (Fukushima et al., 1999a).

The MST contains all the signal components needed to reconstruct target motion in space, including retinal image-slip-velocity, eye-velocity, and even gaze-velocity (Sakata et al., 1983; Kawano et al., 1984; Komatsu and Wurtz, 1988; Newsome et al., 1988; Thier and Erickson, 1992; for review, see Andersen, 1997). Moreover, since the discharge of pursuit cells in the MST is maintained even when the actual target is briefly extinguished, it has been suggested that this area forms an internal positive feedback circuit in the pursuit system for the maintenance of pursuit (Newsome et al., 1988). Since the origin of the eye-velocity signals in the MST is still unknown (Anderson et al., 1999), however, it is still not clear how the MST could be involved in the internal positive feedback circuit for the maintenance of pursuit and possibly also for an estimate of target-velocity (Fig. 1A).

The periarcuate cortex in the caudal part of the FEF contains many pursuit neurons, and the majority of them discharge before the initiation of pursuit with a median (pre-event) latency of -19 to -12 ms (Bruce and Goldberg, 1985; MacAvoy et al., 1991; Gottlieb et al., 1994; Tian and Lynch, 1996; Tanaka and Fukushima, 1998). These neurons also carry retinal image slip-velocity and gaze-velocity signals (Fukushima et al., 2000). This chapter reviews evidence from trained monkeys that the caudal FEF have signals sufficient for estimates of target-velocity-in-space, prediction signals and gaze-velocity commands in 3D space.

Gaze-velocity and retinal image slip-velocity signals in the caudal FEF

Representative discharge characteristics of pursuit neurons in the caudal FEF are illustrated in Fig. 2A–E for an exemplary single cell in its preferred direction.

Figure 2A shows that during smooth-pursuit, the cell discharged with a peak of activity near peak eye-velocity (i.e., eye-velocity was identical to gaze-velocity without chair rotation). Figure 2B shows the cell's response when the monkey, whose head was stabilized to a chair during whole body rotation (Fig. 2G), was required to track a target that moved in space with the same amplitude, direction and phase as the chair. This condition required the monkey to suppress the vestibulo-ocular reflex (VOR) so that the eyes would remain virtually motionless in the orbit, and gaze would move with the chair (Fig. 2B; i.e., VOR suppression). The activity of this cell during smooth-pursuit and VOR suppression was similar (Fig. 2A–B). When the monkey fixated a stationary target in space during chair rotation by generating a perfect VOR with gaze remaining stationary (Fig. 2C; VOR ×1), the neuron showed little modulation. Therefore, the activity of this cell was not related to eye movement per se but rather to gaze movement. The cell also responded, albeit weakly, during the VOR in complete darkness (Fig. 2D), thereby suggesting the effect of vestibular inputs.

Preferred activation directions of individual pursuit cells in the caudal FEF during smooth-pursuit

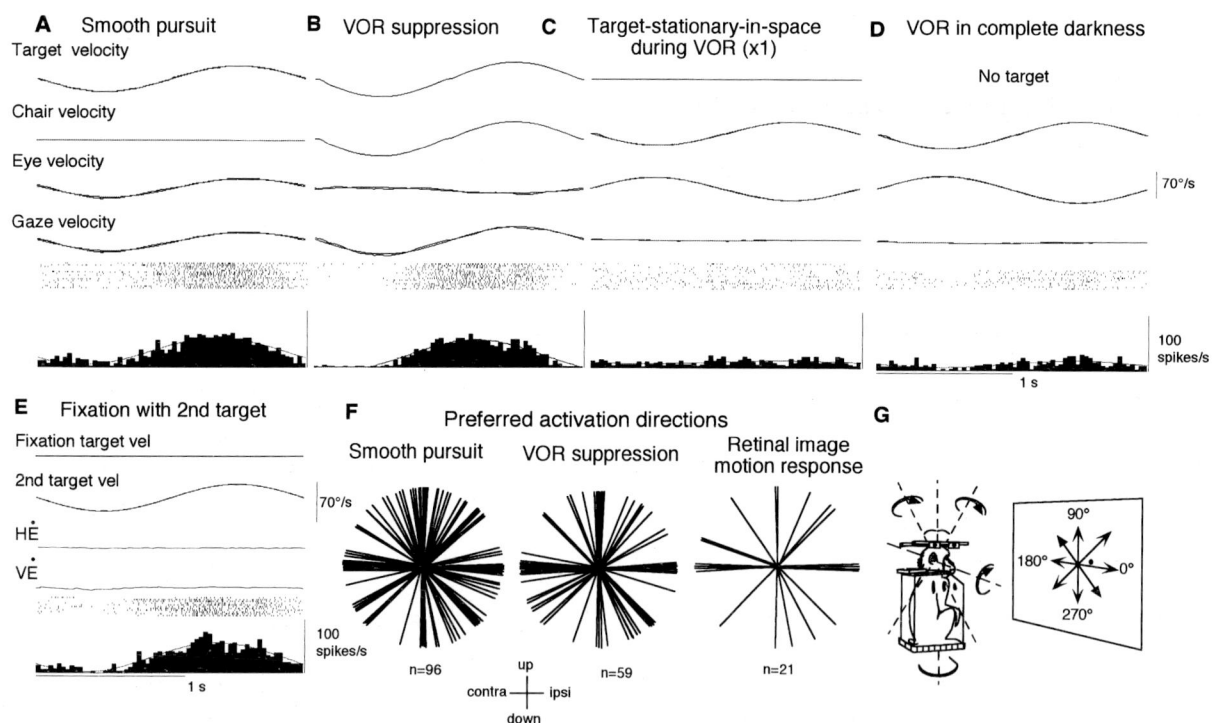

Fig. 2. Discharge characteristics of pursuit neurons in the caudal FEF during different task conditions. (A–E) Responses of a single neuron. Eye-velocity and gaze-velocity traces were de-saccaded and averaged. (F) Preferred activation directions of ensembles of neurons during different task conditions. (G) The stimulus conditions. Abbreviations: HĖ and VĖ, horizontal and vertical eye-velocity, respectively. (Reproduced from Fukushima et al., 2000 with permission from The American Physiological Society.)

and VOR suppression are distributed almost evenly in all directions (Fig. 2F). For the majority of cells, their preferred directions during these two task conditions were similar with similar gains, this being consistent with a gaze-velocity response. Moreover, by plotting amplitudes of cell modulation against peak gaze velocities at different stimulus frequencies, the discharge modulation of these cells was shown to be linearly correlated with gaze-velocity, thereby indicating that they do indeed carry gaze-velocity signals (Fukushima et al., 2000).

Figure 2E shows that the activity of more than half of the pursuit cells tested in the caudal FEF were also modulated in phase with target-velocity when the monkey viewed a target moving on a tangent screen while holding their eyes still by fixating on another stationary spot. Such cells' modulation amplitudes were linearly correlated with the velocity of the moving spot (Fukushima et al., 2000). Preferred directions for such retinal-image-motion responses were distributed in all directions (Fig. 2F), and they were similar to those of single neurons during smooth-pursuit and VOR suppression. These results indicate that pursuit neurons in the caudal FEF carry retinal image-slip-velocity (Fig. 2E), gaze-velocity (Fig. 2A–C) and vestibular signals (Fig. 2D). These signals can provide omnidirectional, target-velocity-in-space and gaze-velocity commands (Fig. 1A).

Predictive response of pursuit neurons in the caudal FEF

Prediction may occur in different ways. On the motor side, it might involve perseverance of ongoing movements. Figure 3A shows representative prediction-related activity of a pursuit neuron in the caudal FEF when a tracking target (moving at 1 Hz) was unexpectedly turned off for a short (700–800 ms) blanking period, and then restarted at 1.5 Hz.

The Fig. 3A cell had a downward preferred direction with the peak modulation close to peak down-eye-velocity and down-target-velocity. When the target was turned off, it still discharged at the time cell activity would have occurred in the presence of the target, thereby reflecting perseverance for the ongoing movement. When the target was

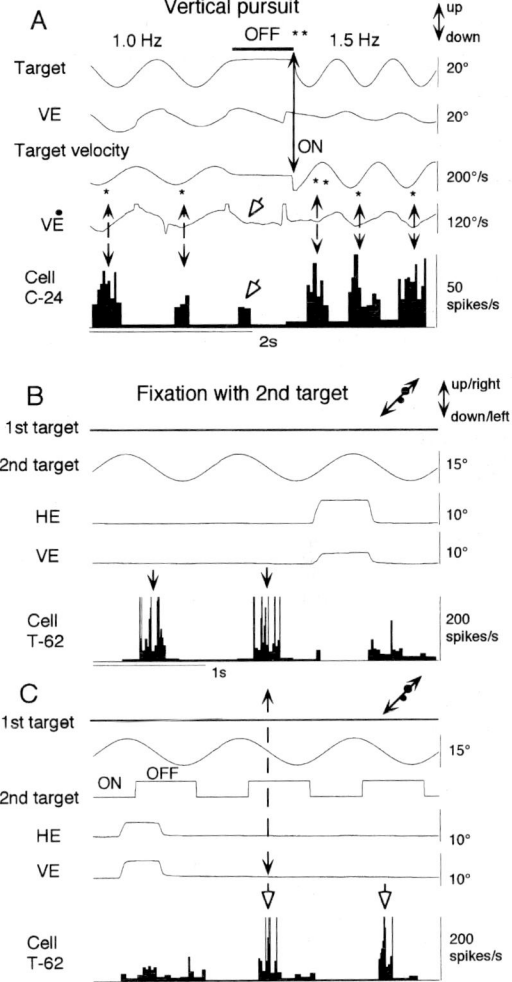

Fig. 3. Prediction-related activity of pursuit neurons in the caudal FEF. (A) Responses of a neuron when a tracking target was unexpectedly turned off and on. Single asterisks and broken lines indicate that this neuron had a downward preferred direction. Open arrows indicate that it still discharged during target blanking (OFF) at the time cell activity would have occurred in the presence of the target. The double asterisk and straight line indicate that the target was abruptly moved downward at 1.5 Hz. Note that this cell discharged during upward target-velocity in response to abrupt downward target motion. (B–C) Response of another neuron (downward arrows in B) when the monkey viewed a target moving on a tangent screen (second target) while holding the eyes still by fixating on another stationary spot (first target). This moving spot was blanked for half of each cycle in C (OFF) and the neuron discharged at the time cell activity would have occurred in the presence of the second target (downward open arrowhead and broken line in C). Abbreviations: HE, horizontal eye position; VE and V̇E, vertical eye position and velocity.

unexpectedly turned on again during downward-target movement at 1.5 Hz, this cell discharged during upward-target velocity, because its response latency to abrupt target motion was 100–110 ms. In subsequent cycles, however, this cell discharged appropriately during downward-eye and downward-target-velocity. Thus, the long delay that occurred, in visual processing when target movement occurred unexpectedly without prediction was compensated for during predictable target motion, and the activity of pursuit cells in the caudal FEF reflected predictive motor activity. We tested a total of 24 neurons in a similar fashion, and by far the majority of them (75%) behaved similarly (Fukushima et al., 2002a).

Prediction might also occur as a visual response that anticipates visual target input (Umeno and Goldberg, 1997). Figure 3B–C shows representative responses of another pursuit neuron with an oblique preferred direction. As a control, its activity was modulated in phase with target-velocity when the monkey viewed a target moving on a tangent screen while holding the eyes still by fixating on a stationary spot (Fig. 3B). To examine predictive visual response, we turned off the moving spot for almost half of each cycle. The monkey continued fixating the stationary spot which was on all the time, and this cell responded clearly during blanking of the moving spot at the time cell activity would have occurred in the presence of the moving spot. The preferred directions remained similar and response magnitudes decreased only slightly during blanking in all 18 of the tested cells (Fukushima et al., 2002a). These results suggest that the predictive responses of pursuit cells in the caudal FEF contain extracted visual components that reflect direction and speed of invisible target motion (i.e., estimates of target-velocity) when its movement is predictable. Clearly, these neurons can convey this information to generate appropriate smooth-pursuit eye movements (Fig. 1A).

Vergence-velocity signals in the caudal FEF

Gamlin and Yoon (2000) reported that neurons in a distinct area rostral to the prearcuate saccade region were selectively modulated during vergence movements, thus supporting the notion that separate frontal cortical pathways control vergence versus version (saccade or smooth-pursuit) eye movements. By examining discharge characteristics of smooth-pursuit neurons ($n = 125$) in the caudal FEF, we obtained quite different results (Fukushima et al., 2002b). To examine the vergence response, Fig. 4A shows that a horizontal screen was positioned at the monkey's nose level.

Figure 4A shows that the monkey was required to track a laser spot moving toward or away from it on a horizontal screen. The spot's movement was sinusoidal in the monkey's view and it moved along the mid-sagittal plane. The caudal FEF was shown to contain both divergence (Fig. 4B) and convergence (Fig. 4C) cells. The majority of them ($\sim 66\%$) responded to both smooth-pursuit in the frontal plane and vergence in depth (Fig. 4A).

Representative discharges are shown for a single cell in Fig. 4D–F. This cell discharged with peak activity near the peak divergence-velocity (Fig. 4D). It also clearly responded to vertical pursuit when the target's movement was large (Fig. 4E; peak velocity, $\sim 31°/s$). Because the horizontal screen was positioned at the monkey's nose level (Fig. 4A), vergence eye movements were accompanied by small ($< 0.8°$) vertical components (Fig. 4B). Such vertical movements alone cannot explain the cell's activity during vergence, however, because its response was negligible during vertical pursuit of even larger amplitude ($\pm 2°$) at the same frequency (Fig. 4F). There was no consistent correlation between preferred directions during smooth-pursuit versus vergence, i.e., these neurons responded to either convergence or divergence, but preferred directions during smooth-pursuit were distributed in all directions (Fig. 2F), thereby suggesting that inputs to caudal FEF neurons are independent for smooth-pursuit and vergence signals. These results further confirm that the cells' activity during vergence cannot be induced by pursuit components, and further indicate that modulation of neurons in the caudal FEF is related to the tracking of 3D target motions common to both smooth-pursuit and vergence (Fukushima et al. 2002b).

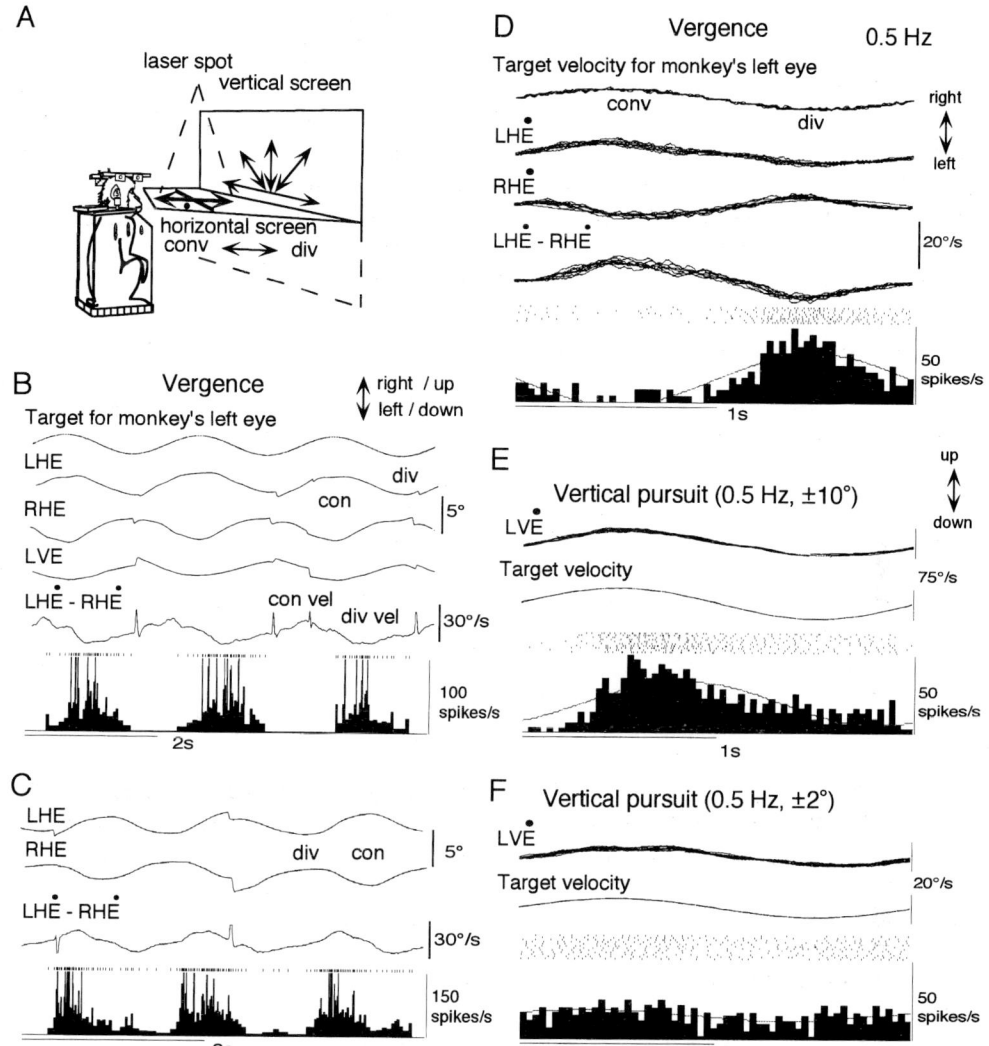

Fig. 4. Vergence-related neuronal activity. Shown are the responses of three pursuit neurons (B, C, D–F) in the caudal FEF. Neurons shown in B and D responded during divergence-velocity, whereas the neuron shown in C responded during convergence-velocity. The divergent neuron shown in D responded during upward pursuit in E (at 0.5 Hz, ±10°) but not in F (at 0.5 Hz, ±2°). Abbreviations: L, R, H, V, E, Ė = left, right, horizontal, vertical, eye position, and eye-velocity, respectively; Con, div and vel = convergence, divergence and velocity, respectively; LHĖ–RHĖ = vergence-velocity.

Chemical deactivation of the caudal FEF impairs smooth-gaze tracking

Pursuit cell activity in the caudal FEF is necessary for smooth-gaze tracking. This was confirmed by injecting a GABA agonist (muscimol) into the region where many gaze-velocity neurons were recorded. Figure 5 (A vs. D) shows that after muscimol infusion, eye-velocity during smooth-pursuit decreased to nearly half and catch-up saccades appeared frequently (Shi et al., 1998; Fukushima et al., 1999b). Vergence tracking was also reduced by a similar amount (Fukushima et al., 2002b), thereby suggesting that this area is necessary for generation of 3D gaze movement.

Figure 5 (B vs. E) shows that the effects of muscimol infusion were even more severe when long-duration (800 ms) blanking was applied before the

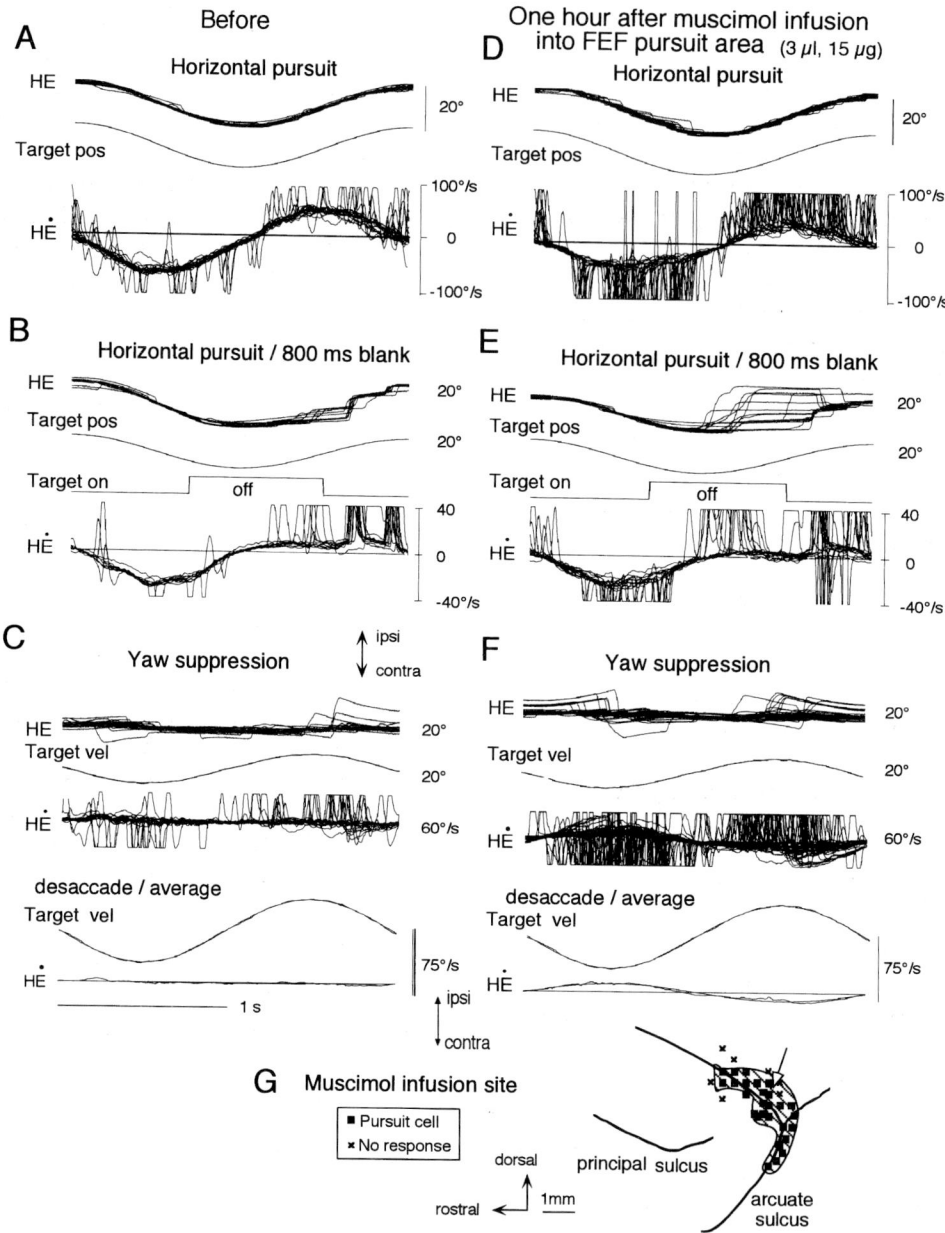

Fig. 5. Effects of muscimol injection on smooth gaze tracking. The tracking is shown before (A–C) and after (D–F) muscimol injection (15 μg) into the caudal FEF during different task conditions, as indicated. (G) The infusion site and the areas where pursuit cell recordings were made. Abbreviations: HE and HĖ, horizontal eye position and velocity, respectively. Pos and vel, position and velocity, respectively. (Reproduced, in part, from Fukushima et al., 1999b with permission from the New York Academy of Sciences.)

target changed its direction. This task condition required an estimate of target movement (Fig. 1A), and the monkey had to perform gaze tracking with an invisible target motion. Before muscimol infusion, the monkey performed smooth eye movements fairly well even during blanking (Fig. 5B). After muscimol infusion, however, the monkey was virtually unable to generate smooth eye movements in the absence of a target. Rather, it performed the task with saccades (Fig. 5E).

The monkey's performance of the VOR suppression task was also severely impaired after muscimol infusion (Fig. 5C vs. F). Before infusion, the VOR was not induced during the suppression task (gain, ~ 0.01). After infusion, however, the animal was unable to suppress the VOR (gain, ~ 0.3), and corrective saccades appeared frequently to compensate for this impairment (Fig. 5F; Fukushima et al., 1999b). In contrast, the monkey's performance of the VOR $\times 1$, which did not require gaze movement (Fig. 2C), was not clearly affected by muscimol infusion. These results indicate that gaze-velocity signals in the caudal FEF are common for both smooth-pursuit and VOR suppression tasks (Fig. 2A–B) and they are necessary, particularly in a task condition that requires prediction of target movement (Fig. 5E, cf. Fig. 1A).

Representation of smooth-gaze movements in 3D space by the caudal FEF

The vergence system has traditionally been considered to have neural origins separate from the smooth-pursuit system. This was suggested more than a century ago by Hering (for review, see Leigh and Zee, 1999). The results indicate, however, that neurons selectively activated during smooth-pursuit or vergence tracking are distinctly in the minority (<34%) in the caudal FEF. Rather, most neurons ($\sim 66\%$) have significant sensitivity to both. Thus, vergence-related signals combine with sensitivity to smooth-pursuit in the frontal plane to give rise to preferred directions widely distributed in 3D space (Fukushima et al., 2002b). Moreover, chemical deactivation of these areas impairs smooth-gaze tracking not only in the frontal plane but also in depth, especially when a prediction of target movement is required (Fig. 5E). This is also consistent with the observations showing that the majority of caudal FEF pursuit neurons carry prediction-related signals (Fukushima et al., 2002a, Fig. 3). These results further support the authors' interpretation that estimates of target-velocity-in-space carried by pursuit neurons in the caudal FEF are necessary for generating appropriate smooth-gaze-velocity commands in 3D space with the utmost of efficiency. How this is done is not yet known, however. It is possible that multiple circuits that include the FEF, supplementary eye fields, and MST are used for such functions (Fig. 1B; for review, see Andersen, 1997; Fukushima, 1997). It is also possible that feedback signals to the above cerebral cortical areas through the cerebellum and/or brainstem may also be used (Fig. 1B; for review, see Leigh and Zee, 1999). Neurons in adjacent regions of the FEF that control small saccades have disparity-tuned visual responses (Ferraina et al., 2000), thereby suggesting that a portion of the saccadic FEF also uses a 3D coordinate frame. Conversely, Gamlin and Yoon (2000) reported separate coding of version and vergence movements in the rostral FEF. This suggests that different areas in the FEF are involved in different aspects of vergence-related activity.

Concluding thoughts

These results indicate that contrary to current understanding, pathways including the caudal FEF are common to the functionally similar eye movement systems, smooth-pursuit and vergence. This suggests that synthesis of these two systems has already taken place in the caudal FEF. This conclusion suggests that Robinson's pursuit-vestibular interaction model (Fig. 1A) may be extended to a 3D coordinate frame for signals in this part of the frontal cortex. This area may function as part of a system that enables primates to track and manipulate objects moving in 3D space with the utmost of efficiency.

Acknowledgments

This work was supported in part by CREST, JAPAN Science and Technology Corporation, Japan Ministry of Science, Culture and Sports (# 12480244, 12878156), and Marna Cosmetics.

Abbreviations

3D	three-dimensional
CNS	central nervous system
DLPN	dorsolateral pontine nucleus

DMPN	dorsomedial pontine nucleus
FEF	frontal eye fields
INC	interstitial nucleus of Cajal
MT	middle temporal visual area
MST	medial superior temporal visual area
NPH	nucleus prepositus hypoglossi
NRTP	nucleus reticularis tegmenti pontis
VOR	vestibulo-ocular reflex

References

Andersen, R.A. (1997). Multimodal integration for the representation of space in the posterior parietal cortex. Phil. Trans. Soc. Lond. B, 352: 1421–1428.

Andersen, R.A., Shenoy, K.V., Snyder, L.H., Bradley, D.C. and Crowell, J.A. (1999). The contributions of vestibular signals to the representations of space in the posterior parietal cortex. Ann. N.Y. Acad. Sci., 871: 282–292.

Barnes, G. (1993). Visual-vestibular interaction in the control of head and eye movement: the role of visual feedback and predictive mechanisms. Prog. Neurobiol., 41: 435–472.

Belton, R. and McCrea, R.A. (2000). Role of the cerebellar flocculus region in cancellation of the VOR during passive whole body rotation. J. Neurophysiol., 84: 1599–1613.

Becker, W. and Fuchs, A.F. (1985). Prediction in the oculomotor system: smooth pursuit during transient disappearance of a visual target. Exp. Brain Res., 57: 562–575.

Bruce, C.J. and Goldberg, M.E. (1985). Primate frontal eye fields. I. Single neurons discharging before saccades. J. Neurophysiol., 53: 603–635.

Ferraina, S., Pare, M. and Wurtz, R.H. (2000). Disparity sensitivity of frontal eye field neurons. J. Neurophysiol., 83: 625–629.

Fukushima, K. (1997). Cortico-vestibular interactions: anatomy, electrophysiology and functional considerations. Exp. Brain Res., 117: 1–16.

Fukushima, K., Kaneko, C.R.S. and Fuchs, A.F. (1992). The neuronal substrate of integration in the oculomotor system. Prog. Neurobiol., 39: 609–639.

Fukushima, K. and Kaneko, C.R.S. (1995). Vestibular integrators in the oculomotor system. Neurosci. Res., 22: 249–258.

Fukushima, K., Chin, S., Fukushima, J. and Tanaka, M. (1996). Simple-spike activity of floccular Purkinje cells responding to sinusoidal vertical rotation and optokinetic stimuli in alert cats. Neurosci. Res., 24: 275–289.

Fukushima, K., Fukushima, J., Kaneko, C.R.S. and Fuchs, A.F. (1999a). Vertical Purkinje cells of the monkey floccular lobe: simple-spike activity during pursuit and passive whole body rotation. J. Neurophysiol., 82: 787–803.

Fukushima, K., Sato, T. and Fukushima, J. (1999b). Vestibular-pursuit interactions: gaze-velocity and target-velocity signals in the monkey frontal eye fields. Ann. N.Y. Acad. Sci., 871: 248–259.

Fukushima, K., Sato, T., Fukushima, J., Shinmei, Y. and Kaneko, C.R.S. (2000). Activity of smooth pursuit-related neurons in the monkey periarcuate cortex during pursuit and passive whole body rotation. J. Neurophysiol., 83: 563–587.

Fukushima, K., Yamanobe, T., Shinmei, Y. and Fukushima, J. (2002a). Predictive responses of peri-arcuate pursuit neurons to visual target motion. Exp. Brain Res., 145: 104–120.

Fukushima, K., Yamanobe, T., Shinmei, Y., Fukushima, J., Kurkin, S. and Peterson, B.W. (2002b). Coding of smooth eye movements in three-dimensional space by frontal cortex. Nature, 419: 157–162.

Gamlin, P.D. (1999). Subcortical neural circuits for ocular accommodation and vergence in primates. Ophthalmic Physiol. Opt., 2: 81–89.

Gamlin, P.D. and Yoon, K. (2000). An area for vergence eye movement in primate frontal cortex. Nature, 407: 1003–1007.

Gottlieb, J.P., MacAvoy, M.G. and Bruce, C.J. (1994). Neural responses related to smooth pursuit eye movements and their correspondence with electrically elicited slow eye movements in the primate frontal eye field. J. Neurophysiol., 72: 1634–1653.

Kawano, K. (1999). Ocular tracking: behavior and neurophysiology. Curr. Opin. Neurobiol., 9: 467–473.

Kawano, K., Sasaki, M. and Yamashita, M. (1984). Response properties of neurons in posterior parietal cortex of monkey during visual-vestibular stimulation. J. Neurophysiol., 51: 340–351.

Keller, E.L. and Heinen, S.J. (1991). Generation of smooth-pursuit eye movements: neuronal mechanisms and pathways. Neurosci. Res., 11: 79–107.

Kettner, R.E., Leung, H.-C. and Peterson, B.W. (1996). Predictive smooth pursuit of complex two-dimensional trajectories in monkey: component interactions. Exp. Brain Res., 108: 221–235.

Komatsu, H. and Wurtz, R.H. (1988). Relation of cortical areas MT and MST to pursuit eye movements. I. Localization and visual properties of neurons. J. Neurophysiol., 60: 580–603.

Krauzlis, R.J. and Lisberger, S.G. (1996). Directional organization of eye movement and visual signals in the floccular lobe of the monkey cerebellum. Exp. Brain Res., 109: 289–302.

Leigh, R.J. and Zee, D.S. (1999). The Neurology of Eye Movements, 3rd ed., Oxford University Press, New York, pp. 4–197.

Lisberger, S.G. and Fuchs, A.F. (1978). Role of primate flocculus during rapid behavioral modification of vestibuloocular reflex. I. Purkinje cell activity during visually guided horizontal smooth-pursuit eye movements and passive head rotation. J. Neurophysiol., 41: 733–763.

MacAvoy, M.G., Gottlieb, J.P. and Bruce, C.J. (1991). Smooth pursuit eye movement representation in the primate frontal eye field. Cereb. Cortex, 1: 95–102.

Mays, L.E. and Gamlin, P.D. (1995). Neuronal circuitry controlling the near response. Curr. Opin. Neurobiol., 5: 763–768.

Mays, L.E., Porter, J.D., Gamlin, D.R. and Tellow, C.A. (1986). Neural control of vergence eye movements: neurons encoding vergence velocity. J. Neurophysiol., 56: 1007–1021.

Miles, F.A. and Fuller, J.H. (1975). Visual tracking and the primate flocculus. Science, 189: 1000–1003.

Miles, F.A., Fuller, D.J., Braitman, D.J. and Dow, B.M. (1980). Long-term adaptive changes in primate vestibulo-ocular reflex. III. Electrophysiological observations in flocculus of normal monkeys. J. Neurophysiol., 43: 1437–1476.

Nagao, S., Kitamura, T., Nakamura, N., Hiramatsu, T. and Yamada, J. (1997). Location of efferent terminals of the primate flocculus and ventral paraflocculus revealed by anterograde axonal transport methods. Neurosci. Res., 27: 257–269.

Newsome, W.T., Wurtz, R.H. and Komatsu, H. (1988). Relation of cortical areas MT and MST to pursuit eye movements. II. Differentiation of retinal from extraretinal inputs. J. Neurophysiol., 60: 604–620.

Rashbus, C. and Westheimer, G. (1961). Independence of conjugate and disjunctive eye movements. J. Physiol. (Lond.), 159: 361–364.

Robinson, D.A. (1981). Control of eye movements. In: Brookhart J.M. and Mountcastle J.M. (Eds.), Handbook of Physiology. American Physiological Society, Bethesda, pp. 1275–1320.

Robinson, D.A. (1982). A model of cancellation of the vestibulo-ocular reflex. In: Lennerstrand G., Zee D.S. and Keller E.L. (Eds.), Functional Basis of Ocular Motility Disorders. Pergamon Press, Oxford, pp. 5–13.

Robinson, D.A., Gordon, J.L. and Gordon, S.E. (1986). A model of the smooth pursuit eye movement system. Biol. Cybern., 55: 43–57.

Sakata, H., Shibutani, H. and Kawano, K. (1983). Functional properties of visual tracking neurons in posterior parietal association cortex of the monkey. J. Neurophysiol., 49: 1364–1380.

Shi, D., Friedman, H.R. and Bruce, C.J. (1998). Deficits in smooth pursuit eye movements after muscimol inactivation within the primate frontal eye field. J. Neurophysiol., 80: 458–464.

Shidara, M. and Kawano, K. (1993). Role of Purkinje cells in the ventral paraflocculus in short-latency ocular following responses. Exp. Brain Res., 93: 185–195.

Stone, L.S. and Lisberger, S.G. (1990). Visual responses of Purkinje cells in the cerebellar flocculus during smooth-pursuit eye movements in monkeys. I. Simple spikes. J. Neurophysiol., 63: 1241–1262.

Suh, M., Leung, H.-C. and Ketter, R.E. (2000). Cerebellar flocculus and ventral paraflocculus Purkinje cell activity during predictive and visually driven pursuit in monkey. J. Neurophysiol., 84: 1835–1850.

Tanaka, K. and Fukushima, K. (1998). Neuronal responses related to smooth pursuit eye movements in the periarcuate cortical area of monkeys. J. Neurophysiol., 80: 28–47.

Thier, P. and Erickson, R.G. (1992). Responses of visual-tracking neurons from cortical area MST-l to visual, eye and head motion. Eur. J. Neurosci., 4: 539–553.

Tian, J. and Lynch, J.C. (1996). Functionally defined smooth and saccadic eye movement subregions in the frontal eye field of Cebus monkeys. J. Neurophysiol., 76: 2740–2771.

Umeno, M.M. and Goldberg, M.E. (1997). Spatial processing in the monkey frontal eye field. I. Predictive visual responses. J. Neurophysiol., 78: 1373–1383.

Yasui, S. and Young, L.R. (1975). Perceived visual motion as effective stimulus to pursuit eye movement system. Science, 190: 906–908.

CHAPTER 38

Role of cross-striolar and commissural inhibition in the vestibulocollic reflex

Yoshio Uchino*

Department of Physiology, Tokyo Medical University, 6-1-1 Shinjuku, Shinjuku-ku, Tokyo 160-8402, Japan

Abstract: In the otolith system, there are two types of neuronal circuitry that can enhance response sensitivity during linear acceleration and tilt of the head. One produces cross-striolar inhibition and the other commissural inhibition. Cross-striolar inhibition can be observed in over 50% of saccular-activated, second-order vestibular neurons. In contrast, it is seen in less than 33% of utricular-activated, second-order vestibular neurons. The majority of vestibular neurons that receive cross-striolar inhibition have axons that project to the spinal cord. Over 50% of the utricular-activated, second-order vestibular neurons received commissural inhibition from the contralateral utricular nerve. On the other hand, almost all the saccular-activated, second-order vestibular neurons exhibit no response to stimulation of the contralateral saccular nerve. The majority of vestibular neurons receiving commissural inhibition also have axons that project to the spinal cord. These results suggest that cross-striolar inhibition is important for increasing the sensitivity of the saccular system, whereas commissural inhibition is more important in the utricular system.

Introduction

Shimazu and Precht (1966) were the first to demonstrate that in the horizontal semicircular canal system, inputs from ipsilateral primary afferents to vestibular neurons are excitatory, whereas contralateral actions mediated by commissural neurons are inhibitory. A similar arrangement prevails in the vertical canal system, where the inhibition is plane-specific (Kasahara and Uchino, 1974). Thus, during angular acceleration, the sensitivity of vestibular neurons to head rotation is increased by the convergence of facilitation and disinhibition, with both effects mediated via commissural fibers.

There was no evidence of commissural inhibition in the otolith system, however, until we recorded intracellularly (IC) from vestibular neurons while stimulating ipsilateral and contralateral utricular and saccular nerves. The results showed that there was

*Corresponding author: Tel.: +81-3-3351-6141 (ext. 322); Fax: +81-3-3351-6544; E-mail: y-uchino@tokyo-med.ac.jp

indeed commissural inhibition in the utricular system, but it was less pronounced than that in the saccular system (Uchino et al., 1999, 2001).

We have also found recently that another mechanism may enhance response sensitivity in the otolith system during linear acceleration and tilt of the head. In the saccular system, vestibular neurons are excited monosynaptically by afferents from one side of the striola, and inhibited disynaptically by afferents from the other side. We have termed the latter 'cross-striolar inhibition' (Uchino et al., 1997b). This circuit also exists in the utricular system, albeit not so definitively (Ogawa et al., 2001).

What follows are some features of our above recent work on commissural and cross-striolar inhibition in the utricular and saccular system.

Experimental features

Each cat was initially anesthetized with ketamine and tracheotomized. Under gaseous anesthesia, it was then decerebrated, paralyzed and artificially

DOI: 10.1016/S0079-6123(03)43038-0

ventilated. A C1 laminectomy allowed placement of electrodes for antidromic stimulation of vestibulospinal neurons. These electrodes were inserted into the medial vestibulospinal tract (MVST) and the left and (sometimes) right lateral vestibulospinal tract (LVST). The caudal part of the cerebellum was aspirated. N1 extracellular field potentials were recorded from the vestibular nuclei. IC recordings were obtained from vestibular neurons in the lateral and descending nuclei. Excitatory (E) and inhibitory (I) postsynaptic potentials (PSPs) and the field potentials were routinely averaged.

To study commissural effects, a pair of fine silver electrodes was used to stimulate the bilateral utricular and saccular nerves (Sasaki et al., 1991; Uchino et al., 1994, 1997a,b). For focal stimulation of the saccular macula (Fig. 1A), two groups of animals had different electrode placements. One group had placements across the striola, with one electrode in the rostro-ventral part of the macula and the other in

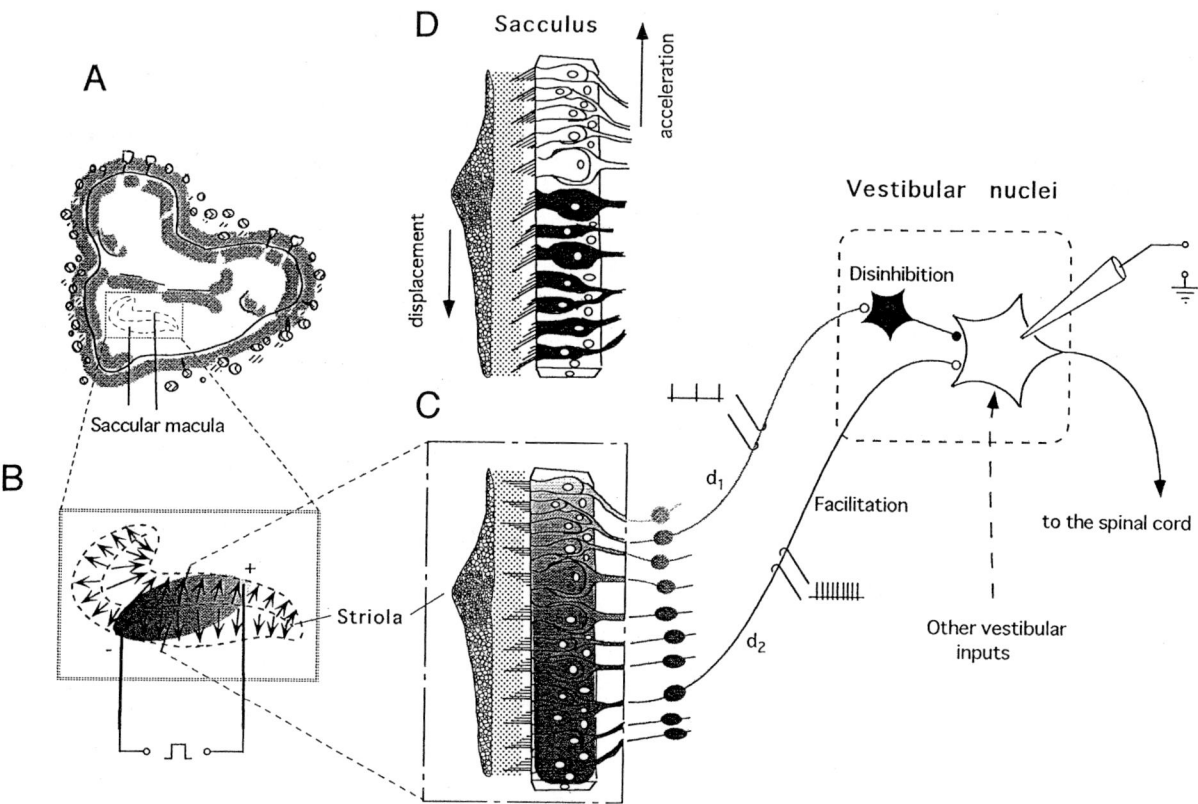

Fig. 1. The experimental arrangement for demonstration of cross-striolar inhibition. (A) Ventro-lateral view of the cat inner ear showing the electrode positions for focal stimulation of the saccular macula. Bipolar electrodes were inserted into the rostro-ventral and caudo-dorsal parts of the saccular macula (for the first group of animals; see text for the other groups). The utricular nerve and the horizontal, anterior and posterior canal nerves were cut. (B) Schematic diagram of the saccular macula showing the morphological polarity of hair cells. Arrows show the direction of kinocilia (see also the thickest/longest hair cells in C–D). Stimulus current was applied such that the rostral–ventral electrode was negative. (C) Cross-section of the saccular receptor at the dashed line in B and its afferent connections to the vestibular nuclei. The experimental arrangement is also shown for demonstration of cross-striolar inhibition (see text). Saccular afferents were depolarized near the negative electrode and hyperpolarized near the positive electrode. Therefore, the firing rate of afferent d_1 decreased, while that of d_2 increased. (D) Cross-section of the saccular receptor showing functional pseudo-equivalence of the focal stimulation in C. If the acceleration is upward, the displacement of hairs is downward, leading to a depolarization of ventral (dark) hair cells and hyperpolarization of dorsal (light) hair cells. (Reproduced, in part, from Uchino et al., 1997b with permission from The American Physiological Society.)

the caudo-dorsal part (Fig. 1B). For the other group, a pair of electrodes was placed on the dorsal edge of the saccular macula (not shown in Fig. 1). A brief, pulse was applied to activate focally a small area of the macula. The activated area could be changed somewhat by reversing the polarity of the stimulus current.

For focal stimulation of the utricular macula, the electrode pairs were inserted in three ways (Ogawa et al., 2000). In the first group of animals, one pair was placed on the lateral edge of the macula and another on the medial edge. In the second group, electrode pairs were placed on each side of the macula, after the utricular nerve had been split sagittally near the lateral 1/3 of the macula. In the third group, one electrode was inserted into the medial edge, and another was placed on the lateral edge. To prevent the spread of stimulus current in the three groups, the nerves and electrodes were covered with a warm semisolid mixture of paraffin and vaseline.

Commissural effects in the utricular system

Commissural effects were studied in 72 utricular-activated, second-order vestibular neurons in the lateral and descending nuclei. These were identified by the monosynaptic latency (≤ 1.4 ms) of their EPSPs (Uchino et al., 2001). EPSP threshold was 4–45 µA (19 ± 12 µA, $n=43$). After contralateral utricular nerve stimulation, 41/72 neurons (57%) received inhibition, and two (3%), excitation, both at a polysynaptic latency (≥ 1.5 ms, most at 1.7–3.5 ms). IPSP threshold was 8–60 µA (32 ± 16 µA, $n=41$). In response to ipsilateral utricular stimulation, EPSP amplitude was 1.3 ± 0.7 mV ($n=38$), and commissural IPSP amplitude was 1.2 ± 1.0 mV ($n=32$). These amplitudes are comparable to those of commissural IPSPs evoked by semicircular canal nerve stimulation (Uchino et al., 1986). Among 34/72 neurons tested for a projection to the spinal cord, 24/34 responded antidromically to spinal stimulation, 15 with axons in i-LVST, eight in MVST, and one in c-LVST.

Among 72 neurons tested for a response to stimulation of the contralateral utricular nerve, 29/72 (40%) exhibited no clear-cut response. Also, among 21/72 neurons tested for a spinal projection, nine had axons in i-LVST and five in MVST. The remaining seven were unresponsive to spinal stimulation.

Commissural effects in the saccular system

To determine such commissural effects, 67 second-order vestibular neurons were tested by selective stimulation of the contralateral saccular nerve (Uchino et al., 2001). Only $\sim 27\%$ of them (18/67) exhibited PSP responses. These were at a polysynaptic latency (≥ 1.5 ms), with 7/67 (10%) receiving inhibition and 11/67 (16%) excitation. The amplitudes of commissural IPSPs evoked by contralateral saccular nerve stimulation were 0.4 ± 0.2 mV; $n=6$). Among 22/67 neurons tested for a spinal projection, eight had a descending axon in i-LVST, five in MVST, and three in c-LVST. The remaining six had no such projection. Among 67 neurons tested for a response to stimulation of the contralateral saccular nerve, 49/67 (73%) exhibited no clear-cut response. An antidromic spike response was tested for in 27/49 neurons. Among these, 21/27 displayed a response; 19 with an axon in i-LVST, and two in MVST. The remaining six were unresponsive to spinal stimulation.

Cross-striolar inhibition in the saccular system

PSPs were recorded in lateral and descending vestibular nucleus neurons in response to focal stimulation of the saccular macula (see Fig. 1; Uchino et al., 1997b). Among the 36 neurons tested with one electrode in the rostro-ventral part of the macula and the other in the caudo-dorsal part, 22/36 (61%) showed the opposite pattern of PSPs when the stimulus polarity was reversed, i.e., they responded with an EPSP to one stimulus polarity and an IPSP to the other polarity. Such EPSPs were monosynaptic (latency, < 1.4 ms), whereas the IPSPs were disynaptic (latency, > 1.5 ms). In the remaining 14/36 neurons, the PSP pattern was not affected by stimulus polarity. These results suggested that a group of saccular afferents had monosynaptic excitatory contacts with vestibular neurons, whereas another group of afferents that originated from hair cells with an opposite 'morphological polarity' had a disynaptic connection with the *same* vestibular neurons by way of an interposed inhibitory interneuron, i.e., they mediated cross-striolar inhibition (see Fig. 1C).

Among the 31 neurons tested with a pair of electrodes placed on the dorsal edge of the saccular macula, either a monosynaptic EPSP (25/31 cells) or a disynaptic IPSP (6/31) was evoked by focal stimulation of particularly the dorsal–lateral edge of the structure. These responses were not affected by a change in stimulus polarity.

Cross-striolar inhibition in the utricular system

Focal stimulation of the medial and/or the lateral part of the utricular macula evoked PSPs in 83 neurons located in the lateral and descending vestibular nuclei (see also Ogawa et al., 2000). Cross-striolar inhibition was observed in 25/83 (30%), i.e., with responses that were of opposite sign (E vs. I, or vice versa) to inputs from the medial versus lateral part of the utricular macula. The majority of such EPSPs were monosynaptic (latency, ≤ 1.4 ms), whereas the IPSPs were usually disynaptic (latency, ≥ 1.5 ms). The percentage of utricular-activated neurons receiving cross-striolar inhibition was much less than that of saccular-activated neurons (30 vs. 61%). Also, the amplitude of utricular-activated disynaptic IPSPs was much smaller than the saccular-activated ones.

Among the remaining vestibular neurons, 32/83 (39%) received the same (E or I) type of inputs from both sides of the utricular macula, while another 26/83 (31%) received input from only one (medial or lateral) side.

Among the 32 utricular-responsive vestibular neurons tested for a spinal connection, 22 were activated antidromically from the C1 segment, 16/22 (73%) with axons in l-LVST and 6/22 (27%) in MVST.

Comment

The substantial difference between the commissural effects evoked by utricular and saccular afferents seems to reflect a difference in the geometry of macula orientation. The two utricular maculae lie approximately in the horizontal plane of the head, and they largely detect accelerations acting parallel to this plane, and head tilts. The striolae separate the maculae mediolaterally, and hair cells on each side of the striola have opposite directions of effective activation. Since the activating directions of hair cells are mainly medio-lateral, hair cells of the corresponding portions of the bilateral utricular maculae (e.g., lateral sides of the striolae in both maculae) have an opposite polarization (see Wilson and Melvill Jones, 1979). This morphological hair-cell arrangement, is the basis for our dual hypothesis that: (1) primary utricular afferents innervating hair cells lateral to the striola terminate monosynaptically on ipsilateral vestibular nucleus neurons; and (2) utricular afferents innervating hair cells in the lateral side in the contralateral macula project di- or polysynaptically to these *same* vestibular neurons by way of intercalated inhibitory interneurons. Figure 2 shows that we also propose that utricular afferents innervating hair cells in the medial sides of the maculae are part of a similar circuit projecting to other vestibular neurons (see also Uchino et al., 2001). As a result, the sensitivity of utricular-activated vestibular neurons to linear acceleration is increased by a disinhibitory mechanism.

In contrast to the utricular system, the bilateral saccular maculae are positioned almost parallel to each other. The striolae separate the saccular maculae mainly dorso-ventrally. Since hair cells in each side of the striola have activating directions almost in the parasagittal plane (mostly upward or downward), hair cells of the corresponding portions of the bilateral saccular maculae (e.g., dorsal to the striola in both maculae) have nearly the *same* polarization direction. This arrangement suggests that in order to increase their sensitivity to linear acceleration, vestibular nucleus neurons should be facilitated by saccular afferents innervating hair cells in the corresponding areas of the bilateral maculae, i.e., they should receive commissural facilitation rather than inhibition. In actuality, we have shown that 16% of the tested vestibular neurons received commissural facilitation following contralateral saccular nerve stimulation, while 10% received commissural inhibition (Uchino et al., 2001). Presumably, commissural inhibition is provided

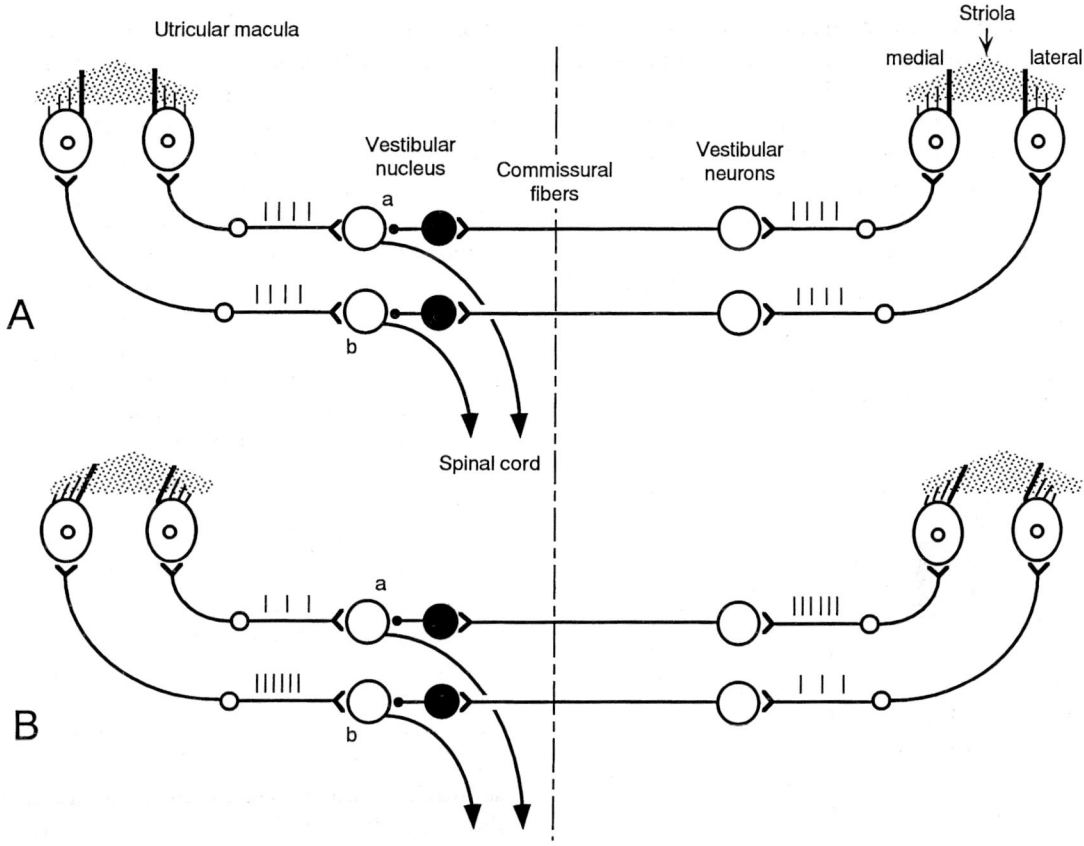

Fig. 2. Hypothetical circuits connecting the utricular maculae to the spinal cord. (A) Schematic diagram of commissural inhibitory pathways showing the 'morphological polarity' of hair cells. The thickest/longest hair in each of the shown cells is a kinocilia. The 'latero-lateral' and 'medio-medial' circuits are hypothetical. The location of inhibitory interneurons (filled circles) is not yet identified, but they are most likely in the contralateral vestibular nuclei. (B) When the acceleration is to the left, the displacement of hairs is to the right. The lateral and medial hair cells in the left macula then depolarize and hyperpolarize, respectively. Simultaneously, medial and lateral hair cells in the right macula depolarize and hyperpolarize, respectively. It follows that the firing rate of afferents from medial hair cells in the left macula is decreased, whereas that from lateral hair cells increase. Thus, the sensitivity of vestibular neuron 'a' to this acceleration would be decreased by the convergence of disfacilitation and inhibition, whereas the sensitivity of vestibular neuron 'b', would be increased by facilitation and disinhibition. (Reproduced, in part, from Uchino et al., 2001 with permission from Springer-Verlag.)

by saccular afferents innervating hair cells in the opposite area across the striola in the contralateral macula, and again via intercalated inhibitory interneurons.

Recently, we used focal stimulation of the macula to reveal another type of cross-striolar inhibition (Uchino et al., 1997b). Vestibular nucleus neurons were excited monosynaptically by otolith afferents from one side of the striola, and inhibited disynaptically by afferents from the other side. About 60% of saccular-activated, second-order vestibular neurons received this type of cross-striolar inhibition, while only 30% of utricular-activated ones were similarly affected (Uchino et al., 1999; Ogawa et al., 2000). Not only were saccular commissural effects much less prevalent than utricular ones, but the amplitudes of commissural IPSPs evoked by contralateral saccular nerve stimulation were smaller than those evoked by contralateral utricular nerve stimulation. These findings suggested that commissural inhibitory circuits may be less prominent in the saccular versus utricular system. It follows that cross-striolar inhibition should be particularly important in increasing the sensitivity to *vertical* acceleration (Uchino et al., 1997b).

What are the targets of these amplified vestibular signals? Utricular-activated, second-order vestibular neurons preferentially innervate spinal motoneurons. More than half of them have projections to the upper cervical cord, whereas only a few are activated antidromically by stimulation in the oculomotor nuclei (Sato et al., 1996). Stimulation of the utricular nerve evokes a disynaptic EPSP in ipsilateral extensor (Bolton et al., 1992), flexor (Ikegami et al., 1994) and contralateral sternocleidomastoid (Kushiro et al., 1999) neck motoneurons. Projections to oculomotor and trochlear (but not abducens) motoneurons are also rare (Sasaki et al., 1991; Uchino et al., 1994, 1996). It is quite likely, however that signals from utricular receptors (enhanced by disinhibition through commissural inhibitory pathways) are transmitted to cervical motoneurons. A similar arrangement prevails for saccular-activated neurons, i.e., a significant projection to the upper cervical cord, and a sparse projection to the oculomotor nuclei, (Sato et al., 1997; Uchino et al., 1997b; Isu et al., 2000; Kushiro et al., 2000). These results indicate that transmission to cervical motoneurons from second-order, saccular-activated neurons are modulated mainly by cross-striolar inhibition, and only to a minor extent by commissural inhibition and/or facilitation. This arrangement results from the morphological geometry of the system, as does the possibility that the sensitivity of utricular-activated vestibular neurons is increased largely by commissural inhibition, whereas that of saccular-activated ones is mainly by cross-striolar inhibition.

Conclusions

We examined whether otolith-activated, second-order vestibular neurons receive commissural inhibition from the contralateral otolith's macula, such as to provide a mechanism for increasing the sensitivity of vestibular neurons to linear acceleration and lateral tilt of the head. For this purpose, we measured the IC-recorded responses of vestibular neurons to stimulation of the ipsi- and contralateral utricular and saccular nerves. Over 50% of the utricular-activated, second-order vestibular neurons received commissural inhibition from the contralateral utricular nerve. In contrast, the majority of saccular-activated, second-order vestibular neurons exhibited no clear-cut response to stimulation of the contralateral saccular nerve. These findings suggest that commissural inhibition is more important in the utricular system than in the saccular system.

We have also shown that primary afferents innervating hair cells on one side of the striola provide monosynaptic excitation to vestibular neurons, whereas the same neurons receive disynaptic inhibition from primary afferents innervating hair cells on the other side of the striola. This inhibition is by way of intercalated inhibitory interneurons in the saccular system. The sensitivity of these neurons to linear acceleration is thereby increased by a combination of facilitation and disinhibition. The neural circuits providing such cross-striolar inhibition were shown to operate on over 50% of the saccular-activated, second-order vestibular neurons. In contrast, such cross-striolar inhibition was observed in less than 30% of the utricular-activated, second-order vestibular neurons, which contribute to utricular afferents. These results suggest that cross-striolar inhibition has a far more prominent excitatory function in the saccular system than in the utricular system.

Acknowledgments

For their collaboration, I would like to thank Drs. Hitoshi Sato, Miridaha Zakir, Keisuke Kushiro, Midori Imagawa, Yasuo Ogawa, Seiji Ono, Hui Meng, Xialing Zhang, Masayuki Katsuta, Mitsuyoshi Sasaki, Naoki Isu and Victor J. Wilson. I also thank Miss Keiko Takayama for her secretarial assistance. This study was supported by a research grant from the Japan Space Forum promoted by NASDA (National Space Development Agency of Japan).

Abbreviations

C1	first segment of the cervical spinal cord
EPSP	excitatory postsynaptic potential
IPSP	inhibitory postsynaptic potential

LVST   lateral vestibulospinal tract
MVST   medial vestibulospinal tract
PSP   postsynaptic potential

References

Bolton, P.S., Endo, K., Goto, T., Imagawa, M., Sasaki, M., Uchino, Y. and Wilson, V.J. (1992). Connections between utricular nerve and dorsal neck motoneurons of the decerebrate cat. J. Neurophysiol., 67: 1695–1697.

Ikegami, H., Sasaki, M. and Uchino, Y. (1994). Connections between utricular nerve and neck flexor motoneurons of decerebrate cats. Exp. Brain Res., 98: 373–378.

Isu, N., Graf, W., Sato, H., Kushiro, K., Zakir, M., Imagawa, M. and Uchino, Y. (2000). Sacculo-ocular reflex connectivity in cats. Exp. Brain Res., 131: 262–268.

Kasahara, M. and Uchino, Y. (1974). Bilateral semicircular canal inputs to neurons in cat vestibular nuclei. Exp. Brain Res., 20: 285–296.

Kushiro, K., Zakir, M., Ogawa, Y., Sato, H. and Uchino, Y. (1999). Saccular and utricular inputs to sternocleidomastoid motoneurons of decerebrate cats. Exp. Brain Res., 126: 410–416.

Kushiro, K., Zakir, M., Ogawa, Y., Katsuta, M., Isu, N., Sato, H., Imagawa, M., Sugihara, Y. and Uchino, Y. (2000). Saccular and utricular inputs to single vestibular neurons in cats. Exp. Brain Res., 131: 406–415.

Ogawa, Y., Sato, H., Kushiro, K., Zakir, M., Imagawa, M. and Uchino, Y. (2000). Neuronal organization of the utricular macula concerned with innervation of single vestibular neurons in the cat. Neurosci. Lett., 278: 89–92.

Sasaki, M., Hiranuma, K., Isu, N. and Uchino, Y. (1991). Is there a three neuron arc in the cat utriculo-trochlear pathway? Exp. Brain Res., 86: 421–425.

Sato, H., Endo, K., Ikegami, H., Imagawa, M., Sasaki, M. and Uchino, Y. (1996). Properties of utricular nerve-activated vestibulospinal neurons in cats. Exp. Brain Res., 112: 197–202.

Sato, H., Imagawa, M., Isu, N. and Uchino, Y. (1997). Properties of saccular nerve-activated vestibulospinal neurons in cats. Exp. Brain Res., 116: 381–388.

Shimazu, H. and Precht, W. (1966). Inhibition of central vestibular neurons from the contralateral labyrinth and its mediating pathway. J. Neurophysiol., 29: 467–492.

Uchino, Y., Ichikawa, T., Isu, N., Nakashima, H. and Watanabe, S. (1986). The commissural inhibition on secondary vestibulo-ocular neurons in the vertical semicircular canal systems in the cat. Neurosci. Lett., 70: 210–216.

Uchino, Y., Ikegami, H., Sasaki, M., Endo, K., Imagawa, M. and Isu, N. (1994). Monosynaptic and disynaptic connections in the utriculo-ocular reflex arc of the cat. J. Neurophysiol., 71: 950–958.

Uchino, Y., Sasaki, M., Sato, H., Imagawa, H., Suwa, H. and Isu, N. (1996). Utriculoocular reflex arc of the cat. J. Neurophysiol., 76: 1896–1903.

Uchino, Y., Sato, H., Sasaki, M., Imagawa, H., Ikegami, H., Isu, N. and Graf, W. (1997a). Sacculocollic reflex arcs in cats. J. Neurophysiol., 77: 3003–3012.

Uchino, Y., Sato, H. and Suwa, H. (1997b). Excitatory and inhibitory inputs from saccular afferents to single vestibular neurons in cats. J. Neurophysiol., 78: 2186–2192.

Uchino, Y., Sato, H., Kushiro, K., Zakir, M., Imagawa, H., Ogawa, Y., Katsuta, M. and Isu, N. (1999). Cross-striolar and commissural inhibition in the otolith system. Ann. N.Y. Acad. Sci., 871: 162–172.

Uchino, Y., Sato, H., Zakir, M., Kushiro, K., Imagawa, M., Ogawa, Y., Ono, S., Meng, H., Zhang, X., Katsuta, M., Isu, N., Wilson, V.J. (2001). Commissural effects in the otolith system. Exp. Brain Res., 136: 421–430.

Wilson, V.J. and Melvill Jones, G. (1979). Mammalian Vestibular. Plenum Press, New York; London.

CHAPTER 39

Functional synergies among neck muscles revealed by branching patterns of single long descending motor-tract axons

Yuriko Sugiuchi, Shinji Kakei, Yoshiko Izawa and Yoshikazu Shinoda*

Department of Systems Neurophysiology, Graduate School of Medicine, Tokyo Medical and Dental University, Tokyo 113-8519, Japan

Abstract: In this chapter, we describe our recent work on the divergent properties of single, long descending motor-tract neurons in the spinal cord, using the method of intra-axonal staining with horseradish peroxidase, and serial-section, three-dimensional reconstruction of their axonal trajectories. This work provides evidence that single motor-tract neurons are implicated in the neural implementation of functional synergies for head movements. Our results further show that single medial vestibulospinal tract (MVST) neurons innervate a functional set of multiple neck muscles, and thereby implement a canal-dependent, head-movement synergy. Additionally, both single MVST and reticulospinal axons may have similar innervation patterns for neck muscles, and thereby control the same functional sets of neck muscles. In order to stabilize redundant control systems in which many muscles generate force across several joints, the CNS routinely uses a combination of a control hierarchy and sensory feedback. In addition, in the head-movement system, the elaboration of functional synergies among neck muscles is another strategy, because it helps to decrease the degrees of freedom in this particularly complicated control system.

Introduction

It had been tacitly assumed, and emphasized in most textbooks, that the pyramidal tract consists of 'private lines', which connect a discrete site in the motor cortex to a single muscle, just as a spinal motor nucleus usually innervates a single muscle. Similarly, corticospinal tract (CST) neurons have long been referred to as 'upper motoneurons' and motoneurons as 'lower motoneurons'. Other long descending motor tracts have generally been considered to be similarly arranged. These assumptions *are no longer tenable*, however, because recent studies have shown that axons of *all* the major long descending motor tracts send axon collaterals to *multiple* spinal segments. This organization was first described by Abzug et al. (1974), who used electrophysiological techniques to show that 50% of the lateral vestibulospinal tract (LVST) neurons which supply axon branches to C6-Th1 segments, are also antidromically driven by stimulation of the lumbar spinal cord. A similar percentage of reticulospinal tract (RTST) neurons were then shown to send branches to the cervical gray matter, as well as to the first lumbar segment (Peterson et al., 1975). Even more unexpectedly, similar electrophysiological experiments have shown that both CST (Shinoda et al., 1986b) and rubrospinal tract (RBST; Shinoda et al., 1977) neurons project to both the cervical gray matter (C4–8) and the thoracic or lumbar spinal cord. Furthermore, virtually all CST neurons examined in the forelimb area of the motor cortex have 3–7 axon collaterals at widely separated segments of the cervical and upper thoracic cord (Shinoda et al., 1976). In addition, multiple axon collaterals of single

*Corresponding author: Tel.: +81-3-5803-5152;
Fax: +81-3-5803-5155; E-mail: yshinoda.phy1@med.tmd.ac.jp

RBST neurons have been found at different spinal segments (Shinoda et al., 1977).

These electrophysiological findings all emphasize that single, motor-tract axons are not simple private lines for connecting the cells of cortical origin to the motoneurons of a single muscle, but, rather, such descending axons concurrently exert *multiple* influences on *different* groups of spinal interneurons and motoneurons at *widely separated* spinal segments.

The above findings clearly have important implications for the descending control of posture and movement. It was not possible until recently, however, to reinforce these electrophysiological findings with relevant morphology, due to the inability to trace long axons from their cell bodies when using classical anatomical techniques. For visualization of the entire axonal morphology of a single neuron, especially one with a relatively long axon, new methods had to be developed, particularly ones which would bring into view the entire morphology of single, *functionally identified* neurons in the central nervous system (CNS). The enzyme, horseradish peroxidase (HRP), was shown to be particularly valuable for this purpose (Jankowska et al., 1976; Kitai et al., 1976; Snow et al., 1976). Admittedly, all of the processes of a single neuron labeled with HRP may not be visualized in a single section. Reconstruction of the axonal trajectory using serial sections, however, reveals the *entire* axonal trajectory of a single-labeled neuron for a distance of 10–30 mm in the mammalian CNS (for review, see Shinoda, 1999).

In this chapter, we summarize our HRP-reconstruction studies on the intraspinal morphology of single axons of various long descending motor tracts in the cat and monkey. We show that the branching patterns of single vestibulocollic and tecto-reticulospinal axons may reveal the functional synergies among neck muscles that the CNS uses for the control of head movements.

Morphology of single CST and other long descending motor-tract axons

Our initial study in this area was undertaken on the cat (Futami et al., 1979). First, we identified CST axons in the lateral funiculus of the cervical spinal cord, by their direct responses to electrical stimulation of the contralateral motor cortex. Next, such an axon was injected iontophoretically with HRP. The trajectory of each single, stained axon was then reconstructed in serial sections of the spinal cord. The stem axon in the lateral funiculus gave rise to multiple axon collaterals at different spinal segments. Primary collaterals ran ventromedially and entered the gray matter. There, they ramified successively to form a delta-like branching pattern in the transverse plane, with terminations in Rexed's laminae V–VII. In the horizontal plane, the axons bifurcated successively, gradually turning their orientation in parallel to the long axis of the spinal cord. Finally, they spread in both ascending and descending directions over a distance of 1.3–3.0 mm. Similar branching patterns were subsequently found for the RBST axons of the cat (Shinoda et al., 1982).

It was long known that in the cat spinal cord, both CST and RBST tracts terminate on interneurons, but not motoneurons (Lloyd, 1941). Therefore, in spite of our above-described demonstration of the existence of multiple axon collaterals at different spinal segments, it did not necessarily indicate that single axons in these tracts participate in the concurrent control of multiple muscles. On the other hand, Preston and Whitlock (1961) had shown that the CST of the monkey does indeed have direct projections onto spinal motoneurons. We took advantage of this morphological connection in an electrophysiological study, which showed that single monkey CST neurons were activated antidromically by stimulation of different motor nuclei in the cervical cord (Shinoda et al., 1979). In a subsequent study (Shinoda et al., 1981), we injected HRP into single CST axons in the lateral funiculus at the level of the cervical cord. These axons originated in the forelimb area of the monkey motor cortex. The stained CST axons had multiple axon collaterals in the cervical cord, and, importantly, such single axons projected to several motor nuclei supplying different forelimb muscles. Terminal boutons of these axons had close (presumed) contacts with dendrites of some of the motoneurons, which we also labeled with HRP, in multiple motor nuclei supplying the forelimb. This latter finding therefore showed that at least some CST neurons of the forelimb area of the motor cortex participate in the simultaneous control of multiple muscles of the forelimb. The question then arose as to which group of muscles is innervated by single, long descending motor-tract axons, i.e., what functional synergies among neck muscles are thereby implemented? The CST system in the monkey is not suitable to address this issue, because the abovementioned experiment is so technically difficult that its experimental yield is too low. Fortunately, it was known (Wilson and Melvill Jones, 1979) that vestibulospinal (VS) neurons have strong monosynaptic connections with neck motoneurons in the cat. Using this vestibulocollic system, we were indeed able to test for the implementation of a functional synergy by even a single, long descending motor-tract neuron.

The axonal trajectories of single vestibulospinal axons

Head position control is an ideal paradigm for studying how the CNS controls a multidimensional motor system (Richmond and Vidal, 1988). Head-movement signals detected by the semicircular canals are mediated through vestibulocollic pathways that link each of the three semicircular canals to a set of neck muscles. For tasks necessitating compensatory head movements, the CNS programs muscle-activation patterns in a *synergy*, i.e., a specific spatial and temporal combination rather than in an infinite variety of patterns. Stimulation of individual semicircular canals produces canal-specific head movements (Suzuki and Cohen, 1964). The plane of the head movement produced by canal stimulation parallels that of the stimulated canal, i.e., they are almost coplanar. Therefore, a signal from each semicircular canal must be distributed to an appropriate set of neck muscles in order to induce compensatory head movement in the same plane as the plane of the stimulated canal. Obviously, this process is more economical when a given descending motor command signal is distributed by a single neuron with divergent branches to multiple target motoneurons that participate in cocontraction of muscles to produce the required movement. The convergence of different canal nerve inputs onto single motoneurons has been extensively analyzed in the vestibulocollic system (Wilson and Melvill Jones, 1979), but its converse, divergent properties of VS axons, have not attracted attention until recently. To address the latter issue, we first analyzed the morphology of single physiologically identified lateral and medial VS tract (LVST and MVST) axons in the cat, by resort to intracellular HRP staining and subsequent three-dimensional reconstruction of axonal trajectories (Shinoda et al., 1986a, 1992a). An example of our approach is shown in Fig. 1.

Axons were penetrated in the cervical cord at C1–8 with an HRP-filled microelectrode. These axons were identified as VS axons by their monosynaptic responses to stimulation of the vestibular nerve, and further classified as either LVST or MVST axons by their responses to stimulation of the LVST and MVST. Figure 1 shows that when stained, these axons could be traced for a rostrocaudal distance of 3–16 mm. Within this length, both LVST and MVST axons were found to have multiple axon collaterals at different segments in the cervical cord. Up to seven collaterals were given off from the stems of both LVST and MVST axons. The LVST axons included neurons that terminated both at the cervical cord level and further caudally in the thoracic or lumbar cord. Each collateral of these LVST axons, after entering the gray matter, ramified successively in a delta-like fashion to terminate mainly in lamina VIII and the medial part of lamina VII. Many boutons of both terminal and en passant type made contact with cell bodies and proximal dendrites of neurons in the ventromedial nucleus (VM). Each collateral had a narrow rostrocaudal extension (0.2–1.6 mm; average, 0.8 mm) in the gray matter in contrast to a much wider intercollateral interval (average, 1.5 mm). This meant that there were gaps, which were free of terminal boutons, between adjacent collateral arborizations.

The morphology of the axon collaterals of MVST axons was very similar to that of the LVST axons. The rostrocaudal extent of their single axon collaterals was very restricted (0.3–2.1 mm) in contrast to the wider spread in the mediolateral and dorsoventral directions (0.9–2.6 mm). MVST axons had dense projections, with multiple axon collaterals, to the upper cervical cord. At the C1–3 levels, 1–7 collaterals of single MVST axons were identified. Their terminals were distributed in laminae VII–IX, including the VM, the nucleus of the spinal accessory

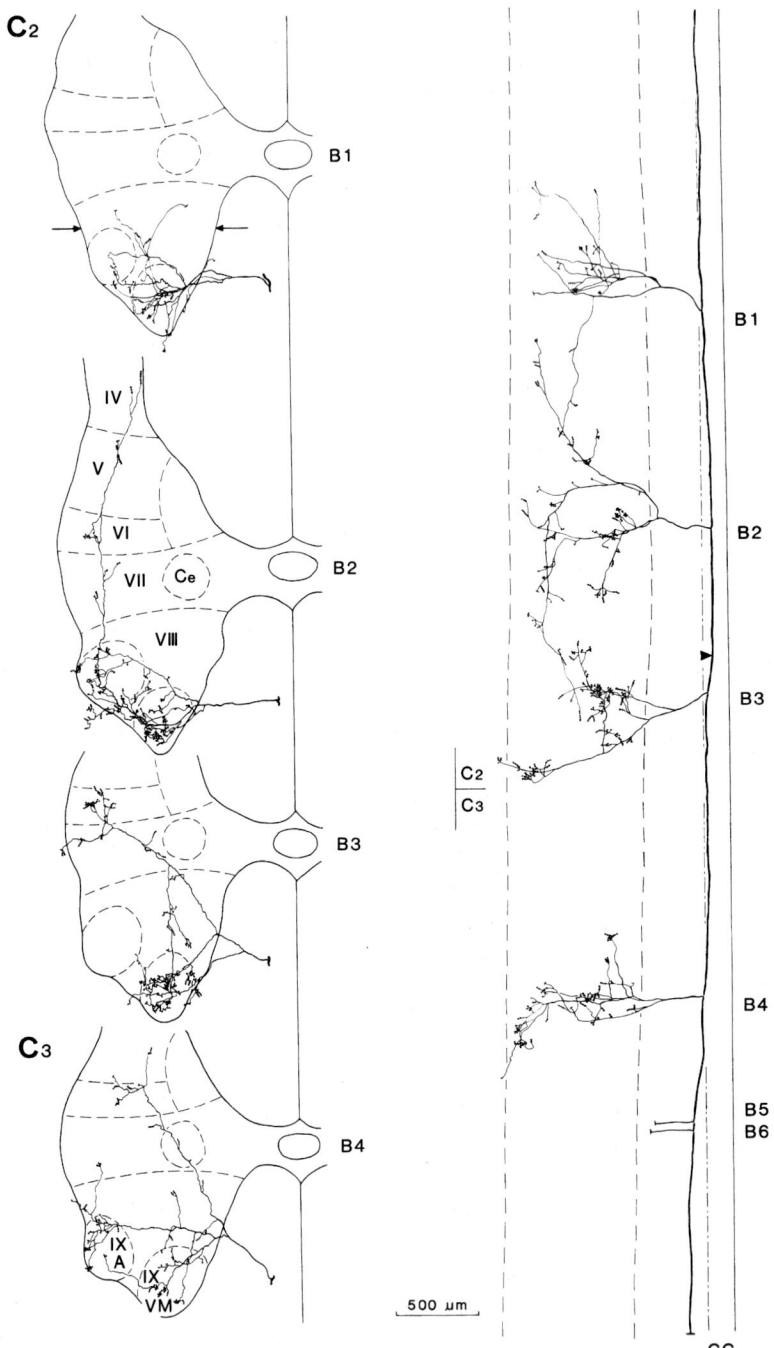

Fig. 1. Spinal distribution of an uncrossed MVST axon. Projection patterns are shown at the C2 and C3 levels in the transverse (left-side) and horizontal (right-side) plane. The overall reconstruction was made from 90 serial sections. The stem axon was traced further rostrally (to 2.8 mm above B1). This axon has terminals in both the ventromedial (VM) and spinal accessory (A) nucleus. Note that these terminals are distributed in Rexed's laminae IV–IX. Note further that many of the terminals of this axon made apparent contact with motoneurons supplying multiple neck muscles. (Reproduced from Shinoda et al., 1992a with permission from Wiley-Liss.)

nerve (A) and the commissural nucleus. Many terminals made contact with retrogradely HRP-labeled motoneurons supplying neck muscles. Both axosomatic and axodendritic contacts were observed on motoneurons of various size. Some collaterals gave rise to terminal arborizations in both the VM and the A. These results suggest that single LVST and MVST axons may help control the excitability of motoneurons to multiple axial muscles concurrent with their control of limb muscles at multisegmental levels.

Convergent patterns of input from the semicircular canals to neck motoneurons

The pattern of connections between different semicircular canals and motoneurons supplying dorsal neck muscles was first investigated by Wilson and Maeda (1974). Since then, the study of VS connections has been dominated by reports describing connections to a specific group of motoneurons that supply large dorsal neck muscles such as the biventer cervicis, complexus and splenius muscles (eg., Uchino and Isu, 1992). It has become increasingly apparent, however, that the large dorsal neck muscles involved in head movement represent a highly specialized group. More than 30 neck muscles must be controlled in an appropriate spatial and temporal pattern to induce a compensatory head movement in the same plane as the stimulated semicircular canal. Hence, it is more likely that there is more than one pattern of input from the six semicircular canals to motoneurons supplying different neck muscles.

To address the above issue, we investigated the patterns of input and the pathways from the six semicircular canals to motoneurons supplying 16 different neck muscles in the anesthetized cat. Intracellularly recorded PSPs from neck motoneurons were obtained during electrical stimulation of the six individual canal nerves (Sugiuchi et al., 1992b; Shinoda et al., 1994, 1997). Their separate stimulation evoked either excitatory (EPSP) or inhibitory (IPSP) postsynaptic potentials in *all* the tested neck motoneurons. Furthermore, virtually all such motoneurons received convergent inputs from the six canal nerves. All the motoneurons supplying a single neck muscle had one of four homogeneous patterns of input from the six semicircular canals. Figure 2 shows that the first of these four patterns was observed in motoneurons to the rectus capitis posterior (RCP) muscle.

In Fig. 2A–C, note that both ipsi- and contralateral stimulation of the anterior canal nerve (ACN) produced an EPSP in the shown (exemplary) RCP motoneuron, whereas analogous stimulation of the posterior canal nerve (PCN) produced an IPSP. Stimulation of the ipsi- and contralateral lateral canal nerve (LCN) produced an IPSP and EPSP, respectively.

To determine the central pathways that link the six semicircular canals to neck motoneurons, the effects of canal nerve stimulation were compared before and after a lesion of the medial longitudinal fasciculus (MLF), ipsilateral to the recorded motoneuron. After the cut, which was at the level of the obex, the EPSP evoked from the contralateral ACN and LCN disappeared, as did the IPSP from the contralateral PCN and the ipsilateral LCN and PCN. In contrast, the EPSP evoked by ipsilateral ACN stimulation remained unchanged. In other experiments, which are not shown in Fig. 2, a lesion was also made in the MLF contralateral to the recording site. In this case, however, no effect of sectioning was observed for bilateral canal inputs to RCP motoneurons.

Most of the EPSPs evoked by stimulation of the bilateral ACNs and contralateral LCN, and the IPSPs evoked by stimulation of the ipsilateral PCN and LCN, had latencies < 1.6 ms, and were therefore considered to be disynaptic. This input and connection pattern was seen in all the tested RCP motoneurons.

The second pattern of input from the six semicircular canals was observed in motoneurons of the obliquus capitis inferior (OCI) muscle. Their EPSPs were evoked by stimulation of the ipsilateral ACN and PCN and the contralateral LCN, whereas IPSPs were evoked by stimulation of the contralateral ACN and PCN, and the ipsilateral LCN. The latency of most of these responses was 0.9–1.6 ms, so they, too, could be classed as disynaptic. In contrast, stimulation of the contralateral ACN induced IPSPs at a latency ≥ 1.8 ms, i.e., ~ 1.0 ms longer than the disynaptic PSPs evoked by stimulation of the other canal nerves (Sugiuchi et al., 1995). These IPSPs were

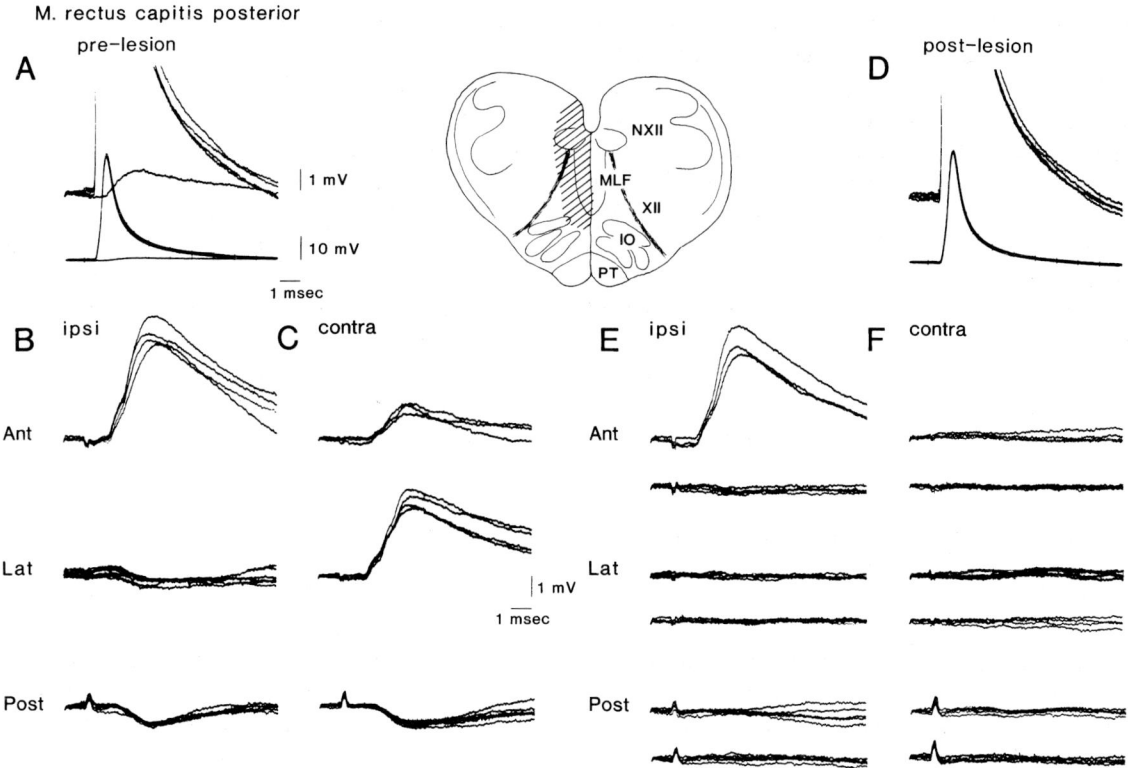

Fig. 2. Postsynaptic responses of an RCP motoneuron to stimulation of semicircular canal nerves before and after ipsilateral section of the MLF. (A–C) Control PSP responses with the MLF intact. (E–F) Analogous test responses after the complete MLF section. (A and D) Antidromic spikes (at two amplification levels) before (A) and after (D) the section. In A, the two smaller responses show monosynaptic EPSPs evoked by stimulation of neck muscle afferents. The upper voltage and the time calibration in A apply to the B–F traces. (B and C) Control PSPs evoked by stimulation of the three ipsilateral (B) and contralateral (C) canal nerves (Ant, anterior; Lat, lateral; Post, posterior). (E and F) Test PSPs recorded after the MLF section. In all cases, the strength of the stimulus pulse was 100 μA. During intracellular recording from this motoneuron, MLF sectioning at the level of the obex was performed in progressive mediolateral steps ipsilateral to the RCP motoneuron. The final (complete) lesion (hatched area) was reconstructed in a montage serial section (top middle sketch). This figure emphasizes that the lesion was well localized in the MLF and its immediate surroundings in the brainstem. Other abbreviations: NXII, hypoglossal nucleus; XII, hypoglossal nerve; IO, inferior olive; PT, pyramidal tract. (Reproduced from Shinoda et al., 1994 with permission from The American Physiological Society.)

classed as trisynaptic from primary vestibular afferents (Sugiuchi et al., 1992a).

The third pattern of semicircular canal input was observed in motoneurons of the muscles innervated by the A. In these motoneurons, stimulation of the three contralateral canal nerves evoked disynaptic EPSPs, whereas stimulation of their three ipsilateral counterparts evoked disynaptic IPSPs. These EPSPs and IPSPs were completely abolished by sectioning the MLF ipsilateral to the recorded motoneurons.

The fourth pattern was observed in motoneurons of the longus capitis (LC) muscle. These motoneurons received EPSPs from the bilateral PCNs and the contralateral LCN, and IPSPs from the bilateral ACNs and the ipsilateral LCN. The latencies of these PSPs were mainly <1.8 ms and therefore considered to be disynaptic. The single exception was the IPSPs evoked by stimulation of the contralateral ACN, which had a latency >1.8 ms. Therefore, these IPSPs were most likely trisynaptic from primary vestibular afferents. In this pattern, section of the ipsilateral MLF eliminated the PSPs evoked by individual canal nerves except those produced by stimulation of the contralateral ACN. In contrast, section of the contralateral LVST eliminated the IPSPs evoked from the contralateral ACN, but left unaffected the PSPs from the other five canal nerves.

Each of the examined 16 neck muscles was supplied by motoneurons with one of the above four patterns of input from the six canal nerves. Input patterns to the motoneurons of six of these muscles are exemplified in Fig. 3. The patterns shown emphasize the convergent nature of input from the six semicircular canals to motoneurons of individual neck muscles.

Note that the same type of data exemplified in Fig. 3 also revealed the divergent output patterns from a particular semicircular canal to motoneurons supplying the 16 selected muscles. For example, an inhibitory output from the LCN is distributed to the six neck muscles shown in Fig. 3, whereas an inhibitory output from the PCN is distributed to five of the same six (i.e., the OCI motoneurons received no such input). By this means, we were able to delineate the divergent output patterns of the six individual semicircular canals to the motoneurons supplying 16 different neck muscles.

To induce a compensatory head movement in the same plane as a stimulated semicircular canal, a signal from the latter must be distributed to an appropriate set of neck muscles. Two models seem possible for explaining the neural mechanisms underlying this spatial transformation in the vestibulocollic pathway. In one model: (1) each single MVST neuron innervates only one motor nucleus supplying a single neck muscle; and (2) primary afferents from a semicircular canal have a divergent projection pattern onto multiple MVST neurons. In the other model: (1) each single MVST neuron has a divergent projection pattern to a specific group of multiple motor nuclei; and (2) primary afferents originating from a semicircular canal innervate those MVST neurons that project to that particular group of multiple motor nuclei. The results presented above exclude the former model and provide indirect support for the latter one, because virtually all our sample of well-stained MVST axons were shown to project to more than one motor nucleus.

Innervation patterns of single MVST axons

The important question arises as to whether the branching pattern of an individual MVST axon reveals if it innervates different motor nuclei randomly or selectively. The assumption is that the latter would involve a combination of motor nuclei to a specific set of functionally relevant neck muscles. To address this issue, we undertook experiments in which we first labeled motoneurons of different neck muscles retrogradely with either HRP or a fluorescent dye, and constructed a map of motor nuclei at the C1 and C2 levels that supplied different neck muscles (Sugiuchi and Shinoda, 1992). Next, MVST axons were impaled by a microelectrode inserted into the ipsilateral or contralateral MLF. They were identified as VS axons by their monosynaptic responses to stimulation of primary vestibular afferents. Each axon was then classified in terms of the semicircular canal input it received, i.e., from the lateral, anterior or posterior canal. This involved determining the axon's maximal response to head rotations on a three-dimensional turntable (Shinoda et al., 1988, 1992b).

As an example of the above experimental approach, we discuss below the properties of MVST axons receiving input from the ipsilateral posterior canal and projecting to ipsilateral neck motor nuclei. After the above-mentioned physiological identification, such axons were injected with HRP in the upper cervical cord where 2–3 motor nuclei were retrogradely labeled with HRP. These axons ran in the ipsilateral MLF, and they commonly exhibited a stereotypic innervation of the motor nuclei of (1) the sternomastoid–cleidomastoid muscles, (2) the semispinalis group (i.e., the biventer, complexus and multifidus cervicis muscles) and (3) the RCP muscle. Each *single* collateral of these axons did not necessarily terminate within *all* of these motor nuclei. Other collaterals arising from the same stem axon, however, terminated in one or more of these nuclei. In other words, single MVST axons innervated the above group of the motor nuclei by way of multiple collaterals. This spatial innervation pattern was found in almost all of the tested posterior canal-related, uncrossed MVST axons. Figure 3 shows that posterior canal-related VS neurons include both excitatory and inhibitory neurons. The excitatory

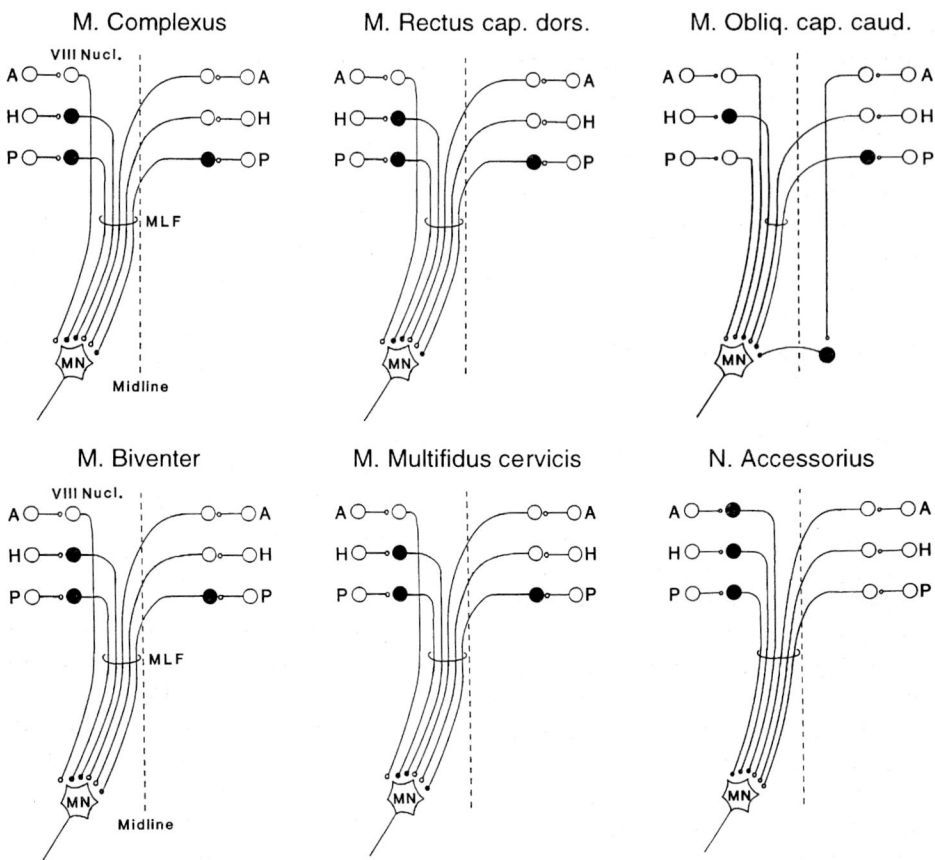

Fig. 3. Convergent input patterns and pathways from the six semicircular canals to motoneurons supplying six different neck muscles. Open and closed circles represent neurons that are excitatory and inhibitory, respectively. The key feature of this figure is that all the motoneurons supplying each neck muscle had a homogeneous pattern of input from the six semicircular canals. Abbreviations: A, H, and P, anterior, horizontal and posterior semicircular canal nerves, respectively; MN, neck motoneuron; MLF, medial longitudinal fasciculus.

neurons projected either contralaterally into the MLF or ipsilaterally in the LVST. Therefore, posterior canal-related VS axons passing in the ipsilateral MLF were most likely inhibitory to their target motoneurons, because posterior canal-related excitatory VS axons that projected ipsilaterally should be LVST axons and innervate motoneurons of the OCI muscle (viz. Fig. 3). Indeed, electron microscopic analysis of some HRP-labeled axon terminals of these VS axons demonstrated that their terminals had morphological characteristics of inhibitory synapses with OCI motoneurons (not illustrated).

We further tried to specify functional synergies among neck muscles that were innervated by single MVST axons, which responded to input from other semicircular canals. The posterior canal on one side is coplanar to the anterior canal on the opposite side. The above-mentioned posterior canal-related MVST axons on the ipsilateral side were inhibitory to the above-mentioned group of muscles. We also found that excitatory MVST axons on the contralateral side receiving input from the anterior semicircular canal terminated in the same group of neck motor nuclei. In other words, this result indicated that single MVST axons receiving input from either one of these coplanar canals exerted their influence on the same functional set of neck muscles, but with opposite effects.

In summary, it was possible to show that single MVST axons receiving input from a particular semicircular canal had a common projection onto the motoneurons of a set of neck motor nuclei that were appropriate for that same semicircular canal. Furthermore, the MVST axons' innervation patterns clearly explained the electrophysiologically determined output patterns from individual semicircular canals to motoneurons supplying functionally related sets of neck muscles.

Role of the tecto-reticulospinal system in the control of neck movements

The superior colliculus (SC) is known to be an important center for the control of coordinated eye–head (orienting) movements. Electrical stimulation of the SC elicits coordinated eye and head movements toward a target, with the orientation direction dependent on the specific site of intra-SC stimulation. Thus, the output signals of the SC should contain specific information on which neck muscles should be coactivated. Collicular influences onto neck motoneurons are conveyed mainly by tectospinal axons (Muto et al., 1996) and more indirect pathways via the brainstem's RTSTs. A key question is whether a single reticulospinal (RS) neuron may implement the same functional synergy among neck muscles as do vestibulocollic axons. To address this issue in the cat, we searched for RS axons in the upper cervical spinal cord of the cat. This involved their monosynaptic activation by stimulation of the contralateral SC, and their subsequent staining with HRP (Kakei et al., 1994). The stem axons of these RS neurons were shown to give rise to multiple axon collaterals which entered laminae VII–IX over a few cervical segments. They made contact with retrogradely HRP-labeled motoneurons supplying different neck muscles. We interpreted these results as meaning that RS axons simultaneously mediate output from the SC to motoneurons supplying functionally different groups of neck muscles. Furthermore, we showed that many RS axons innervate the same group of neck motor nuclei as do the posterior canal-related MVST axons. This finding strongly suggests that branching patterns of single RS axons are similar to those of single MVST axons, and that both single MVST and RS axons may have similar innervation patterns for neck muscles, and thereby control the same functional sets of these muscles.

Summary

In this chapter, we described our recent work on the divergent properties of single, long descending motor-tract neurons in the spinal cord, using the method of intra-axonal staining with HRP, and serial-section, three-dimensional reconstruction of their axonal trajectories. This work provides evidence that single motor-tract neurons are implicated in the neural implementation of functional synergies for head movements. Our results further show that single MVST neurons innervate a functional set of multiple neck muscles, and thereby implement a canal-dependent, head-movement synergy. Additionally, both single MVST and RS axons may have similar innervation patterns for neck muscles, and thereby control the same functional sets of neck muscles. In order to stabilize redundant control systems in which many muscles generate force across several joints, the CNS routinely uses a combination of a control hierarchy and sensory feedback. In addition, in the head-movement system, the elaboration of functional synergies among neck muscles is another strategy, because it helps to decrease the degrees of freedom in this particularly complicated control system.

Acknowledgments

This research was supported by Grants-in-Aid for Scientific Research from the Ministry of Education, Science and Culture of Japan to Y. Shinoda, Y. Sugiuchi and Y. Izawa.

Abbreviations

ACN	anterior canal nerve
CNS	central nervous system
CST	corticospinal tract
EPSP	excitatory postsynaptic potential
HRP	horseradish peroxidase
IPSP	inhibitory postsynaptic potential
LC	longus capitis
LCN	lateral canal nerve

LVST	lateral vestibulospinal tract
MLF	medial longitudinal fasciculus
MVST	medial vestibulospinal tract
A	nucleus of the spinal accessory nerve
OCI	obliquus capitis inferior
PCN	posterior canal nerve
PSP	postsynaptic potential
RBST	rubrospinal tract
RCP	rectus capitis posterior
RS	reticulospinal
RTST	reticulospinal tract
SC	superior colliculus
Th	thoracic
VM	ventromedial nucleus
VS	vestibulospinal

References

Abzug, C., Maeda, M., Peterson, B.W. and Wilson, V.J. (1974). Cervical branching of lumber vestibulospinal axons. J. Physiol. (Lond.), 243: 499–522.

Futami, T., Shinoda, Y. and Yokota, J. (1979). Spinal axon collaterals of corticospinal neurons identified by intracellular injection of horseradish peroxidase. Brain Res., 164: 279–284.

Jankowska, E., Rastad, J. and Westerman, J. (1976). Intracellular application of horseradish peroxidase and its light and electron microscopical appearance in spinocervical tract cells. Brain Res., 105: 552–562.

Kakei, S., Muto, N. and Shinoda, Y. (1994). Innervation of multiple neck motor nuclei by single reticulospinal tract axons receiving tectal input in the upper cervical spinal cord. Neurosci. Lett., 172: 85–88.

Kitai, S.T., Kocsis, J.D., Preston, R.J. and Sugimori, M. (1976). Monosynaptic inputs to caudate neurons identified by intracellular injection of horseradish peroxidase. Brain Res., 109: 601–606.

Lloyd, D.P.C. (1941). The spinal mechanism of the pyramidal system in cats. J. Neurophysiol., 4: 525–546.

Muto, N., Kakei, S. and Shinoda, Y. (1996). Morphology of single axons of tectospinal neurons in the upper cervical spinal cord. J. Comp. Neurol., 372: 9–26.

Peterson, B.W., Maunz, R.A., Pitts, N.G. and Mackel, R. (1975). Patterns of projection and branching of reticulospinal neurons. Exp. Brain Res., 23: 331–351.

Preston, J.B. and Whitlock, D.G. (1961). Intracellular potentials recorded from motoneurones following precentral gyrus stimulation in primate. J. Neurophysiol., 24: 91–100.

Richmond, F.J.R. and Vidal, P.P. (1988). The motor system: joints and muscles of the neck. In: Peterson B.W. and Richmond F.J. (Eds.), Control of Head Movement. Oxford University Press, New York, pp. 1–21.

Shinoda, Y. (1999). Visualization of the entire trajectory of long axons of single mammalian CNS neurons. Brain Res. Bull., 50: 387–388.

Shinoda, Y., Arnold, A. and Asanuma, H. (1976). Spinal branching of corticospinal axons in the cat. Exp. Brain Res., 26: 215–234.

Shinoda, Y., Ghez, C. and Arnold, A. (1977). Spinal branching of rubrospinal axons in the cat spinal cord. Exp. Brain Res., 30: 203–218.

Shinoda, Y., Zarzecki, P. and Asanuma, H. (1979). Spinal branching of pyramidal tract neurons in the monkey. Exp. Brain Res., 34: 59–72.

Shinoda, Y., Yokota, J. and Futami, T. (1981). Divergent projection of individual corticospinal axons to motoneurons of multiple muscles in the monkey. Neurosci. Lett., 23: 7–12.

Shinoda, Y., Yokota, J. and Futami, T. (1982). Morphology of physiologically identified rubrospinal axons in the spinal cord of the cat. Brain Res., 242: 321–325.

Shinoda, Y., Ohgaki, T. and Futami, T. (1986a). The morphology of single lateral vestibulospinal tract axons in the lower cervical cord of the cat. J. Comp. Neurol., 249: 226–241.

Shinoda, Y., Yamaguchi, T. and Futami, T. (1986b). Multiple axon collaterals of single corticospinal axons in the cat spinal cord. J. Neurophysiol., 55: 425–448.

Shinoda, Y., Ohgaki, T., Sugiuchi, Y. and Futami, T. (1988). Structural basis for three-dimensional coding in the vestibulospinal reflex; morphology of single vestibulospinal axons in the cervical cord. Ann. N.Y. Acad. Sci., 545: 216–227.

Shinoda, Y., Ohgaki, T., Sugiuchi, Y. and Futami, T. (1992a). The morphology of single medial vestibulospinal tract axons in the upper cervical spinal cord of the cat. J. Comp. Neurol., 316: 151–172.

Shinoda, Y., Ohgaki, T., Sugiuchi, Y. and Futami, T. (1992b). Spatial innervation patterns of single vestibulospinal axons in neck motor nuclei. In: Berthoz A., Graf W. and Vidal P.P. (Eds.), The Head-Neck Sensory Motor System. Oxford University Press, Oxford, pp. 259–265.

Shinoda, Y., Sugiuchi, Y., Futami, T., Ando, N. and Kawasaki, S. (1994). Input patterns and pathways from six semicircular canals to motoneurons of neck muscles. I. The multifidus muscle group. J. Neurophysiol., 72: 2691–2702.

Shinoda, Y., Sugiuchi, Y., Futami, T., Ando, N. and Yagi, J. (1997). Input patterns and pathways from six semicircular canals to motoneurons of neck muscles. II. The longissimus and semispinalis muscle group. J. Neurophysiol., 77: 1234–1248.

Snow, P.J., Rose, P.K. and Brown, A. (1976). Tracing axons and axon collaterals of spinal neurones using intracellular dye injection. Science, 191: 312–313.

Sugiuchi, Y. and Shinoda, Y. (1992). Organization of the motor nuclei innervating epaxial muscles in the neck and back. In: Berthoz A., Graf W. and Vidal P.P. (Eds.), The Head-Neck

Sensory Motor System. Oxford University Press, Oxford, pp. 235–240.

Sugiuchi, Y., Kakei, S. and Shinoda, Y. (1992a). Spinal commissural neurons mediating vestibular input to neck motoneurons in the cat cervical spinal cord. Neurosci. Lett., 145: 221–225.

Sugiuchi, Y., Futami, T., Ando, N., Kawasaki, T., Yagi, Y. and Shinoda, Y. (1992b). Patterns of connections between six semicircular canals and neck motoneurons. Ann. N.Y. Acad. Sci., 656: 957–959.

Sugiuchi, Y., Izawa, Y. and Shinoda, Y. (1995). Trisynaptic inhibition from the contralateral vertical semicircular canal nerves to neck motoneurons mediated by spinal commissural neurons. J. Neurophysiol., 73: 1973–1987.

Suzuki, J.-I. and Cohen, B. (1964). Head eye, body and limb movements from semicircular canal nerves. Exp. Neurol., 10: 393–405.

Uchino, Y. and Isu, N. (1992). Properties of vestibulo-ocular and/or vestibulo-collic neurons in the cat. In: Berthoz A., Vidal P.P. and Graf W. (Eds.), Head-Neck Sensory-Motor System. Oxford University Press, New York, pp. 266–272.

Wilson, V.J. and Maeda, M. (1974). Connections between semicircular canals and neck motoneurons in the cat. J. Neurophysiol., 37: 346–357.

Wilson, V.J. and Melvill Jones, G. (1979). Mammalian Vestibular Physiology. Plenum Press, New York.

CHAPTER 40

Control of orienting movements: role of multiple tectal projections to the lower brainstem

Alexej Grantyn[1]*, Adonis K. Moschovakis[2] and Toshihiro Kitama[3]

[1]*Laboratoire de Physiologie de la Perception et de l'Action, CNRS-Collège de France, 75005 Paris, France*
[2]*Institute of Applied and Computational Mathematics, F.O.R.T.H., Heraklion, Crete, Greece*
[3]*Department of Physiology, Yamanashi Medical University, Tamaho, Yamanashi 409-38, Japan*

Abstract: In movement neuroscience this past decade, a conceptual approach that puts emphasis on population coding was clearly dominant. The purpose of numerous studies has been to define presumably homogeneous groups of neurons on the basis of the correlation of their discharges with sensory and motor events. The goal of this chapter is to stress the importance of taking into account individual properties of neurons, this being an essential prerequisite for a biologically meaningful definition of neuron populations. Taking as an example the executive limb of the neural network controlling gaze movements, we demonstrate the functional and anatomical *diversity* of tectal and reticular neurons, which are generally considered as homogeneous populations and used, accordingly, as lumped elements in models. We argue that the extraction of effector-specific signals from the global command of gaze displacement is based not on the interplay between discrete neural modules, but rather on a gradual process of signal specification at *all* levels of the executive network. An eventual accurate description of this network will require knowledge of the unique combinations of afferent inputs and efferent connections for as many subsets of its constituent neurons as is conceivably possible.

Introduction

Changing the direction of gaze toward new attracting targets is a well-known pattern of motor behavior. In primates and cats, the most studied component of orienting is eye saccades. Under natural conditions, however, and depending on the eccentricity of targets selected for foveation, gaze displacement is often achieved by coordinated movements of the eyes, head, body and forelimbs. Understanding how neurons located in several brain areas work together to assure a rapid and precise realignment of the line of sight is a fundamental issue in movement neuroscience and one that comprises a number of general conceptual issues. In this chapter we address, in particular, the problems of coordinate transformation, extraction of premotor signals specific to the control of dissimilar effectors, and adjustment of their timing and dynamic properties to optimize the coordination of several groups of muscles. As an example, we use the executive portion of the brainstem that controls horizontal gaze movements. This neural network includes the superior colliculus (SC), its efferent connections and ponto-bulbar cell groups serving as interface between the SC and motor nuclei. The complexity of the underlying connections can be appreciated from Fig. 1, which illustrates extensive and overlapping branching patterns of intra-axonally labeled orienting-related tectal and reticular neurons (Grantyn et al., 1992b; Olivier et al., 1993). The structure and function of the tecto-reticulo-spinal system has been repeatedly reviewed (Hess et al., 1946; Sprague and Meikle, 1965; Wurtz and Albano, 1980; Sparks, 1986; Grantyn et al., 1993; Moschovakis et al., 1996). For this reason, we

*Corresponding author: Tel.: +33-1-4427-1628; Fax: +33-1-4427-1382; E-mail: alexej.grantyn@college-de-france.fr

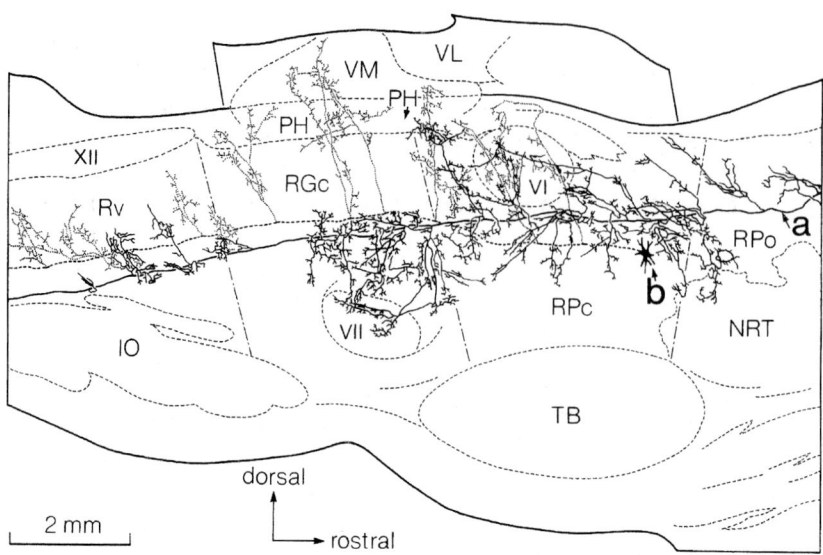

Fig. 1. Overview of connections of the tecto-reticulo-spinal system in the pons and medulla. Parasagittal reconstruction of two orienting-related neurons labeled by intra-axonal injections of HRP in alert cats. Axonal branching of a visuomotor tecto-reticulo-spinal neuron (a, solid lines) is superimposed on a topographically matched reconstruction (b, dotted lines) of an 'eye–neck' reticulospinal neuron. Monosynaptic input from the contralateral SC to this reticular neuron was identified by intra-axonal recording of an attenuated EPSP. Note extensive branching (typical for the two classes of neurons), partial overlap of terminal domains in the caudal pons and differences with respect to the arrays of contacted target areas. Abbreviations: IO, inferior olive; NRT, tegmental reticular nucleus; PH, nucleus prepositus hypoglossi; RGc, gigantocellular reticular nucleus; RPc, RPo, caudal and oral pontine reticular nuclei, respectively; Rv, ventral reticular nucleus; TB, trapezoid body; VM, VL, medial and lateral vestibular nuclei, respectively; VI, VII, XII, abducens, facial and hypoglossal cranial motor nuclei, respectively.

focus here on only those selected aspects of its organization that are closely related to our own experimental data (for allied work, see Peterson, Sasaki, Fukushima et al., Uchino, and Kobayashi et al., Chapters 35–38 and 41 of this volume).

Anatomical substrate for spatiotemporal transformations within the tecto-reticulo-spinal system

The size and velocity of horizontal and vertical saccade components are encoded in the temporal characteristics of discharges emitted by burst neurons of the so-called saccadic burst generators (BGs). Horizontal components are controlled by neurons located in the paramedian pontine reticular formation (PPRF). Cell groups comprising the horizontal saccadic BG have been extensively explored, and the operation of the circuit has been simulated in several models (for review, see Moschovakis et al., 1996). All models agree that in order to produce saccades of larger size, saccade-related bursters in the PPRF must receive a stronger excitatory input from the SC. The discharges of tectal burst neurons projecting to the PPRF, however, are of about the same duration and intensity, irrespective of whether their position on the motor map corresponds to small or large saccades (Sparks and Mays, 1980). These neurons provide a spatial code for saccade metrics, but this code must be changed to a temporal one by the time it reaches the saccade generators. In modeling studies, this spatiotemporal transformation is usually implemented by assuming an increasing weight of projections to the BG from populations of SC neurons located at increasingly larger eccentricities of the motor map (Scudder, 1988; Arai et al., 1994; Moschovakis, 1994; Das et al., 1995). Such models need not specify neural correlates of the postulated differential weighting. The latter can result from appropriately oriented gradients of the concentration of SC neurons projecting to the saccadic generators, or from the differential adjustment of the synaptic efficacy of descending connections. Whereas the efficacy of synaptic transmission cannot be directly explored in experiments requiring the behavioral identification of neurons, two anatomical correlates of connection strength are accessible to experimental verification. First, density gradients of SC neurons projecting to the PPRF have been the object of a few studies in cats but their results are contradictory, thereby ruling out any definite conclusion (reviewed in Olivier et al., 1991). A second possibility is that the size of saccades is encoded in the number of terminals deployed in the region of the saccadic BG. This has not been explored previous to our own recent study.

For this work, we used anterograde tracing with biocytin to evaluate connections between circumscribed zones of the SC and the region of the horizontal saccadic BG in the PPRF (Moschovakis et al., 1998). Tracer injections were combined with the analysis of saccades evoked by microstimulation of the same zones while the cats were alert. As shown in Fig. 2A, the positions of the sites studied were distributed over a substantial portion of the SC's motor map.

Figure 2B illustrates 'characteristic vectors' of saccades evoked from each of the tracer injection sites with standard stimulus parameters. Such vectors correspond to saccades evoked when the eyes are centered in the orbit (McIlwain, 1990). Their horizontal component (β_H in Fig. $2C_1$–E_1 and F) is obtained from the regression lines that link the size of horizontal saccade components to horizontal eye position at the time of stimulation. The calculation of β_H values (the characteristic amplitude of horizontal components of evoked saccades) is necessary to correctly evaluate the metrics of evoked saccades by eliminating the effect of position sensitivity (McIlwain, 1990; Grantyn et al., 1996). In the experiments selected for quantitative evaluation, the zones occupied by biocytin reaction product measured 0.8–1.3 mm (Fig. 2C–E). Their size matched well with the effective spread of stimulation currents (reviewed in Tehovnik, 1996).

The overall distribution of labeled boutons in the pontine tegmentum is shown by composite plots in Fig. $2C_1$–E_1. Counts were confined to the region of the contralateral PPRF corresponding to the location and extent of the horizontal saccadic generator, as defined by previous studies in cats (Curthoys et al., 1981; Kaneko et al., 1981; Sasaki and Shimazu, 1981). The area of counts is indicated by dashed frames in Fig. $2C_1$–E_1. Quantitative data from six experiments are illustrated in Fig. 2F as normalized bouton numbers in sequential 75-μm sections through the PPRF. The normalization consisted of dividing the number of boutons in each section by the number of labeled SC projection axons at their entry in the pons. The final result of the study is given in the inset of Fig. 2F. It demonstrates that average normalized bouton numbers in the region of interest increase linearly with β_H. This relationship was highly significant ($P < 0.0001$). In contrast, significant correlations were not found when either radial saccade amplitude or saccade direction was used as the independent variable. It is important to note that bouton numbers were not correlated to the saccade size predicted from the anatomical location of SC injection sites. A purely anatomical experiment, without simultaneous measurement of saccade parameters, would therefore fail to reveal the gradient of connection strength.

Our study gives the first experimental evidence to support the intuitively attractive idea that the spatiotemporal transformation in the tecto-reticular pathway is based on a positive correlation between the size of saccades and the density of terminations in the region of the saccadic generator. Can this termination gradient account completely for the encoding of saccade size? A definite answer is not yet possible. Worthwhile, however, is a comparison with the gradients of connection weights, suggested by distributed models of the primate SC. For an increment of saccade size from 5° to 10°, which is representative of the saccade range in our study, different optimization algorithms (Arai et al., 1994; Das et al., 1995; Moschovakis, 1994) predict an about 2.5-fold increment of projection strength from the corresponding SC sites. Our normalized bouton counts indicate a two-fold increment for the same change in the size of the horizontal component. The experimentally obtained gradient is somewhat weaker, thereby suggesting that the spatiotemporal transformation is not entirely implemented at the level of synaptic terminals. The second anatomical

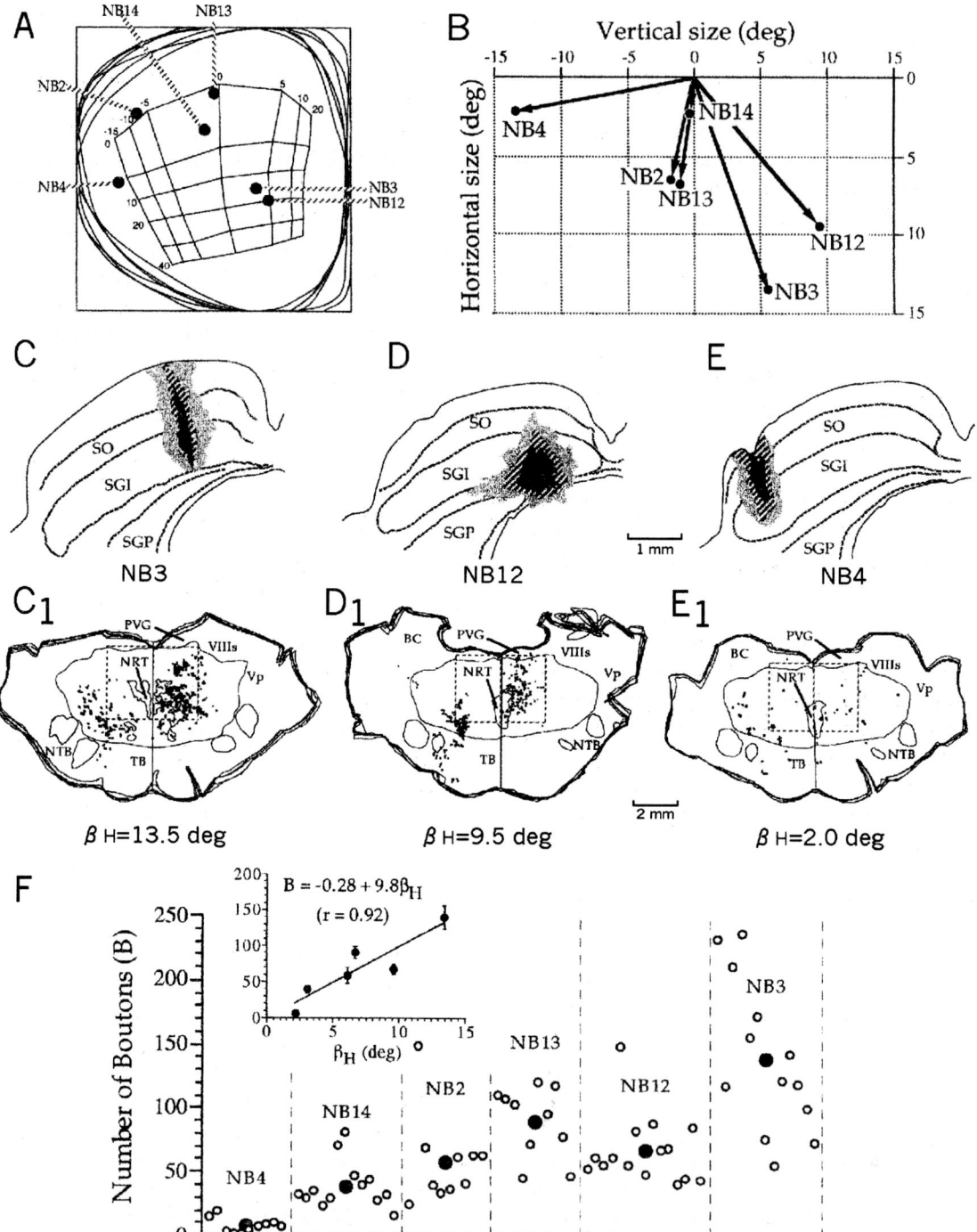

Fig. 2. A combined microstimulation–anterograde-tracing study of SC projections onto the region of the pontine saccadic generator. (A) Sites of stimulation and biocytin injections projected onto the horizontal plane together with the outlines of the SC and its visuomotor map. Labels specifying different experiments (NB2, etc.) are the same in A through F. (B) Characteristic vectors of saccades evoked from the sites shown in A at a stimulus intensity of two times the threshold. (C–E) Examples of serial reconstructions

factor, the nonhomogeneous distribution of saccade-related neurons in the SC, cannot yet be rejected because the results of available studies are inconsistent.

The above straightforward interpretation is based on the implicit assumptions that: (1) SC neurons labeled after biocytin injection participate in saccade generation; and (2) the region of the PPRF selected for bouton counts contains no other neurons but those composing the horizontal saccade generator. It is known, however, that many tectal neurons projecting into the predorsal bundle, as well as their target neurons in the PPRF, do not participate in the control of orienting movements (Grantyn and Berthoz, 1985; Moschovakis et al., 1988; Isa and Naito, 1995; Kitama et al., 1995). By definition, connections of such 'unrelated' cells are unlikely to show a correlation with eye-movement metrics. Their labeling could not therefore produce a spurious gradient of connection strength in our bulk-tracing study. The use of bulk-tracer injections in our and other anatomical studies is a global approach, similar to the use of lumped elements in models of sensorimotor functions. It cannot give a complete description of the premotor network because it does not take into account the functional and anatomical diversity of neurons within populations defined by some global anatomical or behavioral criteria. We know that such populations are heterogeneous and that they are composed of neurons that differ according to their contribution to the control of movement. Neurons do differ also with respect to their individual patterns of projections and, hence, with the destination of their signals to distinct arrays of postsynaptic targets. We provide some examples of the intrinsic heterogeneity of orienting-related cell groups of the tecto-reticular system in the next section.

Anatomical heterogeneity of orienting-related tectal efferent neurons

In alert cats, the axonal architecture of tectal projection neurons has been studied in a small sample of cells identified by their orienting-related discharge patterns as members of a large class of visuomotor tecto-reticulo-spinal neurons (vm-TRSNs; Grantyn and Berthoz, 1985; Olivier et al., 1993). From their quantitative extracellular study in head-unrestrained cats, Munoz et al. (1991) concluded that cells of this class transmit a gaze motor error signal, without specifying the contributions of the eyes and the head to gaze displacement. Antidromic identification and labeling experiments have shown that axons of all tested vm-TRSNs course through the rhombencephalon in the contralateral predorsal bundle and many reach the spinal cord.

The general pattern of axonal branching in the rhombencephalon is shown by solid lines in Fig. 1 for one behaviorally identified vm-TRSN, which was labeled by an intra-axonal HRP injection. We selected this cell to illustrate its exemplary extensive branching, such that the regions receiving its terminations encompass practically all contralateral target areas of the SC reported in anterograde bulk-tracing studies (Huerta and Harting, 1982; Cowie and Holstege, 1992). The general pattern of branching was similar in all labeled vm-TRSNs, but they differed in terms of the combination of their target areas and the numbers of boutons they deployed in each of them. This is illustrated in Fig. 3 for three neurons that displayed a complete labeling of terminals in the caudal pons and rostral medulla (Olivier et al., 1993). Reconstruction of their terminal domains clearly shows an overlap of projections in the reticular formation (RF), including PPRF, and in the abducens nucleus (AbdN). On the other hand,

---

(*Contd.*) of stimulation–injection sites used to estimate the size of the zones that contributed to the tracer uptake. Different degrees of shading for intense (black), moderate (hatched) and weak (stippled) tracer deposits. (C_1–E_1) Composite plots (each five overlaid sections) of labeled boutons in the PPRF. Regions of the saccadic BG are delimited by interrupted lines. Bouton counts were made on the right side, contralateral to the injection sites in the SC. Amplitudes of horizontal components of characteristic saccade vectors (β_H) are given below each bouton plot. (F) Scatter plot of normalized boutons numbers per section (open circles) and their mean values (solid circles) in six experiments, arranged in the order of increasing β_H. Inset: plot of mean bouton numbers as function of β_H and regression line describing this relationship. Abbreviations: BC, brachium conjunctivum; NRT, tegmental reticular nucleus; NTB, nucleus of the trapezoid body; PVG, periventricular gray; SO, SGI, SGP, optic, intermediate gray and deep gray layers of the SC, respectively; TB, trapezoid body; Vp, principal sensory nucleus of the trigeminal nerve; VIIIs, superior vestibular nucleus.

bouton counts reveal that the strength of projections of individual neurons to the AbdN and to areas containing excitatory (Fig. 3A) and inhibitory (Fig. 3C) bursters of the saccadic generator differ by more than three-fold. Bearing in mind that vm-TRSNs control combined eye and head movements, it is important to note that the same areas contain orienting-related reticulospinal neurons (Fig. 3, filled symbols). Several structures outside the medial RF are contacted in a selective manner. For example, the midline region containing omnipause neurons (OPNs) of the saccadic generator receives a substantial projection from only one of the three neurons (Fig. 3A–B). Also the nucleus prepositus hypoglossi (NPH), the site of the oculomotor neural integrator, is contacted only by some of the vm-TRSNs (Figs. 1 and 3C) and it therefore represents a nonobligatory target of orienting-related SC neurons.

Morphological heterogeneity of a functionally distinct population of tectal projection cells has also been clearly demonstrated in monkeys. It concerns neurons named 'tectal long-lead bursters' (TLLBs) by Moschovakis et al. (1988). These generate bursts with a brisk onset about 20 ms before saccades in a certain range of directions and amplitudes. As mentioned above, by their location in the SC, such cells define the metrics of saccades in a vectorial form. Moschovakis et al. (1988) have shown that all these cells belong to the anatomical class of T neurons, so named because, besides projecting in the predorsal

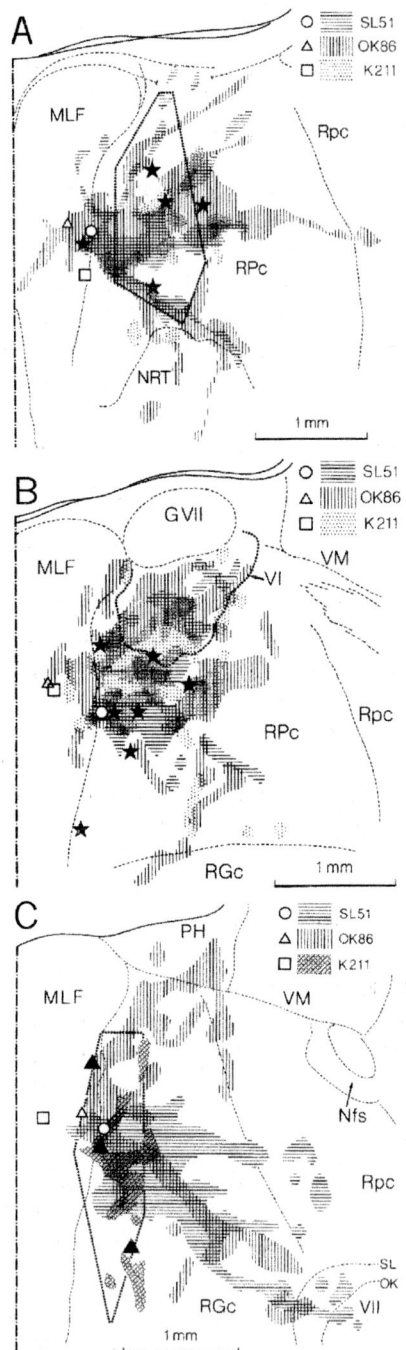

Fig. 3. Terminal fields of three behaviorally identified and intra-axonally labeled visuomotor tecto-reticulo-spinal neurons at three different brainstem levels. (A) Rostral half of the caudal pontine reticular nucleus (RPc). (B) Its caudal half. (C) Rostral portion of the gigantocellular reticular nucleus (RGc). At each level, hatched areas correspond to the locations of boutons collected in 2-mm long segments of the brainstem and projected onto the frontal plane. Insets on the top right provide the key for following terminations of each neuron at different levels. Open symbols: positions of labeled axons in the predorsal bundle. Black stars: positions of phasic-sustained and quasi-tonic 'eye–neck' reticulospinal neurons. Black triangles: positions of phasic reticulospinal neurons. Dotted polygons delimit locations of excitatory (A) and inhibitory (C) burst neurons of the saccadic generator according to Sasaki and Shimazu (1981) and Hikosaka et al. (1978). Abbreviations: GVII, genu of the facial nerve; MLF, medial longitudinal fasciculus; Nfs, nucleus of the solitary tract; NRT, tegmental reticular nucleus; PH, nucleus prepositus hypoglossi; RGc, gigantocellular reticular nucleus; RPc, caudal pontine reticular nucleus; Rpc, parvocellular reticular nucleus; VM, medial vestibular nucleus; VI, VII, abducens and facial nuclei. (Reproduced from Olivier et al., 1993 with permission from Springer-Verlag.)

bundle, they make a commissural connection with the SC on the opposite side. For 12 cells of this class, Scudder et al. (1996) obtained a complete description of their terminations in the ponto-bulbar tegmentum. They noted convergence to a common set of target areas from all members of this population, and individual variations in the distribution of terminal fields. As in the case with the cat vm-TRSN group, the sample was too small for a study of the correlation between individual patterns of connectivity and such functional properties as directional tuning and timing of bursts with respect to saccades.

One way to interpret the individual variations of connectivity is to suppose that they level out at the population level. Alternatively, the nonobligatory connections are weak and may be neglected for a global (coarse-grain) description of the gaze control system. Another point of view maintains that even subtle differences in connections reflect functional heterogeneity of the population. To experimentally support this latter notion, one would need to record and label both the SC projection neurons and their target neurons simultaneously in behaving animals, a task that does not seem currently feasible. As we show in the next section, however, indirect experimental evidence does support the view that differential connections of vm-TRSNs can account for the SC control of orienting movements having different kinematic properties.

Tectal control of movements with different kinematic properties

Figure 4A illustrates motor components of orienting synergy together with the discharge of an exemplary vm-TRSN in a head-restrained cat. Besides saccadic eye movement toward a manually presented visual target, the animal attempts to move the head, as can be recognized from the transient increase in neck muscle EMG activity. We selected this example to show that during the final stage of gaze shifts, cats may generate slow eye movements (postsaccadic drifts; PSD) instead of corrective saccades. When the head is prevented from moving, slow movements may suffice to foveate targets at small eccentricity. Usually, however, such slow movements serve to correct residual gaze errors after the end of saccades

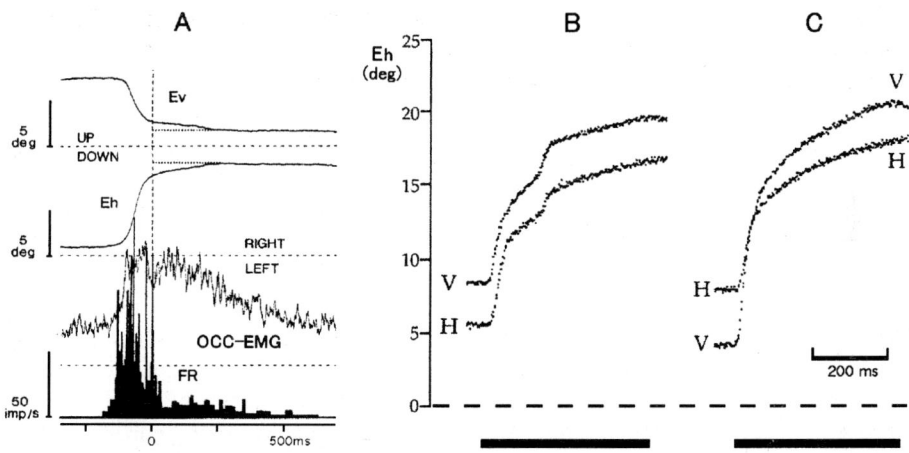

Fig. 4. Combined saccadic and slow eye movements during natural orienting behavior and in response to electrical stimulation of the SC. (A) Averages of visually triggered eye movements, neck EMG activity and associated discharge of a visuomotor tecto-reticulo-spinal neuron. The averages are based on seven trials, which are aligned (vertical dashed line) at the end of saccades and beginning of PSDs. From top to bottom: vertical (Ev) and horizontal (Eh) eye positions, rectified and integrated EMG activity of the obliquus capitis cranialis muscle (OCC-EMG), and instantaneous firing rate (FR) of an exemplary neuron. The cell is a SC projection neuron that generated perisaccadic bursts followed by decremental firing during PSDs and dynamic overshoot of the EMG activity. (B–C) Vertical (V) and horizontal (H) components of eye movements evoked by stimulation of the SC with prolonged (450 ms) 200-Hz pulse trains at three (B) and four times (C) the threshold intensity. Periods of stimulation are indicated by horizontal bars. Note variations of the time course of slow drifts between ramp- (B) or exponential-like (C) waveforms. (Reproduced, in part, from Olivier et al., 1993 with permission from Springer-Verlag; and from Grantyn et al., 1996 with permission from Elsevier Science.)

(Missal et al., 1993; Olivier et al., 1993). The Fig. 4A record of a vm-TRSN's discharge shows a high-frequency, phasic component during the saccade and the rapid buildup of neck muscle activity. About 20% of vm-TRSNs continue to fire at a lower rate after the saccade's offset (Fig. 4A). The duration of postsaccadic discharges tends to be positively correlated with the duration of the PSDs. Based on these observations, Olivier et al. (1993) proposed that the SC is able to control eye movements of quite different velocity. In a subsequent study (Grantyn et al., 1996), we demonstrated that prolonged microstimulation of the SC induces not only the well-known saccadic responses but also PSDs (Fig. 4B–C). We paid particular attention to the properties of the evoked PSDs because we thought that their quantitative analysis might explain how prolonged output volleys of the SC generate a sequence of saccades or, alternatively, a saccade followed by a slow movement.

An obvious difference between saccades and PSDs is their difference in velocity (Fig. 4B–C). Peak velocities of the smallest, and hence the slowest, stimulus-evoked saccades in our study were about 60–80°/s, whereas the fastest PSDs attained only 16°/s. Also, the drifts following natural, visually triggered saccades only rarely reach 20°/s (Olivier et al., 1993). Increasing the duration of the stimulus train, while keeping constant its intensity and pulse rate, has no effect on any of the saccade characteristics, besides eventually giving rise to a sequence of saccades (Fig. 4B). On the contrary, PSDs continue without saturation till the end of the stimulus train. Their amplitude is therefore proportional to train duration, at least within the limits tested in our experiments (450 ms). Varying the pulse rate in the range of 200–500 Hz has no influence on the saccade parameters, but both the amplitude and mean velocity of the PSDs increase with higher rates. All these differences between evoked saccades and PSDs indicate that SC stimulation simultaneously activates more than one neural substrate. As reported by Olivier et al. (1993), bursts of some vm-TRSNs terminate near the end of saccades while others display postsaccadic firing. Such a differentiation is corroborated by the absence of correlation between the amplitudes of saccades and PSDs evoked from different SC sites (Grantyn et al., 1996). Furthermore, the amplitude of saccades and PSDs does not increase at the same rate with increasing stimulus intensity. This can be explained if one assumes that increasing the size of the activated zone changes the ratio of saccade and PSD-related neurons. Such a differentiation at the level of the SC, however, is insufficient to account for stimulus-evoked PSDs because a long-lasting excitatory input to the saccadic BG should produce a sequence of saccades and not a slow movement. An additional factor could be the differential strength of connections between vm-TRSNs and pontine preoculomotor cell groups.

To test the validity of this second factor, we tried to reproduce typical waveforms of electrically evoked PSDs using the 'MSH model' of the saccadic system (Moschovakis, 1994). The experimentally observed time course of PSDs can often be described by either exponential (Fig. 5A) or linear (Fig. 5B) functions. More frequently, the trajectories resemble a superposition of exponential and ramp-like forms (Fig. 4B–C). Fig. 5C–F presents the results of our simulation (Grantyn et al., 1996). Assumptions about the relative strength of direct and indirect SC connections with ocular motoneurons are shown to the left of movements generated by the model in response to a prolonged step-like output from the SC. When SC output is routed to motoneurons exclusively through the saccadic BG (Fig. 5C), the resulting movement is a staircase of saccades. This corresponds to the configuration of connections used in models of the primate saccadic system (for review, see Moschovakis, 1994; Moschovakis et al., 1996). Connections in the cat are different in that some vm-TRSNs make direct collateral connections with abducens motoneurons and with the NPH, the site of the oculomotor neural integrator (see Figs. 1 and 3). These pathways bypass the saccadic BG and are amplified by disynaptic links through orienting-related reticulospinal neurons (Figs. 1 and 3). As expected from the mechanical properties of the eye in the orbit, the output of the model assumes an exponential time course when a step-like SC efferent signal is routed through the direct tecto-oculomotor pathway (Fig. 5D). A selective activation of connections linking the SC to the network of the oculomotor neural integrator would result in a slow movement following a ramp-like trajectory (Fig. 5E). Finally, a more realistic assumption of a simultaneous action

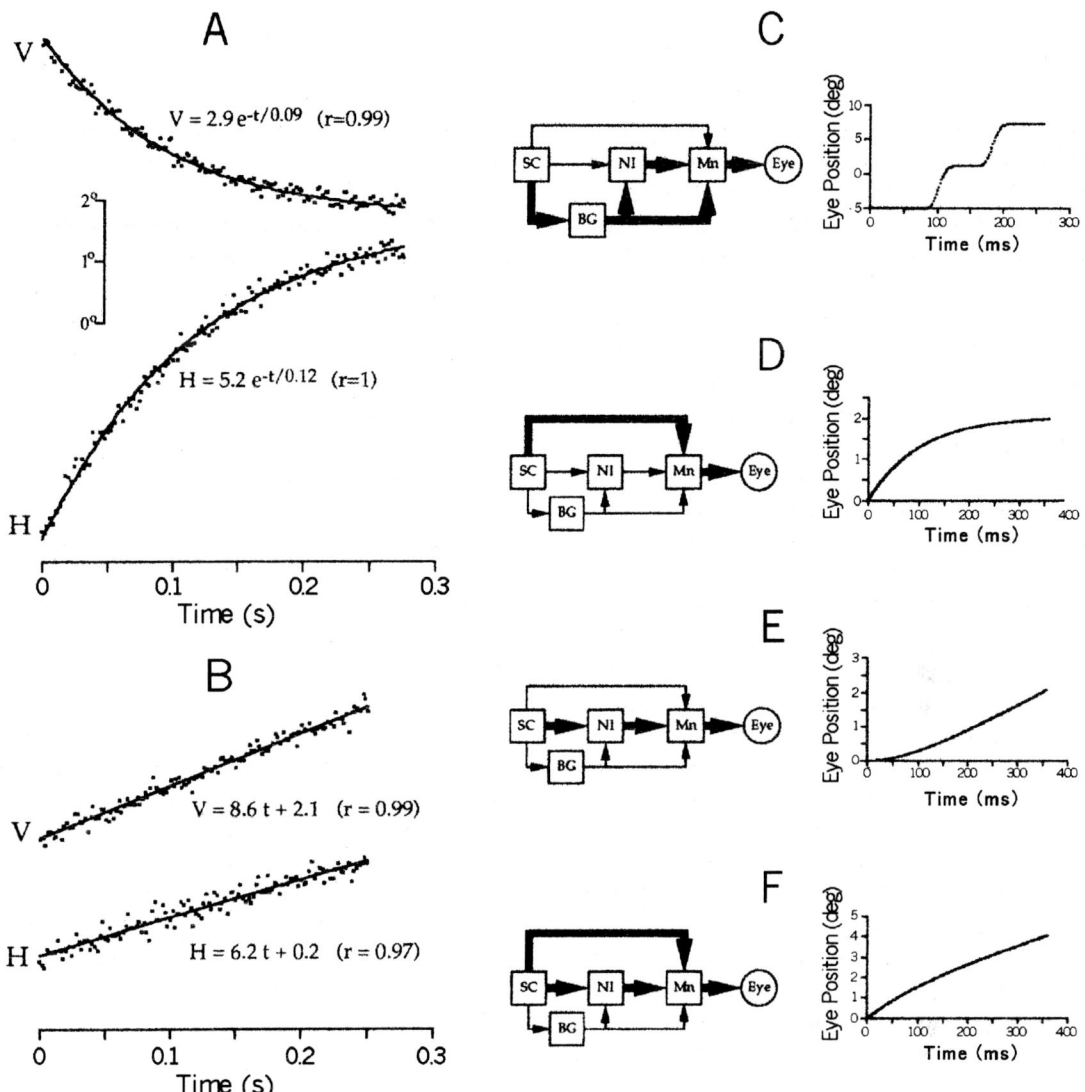

Fig. 5. Model simulations of eye movements evoked by prolonged output activity of the SC. (A–B) Examples of stimulus-evoked PSDs that can be accurately described by exponential (A) or linear (B) functions, as indicated by equations in the insets. Calibration of vertical (V) and horizontal (H) eye positions in A applies also to B. (C–F) Results of simulation with different assumptions about the routing of SC output to the oculomotor plant. Wiring diagrams (left) illustrate the basic network components: the SC, saccadic BG, neural integrator (NI), motoneurons (Mn) and eye (Eye). The impulse response of the Mn and SC elements is a Dirac delta function, the Eye and NI units are first-order systems with time constants of 100 ms and 30 s, respectively. The BG is the same as in the MSH model of Moschovakis (1994). Depending on the pathways activated (thick lines), the output of the model (shown on the right) can appear as a sequence of saccades (C), exponential slow drifts (D), ramp-like slow drifts (E) or combinations of exponential and ramp-like slow drifts (F). (Reproduced, in part, from Grantyn et al., 1996 with permission from Elsevier Science.)

of the two pathways bypassing the saccadic BG predicts the typical time course of stimulus-evoked PSDs, i.e., a combination of exponential and ramp-like waveforms (Fig. 5F).

In this and the preceding sections, we summarized data indicative of the diversity of cat visuomotor (Grantyn and Berthoz, 1985) and 'orientation' (Munoz et al., 1991) tecto-reticulo spinal neurons.

Global criteria defining this population are its cells' burst-like activation during visually triggered gaze shifts and their projection through the predorsal bundle. The population's heterogeneity is expressed at an anatomical level, in that each of its neurons is characterized by an individual combination of target areas in the lower brainstem. Heterogeneity is also expressed at a functional level, in that the population's neurons differ according to dynamic characteristics of their movement-related discharges. Our simulations lead to the important conclusion that realistic patterns of eye movements can only be obtained after taking into account both the anatomical and functional diversity of relevant populations of neurons. Presently available data, however, are inadequate to demonstrate a strict correspondence between morphologically defined and physiologically defined classes of cells. For this, it would be necessary to prove that vm-TRSNs characterized by essentially phasic discharges make quantitatively stronger connections with the saccadic BG in the PPRF, whereas cells capable of prolonged postsaccadic firing tend to bypass the PPRF and preferentially contact AbdN, NPH and orienting-related reticular neurons. This is just one of several frequently encountered problems that cannot be addressed experimentally unless the connections established by individual neurons are elucidated together with their functional properties.

Anatomical and functional heterogeneity of reticular formation neurons receiving tectal projections

Head movements are obligatory components of orienting toward targets at eccentricities larger than 3–5° in cats and 15–20° in primates. Most of the work on the primate SC has focused on the control of eye saccades while the contribution of the cat SC to the control of gaze shifts was traditionally viewed as more versatile. This can be illustrated by referring to two studies, the first affirming that the primary role of the cat SC is to control head movements (Straschill and Schick, 1977), and the second (Peck, 1990) reporting a population of saccade-related bursters, indistinguishable from those in monkeys, and only a few cells loosely coupled to head movements.

In a later study, Munoz et al. (1991) argued that projection neurons of the vm-TRSN class control the sum of eye and head displacements, i.e., gaze. This point of view is supported by a more recent reexamination of cell activity in the SC of head-unrestrained monkeys (Freedman and Sparks, 1997). Because it is difficult to record from cells when the head is moving, the fate of the tectal gaze-displacement signal after it reaches the premotor areas of the brainstem RF remains largely unknown. In this section, we review some of the results relevant to the decomposition of SC output signals in eye and head components. Firing patterns and connections of specific brainstem cell groups that make up the circuits of saccadic BGs have been extensively studied, modeled and summarized in numerous reviews (Robinson, 1975; Hepp et al., 1989; Moschovakis et al., 1996). We therefore limit our discussion to the much less-abundant experimental work in alert animals that has focused on the functional and morphological characteristics of rhombencephalic cell groups comprising the 'generator' of rapid head movements.

Experimental work on alert animals was pioneered by Vidal et al. (1983), using head-restrained cats. The work demonstrated neurons of the caudal pontine RF whose tonic or phasi-tonic discharges are correlated with the intensity of neck EMG activity and, simultaneously, the position of the eyes in the orbit. They hypothesized that their test neurons correspond to reticulospinal neurons controlling head movements during combined eye–head gaze shifts. Subsequent studies refined functional classification of such 'eye–neck' (ENN) neurons and established their morphological identity (Grantyn et al., 1992a,b). The global functional criterion defining the ENN population is a highly reproducible activation during gaze shifts terminating in the ipsilateral hemifield. We later demonstrated that ENNs are specifically activated during active orienting movements, with no contribution to compensatory head movements during passive whole body rotations (Kitama et al., 1995). The locations of ENNs in the caudal pontine and rostral bulbar RF are well within the areas occupied by the saccadic BG and terminal domains of vm-TRSNs (Figs. 1 and 3). Monosynaptic input from the contralateral SC was indeed proven for 92% of the ENN population.

Although antidromic stimulation confirmed the spinal projection of only about 50% of the ENN group (Kitama et al., 1995), axons of all such neurons, when stained with HRP, were followed in the first segment of the spinal cord (Grantyn et al., 1992b). In all, we have defined three classes of ENNs by combining their movement-related discharge properties with their axonal branching patterns (Grantyn et al., 1992a,b; Kitama et al., 1995).

About one-third of the ENNs, the phasic-sustained (PS) type, generate long-lead bursts during saccades, which are synchronous with phasic increments in neck EMG activity. Cell firing continues after termination of the first saccade and overlaps in time with a dynamic overshoot of EMG activity. In contrast to such EMG-related activation, an association of PS-ENN discharges with saccades is not obligatory. This is indicated by the fact that bursts of discharge are generated only with saccades terminating in the ipsilateral hemifield. All labeled PS-ENNs are characterized by an extensive branching in the caudal pons and medulla (Fig. 1; Grantyn et al., 1992b). Projections to oculomotor areas (e.g., AbdN and NPH) overlap the terminations of TRSNs, but the former are more abundant. As discussed above, a combined action of PS-ENNs and TRSNs can underlie the generation of PSDs and explain their time course. It should be emphasized that PS-ENNs are 'oculo-spinal', with a divergent connection to oculomotor structures and to the spinal cord, both directly and via bulbar reticulospinal neurons. Their role in eye–head coupling during combined gaze shifts is therefore very likely. Axons of the two other groups of ENNs lack collateral connections to the oculomotor areas. Rather, they project only into precerebellar nuclei in the medulla. Neurons of the first group, quasi-tonic (QT) ENNs, generate steady discharges that are better correlated with the level of EMG activity during eccentric fixations than with eye position. These reticulospinal cells transmit to the spinal cord a somewhat modified eye position signal, presumably originating in the NPH. Finally, Grantyn et al. (1992a) labeled a few essentially phasic neurons (P-ENNs). They project to the spinal cord without making connections with oculomotor areas. They receive mono- or disynaptic input from the SC and display discharge patterns qualitatively similar to those of TRSNs. They may represent a parallel pathway reinforcing the transmission of the tectal gaze error signal to the spinal cord.

To our knowledge, data from a comparable population of pontobulbar neurons in head-unrestrained cats are available only from an elegant study by Isa and Naito (1995). Among cells discharging in relation to visually guided eye–head gaze shifts, these authors distinguished six groups, which differed in their behavioral properties. Among cells with horizontal directional preference, the most numerous 'directional pretype' neurons closely resembled P- and PS-ENNs recorded in head-restrained cats. Cell discharges of the QT-type were not observed, most probably because eccentric eye positions are usually not maintained after head-unrestrained gaze shifts. Isa and Naito (1995) obtained a positive correlation between their cells' mean firing rate and the amplitude of head movement, as well as between peak firing rate and maximal head velocity. The authors were inclined to conclude that neurons of the 'directional pretype' group control orienting head movements. They conceded, however, that a contribution of these cells to eye movement and total gaze displacement cannot be excluded.

It cannot be overlooked that phasic discharges of P- and PS-ENNs have much in common with bursts of pontine saccade-related long-lead bursters, which are thought to be a specific cell group of the saccadic BG (Kaneko et al., 1981, Fig. 2; Van Gisbergen et al., 1981). It may well be that a strict separation in eye, head and gaze-related classes is an over-simplification due to the fact that it is difficult to interpret the discharge of neurons with intermediate properties. The problem can be illustrated by work on inhibitory saccade-related bursters (IBN) in head-unrestrained cats (Cullen et al., 1993). Bursts of these cells displayed a tight correlation with saccade parameters when the head was restrained but, during combined eye–head gaze shifts, this correlation became insignificant for most of the neurons. The number of spikes and the amplitude of head movement were tightly correlated in some neurons. Nevertheless, the analysis of multiple correlations compelled the authors to conclude that IBN cells encode gaze motor error rather than generate a specific premotor signal of saccade amplitude and velocity. A similar ambiguity has been reported for another class of cells making up the pontine saccadic BG, the omnidirectional pause neurons (OPNs). The duration of their pauses defines the duration of eye saccades in head-restrained cats. However, in head-unrestrained animals the same cells have been shown to pause whenever gaze moves (Paré and Guitton, 1998), even if saccades or other kinds of eye movements are absent.

The problem of a possible functional overlap between cell groups apparently specific for eye, head or gaze movements attracted our attention after we occasionally encountered atypical OPNs in head-restrained cats (Petit et al., 1999) (Fig. 6B). For comparison, Fig. 6A illustrates the commonly known OPN discharge pattern, i.e., the cell stops firing during each of individual saccades in a complex gaze shift. Quantitative analysis reveals a strict correlation between this neuron's pause durations and saccade durations. Firing does resume at the end of each saccade, in spite of slow inter- and postsaccadic eye drifts. The characteristics of such 'saccade' OPNs are well known from previous studies in cats (Evinger et al., 1982; Paré and Guitton, 1994) and monkeys (Keller, 1974). We found, however, that some OPNs interrupt their firing for the total duration of gaze movement when it is composed of a series of saccades and slow eye movements (Fig. 6B). Such 'complex' OPNs are intermingled with the 'saccade' OPN type in the midline region of the caudal pons. To compare these two types, we analyzed the relationships of pauses with individual saccades by selecting those after which the resumption of the background firing was complete. We classified the cells as 'saccade' OPN (S-OPN) if their pause durations had a significant ($P \leq 0.01$) positive correlation with saccade durations. Cells displaying a weaker correlation were pooled in group termed 'complex' OPN

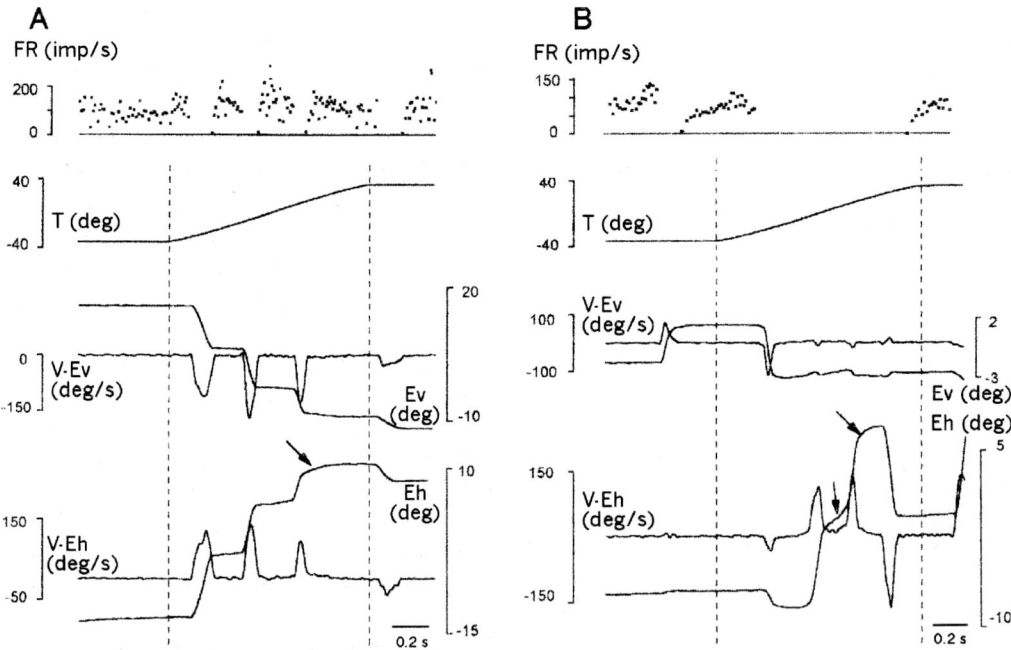

Fig. 6. Contrasting properties of OPNs in alert, head-restrained cats. (A) Recording from a 'saccade' OPN. Note interruption of the discharge during individual saccades of a multisaccadic gaze shift and rapid recovery of activity at the saccade's end. (B) Recording from a 'complex' OPN. Note a prolonged pause corresponding to the total duration of a complex gaze shift and slow recovery of firing rate after the pauses. From top to bottom: instantaneous firing rate (FR), position of a moving visual target (T), vertical eye position (Ev) and its velocity (V-Ev), horizontal eye position (Eh) and velocity (V-Eh). Scale bars: eye position on the right; eye velocity on the left. Arrows point to slow eye movements in between, or after, saccades. Note that firing of the saccade OPN is restored during the PSD (A), but this is not the case for the complex OPN (B). (Reproduced, in part, from Petit et al., 1999 with permission from Springer-Verlag.)

(C-OPN). The lead time of C-OPN pauses with respect to saccade onset was significantly longer than that of the S-OPN group (mean ± SD, 49 ± 39 vs. 15 ± 7 ms). Resumption of firing of the S-OPN group occurred, on average, 13 ± 12 ms before the end of the saccades. In contrast, pauses of the C-OPN group usually terminated after the saccades ended (mean lag, 58 ± 64 ms). Timing parameters of the C-OPN group result in pause durations that are, as a rule, longer than saccade duration. Also, during a rapid succession of saccades in a complex gaze shift, these neurons' firing does not resume until the end of the last saccade and/or the PSD. Unexpectedly, the discharge of the 'complex' OPN group was suppressed by moving visual stimuli, in contrast to the well-documented excitatory visual input to the S-OPN group. Taken together, the peculiar timing of pauses and the presence of visual suppression suggest that the C-OPNs differ from the S-OPN group in terms of the organization of their afferent inputs. Our subdivision of OPNs, based on a single criterion (presence or absence of correlation between pause and saccade durations), may give the impression that the two groups are clearly distinct. It should be emphasized, however, that considerable overlap is seen between the S- and C-OPN groups with respect to their background firing rate, lead time and the timing of their pause end. It is therefore premature to affirm that all pontine OPNs gate the duration of the total gaze shift in both head-restrained and head-unrestrained conditions (Paré and Guitton, 1998). Rather, our data suggest that this population is heterogeneous. Some of this population's neurons (the C-OPN group) may indeed fulfill the role proposed by Paré and Guitton, while at the other extreme of the continuum we find the S-OPN group, whose role is specifically the control of individual saccades.

Conclusions

We have limited this review to a very small fraction of neuronal networks controlling gaze. Taking as an example the executive limb of this motor subsystem, we have illustrated some common methodological and conceptual issues raised by experimental studies addressing the control of goal-directed movements by real neuronal ensembles. The conceptual approach that puts emphasis on population coding can be traced back to the 1930s, and its advantage has been lucidly formulated by McIlwain (1976, p. 227): "... Unlike the experimenter, whose attention is restricted to the activity of the one or two cells discharging near his microelectrode, the brain has immediate access to all the cells responding to a sensory stimulus." At that time, McIlwain felt that in studies on the visual system, the population concept was unjustly "... supplanted by current interest in the stimulus specificities of single... neurons." However, we prefer to opine that an approach focused on neuronal 'specificities' is complementary, rather than antithetical, to analyses that focus on populations. Note further that a population can be correctly defined only after taking into consideration multiple behavior-related and morphological criteria. An appropriate example is given by a recent publication that describes a reconstruction from single-cell recordings of two-dimensional, saccade-related population activity in the primate SC (Anderson et al., 1998). In our opinion, this reconstruction is the most complete and reliable reported to date. Even so, these authors commented that the composite signals they obtained do not necessarily predict "... the relative influence of the neurons... on the downstream structures..." because "... each neuron... projects its own signal to downstream saccadic control structures, and these distributed inputs may be weighted by location of the projecting cells on the map." (p. 814).

During this past decade, the population approach was clearly dominant. Note, for example, the truly large number of extracellular-recording studies, the purpose of which was to define presumably homogeneous groups of neurons on the basis of the correlation of their discharge with sensory and motor events. In this chapter, however, our goal has been to stress the importance of taking into account cell-specific anatomical properties, this being an essential prerequisite for a biologically meaningful definition of a population. By combining microstimulation and anterograde tracing of small populations of the SC projection neurons, we have shown quite clearly that a spatiotemporal transformation can be implemented by the differential weighting of connections to the pontine saccadic generator. We have also conceded that projection neurons differ in the distribution of their collateral connections in the lower brainstem, and that individual weighting of connections to cell groups controlling movements of the eyes, head and other body segments remains unexplored. Work on alert cats has dealt with only one large population of tectal efferent neurons (the vm-TRSN group), with the focus on two criteria: their transient discharge during contraversive orienting movements and their crossed projection in the predorsal bundle. The description of their branching patterns remained limited to a few cells. If we consider all the literature on the firing patterns of cat SC neurons, at least seven functionally defined subclasses can be ascribed. Among them are saccade- and head movement-related cells, further subdivided according to the dynamic properties of their activity, and cells presumably carrying the gaze error signal. A complete description of the tecto-reticulo-spinal system would require the knowledge of the morphological identity and individual connections of these subclasses defined by behavioral criteria. Our modeling experiment has illustrated that a complete sequence of saccadic and slow eye movements in response to SC stimulation can only be reproduced if the dynamic properties of cell firing are combined with their previously established differential connections to the preoculomotor cell groups in the lower brainstem.

We reviewed above the ponto-bulbar cell groups that compose the 'generator' of rapid head movements and its coordination with the oculomotor component of orienting. Here again, only three populations defined by more-or-less similar discharge patterns and some characteristics of axonal branching have been studied. Some observations suggest that previously used functional subdivisions may be too rigid for a realistic description of the network. On the one hand, 'head-bursters' display characteristics of cell classes that would be referred to as saccade-related, if head movements and neck EMG activity were not monitored. On the other hand, our study of OPNs, traditionally viewed as a part of saccadic generator, has demonstrated that this population contains cells with firing patterns appropriate for the gating of complex gaze shifts. If we assume the actually dominating view that the SC sends out the gaze error signal, the above observations suggest that its decomposition in eye and head components is based on a gradual process of specification of premotor signals in the RF. Besides cells clearly related to the one or the other effector, there exist neurons with intermediate properties which should not be discarded for the sake of simplicity. A complete description of the network would require knowledge of specific combinations of afferent inputs and efferent connections for as many subsets of its constituent neurons as possible.

Acknowledgments

Alain Berthoz, Yiannis Dalezios, Antoine Gourdon, Olivier Hardy, Etienne Olivier and Julien Petit participated in the studies reviewed in this chapter. We gratefully acknowledge their contributions. We thank Anne-Marie Brandi, Mireille Chat, Michel Ehrette and Stephanie Lemarchand for their expert technical assistance. This research was supported by the CNRS and grants from the European Community (ERBSC1-CT910643; CHRX-CT 940559).

Abbreviations

AbdN	abducens nucleus
BC	brachium conjunctivum
BG	saccadic burst generator
C-OPN	'complex' omnipause neuron
EMG	electromyogram/electromyography
ENN	'eye–neck' neuron
GVII	genu of the facial nerve
HRP	horseradish peroxidase
IO	inferior olive
MLF	medial longitudinal fasciculus
Nfs	nucleus of the solitary tract
NPH	nucleus prepositus hypoglossi
NRT	tegmental reticular nucleus
NTB	nucleus of the trapezoid body
OPN	omnipause neuron
P-ENN	phasic 'eye–neck' neuron
PH	nucleus prepositus hypoglossi
PPRF	paramedian pontine reticular formation
PS-ENN	phasic-sustained 'eye–neck' neuron
PSD	postsaccadic drift
PVG	periventricular gray

QT-ENN	quasi-tonic 'eye–neck' neuron
RF	reticular formation
RGc	gigantocellular reticular nucleus
RPc	caudal pontine reticular nucleus
Rpc	parvocellular reticular nucleus
RPo	oral pontine reticular nucleus
Rv	ventral reticular nucleus
SC	superior colliculus
SGI	intermediate gray layer
SGP	deep gray layer
SO	optic layer
S-OPN	'saccade' omnipause neuron
TB	trapezoid body
TLLB	tectal long-lead burster
VM	medial vestibular nucleus
vm-TRSN	visuomotor tecto-reticulo-spinal neuron
VL	lateral vestibular nucleus
Vp	principal sensory nucleus of the trigeminal nerve
VI	abducens nucleus
VII	facial nucleus
VIIIs	superior vestibular nucleus
XII	hypoglossal nucleus

References

Anderson, R.W., Keller, E.L., Gandhi, N.J. and Das, S. (1998). Two-dimensional saccade-related population activity in superior colliculus in monkey. J. Neurophysiol., 80: 798–817.

Arai, K., Keller, E.L. and Edelman, J.A. (1994). Two-dimensional neural network model of the primate saccadic system. Neural Netw., 7: 1115–1135.

Cowie, R. and Holstege, G. (1992). Dorsal mesencephalic projections to pons, medulla, and spinal cord in the cat: limbic and non-limbic components. J. Comp. Neurol., 319: 536–559.

Cullen, K.E., Guitton, D., Rey, C.G. and Jiang, W. (1993). Gaze-related activity of putative inhibitory burst neurons in the head-free cat. J. Neurophysiol., 70: 2678–2683.

Curthoys, I.S., Nakao, S. and Markham, C.H. (1981). Cat medial pontine reticular neurons related to vestibular nystagmus: firing pattern, location and projection. Brain Res., 222: 75–94.

Das, S., Gandhi, N.J. and Keller, E.L. (1995). Open-loop simulations of the primate saccadic system using burst cell discharge from the superior colliculus. Biol. Cybern., 73: 509–518.

Evinger, C., Kaneko, C.R.S. and Fuchs, A.F. (1982). Activity of omnipause neurons in alert cats during saccadic eye movements and visual stimuli. J. Neurophysiol., 47: 827–844.

Freedman, E.G. and Sparks, D.L. (1997). Activity of cells in the deeper layers of the superior colliculus of the rhesus monkey: evidence for a gaze displacement command. J. Neurophysiol., 78: 1669–1690.

Grantyn, A. and Berthoz, A. (1985). Burst activity of identified tecto-reticulo-spinal neurons in the alert cat. Exp. Brain Res., 57: 417–421.

Grantyn, A., Berthoz, A., Hardy, O. and Gourdon, A. (1992a). Contribution of reticulo-spinal neurons to the dynamic control of head movements: presumed neck bursters. In: Berthoz A., Graf W. and Vidal P.-P. (Eds.), The Head-Neck Sensory Motor System. Oxford University Press, New York, pp. 318–329.

Grantyn, A., Hardy, O., Olivier, E. and Gourdon, A. (1992b). Relationship between task-related discharge patterns and axonal morphology of brainstem projection neurons involved in orienting eye and head movements. In: Shimazu H. and Shinoda Y. (Eds.), Vestibular and Brain Stem Control of Eye, Head and Body Movements. Japan Scientific Society Press and S. Karger, Tokyo, Basel, pp. 255–273.

Grantyn, A., Olivier, E. and Kitama, T. (1993). Tracing premotor brain stem networks of orienting movements. Curr. Opin. Neurobiol., 3: 973–981.

Grantyn, A., Dalezios, Y., Kitama, T. and Moschovakis, A.K. (1996). Neuronal mechanisms of two-dimensional orienting movements in the cat. I. A quantitative study of saccades and slow drifts produced in response to the electrical stimulation of the superior colliculus. Brain Res. Bull., 41: 65–82.

Hepp, K., Henn, V., Vilis, T. and Cohen, B. (1989). The neural substrates for saccadic eye movements. Brainstem regions related to saccade generation. In: Wurtz R.H. and Goldberg M.E. (Eds.), Neurobiology of Saccadic Eye Movements, Reviews of Oculomotor Research, Vol. 3, Elsevier, New York, pp. 105–212.

Hess, W.R., Bürgi, S. and Bucher, V. (1946). Motorische Funktion des Tektal- und Tegmentalgebietes. Mschr. Psychiat. Neurol., 112: 1–52.

Hikosaka, O., Igusa, Y., Nakai, S. and Shimazu, H. (1978). Direct inhibitory synaptic linkage of pontomedullary reticular burst neurons with abducens motoneurons in the cat. Exp. Brain Res., 33: 337–352.

Huerta, M.F. and Harting, J.K. (1982). Tectal control of spinal cord activity: neuroanatomical demonstration of pathways connecting the superior colliculus with the cervical spinal cord grey. In: Kuypers H.G.J.M. and Martin G.F. (Eds.), Descending Pathways to the Spinal Cord, Prog. Brain Res., Vol. 57, Elsevier, New York, pp. 293–328.

Isa, T. and Naito, K. (1995). Activity of neurons in the medial pontomedullary reticular formation during orienting movements in alert head-free cats. J. Neurophysiol., 74: 73–95.

Kaneko, C.R.S., Evinger, C. and Fuchs, A.F. (1981). The role of cat pontine burst neurons in the generation of saccadic eye movements. J. Neurophysiol., 46: 387–408.

Keller, E.L. (1974). Participation of medial pontine reticular formation in eye movement generation in monkey. J. Neurophysiol., 37: 316–332.

Kitama, T., Grantyn, A. and Berthoz, A. (1995). Orienting-related eye-neck neurons of the medial ponto-bulbar reticular formation do not participate in horizontal canal-dependent vestibular reflexes of alert cats. Brain Res. Bull., 38: 337–347.

McIlwain, J.T. (1976). Large receptive fields and spatial transformations in the visual system. Int. Rev. Physiol., 10: 224–248.

McIlwain, J.T. (1990). Topography of eye-position sensitivity of saccades evoked electrically from the cat's superior colliculus. Vis. Neurosci., 4: 289–298.

Missal, M., Crommelinck, M., Roucoux, A. and Decostre, M.-F. (1993). Slow correcting eye movements of head-fixed, trained cats toward stationary targets. Exp. Brain Res., 96: 65–76.

Moschovakis, A.K. (1994). Neural network simulations of the primate oculomotor system. I. The vertical saccadic burst generator. Biol. Cybern., 70: 291–302.

Moschovakis, A.K., Karabelas, A.B. and Highstein, S.M. (1988). Structure-function relationships in the primate superior colliculus. II. Morphological identity of presaccadic neurons. J. Neurophysiol., 60: 263–302.

Moschovakis, A.K., Scudder, C.A. and Highstein, S.M. (1996). The microscopic anatomy and physiology of the mammalian saccadic system. Prog. Neurobiol., 50: 133–254.

Moschovakis, A.K., Kitama, T., Dalezios, Y., Petit, J., Brandi, A.M. and Grantyn, A. (1998). An anatomical substrate for the spatiotemporal transformation. J. Neurosci., 18: 10219–10229.

Munoz, D.P., Guitton, D. and Pélisson, D. (1991). Control of orienting gaze shifts by the tectoreticulospinal system in the head-free cat. III. Spatiotemporal characteristics of phasic motor discharges. J. Neurophysiol., 66: 1642–1666.

Olivier, E., Chat, M. and Grantyn, A. (1991). Rostocaudal and mediolateral density distributions of superior colliculus neurons projecting in the predorsal bundle and to the spinal cord: a rectrograde HRP study in the cat. Exp. Brain Res., 87: 268–282.

Olivier, E., Grantyn, A., Chat, M. and Berthoz, A. (1993). The control of slow orienting movements by tectoreticulospinal neurons in the cat: behavior, discharge patterns and underlying connections. Exp. Brain Res., 93: 435–449.

Paré, M. and Guitton, D. (1994). Discharge behavior of omnipause neurons in the cat. In: d'Ydewalle G. and Van Rensbergen J. (Eds.), Visual and Oculomotor Functions: Advances in Eye Movement Research, Studies in Visual Information Processing, Vol. 5, Elsevier, Amsterdam, pp. 271–283.

Paré, M. and Guitton, D. (1998). Brain stem omnipause neurons and the control of combined eye-head gaze saccades in the alert cat. J. Neurophysiol., 79: 3060–3076.

Peck, C.K. (1990). Neuronal activity related to head and eye movements in cat superior colliculus. J. Physiol. (Lond.), 421: 79–104.

Petit, J., Klam, F., Grantyn, A. and Berthoz, A. (1999). Saccades and multisaccadic gaze shifts are gated by different pontine omnipause neurons in head-fixed cats. Exp. Brain Res., 125: 287–401.

Robinson, D.A. (1975). Oculomotor control signals. In: Lennerstrand G. and Bach-y-Rita P. (Eds.), Basic Mechanisms of Ocular Motility and their Clinical Implications. Pergamon Press, New York, pp. 337–374.

Sasaki, S. and Shimazu, H. (1981). Reticulovestibular organization participating in generation of horizontal fast eye movements. In: Cohen B. (Ed.), Vestibular and Oculomotor Physiology. The New York Academy of Sciences, New York, pp. 130–143.

Scudder, C.A. (1988). A new local feedback model of the saccadic burst generator. J. Neurophysiol., 59: 1455–1474.

Scudder, C.A., Moschovakis, A.K., Karabelas, A.B. and Highstein, S.M. (1996). Anatomy and physiology of saccadic long-lead burst neurons recorded in the alert squirrel monkey. I. Descending projections from the mesencephalon. J. Neurophysiol., 76: 332–352.

Sparks, D.L. (1986). Translation of sensory signals into commands for control of saccadic eye movements: role of primate superior colliculus. Physiol. Rev., 66: 118–171.

Sparks, D.L. and Mays, L.E. (1980). Movement fields of saccade related burst neurons in the monkey superior colliculus. Brain Res., 190: 39–50.

Sprague, J.M. and Meikle, T.H. (1965). The role of the superior colliculus in visually guided behavior. Exp. Neurol., 11: 115–146.

Straschill, M. and Schick, F. (1977). Discharges of superior colliculus neurons during head and eye movements of the alert cat. Exp. Brain Res., 27: 131–141.

Tehovnik, E.J. (1996). Electrical stimulation of neural tissue to evoke behavioral responses. J. Neurosci. Methods, 65: 1–17.

Van Gisbergen, J.A.M., Robinson, D.A. and Gielen, S. (1981). A quantitative analysis of generation of saccadic eye movements by burst neurons. J. Neurophysiol., 45: 417–441.

Vidal, P.-P., Corvisier, J. and Berthoz, A. (1983). Eye and neck motor signals in periabducens reticular neurons of the alert cat. Exp. Brain Res., 53: 16–28.

Wurtz, R.H. and Albano, J.E. (1980). Visual-motor function of the primate superior colliculus. Annu. Rev. Neurosci., 3: 189–226.

CHAPTER 41

Pedunculo-pontine control of visually guided saccades

Yasushi Kobayashi*, Yuka Inoue and Tadashi Isa

Department of Integrative Physiology, National Institute for Physiological Sciences, Okazaki 444-8585, Japan

Abstract: The cholinergic pedunculopontine tegmental nucleus (PPTN) is one of the major ascending arousal systems in the brainstem, and it is linked to motor, limbic and sensory centers. Despite an abundance of anatomical and physiological data, however, the functional role of PPTN neurons in behavioral control is still unresolved. In this chapter, we hypothesize that the PPTN is implicated in the integrative control of movement, particularly the reinforcement of tasks performed during conscious behavior. We present a new model of the PPTN's involvement in the control of arousal, attention and reinforcement aspects of motor behavior, with a focus on the control of saccadic eye movements.

Introduction

The brainstem's pedunculopontine tegmental nucleus (PPTN) and laterodorsal tegmental nucleus (LDTN) are central components of the reticular-activating system, which complex provides background excitation for several sensory and motor systems, and is essential for perception (Lindsley, 1958) and cognitive processes (Steckler et al., 1994b). The PPTN contains both cholinergic and glutamatergic neurons (Hallanger and Wainer, 1988), and it is one of the major sources of cholinergic projections in the brainstem. The cholinergic system is a key modulatory neurotransmitter system in the brain, controlling, for example, neuronal activity involved in selective attention. There is also anatomical and physiological evidence that supports the idea of a 'cholinergic component' of conscious awareness (Perry et al., 1999).

The PPTN has reciprocal connections with the basal ganglia, including the subthalamic nucleus, globus pallidus and substantia nigra (Edley and Graybiel, 1983; Lavoie and Parent, 1994), and catecholaminergic systems in the brainstem, including the locus coeruleus (LC) and dorsal raphe nucleus (DRN; Koyama and Kayama, 1993). It has been proposed that this basal ganglia–PPTN–catecholaminergic complex is important for gating movement and controlling several attentive behaviors (Garcia-Rill, 1991; see also Takakusaki et al., Garcia-Rill, et al., Skinner et al., and Grantyn et al., Chapters 23, 27, 28 and 40 of this volume). Despite the abundant anatomical findings, however, the functional importance of the PPTN is far from understood. Several anatomical-connection and lesion studies across several species have suggested, however, that the PPTN is involved in the control of attention and motor control (Steckler et al., 1994a) and reinforcement learning (Brown et al., 1999). Saccades are rapid (ballistic) eye movements, which can be triggered by a variety of cues (auditory, visual, memory, etc.). Since their onsets are easy to identify, their preparation and execution processes can be phase-related to neuronal activity. In addition, saccades are altered by attentional (Kustov and Robinson, 1996) and motivational (Kawagoe et al., 1998) states, which are also modifiable by task sequences during experimental sessions. Thus, oculomotor tasks, especially those involving saccades and recording PPTN neuronal activity in awake monkeys, are an optimal cmeans of investigating the neural control of movement, attention and reinforcement

*Corresponding author: Tel.: +81-564-55-7857;
Fax: +81-564-55-7790; E-mail: kobayasi@nips.ac.jp

DOI: 10.1016/S0079-6123(03)43041-0

learning. In this chapter, our group presents a new model of the PPTN's involvement in the control of arousal, attention and reinforcement aspects of motor behavior, with a focus on the control of saccadic eye movements (see also Isa and Kobayashi, and Grantyn et al., this volume).

Execution of saccadic eye movements

There have been reports of the PPTN's involvement in the control of arm movements (Matsumura et al., 1997) and posture and locomotion (Garcia-Rill, 1991). It has also been shown that lesions of the PPTN reduce the frequency of eye movements during rapid eye-movement sleep (Shouse and Siegel, 1992). The frontal eye field (FEF; Bruce and Goldberg, 1985) and substantia nigra pars reticulata (SNr; Hikosaka and Wurtz, 1983) participate in the control of saccadic eye movements and there is evidence that FEF neurons project to the PPTN (Matsumura et al., 2000), which structure receives GABAergic input from SNr (Gerfen et al., 1982; Granata and Kitai, 1991). These findings suggest that the PPTN may well contribute to the control of eye movements. Furthermore, the PPTN strongly innervates the intermediate layer of the superior colliculus (SC) in several mammalian species (Graybiel, 1978; Beninato and Spencer, 1986; Hall et al., 1989; Henderson and Sherriff, 1991; Ma et al., 1991; Schnurr et al., 1992; Jeon et al., 1993). The SC is a key structure for the generation of saccadic eye movements (Sparks and Hartwich-Young, 1989). Neuronal activity in its intermediate layer that precedes saccade execution is influenced by attentional shifts (Kustov and Robinson, 1996), movement selection (Glimcher and Sparks, 1992) and the extent of motor preparation (Dorris and Munoz, 1998). Thus, cognitive influences on saccades and their execution signals may well depend on activation of the PPTN. Accordingly, our group proposes that the PPTN coordinates the relay of visuosensory information to the cerebral cortex and SNr for the execution of saccades.

To clarify the role of cholinergic input to the intermediate layer of the SC for the elaboration of saccades, our group has examined the effect of microinjection of nicotine into the SC on visually guided saccades in the monkey (Aizawa et al., 1999).

After such injection, the frequency of extremely short-latency saccades [express saccades; saccadic reaction time (SRT) < 120 ms; Fischer and Weber, 1993; Pare and Munoz, 1996] increases dramatically. This result suggests that cholinergic PPTN inputs to the SC influence motor preparation signals and the timing of saccade initiation. Our laboratory also recently used slice preparations of the rat SC to show that activation of nicotinic acetylcholine receptors on neurons in the intermediate layer of the SC induces inward currents and depolarization, which may gate signal transmission in the direct visuomotor pathway (i.e., from the superficial to the intermediate layer of the SC; Isa et al., 1998a,b). These results suggest that the neural mechanisms we observed in the rat SC slice also apply to behaving monkeys (Aizawa et al., 1999), i.e., the PPTN's cholinergic input to the SC may control SRT via activation of nicotinic acetylcholine receptors on neurons in the intermediate layer of the SC. Figure 1 shows our model of the PPTN's involvement in the control of saccades.

Neuronal burst activity related to saccade executions may be observed in the PPTN just as it is in the caudal part of the SC. Crossed descending projections from the SC terminate in the PPTN (cat—Huerta and Harting, 1982; rat—Redgrave et al., 1987;

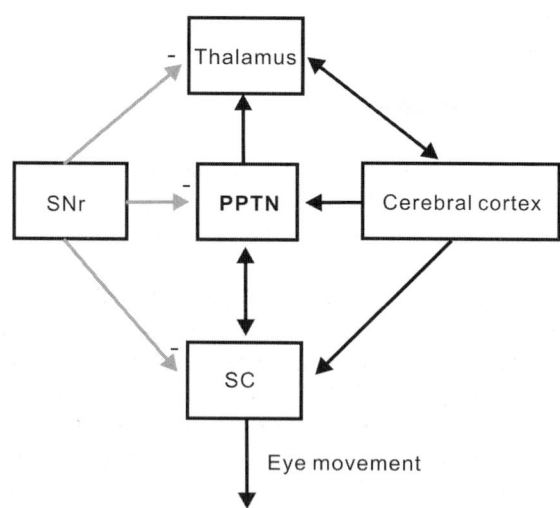

Fig. 1. Model of the PPTN's involvement in controlling saccadic eye movement. Abbreviations: SC, superior colliculus; SNr, substantia nigra pars reticulata; PPTN, pedunculopontine tegmental nucleus. See text for further explanation of this and the subsequent figures.

monkey—May and Porter, 1992), which structure receives ipsilateral projections from motor areas in cerebral cortex in the cat (Edley and Graybiel, 1983) and monkey (Matsumura et al., 2000). In the monkey, the PPTN also receives ipsilateral projections from the infralimbic and prelimbic areas (Chiba et al., 2001) and the FEF (Matsumura et al., 2000). The cause of the saccade burst in the SC may also be explained by disinhibition of GABAergic input from the SNr (Gerfen et al., 1982; Granata and Kitai, 1991). Moreover, a pause in PPTN neuron discharge during a saccade occurs in phase with one in fixation neurons in the rostral pole of the SC. The stimulus for this pause may involve inputs from the SC, basal ganglia and cerebral cortex. Fig. 1 shows that GABAergic inputs from the SNr, glutamatergic inputs from the cerebral cortex, and glutamatergic and cholinergic inputs from the PPTN may all regulate the SC, where multimodal signals converge and interact.

In summary, the PPTN has the capacity to integrate several saccade-related signals, which it receives from the SC, basal ganglia and cerebral cortex. The PPTN-processed signals are then returned to their sites of origin. Clearly, the PPTN is ideally situated to work cooperatively with other brain structures for the initiation and execution of saccades.

Arousal and attention

Many studies suggest that the PPTN and LDTN regulate vigilance levels like sleep and wakefulness. These structures are also involved in inducing an arousal and global attentive state in response to a novel stimulus (Steriade, 1996a,b). Regulation of arousal by the PPTN is coupled with the actions of catecholaminergic systems in the brainstem, the LC (noradrenergic) and DRN (serotonergic; Koyama and Kayama, 1993). A projection from the PPTN to the lateral geniculate nucleus (LGN) forms part of the reticular-activation system, and activation of the PPTN can lead to enhanced visual responses of the LGN in the cat (Uhlrich et al., 1995). Furthermore, in the head-restrained rat, LDTN neurons have heterogeneous properties not only with respect to their spontaneous activity during sleep and wakefulness but also in their response to sensory stimulation. Some of these LDTN neurons may also function to induce a global attentive state in response to a novel stimulus (Koyama et al., 1994).

Figure 2 shows our model for the control of arousal and attention by the PPTN. Arousal-related neuronal activity can be generated by the LC–DRN–PPTN circuit, which is triggered by inputs from the cerebral cortex. Neuronal activity in brainstem cholinergic nuclei (LDTN) is related to tonic-activation processes in thalamo-cortical systems (Steriade et al., 1990). Cholinergic projections (McDonald et al., 1993) from the PPTN to the LGN are also a component of the reticular-activation system. Activation of the PPTN can also regulate visual responses of the LGN in the cat (Uhlrich et al., 1995). The thalamo-cortical projection and cortical projection to the PPTN (cat—Edley and Graybiel, 1983; monkey—Matsumura et al., 2000) form a recurrent PPTN–thalamus–cerebral cortex network. Reverberation of signals in this circuit may regulate sensorimotor processing and attentional modulation within the cerebral cortex.

Scarnati and Florio (1997) found that movement performance in the rat was dramatically reduced by bilateral lesions of the PPTN, with the animal

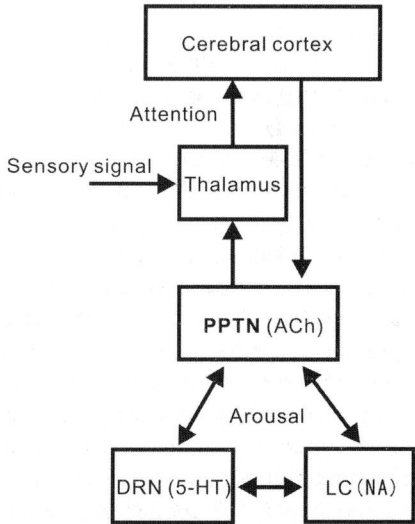

Fig. 2. Model of the PPTN's involvement in controlling attention and arousal. Additional abbreviations: 5-HT, serotonin; ACh, acetylcholine; DRN, dorsal raphe nucleus; LC, locus coeruleus; NA, noradrenaline.

exhibiting significantly increased reaction time and slowed movement duration. In the conditioned cat, reversible inactivation of the PPTN by microinjection of lidocaine or muscimol dramatically slows intertrial intervals and the motor execution of lever release (Conde et al., 1998). These results suggest that the PPTN is involved in selecting the appropriate motor program and attentional state to perform the selected movement, including saccades.

A number of studies have suggested that the PPTN induces a global attentive state in response to a novel stimulus (Steriade, 1996a,b). Recently, context-dependent activity in PPTN neurons was demonstrated in the operantly conditioned cat while performing a lever-release movement (Dormont et al., 1998). In this study, PPTN neurons exhibited excitation at a very short latency after the cue stimulus. Similar PPTN neuronal activity has been observed during a visually guided saccade task (Kobayashi et al., 2002). In this latter case, a fixation point (FP) appeared at an earlier stage of the task.

The proposed PPTN neuronal activity (sustained tonic activity related to task performance or response to the FP) during saccade tasks may enhance responsiveness of the thalamus and thereby modulate sensorimotor processing at the cortical level. Such PPTN activity should improve processing for perception of the next target and the execution of subsequent saccades.

Reinforcement learning

In the monkey, the PPTN receives limbic inputs from the hypothalamus and ventral tegmental area (Semba and Fibiger, 1992) and the limbic cortex (Chiba et al., 2001). Recent studies have emphasized that SNc dopaminergic neurons process the reward-related information necessary for the reinforcement of behavior (Schultz, 1998). Most dopamine neurons exhibit phasic activity after primary liquid and food rewards and conditioned, reward-predicting visual and auditory stimuli (for review, see Schultz, 1998).

A recent computational model predicts that the PPTN is a major source of the excitatory signal to the substantia nigra pars compacta (SNc) and an important neural control component for reinforcement learning (Brown et al., 1999). The model made use of the finding that the PPTN projects to SNc dopaminergic neurons (Beninato and Spencer, 1986), which may encode an error signal sufficient for such learning (Schultz, 1998). Brown et al.'s (1999) model successfully simulated the learning process. Dopamine neurons were activated by rewards during early trials in the learning sessions, when errors were frequent and rewards unpredictable. Activation by rewards was progressively reduced as performance was consolidated and the rewards became more predictable. The suppression of a reward response after learning may be caused by GABAergic input from striosomes (Gerfen, 1992).

Recently, reinforcement-related single-unit activity in the PPTN has also been demonstrated in the cat (Dormont et al., 1998). Glutamatergic and cholinergic inputs from the PPTN make synaptic connections with SNc dopamine neurons in the cat (Futami et al., 1995; Takakusaki et al., 1996), and electrical stimulation of the PPTN induces a time-locked burst in analogous neurons in the rat (Lokwan et al., 1999). Clearly, SNc dopamine neurons can code errors in the prediction of the occurrence of rewards (Hollerman and Schultz, 1998).

Reward-related activity in the PPTN has been observed in the monkey during saccade tasks, even during fully conditioned situations (Kobayashi et al., 2002). Since the PPTN can code primary (direct) reward signals, which are modified by cortical inputs, Fig. 3 shows that our group now hypothesizes that the reward prediction error signal coded by the SNc may be composed of an excitatory reward signal (derived from the PPTN), and inhibitory primary reward prediction signals from the striatum.

Cognitive processes

PPTN neurons have axon collaterals that project to both the LGN and SC (Billet et al., 1999). Accordingly, we hypothesize that the PPTN coordinates the relay of attentive information to the cortex via the thalamus, and the PPTN's effects on the onset and execution of orienting movements, which are mediated via the SC. Neuronal activity related to behavioral performance can improve visual and motor processing via the thalamus, including SC preparation signals in advance of a saccade. Therefore, activity is prominent in those PPTN

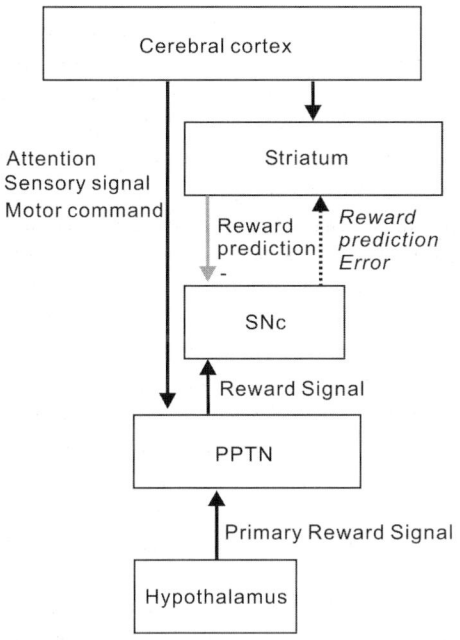

Fig. 3. Model of the PPTN's involvement in controlling reinforcement learning. Additional abbreviation: SNc, substantia nigra pars compacta.

neurons, which respond to both saccade execution and task performance.

PPTN neurons, which have abundant axon collaterals, receive a wide variety of inputs from many areas of the brain. Hence, it is not surprising that some of them are active during both saccades and reward signals. Such neurons may also coordinate the relay of reward information to the SNc when saccades are being executed.

Figure 4 shows our model for the involvement of the PPTN in the control of attention, saccades and reinforcement. The signals related to all three may be impressed on each single PPTN neuron. With regard to movement control, the PPTN can integrate saccade-related signals derived from the SC, FEF and SNr. During movement, the PPTN works cooperatively with the SC by virtue of their reciprocal connectivity. Reward-related signals from the PPTN can affect reinforcement learning within the basal ganglia via the SNc, and signals related to behavioral performance can affect attention at the cerebral cortex via the thalamus.

In final summary, our group's work has shown that by recording neuronal activity in the PPTN

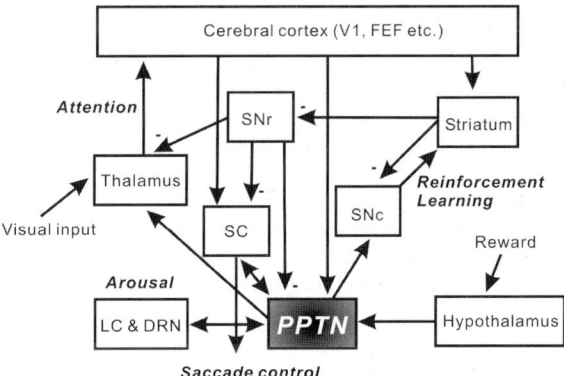

Fig. 4. Model of the PPTN's interactions with brainstem/basal ganglia circuits for the control of saccades. Additional abbreviations: FEF, frontal eye field; V1, primary visual cortex.

during the elaboration of saccades by the conscious monkey serves as a powerful tool for investigating neural mechanisms for motor control, arousal, attention and reinforcement learning, and their interactions.

Acknowledgments

We thank Chika Sasaki, Yasuyuki Takeshima and Junko Yamamoto for their technical assistance. This study was supported by grants from the Ministry of Education, Science, Sports and Culture of Japan to Y.K. (11780599, 13780647 and 13035052).

Abbreviations

5-HT	serotonin
ACh	acetylcholine
DRN	dorsal raphe nucleus
FP	fixation point
FEF	frontal eye field
LC	locus coeruleus
LDTN	laterodorsal tegmental nucleus
LGN	lateral geniculate nucleus
NA	noradrenaline
PPTN	pedunculopontine tegmental nucleus
SRT	saccadic reaction time
SNc	substantia nigra pars compacta
SNr	substantia nigra pars reticulata
SC	superior colliculus
V1	primary visual cortex

References

Aizawa, H., Kobayashi, Y., Yamamoto, M. and Isa, T. (1999). Injection of nicotine into superior colliculus facilitates occurrence of express saccades in monkeys. J. Neurophysiol., 82: 1642–1646.

Beninato, M. and Spencer, R.F. (1986). A cholinergic projection to the rat superior colliculus demonstrated by retrograde transport of horseradish peroxidase and choline acetyltransferase immunohistochemistry. J. Comp. Neurol., 253: 525–538.

Billet, S., Cant, N.B. and Hall, W.C. (1999). Cholinergic projections to the visual thalamus and superior colliculus. Brain Res., 847: 121–123.

Brown, J., Bullock, D. and Grossberg, S. (1999). How the basal ganglia use parallel excitatory and inhibitory learning pathways to selectively respond to unexpected rewarding cues. J. Neurosci., 19: 10502–10511.

Bruce, C.J. and Goldberg, M.E. (1985). Primate frontal eye fields. I. Single neurons discharging before saccades. J. Neurophysiol., 53: 603–635.

Chiba, T., Kayahara, T. and Nakano, K. (2001). Efferent projections of infralimbic and prelimbic areas of the medial prefrontal cortex in the Japanese monkey, Macaca fuscata. Brain Res., 888: 83–101.

Conde, H., Dormont, J.F. and Farin, D. (1998). The role of the pedunculopontine tegmental nucleus in relation to conditioned motor performance in the cat. II. Effects of reversible inactivation by intracerebral microinjections. Exp. Brain Res., 121: 411–418.

Dormont, J.F., Conde, H. and Farin, D. (1998). The role of the pedunculopontine tegmental nucleus in relation to conditioned motor performance in the cat. I. Context-dependent and reinforcement-related single unit activity. Exp. Brain Res., 121: 401–410.

Dorris, M.C. and Munoz, D.P. (1998). Saccadic probability influences motor preparation signals and time to saccadic initiation. J. Neurosci., 18: 7015–7026.

Edley, S.M. and Graybiel, A.M. (1983). The afferent and efferent connections of the feline nucleus tegmenti pedunculopontinus, pars compacta. J. Comp. Neurol., 217: 187–215.

Fischer, B. and Weber, H. (1993). Express saccade and visual attention. Behav. Brain Sci., 16: 553–610.

Futami, T., Takakusaki, K. and Kitai, S.T. (1995). Glutamatergic and cholinergic inputs from the pedunculopontine tegmental nucleus to dopamine neurons in the substantia nigra pars compacta. Neurosci. Res., 21: 331–342.

Garcia-Rill, E. (1991). The pedunculopontine nucleus. Prog. Neurobiol., 36: 363–389.

Gerfen, C.R. (1992). The neostriatal mosaic: multiple levels of compartmental organization. Trends Neurosci., 15: 133–139.

Gerfen, C.R., Staines, W.A., Arbuthnott, G.W. and Fibiger, H.C. (1982). Crossed connections of the substantia nigra in the rat. J. Comp. Neurol., 207: 283–303.

Glimcher, P.W. and Sparks, D.L. (1992). Movement selection in advance of action in the superior colliculus. Nature, 355: 542–545.

Granata, A.R. and Kitai, S.T. (1991). Inhibitory substantia nigra inputs to the pedunculopontine neurons. Exp. Brain Res., 86: 459–466.

Graybiel, A.M. (1978). A stereometric pattern of distribution of acetylthiocholinesterase in the deep layers of the superior colliculus. Nature, 272: 539–541.

Hall, W.C., Fitzpatrick, D., Klatt, L.L. and Raczkowski, D. (1989). Cholinergic innervation of the superior colliculus in the cat. J. Comp. Neurol., 287: 495–514.

Hallanger, A.E. and Wainer, B.H. (1988). Ascending projections from the pedunculopontine tegmental nucleus and the adjacent mesopontine tegmentum in the rat. J. Comp. Neurol., 274: 483–515.

Henderson, Z. and Sherriff, F.E. (1991). Distribution of choline acetyltransferase immunoreactive axons and terminals in the rat and ferret brainstem. J. Comp. Neurol., 314: 147–163.

Hikosaka, O. and Wurtz, R.H. (1983). Visual and oculomotor functions of monkey substantia nigra pars reticulata. I. Relation of visual and auditory responses to saccades. J. Neurophysiol., 49: 1230–1253.

Hollerman, J.R. and Schultz, W. (1998). Dopamine neurons report an error in the temporal prediction of reward during learning. Nat. Neurosci., 1: 304–309.

Huerta, M.F. and Harting, J.K. (1982). Tectal control of spinal cord activity: neuroanatomical demonstration of pathways connecting the superior colliculus with the cervical spinal cord grey. Prog. Brain Res., 57: 293–328.

Isa, T., Endo, T. and Saito, Y. (1998a). Nicotinic facilitation of signal transmission in the local circuits of the rat superior colliculus. Soc. Neurosci. Abstr., 24: 60.13.

Isa, T., Endo, T. and Saito, Y. (1998b). The visuo-motor pathway in the local circuit of the rat superior colliculus. J. Neurosci., 18: 8496–8504.

Jeon, C.J., Spencer, R.F. and Mize, R.R. (1993). Organization and synaptic connections of cholinergic fibers in the cat superior colliculus. J. Comp. Neurol., 333: 360–374.

Kawagoe, R., Takikawa, Y. and Hikosaka, O. (1998). Expectation of reward modulates cognitive signals in the basal ganglia. Nat. Neurosci., 1: 411–416.

Kobayashi, Y., Inoue, Y., Yamamoto, M., Isa, T. and Aizawa, H. (2002). Contribution of pedunculopontine tegmental nucleus neurons to performance of visually guided saccade tasks in monkeys. J. Neurophysiol., 88: 715–731.

Koyama, Y. and Kayama, Y. (1993). Mutual interactions among cholinergic, noradrenergic and serotonergic neurons studied by ionophoresis of these transmitters in rat brainstem nuclei. Neuroscience, 55: 1117–1126.

Koyama, Y., Jodo, E. and Kayama, Y. (1994). Sensory responsiveness of 'broad-spike' neurons in the laterodorsal tegmental nucleus, locus coeruleus and dorsal raphe of awake rats: implications for cholinergic and monoaminergic neuron-specific responses. Neuroscience, 63: 1021–1031.

Kustov, A.A. and Robinson, D.L. (1996). Shared neural control of attentional shifts and eye movements. Nature, 384: 74–77.

Lavoie, B. and Parent, A. (1994). Pedunculopontine nucleus in the squirrel monkey: projections to the basal ganglia as revealed by anterograde tract-tracing methods. J. Comp. Neurol., 344: 210–231.

Lindsley, D.B. (1958). The reticular system and perceptual discrimination. In: Jasper H.H. (Ed.), The Reticular Formation of the Brain. Little Brown & Co, Boston, pp. 513–534.

Lokwan, S.J., Overton, P.G., Berry, M.S. and Clark, D. (1999). Stimulation of the pedunculopontine tegmental nucleus in the rat produces burst firing in A9 dopaminergic neurons. Neuroscience, 92: 245–254.

Ma, T.P., Graybiel, A.M. and Wurtz, R.H. (1991). Location of saccade-related neurons in the macaque superior colliculus. Exp. Brain Res., 85: 21–35.

Matsumura, M., Watanabe, K. and Ohye, C. (1997). Single-unit activity in the primate nucleus tegmenti pedunculopontinus related to voluntary arm movement. Neurosci. Res., 28: 155–165.

Matsumura, M., Nambu, A., Yamaji, Y., Watanabe, K., Imai, H., Inase, M., Tokuno, H. and Takada, M. (2000). Organization of somatic motor inputs from the frontal lobe to the pedunculopontine tegmental nucleus in the macaque monkey. Neuroscience, 98: 97–110.

May, P.J. and Porter, J.D. (1992). The laminar distribution of macaque tectobulbar and tectospinal neurons. Vis. Neurosci., 8: 257–276.

McDonald, C.T., McGuinness, E.R. and Allman, J.M. (1993). Laminar organization of acetylcholinesterase and cytochrome oxidase in the lateral geniculate nucleus of prosimians. Neuroscience, 54: 1091–1101.

Pare, M. and Munoz, D.P. (1996). Saccadic reaction time in the monkey: advanced preparation of oculomotor programs is primarily responsible for express saccade occurrence. J. Neurophysiol., 76: 3666–3681.

Perry, E., Walker, M., Grace, J. and Perry, R. (1999). Acetylcholine in mind: a neurotransmitter correlate of consciousness?. Trends Neurosci., 22: 273–280.

Redgrave, P., Mitchell, I.J. and Dean, P. (1987). Descending projections from the superior colliculus in rat: a study using orthograde transport of wheatgerm-agglutinin conjugated horseradish peroxidase. Exp. Brain Res., 68: 147–167.

Scarnati, E. and Florio, T. (1997). The pedunculopontine nucleus and related structures. Functional organization. Adv. Neurol., 74: 97–110.

Schnurr, B., Spatz, W.B. and Illing, R.B. (1992). Similarities and differences between cholinergic systems in the superior colliculus of guinea pig and rat. Exp. Brain Res., 90: 291–296.

Schultz, W. (1998). Predictive reward signal of dopamine neurons. J. Neurophysiol., 80: 1–27.

Semba, K. and Fibiger, H.C. (1992). Afferent connections of the laterodorsal and the pedunculopontine tegmental nuclei in the rat: a retro- and antero-grade transport and immunohistochemical study. J. Comp. Neurol., 323: 387–410.

Shouse, M.N. and Siegel, J.M. (1992). Pontine regulation of REM sleep components in cats: integrity of the pedunculopontine tegmentum (PPT) is important for phasic events but unnecessary for atonia during REM sleep. Brain Res., 571: 50–63.

Sparks, D.L. and Hartwich-Young, R. (1989). The deep layers of the superior colliculus. Rev. Oculomot. Res., 3: 213–255.

Steckler, T., Keith, A.B. and Sahgal, A. (1994a). Lesions of the pedunculopontine tegmental nucleus do not alter delayed non-matching to position accuracy. Behav. Brain Res., 61: 107–112.

Steckler, T., Inglis, W., Winn, P. and Sahgal, A. (1994b). The pedunculopontine tegmental nucleus: a role in cognitive processes?. Brain Res. Brain Res. Rev., 19: 298–318.

Steriade, M. (1996a). Awakening the brain. Nature, 383: 24–25.

Steriade, M. (1996b). Arousal: revisiting the reticular activating system. Science, 272: 225–226.

Steriade, M., Datta, S., Pare, D., Oakson, G. and Curro Dossi, R.C. (1990). Neuronal activities in brain-stem cholinergic nuclei related to tonic activation processes in thalamocortical systems. J. Neurosci., 10: 2541–2559.

Takakusaki, K., Shiroyama, T., Yamamoto, T. and Kitai, S.T. (1996). Cholinergic and noncholinergic tegmental pedunculopontine projection neurons in rats revealed by intracellular labeling. J. Comp. Neurol., 371: 345–361.

Uhlrich, D.J., Tamamaki, N., Murphy, P.C. and Sherman, S.M. (1995). Effects of brain stem parabrachial activation on receptive field properties of cells in the cat's lateral geniculate nucleus. J. Neurophysiol., 73: 2428–2447.

SECTION X

Higher control mechanisms: basal ganglia, sensorimotor cortex and frontal lobe

CHAPTER 42

Macro-architecture of basal ganglia loops with the cerebral cortex: use of rabies virus to reveal multisynaptic circuits

Roberta M. Kelly[1] and Peter L. Strick[2]*

[1] Johnson and Johnson Pharmaceutical Research and Development, San Diego, CA 92121, USA
[2] Pittsburgh Veterans Affairs Medical Center and Center for the Neural Basis of Cognition, and Departments of Neurobiology, Neurological Surgery, and Psychiatry, University of Pittsburgh, Pittsburgh, PA 15261, USA

Abstract: We have used retrograde transneuronal transport of rabies virus to examine basal ganglia connections with the cerebral cortex. We injected rabies into the primary motor cortex (M1) or into Area 46 of cebus monkeys. A 4-day survival time was long enough to allow transport of rabies from the injection site to 'third-order' neurons in the basal ganglia. After either M1 or Area 46 injections, third-order neurons were found in the external segment of the globus pallidus (GPe), striatum and subthalamic nucleus (STN). In each of these nuclei, the third-order neurons that innervate M1 were spatially separated from those that innervate Area 46. Thus, distinct basal ganglia–thalamocortical circuits innervate M1 and Area 46. Next, we injected a conventional tracer into M1 to define its terminations in the putamen and STN. We found that the regions of the putamen and STN that receive input from M1 are the same as those that contain third-order neurons after M1 injections of virus. On the other hand, virus injections into M1 also labeled a relatively dense group of third-order neurons in a region of the ventral putamen that is not innervated by M1. This region of the putamen is the target of efferents from the amygdala. Thus, the ventral putamen may provide a route for the limbic system to influence motor output. Overall, our results indicate that basal ganglia circuits with the cerebral cortex can be characterized by both open- and closed-loop macro-architectures.

Introduction

Our notions about the structure of basal ganglia circuits with the cerebral cortex have undergone some dramatic modifications in recent years. It was traditionally thought that information from distributed areas of the cerebral cortex was funneled through the basal ganglia in order to gain access to the primary motor cortex (Kemp and Powell, 1971). It has become increasingly clear, however, that the output from the basal ganglia projects to multiple areas of the frontal lobe (for references and review,

*Corresponding author: Tel.: +412-383-9961;
Fax: +412-383-9061; E-mail: strickp@pitt.edu

see Hoover and Strick, 1999; Middleton and Strick, 2000, 2002). These outputs provide the neural substrate for the basal ganglia to influence diverse aspects of motor and nonmotor function (Alexander et al., 1986).

The multisynaptic nature of the interconnections between the basal ganglia and cerebral cortex has made it technically difficult to define the macro-architecture of these circuits with conventional tracing techniques. Selected neurotropic viruses, however, are capable of moving transneuronally through chains of synaptically linked neurons (Fig. 1). The development and use of these viruses as transneuronal tracers in the primate central nervous system (CNS) have provided new opportunities

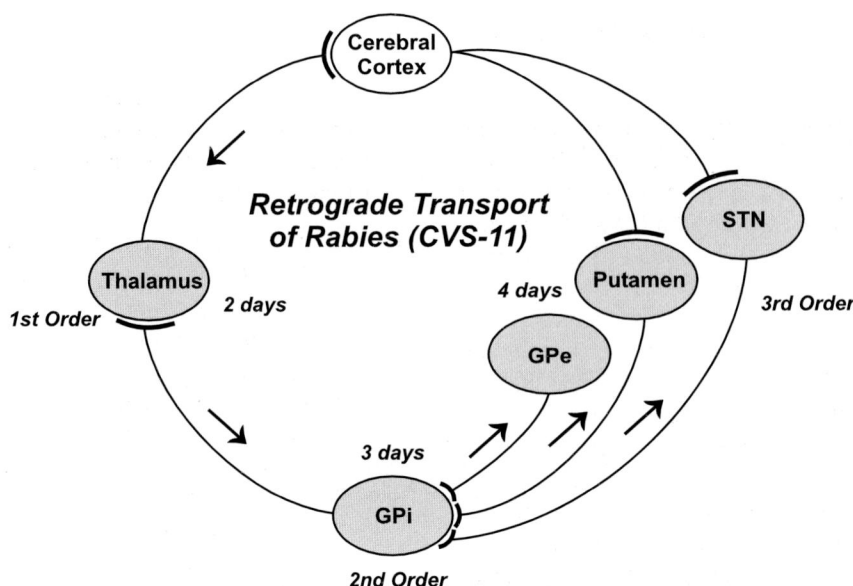

Fig. 1. Retrograde transneuronal transport of rabies virus in basal ganglia loops with the cerebral cortex. This schematic diagram illustrates the time course of infection after an injection of the CVS-11 strain of rabies into the cortex of a cebus monkey. Two days after a cortical injection, virus was transported in the retrograde direction to label first-order neurons in the thalamus that innervate the injection site. Three days postinjection, virus was transported transneuronally in the retrograde direction from first-order neurons in the thalamus to second-order neurons in GPi. After a 4-day survival period, the virus underwent another stage of retrograde transneuronal transport from second-order neurons in GPi to third-order neurons in the external segment of GPe, STN and putamen. Abbreviations: GPe, external segment of the globus pallidus; GPi, internal segment of the globus pallidus; STN, the subthalamic nucleus.

to examine basal ganglia interactions with the cerebral cortex (Zemanick et al., 1991; Strick and Card, 1992; Hoover and Strick, 1993; Kelly and Strick, 2000).

Output channels

We have used retrograde transneuronal transport of herpes simplex virus type 1 (HSV1) to examine the origin of basal ganglia projections to regions of primary motor, premotor, oculomotor, prefrontal and inferior temporal cortex (Hoover and Strick, 1993; Lynch et al., 1994; Middleton and Strick, 1994, 1996; Hoover and Strick, 1999; Middleton and Strick, 2002). In these studies, the survival time was long enough to allow retrograde transneuronal transport of virus to 'second-order' neurons in the output nuclei of the basal ganglia, i.e., the internal segment of the globus pallidus (GPi) and the substantia nigra pars reticulata (SNpr). We found that each of the cortical areas examined is the target of basal ganglia output. Furthermore, the neurons that project to one cortical area were generally clustered together within GPi and/or SNpr to form what we have termed 'output channels' (Hoover and Strick, 1993). In general, such output channels to different cortical areas displayed limited overlap (Hoover and Strick, 1993; Middleton and Strick, 1994, 1996; Hoover and Strick, 1999; Middleton and Strick, 2002). Thus, our studies with HSV1 indicate that the basal ganglia–thalamocortical pathway consists of multiple, spatially separate output channels that innervate a remarkably diverse set of motor and nonmotor cortical areas.

Some technical details on rabies transneuronal transport

In this report, we will describe some of our recent observations using rabies virus to unravel additional levels of the macro-architecture of basal ganglia loops with the cerebral cortex (Kelly and Strick, 1999). Like some strains of HSV1, rabies virus replicates in infected neurons and is transported transneuronally in the retrograde direction in a time-dependent fashion (Fig. 1). Unlike HSV1, however, rabies virus does not cause any cell lysis in the CNS of primates (Kelly and Strick, 2000; see also Ugolini, 1995). The lack of tissue destruction and the absence of behavioral symptoms during the survival period make rabies virus especially useful for revealing chains of synaptically linked neurons.

To reveal multiple stages in basal ganglia loops with the cerebral cortex, we have examined transneuronal transport of rabies after extended survival times (Fig. 1). Overall, we found that a 4-day survival time is long enough to allow transport of the CVS-11 strain of rabies from a cortical injection site to 'third-order' neurons in the basal ganglia. In these studies, we injected the CVS-11 strain of rabies virus ($1 \times 10^{7.7}$ PFU/ml) at multiple sites (0.2–0.4 μl/site) in the primary motor cortex (M1) in six monkeys (*Cebus apella*). Injections were placed throughout the 'arm' area or were localized to sites where intracortical electrical stimulation evoked digit movements at stimulus thresholds <10 μA. After a 2–4 day survival period, immunohistochemical procedures were used to localize virus.

At the 2-day survival time, virus was transported in the retrograde direction from the injection site to first-order neurons in the ventrolateral thalamus, which are known to project to M1 e.g., (Holsapple et al., 1991). In contrast, we found no evidence of anterograde transport of rabies virus to sites innervated by M1 (e.g., striatum, pontine nuclei and red nucleus). At the 3-day survival time, virus was transported transneuronally in the retrograde direction from first-order neurons in the thalamus to second-order neurons in GPi. At the 4-day survival time, virus underwent another stage of retrograde transneuronal transport from second-order neurons in GPi to third-order neurons in the external segment of the globus pallidus (GPe), subthalamic nucleus (STN) and putamen. These results indicate that with careful adjustment of survival time, transneuronal transport of rabies virus can be a powerful tool for mapping basal ganglia–thalamocortical circuitry.

Circuit analysis with rabies transport

We have used transneuronal transport of rabies to examine two key issues concerning the patterns of convergence and divergence in basal ganglia loops with the cerebral cortex.

1. To what extent are the populations of third-order neurons that innervate different cortical areas spatially segregated? For example, we have shown that the output channels in GPi and SNpr to different cortical areas display little evidence of overlap. Does the spatial segregation in these channels extend to the level of third-order neurons in GPe, the striatum and STN?
2. Do the populations of third-order neurons that target a specific cortical area receive input from that area? Two sites of third-order neurons, the striatum and STN, receive direct input from the cerebral cortex. Are the sites of terminations of a cortical area within the striatum and STN in register with the third-order neurons that ultimately project back to the same cortical area?

Third-order projections to M1 and Area 46

Here we report the results of studies in which we examined basal ganglia circuit connections with the arm area of M1 and Area 46 in the prefrontal cortex (Figs. 2–5). Figure 2 shows plots of third-order labeling in GPe and the striatum after cortical injections of rabies. Plots of labeling in STN are shown in Fig. 3. In both figures, labeling after injections into the arm area of M1 is displayed on the left side and that following injections into Area 46 is displayed on the right.

Within the *striatum*, third-order neurons that project to M1 are found most densely in the so-called 'sensorimotor' portion of the putamen (Fig. 2 left, sections 502, 522). In contrast, third-order neurons in the striatum that project to Area 46 are found most densely in rostromedial portions of the caudate (Fig. 2 right, sections 306, 406).

Third-order neurons in *GPe* that project to M1 are found most densely in ventrolateral portions of the nucleus (Fig. 2 left, section 522), whereas third-order neurons that project to Area 46 are found more

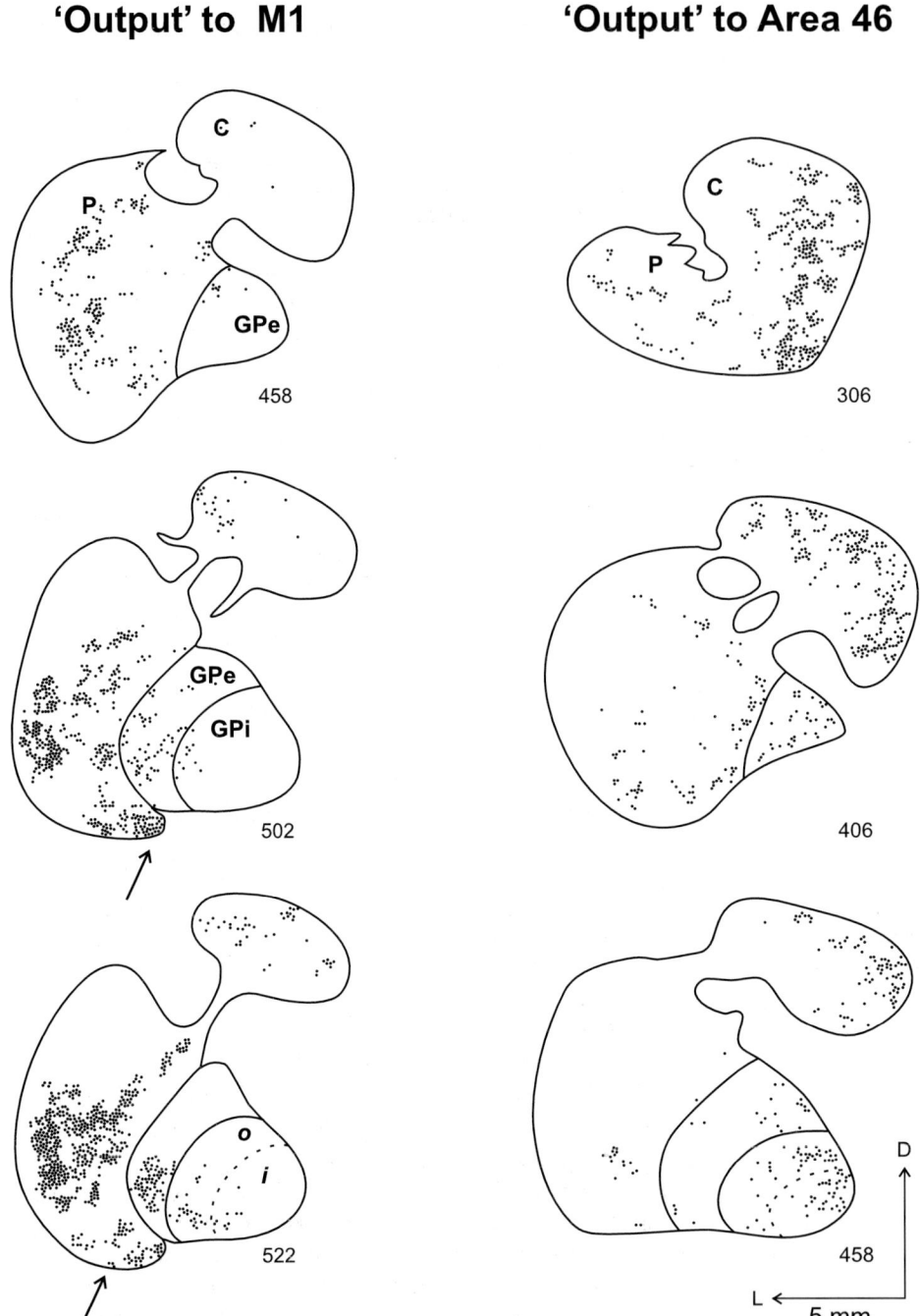

Fig. 2. Third-order neurons in the striatum and GPe. Coronal sections through different levels of the striatum and globus pallidus display the distribution of third-order neurons labeled after either an injection of rabies into M1 (left) or Area 46 (right). The section number is indicated beside each chart. The arrows next to sections 502 and 522 point to a special cluster of labeled neurons in the ventral putamen (see text for details). Abbreviations: C, caudate; D, dorsal; L, lateral; P, putamen; i, inner portion of GPi; o, outer portion of GPi.

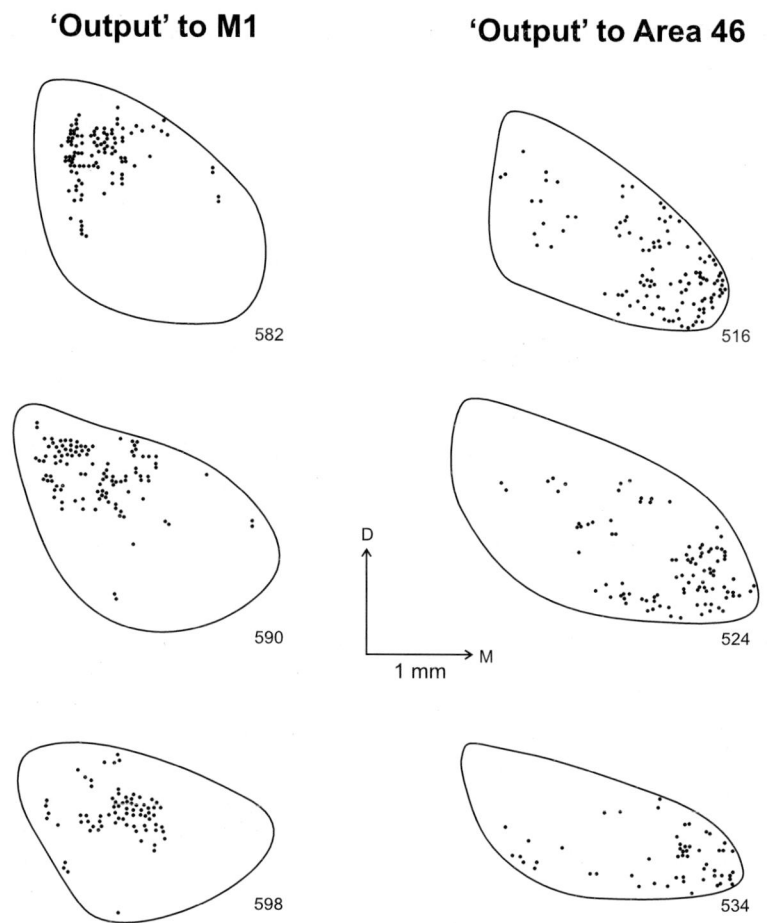

Fig. 3. Third-order neurons in STN. Coronal sections through different levels of STN display the distribution of third-order neurons labeled after either an injection of rabies into M1 (left) or Area 46 (right). The section number is indicated beside each chart.

rostrally in dorsomedial portions of GPe (Fig. 2 right, section 458).

Within *STN*, neurons labeled after M1 injections are located in dorsolateral portions of the nucleus (Fig. 3 left, sections 582, 590). On the other hand, neurons labeled after Area 46 injections are found most densely at more rostral levels of STN in ventromedial portions of the nucleus (Fig. 3 right, sections 516, 524).

These results illustrate that within GPe, the striatum and STN, the third-order neurons that innervate M1 are largely spatially separated from those that innervate Area 46. Thus, the segregation of the output channels found in GPi and SNpr (Hoover and Strick, 1993; Middleton and Strick, 1994; Hoover and Strick, 1999; Middleton and Strick,

2002) extends into earlier stages of basal ganglia processing.

The location of the peak density of third-order neurons in GPe, the putamen and STN following M1 injections of virus corresponds closely to the sites in these nuclei where many neurons with activity related to arm movements have been found in physiological studies (Anderson and Horak, 1985; DeLong et al., 1985; Liles and Updyke, 1985; Bergman et al., 1990; Wichmann et al., 1994; Turner and Anderson, 1997; Baron et al., 2002). On the other hand, few, if any, neurons related to arm movements are found at caudate, GPe and STN sites that project to Area 46. Instead, these regions contain neurons whose activity is modulated by cognitive aspects of behavioral tasks (Hikosaka et al., 1989; Schultz and Romo,

1992; Kawagoe et al., 1998). Thus, the anatomical segregation of third-order neurons in GPe, the striatum and STN fits with the known differential distribution of neurons with motor and nonmotor response properties.

Input–output coupling

A comparison of what is known about cortical inputs to the striatum and STN with the location of third-order neurons that project to M1 and Area 46 suggests some interesting functional linkages. The peak density of third-order neurons in the putamen that project to M1 appears to be located in the region of the putamen that receives the major projection from the sensorimotor cortex (Kunzle, 1975; Jones et al., 1977; Liles and Updyke, 1985; Flaherty and Graybiel, 1993a; Inase et al., 1996; Takada et al., 1998). Likewise, the peak density of third-order neurons in the caudate that project to Area 46 appears to be located in the region of the caudate that receives the major projection from the dorsolateral prefrontal cortex (Selemon and Goldman-Rakic, 1985; Arikuni and Kubota, 1986; Yeterian and Pandya, 1991; Elben and Graybiel, 1995).

A similar input–output arrangement may exist in STN. The dorsolateral region of the nucleus with third-order neurons that project to M1 appears to be the region of STN that receives input from M1 (Hartmann-von Monakow et al., 1978; Nambu et al., 1996). Similarly, the ventromedial region of STN with third-order neurons that project to Area 46 appears to be the region of STN that receives input from Area 46 (Hartmann-von Monakow et al., 1978). These observations suggest that the striatum and STN participate in 'closed loop' circuits with M1 and Area 46.

The anatomical studies of cortical inputs to the striatum and STN cited above were performed in the macaques or squirrel monkey, whereas our studies with rabies virus were performed in the cebus monkey. To further test the closed-loop nature of these circuits, we used anterograde transport of WGA–HRP to define where efferents from the arm area of M1 terminate in STN and the putamen of the cebus monkey. We then compared these sites of termination with the origin of third-order neurons that project to M1 in the same species. The optimal way to do this type of comparison would be to inject both tracers into M1 of the same animal, of course. Such an experiment is not feasible, however, because WGA is known to antagonize rabies infection (Conti et al., 1986). Thus, without further technical developments, this experiment requires comparison of material from separate animals.

In accordance with results from the macaque, we found that efferents from the arm area of M1

'Input' from M1 Arm

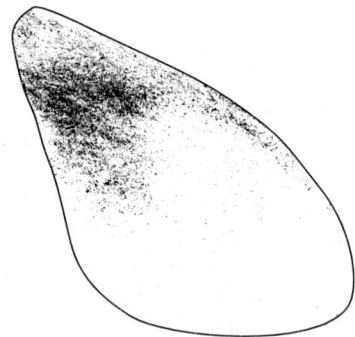

'Output' to M1 Arm

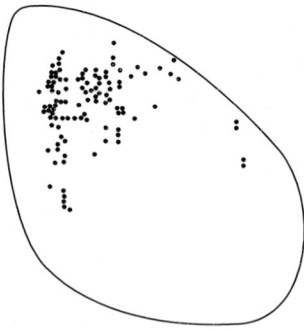

Fig. 4. Input–output organization in STN. We used anterograde transport of WGA–HRP to define where efferents from the arm area of M1 terminate in STN (left). In another animal, we used retrograde transneuronal transport of rabies to define the location of third-order neurons that project to the arm area of M1 (right). The results of these experiments are displayed on coronal sections through approximately the same level of STN. Note that the region of STN that receives input from M1 overlaps the region that projects to M1.

'Input' from M1 Arm   'Output' to M1 Arm

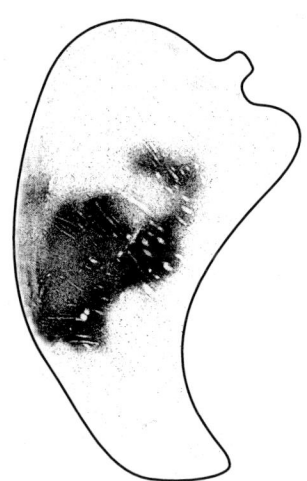 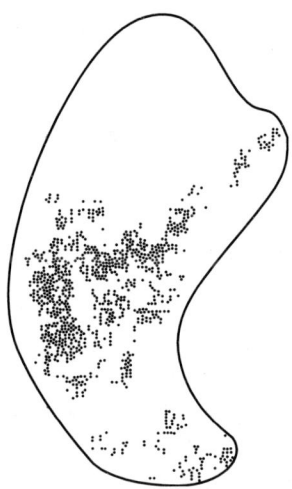

Fig. 5. Input–output organization in the putamen. We used anterograde transport of WGA–HRP to define where efferents from the arm area of M1 terminate in the putamen (left). In another animal, we used retrograde transneuronal transport of rabies to define the location of third-order neurons that project to the arm area of M1 (right). The results of these experiments are displayed on coronal sections through approximately the same level of the putamen. Note that there is extensive overlap in the regions of the putamen that project to and receive from M1.

terminated most densely in a dorsolateral portion of STN in the cebus monkey (Fig. 4 left). This site of termination corresponded closely to the region of STN that contained the peak density of third-order neurons after injections of rabies virus into the arm area of M1 of the same species (Fig. 4 right). Hence, within STN, the region that receives input from M1 overlaps the region that projects to M1.

We found a similar arrangement in the striatum of the cebus monkey. Efferents from the arm area of M1 terminated most densely in a mid-lateral portion of the putamen (Fig. 5 left). This site of termination corresponded closely to the region of the putamen that contained the peak density of third-order neurons after injections of rabies virus into the arm area of M1 (Fig. 5 right). Thus, an important component of the interconnections between M1 and two basal ganglia nuclei, STN and the putamen, can be characterized by a closed-loop architecture.

An open-loop circuit

A detailed examination of the correspondence between M1 input and M1 output in the putamen and STN showed that the match is not always complete. Some M1 terminations in the putamen and STN were scattered outside of the primary focus of labeling. Similarly, some third-order neurons in the putamen and STN were located outside of the primary focus of labeled neurons. These scattered terminations and scattered labeled neurons may or may not be in register.

Perhaps more importantly, a portion of the ventral putamen provides an example where input and output were clearly not in register. Injections of rabies into the arm area of M1 labeled a relatively dense group of third-order neurons in the ventral putamen (Fig. 2 left, sections 502, 522, and Fig. 6). This region of the putamen does not receive input from M1. The third-order neurons in the ventral putamen represented 10–20% of the total number of neurons labeled in the striatum after M1 injections. These results demonstrate that the ventral putamen is the origin of a modest, but consistent 'open-loop' pathway to M1.

The demonstration of an open-loop projection from the ventral portion of putamen to M1 raises some interesting questions about the nature of this circuit. For instance, what are the cortical and/or

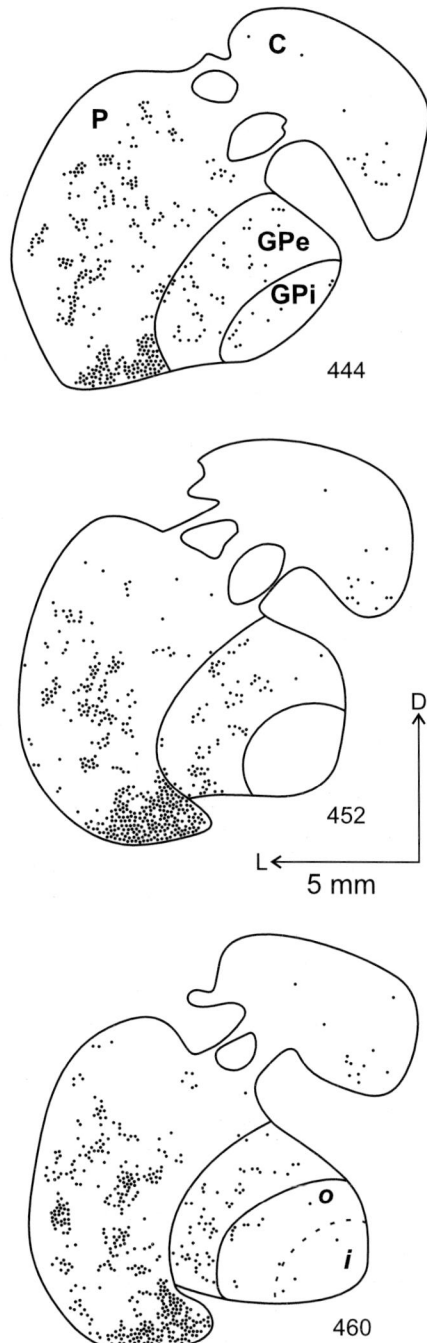

Fig. 6. Third-order neurons in the ventral putamen project to M1. Injections of rabies into the arm area of M1 labeled a relatively dense group of third-order neurons in the ventral putamen (see also Fig. 2 left, sections 502, 522). These coronal sections display levels of the putamen where the ventral labeling was the densest. This region does not receive input from M1.

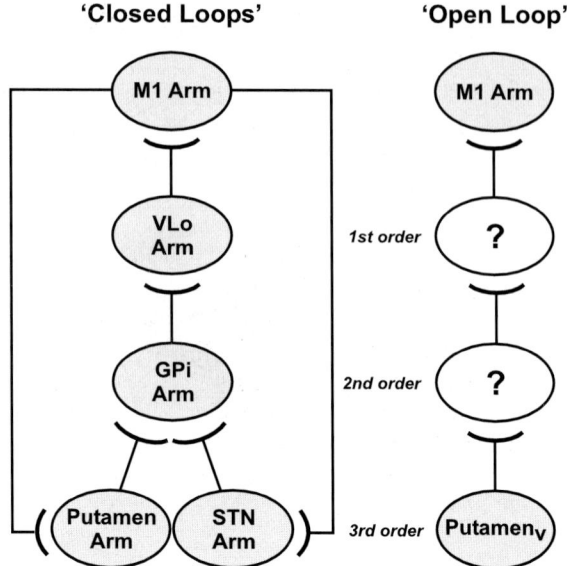

Fig. 7. Macroarchitecture of basal ganglia circuits with M1: 'closed' and 'open' loops. Closed-loop circuits link the arm area of M1 with the arm areas in STN and the sensorimotor portion of the putamen. These loops utilize classical thalamocortical circuitry and operate entirely within the domain of motor control. Our results on Area 46 suggest that similar closed-loop circuits connect the basal ganglia with nonmotor areas in the frontal lobe. In contrast, open-loop circuitry characterizes the connection between M1 and the ventral putamen. The precise route that the ventral putamen uses to project to M1 is unclear. Inputs to the ventral putamen suggest that this circuit provides the limbic system with access to the motor system. See the text for further details. Abbreviations: Putamen$_v$, ventral putamen; VLo, ventralis lateralis pars oralis.

subcortical sources of input to the ventral putamen? There is evidence (Russchen et al., 1985; Fudge et al., 2002) that the amygdala is one source of afferents to this region. In fact, Parent and Hazrati (1995) have designated this portion of the putamen as part of the 'limbic territory' of the striatum. Their observations, together with our present results, suggest that the ventral putamen may provide a means for nonmotor signals from the limbic system to influence the control of movement at the level of M1.

Another important question is how this portion of the ventral putamen gains access to M1. Neurons in the ventral putamen do not appear to project to the region of GPi that contains an output channel to M1 (Flaherty and Graybiel, 1993b; Parent and Hazrati, 1995; Hoover and Strick, 1999). Thus, it is unlikely that the ventral putamen projects to M1 through a traditional basal ganglia–thalamocortical route. Instead, efferents from the ventral putamen are likely to have projection patterns similar to those of other regions of the ventral striatum that include a projection to the ventral pallidum (Haber et al., 1990; Parent and Hazrati, 1995). The precise routes that this system uses to gain access to M1 are unclear. What is clear is that M1 receives two sources of striatal input (Fig. 7). One source originates in the sensorimotor portion of the putamen and uses a classical basal ganglia–thalamocortical circuit to operate in the domain of motor control. The second source originates in the limbic territory of the ventral putamen and uses a nontraditional route to provide M1 with information that is likely to be outside the domain of motor control.

A potential clinical correlate

The existence of the above two sources of striatal input to M1 may have a clinical correlate. In Parkinson's disease, the sensorimotor portion of the putamen is the early site of major dopamine depletion (Kish et al., 1988). Other striatal sites are less affected or unaffected at comparable stages of the disease process. One of the classical symptoms of Parkinson's disease is a disinclination to move. However, Parkinson patients are known to demonstrate an increased mobility in situations of 'fight or flight'. Perhaps the enhanced ability to generate movement in these special circumstances is dependent on the intact function of a pathway from the ventral putamen to M1, and the access this pathway provides for limbic influences to trigger motor output.

Summary and conclusions

In this chapter, we have described some of our recent results from experiments using transneuronal transport of rabies virus to examine multisynaptic circuits in basal ganglia loops with the cerebral cortex. Retrograde transneuronal transport of rabies has enabled us to identify the origin and distribution of third-order neurons in the basal ganglia that project to either the arm area of M1 or Area 46 in dorsolateral prefrontal cortex. In prior studies using HSV1 as a tracer, we have shown that M1 and Area 46 are the target of distinct output channels in GPi and SNpr. Transneuronal transport of rabies after longer survival times demonstrates that much of the spatial segregation of these output channels extends to the level of third-order neurons in GPe, the striatum and STN. In other words, the third-order neurons in GPe, the striatum and STN that influence M1 are largely separate from those that influence Area 46. We find little or no evidence for convergence in the basal ganglia pathways that project to M1 and Area 46.

An examination of cortical input to these third-order neurons has provided some new insights into the macro-architecture of basal ganglia loops with the cerebral cortex. We observed three fundamental circuits between the cerebral cortex and the basal ganglia (Fig. 7). An important aspect of basal ganglia macro-architecture is represented by two closed loops, which link M1 with STN, and M1 with a portion of the striatum. Clearly, these circuits operate entirely within the domain of motor control. The results of our experiments on Area 46 suggest that a similar closed-loop architecture also exists for basal ganglia circuits with nonmotor areas in the frontal lobe. Closed-loop circuits with these latter cortical areas presumably operate outside the domain of motor control.

Our study has also revealed that a distinct open-loop macro-architecture characterizes connections between M1 and the ventral putamen. Although the precise route that the ventral putamen uses to project to M1 is unclear, an input to this circuit from the amygdala has been identified. Thus, this open-loop circuit may provide the limbic system with access to M1. Whether a similar open-loop architecture also exists for other regions of the ventral striatum and nonmotor areas of cortex remains to be determined.

In conclusion, our results indicate that the basal ganglia can participate in two types of functional circuits. An important component of the macro-architecture of basal ganglia loops with the cerebral cortex can be characterized by parallel, segregated and closed-loop circuits. In addition, we have discovered that the basal ganglia may provide for a specific type of convergence that enables the limbic system to influence motor output.

Acknowledgments

We thank Michael Page for the development of computer programs and Wendy Burnette, Karen Hughes and Michelle O'Malley-Davis for their expert technical assistance. We also thank Drs. Charles Rupprecht (CDC, Atlanta, GA) for supplying the CVS-11 strain of rabies, and Alex Wandeler (Animal Diseases Research Institute, Nepean, ONT, CAN) for supplying antibodies to rabies. This work was supported by the Veterans Affairs Medical Research Service and by US Public Health Service grants NS 24328 and MH 56661 (to P.L.S.).

Abbreviations

C	caudate
CNS	central nervous system
GPe	external segment of the globus pallidus
GPi	internal segment of the globus pallidus
HSV1	herpes simplex virus type 1
i	inner portion of Gpi
M1	primary motor cortex
o	outer portion of Gpi
P	putamen
PFU	plaque-forming units
Putamen$_v$	ventral putamen
SNpr	substantia nigra pars reticulata
STN	subthalamic nucleus
VLo	ventralis lateralis pars oralis

References

Alexander, G.E., DeLong, M.R. and Strick, P.L. (1986). Parallel organization of functionally segregated circuits linking basal ganglia and cortex. Annu. Rev. Neurosci., 9: 357–381.

Anderson, M.E. and Horak, F.B. (1985). Influence of the globus pallidus on arm movements in monkeys. III. Timing of movement-related information. J. Neurophysiol., 54: 433–438.

Arikuni, T. and Kubota, K. (1986). The organization of prefrontocaudate projections and their laminar origin in the macaque monkey: a retrograde study using HRP-gel. J. Comp. Neurol., 244: 492–510.

Baron, M.S., Wichmann, T., Ma, D. and DeLong, M.R. (2002). Effects of transient focal inactivation of the basal ganglia in Parkinsonian primates. J. Neurosci., 22: 592–599.

Bergman, H., Wichmann, T. and DeLong, M.R. (1990). Reversal of experimental Parkinsonism by lesions of the subthalamic nucleus. Science, 249: 1436–1438.

Conti, C., Superti, F. and Tsiang, H. (1986). Membrane carbohydrate requirement for rabies virus binding to chicken embryo related cells. Intervirology, 26: 164–168.

DeLong, M.R., Crutcher, M.D. and Georgopoulos, A.P. (1985). Primate globus pallidus and subthalamic nucleus: functional organization. J. Neurophysiol., 52: 530–543.

Elben, F. and Graybiel, A.M. (1995). Highly restricted origin of prefrontal cortical inputs to striosomes in the macaque monkey. J. Neurosci., 15: 5999–6013.

Flaherty, A.W. and Graybiel, A.M. (1993a). Two input systems for body representations in the primate striatal matrix: experimental evidence in the squirrel monkey. J. Neurosci., 13: 1120–1137.

Flaherty, A.W. and Graybiel, A.M. (1993b). Output architecture of the primate putamen. J. Neurosci., 13: 3222–3237.

Fudge, J.L., Kunishio, K., Walsh, P., Richard, C. and Haber, S.N. (2002). Amygdaloid projections to ventromedial striatal subterritories in the primate. Neuroscience, 110: 257–275.

Haber, S.N., Lynd, E., Klein, C. and Grownewegen, H.J. (1990). Topographic organization of the ventral striatal efferent projections in the rhesus monkey: an anterograde tracing study. J. Comp. Neurol., 293: 282–298.

Hartmann-von Monakow, K., Akert, K. and Kunzle, H. (1978). Projections of the precentral motor cortex and other cortical areas of the frontal lobe to the subthalamic nucleus in the monkey. Exp. Brain Res., 33: 395–403.

Hikosaka, O., Sakamoto, M. and Usui, S. (1989). Functional properties of monkey caudate neurons. III. Activities related to expectation of target and reward. J. Neurophysiol., 61: 814–832.

Holsapple, J.W., Preston, J.B. and Strick, P.L. (1991). The origin of thalamic inputs to the 'hand' representation in the primary motor cortex. J. Neurosci., 11: 2644–2654.

Hoover, J.E. and Strick, P.L. (1993). Multiple output channels in the basal ganglia. Science, 259: 819–821.

Hoover, J.E. and Strick, P.L. (1999). The organization of cerebellar and basal ganglia outputs to primary motor cortex as revealed by retrograde transneuronal transport of herpes simplex virus type 1. J. Neurosci., 19: 1446–1463.

Inase, M., Sakai, S.T. and Tanji, J. (1996). Overlapping corticostriatal projections from the supplementary motor area and the primary motor cortex in the macaque monkey: an anterograde double labeling study. J. Comp. Neurol., 373: 283–296.

Jones, E.G., Coulter, J.D., Burton, H. and Porter, R. (1977). Cells of origin and terminal distribution of corticostriatal fibers arising in the sensory-motor cortex of monkeys. J. Comp. Neurol., 173: 53–80.

Kawagoe, R., Takikawa, Y. and Hikosaka, O. (1998). Expectation of reward modulates cognitive signals in the basal ganglia. Nature Neurosci., 1: 411–416.

Kelly, R.M. and Strick, P.L. (1999). Retrograde transneuronal transport of rabies virus through basal ganglia-thalamocortical circuits of primates. Soc. Neurosci. Abstr., 25: 1925.

Kelly, R.M. and Strick, P.L. (2000). Rabies as a transneuronal tracer of circuits in the central nervous system. J. Neurosci. Methods, 103: 63–71.

Kemp, J.M. and Powell, T.P.S. (1971). The connexions of the striatum and globus pallidus: synthesis and speculation. Phil. Trans. R. Soc. Lond., 262: 441–457.

Kish, S.J., Shannak, K. and Hornykiewicz, O. (1988). Uneven pattern of dopamine loss in the striatum of patients with idiopathic Parkinson's disease. Pathophysiologic and clinical implications. N. Engl. J. Med., 318: 876–880.

Kunzle, H. (1975). Bilateral projections from precentral motor cortex to the putamen and other parts of the basal ganglia. An autoradiographic study in macaca fascicularis. Brain Res., 88: 195–209.

Liles, S.L. and Updyke, B.V. (1985). Projection of the digit and wrist area of precentral gyrus to the putamen: relation between topography and physiological properties of neurons in the putamen. Brain Res., 339: 245–255.

Lynch, J.C., Hoover, J.E. and Strick, P.L. (1994). Input to the primate frontal eyefield from the substantia nigra, superior colliculus, and dentate nucleus demonstrated by transneuronal transport. Exp. Brain Res., 1: 181–186.

Middleton, F.A. and Strick, P.L. (1994). Anatomical evidence for cerebellar and basal ganglia involvement in higher cognitive function. Science, 266: 458–461.

Middleton, F.A. and Strick, P.L. (1996). The temporal lobe is the target of output from the basal ganglia. Proc. Natl. Acad. Sci. USA, 93: 8683–8687.

Middleton, F.A. and Strick, P.L. (2000). Basal ganglia and cerebellar loops: motor and cognitive circuits. Brain Res. Rev., 31: 236–250.

Middleton, F.A. and Strick, P.L. (2002). Basal-ganglia 'projections' to the prefrontal cortex of the primate. Cerebral Cortex, 12: 926–935.

Nambu, A., Takada, M., Inase, M. and Tokuno, H. (1996). Dual somatotopical representations in the primate subthalamic nucleus: evidence for ordered but reversed body-map transformations from primary motor cortex and the supplementary motor area. J. Neurosci., 16: 2671–2683.

Parent, A. and Hazrati, L. (1995). Functional anatomy of the basal ganglia. I. The cortico-basal ganglia-thalamocortical loop. Brain Res. Rev., 20: 91–127.

Russchen, F.T., Bakst, I., Amaral, D.G. and Price, J.L. (1985). The amygdalostriatal projections in the monkey. An anterograde tracing study. Brain Res., 329: 241–257.

Schultz, W. and Romo, R. (1992). Role of primate basal ganglia and frontal cortex in the internal generation of movements. I. Preparatory activity in the anterior striatum. Exp. Brain Res., 91: 363–384.

Selemon, L.D. and Goldman-Rakic, P.S. (1985). Longitudinal topography and interdigitation of corticostriatal projections in the rhesus monkey. J. Neurosci., 5: 776–794.

Strick, P.L. and Card, J.P. (1992). Transneuronal mapping of neural circuits with alpha herpesviruses. In: Bolam J.P. (Ed.), Experimental Neuroanatomy: A Practical Approach. Oxford University Press, Oxford, pp. 81–101.

Takada, M., Tokuno, H., Nambu, A. and Inase, M. (1998). Corticostriatal projections from the somatic motor areas of the frontal cortex in the macaque monkey: segregation versus overlap of input zones from the primary motor cortex, the supplementary motor area, and the premotor cortex. Exp. Brain Res., 120: 114–128.

Turner, R.S. and Anderson, M.E. (1997). Pallidal discharge related to kinematics of reaching movements in two dimensions. J. Neurophysiol., 86: 623–632.

Ugolini, G. (1995). Specificity of rabies virus as a transneuronal tracer of motor networks: transfer from hypoglossal motoneurons to connected second-order and higher order central nervous system cell groups. J. Comp. Neurol., 356: 457–480.

Wichmann, T., Bergman, H. and DeLong, M.R. (1994). The primate subthalamic nucleus. I. Functional properties in intact animals. J. Neurophysiol., 72: 494–506.

Yeterian, E.H. and Pandya, D.N. (1991). Prefrontostriatal connections in relation to cortical architectonic organization in rhesus monkeys. J. Comp. Neurol., 312: 43–67.

Zemanick, M.C., Strick, P.L. and Dix, R.D. (1991). Direction of transneuronal transport of herpes simplex virus 1 in the primate motor system is strain-dependent. Proc. Natl. Acad. Sci. USA, 88: 8048–8051.

CHAPTER 43

A new dynamic model of the cortico-basal ganglia loop

Atsushi Nambu*

Department of System Neuroscience, Tokyo Metropolitan Institute for Neuroscience, Tokyo Metropolitan Organization for Medical Research, Tokyo 183-8526, Japan

Abstract: An important issue in the neural control of posture and movement is how motor-related areas of the cerebral cortex modulate the activity of the output nuclei of the basal ganglia. In this chapter, the functional significance of the 'hyperdirect' cortico-subthalamo-pallidal pathway is emphasized, and further a new dynamic model of basal ganglia function is presented. When a voluntary movement is about to be initiated by cortical mechanisms, a corollary signal is conveyed through the 'hyperdirect' pathway to first inhibit large areas of the thalamus and cerebral cortex that are related to both the selected motor program and other competing programs. Next, another corollary signal is sent through the 'direct' cortico-striato-pallidal pathway to disinhibit this second pathway's targets, and ensure activation of only the selected motor program. Finally, a third corollary signal is sent through the 'indirect' cortico-striato-external pallido-subthalamo-internal pallidal pathway to strongly inhibit this third pathway's targets. This sequential information processing ensures that only the selected motor program is initiated, executed and terminated at the appropriate times, whereas other competing programs are canceled.

Introduction

A loop linking the cerebral cortex and the basal ganglia is important for the control of voluntary movement (Kelly and Strick, Chapter 42 of this volume). Information derived from the cortex is processed in the basal ganglia and returns to the cortex via the thalamus (Alexander and Crutcher, 1990). In the current, well-accepted model of the basal ganglia organization, the striatum receives direct excitatory cortical inputs, and projects to two output nuclei; the internal segment (GPi) of the globus pallidus (GP) and the substantia nigra pars reticulata (SNr). This is via two major projection systems, the 'direct' and 'indirect' pathways (Alexander and Crutcher, 1990). The direct pathway arises from GABAergic striatal neurons containing substance P, and it projects monosynaptically to the GPi/SNr. The indirect pathway arises from GABAergic striatal neurons containing enkephalin, and it projects polysynaptically to the GPi/SNr by way of a sequence of connections involving the external segment (GPe) of the GP and the subthalamic nucleus (STN).

To advance understanding of how motor-related cortical areas modulate the activity of the output nuclei of the basal ganglia, we analyzed the response patterns of GPi/SNr neurons to electrical stimulation of the motor-related cortical areas (Nambu et al., 1990; Yoshida et al., 1993; Nambu et al., 2000b). Based on these studies, a new dynamic model of how the basal ganglia function is presented in the next section.

Circuitry

Stimulation of the motor-related cortical areas induces an early, short-latency excitation, followed

---

*Present address: Division of System Neurophysiology, National Institute for Physiological Sciences, Okazaki 444-8585, Japan. Tel.: +81-564-55-7771; Fax: +81-564-52-7913; E-mail: nambu@nips.ac.jp

by an inhibition and a late excitation in pallidal neurons of monkeys (Nambu et al., 1990; Yoshida et al., 1993; Nambu et al., 2000b). What is the origin of each response? The early excitation is considered to be derived from neither the direct nor indirect pathways, but rather, from the cortico-STN-pallidal 'hyperdirect' pathway (Nambu et al., 1996). This assertion is based on six key findings (Nambu et al., 2000b): (1) the monkey STN receives somatotopically organized inputs from the primary motor cortex, supplementary motor area and premotor cortex (Hartmann-von Monakow et al., 1978; Nambu et al., 1996, 1997) and, in turn, it sends outputs to the GPe and GPi/SNr. (2) Simultaneous recordings of neuronal activity in the pallidal complex and STN have shown that the cortical stimulation induces an early, short-latency excitation in STN neurons preceding that in pallidal neurons. (3) Stimulation in the STN through the recording electrode activates pallidal neurons orthodromically. (4) Blockade of neuronal activity in the STN by its injection with muscimol (a $GABA_A$ receptor agonist) abolishes the early and late excitations of pallidal neurons evoked by cortical stimulation. (5) Blockade of the glutamatergic cortico-STN neurotransmission by injection of (±)-3-(2-carboxypiperazin-4-yl)-propyl-1-phosphonic acid (CPP, an NMDA receptor antagonist) into the STN suppresses the early excitation of pallidal neurons. (6) On the other hand, interference with GABAergic pallido-STN neurotransmission by injection of bicuculline (a $GABA_A$ receptor antagonist) into the STN has little effect on the early excitation.

Several studies suggest that the inhibition evoked in pallidal neurons by cortical stimulation is mediated by the net inhibitory cortico-striato-pallidal 'direct' pathway. For example, stimulation of the striatum evokes inhibitory responses in pallidal neurons. The difference between the latency of this inhibition and that evoked by cortical stimulation corresponds well to the cortico-striatal conduction time (Yoshida et al., 1993). Inhibitory postsynaptic potentials (IPSPs), or inhibition evoked by cortical stimulation in pallidal neurons, is abolished by systemic (Kita, 1992) or local (Nambu et al., unpublished observations) injection of a GABAergic blocker, or by blocking cortico-striatal neurotransmission (Maurice et al., 1999).

An open issue is the origin of the late excitation evoked in pallidal neurons by cortical stimulation. It might be via the net excitatory cortico-striato-GPe-STN-GPi 'indirect' or cortico-striato-GPe-GPi pathway. This is suggested by a similar late excitatory response of SNr neurons being markedly reduced after the blockade of cortico-striatal or striato-GPe neurotransmission (Maurice et al., 1999). Injection of CPP into the STN also attenuated the late excitation, however. This suggests an involvement of the cortico-STN-pallidal hyperdirect pathway in the late excitation of pallidal neurons (Nambu et al., 2000b). In either case, the late excitation in pallidal neurons evoked by cortical stimulation is ascribed to the late excitation in STN neurons. This is supported by the observation that blockade of STN activity abolishes the late excitation as well as the early excitation in pallidal neurons (Nambu et al., 2000b). A contribution of rebound firing after IPSPs via the direct pathway cannot yet be ruled out, however, because striatal stimulation causes rebound firing after IPSPs in pallidal neurons in slice preparations (Nambu and Llinás, 1994).

The existence of the cortico-STN-pallidal hyperdirect pathway in the cortico-basal ganglia circuitry is clearly demonstrated by the observations described above. Fig. 1A shows that this hyperdirect pathway conveys powerful excitatory effects from the motor-related cortical areas to the pallidum. It bypasses the striatum, and its effects have a shorter conduction time than those conveyed through the direct and indirect pathways.

The new model

Recent anatomical studies have shown that STN-pallidal fibers arborize more widely and terminate on more proximal neuronal elements than do striato-pallidal fibers (Hazrati and Parent, 1992a,b). These findings suggest a 'center-surround model' of basal ganglia function, which proposes the inhibition of competing motor programs, and the focused release of the selected motor program (Mink and Thach, 1993; Mink, 1996). The hyperdirect pathway exerts powerful excitatory effects on the output nuclei of the basal ganglia, and it is faster in signal conduction time from the cerebral cortex than are the direct and

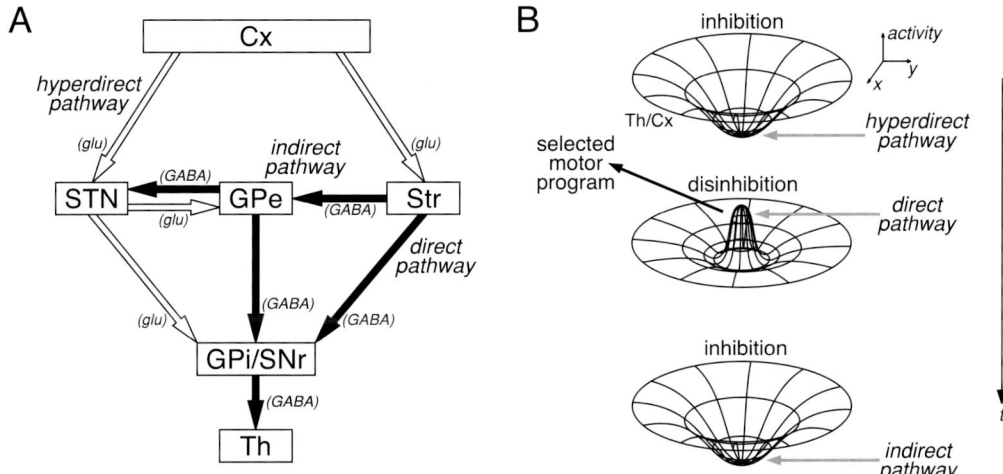

Fig. 1. (A) Schematic diagram of the cortico-STN-GPi/SNr 'hyperdirect', cortico-striato-GPi/SNr 'direct' and cortico-striato-GPe-STN-GPi/SNr 'indirect' pathways. Open and filled arrows represent excitatory glutamatergic (glu) and inhibitory GABAergic (GABA) projections, respectively. Cx, cerebral cortex; GPe, external segment of the globus pallidus; GPi, internal segment of the globus pallidus; SNr, substantia nigra pars reticulata; STN, subthalamic nucleus; Str, striatum; Th, thalamus. (B) Dynamic model of the basal ganglia function explaining the activity changes in the thalamus and/or cortex (Th/Cx) caused by sequential inputs through the hyperdirect (top), direct (middle) and indirect (bottom) pathways. (Reproduced, in part, from Nambu et al., 2000b with permission from The American Physiological Society.)

indirect pathways (viz. Fig. 1A). For the regulation of voluntary limb movements, these findings expand the center-surround model in the temporal domain (Nambu et al., 2000b). The model proposes that when a voluntary movement is about to be initiated by cortical mechanisms, a corollary signal is transmitted simultaneously from the motor cortex to the GPi through the hyperdirect pathway. The corollary signal activates GPi neurons, thereby resulting in the inhibition of large areas of the thalamus and cerebral cortex, which are related to both the selected motor program and other competing programs (Fig. 1B, top). Next, another corollary signal through the direct pathway is conveyed to the GPi. It inhibits a specific population of pallidal neurons in the center area. Such pallidal neurons disinhibit their targets and release only the selected motor program (Fig. 1B, middle). Finally, the third corollary signal through the indirect pathway reaches the GPi to activate neurons therein, thereby extensively suppressing their targets (Fig. 1B, bottom). Throughout the above sequential information processing, only the selected motor program is initiated, executed and terminated at the selected timing. Other competing programs mediated by pallidal neurons in the surrounding area are canceled.

Blockade of STN neuronal activity (Nambu et al., 2000b) or lesions in the STN (Carpenter et al., 1950; Hamada and DeLong, 1992) result in the induction of hemiballismus. Specifically, after blockade of STN neuronal activity, GPi neurons show abnormal activity characterized by pauses and grouped discharges (Nambu et al., 2000b). Motor programs, which are usually inhibited by the constant high-frequency discharges of GPi neurons, may be released randomly by pauses of GPi neuronal activity, thereby resulting in involuntary movements.

This dynamic model of the basal ganglia may explain some features of pallidal activity during voluntary movements. It has always been associated with an increase rather than a decrease in its cells' discharge, and an incidence ratio of spike-frequency increase-to-decrease of 1.6–5.8 (Georgopoulos et al., 1983; Anderson and Horak, 1985; Mitchell et al., 1987; Hamada et al., 1990; Nambu et al., 1990; Mink and Thach, 1991; Turner and Anderson, 1997). In GPi, the movement-related increases in cell discharge tend to occur earlier than decreases (Georgopoulos et al., 1983; Anderson and Horak, 1985). In addition, neurons in the lateral STN (Georgopoulos et al., 1983; DeLong et al., 1985; Wichmann et al., 1994) and STN neurons projecting to the pallidal complex (Jinnai et al., 1990) exhibit movement-related activity. Thus, it is likely that the activity of pallidal neurons during voluntary limb movements is mediated by the net excitatory and faster hyperdirect pathway, as well as by the direct pathway, which has a net inhibitory effect and is slower conducting.

It is also necessary to consider the different nature of the signals conveyed through the hyperdirect, direct and indirect pathways. Cortico-STN and corticostriatal neurons belong to different populations, and they seem to mediate different types of motor information. The cortico-STN projections have been reported to originate from the axon collaterals of pyramidal tract neurons (Giuffrida et al., 1985). This suggests that the STN should receive signals directly related to movements, and this has indeed been demonstrated (Georgopoulos et al., 1983; DeLong et al., 1985; Wichmann et al., 1994). In contrast, corticostriatal neurons transmit signals that are more selective to the parameters of behavioral tasks than do cortical neurons projecting to the spinal cord/brainstem (Bauswein et al., 1989; Turner and DeLong, 2000). Striatal neurons show similar activity to that of corticostriatal neurons (Crutcher and DeLong, 1984; Crutcher and Alexander, 1990). Moreover, striatal neurons display context-dependent (Kimura et al., 1992) and reward-contingent (Kawagoe et al., 1998) activity. Thus, the author's model proposes that when a voluntary movement is about to be initiated by cortical mechanisms, hyperdirect pathway signals widely inhibit motor programs, while direct pathway signals release the motor program that is appropriate for the task at hand.

Future issues

In this chapter, a new dynamic model of basal ganglia function is introduced. Its relevance to voluntary movements has been emphasized, but the supporting evidence has come largely from stimulation studies. Thus, the model should be tested functionally, by recording GPi activity during voluntary movements with selected blocking of the hyperdirect, direct or indirect pathway.

Currently, the known basal ganglia circuitry suggests that if the net inhibitory GPe–STN–GPi and GPe–GPi pathways really work as proposed by the model, then GPe and GPi neurons should display opposite activity. In contrast, the neuronal activity of GPe and GPi has striking similarities, both in response to cortical stimulation (Yoshida et al., 1993; Nambu et al., 2000b) and during the performance of motor tasks (Georgopoulos et al., 1983; Anderson and Horak, 1985; Mitchell et al., 1987; Hamada et al., 1990; Nambu et al., 1990; Mink and Thach, 1991; Turner and Anderson, 1997). These neurons seem to behave as if they are exclusively under the control of the hyperdirect and direct pathways. To resolve this issue, it is necessary to characterize more explicitly the effects of the hyperdirect, direct and indirect pathways on the activity of pallidal neurons during voluntary movements.

Finally, the author points out the relevance of his new model to pathophysiological aspects of the basal ganglia. The mechanisms underlying hypokinetic and hyperkinetic disorders are currently explained as changes in the *static* state of the basal ganglia, i.e., an increase or a decrease in the mean firing rate of GPi/SNr neurons, which is likely caused by an imbalance of activity in the direct and indirect pathways (DeLong, 1990). A reduction of disinhibition in the thalamus and cortex in the *temporal and spatial domains* better explains the akinesia of Parkinson's disease, however (Boraud et al., 2000; Nambu et al., 2000a). This latter finding is accommodated in the author's model, which may prove to be of value for advancing understanding of other basal ganglia disturbances, as well.

Acknowledgments

This study was supported by Grants-in-Aid for Scientific Research (C) and for Scientific Research on Priority Areas (A) from the Ministry of Education, Culture, Sports, Science and Technology of Japan.

Abbreviations

CPP   (±)-3-(2-carboxypiperazin-4-yl)-propyl-1-phosphonic acid

GABA	γ-aminobutyric acid
GP	globus pallidus
GPe	external segment of the globus pallidus
GPi	internal segment of the globus pallidus
IPSP	inhibitory postsynaptic potential
NMDA	N-methyl-D-aspartate
SNr	substantia nigra pars reticulata
STN	subthalamic nucleus

References

Alexander, G.E. and Crutcher, M.D. (1990). Functional architecture of basal ganglia circuits: neural substrates of parallel processing. Trends Neurosci., 13: 266–271.

Anderson, M.E. and Horak, F.B. (1985). Influence of the globus pallidus on arm movements in monkeys. III. Timing of movement-related information. J. Neurophysiol., 54: 433–448.

Bauswein, E., Fromm, C. and Preuss, A. (1989). Corticostriatal cells in comparison with pyramidal tract neurons: contrasting properties in the behaving monkey. Brain Res., 493: 198–203.

Boraud, T., Bezard, E., Bioulac, B. and Gross, C.E. (2000). Ratio of inhibited-to-activated pallidal neurons decreases dramatically during passive limb movement in the MPTP-treated monkey. J. Neurophysiol., 83: 1760–1763.

Carpenter, M.B., Whittier, J.R. and Mettler, F.A. (1950). Analysis of choreoid hyperkinesia in the rhesus monkey: surgical and pharmacological analysis of hyperkinesia resulting from lesions in the subthalamic nucleus of Luys. J. Comp. Neurol., 92: 293–332.

Crutcher, M.D. and Alexander, G.E. (1990). Movement-related neuronal activity selectively coding either direction or muscle pattern in three motor areas of the monkey. J. Neurophysiol., 64: 151–163.

Crutcher, M.D. and DeLong, M.R. (1984). Single cell studies of the primate putamen. II. Relations to direction of movement and pattern of muscular activity. Exp. Brain Res., 53: 244–258.

DeLong, M.R. (1990). Primate models of movement disorders of basal ganglia origin. Trends Neurosci., 13: 281–285.

DeLong, M.R., Crutcher, M.D. and Georgopoulos, A.P. (1985). Primate globus pallidus and subthalamic nucleus: functional organization. J. Neurophysiol., 53: 530–543.

Georgopoulos, A.P., DeLong, M.R. and Crutcher, M.D. (1983). Relations between parameters of step-tracking movements and single cell discharge in the globus pallidus and subthalamic nucleus of the behaving monkey. J. Neurosci., 3: 1586–1598.

Giuffrida, R., Li Volsi, G., Maugeri, G. and Perciavalle, V. (1985). Influences of pyramidal tract on the subthalamic nucleus in the cat. Neurosci. Lett., 54: 231–235.

Hamada, I. and DeLong, M.R. (1992). Excitotoxic acid lesions of the primate subthalamic nucleus result in transient dyskinesias of the contralateral limbs. J. Neurophysiol., 68: 1850–1858.

Hamada, I., DeLong, M.R. and Mano, N. (1990). Activity of identified wrist-related pallidal neurons during step and ramp wrist movements in the monkey. J. Neurophysiol., 64: 1892–1906.

Hartmann-von Monakow, K., Akert, K. and Künzle, H. (1978). Projections of the precentral motor cortex and other cortical areas of the frontal lobe to the subthalamic nucleus in the monkey. Exp. Brain Res., 33: 395–403.

Hazrati, L.-N. and Parent, A. (1992a). Convergence of subthalamic and striatal efferents at pallidal level in primates: an anterograde double-labeling study with biocytin and PHA-L. Brain Res., 569: 336–340.

Hazrati, L.-N. and Parent, A. (1992b). Differential patterns of arborization of striatal and subthalamic fibers in the two pallidal segments in primates. Brain Res., 598: 311–315.

Jinnai, K., Nambu, A., Yoshida, S. and Tanibuchi, I. (1990). Discharge patterns of subthalamo-pallidal projection neurons with input from various areas of the cerebral cortex. Jpn. J. Physiol., Suppl. 40: S204.

Kawagoe, R., Takikawa, Y. and Hikosaka, O. (1998). Expectation of reward modulates cognitive signals in the basal ganglia. Nat. Neurosci., 1: 411–416.

Kimura, M., Aosaki, T., Hu, Y., Ishida, A. and Watanabe, K. (1992). Activity of primate putamen neurons is selective to the mode of voluntary movement: visually guided, self-initiated or memory-guided. Exp. Brain Res., 89: 473–477.

Kita, H. (1992). Responses of globus pallidus neurons to cortical stimulation: intracellular study in the rat. Brain Res., 589: 84–90.

Maurice, N., Deniau, J.-M., Glowinski, J. and Thierry, A.-M. (1999). Relationships between the prefrontal cortex and the basal ganglia in the rat: physiology of the cortico-nigral circuits. J. Neurosci., 19: 4674–4681.

Mink, J.W. (1996). The basal ganglia: focused selection and inhibition of competing motor programs. Prog. Neurobiol., 50: 381–425.

Mink, J.W. and Thach, W.T. (1991). Basal ganglia motor control. II. Late pallidal timing relative to movement onset and inconsistent pallidal coding of movement parameters. J. Neurophysiol., 65: 301–329.

Mink, J.W. and Thach, W.T. (1993). Basal ganglia intrinsic circuits and their role in behavior. Curr. Opin. Neurobiol., 3: 950–957.

Mitchell, S.J., Richardson, R.T., Baker, F.H. and DeLong, M.R. (1987). The primate globus pallidus: neuronal activity related to direction of movement. Exp. Brain Res., 68: 491–505.

Nambu, A. and Llinás, R. (1994). Electrophysiology of globus pallidus neurons in vitro. J. Neurophysiol., 72: 1127–1139.

Nambu, A., Yoshida, S. and Jinnai, K. (1990). Discharge patterns of pallidal neurons with input from various cortical areas during movement in the monkey. Brain Res., 519: 183–191.

Nambu, A., Takada, M., Inase, M. and Tokuno, H. (1996). Dual somatotopical representations in the primate subthalamic nucleus: evidence for ordered but reversed body-map transformations from the primary motor cortex and the supplementary motor area. J. Neurosci., 16: 2671–2683.

Nambu, A., Tokuno, H., Inase, M. and Takada, M. (1997). Corticosubthalamic input zones from forelimb representations of the dorsal and ventral divisions of the premotor cortex in the macaque monkey: comparison with the input zones from the primary motor cortex and the supplementary motor area. Neurosci. Lett., 239: 13–16.

Nambu, A., Kaneda, K., Tokuno, H. and Takada, M. (2000a). Abnormal pallidal activity evoked by cortical stimulation in the parkinsonian monkey. Soc. Neurosci. Abstr., 26: 960.

Nambu, A., Tokuno, H., Hamada, I., Kita, H., Imanishi, M., Akazawa, T., Ikeuchi, Y. and Hasegawa, N. (2000b). Excitatory cortical inputs to pallidal neurons via the subthalamic nucleus in the monkey. J. Neurophysiol., 84: 289–300.

Turner, R.S. and Anderson, M.E. (1997). Pallidal discharge related to the kinematics of reaching movements in two dimensions. J. Neurophysiol., 77: 1051–1074.

Turner, R.S. and DeLong, M.R. (2000). Corticostriatal activity in primary motor cortex of the macaque. J. Neurosci., 20: 7096–7108.

Wichmann, T., Bergman, H. and DeLong, M.R. (1994). The primate subthalamic nucleus. I. Functional properties in intact animals. J. Neurophysiol., 72: 494–506.

Yoshida, S., Nambu, A. and Jinnai, K. (1993). The distribution of the globus pallidus neurons with input from various cortical areas in the monkey. Brain Res., 611: 170–174.

CHAPTER 44

Functional recovery after lesions of the primary motor cortex

Eric M. Rouiller[1],* and Etienne Olivier[2]

[1]Institute of Physiology and Program in Neuroscience, University of Fribourg, CH-1700 Fribourg, Switzerland
[2]Laboratory of Neurophysiology, Faculty of Medicine, Université Catholique de Louvain, B-1200 Brussels, Belgium

Abstract: After a lesion in the motor cortex of an adult primate, are cortical motor maps reorganized? This important question has attracted much interest throughout the past decade. In human subjects, substantial progress has resulted from the use of noninvasive imaging and stimulation techniques. For example, there is recent, well-accepted, albeit indirect evidence that following such a lesion on one side of the human brain, a dramatic reorganization of the hand representation occurs within either the ipsilateral primary motor cortex, nonprimary motor areas or both. The contribution of contralateral motor areas to functional recovery of the paretic hand remains uncertain, however, because of the lack of direct confirmatory evidence obtained from experiments undertaken on nonhuman primates. A better understanding of how the brain selects the optimal strategy for functional recovery following cortical lesions, and the neuronal mechanisms underlying cortical plasticity, will be important challenges for the next decade. To this end, the purpose of the present chapter is to provide an update on what is truly known about the functional recovery that takes place after a lesion in the primary motor cortex of both the nonhuman primate and the human. It bears emphasis that work on these fundamental issues is an essential prerequisite to the development of improved therapeutic and rehabilitation procedures for the brain-injured human.

Introduction: motor areas of the cerebral cortex

Present knowledge on the organization and function of the cortical motor areas in nonhuman primates mainly concerns the control of hand movements (Porter and Lemon, 1993; also Lemon et al., Weber and He, and Wiesendanger and Serrien, Chapters 26, 45 and 46 of this volume). Figure 1 shows that at least four key cortical regions are now known to be involved in such control. (1) The primary motor cortex (M1; i.e., Brodmann's area 4; also termed area F1). (2) The supplementary motor area (SMA), which corresponds to the mesial part of Brodmann's area 6, and is subdivided into a rostral (pre-SMA, area F6) and a caudal (SMA-proper, area F3) part. (3) The premotor cortex (PM), which corresponds to the lateral part of Brodmann's area 6, and is subdivided into a dorsal (PMd) and a ventral (PMv) part. Both the PMd and PMv can be further separated into a rostral (PMd-r, area F7; and PMv-r, area F5) and a caudal (PMd-c, area F2; and PMv-c, area F4) part. (4) The cingulate motor area (CMA) in the cingulate cortex, which is segregated into three subdivisions (CMA-r, CMA-d, CMA-v). The Fig. 1 parcellation of these motor cortical areas is based on several anatomical and functional criteria, which we recently reviewed in depth (Liu et al., 2002).

In M1, neuronal populations that are active during individual movements of different fingers, overlap extensively. The movement of a given finger appears to recruit a set of neurons distributed throughout the *entire* hand area rather than being somatotopically segregated (Schieber and Hibbard, 1993). This finding is in sharp contrast to the somatotopically ordered representation of movements in M1 as originally

*Corresponding author: Tel.: +41-26-300-86-09;
Fax: +41-26-300-96-75; E-mail: eric.rouiller@unifr.ch

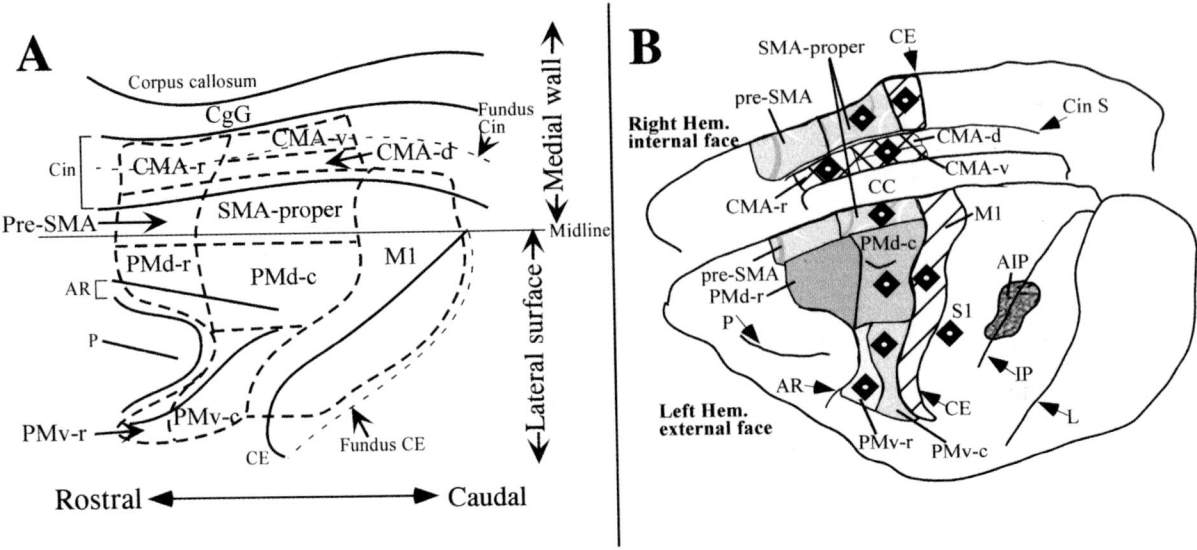

Fig. 1. Schematic representation of the motor areas of the cerebral cortex in the macaque monkey. (A) A portion of the lateral surface of the left hemisphere. The medial wall of the brain is also represented, extending from the midline to the corpus callosum. The central, arcuate and cingulate sulci have been partially unfolded (thick continuous lines), with the fundus depicted by a thin dashed line. The location of the motor cortical areas is shown with thick dashed lines. (B) The same cortical areas are represented on a surface view of the external face of the left hemisphere and the internal face of the right hemisphere. Black diamonds indicate the distribution of CS neurons across the motor cortical areas. See text for most abbreviations. Additional ones include: AR, arcuate sulcus; Cin, cingulate sulcus; CE, central sulcus; P, sulcus principalis.

proposed by Woolsey in the monkey (Woolsey et al., 1952), and Penfield and Jasper (1954) in the human. In the human, the presence of two cortical hand representations has been suggested on the basis of both anatomical and functional data (Geyer et al., 1996). Furthermore, Fig. 1B shows that the cell bodies of neurons at the origin of the corticospinal (CS) pathway are not restricted to M1, but, rather, are also distributed among the nonprimary motor areas (see Dum and Strick, 1991; He et al., 1993, 1995; Rouiller et al., 1996; Morecraft et al., 1997). To be regarded as a 'true' motor area of the cerebral cortex (Fig. 1), a cortical zone has to fulfill one, or several, criteria, including: the presence of CS neurons, a direct projection to M1, muscle activation in response to intracortical microstimulation (ICMS), a connection with the 'motor' thalamus, and neuronal activity tightly linked to the execution of a conditional motor task (Table 1).

The classical description of CS projections is that they are comprised of 90% crossed axons and 10% uncrossed ones. This distribution needs reevaluation. For example, it is now known that some CS axons do not cross the midline at the pyramidal decussation, but, rather, at a segmental level, whereas some of the CS axons that cross at the pyramidal decussation then recross at the spinal level (Galea and Darian-Smith, 1997). The function of these two components of the CS tract remains to be determined. It is likely, however, that they play a role in the functional recovery that occurs after a lesion of the CS system, by actions exerted at either the cortical, medullary or spinal level, or all three.

The development of the CS tract in primates has also been reinvestigated recently. Although CS projections are already present at birth in both the monkey (Armand et al., 1994, 1997) and human (Eyre et al., 2000), the myelination of this tract is much more protracted than originally supposed. It probably continues until the completion of adolescence (Olivier et al., 1997; Paus et al., 1999). The development of corticomotoneuronal connections in primates, however, remains a controversial issue (Armand et al., 1997; Eyre et al., 2000). Finally,

Table 1. List of criteria for definition of motor areas of the cerebral cortex

	Connection with M1	% of CS neurons[1]	ICMS effects	Linked neuronal activity	Connection with motor thalamus
M1 (F1)	–	48%	Y	Y	Y
SMA-proper (F3)	Y	19%	Y	Y	Y
Pre-SMA (F6)	N	N	(Y)	Y	Y
PMd-r (F7)	N	N	(Y)	Y	Y
PMd-c (F2)	Y	7%	Y	Y	Y
PMv-r (F5)	Y	4%	Y	Y	Y
PMv-c (F4)	Y	4%	Y	Y	Y
CMA-r	Y	4%	(Y)	Y	Y
CMA-d	Y	10%	Y	Y	Y
CMA-v	Y	8%	Y	Y	Y
AIP	N	–	(Y)	Y	N

AIP, anterior intraparietal area. See Fig. 1 for schematic location of the motor cortical areas.
[1]Percent distribution of CS neurons based on a retrograde-tracing study (Dum and Strick, 1991).

it deserves mention that in the newborn monkey, the cortical territory occupied by CS neurons is twice as large as that in the adult, and the total number of CS neurons is three times higher (Galea and Darian-Smith, 1995). In parallel, the axons' growth and guidance factors, and the proteins preventing axonal growth, have been the topics of numerous studies in recent years, albeit data on primates are still very scarce (see Schwab and Schnell, 1991; Thallmair et al., 1998; Horner and Gage, 2000; Z'Graggen et al., 2000).

Reorganization of cortical motor areas following a cortical lesion in the monkey

After a lesion of M1, two distinct mechanisms are likely to be responsible for functional recovery: (1) an intrinsic reorganization of the motor map in M1; and (2) a substitution by non-M1 motor cortical areas for the postlesion, deficient function of M1. These two mechanisms are not mutually exclusive, of course.

Figure 2A shows an example of functional recovery after an M1 lesion that was considered due to a remodeling of the motor map within M1. The lesion was made in the *newborn* monkey ($n=6$) and affected mostly the unilateral hand area (Rouiller et al., 1998). When the animal was studied a few years later in adulthood (i.e., at the stage by which there was a significant restoration of manual dexterity), it was shown that ICMS of the *former* M1 hand representation area failed to evoke relevant muscle twitches, thereby confirming the efficacy of the original lesion. Stimulation of M1 territories located medial to the lesion, which normally subserve more proximal segment representations (wrist, elbow, shoulder), were now found to elicit finger movements, however.

The Fig. 2A observations support the hypothesis that the observed shift in hand representation to intact territories of M1 was responsible for the restoration of hand function. To address the issue of the actual contribution of this 'new' hand area to manual control, the same group of lesioned monkeys was trained to perform a prehension task on a modified Brinkman board (Rouiller et al., 1998). Scores for this task, whose successful performance required independent finger movements, were determined for both the ipsilateral (normal) and contralateral (affected) hand. The behavioral data showed that functional recovery of the affected hand was incomplete, albeit substantial, and that the extent of the remaining deficit was associated with the extent of the original lesion (Rouiller et al., 1998). The new hand territory adjacent to the lesion was then reversibly inactivated by an intracortical injection of lidocaine. Ten minutes after the injection, the prehension task score for the affected hand dropped to zero, whereas the score for the normal hand remained unchanged (Fig. 2A). These data showed that the new hand representation contributed significantly to the recovery of fine hand control.

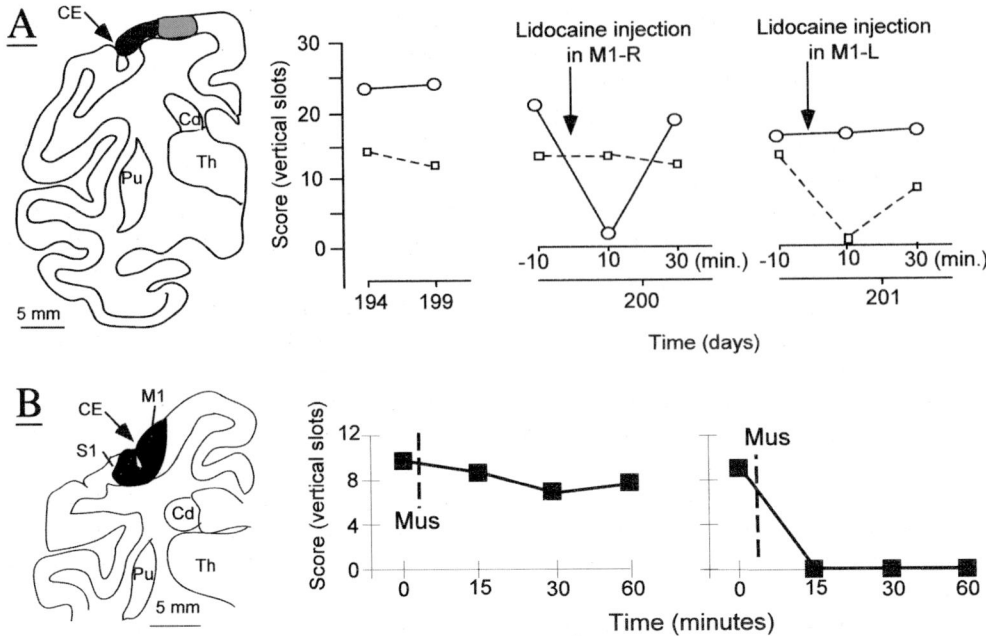

Fig. 2. Prehension responses of the adult monkey after an M1 lesion made at the neonatal versus adult stage. (A) The left-side sketch of a frontal histological section shows at the adult stage the location and extent of a lesion (black spot) in the left M1 (M1-L), which was made when the animal was a neonate (for experimental details, see Rouiller et al., 1998). The right-side plots show behavioral data which consists of scores of prehension performance when grasping for a reward housed in a modified Brinkman board with the normal (open circles) versus affected (open squares) hand. The maximal possible score was 25 vertical slots successfully visited in a given amount of time (60 s). Note that the score for the affected hand is lower than that of the normal hand, but it nonetheless remained substantial, thereby reflecting some (albeit incomplete) preservation of hand dexterity. Time in days is relative to 0, the day of implantation of a chronic chamber above M1 for ICMS mapping. On days 200 and 201 (right-most plots), lidocaine was infused into M1-R versus M1-L. After inactivation of M1-R, the score for the normal hand was dramatically reduced, of course. Inactivation of the zone in M1-L represented by a gray spot on the histological section led to a loss of the preserved behavioral score (see text). (B) As in A, the left-side sketch shows the extent and location of a lesion of M1 made in another monkey at the adult stage (for details, see Liu and Rouiller, 1999). The two right-side plots show similar behavioral scores for two inactivation sessions (10 months postlesion), in which muscimol (Mus) was infused into the intact M1 on the opposite hemisphere (left-side plot) versus the PM (right-side plot) on the same side as the M1 lesion. The score at time 0 is before inactivation and the other time points are scores for 15, 30 and 60 min after Mus infusion. Note a relative lack of effect on the recovered behavioral score in the case of inactivation of the intact M1, whereas inactivation of the PM led to a dramatic loss of the recovery.

Interestingly, a similar observation was made half a century ago in the juvenile (rather than neonatal) monkey, which were subjected to a quite small, thumb-area M1 lesion (Glees and Cole, 1950).

In the Fig. 2A experiment, it was also shown that a reversible inactivation of the contralateral (intact) M1 hand representation failed to modify the behavioral score of the affected hand (Fig. 2A). This result revealed that the contralateral M1 did not contribute significantly to the functional recovery of the ipsilateral, paretic hand. This finding, too, is in accordance with previous observations (Passingham et al., 1983; Sloper et al., 1983). In addition, in the Fig. 2A experiment, the hypothesis was nullified that nonprimary motor cortical areas, such as SMA, can take over the control of a paretic hand. Indeed, a reversible inactivation of SMA, either separately or simultaneously in both hemispheres, did not modify the behavioral score of the affected hand (Rouiller et al., 1998).

In contrast to the above results, a unilateral lesion of M1 in the *adult* monkey (4–5 years old; $n=2$) provided quite different results (Liu and Rouiller, 1999). In this experiment, the M1 hand representation in both hemispheres was identified by ICMS. Then, the hand area in the left M1 was infused with ibotenic acid to produce a chemical lesion, whose extent is shown in Fig. 2B. A few minutes after the infusion, the contralateral hand became completely paretic. The long-term deficit of the paretic hand was more devastating than in the neonatal-lesioned monkey, i.e., the contralateral hand remained paretic for 1–2 months, followed by a slow, progressive recovery over the next 2–3 months. From the sixth month postlesion and onwards, the functional score for the paretic hand reached a plateau at $\sim 30\%$ of the prelesion (control) behavioral score (Liu and Rouiller, 1999). A remapping of the motor cortex performed 9 months after the lesion failed to reveal a reorganization of the territory of M1 adjacent to the lesion. This finding was in sharp contrast to the Fig. 2A observations. Additionally, some territory surrounding the lesioned territory in the Fig. 2B adult monkey (as assessed histologically) was found to be relatively nonreactive to ICMS. Finally, as also assessed by ICMS, the contralateral, intact M1 displayed no reorganization (Liu and Rouiller, 1999).

In the adult-lesioned monkey, we next tested for functional recovery based on substitution by non-primary motor areas. An intracortical injection of muscimol was used to create a reversible inactivation of the PM ipsilateral to the already-lesioned M1. This led to a considerable drop in the behavioral score of the affected hand, i.e., from 30% of the original (control, prelesion) score to 0%. These results support the hypothesis that in the *adult-lesioned* monkey, after a relatively extensive lesion of M1, which covered the whole hand representation, and encroached upon the primary somatosensory cortex (S1) as well, nonprimary motor cortical areas (the PM in this case, but possibly also the SMA and CMA) take over part of the deficient motor control of the contralateral hand (Liu and Rouiller, 1999). Indeed, all three key nonprimary motor zones (PM, SMA, CMA) have appropriate connections for this substitution role because they all contain CS neurons which project to hand motoneurons (Dum and Strick, 1991; He et al., 1993, 1995; Rouiller et al., 1996; Morecraft et al., 1997; Maier et al., 2002). Note also that in the Fig. 2B experiment, a reversible inactivation of M1 in the intact hemisphere did not affect the recovered behavioral score of the paretic hand, thereby showing that the contralateral M1 did not play a crucial role in the recovery process.

Plasticity of the motor cortex in the human

In the human, there is but a modicum of data to suggest that an intrinsic reorganization of M1 may also play a role in functional recovery after a subcortical lesion of the CS tract that was acquired *in adulthood*. For example, Weiller et al. (1993) reported that following an infarct of the posterior arm of the internal capsule, the representation in ipsilateral M1 for voluntary movements of the paretic hand was extended to occupy an area normally devoted solely to facial representation (Weiller et al., 1993). Whether this reorganization was indeed correlated with the functional recovery observed in these patients remains an open question because this extended hand representation cannot be selectively inactivated as in monkeys (see the previous section). Furthermore, such cases of intrinsic reorganization in M1 are rare in human subjects. Rather, numerous studies have supported the view that functional recovery after a unilateral lesion depends on the contribution of other motor cortical areas, in both the ipsilateral and contralateral (lesioned) hemispheres. In hemiplegic patients, for example, functional imaging data have provided evidence for a displacement and/or enlargement of hand representation into most of the nonprimary motor areas (Chollet et al., 1991, 1999; Weiller et al., 1992, 1993; Rossini et al., 1996; Cramer et al., 1997, 2000; Cao et al., 1998; Seitz et al., 1998). It has been also suggested that the ipsilateral S1 may contribute to the substitution for impaired M1 function (Sasaki and Gemba, 1984; Cramer et al., 2000). Again, however, studies in human subjects do not provide evidence on the actual contribution of these different areas to the functional recovery observed in these patients. In particular, the enlargement of activation zones observed in functional imaging studies during movements of the paretic hand in patients may simply reflect the increased difficulty these patients experience in performing a motor task requiring independent finger movements.

In addition to the possible role played by ipsilateral nonprimary motor cortical areas in functional recovery in hemiplegic patients, a contribution of the contralateral M1 has also been suggested. This hypothesis has received some support from PET studies, which have shown a strong activation of the contralateral M1 during movements of a paretic hand (Chollet et al., 1991; Weiller et al., 1993; Chollet and Weiller, 1994). A similar observation is lacking in monkeys following a restricted M1 lesion, however. Nonetheless, such an activation of the contralateral M1 in the human has been confirmed by fMRI studies (Cramer et al., 1997; Cao et al., 1998) and, more recently, by a combined transcranial magnetic stimulation (TMS)-fMRI study (Vandermeeren et al., 2002). The claim is further supported by the finding that in the normal human, the ipsilateral M1 is activated during hand movements (Kim et al., 1993; Kawashima et al., 1993; Cramer et al., 1999). Moreover, a unilateral lesion of M1 has been shown to produce a functional deficit of the ipsilateral forelimb (Jones et al., 1989; Adams et al., 1990; Marque et al., 1997). A correlation between the redistribution of cerebral blood flow in the contralateral M1 and functional recovery (i.e., as would be expected if the contralateral M1 plays a critical role in the control of the paretic hand movements) is still lacking, however. In contrast, the possibility of eliciting a response in the paretic hand after TMS of the ipsilateral M1 is usually associated with a substantial recovery of independent finger movements (Bastings et al., 1997; Binkofski et al., 1998). This observation suggests that the preservation of a (even quite small) group of CS axons originating from the ipsilateral hemisphere is a crucial factor in accomplishing a useful degree of functional recovery in stroke patients. Indeed, in patients with a congenital hemiplegia, the percentage of residual CS neurons originating from the lesioned hemisphere has been found to be closely correlated with the ability to perform a precision grip task (Duqué et al., 2003).

Yet to be determined are the factors underlying the brain's manner of accomplishing an optimal functional reorganization of different motor areas after a unilateral lesion of the CS system. The moment of occurrence of the lesion during development, or even in adulthood, is likely to be one of the key determinants. Different strategies of substitution may also depend on the size and location of the lesion (Friel and Nudo, 1998; Nudo, 1999). After a restricted lesion of M1, for example, functional recovery may depend primarily on an intrinsic reorganization of motor representation within the *ipsilateral* M1 (Glees and Cole, 1950). In the case of a larger lesion of M1, the substitution of M1 appears to involve ipsilateral nonprimary motor cortical areas (Seitz et al., 1998; Liu and Rouiller, 1999). Finally, in case of an even larger unilateral lesion, for example one that covers a wide part of the frontal lobe, the contralateral motor areas may be recruited to insure an effective control of the paretic hand (Vandermeeren et al., 2002).

Rehabilitation and functional recovery

The nature and extent of functional training after a lesion are also crucial aspects of the recovery process. Indeed, in animals subjected to a lesion of the M1 hand area, in the absence of rehabilitative training, no hand representation in the perilesion territory was observed, whereas the representation of more proximal muscles was enlarged (Nudo and Milliken, 1996). In contrast, when lesioned animals underwent an intensive rehabilitative training, a hand representation appeared in the perilesion territory (Nudo et al., 1996b). Similarly, in human subjects, a single physiotherapy session aimed at improving the dexterity of the paretic hand led to a rapid and transient expansion of the hand representation (Liepert et al., 2000a). The persistence of such a motor cortex reorganization varied strongly from one patient to the other, however. Along the same lines, 'constraint-induced movement therapy' has been correlated with a reorganization of motor cortical maps, both in the monkey and human (Nudo et al., 1996a,b; Liepert et al., 1998, 2000b; Taub et al., 1999). In chronic stroke patients, for example, the forced used of the paretic hand for 12 consecutive days led to its M1 representation becoming even larger than that of the normal hand in the contralateral hemisphere (Liepert et al., 2000b). This 'trained' enlargement seemed to parallel a long-lasting functional improvement. After several months, however, the M1 hand representations of the two hemispheres became more equal in size.

In contrast to the above training effects, the *nonuse* of a limb is invariably associated with a decrease in its cortical representation (Liepert et al., 1995; Nudo et al., 1996b). Note, however, that if the instruction to use the paretic hand in the days following the lesion is too demanding, an enlarged lesion is likely ensue (Kozlowski et al., 1996; Humm et al., 1998; Nudo, 1999).

Concluding points and thoughts

There is recent evidence in nonhuman primates that the well-known and well-accepted reorganization of cortical maps following a lesion made in the neonate, is also present after a lesion made in the adult. Different factors affecting such motor area reorganization may come into play in both cases, including: (1) the time of occurrence of the lesion (in both neonate and adult); (2) the precise location and size of the lesion; (3) the nature and extent of postlesion rehabilitative training. For example, after a lesion of the M1 hand area in the neonatal monkey, a reorganization was observed of the perilesion territory in M1. A similar mechanism was observed after a later-occurring lesion in the juvenile monkey, which was restricted to a particularly small (thumb area) M1 territory. Larger lesions in the adult monkey and human, which covered the entire hand area of M1, were followed by an extensive reorganization of the network of ipsilateral nonprimary motor areas. The contribution of the contralateral motor areas to functional recovery of the paretic hand is still debated, because evidence in favor of such a mechanism has been observed indirectly in the human, but not yet directly in the monkey. We would argue that the continual interplay between fundamental experiments undertaken on the nonhuman primate and the human are needed to advance this field of research, which is an essential prerequisite to advancing therapeutic and rehabilitation procedures for brain-injured subjects.

Acknowledgments

We would like to thank Dr. Becky Farley, The University of Arizona, for reviewing a draft of this chapter. This work was supported, in part, by the Swiss National Science Foundation (Grants Nos. 31-43422.95, 4038-43918, 31-61857.00), the Novartis Foundation, and the Swiss National Centre of Competence in Research (an Award on 'Neural Plasticity and Repair').

Abbreviations

AIP	anterior intraparietal area
AR	arcuate sulcus
CE	central sulcus
Cin	cingulate sulcus
CMA	cingulate motor area
CS	corticospinal
fMRI	functional magnetic resonance imaging
ICMS	intracortical microstimulation
Mus	muscimol
M1	primary motor cortex
PM	premotor cortex
PM-d	dorsal part of PM
PM-v	ventral part of PM
SMA	supplementary motor area
TMS	transcranial magnetic stimulation

References

Adams, R.W., Gandevia, S.C. and Skuse, N.F. (1990). The distribution of muscular weakness in upper motor neuron lesions affecting the lower limb. Brain, 113: 1459–1476.

Armand, J., Edgley, S.A., Lemon, R.N. and Olivier, E. (1994). Protracted postnatal development of corticospinal projections from the primary motor cortex to hand motoneurones in the macaque monkey. Exp. Brain Res., 101: 178–182.

Armand, J., Olivier, E., Edgley, S.A. and Lemon, R.N. (1997). Postnatal development of corticospinal projections from motor cortex to the cervical enlargement in the macaque monkey. J. Neurosci., 17: 251–266.

Bastings, E., Rapisarda, G., Pennisi, A., Maertens de Noordhout, A. and Delwaide, P.J. (1997). Mechanisms of hand motor recovery after a stroke: an electrophysiological study of central motor pathways. J. Neurol. Rehab., 11: 97–108.

Binkofski, F., Dohle, C., Posse, S., Stephan, K.M., Hefter, H., Seitz, R.J. and Freund, H.J. (1998). Human anterior intraparietal area subserves prehension: a combined lesion and functional MRI activation study. Neurology, 50: 1253–1259.

Cao, Y., D'Olhaberriague, L., Vikingstad, E.M., Levine, S.R. and Welch, K.M. (1998). Pilot study of functional MRI to assess cerebral activation of motor function after poststroke hemiparesis. Stroke, 29: 112–122.

Chollet, F. and Weiller, C. (1994). Imaging recovery of function following brain injury. Curr. Opin. Neurobiol., 4: 226–230.

Chollet, F., DiPiero, V., Wise, R.J.S., Brooks, D.J., Dolan, R.J. and Frackowiak, R.S.J. (1991). The functional anatomy of motor recovery after stroke in humans: a study with positron emission tomography. Ann. Neurol., 29: 63–71.

Chollet, F., Loubinoux, I., Carel, C., Marque, P., Albucher, J.F. and Guiraud-Chaumeil, B. (1999). Mécanismes de la récupération motrice après accident vasculaire cérébral. Rev. Neurol. (Paris), 155: 718–724.

Cramer, S.C., Nelles, G., Benson, R.R., Kaplan, J.D., Parker, R.A., Kwong, K.K., Kennedy, D.N., Finklestein, S.P. and Rosen, B.R. (1997). A functional MRI study of subjects recovered from hemiparetic stroke. Stroke, 28: 2518–2527.

Cramer, S.C., Finklestein, S.P., Schaechter, J.D., Bush, G. and Rosen, B.R. (1999). Activation of distinct motor cortex regions during ipsilateral and contralateral finger movements. J. Neurophysiol., 81: 383–387.

Cramer, S.C., Moore, C.I., Finklestein, S.P. and Rosen, B.R. (2000). A pilot study of somatotopic mapping after cortical infarct. Stroke, 31: 668–671.

Dum, R.P. and Strick, P. (1991). The origin of corticospinal projections from the premotor areas in the frontal lobe. J. Neurosci., 11: 667–689.

Duqué, J., Sébire, G., Vandermeeren, Y., Cosnard, G., Thonnard, J.-L. and Olivier, E. (2003) Correlation between impaired dexterity and corticospinal tract dysgenesis in congenital hemiplegia. Brain, 126: 732–747.

Eyre, J.A., Miller, S., Clowry, G.J., Conway, E.A. and Watts, C. (2000). Functional corticospinal projections are established prenatally in the human foetus permitting involvement in the development of spinal motor centres. Brain, 123: 51–64.

Friel, K.M. and Nudo, R.J. (1998). Recovery of motor function after focal cortical injury in primates: compensatory movement patterns used during rehabilitative training. Somatosens. Motor Res., 15: 173–189.

Galea, M.P. and Darian-Smith, I. (1995). Postnatal maturation of the direct corticospinal projections in the macaque monkey. Cereb. Cortex, 5: 518–540.

Galea, M.P. and Darian-Smith, I. (1997). Manual dexterity and corticospinal connectivity following unilateral section of the cervical spinal cord in the macaque monkey. J. Comp. Neurol., 381: 307–319.

Geyer, S., Ledberg, A., Schleicher, A., Kinomura, S., Schormann, T., Burgel, U., Klingberg, T., Larsson, J., Zilles, K., Roland, P.E. (1996). Two different areas within the primary motor cortex of man. Nature, 382: 805–807.

Glees, P. and Cole, J. (1950). Recovery of skilled motor functions after small repeated lesions of motor cortex in macaque. J. Neurophysiol., 13: 137–148.

He, S.-Q., Dum, R.P. and Strick, P.L. (1993). Topographic organization of corticospinal projections from the frontal lobe: motor areas on the lateral surface of the hemisphere. J. Neurosci., 13: 952–980.

He, S.-Q., Dum, R.P. and Strick, P.L. (1995). Topographic organization of corticospinal projections from the frontal lobe: motor areas on the medial surface of the hemisphere. J. Neurosci., 15: 3284–3306.

Horner, P.J. and Gage, F.H. (2000). Regenerating the damaged central nervous system. Nature, 407: 963–970.

Humm, J.L., Kozlowski, D.A., James, D.C., Gotts, J.E. and Schallert, T. (1998). Use-dependent exacerbation of brain damage occurs during an early post-lesion vulnerable period. Brain Res., 783: 286–292.

Jones, R.J., Donaldson, I.M. and Parkin, P.J. (1989). Impairment and recovery of ipsilateral sensory-motor function following unilateral cerebral infraction. Brain, 112: 113–132.

Kawashima, R., Yamada, K., Kinomura, S., Yamaguchi, T., Matsui, H., Yoshioka, S. and Fukuda, H. (1993). Regional cerebral blood flow changes of cortical motor areas and prefrontal areas in humans related to ipsilateral and contralateral hand movement. Brain Res., 623: 33–40.

Kim, S.-G., Ashe, J., Hendrich, K., Ellermann, J.M., Merkle, H., Ugurbil, K. and Georgopoulos, A.P. (1993). Functional magnetic resonance imaging of motor cortex: hemispheric asymmetry and handedness. Science, 261: 615–617.

Kozlowski, D.A., James, D.C. and Schallert, T. (1996). Use-dependent exaggeration of neuronal injury after unilateral sensorimotor cortex lesions. J. Neurosci., 16: 4776–4786.

Liepert, J., Tegenthoff, M. and Malin, J.-P. (1995). Changes of cortical motor area size during immobilization. EEG Clin. Neurophysiol., 97: 382–386.

Liepert, J., Miltner, W.H.R., Bauder, H., Sommer, M., Dettmers, C., Taub, E. and Weiller, C. (1998). Motor cortex plasticity during constraint-induced movement therapy in stroke patients. Neurosci. Lett., 250: 5–8.

Liepert, J., Graef, S., Uhde, I., Leidner, O. and Weiller, C. (2000a). Training-induced changes of motor cortex representations in stroke patients. Acta Neurol. Scand., 101: 321–326.

Liepert, J., Bauder, H., Wolfgang, H.R., Miltner, W.H., Taub, E. and Weiller, C. (2000b). Treatment-induced cortical reorganization after stroke in humans. Stroke, 31: 1210–1216.

Liu, Y. and Rouiller, E.M. (1999). Mechanisms of recovery of dexterity following unilateral lesion of the sensorimotor cortex in adult monkeys. Exp. Brain Res., 128: 149–159.

Liu, J., Morel, A., Wannier, T. and Rouiller, E.M. (2002). Origins of callosal projections to the supplementary motor area (SMA): a direct comparison between pre-SMA and SMA-proper in macaque monkeys. J. Comp. Neurol., 443: 71–85.

Maier, M., Armand, J., Kirkwood, P.A., Yang, H.W., Davis, J.N. and Lemon, R.N. (2002). Differences in the corticospinal projection from primary motor cortex and supplementary motor area to macaque upper limb motoneurons: an anatomical and electrophysiological study. Cereb. Cortex, 12: 281–296.

Marque, P., Felez, A., Puel, M., Demonet, J.F., Guiraud-Chaumeil, B., Roques, C.F. and Chollet, F. (1997). Impairment and recovery of left motor function in patients with right hemiplegia. J. Neurol. Neurosurg. Psychiatry, 62: 77–81.

Morecraft, R.J., Louie, J.L., Schroeder, C.M. and Avramov, K. (1997). Segregated parallel inputs to the brachial spinal cord from the cingulate motor cortex in the monkey. Neuroreport, 8: 3933–3938.

Nudo, R.J. (1999). Recovery after damage to motor cortical areas. Curr. Opin. Neurobiol., 9: 740–747.

Nudo, R.J. and Milliken, G.W. (1996). Reorganization of movement representations in primary motor cortex following focal ischemic infarcts in adult squirrel monkeys. J. Neurophysiol., 75: 2144–2149.

Nudo, R.J., Milliken, G.W., Jenkins, W.M. and Merzenich, M.M. (1996a). Use-dependent alterations of movement representations in primary motor cortex of adult squirrel monkeys. J. Neurosci., 16: 785–807.

Nudo, R.J., Wise, B.M., SiFuentes, F. and Milliken, G.W. (1996b). Neural substrates for the effects of rehabilitative training on motor recovery after ischemic infarct. Science, 272: 1791–1794.

Olivier, E., Edgley, S.A., Armand, J. and Lemon, R.N. (1997). An electrophysiological study of the postnatal development of the corticospinal system in the macaque monkey. J. Neurosci., 17: 267–276.

Passingham, R.E., Perry, V.H. and Wilkinson, F. (1983). The long-term effects of removal of sensorimotor cortex in infant and adult rhesus monkeys. Brain, 106: 675–705.

Paus, T., Zijdenbos, A., Worsley, K., Collins, D.L., Blumenthal, J., Giedd, J.N., Rapoport, J.L. and Evans, A.C. (1999). Structural maturation of neural pathways in children and adolescents: in vivo study. Science, 283: 1908–1911.

Penfield, W. and Jasper, H.H. (1954). Epilepsy and the Functional Anatomy of the Human Brain. Little Brown, Boston.

Porter, R.N. and Lemon, R. (1993). Corticospinal Function and Voluntary Movement. Oxford University Press, Oxford.

Rossini, P.M., Rossi, S., Tecchio, F., Pasqualetti, P., Finazzi-Agro, A. and Sabato, A. (1996). Focal brain stimulation in healthy humans: motor map changes following partial hand sensory deprivation. Neurosci. Lett., 214: 191–195.

Rouiller, E.M., Moret, V., Tanne, J. and Boussaoud, D. (1996). Evidence for direct connections between the hand region of the supplementary motor area and cervical motoneurons in the macaque monkey. Eur. J. Neurosci., 8: 1055–1059.

Rouiller, E.M., Yu, X.H., Moret, V., Tempini, A., Wiesendanger, M. and Liang, F. (1998). Dexterity in adult monkeys following early lesion of the motor cortical hand area: the role of cortex adjacent to the lesion. Eur. J. Neurosci., 10: 729–740.

Sasaki, K. and Gemba, H. (1984). Compensatory motor function of the somatosensory cortex for the motor cortex temporarily impaired by cooling in the monkey. Exp. Brain Res., 55: 60–68.

Schieber, M.H. and Hibbard, L.S. (1993). How somatotopic is the motor cortex hand area?. Science, 261: 489–492.

Schwab, M.E. and Schnell, L. (1991). Channeling of developing rat corticospinal tract axons by myelin-associated neurite growth inhibitors. J. Neurosci., 11: 709–721.

Seitz, R.J., Hoflich, P., Binkofski, F., Tellmann, L., Herzog, H. and Freund, H.J. (1998). Role of the premotor cortex in recovery from middle cerebral artery infarction. Arch. Neurol., 55: 1081–1088.

Sloper, J.J., Brodal, P. and Powell, T.P.S. (1983). An anatomical study of the effects of unilateral removal of sensorimotor cortex in infant monkeys on the subcortical projections of the contralateral sensorimotor cortex. Brain, 106: 707–716.

Taub, E., Uswatte, G. and Pidikiti, R. (1999). Constraint-induced movement therapy: a new family of techniques with broad application to physical rehabilitation. J. Rehabil. Res. Dev., 36: 237–251.

Thallmair, M., Metz, G.A., Z'Graggen, W.J., Raineteau, O., Kartje, G.L. and Schwab, M.E. (1998). Neurite growth inhibitors restrict plasticity and functional recovery following corticospinal tract lesions. Nat. Neurosci., 1: 124–131.

Vandermeeren, Y., De Volder, A., Bastings, E., Duqué, J., Thonnard, J.-L., Grandin, C., Cosnard, G., Sébire, G. and Olivier, E. (2002). Functional relevance of the motor cortex reorganization in a child with a unilateral schizencephaly. Neuroreport, 13: 1821–1824.

Weiller, C., Chollet, F., Friston, K.J., Wise, R.J. and Frackowiak, R.S.J. (1992). Functional reorganization of the brain in recovery from striatocapsular infarction in man. Ann. Neurol., 31: 463–472.

Weiller, C., Ramsay, S.C., Wise, R.J., Friston, K.J. and Frackowiak, R.S.J. (1993). Individual patterns of functional reorganization in the human cerebral cortex after capsular infarction. Ann. Neurol., 33: 181–189.

Woolsey, C.N., Settlage, P.N., Meyer, D.N., Sencer, W., Hamuy, T.P. and Travis, A.M. (1952). Patterns of localization in precentral and 'supplementary' motor areas and their relation to the concept of a premotor region. Res. Pub. Ass. Nerv. Ment. Dis., 30: 238–264.

Z'Graggen, W.J., Fouad, K., Raineteau, O., Metz, G.A., Schwab, M.E. and Kartje, G.L. (2000). Compensatory sprouting and impulse rerouting after unilateral pyramidal tract lesion in neonatal rats. J. Neurosci., 20: 6561–6569.

CHAPTER 45

Adaptive behavior of cortical neurons during a perturbed arm-reaching movement in a nonhuman primate

Douglas J. Weber and Jiping He*

Department of Bioengineering, Arizona State University, Tempe, AZ 85287, USA

Abstract: This chapter provides evidence of spatial and temporal changes in the behavior of neurons within Areas 5 and 4 of the sensorimotor cortex of a nonhuman primate while it was executing a perturbed arm-reaching task. Chronically implanted electrode arrays were used to record simultaneously from 37 to 58 neurons. Also measured were the trajectory of arm movement, EMG activity in selected arm muscles and the perturbation force applied to the arm. The adaptation in Area 4 neurons' behavior usually involved a reduction in the latency from the onset of the perturbation to the peak-firing rate of the cell. In contrast, Area 5 neurons exhibited no such adaptive change in this latency. In each cortical area, the adaptation was not uniform across all neurons, and the spatial pattern of neuronal population behavior changed over the period of behavioral adaptation. We also found that the direction of arm movement and its configuration were important in determining which control strategy (predictive trajectory compensation or stiffness control) the animal used to overcome the externally applied perturbation for an improved performance of the reaching task.

Introduction

An active field of current research is the development of cortically controlled neuroprosthetic systems, which derive movement command signals from recorded populations of neurons in the cerebral cortex (Isaacs et al., 2000; Lauer et al., 2000; Schwartz et al., 2001). Such direct brain–machine interfaces will revolutionize the way neuroprostheses are designed and developed for restoring lost motor and sensory functions. This technology has become realizable by several recent advances in (1) the design and fabrication of neural implants for stable chronic recording from brain neurons (Kennedy et al., 1992; Hoogerwerf and Wise, 1994; Nordhausen et al., 1996; Rousche et al., 2001), (2) signal processing for decoding neural signals and control algorithms for

---

*Corresponding author: Tel.: +480-965-0092;
Fax: +480-965-4292; E-mail: jiping.he@asu.edu

adaptation and learning (Lin et al., 1997; Schwartz et al., 2001; Serruya et al., 2002; Taylor et al., 2002), and (3) understanding the fundamental principles of neural control of posture and movement (Loeb et al., 1990; Thoroughman and Shadmehr, 2000). Despite these developments, important problems remain to be solved before such advanced neuroprostheses will achieve clinical feasibility. Information is needed on: (1) how and where to obtain reliable command signals from the brain, both conveniently and consistently; (2) how neural commands and coding are affected by external perturbations; and (3) how neural command signals may change (adapt) when such perturbations or environmental changes are encountered. In this report, we present a new approach to investigate these issues (see also Wiesendanger and Serrien, Chapter 46 of this volume).

The concept of a cortically controlled neuroprosthesis was inspired by the pioneering work of Georgopolous and colleagues (Georgopoulos et al.,

1982, 1988), who demonstrated that populations of neurons in the motor cortex (Rouiller and Olivier, Chapter 44 of this volume) encode information about the direction of an impending movement. Since then, extensive research has been undertaken to study interactions between the neuronal activity patterns of various brain structures and learned behaviors (Kalaska et al., 1989; Caminiti et al., 1991; Crammond and Kalaska, 1996; Gandolfo et al., 2000; Gribble and Scott, 2002). Neuronal signals recorded from selected areas have been used to predict the direction, velocity and force dynamics of arm movements. For example, the Schwartz laboratory has successfully reproduced a monkey's hand trajectories from the neuronal signals recorded in the motor cortex of monkeys making drawing motions (Schwartz, 1992; Moran and Schwartz, 1999). It was shown that the direction and velocity of hand movement were represented in the collective activity of a population of motor cortical neurons that discharged 80–120 ms prior to the movement. Subsequently, researchers have attempted to predict hand trajectories in real-time using small populations of simultaneously recorded neurons. Serruya et al. (2002) used a linear-filter algorithm to demonstrate direct cortical control of a cursor in a planar workspace. To take advantage of the brain's learning ability, Taylor et al. (2002) developed an adaptive, population vector-coding method, which exploits the neural plasticity of cortical neurons to achieve accurate and reliable control of a three-dimensional (3D) cursor in virtual reality. The above experiments demonstrated the feasibility of extracting command signals from cortical neurons, but have not characterized the neural response to dynamic environments, which may disrupt the normal-reach behavior.

A practical requirement for a neuroprosthetic control system is that the device be able to compensate for perturbations that interfere with its normal (control) operation. To this end, we designed a research project to investigate the following questions. Will the activity pattern of cortical neurons change when a well-trained task is disturbed unexpectedly? If the animal is repeatedly subjected to a perturbation, will the neuronal activity change and, if so, by what means? Will changes in such adapted neuronal activity endure after removal of the perturbation?

Many different perturbation paradigms have been used in movement neuroscience and psychophysical investigations. These include constant bias forces (Kalaska et al., 1989; Georgopoulos et al., 1992), complex force fields that depend on the speed and direction of movement (Lackner and Dizio, 1994; Shadmehr and Mussa-Ivaldi, 1994), and transient impulse perturbations (Lacquaniti et al., 1991; Weber et al., 1998). Data from experiments using these approaches have revealed the remarkable capacity of neural-control systems to compensate for various types of external disturbance. In general, such compensation is provided by a combination of sensory feedback and direct cortical control (Hayashi et al., 1990; Lacquaniti et al., 1992; Bhushan and Shadmehr, 1999). The precise nature of their coordination remains an elusive topic, however.

This chapter describes the design of a new apparatus for applying a perturbation while the arm is moving without constraint in a 3D workspace. We have used this apparatus in conjunction with chronic multiunit recording from various cortical areas in the monkey. This has allowed us to investigate: (1) the real-time response of multiple cortical neurons to repeated perturbations of an arm-reaching task; (2) adaptation of the neurons' behavior in the face of the perturbation; and (3) the monkey's development of a control strategy to reduce the perturbation's effect on its arm movement. Our results have important implications for the development of a practical and robust neuroprosthetic system.

Experimental design and data analysis

The Institutional Animal Care and Use Committee of Arizona State University approved the behavioral paradigm and surgical procedures, which followed the mandatory guidelines for general animal care.

Behavioral task

Three rhesus monkeys (*M. mullata*) were trained by operant conditioning to execute a 3D center-out reaching task, which was performed on a behavioral

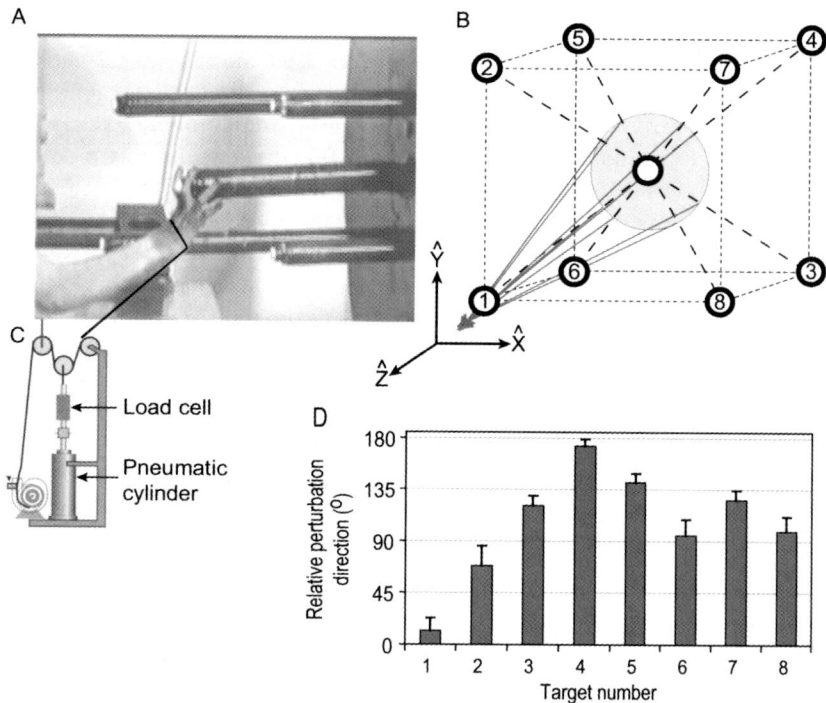

Fig. 1. The experimental arrangement. The top-left photograph (A) shows the apparatus we designed for execution of a 3D, arm-reaching, *center-out* task in which every movement started from the center and ended at one of the eight corners of a cubic space (B). The X–Y–Z diagram between A and B shows the coordinate system (with respect to the subject), with: X, the posterior–anterior direction, Y, the down–up direction; and Z, the left–right direction. A shows that lighted push-button switches (1 cm diameter) were positioned on the ends of two short (14 cm) and two long (27 cm) rods located at the four corners of the 13-cm square that represented four of the eight corners of the cube; as in B. Rotating the plate 180° produced the other four targets. The monkey began each trial by holding the center button, and then reaching for a lighted target, which was randomly selected by a computer program. During perturbation trials, the pneumatic cylinder (C) was actuated after the hand had moved 20 mm from the starting position (i.e., outside of the radius of the gray circle shown in B). The orientation of the perturbation force varied with the target location, and the gray vectors in B represent the approximate direction. The bar plot in D shows that the direction of the perturbation force varied with the reach direction across a wide range of angles. An angle of 0° means that the perturbation and reach directions were coincident, and 180° means that the perturbation directly opposed the reach direction.

apparatus with a perturbation mechanism that can deliver a short instantaneous pulling force. Figure 1 shows that the perturbation force had different effects on movement toward eight targets located at the vertices of a 13-cm cube and a center rod at the middle of the cubic space.

The start position for each movement was at the center of the cube. The eight targets were presented randomly, with five replications of each target, i.e., for a total of 40 successful trials (1 block = 40 trials; accomplished in 0.3–0.6 s for each reaching task). A successful trial required an appropriate sequence of the following actions. Each trial began with the illumination of a center light (*center-on*). The monkey was trained to push and hold the center button for a minimum *center-hold* time (100–500 ms) in order to ensure that the hand was stationary until a target was presented. The center light was then extinguished and one of the eight targets was illuminated (*target-on*) in random order. The reaction-time criterion from *target-on* to *center-release* was 200–750 ms. The monkey had to reach to, and depress, the illuminated target button within the allowed *movement-time* (600 ms). To receive a reward (water), the animal had to hold the target button for a minimum *target-hold* time of 100–500 ms.

Perturbation device

The left-side diagram in Fig. 1(A and C) shows that the perturbation device consisted of a pneumatic cylinder, recoil-mounted string attached to the subject's wrist and a series of pulleys that guided and changed the direction of the perturbation force. The perturbation could be activated at any given time during the movement for a prescribed duration (75 ms was used in the experiments reported here). A strain-gauge force transducer, which was mounted on the pulley attached to the cylinder, measured the tension in the string. The perturbation force was sampled at 500 Hz. The perturbation cylinder was activated after the hand had moved 20 mm from the *center-hold* position.

Kinematic measurements

The arm- and hand-movement trajectories were monitored by a 3D optical motion-capture system in real-time, and subsequently sampled at 200 Hz. One light-emitting diode (LED) was placed on the hand, three on the forearm and two on the upper arm. Another LED was placed on the last pulley of the perturbation system. This latter LED, together with a wrist LED, provided information on the perturbation's direction. At the beginning and end of each recording session, a kinematic calibration record was taken with the arm in a prescribed posture. Intersegmental joint angles for the shoulder, elbow and wrist were computed from the calibrated marker trajectories.

Cortical recording and stimulation procedures

Four arrays of microwire electrodes ($4 \times 16 = 64$ stainless steel or tungsten electrodes; tip diameter, 50 µ) were permanently implanted into the arm region of the left-side Areas 5 and 4 of the sensorimotor cortex. Neural activity was recorded on a 96-channel digital signal-processing system with real-time, multi-unit spike-sorting capability. Adjustable threshold-crossings marked the occurrences of the sorted action potential (spike) discharges. The spike times were recorded in a data file along with associated behavioral events. Sorted waveform samples of the spikes were also recorded throughout the experiment to track the stability of the neural recordings within a single experimental session and across multiple days.

Intracortical microstimulation was used to assess the functional connectivity of each motor cortical array. Brief (<5 s) trains of stimulus current pulses (200 µs, 300 Hz, 50–200 µA) were injected through a single electrode in the array while the arm was visually inspected for muscle twitches in response to this stimulation. The threshold current and location of evoked muscle twitches were recorded on an anatomical map of the dorsal and ventral surface of the arm. We also tested the responsiveness of each cortical cell to passive movement and tactile stimulation. These response properties were compared with the microstimulation results to produce a general somatotopic map for each cortical array.

Experimental paradigm

Our experiment paradigm consisted of four phases completed over a 5-week span: *control*, *naive*, *adaptation* and *extinction*.

(1) We established the baseline data during the *control* phase (unperturbed; 240–640 trials in 6–16 blocks/day; the actual number of days depended on the performance of each animal). The pneumatic cylinder was activated but no perturbation force was applied. This desensitized the animal to the sound emitted by the pneumatic valves and cylinder.

(2) The *naive* or *perturbation-start* phase was performed on the last day of the *control* phase. On this day, the animal experienced the first actual perturbation after completing 2–4 blocks (i.e., after 80–160 successful movements) of unperturbed trials. The perturbation was applied to each consecutive reaching movement until the end of the day, in order to establish the pattern of initial or naive responses to the perturbation. At irregular intervals, a catch trial was performed during which we disengaged the perturbation apparatus so that no perturbation force was generated during a reach to a randomly selected target.

(3) The *adaptation* phase of the experiment consisted of 8–12 perturbation-application days

(depending on the adaptation rate of each animal) on which the perturbation was applied in consecutive trials following completion of daily control trials (i.e., there were 80–240 unperturbed trials/day at the beginning of each day). The number of perturbed trials on each day ranged from 200 to 360 for a total number of unperturbed + perturbed trials of 280–600. The control (unperturbed) trials at the beginning of each day were used to document the daily baseline activity pattern. The latter provided the data set used to detect changes in cortical neuron and arm-movement patterns that might have occurred while the animal was performing unperturbed reaching tasks. The daily baseline activity pattern also provided information to examine (1) the daily adaptation changes, including the rate and robustness of adaptation, and (2) residual effects or memory of previous adaptations. Catch trials were performed throughout the adaptation phase, but the frequency of the catch trials was kept very low (1/40 reaches) in order to preserve an 'expectation' of the perturbation.

(4) On the last perturbation day, we disengaged the pneumatic cylinder after six perturbation blocks (240 trials) were completed. This marked the beginning of the *extinction* phase, which consisted of 2–4 days of trials without the perturbation.

Data analysis

Cortical neuron discharge and arm kinematic data were grouped for analysis according to each of the eight targets and the four experimental phases; *control*, *perturbed*, *adaptation* and *extinction*. We aligned the data at the onset of the perturbation force (time corresponding to 10% of force peak) and set that time as the zero (0-s) reference. This alignment procedure preserved the time scale of the data across trials, which was important for comparing the temporal profile of cortical neuron data with the observed behavioral measurements. For analysis purposes, each trial was subdivided into three epochs of time: *movement-initiation* (center-release to perturbation-on), *perturbation-application* (perturbation-on to perturbation-off) and *target-acquisition* (perturbation-off to target-hit). For the cortical neuron data, we computed average firing rates in each of the above relevant epochs by dividing the number of spikes in each epoch by the epoch's duration. Analyses based on combinations of these epochs were also performed. These data allowed us to compare event-related cortical activity across perturbation conditions and target directions.

Results

Recall from Fig. 1 that the direction of the perturbation pulling force and the direction of the reaching movement toward the eight targets formed a range of angles. As a result, the effect of the perturbation was different for the eight movement directions. When the movement direction was aligned with the perturbation direction, the force either assisted or resisted the movement (targets 1 and 4). This was analogous to a force perturbation or temporary load change. When the movement direction was perpendicular to the perturbation direction (targets 6 and 8), the effect was to divert the movement direction producing a trajectory perturbation. During movements to the remaining targets, a hybrid perturbation effect (combination of force and trajectory perturbation) was produced where angles between the movement-direction and perturbation-force vectors were either acute (target 2) or oblique (targets 3, 5–7). The variety of effects produced by the above perturbations generated a rich set of data for investigation into the nature and adaptation of neural command signals involved in controlling arm movement in the face of external disturbances.

Perturbation effects on the arm-reaching movement

Figure 2 illustrates hand trajectories to four (1, 4, 6, 7) of the eight targets. Movement toward these four targets experienced four different effects from the perturbation as shown in Fig. 1D. The blue traces represent the average hand path from the first and last 10 baseline trials ($n = 20$). The red traces show the hand trajectories during the first (left panel) and last (right panel) 10 perturbation trials. When perturbed,

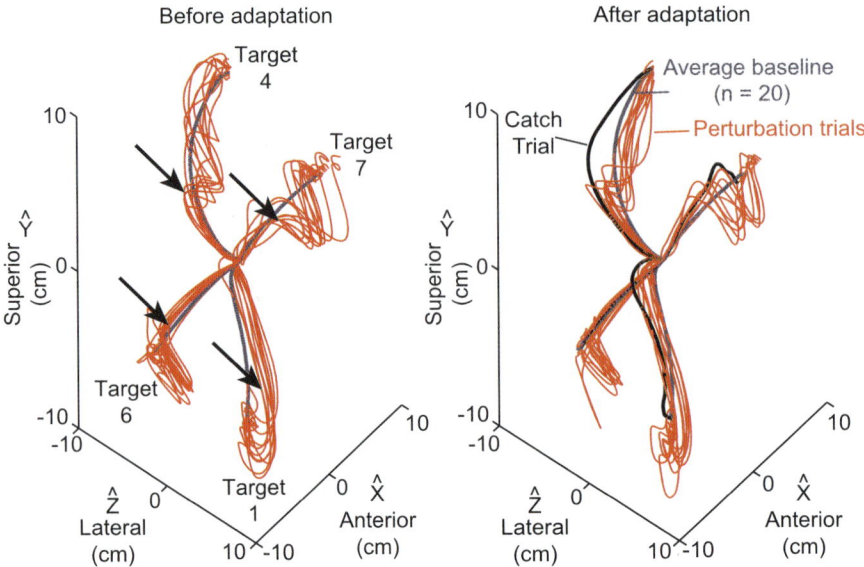

Fig. 2. Hand paths to selected targets under different experimental conditions before versus after adaptation to the repeated perturbation. The average baseline (control) trajectory shown in blue was computed from the first and last 10 baseline trials to each target (1, 4, 6, 7). The thin red lines in each panel represent hand paths from the first 10 (left) and last 10 (right) perturbation trials to each target. The black traces show paths from the final catch trials, which were single, no-perturbation trials interspersed among many repeated perturbations (i.e., to examine the effect of a predictive control strategy when the anticipated perturbation did not occur). The differences between the profile of the trajectories in the two panels gives indication of the range of control strategies used by the CNS to compensate for the perturbations. Note also the lack of a predictive trajectory compensation when the monkey reached to target 6.

the path deviations to different targets were as predicted according to the direction of the movement (viz., Fig. 1). The effect of the perturbation on the movement was transient, however, and the monkeys were able to overcome the perturbation and eventually reach the targets. During the movement-initiation phase of the reach, the hand paths in the control and perturbed trials were similar, but the perturbation created a transient yet significant deviation from the normal path. Presumably during the target-acquisition phase, sensory feedback was used to generate the large corrective movements that often produced an overshoot (especially for targets 4 and 7) before the hand reached the target.

After a few days of exposure to the same perturbation, the animal learned about its timing and dynamic effects. Accordingly, it developed a control strategy to overcome or reduce the disruption produced by the perturbation. This behavior was reflected in an improved success rate for reaching the targets within the allowed time duration and a reduced overshoot in the final correction of the movement trajectory (i.e., in Fig. 2, note the hand paths after the perturbation when moving to targets 4 and 7). Another noticeable difference in hand paths between the 'before' and 'after' adaptation trials was found in the *movement-initiation* direction. There was a clear deviation in its initial phase before the onset of the perturbation, which began after a few days of experiencing the perturbation. The direction of the deviation was opposite to that of the perturbation. Such anticipatory trajectory compensation was observed in hand paths to most targets, and particularly to targets 1, 4 and 7 (i.e., right-side panel in Fig. 2).

To establish clear evidence for an anticipatory correction of the expected disturbance, we inserted a few unperturbed trials to 'catch' the effect of erroneous anticipations. The black traces in the right-side panel represent the last catch trial completed at each target. The largest catch-trial effects were observed at targets 4 and 7, where the hand paths showed predictive trajectory deviations that opposed the displacement caused by the perturbation. Because the perturbation was absent, the movement continued along the initiation direction until (presumably) feedback correction brought the hand back to the target.

Trajectory compensation was not observed during movements to target 6. This suggests that a different control strategy was used to counteract the perturbation experienced at target 6. A possible strategy used by the animal for its arm movement to target 6 was stiffness control by muscle cocontraction or, alternatively, a change of arm posture to increase the limb's resistance to the perturbation. This strategy would not affect the reach trajectory, but would reduce the perturbation-induced arm displacement. Therefore, if a stiffness or impedance control strategy were indeed used, there would be no difference in trajectories between the baseline and catch trials, and a smaller perturbation-induced arm deviation. The movement trajectory to target 6 demonstrated *all* of these characteristics, which were significantly different from the trajectories to the other (1, 4, 7) targets shown in Fig. 2.

Behavior of Area 5 versus 4 neurons during the perturbation trials

Perievent-time histograms

Figure 3 shows perievent-time histograms for an exemplary Area 5 neuron during the control, naive, adapted and extinction phases of the experiment. The total trial time was subdivided into three epochs: *movement-initiation* (between movement onset and perturbation onset), *perturbation-application* (between the perturbation onset and off) and *target-acquisition* (from perturbation off to the acquisition of the target). Note that the neuron consistently began to fire approximately 0.10 s before the onset of movement. During the unperturbed trials, the peak-firing rate occurred soon after the movement initiation. During the perturbation trials, a larger peak was observed in the discharge profile at a 0.06-s latency from the onset of the perturbation force. The postperturbation response presumably represents the sensory feedback to correct the perturbed trajectory.

The timing of this neuron's discharge during the perturbation trials was similar for its naïve and adapted behavior, but its peak spike-frequency was lower after adaptation. During the naïve trials, the cell's spike-frequency reached a peak soon after movement initiation and it then started to taper off. After adaptation, this drop was much less pronounced, or absent, and the spike-frequency maintained a period of activity at the high level as shown by the histograms for adapted and extinction in Fig. 3. Note also that after adaptation and during the initial extinction trials, the cell's firing rate was higher than in the control and naïve states immediately before perturbation onset. This elevated discharge was probably related to the anticipation of a forthcoming perturbation. The difference between the initial and adapted spike-frequency behavior in the presence of the perturbation was mainly reflected in the *increased* adapted discharge rate *before*

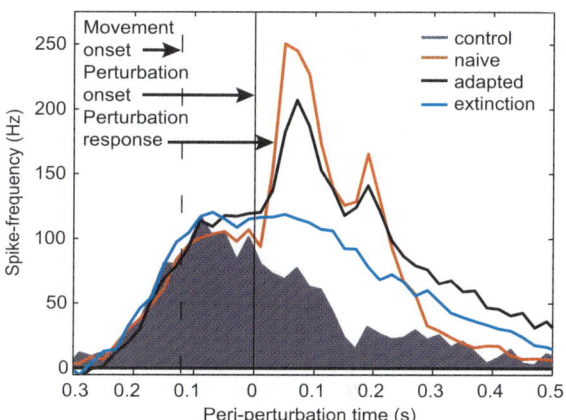

Fig. 3. Perievent time histograms showing the spike-frequency behavior of an Area 5 neuron during control and perturbed arm-reaching movements. The histograms are aligned on the perturbation trigger ($t = 0$; ~110 ms after *movement-initiation*). The blue shaded region indicates the discharge profile of the neuron during control conditions. The red and black traces represent the average discharge profile on the first (naive: before adaptation) and last (adapted) perturbation days. The light-blue trace shows the discharge profile during the extinction phase of the experiment in which the perturbation was removed after adaptation. The period between the first vertical line (dashed) and the second line (solid) represents the duration of the movement-initiation epoch. The period between the perturbation onset and the peak of the neuronal response to the perturbation represents the perturbation-application epoch. The target-acquisition time epoch is the period from the peak to the end of the movement. For further explanation, see the text.

perturbation onset and *decreased* (presumably reflexive) discharge *after* the peak of the perturbation (perturbation off). The Wilcoxan rank-sum test indicated that the increase in firing rate during the *movement-initiation* phase was significant ($\alpha = 0.01$).

Adapted discharge

Overall adaptation. The above-described adaptation in spike-frequency was not observed in all the tested neurons, however. For example, Fig. 4 compares the perievent-time histograms of three exemplary neurons each from Areas 5 and 4, respectively, whose discharges were recorded simultaneously for a 2-week period. Recall that the function of Area 5 cells is predominately, but not exclusively, associated with sensory reception, while Area 4 cells subserve predominately, but again not exclusively, motor output functions (see Rouiller's Chapter 41).

The pattern of adaptation for 12 of the 17 recorded Area 5 neurons was very similar to that shown in Fig. 3. The top three panels of Fig. 4 illustrate the discharge profiles of three such Area 5 neurons for comparison with the three Area 4 neurons in the bottom panels. The overall spike-frequency of the Area 5 neurons increased after adaptation, but the time of occurrence of the peak spike-frequency remained at the same latency relative to the onset of the perturbation force.

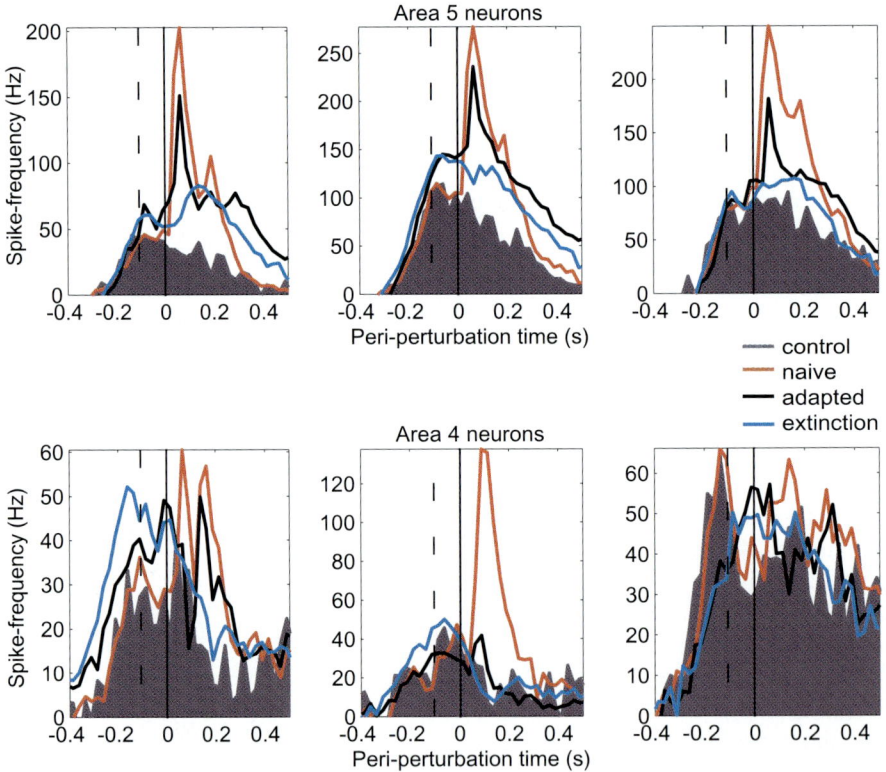

Fig. 4. Perturbation-modulated activity of Area 5 versus 4 neurons. The histograms of three Area 5 and three Area 4 neurons are aligned on the perturbation trigger ($t=0$; ~110 ms after *movement-initiation*). This figure is organized like Fig. 3, with: filled blue-colored area representing the baseline spike-frequency; red-colored line, naïve response; black-colored line, adapted response; and cyan-colored line, catch-trial response. Note that the patterns of adaptation in Area 5 neurons are very similar, with: (1) an increase in spike-frequency during the movement initiation period; (2) a decreased peak spike-frequency following the perturbation; and (3) this peak occurring at a constant latency. The patterns of adaptation were more variable in Area 4 neurons, with: (1) a peak spike-frequency in response to the perturbation that was reduced or absent after adaptation; (2) if present, an earlier peak spike-frequency; and (3) no overall increase in spike-frequency throughout the movement.

The patterns of adaptation of Area 4 neurons were much more variable, as demonstrated by the three neurons shown in the bottom panels of Fig. 4. These three neurons displayed three different patterns of change during adaptation. One (left-side panel) neuron exhibited an increase in spike-frequency during the movement-initiation epoch ($t < 0$). The increase in movement-initiation firing rate between the naive and adapted phases was significant ($\alpha = 0.01$) and lasted into the extinction trials. Another (middle) neuron showed no significant change in its movement-initiation activity pattern, but its peak perturbation response was reduced after adaptation. The third (right-side) neuron exhibited yet another pattern of change after adaptation. Its perturbation-related activity shifted forward in time such that its peak coincided with the onset of the perturbation ($t = 0$). All three of these cells showed changes consistent with an anticipation of the forthcoming perturbation.

Latency adaptation. Latency is here defined as the time interval between the onset of the perturbation (or the equivalent time in the unperturbed trials) and the neurons' peak spike-frequency. Cells that responded to the perturbation were identified by testing for their firing rate changes in the three time epochs between the control and perturbed trials. Those cells that exhibited a significant change in firing rate (Wilcoxan rank-sum test, $\alpha = 0.01$) were selected for measurement of their response latency. Figure 5 shows the average latency for neurons recorded from a microwire array in Area 5 versus those from another microwire array in Area 4. Each array contained 16 recording electrode wires, which normally provided stable recordings from at least one neuron on each of 8–11 wires (i.e., a yield of 50–70%). In the Fig. 5 case, 12 of 17 Area 5 neurons responded to the perturbation as did 13 of 28 neurons in Area 4. For >2 weeks, we continuously monitored the activity patterns of these 25 neurons while the animal performed the *center-out* reaching task under control (no perturbation, days 1–3), repeated perturbation (days 3–10), extinction (no perturbation, days 11–13) and random perturbation (days 14–18).

Latency during control trials. The blue data points in Fig. 5 represent the average onset latency within the group of 12 Area 5 versus 13 Area 4 neurons. These data show that both groups of neurons normally reach their peak discharge rate around $t = -0.05$ s, which was shortly after the initiation of movement. Indeed, the histograms in Figs. 3 and 4

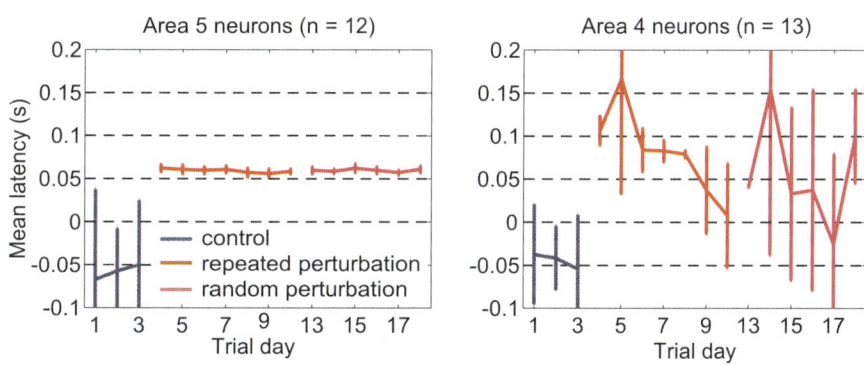

Fig. 5. Latency from perturbation onset to peak firing rate of Area 5 versus 4 cortical cells. Data are for 12 Area 5 neurons, and 13 Area 4 neurons. After 2 control (no perturbation) trial days, the 3rd day featured sequence of control (nonperturbed) movements followed by the first perturbed arm-reaching movements. The repeated perturbations were applied on the latter part of the 3rd day and stopped on the 10th day. Next, there were 3 days of postperturbation trial in which no perturbation was applied (extinction period). Finally, 6 days of random perturbation trials were completed in which case the perturbation was applied during 50% of the trials, selected randomly. The purpose of the random perturbation period was to examine whether the adaptation strategy developed for the repeated (expected) perturbation would also apply to unexpected perturbations. Note that throughout the perturbation days, the latency remained quite constant for Area 5 neurons. In contrast, Area 4 cells exhibited a significant progressive reduction in latency, thereby showing the development of predictive behavior.

showed that the discharge rate began to increase well before the onset of movement for both groups of neurons. Note, however, the large variation in this average value (i.e., large SE bars). This occurred because these cells did not all behave identically. Even though their peak spike-frequency generally occurred before the onset of arm movement, this was not the case for some cells. These observations are consistent with previously reported studies of Area 5 (Mountcastle et al., 1975; Kalaska et al., 1983) and Area 4 (Georgopoulos et al., 1982; Schwartz, 1992) neuronal activity during arm movement. They are also consistent with the observation that not all Area 4 neurons are predominately 'motor' neurons and not all Area 5 neurons are 'sensory' neurons (see Chapter 41). Figs. 3–5 show, however, that Area 5 neurons exhibited a more consistent response pattern to the perturbation than did Area 4 neurons.

Latency during perturbation trials. Perturbations were applied on days 3–10. Figure 5 shows that during these trials the average latency of Area 5 neurons occurred ~65 ms after onset of the perturbation force. This latency remained constant throughout the 7 days of repeated perturbation trials and 6 days of random perturbation trials! In sharp contrast, the Area 4 neurons exhibited *substantially different* behavior. On the first 3 days (i.e., days 3–5 in Fig. 5), the average latency of the Area 4 neurons occurred at least 0.010 s *after* that of the Area 5 cells and as late as 0.08 s, thereby suggesting a sensory-feedback effect being exerted on the Area 4 neurons. Starting on day 6, however, a clear trend was evident for progressively shortening response latency in the Area 4 cells. This suggests a learned expectation of the perturbation. On the last 2 days of the perturbation trials (days 9–10), the average latency of the Area 4 neurons occurred *before* that of the Area 5 cells. This suggests the manifestation of a predictive control action before perturbation-initiated sensory feedback arrives in Area 4.

Latency during random-perturbation trials. After 3 days of extinction trials (days 11–13), during which no perturbations were applied, perturbations were again applied on days 14–18, but with a random occurrence. By then, the animal had prior experience in dealing with the perturbation. At first glance, it would seem that the animal would be inclined to use a predictive control strategy. The unpredictability of perturbation occurrence rendered this strategy tenuous, however. The result was a particularly large variation in the average latency of Area 4 neurons during days 13–18. Again in particularly sharp contrast, there was *no change* in the average latency of Area 5 neurons.

Summary on behavior of Area 5 versus 4 neurons

The similarities in how Area 5 and 4 neurons responded to our experimental paradigm included a clearly identifiable peak spike-frequency, which was induced by a perturbation to the arm-reaching movement, and other more subtle firing pattern changes as the animal adapted to such a perturbation. Adaptation generally involved an increased overall spike-frequency, which was combined with a decrease in perturbation-induced peak spike-frequency. The main difference between the two groups of cortical cells was that the latency to their peak spike-frequency of the Area 5 neurons remained constant throughout the 18 days of perturbation trials, whereas the latency of Area 4 neurons decreased after adaptation. These observations suggest that plasticity occurred in both Area 4 and 5 neurons, but of a different nature. The sensory manifestation of the perturbation event in Area 5 always occurred at the same latency, but the movement-initiation related activity increased after adaptation. In contrast, the firing pattern of Area 4 neurons changed from being responsive to the perturbation to becoming predictive of the perturbation. This change is consistent with the development of a preparatory motor control strategy that compensates for the expected effects of the perturbation.

Trajectory compensation versus stiffness regulation

Figure 2 and its legend showed how we identified at least two different CNS control strategies: anticipatory trajectory compensation and stiffness modulation. The adaptive trajectory compensation can be clearly seen in the shift of hand-trajectory direction before the perturbation's onset (i.e., Fig. 2; movement to targets 1, 4, 7). No significant shift was observed in the hand's trajectory to target 6 even though the reaching performance, as measured by its success rate, was also dramatically improved. The arm-movement trajectory in response to the perturbation to target 6 suggested stiffness control. This can be explained by two observations of the trajectory to target 6: (1) no deviation occurred in catch trials; and (2) the perturbation effect was smaller but it had the same kinematic characteristics.

To search for more supporting evidence about stiffness modulation, we analyzed the arm's configuration during the adaptation process when moving to target 6 versus moving to the other targets. Figure 6 shows that when moving to target 6, the CNS did indeed change the arm configuration during the movement to increase the arm's resistance against the perturbation, though the hand trajectory remained the same as shown in Fig. 2. Whereas the movement to target 7 did not show a significantly different arm configuration during the movement even though the hand trajectory changed as shown in Fig. 2, suggesting the use of trajectory compensation strategy.

Comment

Our experiments featured chronic, multiunit recording from a relatively large population of neurons and use of an experimental paradigm (Fig. 1), which provided the opportunity to analyze the gradual changes that occurred in cortical neuron behavior as learning ensued. The significance of this technical advance is self-obvious.

Our results showed that *M. mullatta* quickly developed an effective control strategy to accommodate a perturbation applied to one of its forearms

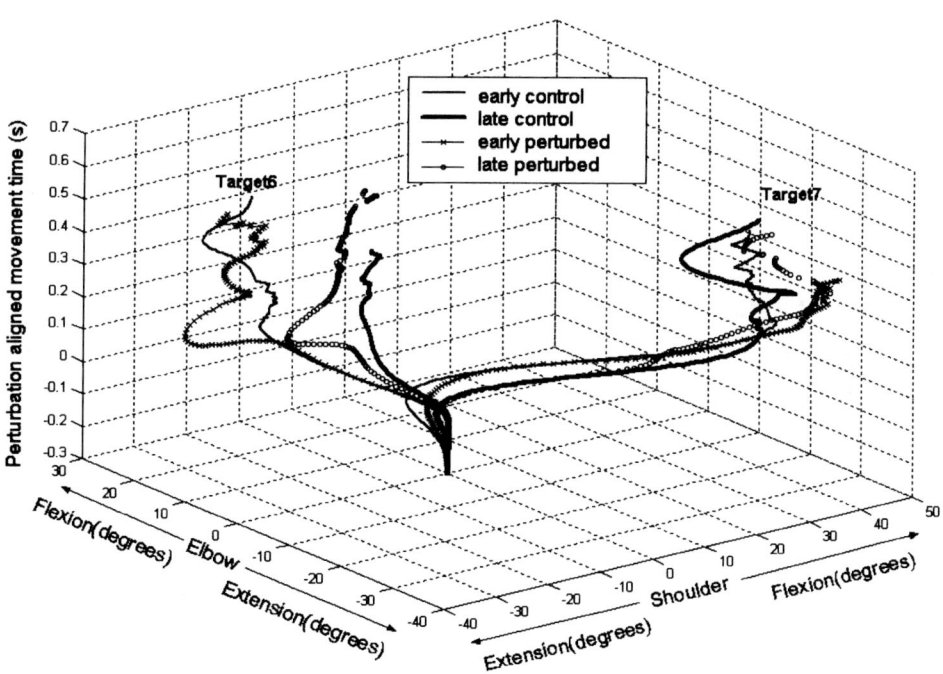

Fig. 6. Joint relationships for the shoulder versus elbow during a control versus perturbed arm-reaching task. Phase plots of elbow flexion (−o)–shoulder flexion (+o) when the animals were reaching toward targets 6 and 7 (see Fig. 1 for direction difference). Thin lines show the arm's configuration on the first day for both the control and perturbation-applied condition. Thick lines show the analogous configurations on the last day of testing. This figure shows that when the arm was moving toward target 7, the changes in its posture (i.e., as represented by the shown joint angles) were small and not even obvious after adaptation to the perturbation. In contrast, when moving toward target 6, a significant change occurred in arm posture during the movement for the perturbed versus control condition. This suggested that a change in arm stiffness might have played a significant role in overcoming the perturbation.

when it was engaged in an operant-trained reaching task. The learning process included changes in the firing-pattern behavior of Area 5 and 4 neurons in the sensorimotor cortex. For each of the three tested animals, this control strategy was not uniform and it depended, in part, on the relation between the perturbation and the direction of movement.

Feedback versus feedforward control

Sensory feedback is used to modulate control commands when a neural controller learns and adapts to a new task and/or environment. The modified control commands then provide feedforward compensation, which is needed to improve task performance. To investigate the development of feedforward-compensation strategies, many recent studies have employed perturbation paradigms to study adaptations in control when subjects are experiencing a standard (nonvarying) perturbation. Human and nonhuman primates have been shown to adapt to such perturbations (Gandolfo et al., 2000). For example, Thoroughman and Shadmehr (2000) found that error-driven feedback produced in a response to a novel force-field perturbation shifted earlier in time as the subject adapted to the perturbation. This indicated that the perturbation's compensation shifted from predominantly feedback to a more feedforward strategy.

In our experiments, the perturbation became a fixture of the task such that the subjects learned its dynamic effect, and could also anticipate its occurrence. As a result, the subjects were able to plan a corrective strategy rather than rely on sensory feedback. The hand paths from the few catch trials exemplified, in particular, the development of a feedforward/predictive strategy to compensate for the expected perturbation (viz., Fig. 2).

Two components of the perturbation response were identified (viz., Evarts and Tanji, 1976). A presumed reflex component appeared at a relatively constant latency across all of the perturbation trials. During adaptation, a volitional component of the perturbation response developed as the animal learned to anticipate the perturbation. It is likely that this latter component was related to the planned corrective movement (e.g., the catch trials).

Behavior of Area 5 versus 4 neurons

It was instructive to compare the neurons' average discharge profile on the first and last days of the perturbation trials. After adaptation, the cells still showed a peak spike-frequency ~80 ms after perturbation onset. There was increased overall activity throughout both the preperturbation and postperturbation epochs. These changes reflected the adaptation in control that had taken place throughout adaptation. When we further separated the analysis of cortical neuron discharge according to the locations of implanted electrodes, we discovered the presence of differential adaptation in the Area 5 versus 4 neurons. The Area 4 (but not 5) neurons' behavior further revealed that the central nervous system (CNS) developed a predictive strategy to accommodate a predictable perturbation. This finding supported the hypothesis that the CNS shifts its control emphasis from sensory feedback to predominantly a feedforward strategy *when the nature of a perturbation is known.*

Trajectory compensation versus stiffness regulation

There are several possible feedforward control strategies to compensate for the perturbation effect on the reaching task, including (1) trajectory compensation to cancel the deviation induced by the perturbation, (2) force compensation to overcome the effect of the perturbation force and (3) stiffness regulation to reduce the effect of the perturbation. We have observed two types of adaptation in hand trajectory that may reflect the adoption of two possible control strategies: trajectory compensation and stiffness regulation. It remains to be determined, however, how these different control strategies are represented in cortical neuronal activity patterns.

Concluding thoughts

The data analysis presented in the previous sections is essentially a preliminary one. Already, however, it shows that our research approach has produced a rich data set for exploring CNS control of arm movement at both the cortical and behavioral levels.

Further analysis is now necessary to evaluate the time course of the adaptation that shifted the mode of perturbation compensation from feedback to feedforward. Changes in the discharge rate of Area 5 versus 4 cortical neurons were of special interest. Already, we have demonstrated two key differences in the firing pattern and time course of adaptation throughout the perturbation trials. In regard to trajectory compensation versus stiffness regulation, we have shown that the CNS adopted each one, as dependent on the direction of the movement and the type of the perturbation. By simultaneously observing a relatively large population of neurons over a relatively long time period (days or months), we had the opportunity to investigate the time course of adaptation and plasticity in cortical control of movement. Clearly, our techniques and experimental paradigm should prove valuable as we extend on these intriguing observations.

Acknowledgments

We would like to thank Dr. Andrew Schwartz for providing technical advice and insightful discussions, Wen-Shan Lin for his daily training and handling of the nonhuman primates used in the research, Xinying Cai for her help with the data analysis and many others in our laboratory. The work was supported in part by grants from the USPHS (NS 37088, NS 62347), DARPA and the Whitaker Foundation.

Abbreviations

3D	three-dimensional
CNS	central nervous system
EMG	electromyogram/electromyographic
LED	light-emitting diode

References

Bhushan, N. and Shadmehr, R. (1999). Computational nature of human adaptive control during learning of reaching movements in force fields. Biol. Cybern., 81: 39–60.

Caminiti, R., Johnson, P.B., Galli, C., Ferraina, S. and Burnod, Y. (1991). Making arm movements within different parts of space: the premotor and motor cortical representation of a coordinate system for reaching to visual targets. J. Neurosci., 11: 1182–1197.

Crammond, D.J. and Kalaska, J.F. (1996). Differential relation of discharge in primary motor cortex and premotor cortex to movements versus actively maintained postures during a reaching task. Exp. Brain Res., 108: 45–61.

Evarts, E.V. and Tanji, J. (1976). Reflex and intended responses in motor cortex pyramidal tract neurons of monkey. J. Neurophysiol., 39: 1069–1080.

Gandolfo, F., Li, C., Benda, B.J., Schioppa, C.P. and Bizzi, E. (2000). Cortical correlates of learning in monkeys adapting to a new dynamical environment. Proc. Natl. Acad. Sci. USA, 97: 2259–2263.

Georgopoulos, A.P., Kalaska, J.F., Caminiti, R. and Massey, J.T. (1982). On the relations between the direction of two-dimensional arm movements and cell discharge in primate motor cortex. J. Neurosci., 2: 1527–1537.

Georgopoulos, A.P., Kettner, R.E. and Schwartz, A.B. (1988). Primate motor cortex and free arm movements to visual targets in three-dimensional space. II. Coding of the direction of movement by a neuronal population. J. Neurosci., 8: 2928–2937.

Georgopoulos, A.P., Ashe, J., Smyrnis, N. and Taira, M. (1992). The motor cortex and the coding of force. Science, 256: 1692–1695.

Gribble, P.L. and Scott, S.H. (2002). Overlap of internal models in motor cortex for mechanical loads during reaching. Nature, 417: 938–941.

Hayashi, R., Becker, W.J. and Lee, R.G. (1990). Effects of unexpected perturbations on trajectories and EMG patterns of rapid wrist flexion movements in humans. Neurosci. Res., 8: 100–113.

Hoogerwerf, A.C. and Wise, K.D. (1994). A three-dimensional microelectrode array for chronic neural recording. IEEE Trans. Biomed. Eng., 41: 1136–1146.

Isaacs, R.E., Weber, D.J. and Schwartz, A.B. (2000). Work toward real-time control of a cortical neural prosthesis. IEEE Trans. Rehabil. Eng., 8: 196–198.

Kalaska, J.F., Caminiti, R. and Georgopoulos, A.P. (1983). Cortical mechanisms related to the direction of two-dimensional arm movements: relations in parietal area 5 and comparison with motor cortex. Exp. Brain Res., 51: 247–260.

Kalaska, J.F., Cohen, D.A., Hyde, M.L. and Prud'homme, M. (1989). A comparison of movement direction-related versus load direction-related activity in primate motor cortex, using a two-dimensional reaching task. J. Neurosci., 9: 2080–2102.

Kennedy, P.R., Bakay, R.A. and Sharpe, S.M. (1992). Behavioral correlates of action potentials recorded chronically inside the Cone Electrode. Neuroreport, 3: 605–608.

Lackner, J.R. and Dizio, P. (1994). Rapid adaptation to Coriolis force perturbations of arm trajectory. J. Neurophysiol., 72: 299–313.

Lacquaniti, F., Borghese, N.A. and Carrozzo, M. (1991). Transient reversal of the stretch reflex in human arm muscles. J. Neurophysiol., 66: 939–954.

Lacquaniti, F., Borghese, N.A. and Carrozzo, M. (1992). Internal models of limb geometry in the control of hand compliance. J. Neurosci., 12: 1750–1762.

Lauer, R.T., Peckham, P.H., Kilgore, K.L. and Heetderks, W.J. (2000). Applications of cortical signals to neuroprosthetic control: a critical review. IEEE Trans. Rehabil. Eng., 8: 205–208.

Lin, S., Si, J. and Schwartz, A.B. (1997). Self-organization of firing activities in monkey's motor cortex: trajectory computation from spike signals. Neural Comput., 9: 607–621.

Loeb, G.E., Levine, W.S. and He, J. (1990). Understanding sensorimotor feedback through optimal control. Cold Spring Harb. Symp. Quant. Biol., 55: 791–803.

Moran, D.W. and Schwartz, A.B. (1999). Motor cortical activity during drawing movements: population representation during spiral tracing. J. Neurophysiol., 82: 2693–2704.

Mountcastle, V.B., Lynch, J.C., Georgopoulos, A., Sakata, H. and Acuna, C. (1975). Posterior parietal association cortex of the monkey: command functions for operations within extrapersonal space. J. Neurophysiol., 38: 871–908.

Nordhausen, C.T., Maynard, E.M. and Normann, R.A. (1996). Single unit recording capabilities of a 100 microelectrode array. Brain Res., 726: 129–140.

Rousche, P.J., Pellinen, D.S., Pivin, D.P., Jr., Williams, J.C., Vetter, R.J. and Kipke, D.R. (2001). Flexible polyimide-based intracortical electrode arrays with bioactive capability. IEEE Trans. Biomed. Eng., 48: 361–371.

Schwartz, A.B. (1992). Motor cortical activity during drawing movements: single- unit activity during sinusoid tracing. J. Neurophysiol., 68: 528–541.

Schwartz, A.B., Taylor, D.M. and Tillery, S.I. (2001). Extraction algorithms for cortical control of arm prosthetics. Curr. Opin. Neurobiol., 11: 701–707.

Serruya, M.D., Hatsopoulos, N.G., Paninski, L., Fellows, M.R. and Donoghue, J.P. (2002). Instant neural control of a movement signal. Nature, 416: 141–142.

Shadmehr, R. and Mussa-Ivaldi, F.A. (1994). Adaptive representation of dynamics during learning of a motor task. J. Neurosci., 14: 3208–3224.

Taylor, D.M., Tillery, S.I. and Schwartz, A.B. (2002). Direct cortical control of 3D neuroprosthetic devices. Science, 296: 1829–1832.

Thoroughman, K.A. and Shadmehr, R. (2000). Learning of action through adaptive combination of motor primitives. Nature, 407: 742–747.

Weber, D.J., Chi, A. and He, J. (1998) Adaptation of arm movement control under repeated perturbations. Int. Conf. IEEE EMBS, Hong Kong, 5: 2354–2357.

CHAPTER 46

The quest to understand bimanual coordination

Mario Wiesendanger* and Deborah J. Serrien

Laboratory of Motor Systems, Department of Neurology, University of Berne, CH-3010 Berne, Switzerland

Abstract: Many skillful manipulations engage both hands for goal achievement. Whereas the goal is planned consciously and achieved quasi-invariantly, the articulators are mobilized automatically, but in a flexible manner (Lashley's principle of motor equivalence). In brain disorders affecting hand functions, adaptive mechanisms are mobilized to improve goal achievement. Thus, chronic cerebellar patients were found to initiate a bimanual drawer task with marked intermanual desynchronization as compared to control subjects. This was partly compensated for, however, by adjusting the kinematics as the individual limbs move toward the goal, thereby improving the initial desynchronization. Adaptive strategies rarely correct deficits completely, however. Bimanual movement patterns, either in-phase or anti-phase are relatively stable in healthy human subjects, whereas brain pathology may preferentially impair the anti-phase pattern. This is the case in patients with acquired pathology of the corpus callosum, thereby suggesting that this structure is important for maintaining temporally independent limb and hand movements.

Introduction: from manual skills of the pebble culture to present artistic skills

Nonhuman primates make use of their hands for many skillful operations, including purposeful cooperative actions of both hands. And yet, from the time of the earliest hominids using an upright gait, it took more than one million years until *Homo habilis* fabricated crude stone tools: this was the 'pebble culture' discovered at the Olduvai site in Tanzania (Leakey, 1981). During this long period, the brain increased from 500 to ~800 cm^3, and further increased enormously until the emergence of *Homo sapiens* with a brain of 1350 cm^3. It is commonly thought that the upright gait freed the hands for manipulatory behavior. This begs the question, however: what is new in the skill performance of the modern human as compared to the already highly developed skillful behavior of monkeys and apes? To cite the physiologist J.B.S. Haldane: ... "The human brain has two superanimal activities, manual skill and logical thought. Manual skill appears to be the earlier acquisition of the two, and the capacity for language and thought has grown-up around it." (cited in Smith, 1959). One might add that human skills are more goal directed, including also goals planned for a more distant future. Furthermore, they include the fabrication of progressively more sophisticated tools. The ultimate achievement in skillful human behavior is perhaps the virtuosic playing of a musical instrument (Fig. 1).

These latter exclusive human faculties depend crucially on the motor-cortical and pyramidal tract as the main executive system. Phillips' pioneering work (Phillips, 1986) led to the appreciation of the role of the pyramidal tract system for skilled manipulatory behavior, particularly its monosynaptic corticomotoneuronal component. He also had the insight that what is transmitted to the spinal cord has already been assembled "... upstream of this corticofugal level" (Phillips, 1966). Currently, much research in

---

*Corresponding author: Tel.: +41-26-402-4528; Fax: +41-31-632-3011; E-mail: neuro@bluewin.ch

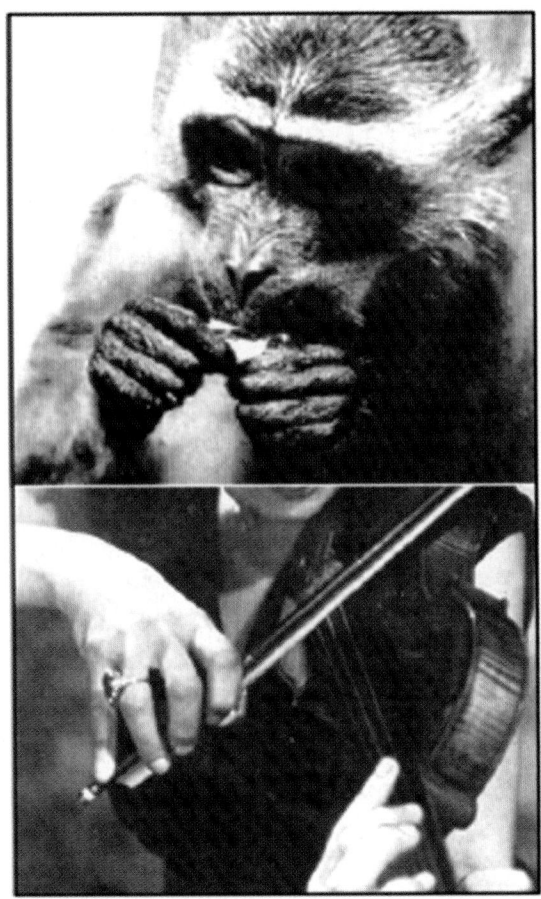

Fig. 1. Bimanual skills of a nonhuman primate and the human. Old world monkeys have acquired considerable manipulative skills, but cannot play music or speak.

motor behavior is dedicated to the role of higher cortical structures, upstream to the primary motor cortex (e.g., Hoshi and Tanji, and Ma et al., this volume). In the *Homo* lineage, the main developmental increase concerned secondary motor areas in the lateral and mesial frontal cortex, together with the particularly large association areas and their interconnections with thalamus, basal ganglia and cerebellum. This is the structural basis for evolving cognitive functions, including the superior forelimb–hand motor behavior of the human. Intentional and goal-oriented actions strongly depend on the predictive capacities of the human brain. They also enable the human to find alternate solutions for achieving the same goal. Goal achievement with variable means is the *principle of motor equivalence* (Lashley, 1933), a mental capacity that makes the human adaptive and innovative. [See also similar principles of '*gliding coupling*' (Bethe, 1931) and of '*equal simplicity*' (Bernstein, 1967).]

Higher-order skills are controlled by a memory-based feedforward model and depend on continuing practice

Familiar skills have usually been learned during childhood and practiced many thousands of times. Over the years, they are eventually performed easily and automatically. The term, 'overtrained', is often used in experimental psychology in the context of operant conditioning when training is continued after the animal reaches an experiment-selected criterion level. The term is misleading, however, in the context of learning skills, since it implies the notion of 'too much'. The gradual acquisition of natural skills means *learning by doing*, whereby conscious attention becomes focused on the goal, rather than on the details of execution. For example, elite musicians have practiced, by the age of 20 years, about 10,500 h on their instrument (Sloboda and Howe, 1991). A high degree of automatization is certainly achieved, and this is absolutely necessary, because it concerns the most basic details of execution. Their automatic mastery allows, in turn, for more freedom in accentuating musical content by slight intended changes in force (loudness) and tempo (Repp, 1999; Drake and Palmer, 2000).

An appealing theory is that learned skills are essentially memorized and represented in neural networks of multiple brain structures. Such brain representation functions, in turn, as a memory-based internal model delivering feedforward commands ('internal forward model'; Wolpert et al., 1998). Since we live in an unstable world with variable relative configurations of the object (goal) and the active motor apparatus, appropriate adjustments are needed to correct for disturbances. It was suggested by von Holst (Holst von and Mittelstaedt, 1950) that a corollary discharge, originally termed 'efference copy', is compared with actual sensory feedback, thereby providing predictive signals about the (sensory) consequences of movement. As a predictor of the consequences of actions, the internal model thus has adaptive power and is capable of stabilizing goal-oriented actions in their actual context (Blakemore et al., 2001).

Synergistic bimanual actions: performing goal- and object-oriented skills

For most purposes, everyday skills require use of both limbs, usually with an asymmetrical engagement of the hands, e.g., when lacing shoes, and eating with fork and knife. Interestingly, the choice of which hand for which task is not arbitrary. For example, most humans use the left hand for lifting and holding a glass while the right hand grasps the bottle for pouring the wine. A division of labor with hand preferences for task components is a hallmark of object-oriented bimanual actions. Rather than the connotation of 'dominance', the division of labor is a term concerning preferential roles: the left hand for object holding and providing a body-related reference frame, and the right hand for manipulating objects (MacNeilage, 1987). Consider again playing an instrument, e.g., the violin. It would make no sense to call the right-bowing hand the dominant one and the left-fingering hand nondominant, because both movements are extremely sophisticated actions. In keeping with the ideas of MacNeilage, however, is the additional postural function of the left hand in supporting the violin at its neck. The emergence of hemispheric asymmetries is a crucial feature in human evolution. Implementation of a goal/object-oriented task imposes spatial as well as temporal coordination of the two hands and the action becomes 'a higher unit of performance' (Welford, 1968). The question then arises about the underlying coordinating mechanisms for these skilled bimanual movements (see also Grillnér and Wallen, this volume).

Interactions between limbs and the occurrence of phase transitions in bimanual rhythmic movements

For many years, there were few studies on the coordination of familiar and object-oriented bimanual actions. Recognizing the need to assess deficits in bimanual coordination, Luria (1969) introduced rhythmic movements as a simple neurological test in cases of intermanual discoordination. He noticed that patients with frontal cortical lesions often had difficulties in keeping the phase constant when sequences had to be performed in an anti-phase (parallel) mode, as compared to an in-phase (mirror) mode. Since Luria's seminal work, a large amount of work on rhythmic bimanual movements ensued. It has demonstrated the strong attraction from the less stable anti-phase pattern to the more stable in-phase pattern. Not only was this paradigm a departure point for explaining biological transition phenomena and attractors in terms of self-organizing, nonlinear systems (Kelso, 1995), but it was also found useful for investigating coordinative mechanisms in both healthy human subjects (Franz et al., 1991; Treffner and Turvey, 1996; Swinnen et al., 1998) and neurological patients (Wyke, 1971; Leonard et al., 1988; Johnson et al., 1998; Serrien et al., 2000). For example, such studies have shown conditions favoring attraction constraints in the spatial–temporal domain.

Recently, the implication of cerebral structures in bimanual coordination of rhythmic hand movements was assessed by use of brain imaging techniques (see the next section). In most such studies, the difference in activation patterns during the in-phase mode and the anti-phase mode was taken as a measure of the extra load imposed on the brain by coordination (i.e., assuming that moving in an anti-phase mode requires more coordination than in an in-phase one). One may argue, however, that any difference found between the two modes may be due largely to the more demanding attention and supervision for the anti-phase mode, rather than to coordination per se.

A number of studies have also focused on the interactions occurring between the two hands when they move in discrete steps with asymmetric assignments. For example, Kelso et al. (1983) have shown that when the two limbs are pointing to separate goals at different distances, they have unequal velocity in order to be matched in time at the goal (Kelso et al., 1979). The authors suggested that the scaling of velocities for goal synchronization was imposed by a 'coordinative structure'. Similar studies have also been fruitful to assess deficits in bimanual coordination (Viallet et al., 1987; Stelmach and Worringham, 1988; Horstink et al., 1990; Alberts et al., 1998; Jackson et al., 2000).

In the following discussion, we focus on investigations which concerned force interactions between forelimbs. In load-lifting tasks, the grip force (normal to the grasped surface) and load force (to overcome gravity and inertial forces) have to be coordinated, with the grip force securing the object from slipping. The ratio of the two forces was found to be fairly constant during the holding of an object with a given friction and shape (Johansson and Westling, 1984; Westling and Johansson, 1984). The ratio was dynamically modulated, however, during object manipulation, such as lifting, tilting, transporting and applying torques for rotating (Flanagan and Wing, 1993, 1997; Johansson et al., 1999).

In our own study of the above issues (Serrien and Wiesendanger, 2001b), the aim was to investigate effects produced by introducing force asymmetries in rhythmic bimanual up-and-down movements. Figure 2 shows that subjects had to move objects grasped at the U-shaped upper end which were instrumented with force transducers to measure grip force. Another force transducer was incorporated in the object to measure the longitudinal load-force changes.

In the unimanual Fig. 2A task, grip and load forces were varied in parallel with the rhythmic vertical movements. These forces decreased during upward movements and increased during downward ones. The force ratio also fluctuated, but in the opposite direction to the above two forces: the maximum ratio was reached at the peak of displacement, and the minimum one at the trough. The fact that grip force diminished less than load force during upward movements means that the safety margin increased transiently, thereby preventing losses of the grasped object when it was nearing the turning point, i.e., when the load force approached zero Newtons. Presumably, this is a learned anticipatory (proactive) adjustment. In the bimanual task, identical loads were moved rhythmically in both the in-phase and anti-phase mode (upper part of Fig. 2B). The weight distribution between the two hands was the same in the in-phase mode, but this was not so in the anti-phase mode. The oscillations were identical to those for the unimanual condition (Fig. 2A). It was discovered, however, that the force ratio maxima

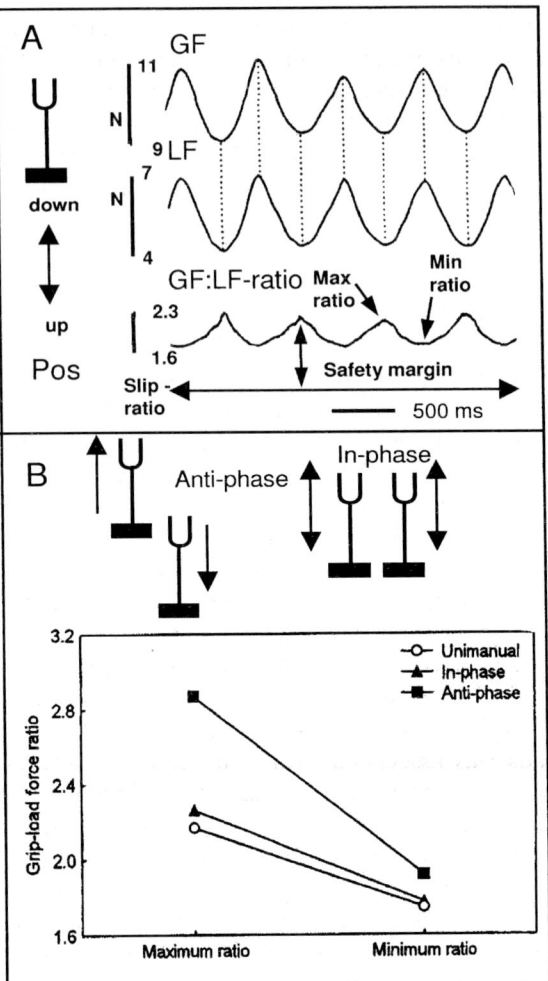

Fig. 2. Grip/load-force coupling and its variation during up-and-down movements. (A) Top-left diagram shows an instrumented object with a U-shaped handle that was to be grasped with a precision grip. Strain gauge was inserted in the vertical rod for measuring load force (LF). Immediately underneath arrows indicate the direction of movement in which the object was moved (Pos, position of the object). Upper two right-side traces show fluctuations in grip force (GF), normal to the surface, and load force (LF) in Newtons (N; vertical bars). The lowest trace shows the fluctuations in the calculated force ratio with arrows at its maxima and minima. The safety margin represents the difference between the slip ratio, the minimum value determined by the friction between skin and object, and the grip–load force ratio employed by the subject. (B) Upper diagrams illustrate the direction of movement of the instrumented objects according to an in-phase and anti-phase mode. Lower plot shows that the maximum force ratio was higher for anti-phase movements.

were scaled to significantly higher values during antiphase movements versus unimanual and in-phase bimanual movements (Fig. 2B). In contrast, the ratio minima did not differ significantly for unimanual versus the two variants of this bimanual task. We suggested that an additional amplification of grip force occurs, and thereby also of the safety margin. This preserves stability in the face of a task that demands more attention for maintaining the coordination pattern when the force oscillations are reciprocal in the two limbs. It also reinforces the notion of a task dependency in the force coordination of bimanual actions and of proactive adjustments of the safety margin in potentially unstable situations.

Another study likewise revealed adjustments of the safety margin when asymmetric assignments were imposed in a bimanual load-lifting task (Serrien and Wiesendanger, 2001a). Objects of a light or heavy weight were lifted for a fixed distance. Subjects performed unimanual and bimanual weight lifts. In the latter condition, they used equal (light–light, heavy–heavy) or unequal (light–heavy, heavy–light) loads. The results showed that the lifting of light loads was associated with a higher force ratio and shorter movement time in the unimanual and bimanual-equal conditions. In contrast, the unequal bimanual conditions (different weights) showed a clear-cut assimilation of the dependent variables, i.e., force ratios and movement times. Table 1 summarizes the significant changes in this study and highlights the results obtained in the bimanual-unequal condition.

When one hand was lifting a light weight simultaneously with a heavy weight in the opposite hand, the movement time for the light lift became longer as compared with the bimanual-equal or unimanual condition. Therefore, the change of the force ratio in the asymmetric loading could be explained by a velocity decrease of the low-weight limb movements, which, in turn, induced a decrease of the force ratio and assimilation between the limbs. Our deduction is that assimilation of force ratio and movement time was imposed by the kinematics of the limb.

Coordinative rules in a bimanual drawer-opening task: experiments on the monkey

These experiments involved a number of collaborators (Wiesendanger et al., 1992; Kazennikov et al., 1994; Wiesendanger et al., 1994a,b, 1996; Kazennikov et al., 1999). Figure 3 shows the bimanual-drawer paradigm that we developed for behavioral and electrophysiological studies in primates. The experience gained in these experiments strongly influenced our further work in healthy human subjects and neurological patients (see the next section).

Monkeys ($n=9$) were trained to reach for and open a drawer with one hand, and to pick up a food morsel with the other. The pull-hand had to keep the spring-loaded drawer open in order to avoid its closing while the pick-hand was still grasping for food. The animals were highly motivated and easily performed ~100 trials per session. The pick-hand arrived at the goal precisely when the drawer was fully opened, with the pull-hand leading by 10–50 ms (Fig. 3A inset). All the monkeys also quickly learned to accomplish a series of trials in complete darkness. By means of electronic signals from a number of

Table 1. Force and time assimilation among hands in the task condition of unequal bimanual load lifting

	Hands		Force ratio	Movement time (MT)
	Left	Right		
Unimanual	Light Heavy	Light Heavy	$Ratio_{light} > ratio_{heavy}$	$MT_{light} < MT_{heavy}$
Bimanual equal	Light Heavy	Light Heavy	$Ratio_{light} > ratio_{heavy}$	$MT_{light} < MT_{heavy}$
Bimanual unequal	Light Heavy	Heavy Light	$Ratio_{light} = ratio_{heavy}$ assimilation	$MT_{light} = MT_{heavy}$ assimilation

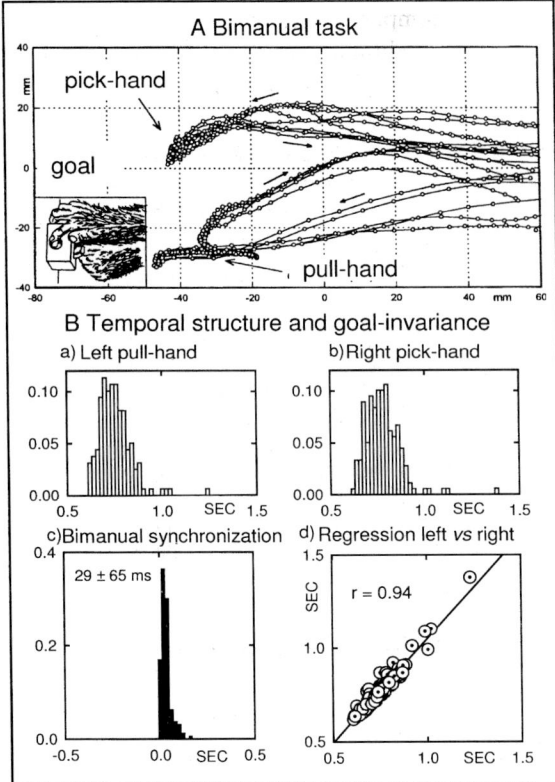

Fig. 3. Execution of a bimanual-drawer task by a monkey without visual guidance. (A) Superimposed trajectories (sagittal view) of the pick- and pull-hands. On the right one can see the almost straight reaching trajectories and the more curved withdrawal trajectories. On the left picking occurs as the hand reaches the opened drawer. Below one can see the pulling and release of the drawer. The inset corresponds to the moment of goal achievement. Movement directions toward the drawer and back to the home position are indicated by arrows. (B) Histograms of arrival times of left (a) and right (b) hands. Also shown are a histogram of right–left intervals (c; synchronization interval) and a regression plot (d) of right- and left-hand arrival times, indicating the correlation coefficient (r). (Reproduced from Wiesendanger et al., 2001 with permission from IOS Press.)

event markers and the displacement profiles of the drawer, it was possible to obtain a quantification of the temporal structure of the task, with and without vision. In addition, trajectories of the two hands were recorded.

Figure 3A, taken from a representative monkey, illustrates superimposed hand trajectories in consecutive trials. The panels in Fig. 3B show the broad histograms of arrival times of the pull- and pick-hand at the goal. Also shown are a histogram of interlimb synchronization intervals at the goal and a regression plot with the correlation coefficient (r) for the right- and left-hand arrival times. The data were taken from experimental series without vision, i.e., when the behavior was memory based. We learned from these experiments that such habitual skills were performed quite reliably provided there was with optimal motivation. None of the monkeys performed the task sequentially, although they would have been free to wait with the pick-hand at the start position until the drawer was opened. Thus, the bimanual synergy was organized as a unitary, goal-directed reach-and-grasp action, with synchronized goal achievement. The bimanual goal invariance contrasted with the high variability of the individual limb components (Fig. 3B, c–d vs. a–b). Goal invariance with variable components is the hallmark of the principle of motor equivalence (Bethe, 1931; Lashley, 1933; Bernstein, 1935). Figure 4 shows that Lashley's principle was also apparent when the displacement traces were aligned and superimposed at the instance the drawer was fully opened. Whereas the variability among trials is obvious in Fig. 4, there is a remarkable spatial–temporal 'attraction' of both hands at the moment of complete drawer opening.

In other monkey experiments using the bimanual drawer task (Kazennikov et al., 1998), we observed that unilateral lesions of the supplementary motor area only transiently affected the contralateral arm that became delayed at movement onset. After a few postoperative sessions, the bimanual goal invariance was again present. This was achieved either by adjusting the ipsilateral limb with an equal delay, or by accelerating the movement of the affected arm. Figure 5 shows that in monkeys with a bilateral lesion, a delayed onset was accompanied by a significant slowing of the two limb components (Fig. 5A–B). After the first few sessions, which featured higher variability in bimanual goal synchronization, the temporal goal invariance was reestablished. Both limbs eventually reexhibited the rule of high covariation (Fig. 5C). The smooth coordination that was reestablished in this 'natural' bimanual-drawer task encouraged us to pursue further studies along these lines in human subjects, including neurological patients.

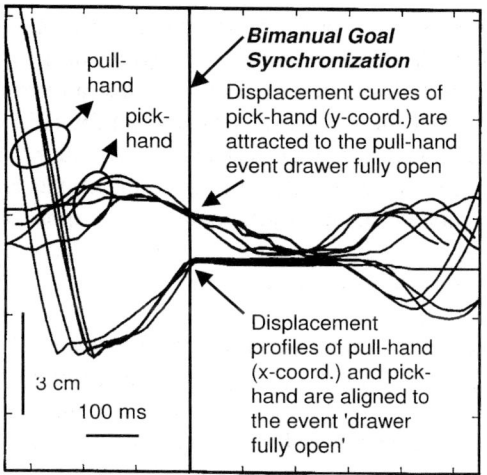

Fig. 4. Further kinematic demonstration of goal synchronization in the bimanual-drawer task by means of bimanual displacement profiles. Goal synchronization occurs at the vertical line, i.e., at the moment of complete drawer opening. All trials were aligned with this pull-hand event (lower arrow). Note the remarkable convergence of pick-hand displacement profiles which are also aligned to the event of the *other* hand's drawer opening. Superimposed displacement profiles of successive trials, time running from left to right.

Bimanual coordination during a drawer-opening task: experiments on the human

We next studied bimanual coordination in the human, using a similar drawer-pulling task with the manipulandum adapted to relevant geometry (Perrig et al., 1999). The task was to open a drawer with one hand while the other hand had to pick up a peg, which was inserted in the drawer's recess (Fig. 6A). After a few trials, the subjects executed the task promptly and consistently, with and without visual input. As in the experiments on monkeys, we observed a goal invariance in terms of synchronization and correlation. Complete drawer opening by the leading left hand was a requisite for the pick-hand to retrieve the peg inside the drawer. As a result, there was a systematic deviation of ~100 ms from perfect synchronization in the no-vision condition. (The Fig. 6B results are presented below.)

The objective in our human experiments was to manipulate some of the task constraints and observe their effects on goal invariance. Intriguingly, synchronization in blindfolded subjects improved

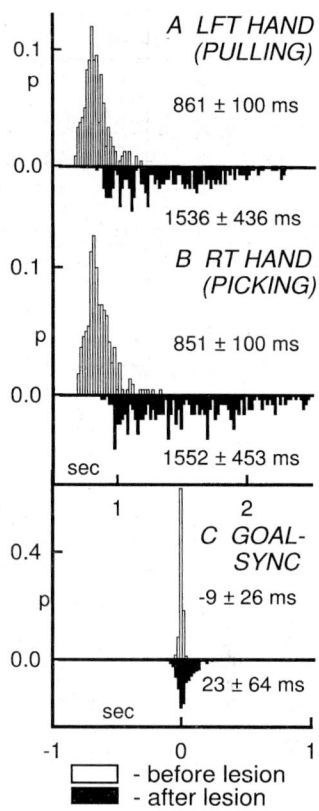

Fig. 5. Execution of the same bimanual-drawer task by a monkey before and after cortical lesion (bilateral supplementary motor area). Latency histograms for goal reaching of the individual limbs (A, B), and interval histograms for bimanual synchronization (C) are plotted on the abscissa. Upward/open bars are for prelesion results, downward/filled bars for postlesion results. Note massive slowing of individual limb movements (A–B), but preserved synchronization, i.e., goal invariance (C). Abbreviations: LFT, left; RT, right; SYNC, synchronization. Numbers indicate mean ± SD.

significantly, as compared to the with-vision condition, possibly because the task relied on proprioceptive signals. Additional loading of the drawer prolonged the total movement time of the pull-hand. This was matched by a covariant slowing of the pick-hand. After local cutaneous anesthesia of the pulling thumb and index finger, the pull-hand was much disturbed in its opening of the drawer and frequent slips occurred. Also, the movement time of successful trials was significantly prolonged. Again, this was accompanied by a covariant prolongation of

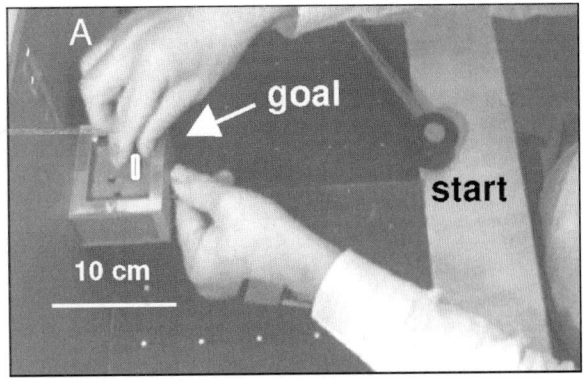

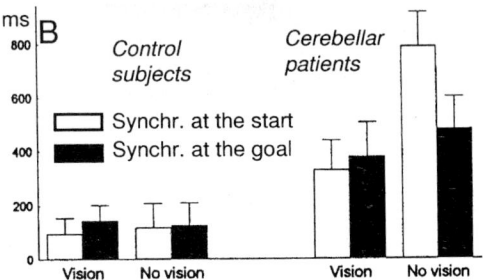

Fig. 6. Execution of the same bimanual-drawer task by a human subject. (A) Note similarity to the monkey's performance (sketch in Fig. 3A inset). Labels: 'start', positions at movement onset (sight on left-hand start position is covered by left arm); 'goal', instance when picking hand is entering the drawer recess of the fully opened drawer. (B) Population mean values of bimanual synchronization (synchr.) at the start and at the goal, in the vision and no-vision condition. Left, age-matched control subjects ($n = 5$); right, patients with cerebellar lesions ($n = 5$). Ordinate shows synchronization intervals in ms.

the pick-hand which did not, however, completely match the synchronization observed in the control series.

We also mimicked a 'split-brain' situation, i.e., the experimenter opened the drawer while the experimental subject reached for the opened drawer under visual control. Whereas goal synchronization was not significantly changed as compared to the bimanual control series, the correlation coefficients between the hand of the experimenter and the subject were significantly lower. This clearly indicates that for normal performance, factors other than vision contribute in adjusting interlimb coordination.

The above experiments in human subjects are in keeping with previous studies (Jeannerod, 1984; Castiello et al., 1993; Weiss and Jeannerod, 1998) that demonstrated a temporal-coordinating rule for goal achievement in bimanual tasks. These and our studies also suggest that in line with the motor equivalence principle, changing task constraints induces adaptations which tend to minimize changes in goal invariance. These studies also indicate that loss of a sensory modality can be compensated for by other modalities, in order to maintain appropriate bimanual coordination.

Disturbed bimanual coordination in neurological pathology

Cerebellar dysfunction

Figure 6B shows that in five patients with bilateral cerebellar pathology (two with a bilateral infarct, two with atrophy, one with a tumor) and moderate symptoms (dysdiadochokinesis in all patients), a pronounced change was revealed in organizing temporal coordination while executing the same bimanual-drawer task (Serrien and Wiesendanger, 2000). Cerebellar patients as compared to age-matched control subjects showed an increased offset for initiating the hand movements. Lack of vision further augmented this degree of desynchronization at the start position. Surprisingly, under the same no-vision condition, the timing at the goal was considerably better than at the start position. This indicates that an adaptive mechanism intervened between start and goal position that was independent of visual guidance. Other factors must be responsible for the relatively preserved goal coordination. The coupling may have been adjusted by proprioceptive signaling and/or learned adjustments in the internal feedforward model (cf. Steyvers et al., 2001). Also, the slower movements provided more time to adjust the limbs on their trajectories to the goal. We therefore concluded that cerebellar circuits are implicated in the temporal structure of the synergy, but chiefly for an initial coupling of the two limbs. Interestingly, the initial desynchronization was, at least in part, compensated for at the goal. The fact that initial desynchronization did not predict desynchronization at the goal, shows that adjustments intervene and contribute to goal synchronization. This finding was also supported by evidence of a high temporal correlation in arrival times of the two limbs at the goal. As discussed previously (Perrig et al., 1999), factors affecting the movements of one limb in healthy subjects (adding weights, cutaneous anesthesia) also resulted in temporal adjustments to preserve, at least partly, the goal invariance.

Congenital and acquired callosal damage

Figure 7 shows results for execution of the drawer task by three patients with congenital agenesis (KB, PH, IT in figure) and three patients with acquired lesions (EC, DM, WN; Serrien et al., 2002b). In the congenital cases, magnetic resonance imaging (MRI) revealed a total absence of the corpus callosum, and yet these patients appeared to have no coordination problems in everyday life. Also, they performed the drawer task as well as the control subjects, with interlimb synchronization within the normal range. Long-term adaptations are likely to intervene in such cases, perhaps by means of interhemispheric transmission via the anterior commissure. All three of these congenital cases had an increased cross-section of the anterior commissure, as revealed by MRI. Among the three patients with acquired callosal lesions, two exhibited desynchronization at task initiation, but only when vision was absent. Nonetheless, they were again synchronized at the goal (Fig. 7A). This again suggests that the coupling mechanisms at task initiation and at the goal are different (recall that this was also observed in the cerebellar patients). This result suggests that the delayed hand was catching up during its course toward the drawer, and that this adjustment was independent of vision. The third patient with an acquired callosal lesion had a large desynchronization at both the start and at the goal, and this was similar in the with-vision and no-vision conditions. In view of his obvious difficulties in adjusting the two limbs for a parallel action, this patient adopted a sequential strategy: first reaching and pulling the drawer, then reaching and grasping the peg. Similar strategies have also been observed in patients with Parkinson's disease (Castiello and Bennett, 1997; Alberts et al., 1998). Moreover, it appears that the mechanisms for goal synchronization are still present in the patients with acquired callosal damage, even though they

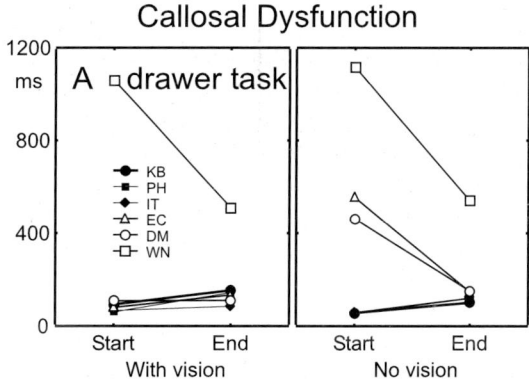

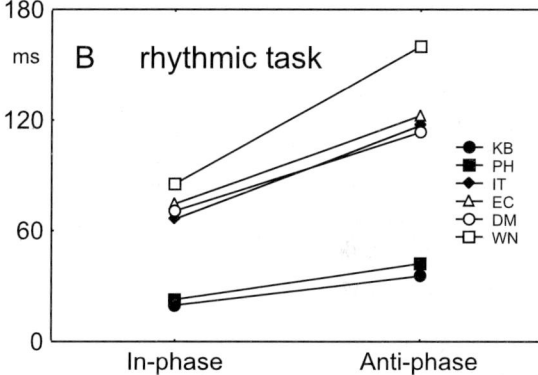

Fig. 7. Execution of bimanual tasks by patients with callosal dysfunction. Subjects included patients with callosal agenesis (filled symbols) and those with acquired lesions of the corpus callosum (open symbols). (A) Execution of the drawer-opening task. On the ordinate are mean values (ms) for synchronization at the start of the synergy and at the goal (end). (B) Execution of in-phase and anti-phase rhythmic circling movements. The mean values are for phase shifts, plotted in ms on the ordinate. See text for further description.

cannot fully compensate for the large initial desynchronization.

What is the performance of the two subgroups of callosal patients when submitted to a rhythmic in-phase and anti-phase paradigm? To test this issue, the patients were asked to trace with each index finger two adjacent circles, using either mirror (=in-phase) or parallel (=anti-phase) circling movements. The measured variable was synchronization, i.e., temporal shifts between the two displacement profiles. For each cycle, the interval between peak positions in the displacement curves of the two tracing index fingers were measured and averaged over a trial of rhythmic movements. Recordings were made by a motion analysis system, with reflecting markers on each index finger. Among the patients with a congenital absence of the corpus callosum, only subject (IT in the Fig. 7B) was impaired in interlimb synchronization. All three patients with acquired lesions had similar difficulties, mostly in the anti-phase mode, with patient WN having the largest temporal shifts. The three patients with an acquired callosal pathology also had an increased score as compared to the congenital cases. The above results suggest that the corpus callosum does not play a major role in the coordination for goal synchronization in familiar skills that have been automatized over years, although split-brain subjects tend to initiate a synergy more sequentially. In chronic callosotomized patients, Sperry and others also noted that familiar skills were performed relatively well whereas the corpus callosum was found to be crucial for acquiring new bimanual skills (Preilowski, 1977; Zaidel and Sperry, 1977; Andres et al., 1999). Although rhythmic bimanual tasks appear to be simpler with regard to the movements per se, they require considerable attention for preserving the phase relationships, or for tracing prescribed asymmetric trajectories. The present results, taken together with other studies on bimanual performance, suggest that the corpus callosum is important for ongoing conscious supervision and attention required by the temporal–spatial task asymmetries.

Recently, it was found that the posterior parietal cortex of the dominant hemisphere might also contribute in the bimanual coupling of anti-phase (parallel) rhythmic tracing of circles (Serrien et al., 2002a). This, too, is in keeping with previous reports (Hécaen, 1978; Fink et al., 1999; Fink, 2001).

Bimanual coordination and its neural representation

Is there a crucial center in the brain that is responsible for bimanual coordination? If so, is it in the left or the right hemisphere, or subdivided in both? In our opinion, these are rather unlikely options since there is considerable evidence that a number of structures are involved, depending on the type of the bimanual task. Rather, taken together, the evidence suggests that widely distributed systems contribute to bimanual coordination, in both a task- and constraint-dependent manner. Exchange of information from one to the other side of the brain and spinal cord is possible via several commissural fiber systems. Early on, both the supplementary motor area (Penfield and Jasper, 1954) and the lateral premotor cortex (Bucy and Fulton, 1933) were found to be bilaterally related structures. Of particular interest is the prominent callosal interconnection of the SMA hand areas (Rouiller et al., 1994), which contrasts markedly with the sparse callosal interconnection between the hand areas of the primary motor cortices (Jenny, 1979).

Each hemisphere has access to both sides of the spinal cord, mostly via subcortical structures with bilaterally descending fiber systems (Kuypers, 1981). Direct corticospinal connections are not restricted to the primary motor cortex, but originate also from the mesial–frontal cortex (Hutchins et al., 1988; Dum and Strick, 1991), with the hand area projecting in the vicinity of motoneurons of the cervical enlargement (Rouiller et al., 1996; Rouiller et al., this volume). Similar anatomical connections have been established for sensory systems. Notably, the SII somatosensory area of both the monkey (Rouiller et al., 1996) and human (Simoes and Hari, 1999) receives considerable input from both sides of the body.

Recent experiments in which the monkey was trained to perform whole-arm movements revealed a considerable amount of neural activity that was related to the ipsilateral limb (Donchin et al., 1996, 1998, 1999). It was also reported that much of the neuronal activity recorded in primary motor cortex and supplementary motor cortex (Kermadi et al., 1998), as well as in premotor, cingulate motor and posterior parietal cortices, had activity patterns associated with a bimanual drawer task that could not be explained as a sum of the patterns seen with separate unimanual performances. The authors therefore argued that during bimanual performance, over 50% of the units in the above areas had activity related specifically to bimanual condition (Kermadi et al., 1997). Note also that hemispheric interactions may be transmitted through basal ganglia (Wiesendanger et al., 1996) and cerebellar circuits (Nirkko et al., 1999; Serrien and Wiesendanger, 2000).

The above list of findings is not comprehensive, but it is sufficient to show that commissural and bilateral descending systems offer many possibilities for coordinating movements on the two sides of the body, including the hands. Indeed, in several of the cortical and subcortical structures, neural activity is related also to ipsilateral movements. Similarly, ipsilateral effects have also frequently been observed in stimulation studies, both in monkeys and humans. This is not surprising because it is also well established that focal movements of one extremity may engage postural adjustments on both sides of the body, e.g., in active bimanual unloading (Hugon et al., 1982; Paulignan et al., 1989) and distal unloading (Kaluzny and Wiesendanger, 1992) tasks. The latter example demonstrates that at the behavioral level, the synergy between action and posture has other constraints than interlimb coordination. Such constraints must come into play in movements like arm swinging during locomotion, pushing a cart, catching a ball and in fine manipulative actions (e.g., dissecting nerve fibers under a microscope, drumming, playing a musical instrument). In a reaction-time situation, the difference in simple finger-response latencies vary only by a few ms, even if a cutaneous stimulus is applied to only one hand (Kaluzny et al., 1994). It begs credulity to imagine that one particular brain structure could control all of these diverse bimanual activities.

Noninvasive brain imaging provides the possibility to determine 'activated' structures in the brain during bimanual actions. In view of the considerable spatial constraints during the measurement of metabolic responses, most functional imaging studies have used the rhythmic paradigm of in-phase versus anti-phase hand/finger movements. Several groups have found more intensive activation of mesial frontal cortex with bimanual anti-phase movements versus in-phase movements (Sadato et al., 1997; Goerres et al., 1998; Stephan et al., 1999; Toyokura et al., 1999), or with 2:1 polyrhythms (Jäncke et al., 2000). In surface-EEG studies, a mesial–parietal 7.8–9.8 Hz EEG was found to be contingent on transitions from uni- to bimanual rhythms (Deiber et al., 2001). The above results tend to confirm the long-standing notion (Penfield and Jasper, 1954) that the mesial–frontal cortex, including the SMA and the cingulate areas, are bilaterally organized structures. It is less clear whether the above newer findings are related to the control of bimanual coordination or whether they reflect the need for increased attention and continuous monitoring while maintaining the stability of anti-phase movements. In experimental bimanual paradigms, the complexity of the task seems also to be an important factor (Cui et al., 2000). In preliminary fMRI work (Nirkko et al., 1999), it was found that performance of a bimanual-drawer task activated many cortical regions bilaterally, including mesial–frontal areas, premotor areas, somatosensory area SII, posterior parietal and prefrontal areas, as well as the overall sensorimotor cortex. Although there is no doubt that complex bimanual tasks engage multiple cortical and subcortical structures, it remains to be seen whether the sum of the unimanual components of the task (pulling alone, picking alone) produce less-activated voxels in the various regions of interest than the combined bimanual task.

Conclusions and future trends

In motor-control research, great efforts have been expended on elucidating neuronal mechanisms underlying bimanually coordinated manipulations. In view of the enormous complexity of the hand, in terms of its anatomy, degrees of freedom, and large and distributed representation in the brain, an investigator needs a high degree of optimism to pursue such research. Indeed, it is unlikely that knowledge about the implication of all elements, from the nerve cells to the articulators, will ever been captured, even for simple hand movements. This would be an inappropriate ambition, however, because it is well known that targeted actions to the same goal are executed differently from one trial to the next (Weiss, 1969; Schmidt, 1988; Newell and Corcos, 1993). A key notion in intentional motor behavior is the principle of motor equivalence, i.e., goal achievement by variable means, as mentioned earlier. Goal specification is the voluntary mental act that induces the mobilization of the articulators according to a learned, *but flexible* scheme. The observed intertrial, spatio-temporal variability suggests considerable flexibility in just how to attain the goal. The general plan of an action, a memorized internal 'feedforward model', is not to produce a machine-like command, but rather to accommodate proactive, context-dependent adjustments (Blakemore et al., 1999). It is this flexible and continuously updated top–down command that imposes the relative invariant goal achievement. Considerable efforts are presently being made to understand how the central 'model' is updated and reflected in the subordinate kinematic changes. For this, the bimanual paradigm has been used extensively, e.g., velocity changes may occur en route to match initiation delays in one hand. In this case, velocity control would be subordinate to invariant goal achievement (Kazennikov et al., 2002). Such a theoretical framework is useful for understanding pathological changes in patients with movement disorders. Postlesion changes can occur in the course of, and contribute to, functional recuperation (Wiesendanger et al., 2001). Functional brain imaging appears to give us additional clues about postlesion changes in the cortical and subcortical representations (Rowe and Frackowiak, 1999) of normal and abnormal movement.

Another research domain for the future is to learn about neural control aspects of the highest motor skills of *H. sapiens*, which were alluded to in Section 1. In spite of the complexity of such research, first steps in this direction have been quite encouraging (Soechting et al., 1996; Engel et al., 1997).

Acknowledgments

We would like to thank Dr. Becky Farley, The University of Arizona, for reviewing a draft of this chapter. The research reported in this review was supported by: Swiss National Science Foundation (NFP-38, Grant No. 4038-044053), Swiss Multiple Sclerosis Society, Novartis Foundation, Roche Foundation, and INTAS.

Abbreviations

EEG	electroencephalogram
fMRI	functional magnetic resonance imaging
GF	grip force
Hz	Hertz
LF	load force
MRI	magnetic resonance imaging
N	Newton
SMA	supplementary motor area
SII	second somatosensory cortical area

References

Alberts, J.L., Tresilian, J.R. and Stelmach, G.E. (1998). The co-ordination and phasing of a bilateral prehension task—the influence of Parkinson's disease. Brain, 121: 725–742.

Andres, F.G., Mima, T., Schulman, A.E., Dichgans, J., Hallett, M. and Gerloff, C. (1999). Functional coupling of human cortical sensorimotor areas during bimanual skill acquisition. Brain, 122: 855–870.

Bernstein, N. (1935). The co-ordination and regulation of movements (in Russian). Archiv biologiceskich nauk, 38: 1–34.

Bernstein, N.A. (1967). The Co-ordination and Regulation of Movements. Pergamon Press, Oxford.

Bethe, A. (1931). Plastizität und Zentrenlehre. In: Bethe A., Bergmann von G., Embden G. and Ellinger A. (Eds.), Handbuch der Normalen und Pathologischen Physiologie. Arbeitsphysiologie II: Orientierung, Plastizität, Stimme und Sprache. Springer, Berlin, pp. 1175–1221.

Blakemore, S.J., Frith, C.D. and Wolpert, D.M. (1999). Spatio-temporal prediction modulates the perception of self-produced stimuli. J. Cogn. Neurosci., 11: 551–559.

Blakemore, S.-J., Frith, C.D. and Wolpert, D.M. (2001). The cerebellum is involved in predicting the sensory consequences of action. Neuroreport, 12: 1879–1884.

Bucy, P.C. and Fulton, J.F. (1933). Ipsilateral representation in the motor and premotor cortex of monkey. Brain, 56: 318–342.

Castiello, U. and Bennett, K.M.B. (1997). The bilateral reach-to-grasp movement of Parkinson's disease subjects. Brain, 120: 593–604.

Castiello, U., Bennett, K.M.B. and Stelmach, G.E. (1993). The bilateral reach to grasp movement. Behav. Brain Res., 56: 43–57.

Cui, R.Q., Huter, D., Egkher, A., Lang, W., Lindinger, G. and Deecke, L. (2000). High-resolution DC-EEG mapping of the Bereitschaftspotential preceding simple or complex bimanual sequential movement. Exp. Brain Res., 134: 49–57.

Deiber, M.-P., Caldara, R., Ibañez, V. and Hauert, C.A. (2001). Alpha band power changes in unimanual and bimanual sequential movements, and during motor transitions. Clin. Neurophysiol., 112: 1419–1435.

Donchin, O., Gribova, A., Bergman, H. and Vaadia, E. (1996). How do the two hemispheres of the cortex communicate in coordinating the actions of the arms: single unit study in a behaving monkey. Soc. Neurosci. Abstr., 22: 795.2.

Donchin, O., Gribova, A., Steinberg, O., Bergman, H. and Vaadia, E. (1998). Primary motor cortex is involved in bimanual coordination. Nature, 395: 274–278.

Donchin, O., De Oliveira, S.C. and Vaadia, E. (1999). Who tells one hand what the other is doing: the neurophysiology of bimanual movements. Neuron, 23: 15–18.

Drake, C. and Palmer, C. (2000). Skill acquisition in music performance: relations between planning and temporal control. Cognition, 74: 1–32.

Dum, R.P. and Strick, P.L. (1991). The origin of corticospinal projections from the premotor areas in the frontal lobe. J. Neurosci., 11: 667–689.

Engel, K.C., Flanders, M. and Soechting, J.F. (1997). Anticipatory and sequential motor control in piano playing. Exp. Brain Res., 113: 189–199.

Fink, G.R. (2001). What the brain needs for managing both hands at the same time. Neuroreport, 12: A69.

Fink, G.R., Marshall, J.C., Halligan, P.W., Frith, C.D., Driver, J., Frackowiak, R.S. and Dolan, R.J. (1999). The neural consequences of conflict between intention and the senses. Brain, 122: 497–512.

Flanagan, J.R. and Wing, A.M. (1993). Modulation of grip force with load force during point-to-point arm movements. Exp. Brain Res., 95: 131–143.

Flanagan, J.R. and Wing, A.M. (1997). The role of internal models in motion planning and control: evidence from grip force adjustments during movements of hand-held loads. J. Neurosci., 17: 1519–1528.

Franz, E.A., Zelaznik, H.N. and McCabe, G. (1991). Spatial topological constraints in a bimanual task. Acta Psychol., 77: 137–151.

Goerres, G.W., Samuel, M., Jenkins, I.H. and Brooks, D.J. (1998). Cerebral control of unimanual and bimanual movements: an H2(15)O PET study. Neuroreport, 9: 3631–3638.

Hécaen, H. (1978). Les apraxies idéomotrices, essai de dissociation. In: Hécaen H. and Jeannerod M. (Eds.), Du Contrôle Moteur à L'Organisation du Geste. Masson, Paris, pp. 343–358.

Holst von, E. and Mittelstaedt, H. (1950). Das reafferenzprinzip (wechselwirkungen zwischen zentralnervensystem und peripherie). Naturwissenschaften, 37: 464–476.

Horstink, M.W.I.M., Berger, H.J.C., Van Spaendonek, K.P.M., Van den Bercken, J.H.L. and Cools, A.R. (1990). Bimanual simultaneous performance and impaired ability to shift attention in Parkinson's disease. J. Neurol. Neurosurg. Psychiatry, 53: 685–690.

Hugon, M., Massion, J. and Wiesendanger, M. (1982). Anticipatory postural changes induced by active unloading and comparison with passive unloading in man. Pflugers Arch., 392: 292–296.

Hutchins, K.D., Martino, A.M. and Strick, P.L. (1988). Corticospinal projections from the medial wall of the hemisphere. Exp. Brain Res., 71: 667–672.

Jackson, G.M., Jackson, S.R. and Hindle, J.V. (2000). The control of bimanual reach-to-grasp movements in hemiparkinsonian patients. Exp. Brain Res., 132: 390–398.

Jäncke, L., Peters, M., Himmelbach, M., Nösselt, T., Shah, J. and Steinmetz, H. (2000). fMRI study of bimanual coordination. Neuropsychologia, 38: 164–174.

Jeannerod, M. (1984). The timing of natural prehension movements. J. Mot. Behav., 16: 235–254.

Jenny, A.B. (1979). Commissural projections of the cortical hand motor area in monkeys. J. Comp. Neurol., 188: 137–146.

Johansson, R.S. and Westling, G. (1984). Roles of glabrous skin receptors and sensorimotor memory in automatic control of precision grip when lifting rougher or more slippery objects. Exp. Brain Res., 56: 550–564.

Johansson, R.S., Backlin, J.L. and Burstedt, M.K. (1999). Control of grasp stability during pronation and supination movements. Exp. Brain Res., 128: 20–30.

Johnson, K.A., Cunnington, R., Bradshaw, J.L., Phillips, J.G., Iansek, R. and Rogers, M.A. (1998). Bimanual co-ordination in Parkinson's disease. Brain, 121: 743–753.

Kaluzny, P. and Wiesendanger, M. (1992). Feedforward postural stabilization in a distal bimanual unloading task. Exp. Brain Res., 92: 173–182.

Kaluzny, P., Palmeri, A. and Wiesendanger, M. (1994). The problem of bimanual coupling: a reaction time study of simple unimanual and bimanual finger responses. Electroencephalogr. Clin. Neurophysiol., 93: 450–458.

Kazennikov, O., Wicki, U., Corboz, M., Hyland, B., Palmeri, A., Rouiller, E.M. and Wiesendanger, M. (1994). Temporal structure of a bimanual goal-directed movement sequence in monkeys. Eur. J. Neurosci., 6: 203–210.

Kazennikov, O., Hyland, B., Wicki, U., Perrig, S., Rouiller, E.M. and Wiesendanger, M. (1998). Effects of lesions in the mesial frontal cortex on bimanual coordination in monkeys. Neuroscience, 85: 703–716.

Kazennikov, O., Hyland, B., Corboz, M., Babalian, A., Rouiller, E.M. and Wiesendanger, M. (1999). Neural activity of supplementary and primary motor areas in monkeys and its relation to bimanual and unimanual movement sequences. Neuroscience, 89: 661–674.

Kazennikov, O., Perrig, S. and Wiesendanger, M. (2002). Kinematics of a coordinated goal-directed bimanual task. Behav. Brain Res., 134: 83–91.

Kelso, J.A.S. (1995). Dynamic Patterns: The Self-Organization of Brain and Behavior. MIT Press, Cambridge, MA.

Kelso, J.A.S., Southard, D.L. and Goodman, D. (1979). On the nature of human interlimb coordination. Science, 203: 1029–1031.

Kelso, J.A.S., Putnam, C.A. and Goodman, D. (1983). On the space-time structure of human interlimb co-ordination. Q. J. Exp. Psychol., 35A: 347–375.

Kermadi, I., Liu, Y., Tempini, A. and Rouiller, E.M. (1997). Effects of reversible inactivation of the supplementary motor area (SMA) on unimanual grasp and bimanual pull and grasp performance in monkeys. Somatosens. Mot. Res., 14: 268–280.

Kermadi, I., Liu, Y., Tempini, A., Calciati, E. and Rouiller, E.M. (1998). Neuronal activity in the primate supplementary motor area and the primary motor cortex in the relation to spatio-temporal bimanual coordination. Somatosens. Mot. Res., 15: 287–308.

Kuypers, H.G.J.M. (1981). Anatomy of descending pathways. In: Brooks V.B. (Ed.), Handbook of Physiology, The Nervous System, Vol. II, American Physiological Society, Bethesda, pp. 597–666 Pt. 1.

Lashley, K.S. (1933). Integrative functions of the cerebral cortex. Physiol. Rev., 13: 1–42.

Leakey, M.D. (1981). Tracks and tools. Philos. Trans. R. Soc. Lond. B Biol. Sci., 292: 95–102.

Leonard, G., Milner, B. and Jones, L.A. (1988). Performance of unimanual and bimanual tapping tasks by patients with lesions of the frontal or the temporal lobe. Neuropsychologia, 26: 79–91.

Luria, A.R. (1969). Frontal lobe syndromes. In: Vinken P.J.B.G.W. (Ed.), Handbook of Clinical Neurology, Vol. 2, Elsevier, Amsterdam, pp. 725–757.

MacNeilage, P.F. (1987). The evolution of hemispheric specialization for manual function and language. In: Wise S.P. (Ed.), Higher Brain Functions. John Wiley, New York, pp. 285–309.

Newell, K.M. and Corcos, D.M. (1993) Variability and Motor Control. Human Kinetics, Champaign, IL.

Nirkko, A.C., Ozdoba, C., Felblinger, J., Schroth, G., Hess, C.W. and Wiesendanger, M. (1999) Human brain areas involved in bimanual coordination. Neuroimage, Fifth Int. Conf. Functional Mapping of the Human Brain: No. 485.

Paulignan, Y., Dufossé, M., Hugon, M. and Massion, J. (1989). Acquisition of co-ordination between posture and movement in a bimanual task. Exp. Brain Res., 77: 337–348.

Penfield, W. and Jasper, H. (1954). Epilepsy and the Functional Anatomy of the Human Brain. Little, Brown, Boston, pp. 1–896.

Perrig, S., Kazennikov, O. and Wiesendanger, M. (1999). Time structure of a goal-directed bimanual skill and its dependence on task constraints. Behav. Brain Res., 103: 95–104.

Phillips, C.G. (1966). Changing concepts of the precentral motor area. In: Eccles J.C. (Ed.), Brain and Conscious Experience. Springer-Verlag, New York, pp. 389–421.

Phillips, C.G. (1986). Movements of the Hand. Liverpool University Press, Liverpool.

Preilowski, B. (1977). Phases of motor-skills acquisition: a neuropsychological approach. J. Mov. Stud., 3: 169–181.

Repp, B.H. (1999). Detecting deviations from metronomic timing in music: effects of perceptual structure on the mental timekeeper. Percept. Psychophys., 61: 529–548.

Rouiller, E.M., Babalian, A., Kazennikov, O., Moret, V., Yu, X.-H. and Wiesendanger, M. (1994). Transcallosal connections of the distal forelimb representations of the primary and supplementary motor cortical areas in macaque monkeys. Exp. Brain Res., 102: 227–243.

Rouiller, E.M., Moret, V., Tanné, J. and Boussaoud, D. (1996). Evidence for direct connections between the hand region of the supplementary motor area and cervical motoneurons in the Macaque monkey. Eur. J. Neurosci., 8: 1055–1059.

Rowe, J.B. and Frackowiak, R.S. (1999). The impact of brain imaging technology on our understanding of motor function and dysfunction. Curr. Opin. Neurobiol., 9: 728–734.

Sadato, N., Yonekura, Y., Waki, A., Yamada, H. and Ishii, Y. (1997). Role of the supplementary motor area and the right premotor cortex in the coordination of bimanual finger movements. J. Neurosci., 17: 9667–9674.

Schmidt, R.A. (1988) Motor Control and Learning. Human Kinetics, Champaign, IL.

Serrien, D.J. and Wiesendanger, M. (2000). Temporal control of a bimanual task in patients with cerebellar dysfunction. Neuropsychologia, 38: 558–565.

Serrien, D.J. and Wiesendanger, M. (2001a). A higher-order mechanism overrules the automatic grip-load force constraint during bimanual asymmetrical movements. Behav. Brain Res., 118: 153–160.

Serrien, D.J. and Wiesendanger, M. (2001b). Regulation of grasping forces during bimanual in-phase and anti-phase coordination. Neuropsychologia, 39: 1379–1384.

Serrien, D.J., Steyvers, M., Debaere, F., Stelmach, G.E. and Swinnen, S.P. (2000). Bimanual coordination and limb-specific parameterization in patients with Parkinson's disease. Neuropsychologia, 38: 1714–1722.

Serrien, D.J., Nirkko, A.C., Loher, T.J., Lövblad, K.-O., Burgunder, J.-M. and Wiesendanger, M. (2002a). Movement control of manipulative tasks in patients with Gilles de la Tourette syndrome. Brain, 125: 290–300.

Serrien, D.J., Nirkko, A.C. and Wiesendanger, M. (2002b). Role of the corpus callosum in bimanual coordination: a comparison of patients with congenital and acquired callosal damage. Eur. J. Neurosci., 14: 1897–1905.

Simoes, C. and Hari, R. (1999). Relationship between responses to contra- and ipsilateral stimuli in the human second somatosensory cortex SII. Neuroimage, 10: 408–416.

Sloboda, J.A. and Howe, M. (1991). Biographical precursors of musical excellence: an interview study. Psychol. Music, 19: 3–21.

Smith, H.W. (1959). From Fish to Philosopher: The Story of Our Internal Environment. Little, Brown, Boston.

Soechting, J.F., Gordon, A.M. and Engel, K.C. (1996). Sequential hand and finger movements: typing and piano playing. In: Bloedel J.R., Ebner T.J. and Wise S.P. (Eds.), The Acquisition of Motor Behavior in Vertebrates. MIT Press, Cambridge, MA.

Stelmach, G.E. and Worringham, C.J. (1988). The control of bimanual aiming movements in Parkinson's disease. J. Neurol. Neurosurg. Psychiatry, 51: 223–231.

Stephan, K.M., Binkofski, F., Halsband, U., Dohle, C., Wunderlich, G., Schnitzler, A., Tass, P., Posse, S., Herzog, H., Sturm, V., Zilles, K., Seitz, R.J., Freund, H.J. (1999). The role of ventral medial wall motor areas in bimanual co-ordination: a combined lesion and activation study. Brain, 122: 351–368.

Steyvers, M., Verschueren, S.M., Levin, O., Ouamer, M. and Swinnen, S.P. (2001). Proprioceptive control of cyclical bimanual forearm movements across different movement frequencies as revealed by means of tendon vibration. Exp. Brain Res., 140: 326–334.

Swinnen, S.P., Jardin, K., Verschueren, S., Meulenbroek, R., Franz, L., Dounskaia, N. and Walter, C.B. (1998). Exploring interlimb constraints during bimanual graphic performance: effects of muscle grouping and direction. Behav. Brain Res., 90: 79–87.

Toyokura, M., Muro, I., Komiya, T. and Obara, M. (1999). Relation of bimanual coordination to activation in the sensorimotor cortex and supplementary motor area: analysis using functional magnetic resonance imaging. Brain Res. Bull., 48: 211–217.

Treffner, P.J. and Turvey, M.T. (1996). Symmetry, broken symmetry, and handedness in bimanual coordination dynamics. Exp. Brain Res., 107: 463–478.

Viallet, F., Massion, J., Massarino, R. and Khalil, R. (1987). Performance of a bimanual load-lifting task by parkinsonian patients. J. Neurol. Neurosurg. Psychiatry, 50: 1274–1283.

Weiss, P.A. (1969). The living system: determinism stratified. In: Koestler A. and Smythies J.R. (Eds.), Beyond Reductionism. Hutchinson, London, pp. 3–55.

Weiss, P. and Jeannerod, M. (1998). Getting a grasp on coordination. News Physiol. Sci., 13: 70–75.

Welford, A.T. (1968). Fundamentals of Skill. Methuen, London.

Westling, G. and Johansson, R.S. (1984). Factors influencing the force control during precision grip. Exp. Brain Res., 53: 277–284.

Wiesendanger, M., Corboz, M., Hyland, B., Palmeri, A., Maier, V., Wicki, U. and Rouiller, E.M. (1992). Bimanual synergies in primates. In: Caminiti R., Johnson P.B. and Burnod Y. (Eds.), Control of Arm Movement in Space, Exp. Brain Res. Series 22. Springer-Verlag, New York, pp. 45–64.

Wiesendanger, M., Kaluzny, P., Kazennikov, O., Palmeri, A. and Perrig, S. (1994a). Temporal coordination in bimanual actions. Can. J. Physiol. Pharmacol., 72: 591–594.

Wiesendanger, M., Wicki, U. and Rouiller, E.M. (1994b). Are there unifying structures in the brain responsible for interlimb coordination?. In: Swinnen S.P., Heuer H., Massion J. and Casaer P. (Eds.), Interlimb Coordination: Neural, Dynamical and Cognitive Constraints. Academic Press, San Diego, pp. 179–207.

Wiesendanger, M., Rouiller, E.M., Kazennikov, O. and Perrig, S. (1996). Is the supplementary motor area a bilaterally organized system?. In: Lüders H.O. (Ed.), Supplementary Sensorimotor Area. Raven Press, New York, pp. 85–93.

Wiesendanger, M., Nirkko, A.C. and Rösler, K.M. (2001). Post-lesion functional restoration of hand dexterity and the principle of motor equivalence. In: Dengler R. and Kossev A.R. (Eds.), Sensorimotor Control. IOS Press, Amsterdam, pp. 141–149.

Wolpert, D.M., Miall, R.C. and Kawato, M. (1998). Internal models in the cerebellum. Trends Cogn. Sci., 2: 338–347.

Wyke, M. (1971). The effects of brain lesions on the learning performance of a bimanual co-ordination task. Cortex, 7: 59–72.

Zaidel, D. and Sperry, R.W. (1977). Some long-term motor effects of cerebral commissurotomy in man. Neuropsychologia, 15: 193–204.

CHAPTER 47

Functional specialization in dorsal and ventral premotor areas

Eiji Hoshi[1] and Jun Tanji[1,2]*

[1]*Department of Physiology, Tohoku University School of Medicine, Sendai 980-8575, Japan and* [2]*CREST, Japan Science and Technology Corporation, Kawaguchi 332-0012, Japan*

Abstract: The premotor cortex (PM) in the bilateral lateral hemisphere of nonhuman primates and the human has been implicated in the sensorial guidance of movements. This is in contrast to more medial motor areas that are involved more in the temporal structuring of movements based on memorized information. The PM is further subdivided into dorsal (PMd) and ventral (PMv) parts. In this chapter, we describe our attempts to find differences in the use of these two areas in a nonhuman primate for programming future motor actions based on visual signals. We show that neurons in the PMv are involved primarily in receiving visuospatial signals and in specifying the spatial location of the target to be reached. In contrast, neurons in the PMd are involved more in integrating information about which arm to use and the target to be reached. Thus, PMd neurons are more implicated than those of the PMv in the preparation for a future motor action.

Introduction

A large body of evidence has accumulated to show that the primate brain has multiple motor areas that individually are further subdivided into subareas (Matelli et al., 1985; Barbas and Pandya, 1987; Dum and Strick, 1996). The supplementary motor area (SMA) and rostrally adjacent presupplementary motor area (pre-SMA) and supplementary eye field (SEF) are located in the medial part of the frontal cortex, rostral to the primary motor cortex (MI). The cingulate motor areas (CMA), with their rostral and caudal subdivisions, are in the banks of the cingulate sulcus. Functional specializations in all these areas have been thoroughly reviewed (Passingham, 1993; Tanji, 1994, 1996; Rizzolatti et al., 1998; Tanji, 2001; Rouiller et al., this volume). In addition, at least two limb-motor areas are known to exist rostral to MI in the premotor cortex (PM), which is located in the lateral surface of the frontal cortex. Each area,

termed the dorsal and ventral premotor cortex (PMd and PMv), has been suggested to be further divided into subareas (Gentilucci et al., 1988; Rizzolatti et al., 1988; Kurata, 1994; Fujii et al., 2000). The role played by these two areas has been discussed previously (Wise, 1985; Wise et al., 1997; Rizzolatti et al., 1998). Their characteristics are far from understood, however. In this report, we briefly review previous lesion and unit-recording studies, which compared the motor function of these two areas. Finally, we present our view on the selective use of the PMd and PMv in the planning and execution of visually guided limb movements (see also Weber and He, Chapter 45 of this volume).

Overview of studies characterizing the PMd and PMv

In his pioneering lesion study, Passingham (1985) showed that the PM is essential for motor selection based on visual cues. Subsequent lesion studies have revealed functional deficits selective to lesions restricted to the PMd versus PMv. PMv lesions

*Corresponding author: Tel.: +81227178071;
Fax: +81227178077; E-mail: tanjij@mail.cc.tohoku.ac.jp

DOI: 10.1016/S0079-6123(03)43047-1

have revealed attentional deficits in the peripersonal space within a reaching distance (Rizzolatti et al., 1983), and deficits in shift-prism adaptation (Kurata and Hoshi, 1999) and hand-preshaping for grasping objects (Fogassi et al., 2001). A tendency to select an object ipsilateral to the lesioned PMv has also been reported (Schieber, 2000). On the other hand, lesions in the PMd were shown to lead to an inability in planning adequate wrist movements based on visual conditional cues (Kurata and Hoffman, 1994). Importantly, the above deficits were observed in the absence of difficulties in perceiving visual information, or in executing limb movements per se (Passingham, 1985; Kurata and Hoshi, 1999).

Neurons in the PMd and PMv respond to the appearance of visual signals instructing future movements and they exhibit sustained activity during the subsequent motor-set period. Their discharge is related to different aspects of forthcoming motor behavior, however. Activity of PMd neurons reflects forthcoming movement parameters like direction and amplitude (Fu et al., 1993; Kurata, 1993; Crammond and Kalaska, 2000), movement trajectory (Hocherman and Wise, 1990), and a combination of target location and movement trajectory (Shen and Alexander, 1997). On the other hand, activity of PMv neurons reflects target location (Gentilucci et al., 1988; Mushiake et al., 1997), the three-dimensional shape of motor targets (Murata et al., 1997) and peripersonal space (Graziano et al., 1997). Boussaoud and Wise (1993a,b) found differences in the firing patterns PMd and PMv neurons in monkeys performing a behavioral task. Most PMd neurons were more active when the direction of a future movement was instructed, whereas most PMv neurons were more active when a locus of attention was indicated. Furthermore, PMd neuronal activity reflected the direction of movements rather than the characteristics of the visual cues.

In summary, both lesion and unit-recording studies have suggested differences in the use of PMd and PMv for the visual guidance of movements and for instructing future actions by use of visual information. We describe in the next section, the extension of these findings by defining more precisely how each of the two areas is involved in retrieving information out of visual signals and in formulating plans for future actions.

Specification of target-choice and effector-choice in PMd versus PMv

When intending to initiate an action, an object in the outside world is selected and a decision made concerning which part (effector) of the body is to be used for the action. For example, to plan an arm-reach movement, different sets of information are required for the selection (choice) of both the target to reach and the arm to use. These sets must then be integrated in order to specify an appropriate motor program for the arm-reach. If the source of information is visual signals, then the most plausible site of information processing for the motor planning is neuronal circuitry within the PM. With this construct in mind, we designed a series of experiments to investigate the role of PMd versus PMv in retrieving, processing and integrating information to plan arm-reaching movements.

We trained two monkeys (*Macaca fuscata*) to plan and prepare future movements following two instruction cues (Hoshi and Tanji, 2000). One cue indicated the arm-choice (right vs. left) for the reaching movement, and the other indicated the target-choice (again right vs. left). Each instruction consisted of a color cue and a white square. The color cue indicated whether the instruction was for arm-choice or for target-choice, and the location of the white square indicated the side (right vs. left) for arm-choice or target-choice. Since the two instructions were presented separately and sequentially in a randomized order, the monkeys were required to (1) retrieve information for arm-choice versus target-choice indicated by the first cue, (2) retain this information, (3) retrieve information indicated by the second cue, and (4) integrate the first and second set of information in order to plan and prepare future movements.

Neuronal activity was recorded in the PMd and PMv while the monkey was performing the task. After the appearance of the first cue, different groups of PMd neurons extracted and retained specific information about arm-choice and target-choice. For example, Fig. 1 shows that the activity of Cell 1 (in the PMd) was most active when the first cue instructed right-arm choice. On the other hand, activity of Cell 2 (also in the PMd) was prominent

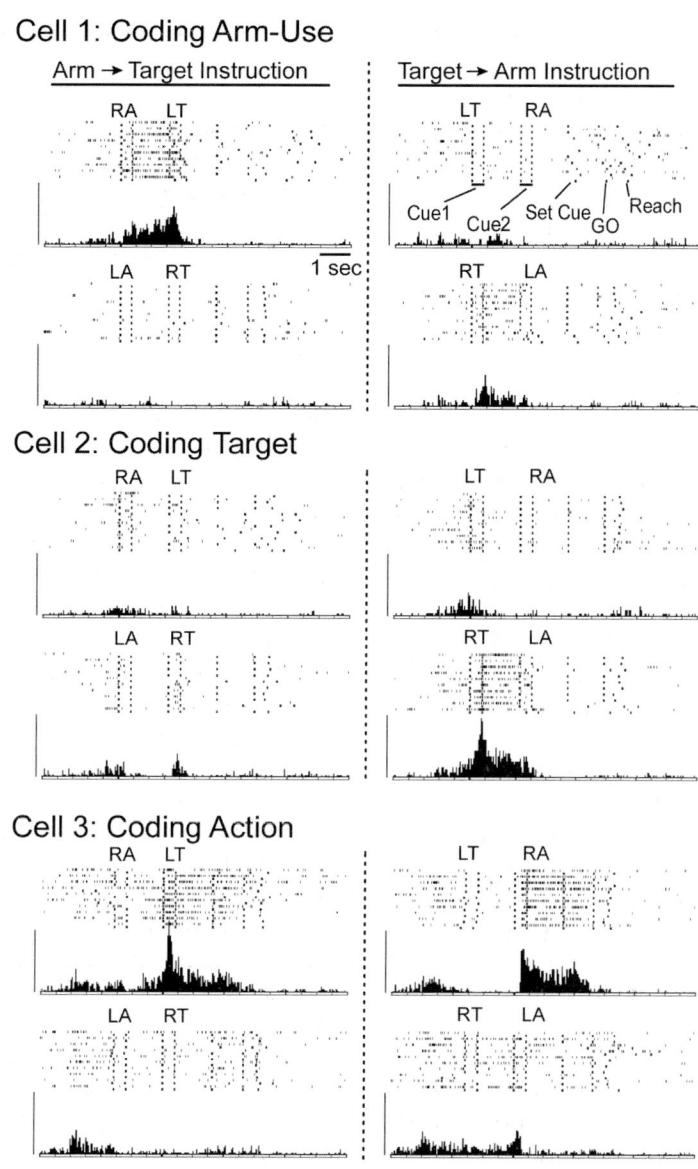

Fig. 1. Three examples of neuronal activity in the PMd. Cell 1, arm-choice activity. In the rasters, each row represents a trial and each dot shows when the cell discharged. The seven marks under each trial represent when the task events appeared, as denoted in the left panel. The first and second instructions in each panel are shown above the rasters (RA, right arm; LA, left arm; RT, right target; LT, left target). In the perievent histograms (bin width, 40 ms) below the rasters, the ordinates represent the number of discharges per second. Scale bar, 1 s. Cell 2, target-choice activity. Cell 3, action-selective activity. This latter neuron responded to the appearance of the second cue if the two instructions specified 'right arm' and 'left target' (top row). (Reproduced, in part, from Hoshi and Tanji, 2000 with permission from Macmillan Magazines Ltd.)

when the first cue instructed reaching to the correct target. The existence of these two types of activity suggested that PMd neurons selectively collect arm-choice and target-choice information, and retain this information before the second information (instruction) is provided.

In contrast to PMd cells, those tested in PMv were mainly selective during the first cue for the physical location of the white square, with far fewer encountered that selectively represented arm-choice or target-choice information. The second-cue responses were also different for PMd versus PMv cells. Many neurons in PMd, but not PMv, integrated the two sets of information and specified future action. For example, Fig. 1 shows that Cell 3 (another PMd cell) was active in the second delay if the two cues indicated right-arm choice and left-target choice, regardless of the order of instruction presentation. This result suggested that the PMd is more involved than PMv in integrating the arm-choice and target-choice information. Furthermore, during the motor-set period, while the monkey prepared a future movement just before the appearance of the GO signal, the two factors were differently represented in PMd versus PMv. Selectivity for the target was more represented in the PMv. In contrast, selectivity for the arm-choice was more represented in the PMd. Interestingly, the number of PMd neurons encountered that were selective for arm-choice versus target-choice was quite similar, thereby possibly further supporting the concept that the integration of arm-choice and target-choice occurs in PMd. In contrast, PMv neurons mainly represented target-choice information rather than arm-choice.

The above results indicate that neurons in PMd have a capacity to selectively represent information about arm-choice versus target-choice. Once the two sets of information become available in PMd, they are quickly integrated. During the motor-set period, PMd neurons also achieve a representation of future action (i.e., a hybrid of arm-choice and target-choice information). In contrast, neurons in PMv mainly represented right versus left aspects of the two instruction cues, and, during the motor-set period, they tend to represent location of the target, regardless of the effector (arm) to be used.

Functional specialization in PMd and PMv

As reviewed in the previous section, many lines of evidence now point to functional specificity within the PMd and PMv. Based on these results, we propose some functional differences between PMd and PMv neurons and their circuitry. PMd retrieves and holds both target information and body-part information, and integrate the two sets of information to plan an action. PMv mainly represents the nature of targets, regardless of the specific effectors required to reach them.

How does the above concept relate to the broad array of previous findings on these two structures? The loss of sensorimotor integration in planning movements after a PMd lesion can be viewed as a deficit in retrieving information and/or in integrating multiple sets of information, such as the choice of an effector to reach a specific target. Many aspects of neuronal activity found in the PMd can be viewed as collecting and integrating component information of movements, and planning and preparing future movements. On the other hand, a PMv-lesion-evoked deficit in locating a target or in specifying a three-dimensional shape can be explained as a failure in representing properties of targets that are relevant to an intended action. It has been shown that neurons in PMv represent the nature of targets for action, such as the location of motor targets or their three-dimensional shape (Rizzolatti et al., 1998; Hoshi and Tanji, 2001). These functional properties, which are unique to the PMd and PMv, seem to be provided by the wealth of anatomical connections with other areas in the CNS, such as the parietal and prefrontal cortex. In future research, it will be necessary to clarify how the PMd and PMv operate as part of global networks for the control of movement.

Acknowledgments

This work was supported by Japan Science and Technology Corporation.

Abbreviations

CMA	cingulate motor area
MI	primary motor cortex
PM	premotor cortex
SEF	supplementary eye field
SMA	supplementary motor area
Pre-SMA	presupplementary motor area
PMd	dorsal premotor cortex
PMv	ventral premotor cortex

References

Barbas, H. and Pandya, D.N. (1987). Architecture and frontal cortical connections of the premotor cortex (area 6) in the rhesus monkey. J. Comp. Neurol., 256: 211–228.

Boussaoud, D. and Wise, S.P. (1993a). Primate frontal cortex: effects of stimulus and movement. Exp. Brain Res., 95: 28–40.

Boussaoud, D. and Wise, S.P. (1993b). Primate frontal cortex: neuronal activity following attentional versus intentional cues. Exp. Brain Res., 95: 15–27.

Crammond, D.J. and Kalaska, J.F. (2000). Prior information in motor and premotor cortex: activity during the delay period and effect on pre-movement activity. J. Neurophysiol., 84: 986–1005.

Dum, R.P. and Strick, P.L. (1996). Spinal cord terminations of the medial wall motor areas in macaque monkeys. J. Neurosci., 16: 6513–6525.

Fogassi, L., Gallese, V., Buccino, G., Craighero, L., Fadiga, L. and Rizzolatti, G. (2001). Cortical mechanism for the visual guidance of hand grasping movements in the monkey: a reversible inactivation study. Brain, 124: 571–586.

Fu, Q.G., Suarez, J.I. and Ebner, T.J. (1993). Neuronal specification of direction and distance during reaching movements in the superior precentral premotor area and primary motor cortex of monkeys. J. Neurophysiol., 70: 2097–2116.

Fujii, N., Mushiake, H. and Tanji, J. (2000). Rostrocaudal distinction of the dorsal premotor area based on oculomotor involvement. J. Neurophysiol., 83: 1764–1769.

Gentilucci, M., Fogassi, L., Luppino, G., Matelli, M., Camarda, R. and Rizzolatti, G. (1988). Functional organization of inferior area 6 in the macaque monkey. I. Somatotopy and the control of proximal movements. Exp. Brain Res., 71: 475–490.

Graziano, M.S., Hu, X.T. and Gross, C.G. (1997). Visuospatial properties of ventral premotor cortex. J. Neurophysiol., 77: 2268–2292.

Hocherman, S. and Wise, S.P. (1990). Trajectory-selective neuronal activity in the motor cortex of rhesus monkeys (Macaca mulatta). Behav. Neurosci., 104: 495–499.

Hoshi, E. and Tanji, J. (2000). Integration of target and body-part information in the premotor cortex when planning action. Nature, 408: 466–470.

Hoshi, E. and Tanji, J. (2001). Contrasting neuronal activity in the dorsal and ventral premotor areas during preparation to reach. J. Neurophysiol., 87: 1123–1128.

Kurata, K. (1993). Premotor cortex of monkeys: set- and movement-related activity reflecting amplitude and direction of wrist movements. J. Neurophysiol., 69: 187–200.

Kurata, K. (1994). Information processing for motor control in primate premotor cortex. Behav. Brain Res., 61: 135–142.

Kurata, K. and Hoffman, D.S. (1994). Differential effects of muscimol microinjection into dorsal and ventral aspects of the premotor cortex of monkeys. J. Neurophysiol., 71: 1151–1164.

Kurata, K. and Hoshi, E. (1999). Reacquisition deficits in prism adaptation after muscimol microinjection into the ventral premotor cortex of monkeys. J. Neurophysiol., 81: 1927–1938.

Matelli, M., Luppino, G. and Rizzolatti, G. (1985). Patterns of cytochrome oxidase activity in the frontal agranular cortex of the macaque monkey. Behav. Brain Res., 18: 125–136.

Murata, A., Fadiga, L., Fogassi, L., Gallese, V., Raos, V. and Rizzolatti, G. (1997). Object representation in the ventral premotor cortex (area F5) of the monkey. J. Neurophysiol., 78: 2226–2230.

Mushiake, H., Tanatsugu, Y. and Tanji, J. (1997). Neuronal activity in the ventral part of premotor cortex during target-reach movement is modulated by direction of gaze. J. Neurophysiol., 78: 567–571.

Passingham, R.E. (1985). Premotor cortex: sensory cues and movement. Behav. Brain Res., 18: 175–185.

Passingham, R.E. (1993). The Frontal Lobes and Voluntary Action. Oxford University Press, Oxford.

Rizzolatti, G., Matelli, M. and Pavesi, G. (1983). Deficits in attention and movement following the removal of postarcuate (area 6) and prearcuate (area 8) cortex in macaque monkeys. Brain, 106: 655–673.

Rizzolatti, G., Camarda, R., Fogassi, L., Gentilucci, M., Luppino, G. and Matelli, M. (1988). Functional organization of inferior area 6 in the macaque monkey. II. Area F5 and the control of distal movements. Exp. Brain Res., 71: 491–507.

Rizzolatti, G., Luppino, G. and Matelli, M. (1998). The organization of the cortical motor system: new concepts. Electroencephalogr. Clin. Neurophysiol., 106: 283–296.

Schieber, M.H. (2000). Inactivation of the ventral premotor cortex biases the laterality of motoric choices. Exp. Brain Res., 130: 497–507.

Shen, L. and Alexander, G.E. (1997). Preferential representation of instructed target location versus limb trajectory in dorsal premotor area. J. Neurophysiol., 77: 1195–1212.

Tanji, J. (1994). The supplementary motor area in the cerebral cortex. Neurosci. Res., 19: 251–268.

Tanji, J. (1996). New concepts of the supplementary motor area. Curr. Opin. Neurobiol., 6: 782–787.

Tanji, J. (2001). Sequential organization of multiple movements: Involvement of Cortical Motor Areas. Annu. Rev. Neurosci.: 631–651.

Wise, S.P. (1985). The primate premotor cortex: past, present, and preparatory. Annu. Rev. Neurosci., 8: 1–19.

Wise, S.P., Boussaoud, D., Johnson, P.B. and Caminiti, R. (1997). Premotor and parietal cortex: corticocortical connectivity and combinatorial computations. Annu. Rev. Neurosci., 20: 25–42.

CHAPTER 48

Spatially directed movement and neuronal activity in freely moving monkey

Yuan-Ye Ma[1,2], Jae-Wook Ryou[2,3], Byoung-Hoon Kim[2,3] and Fraser A.W. Wilson[2-4]*

[1]*Kunming Institute of Zoology, Kunming, China*
[2,3]*ARL Division of Neural Systems, Memory and Aging, and Department of Psychology, University of Arizona, Tucson, AZ 85721, USA*
[4]*Department of Neurology, University of Arizona College of Medicine, Tucson, AZ 85724, USA*

Abstract: The abilities to plan a series of movements and to navigate within the environment require the functions of the frontal and ventromedial temporal lobes, respectively. Neuropsychological studies posit the existence of egocentric (prefrontal) and allocentric (ventromedial temporal) spatial frames of reference that mediate these functions. To examine neural mechanisms underlying egocentric and allocentric guidance of movement, we have developed behavioral and neurophysiological techniques for freely moving monkey. In this chapter, we provide evidence that the dorsolateral prefrontal cortex is important for egocentric spatial tasks in both the visual and tactile modalities, but it does not contribute to performance of an allocentric spatial task. Moreover, neurophysiological recordings indicate that prefrontal neurons are involved in monitoring the spatial nature of behavioral sequences in an egocentric memory task. In contrast, hippocampal neurons are active during spatially directed locomotion, apparently reflecting the monkey's location in a testing room. This discharge is independent of the task's contingencies.

Introduction

The focus of our research is the planning and execution of spatially directed movements. Seminal studies (Butters and Pandya, 1969; Goldman and Rosvold, 1970) showed that the dorsolateral prefrontal cortex (dlPFC) is essential for performing tasks in which arm movements are directed to remembered spatial locations. Neurophysiologists (Fuster and Alexander, 1971; Kubota and Niki, 1971) recorded dlPFC neurons that were active in delayed response tasks, indicative of memory for the location of a target or direction of a response.

Many questions arise from these observations. Here, we identify two. First, is the delay-related neuronal activity due to the impending movement or to the spatial location? Wilson et al. (1993) showed that both kinds of neurons exist. This observation provides the basis to probe the spatial memory mechanism and its relationship to planned action, the domain of frontal lobe function (Fuster, 1997; Goldman-Rakic, 1987). Second, does dlPFC participate in tasks requiring memorized spatially directed body movements? This chapter describes evidence for such participation when the movements are part of a planned sequence.

We begin by noting that knowledge about planned sequences of spatially directed movement is limited compared to information on ballistic, pursuit and locomotor movements. When a movement occurs, its spatial goal can be defined in several ways: (1) a set of coordinated muscle movements of particular amplitude, timing, etc.; (2) the current position of the effector (eye-centered, arm-centered, shoulder-centered); and (3) an absolute set of coordinates

*Corresponding author: Tel.: +520-626-2374; Fax: +52-621-9306; E-mail: fraser@u.arizona.edu

centered on the body. This list is not exhaustive. Recent research provides evidence for brain mechanisms underlying the execution of sequences of arm movements (Funahashi et al., 1997; Hikosaka et al., 1999), though little is known about how a subject organizes movement sequences such as planning a route to the coffee room after attending a scientific session. Neuropsychologists, however, have long recognized that damage to certain brain regions impairs the planning of sequences and route-finding abilities. Pohl (1973) noted that the spatial position of a target may be defined by reference to the position of the observer, or relative to other objects functioning as landmarks—the egocentric and allocentric dichotomy. Pohl's research indicated that egocentric and allocentric mechanisms were mediated by frontal and parietal cortices. Although primate studies have not subsequently addressed this topic, human neuropsychologists often use the egocentric/allocentric dichotomy to describe disorders resulting from brain damage.

Update on the egocentric/allocentric dichotomy

Recent research indicates that the temporal, parietal and frontal lobes play different roles in spatially directed movement. Accumulating evidence indicates that the hippocampus, and the ventromedial parahippocampal and caudal temporal gyri, play a role in the ability to find one's way in the environment. Neuropsychological studies have identified several disorders of allocentric space: place agnosia, memory impairments for landmarks, and dysfunction of the memory for spatial relationships between buildings and landmarks (Maguire et al., 1996). Adding weight to these observations, functional imaging studies have shown that hippocampal activity is recruited specifically for navigation in large-scale environments (Maguire et al., 1997). These observations provide evidence for the existence of an allocentric representation of space, which does not require processing in the inferior frontal region.

Disorders (neglect) of visual and tactile space are the hallmark of damage to the right parietal cortex. Studies indicate a role in spatially directed attention, movements and memory. It is often difficult to separate these functions (limited damage is rare) but certain studies have found it possible, presumably because brain damage was fortuitously circumscribed. For example, Bisiach et al. (1993) cite cases in which subjects attempt to describe the spatial organization of a familiar visual scene, such as a town. They omit description, however, of objects in the left half of visual space, contralateral to the damage. Remarkably, when asked to visualize the scene from the opposite vantage point, they again describe the objects on the right, objects that were omitted in the previous description. Evidently, allocentric representation of space exists, but *the ability to use that information is limited to one-half of visual space, i.e., an egocentric perspective*. Bisiach et al. (1993) also showed that such patients are deficient in noting left turns when describing a route between two locations.

According to Stark et al. (1996), parietal damage "appears to disrupt an egocentric spatial representation system, or mastermap of locations". Their patient was severely impaired in reaching to the remembered location of visual targets using proprioceptive information, although was able to do so with visual guidance. This comprehensive study provides compelling evidence of a memory mechanism for spatial location; indeed, the subject's memory of the spatial layout of her home was distorted. Other studies (Denes et al., 2000) have found specific deficits in encoding body position after parietal damage.

Finally, research into the frontal lobe has demonstrated neuronal activity occurring when monkeys are memorizing the location of a spatial target (Funahashi et al., 1989). But what exactly does this memory-related activity reflect? It is clear that parietal and frontal lobes (Stark et al., 1996; Walker et al., 1998; Wojciulik et al., 2001) contribute to spatially directed behavior as evidenced by the effects of lesions. But what then does the frontal lobe contributes that is different from the parietal lobe? Reviews of many studies indicate that the frontal lobe is involved in the planning of complex sequences of behavior (Stuss and Benson, 1986; Fuster, 1997). From this position, we formulated a program to examine the notion of egocentric and allocentric spatially directed mechanisms. We developed variants of spatial delayed response (SDR) tasks that are known to involve both prefrontal cortex and the hippocampal system (Stamm, 1987; Murray et al.,

1989). The monkeys were trained both in their cages and in a freely moving condition. These two conditions were designed to promote the use of egocentric and allocentric forms of spatial location, respectively. Note that cage training always provides a single perspective, encouraging the use of egocentric mechanisms. In the freely moving condition, however, the testing encourages the encoding of spatial location in relation to environmental cues.

Prefrontal lesions selectively disrupt egocentric performance

Monkeys were trained in three versions of the SDR task (Ma et al., 2003). First, in the Wisconsin spatial delayed response (W-SDR) task, each trial required memory for a spatial cue: the left or right position of a reward hidden on a food board placed in front of their cages. At the end of a 25-s delay, they retrieved the food from the remembered location. This task is virtually identical to those used previously in classical studies. Second, a tactile version of the task was learned: monkeys extended both hands to the experimenter, who then touched the left or the right hand. This cue had to be remembered over a delay and the cued hand was used to make the response.

Third, in the allocentric (A-SDR) task, monkeys had to remember the spatial location of a reward hidden between a variety of large objects (buildings, trees, etc.) located outside the buildings of a primate housing facility. Monkeys were leashed, and taken into a compound (Fig. 1) where rewards were placed on the ground while they watched. The monkey was then walked to a location out of sight of the reward, involving at least one turn of a corner so that the reward was hidden from sight. The monkey was then allowed to find the hidden reward. This task required spatial memory for periods of 25 s.

Monkeys were tested pre- and postoperatively after aspiration lesions of cortex dorsolateral to the principal sulcus (Fig. 1B). *The lack of impairment in performing the A-SDR task was striking.* Preoperatively, monkeys averaged 81% correct; postoperatively, monkeys averaged 84%, and the scores did not differ significantly. The lesioned monkeys had excellent retention of the location of the reward sites, even though experiencing substantial delays, the distraction of events in the compound, and continual locomotion both away from and returning to the reward sites. Moreover, in order to locate the reward, the monkey's route resulted in the experience of different sets of visual cues on the outward versus inward journey, as well as changes in direction that differed by 180°. This route-finding behavior suggests that the monkeys performed the task using intact memories of a mental map of the compound, rather than solving the task based on a set of body turns, or navigation using an invariant set of visual cues which, in fact, differed on the outward versus inward path.

In contrast to the lack of impairment in the A-DR task, prefrontal damage severely impaired the W-SDR task for delay intervals averaging 25 s. Although the monkeys had only two possible choices, a constant visual environment, and were not moved during the delay periods, their postoperative average score was 63% correct, compared to the preoperative scores of 90% correct. These differences were all highly significant. Moreover, the W-SDR task was tested on the same days as the A-DR, but prior to it, and thus the monkey must have been particularly motivated to work on the egocentric task. Evidently, the lesions destroyed brain tissue that was essential for performance of the W-SDR, thereby confirming the importance of dlPFC in this performance. We consider this task to be egocentric because the subject's viewing perspective is constant, and because there are no substantial environmental cues to enable the monkey to perform the left/right choice.

The tactile version of the SDR is also an egocentric task. It requires memory for a tactile cue to the hand, and in particular, whether the touch is to the left versus right hand. This tactile cue has to be memorized and appropriately related to the body schema. Monkeys were severely impaired in this task, with average delays of 25 s. Their preoperative scores averaged 86% correct, compared to a significantly different postoperative average of 65% correct. Although the tactile cue is a sensory stimulus, there is nothing in the environment that serves to aid (or interfere with) the task of remembering the cue during the delay period, once the cue is presented.

It is apparent, then, that the W-SDR task requires spatial abilities that are fundamentally different from

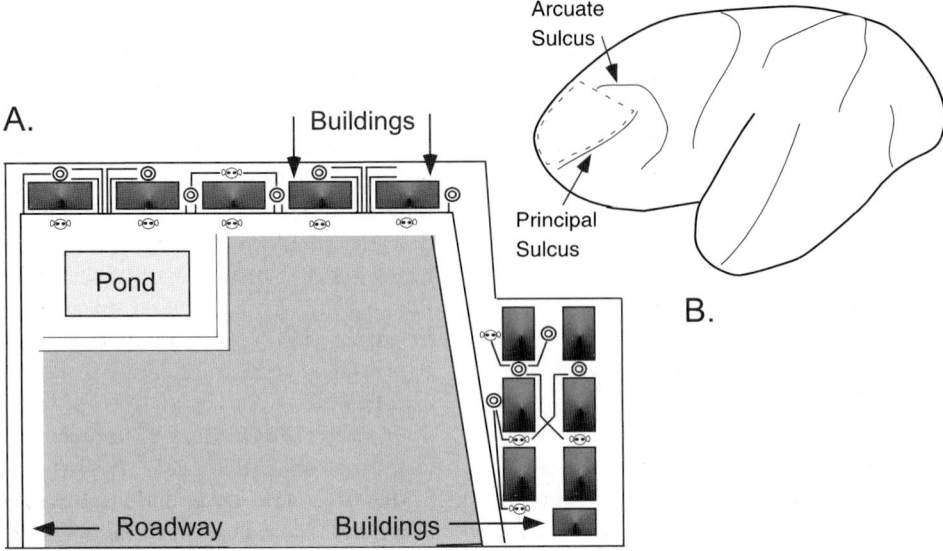

Fig. 1. Testing site for the A-SDR task in FMMs. (A) Food (bull's eyes) was hidden around the 12 buildings in view of the monkeys. After watching, monkeys (small faces) were walked to a new location and then released to find the food. (B) Lesions of the mid-dlPFC. Cortical removal (the region within the dotted lines) includes the dorsal lip of Area 46, Areas 9–46 (Petrides and Pandya, 1999), and medial Area 9.

those required in the A-SDR task. Also, impairment in performing the W-SDR task is not due to a fundamental loss of allocentric spatial memory (i.e., the ability to remember spatial relations between objects). In that the A-DR task requires memory for 'allocentric spatial location'—which is clearly intact after the lesions—it follows that the performance of the W-SDR task does not depend upon memory for location as determined by geometrical relations between stimuli. What, then, is the spatial frame of reference represented in the operation of prefrontal cortex? One possibility is that dlPFC mediates the W-SDR task by relating the spatial position of the current reward site (left vs. right) to the body image, i.e., egocentric spatial memory.

Our lesion study conclusions are tentative. For example, it is arguable that the allocentric and egocentric tasks are not directly comparable in terms of distractability. Some objections can be discarded, however, e.g., that the allocentric task is easier. Rather, preoperative performance was worse on the A-SDR (80%) than on W-SDR (90%) task. It would be interesting to carry out these tests on monkeys with lesions of the hippocampus and parahippocampal gyrus.

Neuronal activity in the dorsolateral prefrontal cortex

Our lesion study indicates that dlPFC mediates spatially directed movement referenced to the body, not to environmental cues. We explored this hypothesis in more detail by developing techniques for neuronal recordings in freely moving monkeys (FMMs), who walked between four food dispensers located close to the walls of a 15′ × 15′ room (see Fig. 3).

The delayed alternation (DA) task was selected because it requires (1) movements in different directions, (2) memory for a previously performed response, and (3) planning and execution of multiple consecutive responses. These are conditions that elicit impairments in patients with prefrontal damage (Petrides and Milner, 1982; Lepage and Richer, 1996) who have difficulty in initiating movements, completing motor sequences and inhibiting inappropriate responses (Kertesz, 1994). Finally, performance of the DA task is impaired in FMMs with prefrontal and hippocampal system lesions (Stamm, 1987; Murray et al., 1989).

The task requires a sequence of movements between the west–north–east–north–west dispensers (north task) and between the west–south–east–south-west dispensers (south task). The tasks are switched after 30 trials. On arrival at the dispenser, the monkeys lift a lid and retrieves food. Two light-emitting diodes are placed on the head and used to track head position and head direction. A head-mounted videocamera provides a view of the monkey's head direction.

Recordings take place in the prefrontal cortex in a region 22–34 mm anterior to the interaural line, and from 4–16 mm lateral to the midline. Based on MRI images and X-radiographs, this region includes Areas 9 and 46.

A variety of neuronal types have been recorded, but we have not found neuronal activity related to simple movements, during locomotion, nor to place information that could directly contribute to allocentric representations. In contrast, hippocampal place fields are typically found dispersed throughout a testing apparatus (O'Keefe, 1999), rather than focused on landmarks (e.g., food dispensers).

We characterize dlPFC neurons as being responsive at different points in the task sequence, and usually in relation to the dispensers. Two examples illustrate these responses. Figure 2 shows neuronal activity that is maximal shortly before the monkey opens the lid of the north dispenser. This activity continues until the monkey has turned and begun to move toward the west dispenser. Monkeys spend approximately 3 s at each dispenser: their heads reach the side of the dispenser, they sit, open the lid with the hand on the side of the dispenser which they approached, use the other hand to retrieve the food, and then their heads rotate to identify the next target feeder. In making a turn to locate the next dispenser, they face the walls, and not the center of the room. The responses of the neuron shown in Fig. 2 are best correlated with food retrieval, but the activity is selective for one of the four dispensers (food retrieval occurs at each dispenser). Such selective responses are observed routinely.

There are many possible explanations for the highly selective activity shown in Fig. 2. The explanation we favor is that it reflects monitoring of the behavioral sequence underlying the task (Petrides, 1995). This contention is supported by the finding that the selectivity markedly declines when target selection is done randomly under visual guidance, rather than as an internal sequence. We can also discard other explanations like vision, because all the feeding dispensers, their lids, movement of the carousels, etc., are identical, so there is little to cue

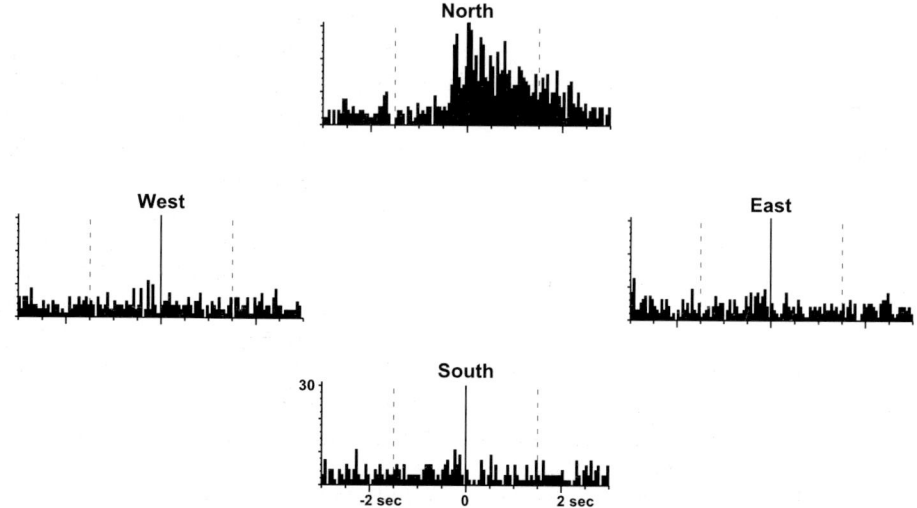

Fig. 2. Responses of a prefrontal neuron during performance of the DA task by a FMM. The cell was selectively responsive in proximity to the north dispenser. Neuronal firing is aligned with the monkey's lid opening at each of the four dispensers. Shown are trials in which the approach is from a single direction; north, from the east; south, from the east; west, from the north; and east, from the north. Apart from the approach to the north dispenser, the neuron was inactive. Trials were done in sequential order. Time 0 represents the opening of the dispenser. Dashed lines represent approximate times that the monkey was within one foot of the target.

the monkey as to the identity of individual dispensers. Note also that (1) head turns to the right or left occur at all dispensers, (2) arm movements using left and right hands occur at all dispensers, and (3) posture is the same in all dispensers. What is different is that the monkeys always start at the west dispenser, perform the north task for 30 trials and then switch to the south task, etc. Thus, the task has an important behavioral asymmetry (i.e., start point), which presumably cues the monkeys as to position within the sequence that forms the structure of the SDR task.

Another example is shown in Fig. 3. In this case, the neuron was active for movements in the southerly direction, and was most active as the monkey arrived at the south dispenser from either the west or the east. Neuronal activity was suppressed by movement toward the north, northwest or northeast. One explanation is that this activity resembles that of 'head-direction' cells (Taube et al., 1990; see also chapters of Grantyn et al., Sasaki, and Peterson, this volume). It is evident that information about the direction of the monkey's movement is reflected in the dlPFC, and it is possible that allocentric information influences these neurons.

Neuronal activity in the hippocampus

Recordings were made in the hippocampus as monkeys walked to different locations within the FMM room, directed by visual cues on the feeding dispensers or in the DA memory task. Computer screens in each corner displayed allocentric visual cues. As in our earlier studies (Wilson et al., 1990), neuronal indicated the importance of spatially directed behavioral responses. Our previous work was carried out in chaired monkeys, however, while the present study used freely moving animals. Our current hippocampal recordings target a region from 2 to 14 mm anterior to the interaural line, and from 6 to 18 mm lateral to the midline.

Figure 4 illustrates a responsive neuron. This cell was maximally active as the monkey approached the east dispenser from the south. Neuronal firing increased as the monkey approached the dispenser and decreased before arrival at the target. These responses were directional as firing rate did not increase when the monkey went from east to south, though other neurons were not directional. Like studies in the rat (O'Keefe, 1999), we have found neurons that fire maximally at random (with respect to feeding dispensers) locations within the testing room (see also Georges-François et al., 1999). Thus, while some responsive hippocampal neurons are goal directed, with discharge terminating on the monkey's arrival or leaving the feeding dispenser, other cells' discharge appears to be spatially selective with respect to the testing room and independent of the task. These hippocampal responses are unlike those observed in dlPFC.

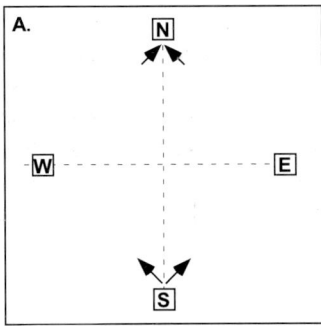

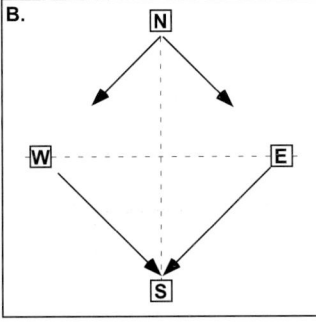

Fig. 3. Directional neuronal responses during performance of the DA task. (A) Neuronal activity is suppressed during northward movements. (B) Activity increases during all southward movements, and is maximal when the monkey is heading for the south dispenser. It decreases when the monkey leaves. Arrows are vectors, representing neuronal activity for particular directions with the strength of the response indicated by the length of the arrows.

Conclusions

We have taken preliminary steps in investigating neural mechanisms underlying spatially directed behavior in FMMs. The clearest data are from our lesion study—there is a clear dissociation indicating

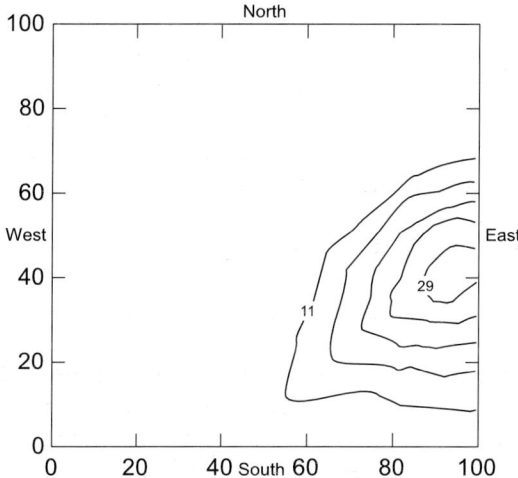

Fig. 4. Hippocampal neuron activity during the DA task. The cell became active as the monkey approached the east dispenser from the south. X and Y axes represent a 100 × 100 spatial grid. Contour lines represent firing rates, ranging between 29 and 11 spikes/pixel within a specific region of the testing room in which the neuron was inactive. There was no discharge when the monkey was in other regions.

the involvement of the dlPFC in egocentric, but not allocentric spatial tasks.

Neurophysiological recordings in dlPFC reveal striking correlates between spatially directed behavior and neuronal activity. The dlPFC neurons are responsive in relation to the spatial and sequential components of a task. One set of neurons is characterized by selectivity during actions in the task—each neuron is most responsive at one, some but not all of the dispensers. One explanation is that these cells reflect a mechanism for monitoring sequences of recent actions, which is impaired in both the monkey and man (Petrides and Milner, 1982; Petrides, 1995). This interpretation is reasonable at this stage because it is consistent with the dorsal location of the recordings, in brain regions that are usually quiescent in monkeys performing tasks while seated (Wilson et al., 1993; ÓScalaidhe et al., 1999). An important future experiment is to compare responses in both the chaired and freely moving condition.

Other dlPFC neurons reflect the orientation of movement within the room (Fig. 3). Similar observations have been made in the postsubiculum (Taube et al., 1990), and recently described connections between Area 9 and the presubicular region (Morris et al., 1999) could provide pathways carrying spatial information. Experiments are needed to identify the cues (including allocentric ones) that enable the brain to derive the signal for orientation, and examine the relationship between neuronal firing and head angle within the room.

Hippocampal neurons can show directional activity, but of a different form to the dlPFC. Certain neurons are active in movements on the monkey's approach and departure relative to the feeding dispensers. Other neurons are active at random locations within the testing room, rather than to monkey actions around the dispenser, per se. Such responses resemble those of place-field neurons found in the rat hippocampus (O'Keefe, 1999). A next step will be to determine the importance of allocentric cues for this neuronal activity.

Collectively, neuropsychological and anatomical studies are providing insights into neural mechanisms underlying spatially directed behavioral sequences. Based on neuropsychological studies, it appears that topographical knowledge is developed within the ventromedial temporal lobes. Parietal lobes provide a schema of the body, and the frontal lobes develop plans of action within the framework provided by the temporal and parietal lobes.

Acknowledgments

We greatly appreciate the support of the Whitehall Foundation (A98-04) and the NIH (MH 58415).

Abbreviations

DA   delayed alternation task
dlPFC   dorsolateral prefrontal cortex
FMM   freely moving monkey
SDR   spatial delayed response

References

Bisiach, E., Brouchon, M., Poncet, M. and Susconi, M.L. (1993). Unilateral neglect in route description. Neuropsychologia, 11: 1255–1262.

Butters, N. and Pandya, D. (1969). Retention of delayed-alternation. Science, 165: 1271–1273.

Denes, G., Cappelletti, J.Y., Zilli, T., Dalla Porta, F. and Gallana, A. (2000). A category-specific deficit of spatial representation: the case of autotopagnosia. Neuropsychologia, 38: 345–350.

Funahashi, S., Bruce, C.J. and Goldman-Rakic, P.S. (1989). Mnemonic coding of visual space in the monkey's dorsolateral prefrontal cortex. J. Neurophysiol., 61: 331–349.

Funahashi, S., Chafee, M.V. and Goldman-Rakic, P.S. (1993). Prefrontal neuronal activity in rhesus monkeys performing a delayed anti-saccade task. Nature, 365: 753–756.

Funahashi, S., Inoue, M. and Kubota, K. (1997). Delay-period activity in the primate prefrontal cortex encoding multiple spatial positions and their order of presentation. Behav. Brain Res., 84: 203–223.

Fuster, J.M. (1997). The Prefrontal Cortex. Anatomy, Physiology and Neuropsychology of the Frontal Lobe, 3rd ed., Lippincott-Raven, Philadelphia.

Fuster, J.M. and Alexander, G.E. (1971). Neuron activity related to short-term memory. Science, 173: 652–654.

Georges-François, P., Rolls, E.T. and Robertson, R.G. (1999). Spatial view cells in the primate hippocampus: allocentric view not head direction or eye position or place. Cereb. Cortex, 9: 197–212.

Goldman-Rakic, P.S. (1987). Circuitry of the primate prefrontal cortex and the regulation of behavior by representational memory. In: Plum F. (Ed.), Handbook of Physiology, The Nervous System, Vol. V, Higher Functions of the Brain. American Physiological Society, Bethesda, pp. 373–417 Part 1.

Goldman, P.S. and Rosvold, H.E. (1970). Localization of function within the dorsolateral prefrontal cortex of the rhesus monkey. Exp. Neurol., 27: 291–304.

Hikosaka, O., Nakahara, H., Ran, M.K., Sakai, K., Lu, X., Nakamura, K., Miyachi, S. and Doya, K. (1999). Parallel neural networks for learning sequential procedures. Trends Neurosci., 22: 464–471.

Kertesz, A. (1994). Frontal lesions and function: Localization and Neuroimaging in Neuropsychology. Academic Press, San Diego, pp. 567–598.

Kubota, K. and Niki, H. (1971). Prefrontal cortical unit activity and delayed alternation performance in monkeys. J. Neurophysiol., 34: 337–347.

Lepage, M. and Richer, F. (1996). Inter-response interference contributes to the sequencing deficit in frontal lobe lesions. Brain, 119: 1289–1295.

Ma, Y.-Y., Tian, B.P. and Wilson, F.A.W. (2003) Dissociation of egocentric and allocentric spatial processing in prefrontal cortex. Neuroreport, 14: in press.

Maguire, E.A., Burke, T., Phillips, J. and Staunton, H. (1996). Topographical disorientation following unilateral temporal lobe lesions in humans. Neuropsychologia, 34: 993–1001.

Maguire, E.A., Frackowiak, R.S. and Frith, C.D. (1997). Recalling routes around London: activation of the right hippocampus in taxi drivers. J. Neurosci., 17: 7103–7110.

Morris, R., Petrides, M. and Pandya, D.N. (1999). Architecture and connections of retrosplenial area 30 in the rhesus monkey (*Macaca mulatta*). Eur. J. Neurosci., 11: 2506–2518.

Murray, E.A., Davidson, M., Gaffan, D., Olton, D.S. and Suomi, S. (1989). Effects of fornix transection and cingulate cortical ablation on spatial memory in rhesus monkeys. Exp. Brain Res., 74: 173–186.

O'Keefe, J. (1999). Do hippocampal pyramidal cells signal non-spatial as well as spatial information?. Hippocampus, 9: 352–364.

ÓScalaidhe, S.P., Wilson, F.A.W. and Goldman-Rakic, P.S. (1999). Face-selective neurons during passive viewing and working memory performance of Rhesus monkeys: evidence for intrinsic specialization of neuronal coding. Cereb. Cortex, 9: 459–475.

Petrides, M. (1995). Impairments on non-spatial self-ordered and externally ordered working memory tasks after lesions of the mid-dorsal part of the lateral frontal cortex in the monkey. J. Neurosci., 15: 359–375.

Petrides, M. and Milner, B. (1982). Deficits on subject-ordered tasks after frontal and temporal-lobe lesions in man. Neuropsychologia, 20: 249–262.

Petrides, M. and Pandya, D. (1999). Dorsolateral prefrontal cortex: comparative cytoarchitectonic analysis in the human and the macaque brain and corticocortical connection patterns. Eur. J. Neurosci., 11: 1011–1036.

Pohl, W. (1973). Dissociation of spatial discrimination deficits following frontal and parietal lesions in monkeys. J. Comp. Physiol. Psychol., 82: 227–239.

Stamm, J.S. (1987). The riddle of the monkey's delayed-response deficit has been solved. In: Perecman E. (Ed.), The Frontal Lobes Revisited. IRBN Press, New York, pp. 73–89.

Stark, M.E., Coslett, H.B. and Saffran, E.M. (1996). Impairment of an egocentric map of locations: implications for perception and action. Cognit. Neuropsychol., 13: 481–523.

Stuss, D.T. and Benson, D.F. (1986). The Frontal Lobes. Raven Press, New York.

Taube, J.S., Muller, R.U. and Ranck, J.B. (1990). Head-direction cells recorded from the postsubiculum in freely moving rats. 1. Description and quantitative analysis. J. Neurosci., 10: 420–435.

Walker, R., Husain, M., Hodgson, T.L., Harrison, J. and Kennard, C. (1998). Saccadic eye movement and working memory deficits following damage to human prefrontal cortex. Neuropsychologia, 36: 1141–1159.

Wilson, F.A.W., Riches, I.P. and Brown, M.W. (1990). Hippocampus and medial temporal cortex: neuronal activity related to behavioural responses during the performance of memory tasks by primates. Behav. Brain Res., 40: 7–28.

Wilson, F.A.W., ÓScalaidhe, S.P. and Goldman-Rakic, P.S. (1993). Dissociation of object and spatial processing domains in primate prefrontal cortex. Science, 260: 1955–1958.

Wojciulik, E., Husain, M., Clark, K. and Driver, J. (2001). Spatial working memory deficit in unilateral neglect. Neuropsychologia, 39: 390–396.

Subject Index

Subjects listed are discussed in the chapters that start on pages referenced here

5-hydroxytryptamine; *see* Transmitter
Acetylcholine; *see* Transmitter
Acquired callosal damage; *see* Patient (with motor disorder)
Action selection (by brain) 507
Adaptation (of firing rate) 319, 353
Adaptive control; *see* Control mechanism
Adaptive plasticity; *see* Plasticity (neuronal or network)
Afterhyperpolarization (AHP) 77
Allocentric space 513
Amygdala 39
Animal model
  cat (intact and/or decerebrate) 3, 39, 67, 77, 105, 115, 123, 163, 175, 231, 239, 251, 263, 283, 291, 309, 319, 341, 353, 369, 383, 403, 411, 423
  cat (spinal, spinal-cord-injured) 77, 133, 147, 155, 163, 263, 353
  in vitro 49, 57, 67, 97, 299
  in vivo vs. in vitro, 57, 67
  lamprey 3, 123
  locust 123
  mouse (neonatal) 97
  nonhuman primate 3, 183, 191, 263, 299, 309, 341, 353, 369, 391, 411, 423, 439, 449, 461, 467, 477, 491, 507, 513
  nonhuman primate (juvenile) 183, 191, 341
  nonhuman primate (neonatal) 467
  rabbit 319, 331
  rat 67, 77, 283, 291, 309
  rat (brainstem slice) 97, 299
  rat (fetal) 49,
  rat (neonatal) 49, 57, 67, 97
  rat (neonatal, spinal-cord-injured) 57, 133, 147, 155, 163
  rat (neonatal, suckling), 97
  rat (spinal-cord injured) 77, 133, 147, 155, 163
  salamander 221

Anterograde neural tracing (labeling); *see* Experimental/analytical approach
Anticipation (of a stimulus) 191, 251, 319, 331, 353, 477, 491, 507
Anticipatory postural adjustment; *see* Posture
AP-5; *see* Transmitter antagonist
Apoptosis 133
Arm reach; *see* Movement (type/neural control)
Arousal (of central nervous system) 283, 291, 439
Associative learning; *see* Learning
Astasia–abasia; *see* Motor disorder
Astrocytes 133, 147
Auditory mid-latency evoked potential 291
Autoradiography; *see* Experimental/analytical approach
Axonal guidance system 133, 147, 155

Basal ganglia 207, 231, 449, 461
  direct, hyperdirect, indirect pathways 461
BBB scale; *see* Experimental/analytical approach
Behavior; *see also* Movement (type/neural control)
  female receptive 39, 105
  motor 3, 39, 57, 105, 115, 133, 207, 263, 319, 467, 477, 491
  orienting 383, 411, 423
  sexual 39, 105
Bimanual coordination 491
  pathology of 491
Biomechanical perspective 13, 183, 207
Body schema 13, 353
Brain imaging (non-invasive); *see* Experimental/analytical approach
Brainstem 67, 97, 105, 221, 231, 251, 353, 403, 411, 423
Brainstem cholinergic system 439
Brainstem reticular formation; *see* Reticular formation
Bulbospinal projection 105, 383

Cat; *see* Animal model
Cat (spinal, spinal-cord-injured); *see* Animal model
Central pattern (rhythm) generator (CPG) 3, 49, 57, 67, 97, 105, 115, 123, 163, 231, 239, 353
  chewing, 97
  female receptive behavior 105
  locomotion 3, 49, 57, 105, 115, 123, 163, 239, 341, 353
  respiratory 67, 105
  spinal 3, 49, 115, 123, 163, 353
Central respiratory drive potential 105
Cerebellar locomotor region (CLR); *see* Locomotor region
Cerebellar EPSP (in motor cortex cell) 309
Cerebellar patient; *see* Patient (with motor disorder)
Cerebellum 13, 309, 319, 331, 341, 353
Cerebral cortex
  dorsolateral prefrontal 513
  frontal 39, 391
  motor (primary, M1) 251, 263, 309, 461, 467, 477
  prefrontal 39, 449, 507
  premotor 461, 467, 507
  premotor lateral 199
  sensorimotor 467, 477
  sensory (primary, S1) 477
  supplementary motor 461, 467, 491
Chondroitin sulfate proteoglycans 147
Circuit gating; *see* Control strategy
Classical (Pavlovian) conditioning; *see* Conditioning (behavioral)
Clonidine; *see* Transmitter agonist
Commissural inhibition; *see* Inhibition
Conditioning (behavioral)
  classical (Pavlovian) 319, 331
  operant 183, 191, 319, 353
Congenital agenesis; *see* Patient (with motor disorder)
Control mechanism
  adaptive 3, 353, 477, 491
  anticipatory (feed-forward) response 13, 191, 251, 263, 207, 319, 331, 341, 353, 383, 477, 491, 507
  cortical coding 477
  reactive (sensory feedback) response 13, 29, 123, 191, 207, 341, 353, 477
  spatiotemporal transformation 411, 423
Control strategy (principle)
  cerebellar 309, 319, 331, 341, 353

  context dependency 319
  circuit gating 251, 299, 423
  direct and inverse dynamics 13, 207, 353
  hierarchical 13, 29, 207
  locomotion integration center 341, 353
  multifunctional 105, 423
  multijoint 13, 207, 353, 411, 477
  multimodal signaling 439
  neural integration 263, 391
  optimal control 369
  parallel processing 199, 263, 423
  population coding 423
  prediction 391, 477, 491
  referent posture 13
  signal decomposition 423
  stiffness (joint) control 207, 477
  task dependency 319, 491, 507
Coordination; *see* Motor coordination
Corpus callosum 491
Cortical coding; *see* Control mechanism
Cortical lesion; *see* Patient (with motor disorder)
Cortical neuron; *see* Neuron (type)
Corticospinal neuron; *see* Neuron (type)
Corticospinal tract; *see* Tract (pathway, projection)
Conditioning lesion effect 133, 147
Corticoreticular projection; *see* Tract (pathway, projection)
Cross-striolar inhibition; *see* Inhibition

Dependency-context; *see* Control stategy (principle)
Dependency (task); *see* Control stategy (principle)
Design feature (of central nervous system)
  axon branching (multiple) 239, 383, 411, 423
  divergence (of neural connections) 251, 411, 423
  overcomplete motor system 369
  population (neuronal) heterogeneity (anatomical) 239, 423
  population (neuronal) heterogeneity (functional) 423
  weighting of connections (anatomical) 423
Development
  excitation (in fetus) 49
  fetal motor system 49
  inhibition (in fetus) 49
  neonatal 49, 57, 67, 467
  neural 49, 57, 67
  swimming 57
  walking 57

Dexterity 263, 467, 491
Disinhibition 299, 353, 403, 461
Dorsal root entry zone 147
Dorsolateral prefrontal cortex; *see* Cerebral cortex
Dynamics; *see* Experimental/analytical approach

Effector (body part as)
  arm 207, 319, 353, 467, 507
  eye 319, 369, 383, 391, 403, 423, 439
  head 369, 383, 403, 411, 423
  trunk 13, 29, 353
Egocentric space 513
Eigenmovement
  hip and ankle 13
Ejaculation
  (by) human male 39
Electromyography (EMG); *see* Experimental/analytical approach
Electrophysiological techniques; *see* Experimental/analytical approach
Embryonic repair graft 133, 147, 155
Emotional motor system 39, 105
Equilibrium control 13, 191, 353, 411
Estrogen 39
Estrus 39, 105
Evoked potential 309
Experimental/analytical approach
  anterograde neural tracing (labeling) 39, 239, 423
  antidromic latency (variation) 105
  autoradiography 163
  BBB scale 133, 155
  dynamics (of movement) 13, 29, 207, 369, 423, 477
  electromyography (EMG) 29, 49, 163, 175, 207, 353, 423
  experimental neuroanatomy 133, 147, 155, 191, 199, 239, 251, 353, 391, 411, 423, 449, 467
  extracellular recording 57, 105, 251, 319, 331, 353, 391, 423, 461, 477, 507
  glucose metabolism 191
  intraaxonal labeling /staining 239, 411, 423
  intraaxonal recording 239, 383, 411, 423
  intracellular labeling/staining 411
  intracellular recording 77, 105, 115, 263, 309, 411
  kinematic analysis 13, 29, 133, 163, 175, 183, 191, 207, 319, 353, 369, 423, 477, 491
  kinetic analysis 13, 133, 207, 353, 491
  microstimulation (in brainstem) 239, 423
  microstimulation (in spinal cord) 105, 115

  microstimulation (intracortical) 467
  multichannel unitary cortical recording 319, 477
  neurectomy 163
  non-invasive imaging 39, 191, 199
  positron emission tomography (PET) 39, 191, 199
  rabies virus transport 449
  retrograde neural tracing (labeling) 155, 449
  reversible inactivation (of brain tissue) 319, 331, 467
  single neuron recording 423, 461, 507
  single photon emission computed tomography (SPECT) 199
  torque analysis 207
  transneuronal transport 507
Express saccade; *see* Movement (type/neural control)
Eye movement, *see* Movement (type/neural control)

Fastigial nucleus, *see* Nucleus
Feedback
  negative 123
  positive 123
  sensory 29, 123, 163, 207, 477
  sensory servo control 29
  sensory triggering 29
Female receptive behavior; *see* Behavior
Fetal motor development; *see* Development
Fetal spinal cord transplant 133, 147, 155, 163
Fictive locomotion 115, 239
Fight-vs.-flight response 283
Flying; *see* Movement (type/neural control)
Forelimb; *see* Movement (type/neural control)
Frontal cortex; *see* Cerebral cortex
Frontal cortical lesion; *see* Patient (with motor disorder)
Frontal eye field 391
Functional (non-invasive) brain imaging; *see* Experimental/analytical approach

Gait, of cerebellar patient 353
Gamma-amino butyric acid; *see* Transmitter
Gaze (velocity) 391, 411, 423
Gaze shift; *see* Movement (type/neural control)
Gene therapy 133, 147
Glucose metabolism; *see* Experimental/analytical approach
Glutamate; *see* Transmitter
Glycine; *see* Transmitter

Globus pallidus 461
Goal-directed actions; *see* Movement (type/neural control)

Hand (neural representation of) 263, 467, 491
Head movement; *see* Movement (type/neural control)
Hierarchical control; *see* Control strategy
Hindbrain neurons; *see* Neuron(s)
Hindlimb joint; *see* Joint (measurement/movement)
Hip and ankle strategy; *see* Posture
Hippocampus 513
Human (as experimental subject) 3, 13, 29, 39, 77, 199, 207, 263, 353, 507
  spinal-cord-injured 133, 147, 155, 163, 263
Hypoglossal nerve; *see* Nerve
In vitro preparation; *see* Animal model
In vivo vs. in vitro preparation; *see* Animal model

Inhibition 147
  commissural 49, 403
  cross-striolar 403
  development of 49, 77
  feed-forward 263
  recurrent 77
  Renshaw 77
Integration of information (by prefrontal cortex) 507
Interactive torques 13, 207, 353
Interlaminar connection 299
Interneuron; *see* Neuron (type)
Interposed (interpositus) nucleus: *see* Nucleus
Intraaxonal labeling/staining; *see* Experimental/analytical approach
Intracortical microstimulation; *see* Experimental/analytical approach
Intracellular labeling/staining; *see* Experimental/analytical approach
Intracellular recording; *see* Experimental/analytical approach
Joint (measurement/movement)
  arm 207, 353
  axis 29
  hindlimb 13, 163, 183, 353

Kinematic analysis; *see* Experimental/analytical approach
"Kinésie paradoxale," *see* Motor disorder
Kinetic analysis; *see* Experimental/analytical approach

Labyrinth 403, 411
Lamprey; *see* Animal model
Learning
  associative 319, 331, 353
  locomotor 341, 353
  overall mechanisms 331
  reinforcement 439
  spatial and motor sequence 353, 513
Lesion (effects on motor function) 77, 133, 147, 155, 163, 263, 353, 467, 491
Linear acceleration (increased labyrinthine sensitivity to) 403
Local circuit; *see* Neuronal network (type) 299
Locomotion; *see* Movement (type/neural control)
  bipedal vs. quadrupedal 183, 191, 353
  in human 199, 353
  in robot 29
Locomotor control; *see* Control mechanism
Locomotor integration center; *see* Control strategy (principle)
Locomotor performance (evaluation of) 133, 147, 155, 163, 353
Locomotor region (of brain)
  cerebellar (CLR) 341, 353
  mesencephalic (midbrain) (MLR) 3, 57, 221, 231, 239, 341
  subthalamic (SLR) 3, 341
Locomotor rhythm; *see* Motor rhythm
Locust; *see* Animal model
Lumbar commissural neurons; *see* Neuron(type)
Lumbar lift 29

Medulla oblongata 67, 105, 423
Memory
  locomotor 341, 353
  motor 319, 331, 341
  spatial 513
Mesencephalic (midbrain) locomotor region (MLR); *see* Locomotor region (of brain)
Mesodiencephalic junction 39
Mastication; *see* Movement (type/neural control)
Microstimulation; *see* Experimental/analytical approach
Modeling
  basal ganglia operation 461
  biomechanical 13
  descending control of locomotion 251
  internal represenation (of posture) 13

multijoint torque 207, 353
neuronal circuits 423
Monkey; *see also* Animal model (non-human primate)
   Macaca fascicularis 467, 491
   Macaca fuscata 183, 191, 309, 341, 391, 507
   Macaca mulatta 263, 353, 477
   Saimiri sciureus 263, 309
Motivation 439
Motoneuron; *see* Neuron (type)
Motoneuron recruitment; *see* Recruitment
Motor activity; *see* Neuronal activity
Motor (action) planning 207, 353, 507, 513
Motor control
   generic issues 3, 29, 123, 207, 341, 353, 423, 439
Motor command
   associative 29
   primary 29
Motor coordination
   bimanual 491
   broad (generic) issues 3, 13, 29, 207, 341, 353, 507
   eye–head 3, 369, 383, 391, 403, 423, 439
   grip/load force 467, 491
   head–neck 369, 403, 411
   hindlimb–forelimb 155
   left–right limb 49
   posture with movement 3, 13, 29, 105, 183, 191, 251, 341, 353, 383
Motor cortex (primary, M1) 3, 239, 251, 263, 309, 411, 449, 461, 467, 477, 491
Motor disorder; *see also* Patient (with motor disorder)
   astasia–abasia 353
   asynergia 13, 29, 353
   ataxia 353
   hemiballismus 461
   incoordination 353
   "kinésie paradoxale" 199
Motor equivalence 491
Motor incoordination; *see* Motor disorder
Motor learning 3, 163, 183, 191, 319, 331, 341, 353, 439, 491, 513
Motor recovery (from damage to central nervous system); *see* Plasticity
Motor rhythm
   abdominal straining 105
   locomotor 3, 49, 163, 239, 353
   respiratory 29, 67, 105
   retching 105

Motor segments (control of) 13, 49, 183, 341, 411
Motor tract; *see* Tract (pathway, projection)
Mouse (neonatal); *see* Animal model
Movement (type/neural control)
   abdominal straining 105
   arm reach 3, 207, 251, 263, 319, 467, 477, 507
   associated 29
   axial 29
   bipedal locomotion 183, 191, 199, 353
   bimanual 491
   complex 29, 207
   eigen 13
   express saccade 299
   eye 3, 369, 383, 391, 403, 411, 423
   flying (by locust) 123
   forelimb 115, 251, 467
   gaze shifts 391, 423
   goal-directed actions 207, 491
   grasping 263, 467, 491
   head 369, 383, 403, 411, 423
   innate 3
   integration with posture 13, 183, 191, 251, 341, 353, 383
   internal representation 13, 263
   learned; *see* Motor learning
   locomotion 3, 49, 57, 115, 123, 133, 163, 175, 183, 199, 221, 231, 239, 251, 283, 341, 353
   lordosis 39, 105
   mastication 97
   mobility vs. stability 29
   orienting 383, 411, 423
   purposive vs. non-purposive
   quadrupedal locomotion 183, 353
   reach and grasp 133, 263
   reflex 3, 411
   respiration 3, 67, 105
   saccade 3, 299, 383, 391, 403, 411, 423, 439
   sit-up 29
   smooth (eye) pursuit 391
   smooth tracking (by eyes) 391, 423
   stepping over obstacles 191, 251
   swimming 57, 123
   tongue, 97
   transition from rest to locomotion 221
   treadmill locomotion 133, 183, 191

Movement (type/neural control) (*Cont.*)
  trunk 175, 383
  trunk bending 13
  upslope locomotion 175
  vergence (of eyes) 391
  vertebral column (during walking) 175
  visual motion 391
  voluntary 3
  voluntary vs. reflex 3
  vomiting 105
  walking (as a specific gait) 123, 183, 191, 199
Multichannel unitary cortical recording; *see* Experimental/analytical approach
Multifunctionality; *see* Control strategy (principle)
Multijoint control (of movement); *see* Control strategy (principle)
Multimodal signaling; *see* Control stategy (principle)
Multiple axon branching; *see* Design (of organism) feature
Muscle
  longissimus 175, 411
  postural tone 13, 231, 283
  roles (functions) 29, 175, 207
  synergies 29, 353, 411, 423, 491
  tone 231
Myelin 133, 147, 155
Myelin-associated glycoprotein 133, 147

NI250/Nogo 147
Negative feedback; *see* Feedback
Neck motoneuron; *see* Motoneuron (type/neural control)
Nerve
  hypoglossal 97
Neural development; *see* Development
Neural integration; *see* Control strategy (principle)
Neurectomy; *see* Experimental/analytical approach
Neuroanatomical techniques; *see* Experimental/analytical approach
Neuron (single recording from/type)
  cortical 251, 477, 507
  corticospinal 3, 263, 411
  excitatory burst 423
  forelimb motoneuron 115, 263
  generic motoneuron 57, 77, 263
  head motoneuron 369, 411
  hindbrain 221
  hindlimb motoneuron 77, 105
  inhibitory burst 423
  interneuron 115, 263
  lumbar commissural 239
  neck motoneuron 369, 411
  olfactory ensheathing 147
  omnipause (complex) 423
  omnipause (saccadic) 423
  propriospinal 105, 263
  Purkinje 319, 341
  pyramidal 251, 263, 309
  respiratory 67, 105
  reticular 423
  reticulospinal 3, 251, 341, 383, 411, 423
  tectal 423
  tectoreticulospinal 411, 423
  vestibular 341, 391, 403, 411
Neuronal activity
  cortical 3, 251, 263, 477, 507
  motor 49, 263
  premotor 423
Neuronal network 3, 57, 67, 331
  brainstem 3, 67, 105, 403
  local circuit 299
  spinal 3, 49, 57, 115
  tectoreticular 411, 423
  vestibulospinal 403, 411
  vs. pacemaker circuit 67
Neuronal regrowth (capacity for) 133, 147, 155
Neurotransmitter; *see* Transmitter
Neurotrophic factor 133, 147, 155
Newborn; *see also* Animal model 49, 57, 67, 97, 467
NMDA; *see* Transmitter
Nonhuman primate; *see* Animal model
Nucleus
  abducens 423
  fastigial 341, 353
  interposed (interpositus) 319, 331, 353
  lateral reticular (LRN) 263
  pedunculopontine (tegmental) (PPN, PPTg, or PPTN) 231, 283, 291, 299, 439
  pontomedullary reticular formation (PMRF) 251, 383
  prepositus hypoglossi 423
  red 411
  reticularis gigantocellularis (NRGc, RGc) 239, 383, 423
  reticularis magnocellularis (NRMc) 239

reticularis pontis caudalis (NRPc, RPc) 239, 383, 423
reticularis pontis oralis (NRPo) 239
retroambiguus (NRA) 39, 105
vestibular 403, 411

Olfactory ensheathing cell; *see* Neuron (type)
Oligodendrocyte-myelin glycoprotein 147
Omnipause neuron; *see* Neuron (type)
Operant conditioning; *see* Conditioning (behavioral)
Orgasm and reward 39
Orienting; *see* Movement (type/neural control)

P13 mid-latency potential 291
P50 mid-latency potential 291
Parallel processing; *see* Control strategy (principle)
Parkinson's disease; *see* Patient (with motor disorder)
Path and target finding (neuronal) 133, 147, 155
Patient (with motor disorder)
   cortical lesion 467, 491
   acquired callosal damage 491
   cerebellar lesion 13, 29, 319, 331, 353, 491
   congenital agenesis 491
   elderly 207
   frontal cortical lesion 467, 491
   Parkinson's disease 199, 207, 353, 461, 491
   spasticity 77
   spinal-cord-injured 133, 147, 155, 163, 263
   stroke 467
Pedunculopontine (tegmental) nucleus; *see* Nucleus
Periaqueductal gray (PAG) 39
Persistent inward current 77, 105
Perturbation (of)
   forelimb 319, 477
   locomotion 191, 251, 353
   lower limb 191
   posture 13, 353
Pharmacotherapy 133, 147
Phylogenetic comparison 309, 263, 467
Place field 513
Plasticity (neuronal or network) 3, 39, 57, 105, 123, 467, 491
   recovery from motor cortex injury 467
   recovery from spinal cord injury 133, 147, 155, 163, 263
Plateau potential 77, 105

Polysynaptic propagation 221
Positive feedback; *see* Feedback
Positron emission tomography (PET); *see* Experimental/analytical approach
Postural control
   against gravity 13, 341
   integration with movement control 13, 341, 383
   parallel control (of posture and movement) 13
   role of cerebellum 13, 341, 353
Postural reactions; *see* Posture
Postural stabilization; *see* Posture
Posture
   anticipatory adjustment 13, 29, 191, 341, 353
   body orientation 13
   during bipedal locomotion 183, 191, 353
   during orienting 383
   female receptive behavior 105
   flexibility 13
   genetic model 13
   hierarchical model 13
   hip and ankle strategy 13
   hip and knee stategy 191
   integration with movement 3, 13, 183, 191, 251, 33, 353
   interface between perception and action 13
   segmental 341
   stabilization 29, 369
Prediction; *see* Control strategy (principle)
Prefrontal cortex; *see* Cerebral cortex
Premotor cortex; *see* Cerebral cortex
Premotor lateral cortex; *see* Cerebral cortex
Presaccadic burst activity 299, 423
Primate (non-human); *see* Animal model
Processing of midbrain volleys 221
Proprioceptor; *see* Sensory receptor
Purkinje cell; *see* Neuron (type)
Pyramidal tract; *see* Tract (pathway, projection)
Pyramidal tract neuron; *see* Neuron (type)

Rabbit; *see* Animal model
Rabies virus transport; *see* Experimental/analytical approach
Rat ; *see* Animal model
Rat (brainstem slice, fetal, neonatal,neonatal–spinal-cord-injured, neonatal-suckling, spinal-cord-injured); *see* Animal model
Reaching (by arm or forelimb); *see* Movement (type/neural control)

Receptor (of transmitter)
  AMPA 163
  Noradrenergic 163
  Serotonergic 163
Recovery of motor function 133, 147, 155, 163, 263, 319, 331, 353, 467, 491
Recruitment
  motoneuron 77
  motor unit 77
Recurrent inhibition; *see* Inhibition
Red nucleus; *see* Nucleus (type)
Reflex
  eye-blink 319, 331
  mating 39
  modulation 369
  sexual 39
  spinal 3, 29
  vestibulo-collic 13, 369, 403, 411
  vestibulo-ocular 391, 403
  vestibulo-spinal 369, 403
  withdrawal 331
Reinforcement learning; *see* Learning
Renshaw inhibition; *see* Inhibition
Respiratory center 3, 67, 105
Respiratory control; *see* Movement (type/neural control)
Respiratory neuron; *see* Neuron (type)
Respiratory rhythm generation; *see* Motor rhythm
Reticular activating system 283, 291
Reticular formation 231, 239, 251, 383, 423
  brainstem 251
  paramedian pontine (PPRF) 423
  gigantocellular tegmental field (FTG) 239
  magnocellular tegmental field (FTM) 239
Reticulospinal neuron; *see* Neuron(type)
Reticulospinal projection; *see* Tract (pathway,projection)
Retroambiguus nucleus; *see* Nucleus
Retrograde neural tracing (labeling) ; *see* Experimental/analytical approach
Reverberation (in a neuronal network) 115
Reversible inactivation (of brain tissue); *see* Experimental/analytical approach

Saccade; *see* Movement (type/neural control)
Saccadic burst generator 423
Saccadic reaction time 439
Sacculus 403

Salamander; *see* Animal model
Selection of body part (for movement) 29, 207, 507
Semicircular canal 391, 403, 411
Sensorimotor cortex: *see* Cerebral cortex
Sensory cortex (primary, S1); *see* Cerebral cortex
Sensory feedback (input); *see* Feedback
Sensory receptor
  proprioceptor 123
Sexual behavior; *see* Behavior
Sherrington, Charles 123, 341
Single cortical neuron recording; *see* Experimental/analytical approach
Single photon emission computed tomography; *see* Experimental/analytical approach
Sleep–waking (control of) 283, 291
Smooth (eye) pursuit; *see* Movement (type/neural control)
Smooth (eye) tracking; *see* Movement (type/neural control)
Somatotopy 353, 467
Spatial and motor sequence learning; *see* Learning
Spatial memory; *see* Memory
Spinal cat; *see* Animal model
Spinal cord 3, 39, 49, 57, 77, 133, 147, 155, 163, 239, 251, 263
  cervical 115, 411
  midlumbar segments 163
  replacement 155
Spinal cord injury
  cat 77, 133, 163
  human 133, 147, 155, 163, 263
  rat 77, 133, 147, 155
  rat (neonatal) 133, 147, 155, 163
Spinal plasticity; *see* Plasticity
Spinal segment
  intersegmental 49
  midlumbar 163
Sprouting (of axon) 133, 147
Stance (of cerebellar patient) 353
Startle response 283
Stem cell 133, 147
Stepping parameter 183
Stroke; *see* Patient (with motor disorder)
Substantia nigra pars reticulata 231
Subthalamic locomotor region (SLR); *see* Locomotor region (of brain)
Subthalamic nucleus 461

Suckling; *see* Animal model
Superior colliculus 299, 383, 411, 423, 439
Survival of the species 39
Synergies; *see* Muscle

Target choice 507
Target selection 507
Tectoreticulospinal system 411, 423
Thalamocortical response 309
Torque analyis; *see* Experimental/analytical approach
Tract (pathway, projection)
  cortico-basal ganglia loop 461
  corticoreticular 239
  corticospinal 263, 369, 467
  descending 39, 57
  fastigiospinal 341
  motor 411
  propriospinal 263
  pyramidal 155, 251, 263, 309, 467
  reticulospinal 231, 239, 251, 341, 369, 383
  tectoreticulospinal 341, 423
  vestibulospinal 341, 369, 403, 411
Transition (from rest to locomotion); *see* Movement (type/neural control)
Transmitter
  5-hydroxytryptamine (5-HT) 49, 57, 77
  acetylcholine 299
  gamma-amino butyric acid (GABA) 49, 231, 461
  glycine 49
  glutamate 49, 461
  neurotransmitters 163
  NMDA 97, 163, 299, 461
Transmitter agonist
  clonidine(noradrenergic) 163
  muscimol (GABAA) 461
Transmitter antagonist
  AP-5 (NMDA) 163
  bicuculline (GABAA) 461
  CPP (NMDA) 461
  yohimbine (noradrenergic) 163
Transneuronal transport; *see* Experimental/analytical technique
Training (motor) 133, 163, 183, 191, 341, 353
Transplantation (of spinal tissue) 133, 147, 155
Trunk movement; *see* Movement (type/neural control)

Utriculus 403

Vergence (of eye); *see* Movement (type/neural control)
Vertebral column; *see* Movement (type/neural control)
Vestibular neuron; *see* Neuron (type)
Vestibular nucleus; *see* Nucleus (type)
Vestibular system 391, 403, 411
Vestibulospinal tract; *see* Tract (pathway, projection)
Visual motion; *see* Movement (type/neural control)

Yohimbine; *see* Transmitter antagonist